8th International Conference on Computer and Automation Engineering (ICCAE 2016)

MATEC Web of Conferences Volume 56 (2016)

Melbourne, Australia
3 - 4 March 2016

Editors:

Manolya Kavakli
Momoh-Jimoh Eyiomika Salami
Abbas Amini
Mohd Ariffanan Bin Mohd Basri
Azman Bujang Masli
Stephanie Chua Hui Li
Mayuka Pal

ISBN: 978-1-5108-2334-1

Printed from e-media with permission by:

Curran Associates, Inc.
57 Morehouse Lane
Red Hook, NY 12571

Some format issues inherent in the e-media version may also appear in this print version.

Printed by Curran Associates, Inc. (2016)

For additional information, please contact EDP Sciences – Web of Conferences at the address below.

EDP Sciences – Web of Conferences
17, Avenue du Hoggar
Parc d'Activité de Courtabœuf
BP 112
F-91944 Les Ulis Cedex A
France

Phone: +33 (0) 1 69 18 75 75
Fax: +33 (0) 1 69 28 84 91

contact-edps@webofconferences.org

Additional copies of this publication are available from:

Curran Associates, Inc.
57 Morehouse Lane
Red Hook, NY 12571 USA
Phone: 845-758-0400
Fax: 845-758-2633
Email: curran@proceedings.com
Web: www.proceedings.com

TABLE OF CONTENTS

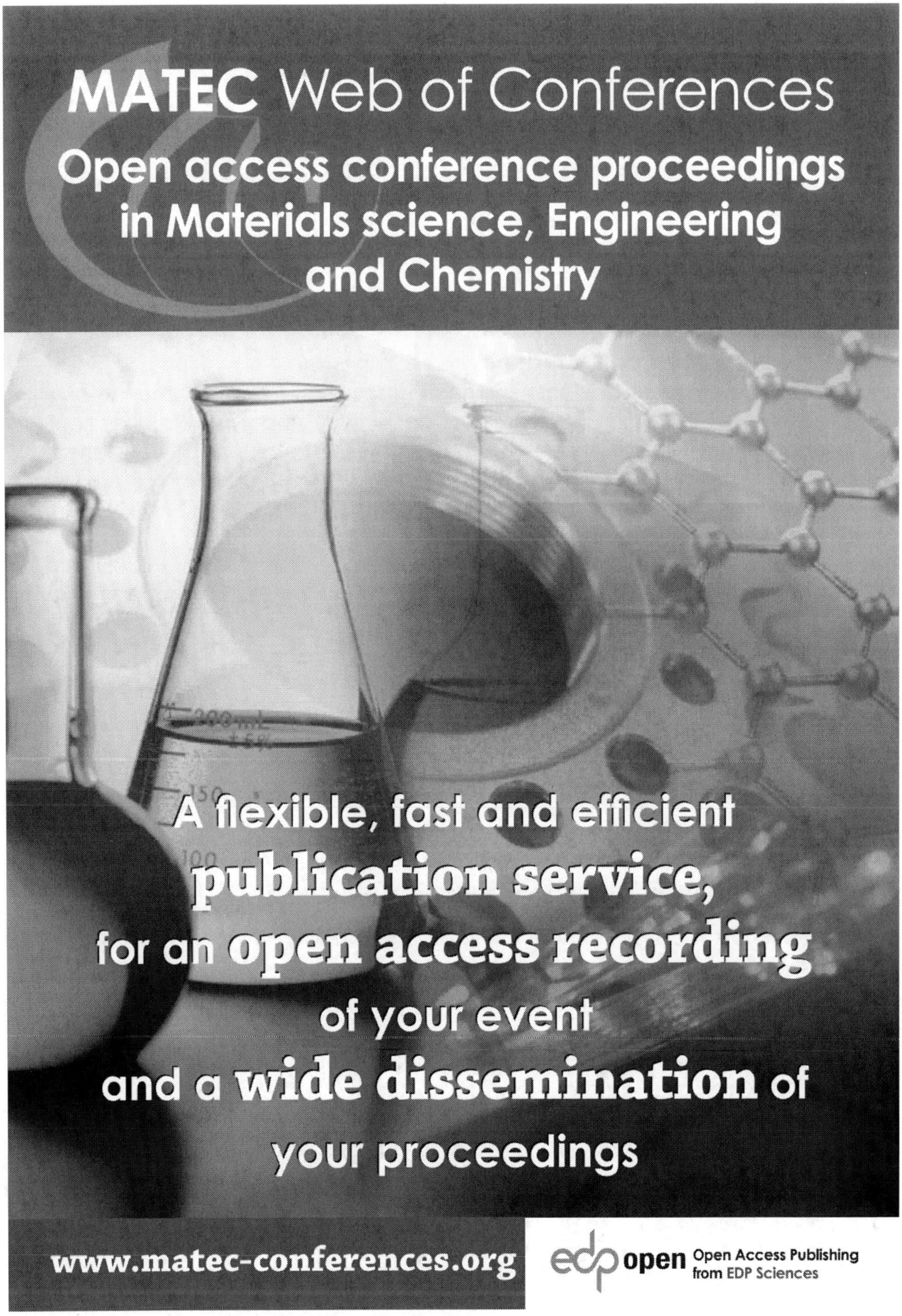
MATEC Web of Conferences
Open access conference proceedings in Materials science, Engineering and Chemistry
A flexible, fast and efficient publication service, for an open access recording of your event and a wide dissemination of your proceedings
www.matec-conferences.org
edp open Open Access Publishing from EDP Sciences

ISSN (electronic edition): 2261-236X

Preface

Nowadays, computer and automation engineering has played a more and more important role in people's modern life. And it has been changing all aspects of our lives in such diverse areas and makes our lives easier. In light of the fast-paced advancements in computer and automation engineering taking place all over the world, it is of interest to keep an eye on the state-of-the-art research and development and to facilitate collaboration in multidisciplinary research areas. With this end in view, 2016 8th International Conference on Computer and Automation Engineering was held once again this year in Melbourne, Australia during March 3-4, 2016.

The aim of the conference was to address and deliberate on the latest technical status and recent trends in the research and applications of computer and automation engineering. This conference has been so designed with the view to provide an opportunity for the scientists, engineers, industrialists, students and other professionals from all over the world to interact and exchange their new ideas and research outcomes for future collaboration. The conference was also aimed at motivating the next generation of researchers to promote their interests in computer and automation engineering and their applications.

The proceedings of ICCAE 2016 contain 63 selected papers from the conference which are presented orally at the conference or via poster sessions, and provide up-to-date, comprehensive and worldwide state-of-the art knowledge in this field. Each contributed paper was rigorously peer-reviewed by international reviewers who were drawn from the organizing and advisory committee members as well as other experts in the field from all over the world. The proceedings are divided into 8 chapters based on the different topics of the papers, namely, Computer and information technology; Image processing and Application; Signal analysis and processing technology; Bioinfomatics; Modern communication technology and application; Control theory and Engineering; Machine vision and robot; Electronic technology and power engineering.

We'd like to express our genuine gratitude to everyone who has contributed to ICCAE 2016. Especially the keynote speakers, Prof. Saman Halgamuge, the program chairs, Asst. Prof. Abbas Amini and Dr. Lihong Zheng as well as invited speakers, Scientific Editor, Dr. Mohd Ariffanan Bin Mohd Basri, Dr. Azman Bujang Mas, Dr. Abbas Amin, Dr. Mayukha Pal and all the technical committee members. Let's wish ICCAE the same success for next year.

Assoc. Prof. Manolya Kavakli,
Macquarie University, Faculty of Science Department of Computing, Australia

Conference Co-Chairs

Assoc. Prof. Manolya Kavakli, Macquarie University, Faculty of Science Department of Computing, Australia

Program Co-Chairs

Asst. Prof. Abbas Amini, University of Western Sydney, Australia

Dr. Lihong Zheng, Charles Sturt University, Australia

Technical Committee

Prof. Gang Su, Huazhong University of Science and Technology, China

Prof. Gee-Sern (Jison) Hsu, National Taiwan University of Science and Technology, Taiwan

Prof. Nahrul Khair bin Alang Md Rashid, International Islamic University Malaysia

Prof. Momoh -Jimoh Eyiomika Salami, International Islamic University Malaysia

Prof. Michael Kraft, University of Liege, Belgium

Prof. Chyi-Yeu Lin, National Taiwan University of Science and Technology, Taiwan

Asst. Prof. Sungho Kim，Yeungnam University, The Republic of Korea

Prof. Dr. Zurni B Omar, Northern University, Malaysia

Dr. Fatimah Sham Ismail, Universiti Teknologi Malaysia, Malaysia

Dr. Rosemizi Abd. Rahim，Universiti Malaysia Perlis, Malaysia

Ms.Noor Hazlini bt Hj Borhan, Universiti Malaysia Sarawak, Malaysia

Dr. Mohd Ariffanan Bin Mohd Basri, Universiti Teknologi Malaysia, Malaysia

Dr. Azman Bujang Masli, Universiti Malaysia Sarawak, Malaysia

Hessam Jahani Fariman, Macquarie University, Australia

Hasan alyamani, Macquarie University, Australia

Mr. Majed Abdullah Alrowaily, Macquarie University, Australia

Dr. Stephanie Chua Hui Li, Universiti Malaysia Sarawak, Malaysia

Dr. Iwan Kelaiah, The University of New South Wales, Australia

Statement of Peer review

In submitting conference proceedings to MATEC Web of Conferences, I certify to the Publisher that I adhere to the **Policy on Publishing Integrity** of the journal in order to safeguard good scientific practice in publishing.

1. All articles have been subjected to peer review administered by the proceedings editors.
2. Reviews have been conducted by expert referees, who have been requested to provide unbiased and constructive comments aimed, whenever possible, at improving the work.
3. Proceedings editors have taken all reasonable steps to ensure the quality of the materials they publish and their decision to accept or reject a paper for publication has been based only on the merits of the work and the relevance to the journal.

Title, date and place of the conference

2016 8th International Conference on Computer and Automation Engineering (ICCAE 2016)

March 3-4, 2016

Melbourne, Australia

Proceedings editor(s):

Assoc. Prof. Manolya Kavakli, Macquarie University, Faculty of Science Department of Computing, Australia
Prof. Momoh -Jimoh Eyiomika Salami, International Islamic University Malaysia
Asst. Prof. Abbas Amini-University of Western Sydney, Australia
Dr. Mohd Ariffanan Bin Mohd Basri, Universiti Teknologi Malaysia, Malaysia
Dr. Azman Bujang Masli, Universiti Malaysia Sarawak, Malaysia
Dr. Stephanie Chua Hui Li, Universiti Malaysia Sarawak, Malaysia
Dr. Mayukha Pal, C R Rao Advanced Institute of Mathematics Statistics, and Computer Science, India

Date and editor's signature

MATEC Web of Conferences **56**, 01001 (2016) DOI: 10.1051/matecconf/20165601001
ICCAE 2016

Context Quantization based on Minimum Description Length and Hierarchical Clustering

Hui Chen , Jianhua Chen

Department of Electronic Engineering, Dianchi College, Kunming, China

Abstract. The code length of a source can be reduced effectively by using conditional probability distributions in a context model. However, the larger the size of the context model, the more difficult the estimation of the conditional probability distributions in the model by using the counting statistics from the source symbols. In order to deal with this problem, a hierarchical clustering based context quantization algorithm is used to combine the conditional probability distributions in the context model to minimize the description length. The simulation results show that it is a good method for quantizing the context model. Meanwhile, the initial cluster centers and the number of classes do not need to be determined in advance any more. Thus, it can greatly simplify the quantizer design for the context quantization problem.

1 Introduction

Entropy coding based on the probability distribution of an information source is used in the lossless compression of the source. The set of probability distributions of the current symbol which are conditioned on the past observations is called the context model. Clearly, it is the model that determines the rate at which we can encode the symbol sequence. According to the information theory [1], we know that the conditional entropy is less than or equals to the unconditional entropy and the more conditions may result in the lower conditional entropy, which is described by

$$H(Z \mid Z_1 Z_2) \leq H(Z \mid Z_1) \leq H(Z) \qquad (1)$$

The reduction of the entropy of a source, which is the lower bound of the average code length, will increase the possibility of the compression of the source. However, too large a modeling context spreads the counting statistics too thin among all possible modeling samples to achieve a good conditional probability estimate. This phenomenon is commonly called "context dilution" [2]. Context quantization is an approach to tackle this problem, where the total number of conditional probability distributions is reduced by combine similar distributions with some kind of clustering algorithm. In [3], the authors designed a vector quantization like context quantization algorithm which is called the minimum conditional entropy context quantization (MCECQ) algorithm and the conditional probability distributions are merged with K-means clustering algorithm. A similar method is proposed in [4] to which the optimization objective is the maximum mutual information (MMI) between the current symbol and the contexts. In these algorithms, K-means clustering is used to implement the context quantization. However, the number of classes and initial cluster centers should be set in advance and the algorithm is easy to be trapped into local optima. In [5], Forchhammer et. proposed the minimum adaptive code length context quantization (MCLCQ) algorithm which is implemented with the dynamic programming algorithm. The most important advantage of this algorithm is that it does not need the predetermined number of classes and the initial cluster centers. Meanwhile, the optimality is ensured by using the dynamic programming. Another way to implement this context quantization technique is to use the shortest path algorithm [6]. However, these two context quantization methods can only be applied to binary source coding.

In the light of the above observation, we want to find a clustering algorithm that does not need to know the optimal number of classes and the initial cluster centers in advance. In fact, the clustering algorithm is divided into partitional clustering algorithms and hierarchical clustering algorithms. K-means algorithm is the most well-known partitional clustering algorithm, where the K initial cluster centers can be determined by using the randomly selected K objects. An iterative procedure is used to assign the objects to their most appropriate classes according to a distance measure between an object and its cluster centers until a given criterion is met. The computational complexity of K-means is moderate and generally, the similarity among the classes is low.

The hierarchical clustering algorithms can be divided into agglomerative hierarchical clustering and split hierarchical clustering algorithms [7]. The most commonly used are the agglomerative hierarchical clustering algorithms. In this kind of algorithms, each object is treated as a class at first. And then, they are merged one by one, according to some distance measure, until all objects are merged into one class or the algorithm can be stopped by a given termination condition. The agglomerative hierarchical clustering is used in statistical classification with a large data sample set initially. In [8], it is used in the segmentation of image pixels and in [9] it is applied to the fingerprint recognition.

In this paper, a hierarchical clustering based context quantization algorithm is proposed. The optimal number of classes and the initial cluster centers do not need to be determined in advance and good context quantization results can still be obtained.

2 Hierarchical Clustering

It is known that the agglomerative hierarchical clustering algorithm is a popular method that is mostly used in the hierarchical clustering. The various similarity measures between the classes can be applied in practice, the four mostly used similarity measures are as follows:

(1)The minimum distance

$$d_{\min}(c_i, c_j) = \min_{p \in c_i, p' \in c_j} |p - p'| \quad (2)$$

(2)The maximum distance

$$d_{\max}(c_i, c_j) = \max_{p \in c_i, p' \in c_j} |p - p'| \quad (3)$$

(3) The average distance

$$d_{mean} = |m_i - m_j| \quad (4)$$

(4) The average weighted distance

$$d_{mg}(c_i, c_j) = \frac{1}{n_i n_j} \sum_{p \in c_i} \sum_{p' \in c_j} |p - p'| \quad (5)$$

Where $|p - p'|$ is the distance between the classes p and p'; m_i is the mean of class c_i, m_j is the mean of class c_j, n_i is the number of objects in class c_i, n_j is the number of objects in class c_j.

With a properly defined distance measure between two classes, the agglomerative hierarchical clustering algorithm which minimizes the given distance measure at each level can be implemented as follows:

- Step1. Initialize each object as a class, calculate the distance between each pair of objects.
- Step2. Merge the two classes into a new class with the minimum distance.
- Step3. Recalculate the distances among the new class and all other classes.
- Step4. If the remaining number of classes is not equal to one, go back to Step 2 for the next iteration.

The most important advantage of the hierarchical clustering is that the number of classes and the corresponding clusters are directly obtained at each level of the tree structured clustering procedure. However, the computational complexity of hierarchical clustering is higher than that of partitional clustering.

A basic hierarchical clustering algorithm will eventually combine all objects into one class. However, if the optimal number of classes is reached, we need to terminate the hierarchical clustering procedure with a reasonable criterion. The description length introduced by Rissanen in [10] can be used as the criterion. Since it reflects not only the complexity of a statistical model, but also the average code length of a source sequence coded based on this model.

3 Description Length

For an I-ary source sequence $x_1, \ldots, x_t$, $x_t \in \{1, 2, \ldots, I\}$,the symbols $x_{t-1}, x_{t-2}, \cdots, x_{t-k}$ $(t > k)$ (with $K = I^k$ possible combinations) observed before the time t are referred to as the context of x_t. Let c_n, $n \in \{1, 2, \ldots, K\}$ denote a context event of x_t. It is actually a specific combination of the context symbols $x_{t-1}, x_{t-2}, \cdots, x_{t-k}$. The conditional probability $P(x_t = i \mid c_n)$ can be estimated by

$$P(x_t = i \mid c_n) = \frac{N_i^t + 1}{N^t + I} \qquad N^t = \sum_{i=1}^{I} N_i^t \quad (6)$$

Where N_i^t denotes the number of those symbols, in the sequence $x_1, \ldots, x_t$, with context event c_n and taking on value i. Apparently, N^t is the sum of the counts in the vector $\{N_1^t, \ldots, N_I^t\}$ on which the estimate of the conditional probability distribution $P(X_t \mid c_n)$ is based.

According to [5], let L_n denote the description length of $P(X_t \mid c_n)$ which is estimated from the vector of counts $\{N_1^t, \ldots, N_I^t\}$. Then, L_n can be calculated by the following formula:

$$L_n = \sum_{s=1}^{N^t} \log(s + I - 1) - \sum_{i=1}^{I} \sum_{j=0}^{N_i^t - 1} \log(j + 1) \quad (7)$$

In fact, (7) can be calculated in the form of the factorial operation

$$L_n = \log \frac{(N^t + I - 1)!}{(I-1)! N_1^t ! N_2^t ! \ldots N_I^t !} \quad (8)$$

However, the computation of (8) is inefficient due to the high cost of the factorial calculation. Instead, the Striling formula (9) is used in our work to calculate the factorials in (8) such that the computational complexity can be alleviated.

$$\ln(x!) \approx (x+0.5)\ln x - x(2x\sin\frac{1}{x})^2 + \ln\sqrt{2\pi} \tag{9}$$

The total description length of the sequence $x_1,\ldots,x_t$ can then be calculated by adding together the description lengths for all possible context events: $L=\sum_{n=1}^{K}L_n$. From [10], the cost for transmitting the specific subsequence in the source sequence $x_1,\ldots,x_t$, in which all symbols have the same context event c_n can be represented equivalently by

$$L_n = N_t H(X\mid c_n) + \frac{I\text{-}1}{2}\log N_t \tag{10}$$

where the conditional entropy $H(X\mid c_n)$ is the average code length for coding each symbol x_t in the subsequence based on the conditional probability distribution $P(X_t\mid c_n)$.

From the above analysis, it is apparent that the cost for encoding a subsequence includes two parts, i.e., the cost for encoding the symbols in the subsequence: $N_t H(X\mid c_n)$ and the cost for encoding the conditional probability distribution $P(X_t\mid c_n)$: $\frac{I\text{-}1}{2}\log N_t$.

4 Context Quantization based on the Minimum Description Length

As stated in section II, a suitable similarity measure between classes should be defined to enable the hierarchical clustering. Here, we propose a description length based similarity measure to fulfill this requirement.

If L_m and L_n are the description lengths of two conditional probability distributions $P(X_t\mid c_m)$ and $P(X_t\mid c_n)$, let L_{mn} denote the description length of the merged probability distributions estimated from the counts vector obtained by adding together the two counts vectors for estimating $P(X_t\mid c_m)$ and $P(X_t\mid c_n)$, an increment of the description length can be observed when these two conditional probability distributions are merged:

$$\Delta L_{mn} = L_{mn} - (L_m + L_n) \tag{11}$$

In fact, this increment reflects the difference of the description length before and after the merging of the two probability distributions. According to (10), this increment can be represented by

$$\begin{aligned}\Delta L_{mn} &= N_t D[P(X_t\mid c_n)\mid P(X_t\mid c_{mn})]\\ &+ M_t D[P(X_t\mid c_m)\mid P(X_t\mid c_{mn})]\\ &- \frac{I\text{-}1}{2}\log\frac{N_t M_t}{N_t+M_t}\end{aligned} \tag{12}$$

where $D[\bullet\mid\bullet]$ is the relative entropy between two probability distributions, M_t is the number of symbols in the sequence $x_1,\ldots,x_t$ with the same context event c_m , $P(X_t\mid c_{mn})$ is the probability distribution estimated by merging the two vectors of counts on which $P(X_t\mid c_m)$ and $P(X_t\mid c_n)$ are estimated.

Some of the interesting properties of this increment can be summarized as follows: Firstly, it is symmetric since $\Delta L_{mn}=\Delta L_{nm}$. Secondly, it also reflects a kind of weighted average distance between $P(X_t\mid c_m)$ and $P(X_t\mid c_n)$ since $D[\bullet\mid\bullet]$ can be used as some kind of distance between two probability distributions.

These two characteristics of ΔL_{mn} means that it can be used as the similarity measure between two conditional probability distributions and can be applied in our hierarchical clustering algorithm to evaluate the similarity between two classes.

Thirdly, it may take a negative value. This feature means that in the context quantization procedure for a given source sequence, the optimal number of classes can be obtained by minimizing the total description length for the sequence since during the evaluation of the fitness of the merging of two probability distributions, a negative increment means the merging will reduce the total description length while a positive one will not.

In this work, the context quantization is actually implemented by merging the conditional probability distributions $P(X_t\mid c_n)$ (for all possible c_n) by the hierarchical clustering algorithm and every conditional probability distribution can be viewed as a class at the beginning of the clustering. Thus, we use the above mentioned increment ΔL_{mn} as the distance measure between two conditional probability distributions in the hierarchical clustering operations. During the clustering procedure, a probability distribution $P(X_t\mid c_n)$ is tested against all other probability distributions $P(X_t\mid c_m)$ for the possible merging and is finally merged with the one that can minimize ΔL_{mn}. In this way, the clustering operation will be stopped when no merging of two distributions can result in a negative ΔL_{mn} . The total description length of the remaining conditional probability distributions will reach the minimum at this time. These remaining classes actually characterize the final clustering results and the optimal number of classes is found accordingly.

The context quantization based on hierarchical clustering and the minimum description length is listed as follows:

- Step1. Initialize the conditions probability distributions by counting from the observed source sequence and treat each of them as an initial class.
- Step2. Calculate ΔL_{mn} for each pair of classes, if $\Delta L_{mn} < 0$, the two classes are viewed as a candidate pair of classes for merging at this level.
- Step3. Find out all candidate pairs, the one that minimize ΔL_{mn} , i.e. , $\Delta L_{m'n'} = \min_{m,n} \{\Delta L_{mn}\}$ (corresponding to $P(X_t \mid c_{m'})$ and $P(X_t \mid c_{n'})$) is merged, the number of classes is reduced by one.

Step4. If no candidate pair is found, stop the algorithm and output the quantization result. Otherwise, go back to Step 2 for the next iteration.

5 Simulation

In our simulations, the original 256 by 256 gray scale images with 8 bits per pixel are quantized to images of the same size but with 3 bits per pixel (8 gray levels) and we quantize 7 gray scale images (8 bits per pixel) into 8 gray level per pixel images as the testing source sequence. Using the quantized images as the source sequences is to simplify the simulation. Among the 7 quantized images, two images (Girl and Barb) are used as the training sequences to estimate the 2-order conditional probabilities $p(x_i \mid x_{i-1}x_{i-2})$, where $x_{i-1}x_{i-2}$ have 64 possible combinations.

Context quantization is implemented using the proposed algorithm to merge these 64 conditional probability distributions. After context quantization, we can obtain the mapping scheme in which each conditional probability distribution is mapped into its corresponding cluster. We use this mapping scheme to help the coding of the test images. Three images (Lena, Woman and Baby) are used as the test sequences and the conditional probability distributions in the quantized context model are applied to drive an arithmetic encoder to implement the compression. The resulting total numbers of bits in all test images are listed in Table I. For comparison, the coding results based on context quantization implemented by the K-means clustering algorithm (MCECQ) are listed in Table II.

Table 1. The Results For The Test Images By Hierarchical Quantization (bits)

Number of classes	Lena	Woman	Baby
27	117376	170259	185982

Table 2. The Results By K-Means Quantization On Test Images (bits)

Number of classes	Lena	Woman	Baby
5	164601	201254	234216
10	143754	216534	227311
27	118928	173287	195431
30	153864	216543	217512
40	175489	223876	223754
45	176236	230874	237095
50	183765	236892	248541

From Table I , The optimal number of classes of the context quantizer by the proposed algorithm is 27 based on the given training sequences. In Table II , various numbers of classes are tested and the related context quantizers are designed accordingly by using the MCECQ algorithm. It can be found that the best coding results for all test images are also obtained when the number of classes is set to 27 for the same training sequences. The coding results for the three test images are almost the same as those in Table I . However, due to the fact that initial cluster centers are randomly selected, the context quantizers are designed by multiple executions of the algorithm for each given number of classes and the one with the best coding results is chosen. For the proposed algorithm, only one execution is needed although its computational complexity is higher than that of the MCECQ algorithm. That means it takes a longer time for the MCECQ algorithm to design a good context quantizer than the proposed algorithm. Therefore, the good coding results indicate that the design objective is achieved by the proposed hierarchical clustering based context quantization algorithm.

6 Conclusion

Context quantization in high order entropy coding is applied more and more widely. With the employment of hierarchical clustering and the description length, the proposed context quantization algorithm successfully address the problems that the optimal number of classes and the initial cluster centers have to be given in advance, which are faced in the MCECQ algorithm. In addition, simulation results show that better coding results can be obtained with the context quantizer designed with the proposed algorithm.

Acknowledgment

This work is supported by a grant from the National Natural Science Foundation of China under Grant 61062005.

References

1. T. M. Cover, J. A. Thomas, Elements of Information Theory, Wiley, (2006)
2. M. J. Weinberger, J J. Rissanen, R. B. Arps, Applications of universal context modelling to lossless compression of gray-scale images, IEEE Transactions on Image Processing, 5 (4), (1996), 575-586.
3. J. Chen, Context modelling based on context quantization with application in wavelet image coding, IEEE Transactions on Image Processing, 13 (1), (2004), 26-32.

4. M. Cagnazzo, Mutual information-based context quantization, Signal Processing Image Communication, 25, (2010), 64-74.
5. S. Forchhammer, X.Wu, Optimal context quantization in lossless compression of image data sequences, IEEE Trans on Image Processing, 13(4) , (2004), 509-517.
6. S. Forchhammer, Context quantization by minimum adaptive code length, in: Proceedings of IEEE International Symposium on Information Theory, Nice, France, June (2007), pp.246-251.
7. A. K. Jain, R. C. Dubes, Algorithms for Clustering Data, Prentice Hall, (1998).
8. X. Hu, The image segmentation based on hierarchical clustering of LS-SVM, Computer and Digital Engineering, 38 (1), 2010, 143-147.
9. Anil Jain, Lin Hong and Ruud Bolle, On-Line Fingerprint Verification, IEEE Trans on Pattern Analysis and Machine Intelligence, 19 (4), 1997, 302-313.
10. J. Rissannen, Universal coding, information, prediction and estimation, IEEE Trans. Inform. Theory, 30, 1984, 629-636.

MATEC Web of Conferences 56, 01002 (2016) DOI: 10.1051/matecconf/20165601002
ICCAE 2016

Efficient Multi-keyword Ranked Search over Outsourced Cloud Data based on Homomorphic Encryption

Mengxi Nie[1,2], Peng Ran[1] and HaoMiao Yang[1,2]

[1]*University of Electronic Science and Technology of China, 611731 Chengdu, China*
[2]*Science and Technology on Communication Security Laboratory, 610041 Chengdu, China*

Abstract. With the development of cloud computing, more and more data owners are motivated to outsource their data to the cloud server for great flexibility and less saving expenditure. Because the security of outsourced data must be guaranteed, some encryption methods should be used which obsoletes traditional data utilization based on plaintext, e.g. keyword search. To solve the search of encrypted data, some schemes were proposed to solve the search of encrypted data, e.g. top-k single or multiple keywords retrieval. However, the efficiency of these proposed schemes is not high enough to be impractical in the cloud computing. In this paper, we propose a new scheme based on homomorphic encryption to solve this challenging problem of privacy-preserving efficient multi-keyword ranked search over outsourced cloud data. In our scheme, the inner product is adopted to measure the relevance scores and the technique of relevance feedback is used to reflect the search preference of the data users. Security analysis shows that the proposed scheme can meet strict privacy requirements for such a secure cloud data utilization system. Performance evaluation demonstrates that the proposed scheme can achieve low overhead on both computation and communication.

1 Introduction

Cloud computing, a revolution of information technology, can significantly change people's life in the foreseeable future. However, the problem of cloud security hasn't been solved well in practical applications. As we know, cloud computing is based on demand. So how to store data in the cloud is non-transparent for most of data users, which will cause users to worry about the data stored in the cloud be misused or leaked. Though some traditional methods can be used to guarantee the security of cloud, such as authentication, authorization, access control and so on, all of these solutions are based on trusted cloud service providers.

Unfortunately, cloud service providers are not as honest as we wish. A prism door event in 2013 was a typical case that the providers' activities threatened the security of the cloud. According to the report, nine IT giants including Google, Microsoft and Apple open their servers to the United State Security Agency, which makes monitoring worldwide e-mails, calls, and stored dates easily. Because the data are stored on the invisible "cloud" supplied by IT giants, it will undoubtedly give the national security and social stability of one's country an enormous threat.

To ensure privacy, users usually encrypt the data before outsourcing it onto the cloud. However, data encryption makes effective data utilization a very challenging task. It restricts user's ability to search the keywords and fails to satisfy further demands of keyword privacy. Due to these reasons, searchable encryption schemes appear.

Although a series of traditional searchable encryption schemes have been proposed to enable search on ciphertexts, there are still many problems remaining to be solved, e.g. low efficiency, high communication and computing expenditure. In order to address these issues, Yu et al. [1] recently proposed a two-round searchable encryption (TRSE) scheme using vector space model and homomorphic encryption. However, in this scheme, data owners have to encrypt each dimension of the searchable vector which lowers the efficiency. For example, for a vector of dimension n, the data owner will take n encryption operations and output n ciphertexts. In addition, in TRSE, the query vector is generated where each dimension is equal to 0 or 1, which doesn't relate to the search weight of a term.

To achieve low overhead, we propose an efficient multi-keyword ranked searchable scheme (EMRS) over outsourced cloud data based on homomorphic encryption that can pack a vector into only one ciphertext, and thus alleviate the burden of computation and communication. Our contributions can be summarized as follows:

- We adopt a new homomorphic encryption scheme which supports the packing method. Therefore, we can encrypt a vector as a whole in a batch manner. As a result, for a vector of dimension n, we only take one encryption operation, and thus reduce the computation and communication expenditure.

- We put the search weight of a term into the query vector, i.e., we use values instead of 0 or 1 to fill the query vector. Using this method, the query vector can reflect the data user's search preference.

The rest of this paper is organized as follows: We give the scenario and related background in Section 2. In Section 3, we give the main techniques used in our scheme. In Section4, we present the detailed description of the proposed searchable encryption scheme. Section 5 gives the security analysis of our scheme. Related performance evaluations for both searchable encryption and secure result ranking are discussed in Section 6. Finally, Section 7 gives the concluding remark of the paper.

2 Problem statements

2.1 Scenario

An encrypted cloud data hosting service involves three different entities as illustrated in Fig. 1: the data owner, the data user, and the cloud server.

The cloud server takes the responsibility for holding the third-party data storage and retrieve services. Since there will be some sensitive information in the data and the server cannot be fully entrusted in protecting data, the data must be encrypted before outsourcing. And the leakage of information is unacceptable.

The data owners are group of people who have a collection of n files $C = \{f_1, f_2, \dots, f_n\}$ to outsource onto the cloud in encrypted form while still keeping the capability to search through them for effective data utilization reasons. To achieve this, the data owner has to build a searchable index I from a set of l distinct keywords $W = \{w_1, w_2, \dots, w_l\}$ extracted from C, and then outsource the index I in encrypted form I' and encrypted files onto the cloud server.

The authorized data user processes multi-keyword retrieval over the outsourced data. Firstly, the data user should generate a query REQ. Secondly, the user encrypts the query and sends it to the cloud server. Finally, the data user gets the relevant files from the cloud server, decrypts it, and then makes use of these files. Note that because of the limitation of user's computing power, operations on the user side should be simplified.

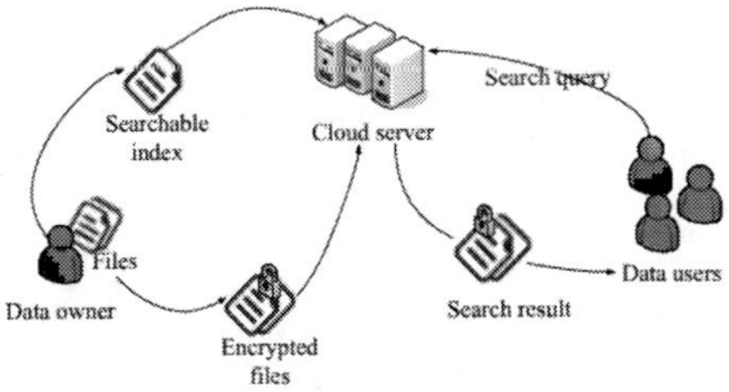

Figure 1. Scenario of retrieval of encrypted cloud data.

2.2 Design Goals

As analyzed above, to realize our efficient and privacy-preserving multi-keyword ranked search over outsourced encrypted cloud data, the following goals should be achieved:

- Confidentiality: The cloud server should not learn plaintexts of data files, index, and searched keywords.
- Unlinkability of trapdoors: The cloud server should not learn whether two searches are performed for the same keywords.
- Preference search: In multi-keyword ranked search, multiple keywords that a data user input are treated as the different weights to reflect the data user' preference.
- Low cost of computation and communication.

3 Preliminaries

3.1 Inner Product Similarity

In our scheme, we use the vector space model to score a file on multi-keywords. A file can be represented as vector $\vec{v_f}$ and each dimension of the vector stands for a separate term, i.e., if this file contains this term, the corresponding value is nonzero, otherwise is zero. Meanwhile, a query can be represented as a vector $\vec{q}$ and each dimension of the vector is padded based on weight. Then we have $score_{f,q} = \vec{v_f} \cdot \vec{q}$. As a result, files can be ranked in order by the scores and the most relevant files can be picked out.

3.2 Relevance Scoring

Here, we adopt the most widely used tf-idf weighting, which involves two attributes: Term Frequency and Inverse document frequency. Term frequency is simply the number of times a given term or keyword appears within a file, and Inverse document frequency is obtained by dividing the number of files in the whole collection by the number of files containing the term.

3.3 Relevance Feedback

Using a Boolean vector to create a query is very imprecise. We use the technique of relevance feedback to add some features to the search query to reflect the search preference of data user [2]-[4].

3.4 BGV Homomorphic Encryption

In this section, we describe a scheme proposed by Brakerski, Gentry, and Vaikuntanathan (BGV) [5], which is regarded as a stripped down vision of FHE scheme and can meet our demand. In our scheme, we use one version of the BGV scheme which deals with integer polynomials and the security of this cryptosystem is linked with the hardness of decisional R-LWE (Ring-Learning with Errors) problem [6]. The BGV scheme is described as follows:

(1)BGV.ParamGen(λ).

The message space is the ring $R_p := \mathbb{Z}_p[X]/\Phi_m(X)$, where $\Phi_m(X)$ is the mth cyclotomic polynomial in $\mathbb{Z}_p[X]$, of degree $n = \varphi(m)$. The ciphertext space is described as follows: pick a large q, which is co-prime to p, and define $R_p := \mathbb{Z}_p[X]/\Phi_m(X)$. The algorithm defines the following distributions:

$D_{\mathbb{Z}^n,\sigma}$: The discrete Gaussian with parameter σ: it is the random variable over $\mathbb{Z}^n$ obtained from sampling x ∈ $\mathbb{Z}^n$ with probability $e^{-\pi\|x\|/\sigma^2}$ and then rounding at the nearest lattice point.

ZO_n : Sample $x = (x_1, \dots, x_n)$ with $x_i \in \{-1, 0, +1\}$ and Pr[xi = −1] = 1/4; Pr[xi = 1] = 1/4; Pr[xi = 0] = 1/2.

In the following, we assume that the parameters generated here are inputs of any subsequent algorithm.

(2)BGV.KeyGen( )→(*pk*,*dk*).

We obtain a public key encryption scheme, so we need the ability to generate the secret key and the public key. Sample $a \leftarrow R_q$, and s,e $\leftarrow D_{\mathbb{Z}^n,\sigma}$. Let $b \leftarrow as + pe$. We set dk$\leftarrow s$ and $pk \leftarrow (a, b)$.

(3)BGV.Enc$_{pk}$(*m*)→*c*.

Given $m \in R_p$, and sample $u \leftarrow ZO_n$, and $v, w \leftarrow D_{\mathbb{Z}^n,\sigma}$. The output is $c = (c_0, c_1)$, where $c_0 \leftarrow bv + pw + m$ and $c_1 \leftarrow av + pu$.

(4)BGV.Dec$_{dk}$(*c*)→*m*.

Compute $t \in R_q$ as $t \leftarrow c_0 - c_1 s$. The output is then $m = t \bmod p$.

3.5 Smart-Vercauteren ciphertext packing (batch) techniques

One appealing optimization that applies to the protocol in this paper is to use "batch homomorphic encryption" where a single ciphertext represents a vector of encrypted values and a single homomorphic operation on two such ciphertexts applies the homomorphic operation componentwise to the entire vector. In this way, at the cost of a single homomorphic operation, we get to compute on an entire vector of encrypted plaintexts. This is a cryptographic analogue of the Single Instruction Multiple Data (SIMD) architecture and is supported by recent fully homomorphic encryption systems [7].

4 Scheme Design

4.1 Secure Computation of Inner Product

Using homomorphic encryption based on RLWE in our encryption model makes it efficient and secure to calculate the inner production of two vectors. The basic processes are described as follows:

(1) The data user uses the techniques of homomorphic encryption and packing to encrypt all coordinates into a ciphertext. The plaintext space $R_2 = \mathbb{Z}_2[x]/(x^n + 1)$ can be decomposed into *l* plaintext slots: $\mathbb{Z}_2[x]/(f_1(x)), \dots, \mathbb{Z}_2[x]/(f_l(x))$. $x^n + 1 = f_1(x) \cdots f_l(x) \bmod 2$, and each $f_i(x)$, $i = 1, \dots, l$,has the same degree *d*=*n*/*l* only if the system could choose the parameters *n* and *l* properly. Suppose that the data user *A* has a vector $V = (V_1, \dots, V_l) \in \mathbb{Z}^l$ and each dimension of *V* can be encoded into $V_i(x)$, $i = 1, \dots, l$. Then $V_i(x)$, $i = 1, \dots, l$, will be packed into $m_v(x)$ through the Chinese Remainder Theorem (CRT), and an encrypted operation is invoked:$BGV.Enc_{pk}(m_v(x)) \rightarrow c_v(x)$. Similarly, the data user *B* can encrypt another vector *W* into $c_w(x)$.

(2) Through calling one operation of homomorphic multiplication $c(x) = c_v(x) \cdot c_w(x)$, the service provider can compute $C_i(x) = V_i(x) \cdot W_i(x)$, i=1,…,l, parallelly, which is similar to using SIMD.

(3)We call Frobenius isomorphism mapping for *l*-1 times and *l* times of homomorphic addition to deal with *c*(*x*). In details, by using the mapping ϕ to $c^{(i)}(x)$:

$$\phi: x \rightarrow x^2$$

$c^{(i)}(x)$ becomes $c^{(i+1)}(x)$, *i*=0,…,*l*-2, and the corresp-onding plaintext is cyclic shifted for one shot. Then, through calling the operation of homomorphic addition, we can get $c_{inner} = \sum_{i=1}^{l-1} c^{(i)}(x)$, where $c^{(0)}(x)$ is the original $c(x)$.

(4) Finally, by calling $BGV.Dec_{dk}(c_{inner})$ for decrypting, we can get the plaintext of the inner product $m(x) = \sum_{i=1}^{l} V_i(x) \cdot W_i(x)$.

4.2 Relevance Feedback

In our scheme, we use the technique of relevance feedback [2]-[4] instead of a binary vector to create a query aiming at improving the accuracy. Relevance feedback is a process of automatically adjusting an existing query using information feedback by the search entity about the preference of previously retrieved documents. By using the relevance feedback technique, we can optimize the query vector by replacing each dimension with the search weight of a term without change any phase of previous scheme. One of the methods is Rocchio [8] and the main idea is abstracted as follows:

$$Q_{opt} = Q + \beta\left(\frac{1}{N_R}\sum_{j \in U_R} P_j\right) - \gamma\frac{1}{N_N}\sum_{j \in U_N} P_j,$$

Where N_R is the number of relevant data users, U_R is the returned result set and N_N is non-relevant ones U_N; P_j is feature vector of data user U_j; β and γ are suitable constants [3][4].The search entity can define its preference weights according to the feature vectors retrieved in advance in method. The optimal query can be generated by putting more weights on the relevant features and less weights on the non-relevant features.

Therefore, the cloud server should not only send the identities of the most matching data users, but also return the feature vectors of them in order to define the preference weight using the historical information.

4.3 Privacy-Preserving and Efficient

The framework of EMRS consists of two phases: *Initialization* and *Retrieval*.

The steps of *Initialization* are described as follows:

(1) The data owner calls ***BGV.keygen(λ)*** to get *pk* and *dk*, then assigns *dk* to the authorized data users.

(2) The data owner extracts the collections of l keywords, $W = \{w_1, w_2, \dots, w_l\}$ from n files, $C = \{f_1, f_2, \dots, f_n\}$ and compute their TF and IDF values. Then, the data owner builds a (l+1)-dimensional vector $v_i = \{id_i, t_{i,1}, \dots, t_{i,j}\}$, where $t_{i,j}$ = $tf\text{-}idf_{i,j}$, for each f_i of C. The searchable index $I = \{v_i | 1 \le i \le n\}$.

(3) The owner uses the smart-vercauteren ciphertext packing technique to pack the index I and then calls $BGV.Enc_{pk}(I)$ to encrypt index I into $I' = \{v_i' | 1 \le i \le n\}$.

(4) The owner encrypts C into C' with cryptology schemes and outsources C' and I' to the cloud server.

The *Retrieval* phase includes the following steps:

(1) The data user generates a set of query keywords $REQ = \{w_1', w_2', \dots, w_s'\}$, and then the query vector $T_w = \{m_1, \dots, m_l\}$ is built in which m_i is the weight of a query keyword belongs to REQ; Othewise, $m_i = 0, 1 \le j \le l$. Then, the user packs the T_w and encrypts it into a trapdoor T_w' and sends T_w' to the server.

(2) For $v_i'(1 \le i \le n)$ in I', the cloud server evaluates $s_i' = v_i'[1{:}l] \cdot T_w$ homomorphically on ciphertexts and returns $N' = \{(id_1', s_1'), (id_2', s_2'), \dots, (id_n', s_n')$ } to the user.

(3) The user decrypts N' into $N' = \{(id_1, s_1), \dots, (id_n, s_n)\}$, where $s_i = BGV.Dec_{dk}(s_i')$. Note s_i are the relevance scores. Then the data user chooses and sends top-k highest-scoring files' identifies $\{i_1, \dots, i_k\}$ to the server.

(4) The server returns the encrypted k files $\{f_{i1}, f_{i2}, \dots, f_{ik}\}$ to the user.

5 Security Analysis

5.1 Confidentiality

Due to the security strength of the file encryption scheme, the file content is clearly well protected. Thus, we only need to focus on the index and the searched keywords. Because the index on the data owner side and the trapdoor on the data user side are encrypted by homomorphic encryption scheme, the privacy of the index and the trapdoor are reduced to the security of the BGV encryption and the packing technique. As showed in [9], the security of BGV and packing is guaranteed by their algorithms, the cloud server can detect nothing about the original data. Namely, the confidentiality will be guaranteed if the secret key dk can be kept properly. It is impossible that the cloud server gets the index vector and queried vector of original files, then infers the TF or ICF of the data files through the secure index vector, secure queried vector and the similarity values. Hence from the analysis, the multikeyword secure ranking algorithm protects the privacy of index, query, and the relevance scores very well.

Additionally, the BGV encryption builds its security on the RLWE hardness problem, the confidentiality in our scheme has a higher level of security than the schemes based on integer[6], e.g. the scheme proposed by Yu. et al [1].

5.2 Unlinkability of trapdoor

The queried vector is encrypted every time before search. The public key *pk* is randomly generated in every encryption, so the searches are independent, i.e., have no relevance. The cloud server cannot mining association rules of the two same secure queried vectors or infer the frequency of keywords in different secure queried vectors through quantities of queried vectors it has received. Namely, in terms of search pattern, the same keywords in different queries, as well as the different keywords in same queries are encrypted independently, e.g., if the same keyword w is requested in different queries, REQ_1 and REQ_2, then the w is encrypted into different ciphertexts by calling $BGV.Enc_{pk}()$. As a result, the unlinkability of trapdoor can be achieved. The cloud server cannot mine association rules of the two same secure queried vectors or infer the frequency of keywords in different secure queried vectors through quantities of queried vectors it has received. As a result, the keywords that the data users search can be protected very well.

5.3 Preference search

In multi-keyword ranked search, multiple keywords that a data user input are treated as the different weights to reflect the data user' preference. The vector is encrypted by the secure BGV encryption scheme and its security can refer to Ref. [6]. As a result, the query vector can reflect the data user's preference and its privacy can be protected very well.

6 Performance Evaluation

In this section, we evaluate the performance of the proposed scheme in terms of computation and communication overheads. We denote an addition in G_A, a multiplication in G_M, an encryption in ***Enc***, and an isomorphic mapping in G_I. Here, the other operations are neglected, as the cost is very small. Meanwhile, we suppose that the data owner extracts the collections of *l* keywords.

6.1 Time for encrypting the index

In *Initialization* Phase, the data owner needs to encrypt the searchable index *I* into *I'*. The technique of packing makes our scheme has low computation and communication overheads. As shown in Table 1, by using the packing technique, the data owner is able to encrypt one vector for just one time which increases the efficiency.

Table 1. Time for encrypting the index

	TRSE	EMRS
Computing Overhead	The cloud server: $(l-1)G_A+lG_M$ Data User: l *Enc*	The cloud server: $1G_M + lG_A + l\,G_I$ Data User: 1*Enc*
Communication Overhead	Number of ciphertexts: l	Number of ciphertexts: 1

6.2 Performance of the Retrieval Phase

The *Retrieval* phase consists of generating trapdoor, calculating inner production and ranking. We compare the efficiency of this phase between our approach and the approach from TRSE [1].The result is shown in Table 2.

Table 2. Performance of the Retrieval Phase

Scheme	Time
TRSE	$(l+1)$ *Enc*
EMRS	1*Enc*

Due to the small cost of isomorphic mapping, we can neglect it. Through comparison, we can conclude that our scheme has high efficiency and low computing cost. Besides, the number of ciphertexts in our scheme is extremely small which makes the communication overhead very low.

7 Conclusion

In this paper, we propose an efficient and privacy-preserving multi-keyword searchable scheme over outsourced cloud data based on homomorphic encryption. We use "inner product similarity" to quantitatively evaluate such similarity measure. For reducing the burden of communication and computation, we choose the efficient measure of encryption and inner product calculation which can help us packing the message before dealing with them. Moreover, we put the search weight of a term into the query vector which can reflect the relevance more precisely but not only existence.

Thorough analysis investigating privacy and efficiency guarantees of proposed schemes is given, and experiments on the real-world data set show our proposed schemes introduce low overhead on both computation and communication.

Acknowledgement

This work is supported by the National Natural Science Foundation of China under Grants U1233108 and U1333127, the International Science and Technology Cooperation and Exchange Program of Sichuan Province, China under Grant 2015HH0040, and China Postdoctoral Science Foundation funded project under Grant 2014M562309.

References

1. J.D Yu,P Lu,Y.M Zhu,G.T Xue and M.L Li ,Toward Secure Mulitikeyword Top-k Retrieval over Encrypted Cloud Data,,*IEEE TRANSACTIONS ON DEPENDABLE AND SECURE COMPUTING*, VOL.**10**,NO.**4**,7/8 (2013)
2. G.Salton,M.J.McGill,Introduction to modern information retrieval,McGraw-Hill NewYork(1983)
3. W.M.Shaw Jr,Term-relevance computations and prefect retrieval performance,*Information Processing&Management* 31 (**4**) (1995) 491-498
4. C.Buckley,G.Salton,Optimization of relevance feedback weights,in: Proceedings of the 18th annual international ACM SIGIR conference on Research and development in information retrie-val,351-357(1995)
5. Z. Brakerski, C. Gentry, and V. Vaikuntanathan. (leveled) fully homomorphic encryption without bootstrapping.In Goldwasser [**32**], pages 309-325.
6. V. Lyubashevsky, C. Peikert, and O. Regev. On ideal lattices and learning with errors over rings. In EUROCRYPT, pages 1–23, (2010)
7. N. P. Smart and F. Vercauteren. Fully homomorphic SIMD operations. Des. Codes Cryptogra-phy, 71(**1**):57/81(2014) .Early verion at http://eprint.iacr.org/2011/133.
8. J.J.Rocchio,Relevance feedback in information retrieval, Prentice-Hall,Englewood Cliffs NJ(1971).
9. Zvika Brakerski and Vinod Vaikuntanathan. Fully homomorphic encryption from ring-LWE and security for key dependent messages. Manuscript, to appear in CRYPTO (2011).

MATEC Web of Conferences 56, 01003 (2016) DOI: 10.1051/matecconf/20165601003
ICCAE 2016

Spectral Clustering based Active Learning with Applications to Text Classification

Wenbo Guo, Chun Zhong ,Yupu Yang

Department of Automation, Shanghai Jiao Tong University, Shanghai, China

Abstract. Active learning is a kind of machine learning algorithms that spontaneously choose data samples from which they will learn. It has been widely used in many data mining fields such as text classification, in which large amounts of unlabelled data samples are available, but labels are hard to get. In this paper, an improved active learning algorithm is proposed, which takes advantages of the distribution feature of the datasets to reduce the labelling cost and increase the accuracy. Before the active learning process, spectral clustering algorithm is applied to divide the datasets into two categories, and instances located at the boundary of two categories are labelled to train the initial classifier. In order to reduce the calculation cost, an incremental method is added in the present algorithm. The algorithm is applied to several text classification problems. The results show it is more effective and more accurate than the traditional active learning algorithm.

1 Introduction

In the last few years, active learning [1] has become more and more popular because of its effectiveness, especially when dealing with the kind of learning tasks where class labels of each data sample are difficult to get and unlabeled data are sufficient or easy to collect. By applying active learning algorithms, the most informative samples are selected in order to learning the correct classifier with less labeled data samples. Another advantage of active learning is that by selecting the most informative data sample, the iteration times of the learning process can be reduced, thus the training time can be reduced. Also active learning introduces a select engine to select data samples, which makes active learning has a better generation performance, due to the violation of the assumption that training set and the test set are identically distributed.

In most situations of pool-based active learning scenarios, active learning deals with a small set of labeled data L and a large pool of unlabeled data U. With the use of few labeled instances, an initial model can be trained. Then in order to keep improving the model, the active learner selects a few unlabeled instances during each consecutive iteration which contain most information in term of the current model. The selected instances are used to update the classification model. This active learning process stops when some certain criterion is met. In this way, unnecessary and redundant samples are less likely to be included in the training set, thus greatly reducing the labeling cost and potentially the computational cost.

Active learning is an iteration process [2]. Each time a new instance is selected, the classification model will be updated until the stopping criterion is met. This process needs large amount of calculation. It is possible that after the first several iterations, the result changes little with the conduction of following iterations. What's more, because the instances are selected according to the current model, it happens that with the conduct of active learning, the selected instances may become useless, which increases the labeling cost. For example, if the current hyper-plane lies far away from the optimal one, the instances selected according to the current model will be useless for updating of the model and getting the correct hyper-plane. This cost dues ignoring the distribution feature of the training data.

Incremental learning method is applied, in this paper. When an instance (x, y) is labeled, the new classification model is updated based on only the new labeled instance (x, y), instead of training the former classification model. In order to cut down the labeling cost and make use of the information about the data distribution feature, a spectral clustering based active learning algorithm is proposed. Before the start of active learning, the whole datasets is clustered into two categories, and the instances located on the border of the two categories are picked to be the initial support vectors, and during the learning process, the points closest to the hyper plane will be chosen to be the new instance of the training set. The effect of this algorithm is show in the results of applying it to several text classification problems.

2 Active learning based on spectral clustering

2.1 Active learning with support vector machine

Active learning refers to machine learning algorithms, which autonomously select data samples. Thus any passive supervised learning method can be applied to train classification model. This paper focus on support vector machine (SVM), which has the only optimal solution with high accuracy. SVM learns a linear classifier, typical in a kernel-based feature space, which can be used to estimate the informativeness of data samples. This property make it perfectly suitable for active learning, because the data sample that is closest to the boundary is the most informative sample which will be labelled for the next iteration of learning.

2.1.1 Support vector machine

Support vector machine aims to learning a separating hyper plane classifier based on the idea of maximum margin. A SVM algorithm for binary classification has the input of N training data samples $\{(x_1, y_1),(x_2, y_2),\ldots,(x_N, y_N)\}$, where $x_i \in R^n$ and $y_i \in \{\pm 1\}$. The classification decision function of SVM is as followed.

$$f(x)=sign(w^T \cdot x+b) \tag{1}$$

where w and b determine the classification hyper plane.

As to linear support vector machine, the decision function can be obtain by solving the following optimization problem.

$$\min_{w,b,\xi_i} \frac{1}{2}\|w\| + C\sum_{i=1}^{\ell} \xi_i \tag{2}$$

$$s.t. \quad y_i(w \cdot \Phi(x)+b) \geq 1-\xi_i, \ \forall i \tag{3}$$

$$\xi_i \geq 0, \ \forall i \tag{4}$$

where $\Phi(\cdot)$ is a mapping from input space to feature space, $C > 0$ is penalty coefficient, which controls the penalty degree of misclassification and ξ_i is relaxation coefficient. The aim of the introduction of these coefficients is to deal with the problem, whose training set is linear inseparable.

In practice, this problem is a convex quadratic problem, which can be solved by optimizing its dual problem according to the Lagrange duality.

$$\max_{\alpha} \sum_{i=1}^{N} \alpha_i - \frac{1}{2}\sum_{i,j} \alpha_i \alpha_j y_i y_j \Phi(x_i)\Phi(x_j) \tag{5}$$

$$s.t. \quad 0 \leq \alpha_i \leq C, \ \forall i \tag{6}$$

$$\sum_{i=1}^{N} \alpha_i y_i = 0 \tag{7}$$

where α_i is Lagrange multiplier, and C is a constant chosen by user.

In order to solve the non-linear classification problem, the kernel trick is introduced to linear support vector machine. The widely used inner product as the mapping from input space to feature space is replaced by kernel function in non-linear support vector machine. The kernel function chosen by this paper is Gaussian kernel [3].

$$K(x_i, x_j) = e^{-\gamma\|x_i - x_j\|^2} \tag{8}$$

where x_i and x_j are two data samples and γ is the parameter of kernel function.

The classification decision function can be written by:

$$f(x) = sign(\sum_{i=1}^{N} \alpha_i y_i K(x_i, x) + b) \tag{9}$$

This decision function can be obtained by changing the $\Phi(x_i)\ \Phi(x_j)$ in equation (5) with equation (8) and solving the dual problem (5) ~ (7). Thus active learning can use this non-linear SVM model as classification model and the follow function can be used to select the most informative data sample.

$$x = \arg\min_{x \in U} \left| \sum_{i=1}^{N} \alpha_i y_i K(x_i, x) + b \right| \tag{10}$$

where U is unlabeled datasets.

2.2 Spectral clustering

Spectral clustering is a cluster algorithm based on graph theory. Datasets are classified into two categories based on the similarities among samples. The graph is $G = (V, E)$, where the V denotes the data samples, the E represents the similarities among two samples. Then clustering problem is transformed into a graph cut problem. To solve this problem means finding a method who can cut graph G into two categories, by maximizing the value of E inside the categories and minimizing the value of E between two categories.

The spectral clustering algorithm used in this paper is proposed in Table 1:

Table 1: Spectral clustering algorithm

Algorithm 1: Spectral clustering algorithm [3]
Input: Unlabeled datasets $X=\{x_i\}$.
Output: Label vector Y, where $y_i \in \{\pm 1\}$.
Step 1: Build the graph $G=(V, E)$, and calculate adjacent matrix W, where $W_{ij} = K_{ij}(x_i, x_j)$.
Step 2: Calculate the degree of each point, $$d_i = \sum_j W_{ij} \tag{11}$$
Step 3: Calculate diagonal matrix D, $$D_{ij} = \begin{cases} 0, i \neq j \\ d_i, i = j \end{cases} \tag{12}$$
Step 4: Calculate the Laplace matrix of G: $$L_{sym} = I - D^{-1/2} W D^{1/2} \tag{13}$$
Step 5: Apply eigenvalue decomposition to L_{sym} and the second less feature of L_{sym} can be obtained, written as v_2.
Step 6: $\hat{y} = sgn(v_2)$, and return $\hat{y}$.

2.3 Spectral clustering based active learning

Suppose a classification task on a datasets containing samples pertaining to two categories, In order to explore its distribution feature, reduce redundancy and increase the accuracy, the spectral clustering is applied on the datasets.

Prior to active learning, the spectral clustering method is applied to the datasets U, then the data points located at the border of the two categories are chosen to be the initial support vectors. During the learning process, the data points near the hyper plane are selected to be labeled and applied to the incremental learning. The final classification model can be obtained after consecutive iterations.

Details of the algorithm is proposed in Table 2.

Table 2: Active learning with spectral clustering

Algorithm 2: Active learning with spectral clustering

Input: Unlabeled datasets U, the number of initial labeled data instance k,

Output: An SVM classifier

Step 1: Divide U into two categories with the use of spectral clustering algorithm, where V_i is the index of which category X_i belongs to, if $V_i<0$, X_i belongs to the first category, if $V_i>0$, X_i be- longs to the second category, if $|V_i|\to 0$, X_i is located at the border of two categories.

Step 2: Choose k data points which have the least value of V for labeling, save them as datasets L.

Step 3: Learn a classification model $f^{(0)}$ from the labeled datasets L.

Step 4: Set t=0.

Step 5: Actively select the unlabeled instance X^* which is located nearest to the hyper plane, delete X^* from U and let:

$$L = L\cup X^* \quad (14)$$

Step 6: Learn the new classification model $f^{(t+1)}$ from the former classifier $f^{(t)}$ and new labeled data X^*, that is:

$$f^{(t)} + X^* \xrightarrow{S.V.M} f^{(t+1)} \quad (15)$$

Step 7: $t = t+1$.

Step 8: Repeat step 5 ~ step 8 until the stopping criterion is met.

Step 9: Return final classification model $f^{(t)}$.

3 Application to text classification

In this paper, we apply this method to text classification problem, especially the widely used datasets: news20 and w2a. These two datasets come from the LIBSVM [4]. By comparing the result of Algorithm 2 with traditional active learning method, the effectiveness of it can be shown. The news20 contains 20,000 instances, each representing an item of news written in English. There are 20 different kinds of news. The w2a datasets has 3470 instances in training set and 46279 instances in testing set.

Linear support vector machine was applied to news20, for the instances in this datasets are linear separable. A single SVM can deal with binary classification task. SVM with Gaussian kernel was employed to be the classifier for w2a.In this experiment, γ is set to be 0.01 and the other parameter of SVM is set as C=100.

The traditional active learning support vector machines (ALSVM) are used for comparison. They are applied on the two datasets respectively. The dimension of the datasets is fixed during the active learning process.

Each experiment is repeated 100 times, and each time initialized with 5 selected instances. Both the traditional ALSVMs and proposed method use the same set of instances for initialization. In each active learning iteration, the data samples nearest to the current separating hyper plane are selected according to equation (10).

Average classification error rates of the datasets news20 with respect to the number of iterations are shown in the following two figures. In each figure, the blue line represents the traditional linear ALSVMs, the red line means the proposed algorithm. The results of iteration range from 20 to 51 times are shown, in which the difference of two method can be clearly distinguished.

Figure 1 shows that correct rates of proposed algorithm always beat the traditional ALSVMs. When it comes to the iteration 51, error rates of proposed algorithm are relatively 33% lower than the traditional ALSVMs

Figure 2 exhibits that error rates of proposed algorithm is lower than the traditional ALSVMs at the beginning, and as the process goes on, the traditional active learning method reaches almost the same correct rates with the proposed method.

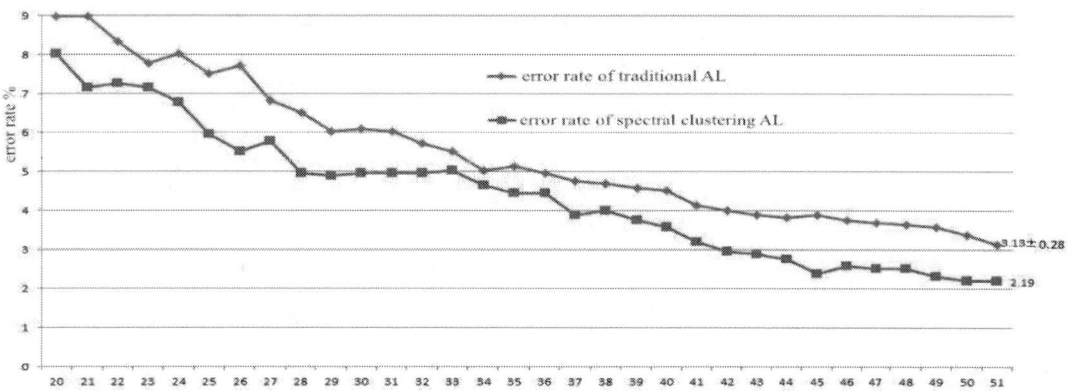

Figure 1. Classification error rates of No.11 and No.12 news in news20.

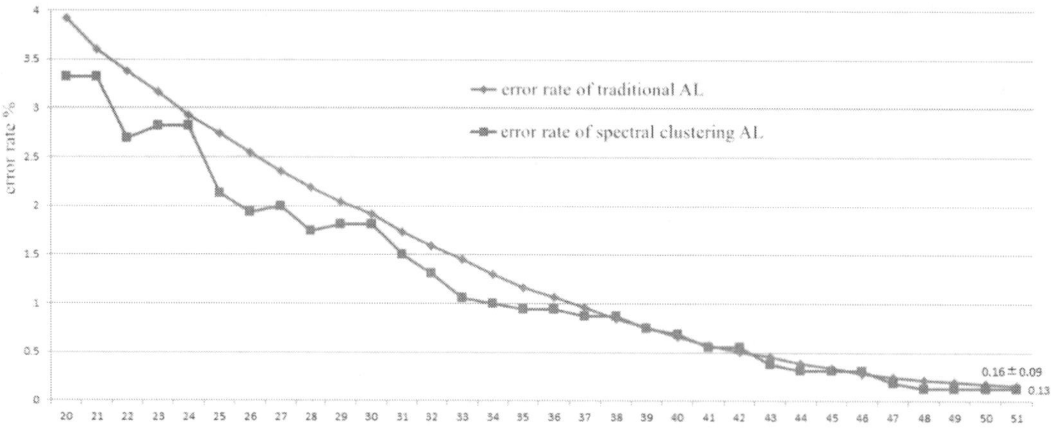

Figure 2. Classification error rates of No.7 and No.15 news in news20.

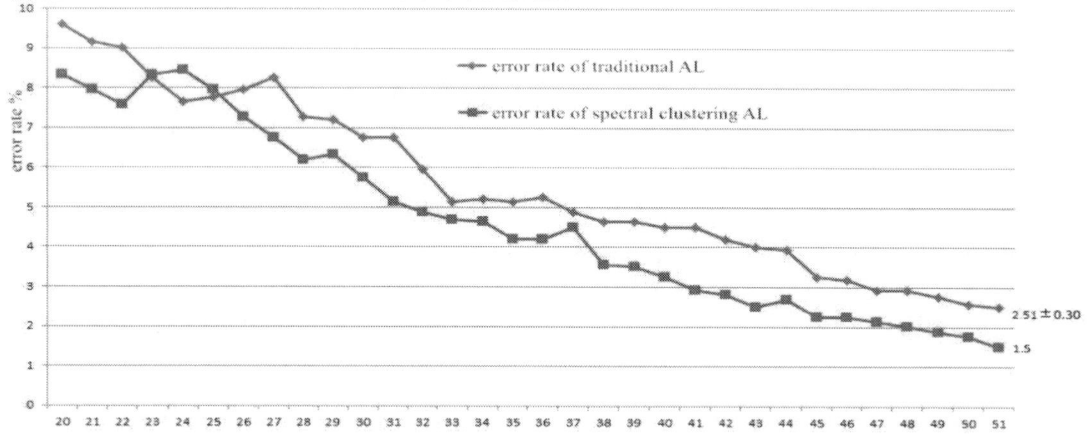

Figure 3. Classification error rates of No 19 and 20 news in news20.

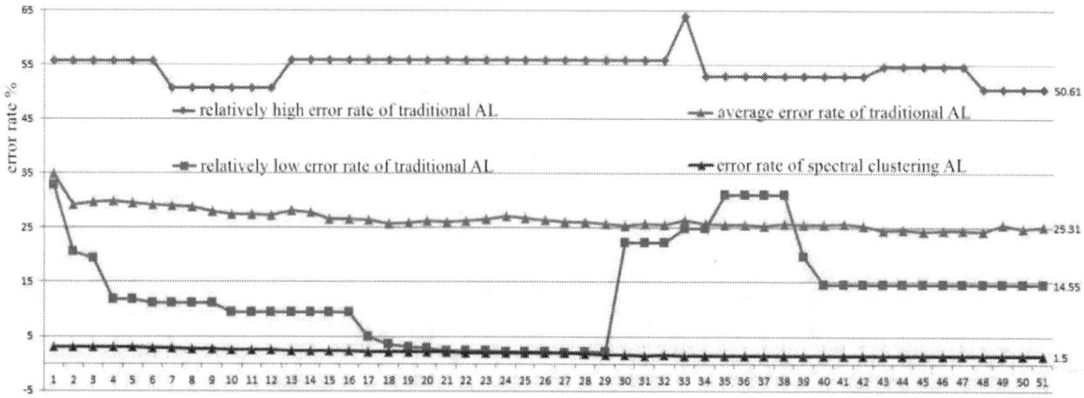

Figure 4. Classification error rates of datasets w2a.

Figure 3 shows that despite the lower error rates of traditional ALSVMs at the early iterations, the proposed algorithm eventually beats the traditional method. Also from the former three figures, the fluctuations of the results are caused by the shortage of active learning that the current selected instances may have less influence to the model as iteration goes on.

The traditional ALSVMs was applied to datasets w2a 100 times in this experiment, and a result with relatively high error rate, a result with relatively low error rate and average error rate are shown in the following figure, with the average error rates of proposed algorithm. The results of iteration range from 1 to 51 times are shown in Figure 4.

In the Figure 4, we can see that the error rates of spectral clustering active learning are lower than the traditional method and it has a faster convergence speed. The error rates of news20 are relatively lower than the w2a for news20 was applied with linear SVM and has less noise. The Figure 4 also shows that the results of traditional active learning algorithm can reach an error rate up to 65%, and vary a lot, which indicate that traditional active learning is affected by initial states and has strong randomness. However, spectral clustering based active learning is less affected by initial states, and the results are more stable.

4 Conclusion

The experiment results indicate that the proposed spectral clustering active learning algorithm is superior to traditional ones, since the error rates are lower and the results are more stable. Taking advantages of the distribution feature of the datasets before selecting instances, the performance of active learning can be promoted.

Actually, the proposed method acts as a framework, spectral clustering is used for digging out the distribution feature of the datasets. As the datasets grows more complicated, more advanced clustering algorithm can be applied, such as Greedy Gradient Max-Cut (GGMC) [5], Sparse Subspace Clustering (SSC) [6], Spectral multi-manifold Clustering (SMMC) [7] and so on.

References

1. Schohn G, Cohn D, Less is more: Active learning with support vector machines, *International Conference on Machine Learning*, 839-846, (2000).
2. Tong S, *Active learning: theory and application*, Stanford University, (2001)
3. Zhou D, Bousquet O, Lal T N, et al, Learning with local and global consistency, *Advances in neural information processing systems*, **16**(16): 321-328, (2004)
4. Chang C C, Lin C J, LIBSVM: A library for support vector machines, ACM TIST , **2**(3): 27, (2011)
5. Wang J, Jebara T, Chang S F, Semi-supervised learning using greedy max-cut, J. Mach. Learn. Res. (JMLR), , **14**(1): 771-800, (2013)
6. Elhamifar E, Vidal R, Sparse subspace clustering: Algorithm, theory, and applications, " IEEE Trans. Pattern Anal. Mach. Intell.(TPAMI), **35**(11): 2765-2781, (2013)
7. Wang Y, Jiang Y, Wu Y, et al, Spectral clustering on multiple manifolds, Neural Networks, IEEE Trans. Neural Netw., **22**(7): 1149-1161, (2011)

Recursive Approach for Evaluation of Time Intervals between Transactions in Polling Procedure

Eugene Larkin[1], Alexey Ivutin[2] , Dmitry Esikov[2]

[1]*Tula State University, Department of Robotics and Industry Mechanization, Tula, Russia*
[2]*Tula State University, Department of Computer Technology, Tula, Russia*

Abstract. An ergodic semi-Markov process with the structure represented by the full graph with loops, which simulates a digital control algorithm that generated transactions onto an object, is investigated. Elementary simplifications for reduction of semi-Markov processes are defined. Recursive procedure for reduction of initial semi-Markov process structure till the model, which includes selected states with its links only, is proposed. Formulae for recalculation of probabilities, weighted densities and expectations of time of switching to linked states are obtained. It is shown that recursive procedure may be used also for calculation of time expectation of return the process to one of selected states that simplified the task of evaluation of time intervals between transactions in polling procedure.
Keywords. Polling, semi-Markov process, recursive procedure, expectation, dispersion, time interval.

1 Introduction

Polling is one of the most common procedure for transactions organization in real time, embedded, swarm etc., control systems [1, 2]. In such a systems control computer algorithmically organizes a cycle of control of peripheral equipment, so there are rigid restrictions for time period between transactions.

For evaluation of time intervals the model of data processing must be worked out. Taking into account the number of specific features of control algorithms, such as cyclic recurrence, quasi-stochastic way of continuation in places of branching, quasi-probabilistic of time of execution of operators, there is the only formalism for modeling the computation, ergodic semi-Markov process [3-6], and interpretation of algorithm may be represented as a wandering through states of semi-Markov process [7, 8].

In [9, 10] matrix expressions for evaluation of time intervals were obtained, but they are cumbersome enough for practice use. So, the known approach needs high computational resources, commensurable with compilation of program and experimental evaluation of time intervals under consideration [11, 12]. Recursive procedure proposed below permits substantially decrease time complexity of analysis of ergodic semi-Markov processes.

2 Initial semi-Markov process

In most general case semi-Markov process, which simulates a polling generator of transactions, is the next [3-6]:

$$ {}^{S+J}M = \left\{ {}^{S+J}A, {}^{S+J}\boldsymbol{r}, {}^{S+J}\boldsymbol{h}(t) \right\}, \qquad (1) $$

where ${}^{S+J}A$ - is the set of states; ${}^{S+J}A = {}^{S}A \cup {}^{\bar{S}}A$; ${}^{S}A \cap {}^{\bar{S}}A = \varnothing$; ${}^{S}A = \{a_1, \ldots, a_s, \ldots, a_S\}$ - is the subset of selected states, which simulate operators of generation of transactions; ${}^{\bar{S}}A = \{a_{S+1}, \ldots, a_{S+j}, \ldots, a_{S+J}\}$ - is the subset of other states; ${}^{S+J}\boldsymbol{r} = \left({}^{S+J}r_{mn} \right)$ - is the adjacency matrix of size $(S+J)\times(S+J)$; ${}^{S+J}\boldsymbol{h}(t)$ - is the semi-Markov matrix of size $(S+J)\times(S+J)$; t - time;

$$ {}^{S+J}\boldsymbol{h}(t) = \left[{}^{S+J}\boldsymbol{p} \otimes {}^{S+J}\boldsymbol{f}(t) \right]; \qquad (2) $$

${}^{S+J}\boldsymbol{p}$ - is the stochastic marix; ${}^{S+J}\boldsymbol{f}(t)$ - is the matrix of densities which specifies time intervals; ${}^{S+J}r_{mn} = 1$ for $1 \le m, n \le J+S$.

Typical graph [13-14] which describes structure of process is shown on figure 1.

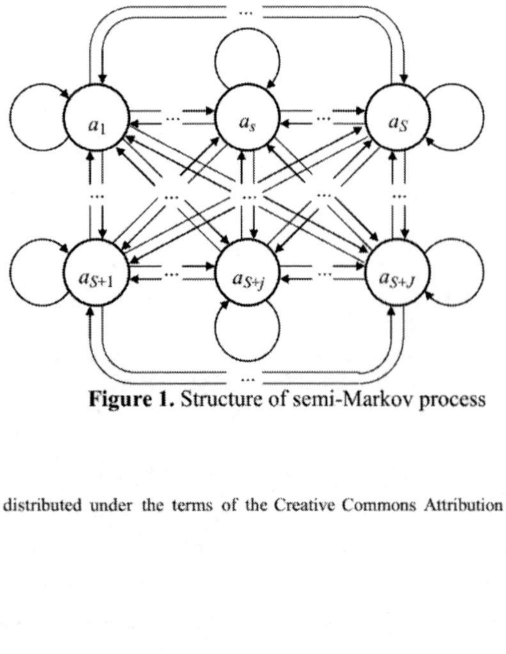

Figure 1. Structure of semi-Markov process

The process (1) generates a transaction in one of the next cases:

1) when the direct switch from $a_m \in {}^S A$ to $a_n \in {}^S A$ occurs;

2) when occurs switch from $a_m \in {}^S A$ to $a_k \in {}^{\bar S} A$ and subsequent wandering through states of subset ${}^{\bar S} A$ from $a_k \in {}^{\bar S} A$ to $a_n \in {}^S A$.

For evaluation of time intervals it is necessary to replace the walk through states of subset ${}^{\bar S} A$ to direct switch, i.e. reduce (1) till the semi-Markov process

$$ {}^S M = \left\{ {}^S A, {}^S \boldsymbol{r}, {}^S \boldsymbol{h}(t) \right\}, \tag{3} $$

where ${}^S \boldsymbol{r} = \left({}^S r_{mn} \right)$ - is the adjacency matrix of size $S \times S$; ${}^S r_{mn} = 1$, $1 \le m, n \le S$; ${}^S \boldsymbol{h}(t)$ - is the semi-Markov matrix of size $S \times S$.

When reducing, states of subset ${}^{\bar S} A$ are eliminated, so every switch of semi-Markov process (3) generates one transaction. For the reducing may be used matrix method, described in [10], however similar result may be obtained with the aid of recursive procedure of simplification of (1).

3 Elementary operations of reducing

For realization of recursive procedure let us to introduce three elementary operations of reducing of semi-Markov process, which are shown on figure 2 [15-16].

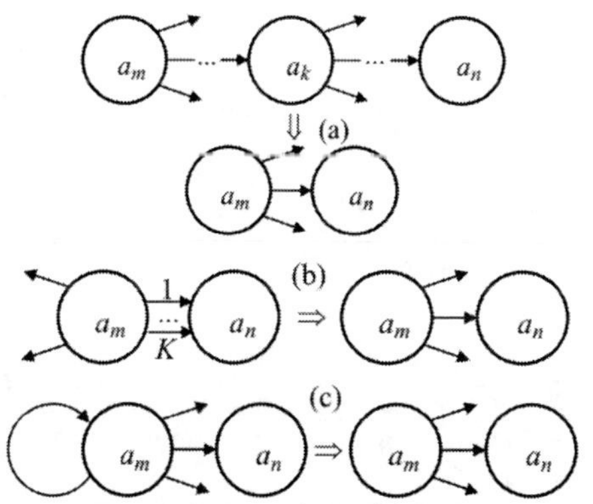

Figure 2. Operations of reducing

On the fig. 2 the union of sequential states is shown. Probability, weighted density, expectation and dispersion of time of wandering from a_m to a_n are determined by the next dependencies:

$$ \tilde p_{m,n} = \prod_{k=m}^{n-1} p_{k,k+1}; $$

$$ \tilde h_{m,n}(t) = L^{-1}\left[\prod_{k=m}^{n-1} L\left[h_{k,k+1}(t) \right] \right]; $$

$$ \tilde T_{m,n} = \sum_{k=m}^{n-1} T_{k,k+1}, $$

$$ \tilde D_{m,n} = \sum_{k=m}^{n-1} D_{k,k+1} \tag{4} $$

where $p_{k,k+1}$ - is the probability of switch to linked state; $h_{k,k+1}(t)$ - weighted density of time of switching to linked state; $T_{k,k+1}$ - is the expectation of time of switching; $L[..]$, $L^{-1}[..]$ - are direct and inverse Laplace transforms.

On the fig. 2 (b) the union of parallel arcs is shown. Probability, weighted density, expectation and dispersion of time of switching from a_m to a_n are determined by the next dependencies:

$$ \tilde p_{m,n} = \sum_{k=1}^{K} p_{m,n,k}; $$

$$ \tilde h_{m,n}(t) = \sum_{k=1}^{K} h_{m,n,k}(t); $$

$$ \tilde T_{m,n} = \frac{\sum_{k=1}^{K} p_{m,n,k} \cdot T_{m,n,k}}{\tilde p_{m,n}}, $$

$$ \tilde D_{m,n} = \frac{\sum_{k=1}^{K} p_{m,n,k} \cdot \left(D_{m,n,k} + T_{m,n,k}^2 \right)}{\tilde p_{m,n}} - \tilde T_{m,n}^2 \tag{5} $$

where k - is the number of arcs from a_m to a_n; $p_{m,n,k}$, $h_{m,n,k}(t)$, $T_{m,n,k}$, $D_{m,n,k}$ - probability, weighted density and expectation of time of switching from a_m to a_n on k-th arc.

On the fig. 2 (c) elimination of loop is shown. Probability, weighted density, expectation and dispersion of time of switching from a_m to a_n are determined by the next dependencies:

$$ \tilde p_{m,n} = \frac{p_{m,n}}{1 + p_{m,m}}; $$

$$ \tilde h_{m,n}(t) = L^{-1}\left[\boldsymbol{I}_r \cdot \sum_{k=1}^{\infty} \left\{ L\left[\boldsymbol{h}_c(t) \right] \right\}^k \cdot \boldsymbol{I}_c \right]; $$

$$ \boldsymbol{h}_c = \begin{pmatrix} h_{m,m}(t) & h_{m,n}(t) \\ 0 & 0 \end{pmatrix}; \ \boldsymbol{I}_c = \begin{pmatrix} 0 \\ 1 \end{pmatrix}; \ \boldsymbol{I}_r = (0, 1); $$

$$ \tilde T_{m,n} = T_{m,n} + \frac{T_{m,m} \cdot p_{m,m}}{1 - p_{m,m}}; $$

$$ \tilde D_{m,n} = D_{m,n} + \frac{D_{m,m} \cdot p_{m,m}}{1 - p_{m,m}} + \frac{T_{m,m}^2 \cdot p_{m,m}}{\left(1 - p_{m,m} \right)^2}. \tag{6} $$

4 The recursive procedure

The recursive procedure presupposes sequential elimination of states from (S+J)-th, to (S+1)-th. Let us assume that at the previous simplifications states from (S+J)-th till (S+I+1)-e were eliminated, i.e. semi-Markov process (1) was reduced till the process

$$ {}^{S+j} M = \left\{ {}^{S+j} A, {}^{S+j} \boldsymbol{r}, {}^{S+j} \boldsymbol{h}(t) \right\}, \tag{7} $$

where ${}^{S+j} \boldsymbol{r} = \left({}^{S+j} r_{mn} \right)$ and ${}^{S+J} \boldsymbol{h}(t) = \left[{}^{S+J} h_{mn}(t) \right]$ are of size $S + j \times S + j$; ${}^S r_{mn} = 1$, $1 \le m, n \le S + j$.

The first operation of reducing is elimination of loop on the state a_{S+j} due to dependencies (6):

$$ {}^{S+j}\tilde{p}_{S+j,n} = \frac{{}^{S+j}p_{S+j,n}}{1+{}^{S+j}p_{S+j,S+j}}; $$

$$ \tilde{h}_{S+j,n}(t) = L^{-1}\left[\boldsymbol{I}_r \cdot \sum_{k=1}^{\infty} \left\{ L\left[{}^{S+j}\boldsymbol{h}_c(t) \right] \right\}^k \cdot \boldsymbol{I}_c \right]; $$

$$ {}^{S+j}\boldsymbol{h}_c = \begin{pmatrix} {}^{S+j}h_{S+j,S+j}(t) & {}^{S+j}h_{S+j,n}(t) \\ 0 & 0 \end{pmatrix}; $$

$$ {}^{S+j}\tilde{T}_{S+j,n} = {}^{S+j}T_{S+j,n} + \frac{{}^{S+j}T_{S+j,S+j} \cdot {}^{S+j}p_{S+j,S+j}}{1-{}^{S+j}p_{S+j,S+j}}; $$

$$ {}^{S+j}\tilde{D}_{S+j,n} = {}^{S+j}D_{S+j,n} + \frac{{}^{S+j}D_{S+j,S+j} \cdot {}^{S+j}p_{S+j,S+j}}{1-{}^{S+j}p_{S+j,S+j}} + $$

$$ + \frac{{}^{S+j}T^2_{S+j,S+j} \cdot {}^{S+j}p_{S+j,S+j}}{\left(1-{}^{S+j}p_{S+j,S+j}\right)^2}, \tag{8} $$

The next operation of the recursive cycle is splitting the state a_{S+j} onto $(S+j-1)^2$ states as it is shown on fig. 3.

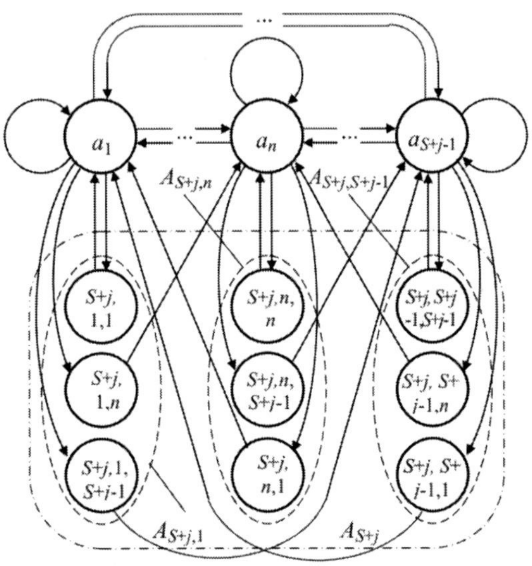

Figure 3. Splitting the state a_{S+j}

At first the state a_{S+j} (is circled with dash-dotted line) is splitting onto set $A_{S+j} = \{a_{S+j,1}, \ldots, a_{S+j,n}, \ldots, a_{S-j,S+j-1}\}$ (every element of set is circled with dashed line), and then states $a_{S+j,n}$, $1 \le n \le S+j-1$ are splitting onto sets $A_{S+j,n} = \{a_{S+j,n,1}, \ldots, a_{S+j,n,n}, \ldots, a_{S-j,n,S+j-1}\}$. In other to ensure equivalency of transforms probabilities, weighted densities and expectations must be recalculated in accordance with dependencies (4):

$$ \hat{p}_{n,m} = {}^{S+j}p_{n,S+j} \cdot {}^{S+j}\tilde{p}_{S+j,m}; $$

$$ \hat{h}_{n,m}(t) = L^{-1}\left[L\left[{}^{S+j}h_{n,S+J}(t) \right] \cdot L\left[{}^{S+j}\tilde{h}_{S+J,m}(t) \right] \right]; $$

$$ \hat{T}_{n,m}(t) = {}^{S+j}T_{n,S+J} + {}^{S+j}\tilde{T}_{S+J,m}; $$

$$ \hat{D}_{n,m}(t) = {}^{S+j}D_{n,S+J} + {}^{S+j}\tilde{D}_{S+J,m}, $$

$$ 1 \le n,m \le S+j-1 \tag{9} $$

On the next step in accordance with (5) must be recalculated probabilities, weighted densities and expectations of semi-Markov process ${}^{S+j-1}M = \{{}^{S+j-1}A, {}^{S+j-1}\boldsymbol{r}, {}^{S+j-1}\boldsymbol{h}(t)\}$, which have already $S+j-1$ states.

$$ {}^{S+j-1}p_{n,m} = {}^{S+j}p_{n,m} + \hat{p}_{n,m}; $$

$$ {}^{S+j-1}h_{n,m} = {}^{S+j}h_{n,m}(t) + \hat{h}_{n,m}(t); $$

$$ {}^{S+j-1}T_{n,m} = \frac{{}^{S+j}p_{n,m} \cdot {}^{S+j}T_{n,m} + \hat{p}_{n,m} \cdot \hat{T}_{n,m}}{{}^{S+j}p_{n,m} + \hat{p}_{n,m}}; $$

$$ {}^{S+j-1}D_{n,m} = -{}^{S+j-1}T^2_{n,m} + $$

$$ + \frac{{}^{S+j}p_{n,m} \cdot \left({}^{S+j}D_{n,m} + {}^{S+j}T^2_{n,m} \right) + \hat{p}_{n,m}\left(\hat{D}_{n,m} + \hat{T}^2_{n,m} \right)}{{}^{S+j}p_{n,m} + \hat{p}_{n,m}}. $$

Recursion procedure repeated from $j = J$ till $j = 1$. As a result of such a procedure elements of matrices ${}^{S}\boldsymbol{p} = \left[{}^{S}p_{mn} \right]$, ${}^{S}\boldsymbol{h} = \left[{}^{S}h_{mn}(t) \right]$ and ${}^{S}\boldsymbol{T} = \left({}^{S}T_{mn} \right)$ are calculated for the semi-Markov process (3).

5 Time intervals between transactions

Process (3) still stay the ergodic one, so density of time intervals between transactions may be defined in accordance with [3, 4] the next expression:

$$ g(t) = \sum_{m=1}^{S} \frac{\sum_{n=1}^{S} {}^{S}p_{m,n} \, {}^{S}T_{m,n}}{{}^{1}T_{m,m}} \left[\sum_{n=1}^{S} {}^{S}h_{m,n} \right], \tag{10} $$

where $\sum_{n=1}^{S} {}^{S}h_{m,n}$ and $\sum_{n=1}^{S} {}^{S}p_{m,n} \, {}^{S}T_{m,n}$ - density and expectation of time of stay of semi-Markov process (3) in the state $a_m \in {}^{S}A$; ${}^{1}T_{m,m}$ - expectation of returning of the ergodic process (3) to state a_m.

For determining ${}^{1}T_{m,m}$ one may use recursive procedure, described above. Without loss of community let us describe it for determining ${}^{1}T_{1,1}$. For other nominations of $1 \le m \le S$ procedure may be applied to process (3) after a changing of numeration of states of the process.

Procedure for decrement of m, from S to 1, is the next:

the elimination of loop

$$ {}^{\hat{o}}\tilde{p}_{m,n} = \frac{{}^{m}p_{m,n}}{1+{}^{m}p_{m,m}}; \quad {}^{m}\tilde{T}_{m,n} = {}^{m}T_{m,n} + \frac{{}^{m}T_{m,m} \cdot {}^{m}p_{m,m}}{1-{}^{m}p_{m,m}}; $$

the splitting of set and union of sequential states

$$ \hat{p}_{n,k} = {}^{m}p_{n,m} \cdot {}^{m}\tilde{p}_{m,k}; \quad \hat{T}_{n,k}(t) = {}^{m}T_{n,m} + {}^{m}\tilde{T}_{m,k}; $$

$$ 1 \le n,k \le m-1; $$

union of parallel arcs

$$ {}^{m-1}p_{n,k} = {}^{m}p_{n,k} + \hat{p}_{n,k}; $$

$$ {}^{m-1}T_{n,k} = \frac{{}^{m}p_{n,k} \cdot {}^{m}T_{n,k} + \hat{p}_{n,k} \cdot \hat{T}_{n,k}}{{}^{m-1}p_{n,k}}. $$

Procedure must be repeated for different states, which must be placed onto the first position to achieve all parameters necessary to use the formula (10).

6 Experimental verification of method

For verification of proposed method direct computer experiment was executed with use the Monte-Carlo method. Under verification was homogeneous semi-Markov process ${}^{3}M = \{{}^{3}A, {}^{3}\boldsymbol{r}, {}^{3}\boldsymbol{h}(t)\}$ which is shown on figure 4.

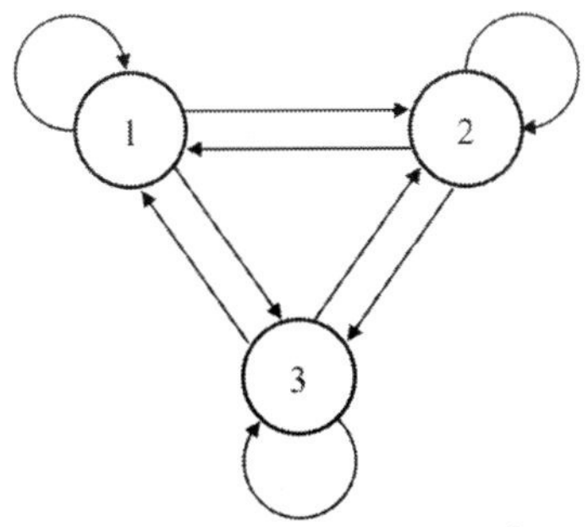

Figure 4. Semi-Markov process ${}^{3}M$

In semi-Markov process ${}^{3}M$:

$$ {}^{3}A = \{1, 2, 3\}; $$

$$ {}^{3}\boldsymbol{r} = \begin{pmatrix} 1 & 1 & 1 \\ 1 & 1 & 1 \\ 1 & 1 & 1 \end{pmatrix}; $$

$$ {}^{3}\boldsymbol{h}(t) = \begin{bmatrix} 1/3 \cdot f(t) & 1/3 \cdot f(t) & 1/3 \cdot f(t) \\ 1/3 \cdot f(t) & 1/3 \cdot f(t) & 1/3 \cdot f(t) \\ 1/3 \cdot f(t) & 1/3 \cdot f(t) & 1/3 \cdot f(t) \end{bmatrix}; $$

$$ f(t) = \begin{cases} 5/3, \text{ if } 0{,}7 \le t \le 1{,}3; \\ 0, \text{ if } t < 0{,}7, \text{ or } t > 1{,}3. \end{cases} $$

Expectation and dispersion of $f(t)$ are equal $T_{m,n} = 1\,s$, $D_{m,n} = 0{,}03\,s^2, 1 \le m, n \le 3$.

Calculation with use the proposed method gives the following result: ${}^{1}T_{1,1} = 3\,s$; ${}^{1}D_{1,1} = 6{,}123\,s^2$.

Computer experiment was carried out in accordance with the next classical method.

1) Reset the counter of number of realizations, $l = 0$.

2) Assignment the status of current to the first state, $a_l = 1$.

3) Reset the timer $t_l = 0$.

4) Receiving a random value $1 \le \pi \le 1$ with uniform distribution.

5) If $0 \le \pi \le 0{,}333$, then $a_l = 1$; if $0{,}333 < \pi \le 0{,}666$, then $a_l = 2$; if $0{,}666 < \pi \le 1$, then $a_l = 3$.

6) Receiving a random value $1 \le \pi \le 1$.

7) Calculation of time increment on the formula $\Delta_t = 0{,}6\pi + 0{,}7\,s$

8) Calculation of current time $t_l := t_l + \Delta_t$;

9) If $a_l \ne 1$, then 4.

10) Unloading t_l to array of results.

11) $l := l + 1$. If $l < L$, then 3).

12) End of experiment.

On figure 5 the histogram of time of returning to state 1 is shown, which was constructer after 10000 realizations of the process ${}^{3}M$.

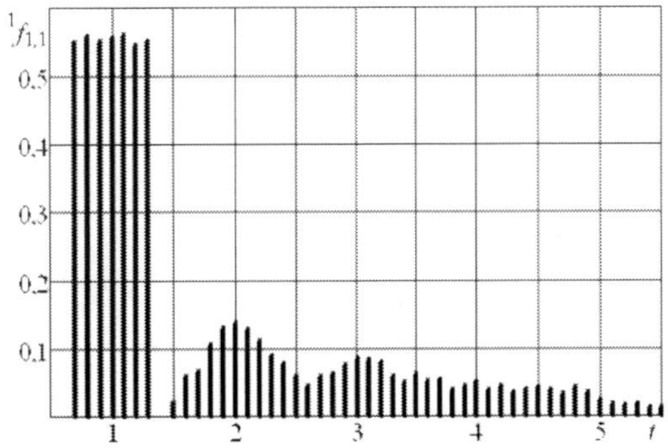

Figure 5. Histogram of ${}^{1}f_{1,1}(t)$

Expectation and dispersion of density ${}^{1}f_{1,1}(t)$, resulting from the processing of experimental data, are equal, correspondingly ${}^{1}T_{1,1} = 2{,}996\,s$, with error 0,13%, and $\sqrt{{}^{1}D_{1,1}} = 2{,}493$ with error 0,75 %. Experimental parameters rather precisely correspond to theoretical settlements.

Conclusion

In this article relatively simple universal procedure of recursive simplification of semi-Markov process was presented to determine time intervals between transactions at polling. Time characteristics were obtained for most common structure of semi-Markov generator of transactions (full graph with loops). Time intervals are significantly important parameter from point of view insertion the computer into control circuit as the feedback element. Optimization of time under consideration permits to improve quality characteristics of control. Further research in this area may be directed onto working out the method of optimization of time intervals with use recursive procedure.

References

1. Olsson, G., Piani, G. *Computer Systems for Automation and Control* (2000)
2. Ivutin, A., Larkin, E., Kotov, V. *LNCS,* 415 (2015)
3. Janssen, J. *Semi-Markov models: theory and applications* (2013)
4. Meerschaert, M., Straka, P. The Annals of Probability, **42**, 1699 (2014)
5. Barbu V.S., Limnios N. Algebraic Methods in Statistics and Probability, **2**, 19 (2010)
6. Barbu V.S., Limnios N. Lecture Notes in Statistics, **191** (2008)

7. Howard, Ronald A. *Dynamic Probabilistic Systems, Volume I: Markov Models. Vol. 1* (2012)
8. Howard, Ronald A. *Dynamic Probabilistic Systems, Volume II: Semi-Markov and Decision Processes. Vol. 2* (2013)
9. Ivutin, A., Larkin, E. *MECO*, 236 (2014)
10. Ivutin, A., Larkin, E., Lutskov, Y. *MECO*, 250 (2015)
11. Zaroliagis, Ch., Grammati P., Kontogiannis S. Alg., Prob., Nets and Games, **9295** (2015)
12. Iverson, M., Ozguner, F., Follen, G. *HPDC Proc.* 263 (1996)
13. Godsil C., Royle G. *Algebraic graph theory* (2013)
14. Bollobas B. *Modern graph theory* (2013)
15. Beizer, B. *Micro-Analysis of Computer System Performance* (1978)
16. Ferrari, B., *Computer Systems Performance Evaluation* (1978)

Incremental Learning Algorithm of Least Square Twin KSVC

Wang Yaru , Yang Ling , Chen Mohan , Xi Jikui

School of Information Science and Engineering, Lanzhou University, Lanzhou 730000, China

Abstract. In view of the batch implementations of standard support vector machine must be retrained from scratch every time when the training set is incremental modified, an incremental learning algorithm based on least squares twin multi-class classification support vector machine (ILST-KSVC) is proposed by solving two inverse matrix. The method will be applied on online environment to update initial data, which avoided cumbersome double counting. ILST-KSVC inherited the advantages of the basic algorithm and has some merits of Least square twin support vector machine for excellent performance on training speed and support vector classification regression for K-class's well classification accuracy. The result will be confirmed no matter in low dimension or in high dimension in UCI datasets.

1 Introduction

Support vector machine is a type of machine learning algorithm, which aims at finding one optimal hyperplane to make two kinds of training data as far as possible away from the hyperplane in the high dimensional space[1].It has great advantages of solving small sample, nonlinear and high dimensional pattern recognition problem, therefore, its been successfully applied to many aspects, such as, medical image detection[2], network information classification[3,4], pattern recognition[5], etc. Classical SVM needs to solving a Quadratic Programming Problem (QPP), which computational complexity is O(l3)(l refers to the total number of training data), so when dealing with large-scale data is restricted. Twin support vector machines (TSVM) been proposed by Jayadeva, etc in 2007. It divides one QPP into two smaller QPPs from SVM. By solving two smaller QPPs to construct two non-parallel hyperplanes, and makes each hyperplane as far as possible close to the kind of sample data and away from the other. TSVM simplifies the computational complexity of SVM and theory has been shown that its computational complexity is O(l3/4), about a quarter of the SVM[6]. Least squares twin support vector machine (LS-TSVM)[7] greatly simplify the computational complexity of TSVM. It constructs a squared loss function modifying with convex linear system to replace the TSVM convex QPPs and substitutes equality constraints for inequality constraints. Thus, solving two equations improves the solving efficiency and accuracy of classification.

Classical SVM and TSVM are designed for binary classification problems; however, we often encounter many classification problems in real life. We usually use '1-versus-1'[8] and '1-versus-1'[9] to dealing with muti-class classification problems in these methods. For K-class, the first method is to construct K two classifiers, each class corresponding to one of them been separated from other classes, the method is faster but easy to cause the imbalance of classifier. The second method construct all the possible binary classifiers in K class of training samples , so need to construct K(K-1)/2 binary classifiers. Each class training data only taken from the corresponding two classes, without considering other sample information, which leading to classification's result is not ideal.2003, Angulo proposed the '1-versus-1-versus-rest 'multi-class classification algorithm---support vector classification regression for K-class classification(K-SVCR)[10]. It constructs K(K-1)/2 K-SVCR triple classifier to deal with K class classification problem, this structure improve accuracy of muti-class problems, but the computational complexity is high. Twin multi-class classification support vector machine (Twin-KSVC) has been come up with, based on TSVM and K-SVCR [11]. It takes advantages of TSVM and SVCR that running time is close to TSVM but far less than K-SVCR. In order to reduce the computational complexity and simplified the structure of classifier further, Jalal a. etc. combined with the LS-TSVM and K-SVCR put forward Least squares twin multi-class classification support vector machine (LST-KSVC)[12].LST-KSVCR inherited the faster training speed of LS-TSVM and high accuracy of KSVC.

In many online scenario, the new data will be continuously added to the data set, while the standard SVM, TSVM will no longer be applicable to the classification of the real-time data. In order to solve the classification problem of the real-time data, many incremental learning algorithm based on SVM and TSVM [13]-[16] has been proposed. In this paper, we propose an incremental learning algorithm suitable for an online environment called Incremental Learning

Algorithm for LST-KSVC; we also call its ILST-KSVC, briefly. The Algorithm introduces SMW (Sherman - Morrison - Woodbury) formula to solve the inverse of the matrix, and every time a new sample data increases, the only need to update inverse matrix on the basis of original weight matrix, without calculating weight matrix inverse matrix, the process will not have an impact on the result of the classification. Decremental is the inverse process of incremental, therefore, the algorithm also applies to the decremental process.

2 Least square twin K-SVCR

2.1 1-versus-1-versus-rest

'one-versus-one-versus-rest' method is a new breakthrough to solve the multiple classification problems. It combines the standard support vector classifier and support vector regression machine together to solve the triple class directly.

Set training set $Q=\{(x_p,y_p)\mid p=1,2,..k\}$

$$\text{Here, } y_p=\begin{cases}1 & p=1,\ldots,l_1\\ -1 & p=l_1+1,\ldots,l_1+l_2\\ 0 & p=l_1+1,\ldots,l_1+l_2,\ldots,l\end{cases}$$

Let matrix $A\subset R^{l_1\times n}$ indicates left class training data labeled '1', matrix $B\subset R^{l_2\times n}$ indicates right class training data labeled '-1' and matrix $C\subset R^{l_3\times n}$ is the other class training data labeled '0'. l is the total training data and $l=l_1+l_2+l_3$.

For K-class classification, the method chooses each i, $j(i\neq j)$ as left class and right class, the rest K-2 classes as class 0, called 'one-versus-one-versus-rest'. Training each i, j class to get K (K - 1) / 2 ternary classifier, corresponding to get K (K - 1) / 2 the output of the decision function. In order to determine the category of the test sample, using the voting as follows: when the decision function to output 1,cast class i one vote; When the decision function output of -1, cast class j one vote; When the decision function output is 0, do not vote. Inspection after all the K (K - 1) / 2 the output of the decision function, put the test sample into the category with the most votes

2.2 linear LST-KSVC

Twin-KSVC needs to address two QPPs, which costs high computational. And LST-KSVCR reduced computation cost by solving two reversible matrix, make it can be used for processing data of high dimension matrices. LST-KSVC change Twin-KSVC's inequality constraints to the linear equality constraint conditions, thus, the original optimization problem as follows.

$$\begin{aligned}&\min_{w_1,b_1}\frac{1}{2}\|Aw_1+e_1b_1\|^2+\frac{c_1}{2}y^Ty+\frac{c_2}{2}z^Tz\\ &\text{s.t. }-(Bw_1+e_2b_1)+y=e_2\\ &\quad -(Cw_1+e_3b_1)+z=e_3(1-\varepsilon)\end{aligned}\tag{1}$$

$$\begin{aligned}&\min_{w_2,b_2}\frac{1}{2}\|Bw_2+e_2b_2\|^2+\frac{c_3}{2}y^Ty+\frac{c_4}{2}z^Tz\\ &\text{s.t. }(Aw_2+e_1b_2)+y=e_1\\ &\quad (Cw_2+e_3b_2)+z=e_3(1-\varepsilon)\end{aligned}\tag{2}$$

Where, c_1, c_2, c_3, c_4 are penalty parameter, y>0, z>0 is slack variable, ε is between 0 and 1. w, b is weights and bias for solving. Substituting the equality constraint into the objective function in formula (1), we obtain

$$\min\frac{1}{2}\|Aw_1+e_1b_1\|^2+\frac{c_1}{2}\|Bw_1+e_2b_1+e_2\|^2+\frac{c_2}{2}\|Cw_1+e_3b_1+e_3(1-\varepsilon)\|^2\tag{3}$$

Setting the gradient of (3) with respect to w_1, b_1 to zeros gives

$$\begin{bmatrix}w_1\\ b_1\end{bmatrix}=-[c_1\begin{bmatrix}B^T\\ e_2^T\end{bmatrix}[B\ \ e_2]+\begin{bmatrix}A^T\\ e_1^T\end{bmatrix}[A\ \ e_1]+c_2\begin{bmatrix}C^T\\ e_3^T\end{bmatrix}\times[C\ \ e_3]]^{-1}\times[c_1\begin{bmatrix}B^T\\ e_2^T\end{bmatrix}e_2+c_2\begin{bmatrix}C^T\\ e_3^T\end{bmatrix}e_3(1-\varepsilon)]\tag{4}$$

In brief, let E=[A e_1], F=[B e_2], G=[C e_3], then,

$$[w_1,b_1]^T=-(c_1F^TF+E^TE+c_2G^TG)^{-1}(c_1F^Te_2+c_2G^Te_3(1-\varepsilon))\tag{5}$$

In the same way, we get

$$[w_2,b_2]^T=(c_3E^TE+F^TF+c_4G^TG)^{-1}(c_3E^Te_1+c_4G^Te_3(1-\varepsilon))\tag{6}$$

By formula (5) and formula (6), we can obtain two linear separable non parallel hyperplanes.

$$xw_1^T+b_1=0\text{ and }xw_2^T+b_2=0\tag{7}$$

LST-KSVC use 'one-versus-one-versus-rest' method to training dataset. Linear LST-KSVC use the following decision function in formula (8) to classify the class with output {1, -1, 0}.

$$f(x_i)=\begin{cases}+1 & \text{if } x_iw_1+eb_1>-1+\varepsilon\\ -1 & \text{if } x_iw_1+eb_1<1-\varepsilon\\ 0 & \text{otherwise}\end{cases}\tag{8}$$

For K class classification problems, choose section 2.1 of the classified voting method.

2.3 non-linear LST-KSVC

By introducing a gaussian kernel function, linear LST - K SVC can be extended to non-linear. The original optimiza tion problem of non-linear LST-KSVC as follows.

$$\begin{aligned}&\min_{u_1,\gamma_1}\frac{1}{2}\|K(A,D^T)u_1+e_1\gamma_1\|^2+\frac{c_1}{2}y^Ty+\frac{c_2}{2}z^Tz\\ &\text{s.t. }-(K(B,D^T)u_1+e_2\gamma_1)+y=e_2\\ &\text{s.t. }-(K(C,D^T)u_1+e_3\gamma_1)+z=e_3(1-\varepsilon)\end{aligned}\tag{9}$$

$$\begin{aligned}&\min_{u_2,\gamma_2}\frac{1}{2}\|K(B,D^T)u_2+e_2\gamma_2\|^2+\frac{c_3}{2}y^Ty+\frac{c_4}{2}z^Tz\\ &\text{s.t. }(K(A,D^T)u_2+e_1\gamma_2)+y=e_1\\ &\text{s.t. }(K(C,D^T)u_2+e_3\gamma_2)+z=e_3(1-\varepsilon)\end{aligned}\tag{10}$$

Here, $K(\cdot)$ is arbitrary kernel function. D=[A; B; C], u,γ is weights and bias for solving. Similarly in the solution process of linear LST-KSVC, the solution of formula (9) and formula (10) are as follows.

$$[u_1,\gamma_1]^T=-(c_1N^TN+M^TM+c_2O^TO)^{-1}(c_1N^Te_2+c_2O^Te_3(1-\varepsilon))\tag{11}$$

$$[u_2,\gamma_2]^T=(c_3M^TM+N^TN+c_4O^TO)^{-1}(c_3M^Te_1+c_4O^Te_3(1-\varepsilon))\tag{12}$$

Here, M=[K(A, D^T) e_1], N=[K(B, D^T) e_2], O=[K(C, D^T) e_3].

By formula (11) and formula (12), we can obtain two linear separable non parallel hyperplanes.

$$K(x^T, D^T)u_1 + \gamma_1 = 0 \quad \text{and} \quad K(x^T, D^T)u_2 + \gamma_2 = 0 \quad (13)$$

Non-linear LST-KSVC use this decision function in formula (14).

$$f(x_i) = \begin{cases} +1 & \text{if } K(x_i, D^T)u_1 + e\gamma_1 > -1 + \varepsilon \\ -1 & \text{if } K(x_i, D^T)u_2 + e\gamma_2 < 1 - \varepsilon \\ 0 & otherwise \end{cases} \quad (14)$$

For K class classification problems, also choose section 2.1 of the classified voting method.

3 Incremental learning algorithm of LST-KSVC (ILST-KSVC)

In this section, we extend the LS-TSVCR to incremental learning and decremental learning version from linear to non-linear as well as LS-TSVCR. First, we need to simplify some symbols for convenience. Let w_1 replace $[w_1, b_1]^T$ in formula (5), w_2 replace $[w_2, b_2]^T$ in formula (6), so $\hat{w}_1$ is the update of w_1, and $\hat{w}_2$ is the update of w_2.

3.1 linear ILST-KSVC

In this paper, we adopt a kind of split matrix inverse, and reorganization of the inverse of the matrix. Finally obtain incremental learning algorithm which has a simple structure and small amount of calculation.

Adding new positive training data $S_i=[x_i^T, 1]^T$, for example. Then the matrix E can be extended to $\widetilde{E}$.

$$\widetilde{E}^T\widetilde{E} = \begin{bmatrix} E \\ S \end{bmatrix}^T \begin{bmatrix} E \\ S \end{bmatrix} = [E^T \quad S^T] \begin{bmatrix} E \\ S \end{bmatrix} = E^T E + S^T S \quad (15)$$

According to formula (5) and formula (6), we get

$$\hat{w}_1 = -(c_1 F^T F + E^T E + c_2 G^T G + S^T S)^{-1} \times (c_1 F^T e_2 + c_2 G^T e_3 (1-\varepsilon)) \quad (16)$$

$$\hat{w}_2 = (c_3 E^T E + F^T F + c_4 G^T G + S^T S)^{-1} \times (c_3 E^T e_1 + c_4 G^T e_3 (1-\varepsilon)) \quad (17)$$

For solving formula (16) and formula (17), we need to solve two inverse matrices by introducing the SMW theorem.

Let

$$H_1 = c_1 F^T F + E^T E + c_2 G^T G \quad (18)$$

$$H_2 = c_3 E^T E + F^T F + c_4 G^T G \quad (19)$$

Thus,

$$\begin{aligned} \hat{H}_1^{-1} &= (H_1 + S^T S)^{-1} = H_1^{-1} - \frac{H_1^{-1} S^T}{1 + S H_1^{-1} S^T} S H_1^{-1} \\ &= H_1^{-1} - \frac{\alpha_1}{1 + S\alpha_1} S H_1^{-1} = H_1^{-1} - \beta_1 S H_1^{-1} \\ &= (I - \beta_1 S) H_1^{-1} = \eta_1 H_1^{-1} \end{aligned} \quad (20)$$

$$\begin{aligned} \hat{H}_2^{-1} &= (H_2 + S^T S)^{-1} = H_2^{-1} - \frac{H_2^{-1} S^T}{1 + S H_2^{-1} S^T} S H_2^{-1} \\ &= H_2^{-1} - \frac{\alpha_2}{1 + S\alpha_2} S H_2^{-1} = H_2^{-1} - \beta_2 S H_2^{-1} \\ &= (I - \beta_2 S) H_2^{-1} = \eta_2 H_2^{-1} \end{aligned} \quad (21)$$

Where, $\alpha_1 = H_1^{-1} S^T$, $\alpha_2 = H_2^{-1} S^T$, $\beta_i = \frac{\alpha_i}{1 + S\alpha_i}, (i = 1,2)$, $\eta_i = I - \beta_i S, (i = 1,2)$.

H_1^{-1} and H_2^{-1} can be gained by solving inverse matrixs. Therefor, we only need to calculate η_1 and η_2 to get the after updating matrixs $\hat{H}_1^{-1}$ and $\hat{H}_2^{-1}$ of $H_1^{-1}$and H_2^{-1}. Finally, we get the updating $\hat{w}_1$ and $\hat{w}_2$.

$$\hat{w}_1 = -\hat{H}_1^{-1}(c_1 F^T e_2 + c_2 G^T e_3 (1-\varepsilon)) \quad (22)$$

$$\hat{w}_2 = \hat{H}_2^{-1}(c_3 E^T e_1 + c_4 G^T e_3 (1-\varepsilon)) \quad (23)$$

3.2 Non-linear ILST-KSVC

Getting Nonlinear LST - KSVC incremental algorithm we also add a right class training sample, for instance. Mapping training sample to high-dimensional $S_i' = [K(x_i, D^T) \quad e]$.

According to formula (11) and formula (12), in nonlinear LST-KSVC, we can easily get the new weight and bias after updating, like formula (24) -(25).

$$\hat{u}_1 = -(c_1 N^T N + M^T M + c_2 O^T O + S'^T S')^{-1}(c_1 N^T e_2 + c_2 O^T e_3 (1-\varepsilon)) \quad (24)$$

$$\hat{u}_2 = (c_3 M^T M + N^T N + c_4 O^T O + S'^T S')^{-1}(c_3 M^T e_1 + c_4 O^T e_3 (1-\varepsilon)) \quad (25)$$

Now we found two $(l+1)\times(l+1)$ dimension invertible matrices in the above formula. In order to reduce the computation cost, we use SMW method theorem to solve the inverse matrix.

Let

$$L = c_1 N^T N + M^T M + c_2 U^T O \quad (26)$$

$$R = c_3 M^T M + N^T N + c_4 O^T O \quad (27)$$

Then

$$\begin{aligned} \hat{L}^{-1} &= (L + S'^T S')^{-1} = L^{-1} - L^{-1} S'^T (1 + S' L^{-1} S'^T) S' L^{-1} \\ &= L^{-1} - \frac{L^{-1} S'^T}{1 + S' L^{-1} S'^T} S' L^{-1} = L^{-1} - \frac{\varphi_1}{1 + S'\varphi_1} S' L^{-1} \\ &= L^{-1} - \phi_1 S' L^{-1} = (I - \phi_1 S') L^{-1} = \xi_1 L^{-1} \end{aligned} \quad (28)$$

$$\begin{aligned} \hat{R}^{-1} &= (R + S'^T S')^{-1} = R^{-1} - R^{-1} S'^T (1 + S' R^{-1} S'^T) S' R^{-1} \\ &= R^{-1} - \frac{R^{-1} S'^T}{1 + S' R^{-1} S'^T} R L^{-1} = R^{-1} - \frac{\varphi_2}{1 + S'\varphi_2} S' R^{-1} \\ &= R^{-1} - \phi_2 S' R^{-1} = (I - \phi_2 S') R^{-1} = \xi_2 R^{-1} \end{aligned} \quad (29)$$

Where, $\varphi_1 = L^{-1} S'^T$, $\varphi_2 = R^{-1} S'^T$, $\phi_i = \frac{\varphi_i}{1 + S'\varphi_i}, (i = 1,2)$, $\xi_i = I - \phi_i S', (i = 1,2)$ .Then the formula (25) and formula (26) can be

$$\hat{u}_1 = -\hat{L}^{-1}(c_1 N^T e_2 + c_2 O^T e_3 (1-\varepsilon)) \quad (30)$$

$$\hat{u}_2 = \hat{R}^{-1}(c_3 M^T e_1 + c_4 O^T e_3 (1-\varepsilon)) \quad (31)$$

Thus, we just need to solve the L^{-1} and R^{-1} to get these inverse matrixes updating matrixes.

The solving process of L^{-1} is as follows.

Let $J=(M^TM+c_2O^TO)^{-1}$, $P=(N^TN+c_1O^TO)^{-1}$, using SMW method theorem, there is

$$L^{-1}=J-JN^T(\frac{I}{c_1}+NJN^T)^{-1}NJ \quad (32)$$

Similarly to L^{-1} , R^{-1} can be expressed as

$$R^{-1}=P-PM^T(\frac{I}{c_3}+MPM^T)^{-1}MP \quad (33)$$

3.3 Decremental learning of LST-KSVC

When constantly add new data to a certain degree, storage space will be occupied larger. Besides the original data is no longer useful, so need to delete the data to reduce the storage space. Decremental learning algorithm delete data without affecting the original classification results and is the inverse process of incremental learning. In this section, we imitate the incremental learning algorithm to conduct decremental learning algorithm. In the case of nonlinear, removing a class $S_i'=[K(x_i,D^T)\ \ e]^T$, for example. Similarly to formula (28)-(29), the derementrtal expression is

$$\begin{aligned}\hat{L}_-^{-1}&=(L-S'^TS')^{-1}=L^{-1}+L^{-1}S'^T(1-S'L^{-1}S'^T)S'L^{-1}\\&=L^{-1}+\frac{L^{-1}S'^T}{1-S'L^{-1}S'^T}S'L^{-1}=L^{-1}+\frac{\varphi_{-1}}{1-S'\varphi_1}S'L^{-1}\\&=L^{-1}+\beta_{-1}S'L^{-1}=(I+\beta_{-1}S')L^{-1}=\phi_{-1}L^{-1}\end{aligned} \quad (34)$$

$$\begin{aligned}\hat{R}_-^{-1}&=(R-S'^TS')^{-1}=R^{-1}+R^{-1}S'^T(1-S'R^{-1}S'^T)S'R^{-1}\\&=R^{-1}+\frac{R^{-1}S'^T}{1-S'R^{-1}S'^T}S'R^{-1}=R^{-1}+\frac{\varphi_{-2}}{1-S'\varphi_2}S'R^{-1}\\&=R^{-1}+\beta_{-2}S'R^{-1}=(I+\beta_{-2}S')R^{-1}=\phi_{-2}R^{-1}\end{aligned} \quad (35)$$

Thus, we only need to recalculate ϕ_{-1} and ϕ_{-2} , and make use of L^{-1} and R^{-1}, which has been required in the incremental learning process to gain the updating result.

4 Experiment and result

The validity of this algorithm has been tested in two experiments. All experiments are based on a 2.5 GHz pentium dual-core processor, 2GB memory hardware platform and using MATLAB 2013b to write programs.

Table 1. datasets in experiment.

datasets	Data scale	class
Hayes-roth	132×6=792	3
Iris	150×5=450	3
wine	178×14=2492	3
Soybean	47×35=1645	4
Ecoli	361×7=2289	5
Glass	214×9=1926	6
Determatology	357×34=12138	6

Experiment 1: selecting 10 group on UCI datasets, which is widely used for classification as showed in Table 1. Noting the dimensions of the car dataset is very high. We will leave it in the second experiments. Article in reference [12] compared LST - KSVC, K - SVCR, TSVM, Twin - KSVC and one-to-rest TSVM algorithms, the results show that the LST - KSVC has a great advantage in classification accuracy and the algorithm running time. Therefore, the first experiment mainly compares for ILST - KSVC and LST - KSVC training time and recognition accuracy.

We use 10-fold cross-validation and Gaussian kernel function $K(x_i,x_j)=\exp(-\|x_i-x_j\|^2/p)$ to convey the effect of different datasets' classification. In this paper, we choose penalty parameters c_1=c_3, c_2=c_4, c_1, c_2 are belong to $\{2^i \mid i=-4,-2,0,1,2,4,6,8\}$, ε is belonging to $\{0.1, 0.2, 0.3, 0.4, 0.5\}$, choose p in $\{2^i \mid i=-4,-2,0,1,2,4,6,8\}$.

Table 2 shows the LST - KSVC and ILST - KSVC classification results. Analysis of Table 2, on the left side of the ternary classification data, ILST KSVC can always get with LST - KSVC compare to the classification accuracy and didn't take too much running time.Table 2 on the right is the training data, which category number is more than 3. In the same way, it's easy to observe and obtain the same conclusion as the left side. Experiment show that ILST KSVC recognition accuracy can match with LST - KSVC, and training time is not that long for multiple classification.

Table 2. Muti-class classification.

dataset	LST-KSVC acc% ±std time/s(×10-2)	ILST-KSVC acc% ±std time/s(×10-2)
Hayes-roth	90.0±8.15 0.028	91.5±5.67 0.96
Iris	98.0±3.22 0.024	98.5±4.51 2.50
wine	98.86 ±4.80 0.03	98.86 ±4.11 1.49
Soybean	100±0.00 0.08	100±0.00 1.50
Ecoli	88.93±7.91 0.032	88.21±4.05 8.30
Glass	68.40±6.40 0.040	57.14±1.96 8.20
Determatology	95.67±2.89 0.17	95.07±3.83 34.57

Experiment 2: the increment algorithm of binary classification time-consuming is very big when dealing with high-dimensional data, this paper puts forward the incremental algorithm based on ternary classification problem to do with thousands of dimensional data, which only need a short period of time.This Experiment will increase the dimension from low to high gradually on a data named car which dimension is 1728 * 6.Figure 1 and Figure 2 compare LST KSVC, which apply to offline environment for ILST KSVC that suitable for the online environment in classification accuracy and running time, respectively.From figure 1 we can see: with increasing of data dimension, the accuracy of both methods is reduced, but always stay above 90%, and has little difference. Figure 2 is the contrast of these two algorithms, it can be

seen that with the continuous improvement of data dimension, training time also increases, but still at a low level.Thus, the incremental algorithm proposed in this paper can be employed for a number of high-dimensional data processing without affecting the classification results.

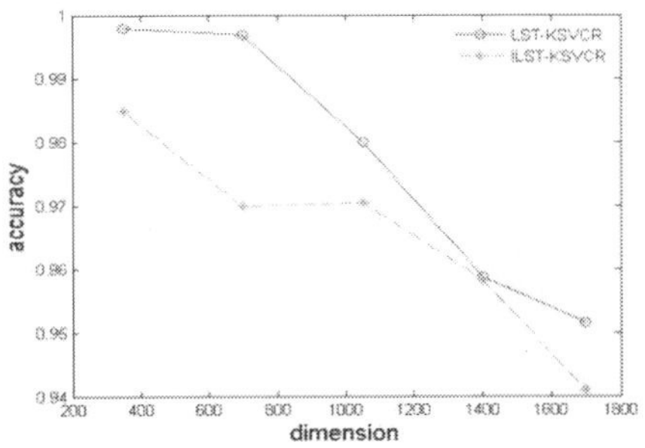

Figure 1. Contrast on accuracy.

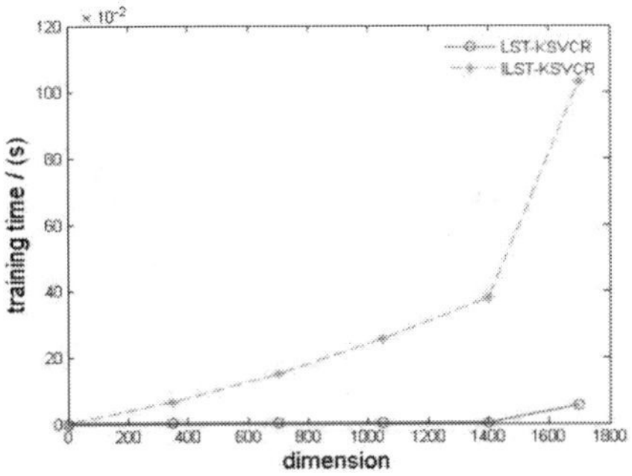

Figure 2. Contrast on training time.

5 Ending

This paper proposed an incremental learning algorithm (ILST - KSVC) suitable for the online environment by solving inverse matrixs based on least squares twin multi-class classification support vector machine (LST - KSVC). ILST - KSVC use historical data to update training data avoiding the repeated calculation, and reducing computational complexity. Selecting 10 groups of UCI data sets, based on MATLAB simulation platform for the two groups of experiments, the experimental results show that the ILST - KSCV inherits the advantages of LST - KSVC that has faster training speed and a high classification recognition rate when dealing with multiple classification data not only suitable for low dimensional data processing, also can be applied to the processing of high-dimensional data.

References

1. C. Cortes, V. Vapnik. M. L., *Support vector networks* **20**, 273-297 (1995)
2. W. Qingzhu, Z. Wenchao, W. Bin, Journal of Medical Systems **39**, 171 (2015)
3. Z. Ruijie, L. Bicheng, W. Fushan, Chinese Jounal of Electronics **42,** 646-652 (2014)
4. G. Farong, W. Jiajia, X. Xu-gang, et al., Jounal of Electronics & Information Technology **37**, 1154-1159 (2015)
5. D. K. Renuka, and P. Visalakshi, Journal of Scientific & Industrial Research **73**, 437-442 (2014)
6. K. R. Jayadeva, S. Chandra, Pattern Analysis & Machine Intelligence IEEE Transactions on **29**, 905-910 (2007)
7. K. M. Arun, M. Gopal, Expert Systems with Applications **36**, 7535-7543 (2009)
8. L. Bottou, C. Cortes, J. S. Denker, et al., *International Conferenceon Pattern Recognition, IEEE Computer Society Press*, 77-82(1994)
9. H. G. Kreßel, *Advances in Kernel Methods*, 255-268 (1999)
10. C. Angulo, X. Parra, A. Catala, Neurocomputing, KSVCR: a support vector machine for multi-class classification 55, 57-77 (2003)
11. X. Yitian, G. Rui and W. Laisheng, Cognitive computation **5**, 580-588 (2013)
12. J. A. Nasiri, N. M. Charkari and S. Jalili, Pattern Recognition **48**, 984-992 (2015)
13. C. P. Diehl, and G. Cauwenberghs, *International Joint Conference on Neural Networks*, 2685-2690 (2003)
14. Z. Guanhua, and H. Min, *16th International Conference on Management Science and Engineering*, 95-100 (2009)
15. F. Orabona, C. Castellini, B. Caputo, et al., Pattern Recognition On-line independent support vector machines **43**, 1402-1412 (2010)
16. L. Yang, L. Kai, et al., ICACI, 18-20 (2012)

A Data-origin Authentication Protocol Based on ONOS Cluster

Hua Qin and Na Wang

College of Computer Science, Beijing University of Technology, Beijing, China.

Abstract. This paper is aim to propose a data-origin authentication protocol based on ONOS cluster. ONOS is a SDN controller which can work under a distributed environment. However, the security of an ONOS cluster is seldom considered, and the communication in an ONOS cluster may suffer from lots of security threats. In this paper, we used a two-tier self-renewable hash chain for identity authentication and data-origin authentication. We analyse the security and overhead of our proposal and made a comparison with current security measure. It showed that with the help of our proposal, communication in an ONOS cluster could be protected from identity forging, replay attacks, data tampering, MITM attacks and repudiation, also the computational overhead would decrease apparently.

1 Introduction

Recently, there has been wide interest in Software-defined networking and how to secure it. A SDN controller, ONOS [1], has been designed to work under a distributed environment, and secured east-westbound protocol for the first time. There are two types of communication scenario in an ONOS cluster, where different kinds of information that need to be synchronized. One of the scenario synchronize information that needs to be strong consistent using RAFT protocol[2], and another one synchronize information that needs to be eventually consistent using gossip based anti-entropy protocol[3]. Table 1 shows the different between two communication scenarios.

Table 1. Communication scenario comparing [1]

	Communication Scenario	
	Strong Consistency	**Eventually Consistency**
Data Structure	ConsistentMap	EventuallyConsistentMap
Algorithm	RAFT	Gossip
Update Period	instantly	periodically (5ms)

In ONOS, data transfer between controllers in a cluster is default in plain text and suffer lots of vulnerability. To deal with the security challenges on control plane, ONOS has implemented TLS tunnel as a safety measure between controllers. It makes information transfer between two controllers confidentially. But it still has some problems, and it can be described as follows:

First of all, current TLS transmission between ONOS controllers is fully encrypted. One cannot implements different security policy for different communication scenarios.

In the strong consistency communication scenario, controllers keep information consistent with the leader controller. Attacker can hijack information sent by leader and make attacks which may result in fraudulent data be preserved in every controller in the cluster and even may cause the data plane impossible to use. Thus in this scenario encryption is a must. By contrast, in the eventually consistency communication scenario, where each ONOS controller has an equal role, and sends information more frequently than strong consistency communication (every 5ms). In this scenario, encryption is not necessary because of the characteristic of gossip protocol. Even if an attacker tampers information and sends to an ONOS controller, correct information can be received from other controller. If TLS is implemented, information will become consistent once it is received, but the encryption and decryption will become the performance bottleneck of the controller when a cluster has a large scale. On the contrary, transfer information in plain text in this scenario will greatly decrease the system cost, while the time when information become consistent will be longer. Due to the characteristic of gossip protocol, information transferred in this scenario is not time sensitive, lower cost security methods like integrity verification and digital signature are enough to protect information from attacks. Therefore in an ONOS cluster, different communication scenarios need different security policies.

Secondly, communication between ONOS controllers doesn't have the mechanism of data-origin authentication. An ONOS controller keeps listening a specific port, and with no authentication an attacker can impersonate an controller in a cluster and send fraudulent data to the cluster easily, which would cause severe problem on the data forwarding plane. For this reason, a light-weight authentication is needed for east-westbound communication, which secure ONOS controllers from malicious attacks.

What's more, TLS [4] tunnel itself is prone to security threats. In our scenario, MITM attacks would take place and both parties of communication could deny the message they have sent [5]. During the handshaking phase of TLS session, all message is sent in plain text, so an attacker can get these information easily, and use them to build a TLS tunnel with a valid controller. Moreover, the secret session key for TLS tunnel is shared by both parties, TLS tunnel cannot assure non-repudiation.

In order to prevent these problem, we proposed a light-weight data origin authentication protocol based on a two-tier hash chain on ONOS controllers. The main purpose of our proposal is to secure information in an ONOS cluster from eavesdropping, tampering and malicious controller. This protocol can be a supplemental security measure for the current TLS tunnel between ONOS controllers, also can provide sufficient security for eventually consistency communication scenario when message transfers in plain text.

We have organized the rest of this paper in the following way. In section two, we introduce the background and related work of our proposal. Section three describes the protocol in detail. And in section four, we analyze the security and overhead of our proposal. Finally in section five, we present our conclusion.

2 Background and Related Work

Digital signature can be used for data-origin authentication. The key technology of digital signature are: public key cryptography, shared key cryptography and Hash algorithms. Compare with the first two technology, hash algorithms demonstrated high reliability and high efficiency.

Due to the frequent traffics between ONOS controllers in a cluster and high computation overhead caused by public key and shared key cryptography, an authentication method with low computation cost and high efficiency is needed. It follows that Hash algorithm is more suitable for communication in an ONOS cluster and capable to meet the security requirements. Hash chain is a method widely used for data-origin authentication [6], and is capable of securing information in an ONOS cluster from tampering, data replay and repudiation.

To work around the length limitation of tradition hash chain [7], the Self-Renewal Hash Chain [6] has been proposed. This new kind of hash chain can be self-renewing without addition protocols and re-initialization smoothly. Another approach, the Two-tier One-way Hash Chain [8], have gave significant performance improvement over previously proposal. This proposal has a single first-tier chain and multiple second-tier chains. The first-tier hash chain offers seeds for second-tier chains which is computed by the first-tier hash function. And second-tier chains are used for authenticating the transaction between the server and clients. These methods based on hierarchical hash chains are more suitable for the communication scenarios.

Recent research on SDN has been focusing on securing east-westbound communication [9, 10]. Jun-Huy Lam et al proposed a scheme of securing the distributed SDN communication with a multi-domain capable Identity-Based Cryptography (IBC). This scheme uses symmetric session key to encrypt communication between two controllers. Study in [10] also used IBC to authenticating the identities of controllers and shared a session key for east-westbound communication. Both of studies in studies in [9, 10] provide a way to communicate encrypted in a SDN cluster. But they still cannot protect SDN controller from illegal controller which will eavesdrop, tamper or replay the traffic in a cluster. Moreover, these schemes depends on symmetric encryption which will cause massive computation cost.

3 A Data-origin Authentication Protocol Based on ONOS Cluster

In this section, we propose a two-way data-origin authentication protocol for ONOS cluster. Our proposal is based on two-tier self-renewable hash chain introduced by Alabrah et al [8] and Zhang et al [6] which are not meant for SDN. In our proposal, two tiers of hash chain, as shown in Fig.1, are used for different purposes. The first tier of hash chain is used for identity authentication between two ONOS controllers and also for providing seed for second tiers at the beginning of the communication in an ONOS cluster. And second tiers of hash chain are used for data-origin authentication during the communication.

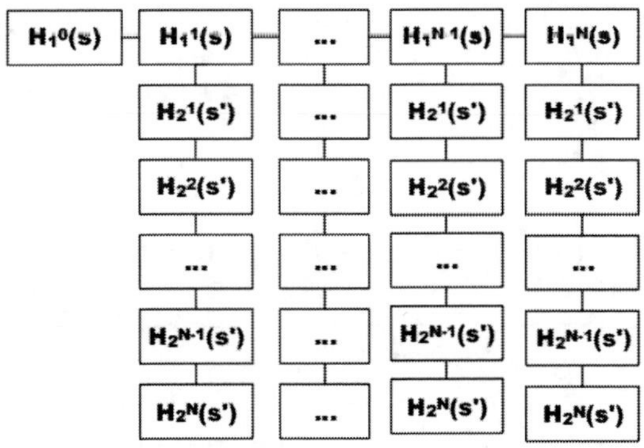

Figure 1. Two-tier self-renewal hash chain

The whole protocol is done in three routines: identity authentication, data-origin authentication and seed update. For eventually consistency communication scenario, all three steps are needed. For strong consistency communication scenario, only data-origin authentication and seed update are needed because TLS already offers identity authentication. Using the notation presented in Table 2, we describe the main components of the proposal in order to communicate.

Table 2. Notation

h_x	Hash function of the x tier hash chain
S_X	Seed of ONOS controller X
PKX, SKX	Private and public key of ONOS controller X
MAC	Integrity check code
N_{XY}	Length of the y tier of hash chain of ONOS controller X
E	Encryption
rX	Random number produced by ONOS controller X
s	Serial number of packets
T_X	Data-origin authentication token from ONOS controller X

3.1 Identity Authentication

The identity authentication routine contains three steps: *Initialization*, *Message transaction* and *Verification*, and two participants which are denoted by controller A and controller B, as shown in Fig. 2.

Initialization is used to distribute the information for identity authentication and seed of the first tier hash chain. Before cluster deployment, the deployment server which works as a Trusted Third Party (TTP) distributes the private and public key for identity authentication and the seed of first tier hash chain. Also the deployment server of the ONOS cluster will send a cluster description file to every ONOS controller in the cluster which contains all ONOS controllers' ID and their states. After ONOS controllers receive everything mentioned above, each controller computes the first hash chain and broadcasts the tail of first hash chain to every controllers in the cluster. In what follows, we give the notations of the identity authentication routine.

ID_A = Identity code for controller A

ID_B = Identity code for controller B

N_1 = length of first tier hash chain

N_2 = length of second tier hash chain

IDSet = controllers' ID set in the cluster description file

mac = hash function

M = message payload between hash chain node and MAC

H = authentication token

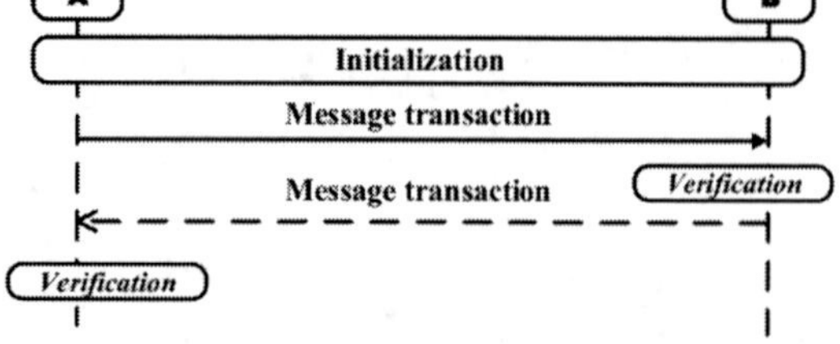

Figure 2. Identity authentication steps

Then, every ONOS controller begins *Message transaction* and *Verification*. Fig. 3 illustrates the messages transform between controllers in detail. During *Message transaction*, controllers select and exchange a random number to authenticate the identity of each other and exchange the tail of second hash chain for later data-origin authentication after they pass the identity authentication.

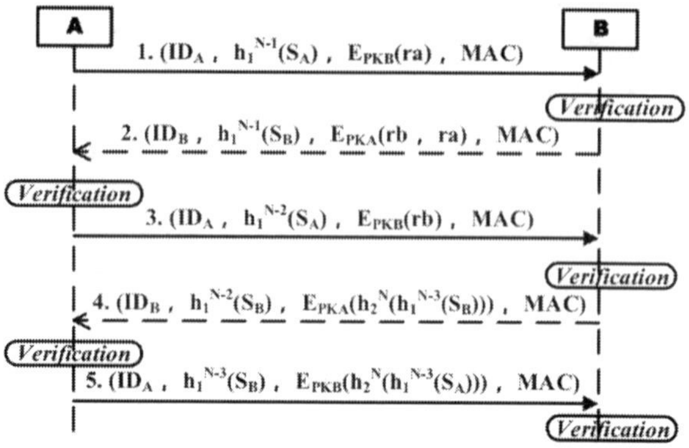

Figure 3. Message transaction step

After receiving message from other controllers, every controller needs to do *Verification* to check the massage is valid or not. Follows are the *Verification* routine, in which payload is all encrypted information during *Message transaction*.

For the kth verification on controller B:

Verification (ID_A,*H*,*Payload*,*MAC*)
Begin
If (ID_A is in IDSet) then
    $T := TA$
    If ($T == h_1(H)$) then
        $MAC' := mac(ID_A, H, Payload)$
        If ($MAC' == MAC$) then
            $T_A := H$
            $N_{A1} := N_{A1}$-1; end_if
    end_if
end_if
if(N_{A1}==0) then
    T_A:= Call *Seed_Update()*
    N_{A1}:=*N1*; end_if
Return (*TRUE*);
End

3.2 Data-origin Authentication

When controllers finish identity authentication, consistency communication begins. And each controller need to sign on the message it sends and let others do data-origin authentication to check message sender is valid or not. Data-origin authentication routine is shown in Fig. 4, where S'_A and S'_B is the next token of the first hash chain from the sender. Each time a controller receives a message, it checks serial number and token of the second tier hash chain by *Verification*. The *Verification* routine is as same as we mentioned above, and the Payload is the serial number of the message and payload it carried. What's different is that at beginning of data-origin routine T_A and T_B equals to the tail of each controller's second hash chain.

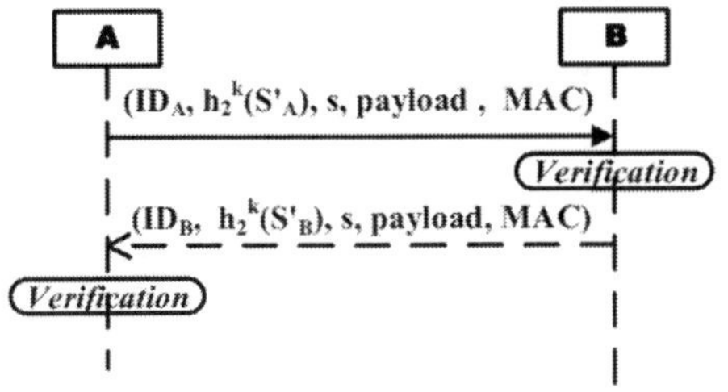

Figure 4. Data-origin authentication routine

3.3 Seed Update

When the hash chain is going to use up, the seed update routine is needed. For the first tier hash chain, the seed is updated by a right shift operation. And for the second tier hash chain, the new seed equals to the first unused token of the first hash chain. For example, controller B is going to do the kth time of seed update, the new seed is h_1^k+5(S_B). After update the seed, new tail of hash chain is carried by the Message transaction to the controllers in the cluster for later authentication. Fig. 5 shows the message that carries new tail of hash chain, where S'new and Snew are respectively the new seeds of second tier hash chain and first tier hash chain.

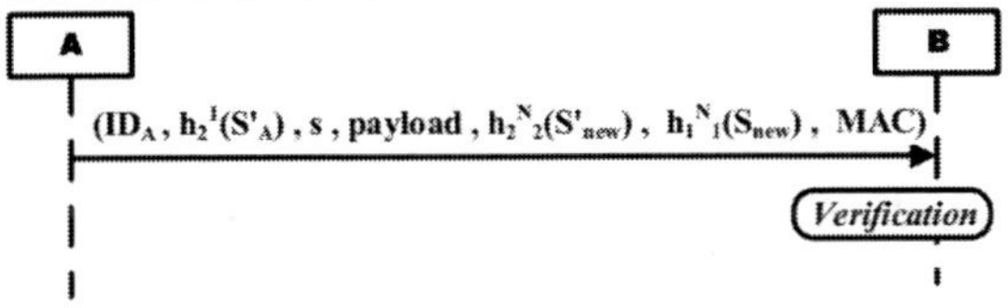

Figure 5. Message carries new tail of hash chain

4 Analysis

4.1. Security analysis

In this section, we analysis the security of the data-origin authentication protocol we proposed. The proposed scheme is a light-weight protocol and can provide several security assurances which are analysis in detail.

Unforgeability of controller's identity: In this proposal, the seed which is used to calculate the hash chain is securely sent from the TTP. Thus only a valid controller owns the seed and the hash chain generated by the seed. And because of the irreversible of the hash function, attacker who captures message between two valid controllers and get the token, cannot deduce the seed from a token. Therefore the identity of a valid controller cannot be forged.

Protection against MITM attacks: Suppose an attacker uses MIMT attack to get the certificates of valid controllers and creates a TLS tunnel with them. Because of lacking of the token, message sent by the attacker cannot pass the data-origin authentication, and a valid controller will notice that there is an attack, then cuts the connection.

Availability of data: The proposal provides serialization and integrity verification for every message transferred between controllers. Consequently, replay attacks and data tampering are avoided.

Non-repudiation: In this proposal, only valid controllers can provide tokens, thus a controller cannot deny the message it sent.

4.2. Overhead Analysis

We analysis performance overhead of our protocol in detail, and compare with tradition TLS protocol. We present a detailed analysis of the overhead associated with the identity authentication routine and data-origin authentication. T_{ASYM} and T_{SYM} are computational overhead for encryption or decryption using asymmetric algorithm and symmetric algorithm respectively. Thash is computational overhead for hash function. And T_{CERT} is computational overhead for certificate verification.

During the identity authentication routine, each controller costs $5T_{ASYM}+3T_{hash}$ computational overhead. And during data transfer, for each transaction, a controller only costs 2Thash computational overhead because of data-origin authentication. Compare with traditional TLS protocol, which will cost $T_{CERT}+2T_{ASYM}$ during TLS handshake for two-way identity authentication, and $T_{SYM}+T_{hash}$ computational overhead during data transfer. Table 3 shows the overhead comparison between our protocol and traditional TLS protocol.

Table 3. Comparison between our protocol and traditional TLS protocol

Protocol	Overhead during identity authentication	Overhead during data transfer
Our Protocol	$5T_{ASYM}+3T_{hash}$	$2T_{hash}$
TLS Protocol	$T_{CERT}+2T_{ASYM}$	$T_{SYM}+T_{hash}$

According to the analysis in studies [11, 12], the computational overhead of encryption or decryption using asymmetric algorithm or symmetric algorithm, which are T_{ASYM} and T_{SYM}, is much higher than the computational overhead of hash function, which is T_{hash}. In addition, Thash of the commonly used hashing algorithms, such as MD5 and SHA, is usually quiet low. Thus, when we use the proposed protocol as a supplemental security measure for the current TLS tunnel, the computational overhead would rise slightly. Also, when we use is as the security measure for eventually consistency communication scenario, the computational overhead would decrease apparently than current TLS tunnel.

5 Conclusion

We proposed a two-way light-weight data-origin authentication protocol for ONOS cluster, which is based on two-tier self-renewable hash chain. The proposal we designed can be used as a supplemental security measure for the current TLS tunnel between controllers in an ONOS cluster, also can provide sufficient security for eventually consistency communication scenario. The protocol can protect controllers from attacks such as MIMT, data tampering, replay, and identity forge. Also the protocol can provide non-repudiation for controllers, which controllers cannot deny the messages they send. As future work, we envision an improvement of the protocol from inter-domain to multi-domain.

References

[1] Information on: https:/wiki.onosproject.org
[2] Information on: http://thesecretlivesofdata.com/raft/
[3] Information on: https://wiki.onosproject.org/display/ONOS/Network+Topology+State#NetworkTopologyState-FailureHandlingandAnti-Entropy
[4] Dierks T, Rescorla E. The Transport Layer Security (TLS) Protocol Version 1.1. In: E R, editor. (2006)
[5] Information on: http://www.ntu.edu.sg/home/ehchua/programming/webprogramming/HTTP_SSL.html
[6] Haojun Z, Xiaoxue L, Rui R. A Novel Self-Renewal Hash Chain and Its Implementation. 2008 IEEE/IFIP International Conference on Embedded and Ubiquitous Computing. Shanghai, p.144-149(2008)
[7] Xiao-Yuan Y, Jia-Yong C, Jing-Jing W. A Self-Renewal Hash Chain Scheme Based on Fair Exchange Idea(SRHC-FEI). 3rd IEEE International Conference on Computer Science and Information Technology. Sichuan, p.152-156(2010)
[8] Alabrah A, Bassiouni M. A Hierarchical Two-Tier One-way Hash Chain Protocol for Secure Internet Transactions. Anaheim, CA, p.868-873(2012)
[9] Jun-Huy L, Sang-Gon L, Hoon-Jae L, Eko OY. Securing Distributed SDN with IBC. ICUFN 2015. Sapporo, p.921-925(2015)
[10] Santos MAS, Nunes BAA, Obraczka K, Turletti T, de Oliveira And Cintia B. Margi BT. Decentralizing SDN's Control Plane. 39th Annual IEEE Conference on Local Computer Networks. Edmonton, AB, p.402-405(2014)
[11] PAULSON LC. Inductive Analysis of the Internet Protocol TLS. ACM Transactions on Information and System Security. **2**, 3, p.332-351(1999)
[12] Apostolopoulos G, Peris V, Saha D. Transport Layer Security: How much does it really cost. INFOCOM '99. Eighteenth Annual Joint Conference of the IEEE Computer and Communications Societies. New York, p.717-725(1999)

2-D Layout for Tree Visualization: a survey

Guanqun Wang[1], Tsuneo Nakanishi[2] and Akira Fukuda[1]

[1]*Faculty of Information Science and Electrical Engineering, Kyushu University, Fukuoka, Japan 819-0395*
[2]*Department of Electronics Engineering and Computer Science, Faculty of Engineering, Fukuoka University, Fukuoka, Japan 814-0180*

Abstract. This is a survey of recent researches on 2D tree visualization approaches. The whole paper will focus on 2D layouts for general trees including different styles of node-link diagram, main variations of Treemap, some solutions designed especially for mobile devices, and several hybrid approaches. We are not trying to give an exhaustive list of tree layouts here. Instead, we will introduce new ideas in the main categories of general tree layouts, which may inspire more efficient and creative designs.

1 Introduction

Since a major part of real world information is hierarchically structured, and tree is a perfect data structure for the storage of hierarchy information, an effective method of tree visualization becomes necessary in a lot applications. This paper will introduce new tree visualization ideas proposed recently. Although there are already some researches on 2.5D[1, 2] and 3D[3, 4] visualization of trees, to narrow down the topic, this paper will only focus on 2D tree visualization methods. For the same reason, we mainly introduce layouts in this paper and very little information about navigation or interaction techniques will be included.

The researches on tree visualization started about half a century ago [5], and there have been a lot of developed approaches to visualize trees in different situations. Based on how the parent-child relationship is represented, all tree visualization methods can be roughly divided into two large groups, connection methods and enclosure methods [6, 7]. In the following part of this paper, Section 2 will introduce some widely used connection methods; Section 3 will introduce some famous enclosure methods; Section 4 will discuss about other tree visualization methods designed especially for mobile devices; Section 5 will talk about some hybrid methods.

2 Connection Methods

Connection methods uses node-link diagram to explicitly show the relationships of nodes. They make it very easy for the user to get an immediate perception of the structure. However, usually they are not efficient in display space utilization [6]. The following subsections will introduce some main categories of connection methods.

2.1 Level-Based Drawing

In a level-based drawing, all nodes are placed on a set of layers which are several parallel lines in the display area, and typically all nodes from the same level are placed on one unique layer. The structure is pretty explicit in such drawings. However, it usually takes a lot of efforts to improve the space utilization.

Buchheim, Jünger and Leipert summarized 5 aesthetic properties which are commonly required when drawing a rooted tree [8]:

- The y-coordinate of each node corresponds its level.
- The edges do not cross.
- The drawing of a subtree is independent of its position in the tree.
- If the tree is ordered, the ordering information is shown in the drawing.
- The drawing of the reflection of a tree is the reflected drawing of the original tree.

In 1981, Reingold and Tilford proposed the first linear time layout algorithm satisfying all 5 aesthetics, thus it is one of the most famous tree layout algorithms and usually called Reingold-Tilford algorithm now [9]. This algorithm, in a bottom-up direction, recursively draws subtrees independently and puts subtrees with the same parent as close as possible. After all nodes have fixed coordinates, the edges are then inserted accordingly. The Reingold-Tilford algorithm was originally proposed to draw ordered binary trees, but Walker modified it later and successfully made a variation which works for general trees without violating any of the aesthetics [10]. This variation is known as Walker's algorithm in a lot of researches. In 2002, Buchheim, Jünger and Leipert proved that Walker's algorithm does not run in linear time like Walker claimed and presented a modification of Walker's algorithm with linear runtime [8]. So far all variations of Reingold-Tilford algorithm place a parent at the centre of its children, which in fact, according to the research of Marriott and Sbarski, can lead to an unnecessarily large width. By relaxing the requirement of placing a parent exactly in the centre of its children and taking edge lengths into consideration during drawing, Marriott and Sbarski presented a compact version of Walker's algorithm [11].

When the size varies from node to node, layered drawing may use more space than necessary. In such situations, a non-layered drawing method which places children at a fixed distance from the parent may be more practical. A lot of work on non-layered drawing can be found in the existing literature [12–16], however, none of them gave a linear algorithm for the general case. In 2014, Ploeg proposed a non-layered variation of Reingold-Tilford algorithm and proved it can run in linear time [17].

When we allow nodes from different levels to be placed on the same layer, we can obtain a shorter drawing and the minimum-layer drawing becomes an interesting problem. Some researches have been done on drawing graphs on two or three layers [18]. Suderman have proved optimal upper and lower bounds for h-layer drawing of trees and proposed a linear layout algorithm matching the upper bound [19]. Alam, Samee, Rabbi, and Rah- man proposed a layout algorithm for minimum-layer up- ward drawing which can run in linear time [20]. One year later, Mondal, Alam, and Rahman demonstrated a linear algorithm for general minimum-layer drawing [21]. In- stead of the pathwidth [22] used in a lot of layered drawings, these two algorithms use a new parameter called line- labelling to divide the tree, and the results are satisfying.

2.2 Non-Level-Based Drawing

Non-level-based drawings are not as good at showing structure information as level-based drawing. However, given more freedom of nodes placements, they can achieve improvements in some metrics such as angular resolution, space utilization, aspect ratio, algorithm runtime and so on.

The H-V (Horizontal-Vertical) approach is one of the most well-known none-level-based tree drawing methods. The basic idea of this approach is very simple. It recursively draws the root in the top-left corner and subtrees in horizontal composition (one next to the other) or vertical composition (one below the other), which is shown in Fig.1 [23]. One algorithm based on this approach has been developed for binary trees [24], and it can be easily extended to general trees [23].

Hyperbolic tree is a famous "Focus+Context" visualization method for large hierarchies. It was first proposed by Lamping and Rao [25, 26]. This technique lays out the tree uniformly on a hyperbolic plane which is mapped onto a circular display area. As shown in Fig.2 [25], this kind of lay- out assigns more space to a specific portion of the tree (focus) without sacrifice of context presentation so that it can effectively visualize a much larger tree than the conventional methods.

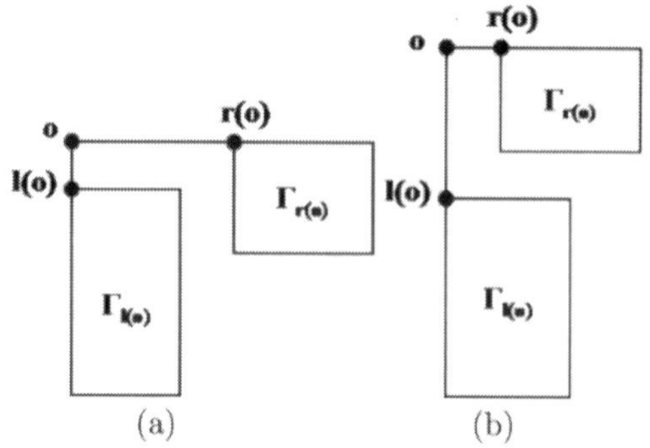

Figure 1. (a) Horizontal composition (b) Vertical composition

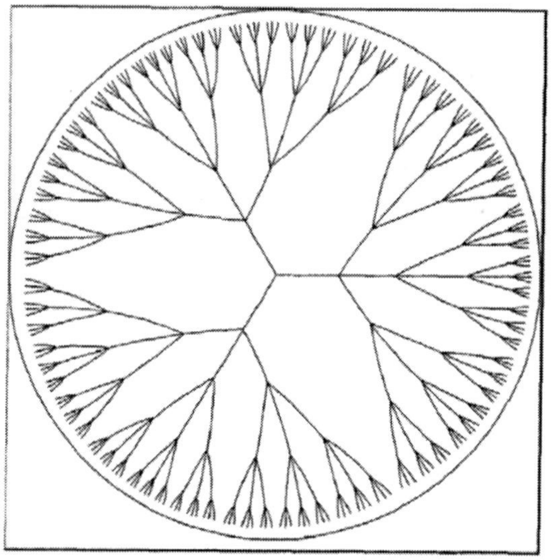

Figure 2. A uniform tree of depth 5 and branching factor 3 (364 nodes)

Radial drawing is another popular category of graph visualization. Fig.3 [24] shows a transformation from layered drawing to radial drawing. Radial drawing can be considered as a special category of level-based drawing which uses polar coordinates instead of Cartesian coordinates. All nodes are placed on several concentric circles while the root is placed at the centre, and nodes from the same level are places on the same circle. Bachmaier presented a radial version of the Sugiyama Framework [27] by placing nodes on concentric circles instead of horizontal lines and drawing edges as outward monotone segments of spirals instead of straight lines. The experiments showed that this drawing style allows more flexible edge and the number of crossings is reduced by about 30% [28].

Duncan, Eppstein, Goodrich, Kobourov, and Nollenburg have demonstrated how to draw trees with perfect angular resolution and polynomial area [29]. As shown in

Fig.4 [29], the result is a little similar to radial drawing. They also found that it's not always possible to achieve perfect angular resolution and polynomial area if the tree is ordered. However, if we use Lombardi drawing style [30], which means we draw edges with arcs instead of straight lines, both perfect angular resolution and polynomial area can be guaranteed as shown in Fig.5 [29].

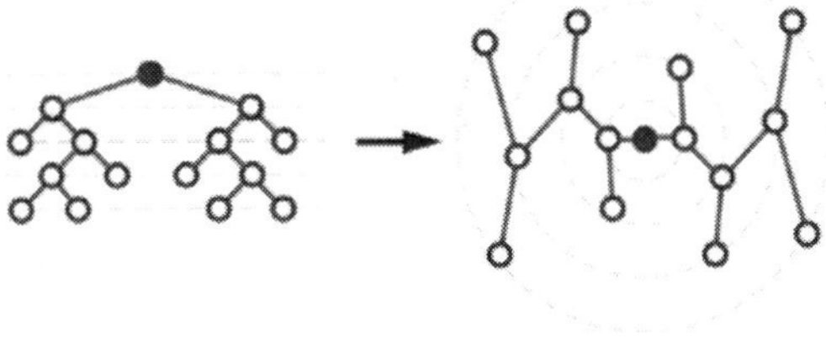

Figure 3. A transformation from layered drawing to radial drawing

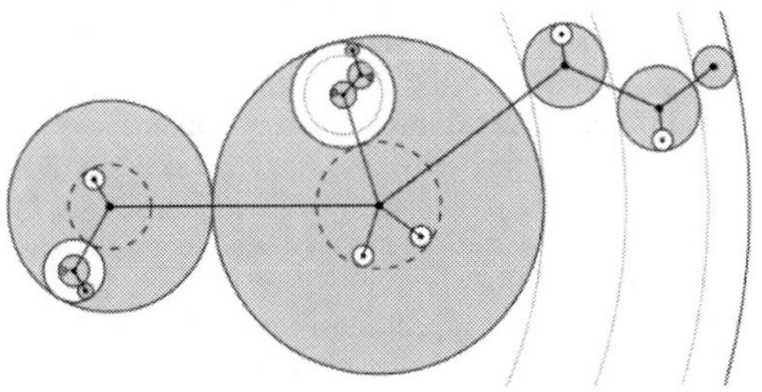

Figure 4. Straight-line drawing for an unordered tree

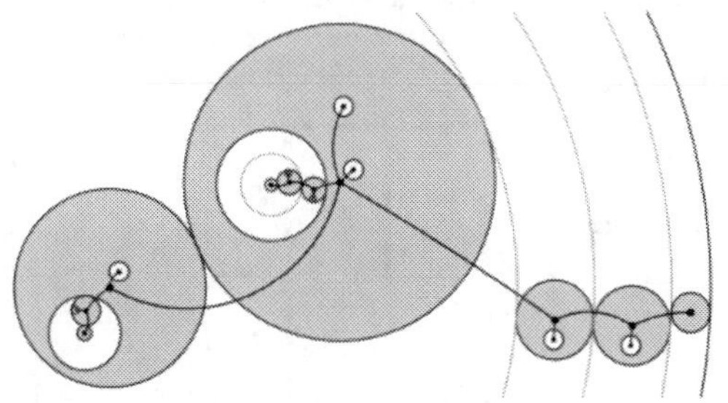

Figure 5. Lombardi drawing for an ordered tree

Ring-based drawing is a kind of drawing style usually used for trees with high degrees. In a ring-based drawing, children are typically placed on the circumference of a circle centred at their parents as shown in Fig.6 [24]. Rusu, crowell, Petzinger, and Fabian also presented an edgeless ring-based tree drawing method called PieVis, one example of which can be found in Fig.7 [31]. All nodes are rep- resented by a circle in this kind of drawing, and children are placed inside of the circle of the corresponding parent. It's important to notice that this PieVis method can only partially visualize the tree by defining a parameter called "degree of interest".

Force-directed layout is another very interesting visualization method. As early as in 1984, Eades modelled a graph as a physical system with rings and springs. With decent parameters, the optimal drawing should occur when the system stays in a stable state [32]. Although there is springs in the system, Eades didn't use Hooke's law to model the force exerted by springs. Instead, he chose his own formula. In fact, it was found that it's very difficult to overcome local minima when using Hooke's law [33]. Fruchterman and Reingold presented a modification of the approach of Eades in 1991. They used a simpler but effective equation to calculate the virtual force and create a faster algorithm. Recently, efforts on graph drawing by physical system simulation and quadratic programming can also be found [34–36]. Even for the Lombardi drawing [30], a relatively new style, there are already some work done on applying force-directed algorithm to generate the optimal drawing [37].

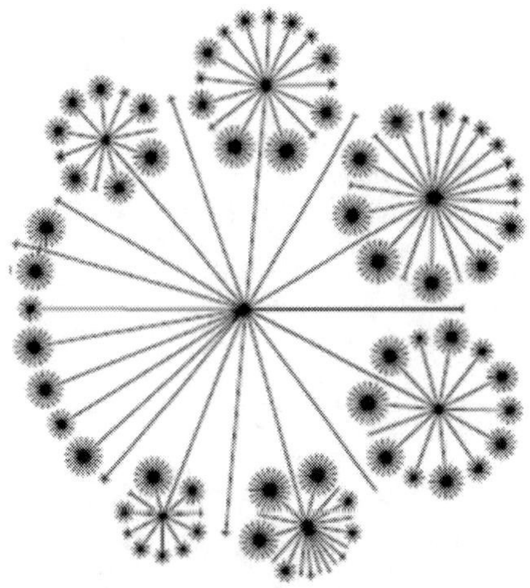

Figure 6. An example of general ring-based drawing

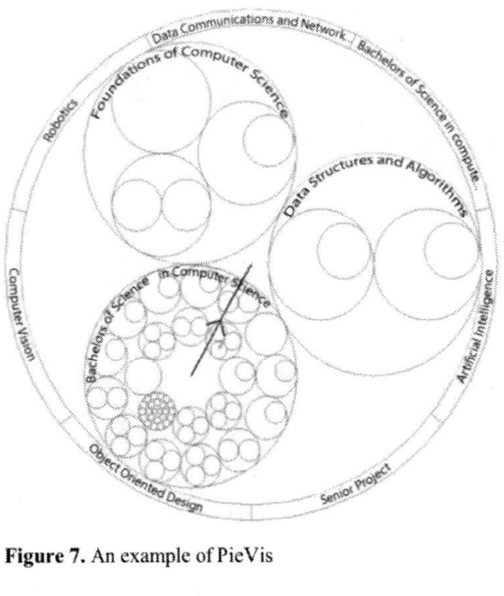

Figure 7. An example of PieVis

3 Enclosure Methods

Enclosure methods use the space-filling concept to map the hierarchical information to a display area and usually can achieve 100% space utilization [6]. Treemap is the most famous and widely used category of enclosure methods and has a lot of variations, so the rest part of this section will focus on Treemap.

Treemap is a widely used tree visualization tool. Application based on it can be found in a lot of fields including financial solution [38], sport information analysis [39, 40], browsing software [41], and even in usenet activity research [42, 43]. The following subsections will introduce the original treemap at first and then some variations based on it.

3.1. Original Idea

In 1990, Shneiderman introduced the concept of Treemap [44, 45]. It is a layout method that represents hierarchical structures with enclosing rectangles while fully utilizing space. In Treemap, each node is represented by a rectangle. The sizes of the rectangles are proportional to the number of children of the node to be partitioned. The general algorithm is usually called "slice and dice". It recursively uses parallel lines to divide a rectangle representing a node into smaller rectangles representing its children nodes. At each level of hierarchy, the orientation of the lines, which is either vertical or horizontal, is switched. The whole process stops when all leaf nodes are reached. Treemap of a tree shown in Fig.8 [44] is shown in Fig.9 [44].

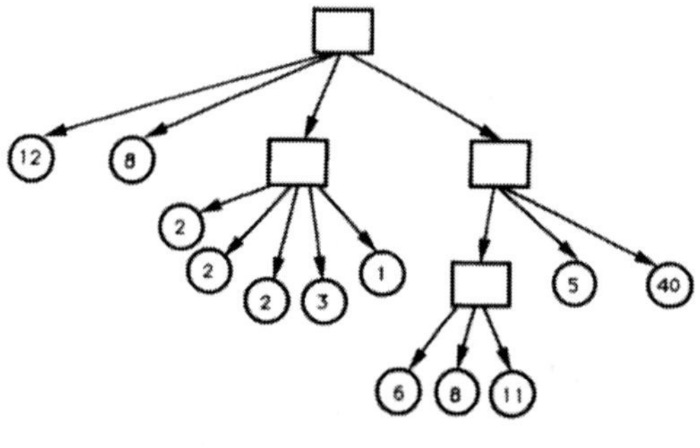

Figure 8. A 3-level tree with a number in each leaf indicating the weight

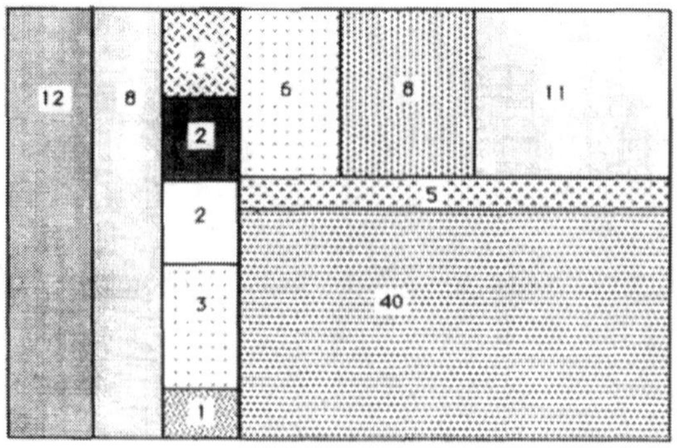

Figure 9. Treemap of the tree shown in Fig.8

3.2 Clustered Treemap

Wattenberg pointed out that there are two problems in the original treemap visualization approach proposed by Shneiderman. The first problem is that this approach may lead to a layout containing rectangles with extremely high aspect ratio, which is very hard to read or label. The second problem is that the original approach can only visualize hierarchical information, which means it cannot show relationships between some special vertices. To solve these two problems, Wattenberg proposed Cluster Treemap [46]. In Cluster Treemap, both vertical and horizontal partitions are employed at each level of the tree to avoid high-aspect-ratio rectangles. Although this new partition method doesn't eliminate extreme aspect ratios theoretically, very obvious improvement in readability was observed in the experiment with real-world data. In all possible layouts, one is chosen to make sure neighbouring rectangles have as much similarity as possible. The more similarity there are between two rectangles, the more similarly how their values have changed in the past. A Cluster Treemap of the original treemap shown in Fig.10 [46] can be found in Fig.11 [46].

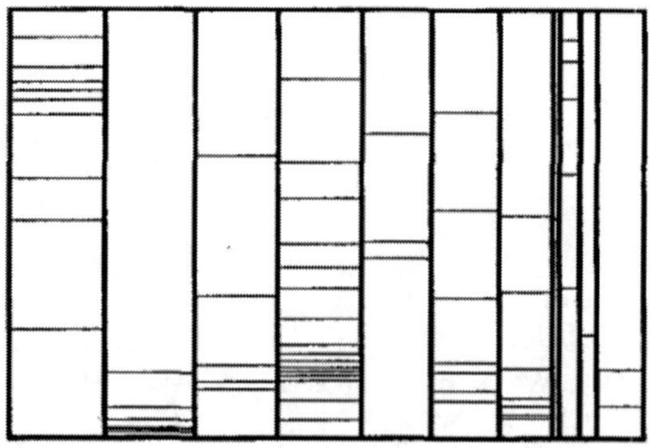

Figure 10. A treemap generated by slice and dice algorithm

Figure 11. Cluster Treemap of the tree shown in Fig.10

3.3 Cushion Treemap

Wetering and Wijk argued that the original Treemap is not good at visualizing the structure of the tree. Especially when the tree is a balanced tree, the corresponding Treemap turns to a regular grid. To solve this problem, they presented a new visualization method called Cushion Treemap [47]. Cushion Treemaps use intuitive shading to provide insight in the hierarchical structure. A ridge rendered with a simple shading model is added when drawing the rectangle for each node. "Valleys" between adjacent rectangles can be observed after shading, and the depth of the "valleys" is proportional to the distance between the corresponding nodes in the tree. Fig.12 [47] shows one example of Cushion Treemap.

Figure 12. Cushion Treemap example

3.4 Squarified Treemap

After the Cushion Treemap, Bruils, Huizing, and Wijk also proposed another visualization method, Squarified Treemap [48], to solve the large aspect ratio problem of the original Treemap, which was pointed out by Wattenberg before, too. The general algorithm is a little similar to Cluster Treemap. Squarified Treemap also lay out rectangles both horizontally and vertically for nodes in the same level. When drawing rectangles of a set of siblings, the algorithm lays out the new rectangle in either the same direction as the previous one or a new direction. The algorithm will choose whichever direction that has a better worst case of aspect ratio. Although there is no theoretical evidence to show that this algorithm is the best solution to improve the aspect ratios, Bruls, Huizing, and Wijk claimed that this method empirically turned out to give the best results. The Squarified Treemap of the Treemap shown in Fig.13 [48] can be found in Fig.14 [48].

3.5 Ordered and Strip Treemap

In 2001, Shneiderman, and Wattenberg pointed out that Cluster Treemap and Squarified Treemap share several drawbacks [49]. At first, changes in data set can lead to dramatic discontinuous changes in the layouts, which may be very hard for users to follow. In fact, Heer and Roberston have proved by experiments that smooth transitions can help users track changes in data [50]. The other problem is that no ordering information is included, which is sometimes very useful for readers to see patterns or find particular items.

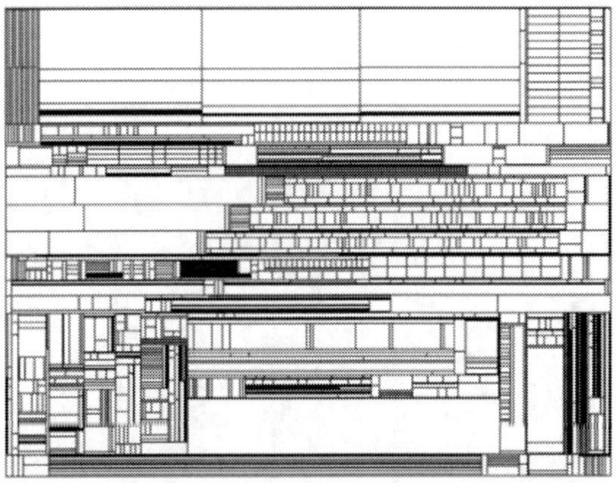

Figure 13. A treemap generated by slice and dice algorithm

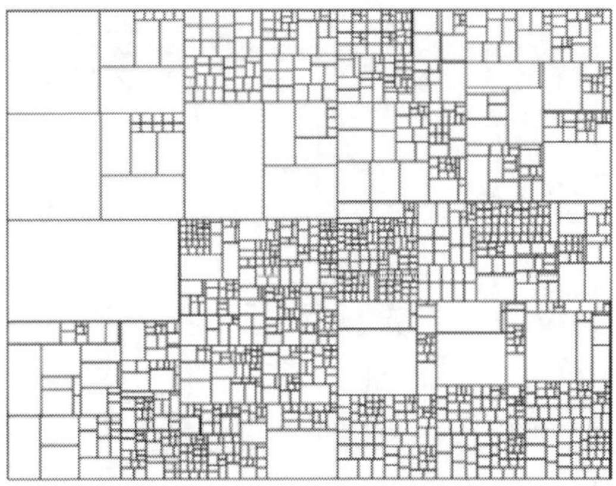

Figure 14. Squarified Treemap of the tree shown in Fig.13

DOI: 10.1051/matecconf/20165601007

In order to solve these two problems, they presented Ordered Treemap [49]. In Ordered Treemaps, items that are next to each other in the given order are also adjacent. The basic idea of Ordered Treemap algorithm is similar to the conventional Quicksort algorithm. It recursively divides the display area (a rectangle) into 4 sub-areas in a fixed way, divide the items to display into 3 groups and one single item while all items in each group are continuous in the given order, and then assign the 3 groups and one item to 4 sub-areas until every single item has its own rectangle. The single item apart from 3 groups is selected based on some assumptions and there are several decent options to make the decision.

An alternative solution for Ordered Treemap, Strip Treemap, was also introduced by Bederson, Shneiderman, and Wattenberg [51]. Strip Treemap is a modification of Squarified Treemap [48]. It divides the whole display area into several strips with different widths. Each strip is divided again into several rectangles, each of which represents a node in the tree. The general algorithm recursively decide to either draw the new rectangle in the current strip or a new strip. The decision is made based on the how the average aspect ratio changes respectively. Fig.15, Fig.16, and Fig.17 [51] shows visualization results of a tree with original treemap algorithm, Ordered Treemap algorithm, and Strip Treemap respectively.

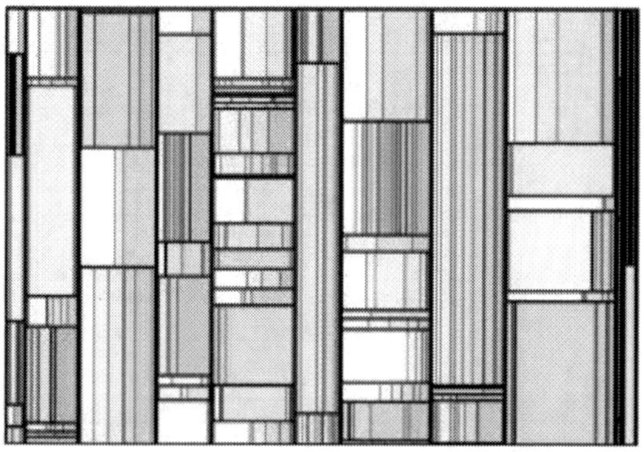

Figure 15. A treemap generated by slice and dice algorithm

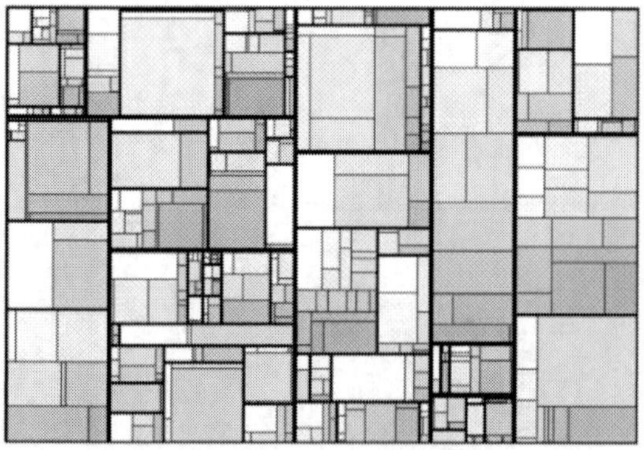

Figure 16. Ordered Treemap of the tree shown in Fig.15

Figure 17. Strip Treemap of the tree shown in Fig.15

3.6 Quantum Treemap

Bederson, Shneiderman, and Wattenberg also proposed Quantum Treemap in the paper introducing Strip Treemap [51]. When images with fixed aspect ratios need to be dis- played in each rectangles of treemaps, some images may have to be scaled down to fit their own rectangles, which may finally make the images very hard to read. Quantum Treemap was designed to solve this problem exactly. All widths and heights of rectangles in Quantum Treemaps are multiples of a step size decided by the user. In this way, all images displayed in the treemap can have exactly the same size and perfectly aligned. Any treemap can be modified to a Quantum Treemap easily by adding one quantization step at the end of regular algorithm.

3.7 Voronoi Treemap

According to [52, 53], various shapes in treemap may help the user with pattern recognition and change tracking. However, all conventional treemaps can only rep- resent each node with a rectangle and the entire display area has to be a rectangle, too. Several researches were done to break through this limitation. In 2005, Balzer and Deussen proposed Voronoi Treemap [54, 55]. In Voronoi Treemap, one display area is divided based on centroidal Voronoi tessellations [56] which allows subareas with different shapes. Fig.18 [54] shows an example of Voronoi Treemap where PW is a distance function chosen in Voronoi tessellation. Finding an optimal solution is a NP-complete problem, so the layout algorithm actually tries different parameters to look for an approximation good enough for treemaps [54], which means a lot of computation power is required. In 2010, Sud, Fisher, and Lee proposed a GPGPU-based technique to speed up the layout algorithm [57].

Figure 18. An example of Voronoi Treemap

3.8 Spiral Treemap

In 2007, Tu and Shen presented Spiral Treemap to reduce abrupt layout changes and produce smooth transitions in treemaps [58]. Ordering information is also stored in Spiral Treemap, so it can be considered as an alternative of Ordered Treemap. The main idea of Spiral Treemap is to use 'spirals' for area partition. Items with the same parent are organized as a spiral in the treemap, and the order is reserved. Since usually there are many possibilities for the layout, the layout algorithm use a similar technique to Strip Treemap. The algorithm only starts drawing a rectangle on the next side if the average aspect ratio decreases after adding this rectangle on the current side. One example of Spiral Treemap is shown in Fig.19 [58].

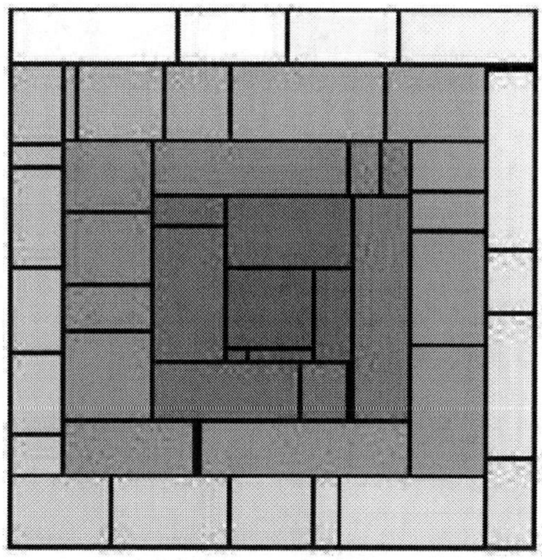

Figure 19. An example of Spiral Treemap

3.9 D&C Treemap

In 2013, Liang and Nguyen proposed D&C Treemap as an alternative solution to break through the "rectangle limitation", and this kind of treemap was claimed to be better than Voronoi Treemap because of lower computation power requirement [59, 60]. D&C Treemap visualizes trees just like other treemaps except that it can partition a polygonal display space into smaller polygons. The algorithm recursively divides the nodes at the same level into two groups with similar total weights and divide the whole display area into two polygons, each of which is as- signed to one group. This process stops when every node has its own polygon. When partitioning the display area, three kinds of algorithms (polygonal/angular/rectangular) are available. Each algorithm can be applied alone, and multiple ones can also be applied together. One example of D&C Treemap is shown in Fig.20 [60].

3.10 Hilbert and Moore Treemap

To achieve smooth transition and store ordering information during data change like Ordered Treemap and Spiral Treemap did, Tak and Cockburn proposed Hilbert and Moore Treemaps, which are two kinds of treemaps based on Hilbert and Moore space-filling curves respectively [61]. The layout algorithm recursively divides the data into four quadrants until each quadrant has four or fewer items. All quadrants have similar weights, and the order or items is reserved. Each level of four quadrants are then laid out by Hilbert or Moore curve. In quadrants at the lowest level, items are laid out by a brute-search method to achieve the lowest aspect ratio. Fig.21 [61] gives an example of Hilbert and Moore Treemap.

Figure 20. An example of D&C Treemap

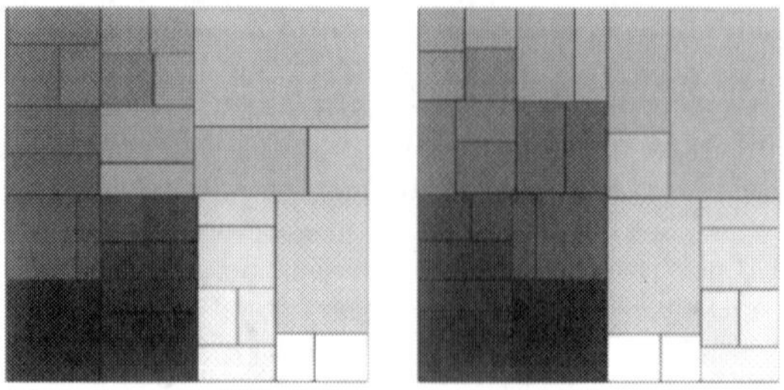

Figure 21. An example of Hilbert (left) and Moore (right) Treemap

4 Methods for Mobile Devices

When mobile devices are used, the display area is usually extremely small and some approaches other than conventional connection and enclosure methods have been pro- posed especially for this situation.

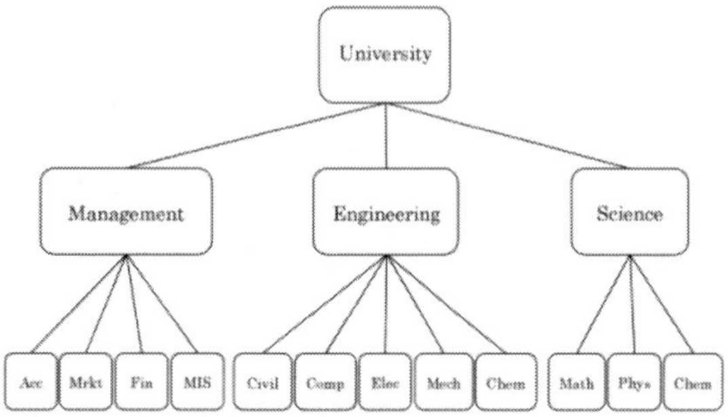

Figure 22. A university structure

Hao, Zhang, and Huang demonstrated "Radial Edge-less Tree" (RELT), a tree visualization method based on space-fill idea for extremely small screen [62, 63]. In RELT, the whole display area is partitioned into several polygons defined by three or four edges. Each polygon represents one node in the tree, and boundary sharing is used to represent the relationship between nodes. Children of one node share the same boundary of this node, and adjacent siblings share one entire boundary. The root is typically placed on the top-left corner of the screen, and other nodes are hierarchically drawn from the top-left to bottom-right, which finally leads to a radial shape. One example of visualization of a university structure by RELT is shown in Fig.22 and Fig.23 [63]. By using boundary sharing instead of links between nodes, a high space utilization is achieved so that a small screen can display more. A proto- type was implemented in Android and compared with the traditional user interface for showing hierarchy information. The results showed that RELT has a relatively long learning curve, which means the compact layout of RELT may hurt its readability.

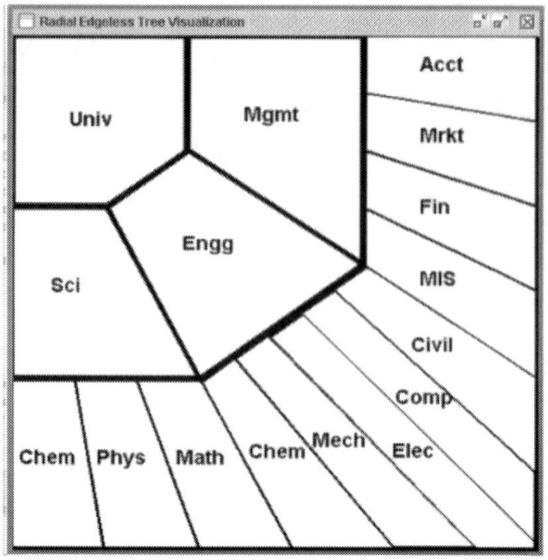

Figure 23. RELT of the university structure shown in Fig.22

Tablorer is another tree visualization system designed especially for mobile devices, which is proposed by Shin, Park, and Han in 2011[64]. In Tablorer, all children of a node are drawn in one rectangle called sibling block. A sibling block is placed either horizontally or vertically, and the orientation alternates according to the level of sib- ling block. In the tree represented by Tablorer, one node is chosen as focus, and all its descendants are defined as the "focus area" while all nodes with higher levels as well as siblings of the focus are defined as the "context area". Three examples of Tablorer can be found in Fig.24 [64]. As Fig.24 (b) and (c) show, all nodes that doesn't belong to focus area or context area are not displayed in Tablorer so that a compact layout can be obtained. Some experiments were also performed to compare Tablorer and other conventional visualization methods and the results showed that it takes 22% less time for users to search information in Tablorer [64].

5 Hybrid Methods

In the existing literature, there are also some examples of tree visualization using hybrid method which is a combination of conventional enclosure approach and connection methods.

In 2005, Zhao, McGuffin, and Chignell proposed Elastic Hierarchies, which is a visualization method combining treemap and node-link diagram [65]. They numerated every possible combination in Fig.25 [65], where **A** is the pure node-link diagram, and **F** is the pure treemap. As shown in Fig.25, **B** is using treemap as a part of the node-link diagram to show higher-level nodes in the tree; **C** and **D** use treemap for the overall structure, and represent some children outside of their parents; **E** uses node-link diagram as a part of treemap to show some higher-level nodes. Zhao, McGuffin, and Chignell implemented a prototype which supports every kind of Elastic Hierarchy shown in Fig.25 except **E**.

Another hybrid approach, EncCon, was also demonstrated in 2005 by Nguyen and Huang [6]. Instead of directly combining treemap and node-link diagram, EncCon uses the basic idea of treemap, space-filling technique, to layout a node-link diagram for maximization of space utilization. EncCon recursively divide the display area into several rectangles and assign each rectangle to a node. Each node is drawn at the centre of its rectangle and edges connecting nodes are added later accordingly. The rectangular division naturally guarantees very few node over- lapping or edge crossing and high space utilization. Be- cause children will finally be around their parent like a circle, a rectangular division algorithm similar to Squarified Treemap [48] is applied so that each node's rectangle can have an aspect ratio close to 1, which will lead to better readability. One example of EncCon is shown in Fig.26 [6], where the numbers are only names of nodes.

In 2011, Marriott, Sbarski, Gelder, Prager, and Bulka proposed a new tree visualization method, Hi-Tree [66]. Nodes in Hi-Tree are categorized into two groups, simple and compound. Children of a simple node are connected to the node by edges, which is similar to node-link diagrams. Children of a compound node are contained within the node, which is similar to treemaps. A compound node is required to have at least one child and its children must be simple nodes. One algorithm for compact layout was also demonstrated in [66] so that Hi-Tree can be applied in small display area. One example of Hi-Tree is shown in Fig.27 [66].

Acknowledgement

This work was partially supported by JSPS KAKENHI Grant Number 15H05708.

References

1. D. Turo, B. Johnson, "Improving the visualization of hierarchies with treemaps: design issues and experimentation." *Proceedings of the 3rd conference on Visualization '92*, 124-131 (1992)
2. H. Lü, J. Fogarty, "Cascaded treemaps: examining the visibility and stability of structure in treemaps." *Proceedings of graphics interface 2008.*, (2008)
3. D. Urribarri, S. M. Castro, S. Martig. "Gyrolayout: A Hyperbolic Level-of-Detail Tree Layout." Journal of universal computer science, **19.1**, 132-156, (2013)
4. G. G. Robertson, J. D. Mackinlay, S. K. Card, "Cone trees: animated 3D visualizations of hierarchical information." *Proceedings of the SIGCHI conference on Hu- man factors in computing systems.*, 189-194 (1991)
5. D. E. Knuth, *The Art of Computer Programming* (Addison-Wesley, Reading, MA, 1968) Chapter1
6. Q. V. Nguyen, M. L. Huang, "EncCon: an approach to constructing interactive visualization of large hierarchical data." Information Visualization, **4.1**, 1-21(2005)
7. M. J. McGuffin, R. Balakrishnan, Interactive "Visualization of Genealogical Graphs." *Proc. IEEE Symp. Information Visualization*, 16-23 (2005)
8. C. Buchheim, M. Jünger, S. Leipert, "Improving Walker's algorithm to run in linear time." *Graph Drawing*, 344-353 (2002)
9. E. M. Reingold, J. S. Tilford, "Tidier drawings of trees." IEEE Transactions on Software Engineering, **2**, 223-228 (1981)
10. J. Q. Walker, "A node-positioning algorithm for general trees." Software: Practice and Experience, **20**, 685–705 (1990)
11. K. Marriott, P. Sbarski, "Compact layout of layered trees." *Proceedings of the thirtieth Australasian conference on Computer science*, **62**, 7-14 (2007)
12. A. Bloesch, "Aesthetic layout of generalized trees." Software: Practice and Experience, **23**, 817–827 (1993)
13. M.Hasan, Md. S. Rahman, T.Nishizeki, "A linear algorithm for compact box-drawings of trees." Networks, **42.3**, 160–164 (2003)
14. Y. Miyadera, K. Anzai, H. Unno, T. Yaku, "Depth-first layout algorithm for trees." Information Processing Letters, **66**, 187–194 (1998)
15. B. Stein, F. Benteler, "On the generalized box-drawing of trees: Survey and new technology." *Proceeding of I-KNOW '07*, (2007)
16. L. Xiaohong, H. Jingwei, "An improved generalized tree layout algorithm." *Proceedings of the 2nd International Asia Conference on Informatics in Control, Automation and Robotics*, **2**, 163–166 (2010)
17. A. Ploeg, "Drawing non-layered tidy trees in linear time." Software: Practice and Experience, **44.12**, 1467- 1484 (2014)
18. S. Cornelsen, T. Schank, D. Wagner, "Drawing graphs on two and three lines." *Graph Drawing*, 31-41 (2002)
19. M. Suderman, "Pathwidth and layered drawings of trees." International Journal of Computational Geometry & Applications, **14.03**, 203-225 (2004)
20. M. J. Alam, Md. A. H. Samee, M. Rabbi, Md. S. Rahman, "Minimum-Layer Upward Drawings of Trees." Journal of Graph Algorithms and Applications, **14.2**, 245-267 (2010)
21. D. Mondal, M. J. Alam, Md. S. Rahman, "Minimum-layer drawings of trees." *WALCOM: Algorithms and Computation*, 221-232 (2011)
22. P. Scheffler, "A linear algorithm for the pathwidth of trees." Topics in combinatorics and graph theory, (Physica-Verlag HD, 1990) 613-620
23. R. Tamassia(editor), "Handbook of Graph Drawing and Visualization." (CRC Press, 2013) 160
24. P. Crescenzi, G. Di Battista, A. Piperno, "A note on optimal area algorithms for upward drawings of bi-nary trees." Comput. Geom. Theory Appl., **2**, 187-200 (1992)
25. J. Lamping, R. Rao, "Laying out and visualizing large trees using a hyperbolic space." *Proceedings of the 7th annual ACM symposium on User interface software and technology*, 13-14 (1994)
26. J. Lamping, R. Rao, P. Pirolli, "A focus+context technique based on hyperbolic geometry for

visualizing large hierarchies." *Proceedings of the SIGCHI conference on Human factors in computing systems*, 401-408 (1995)
27. K. Sugiyama, S. Tagawa, M. Toda, "Methods for Visual Understanding of Hierarchical System Structures." IEEE Trans. Systems, Man, and Cybernetics, **11. 2**, 109-125 (1981)
28. C. Bachmaier, "A radial adaptation of the Sugiyama framework for visualizing hierarchical information." IEEE Transactions on Visualization and Computer Graphics, **13.3**, 583–594 (2007)
29. C. A. Duncan, D. Eppstein, M. T. Goodrick, S. G. Kobourov, M. Nollenburg, "Drawing trees with perfect angular resolution and polynomial area.", *Graph Drawing*, 183-194 (2010)
30. C. A. Duncan, D. Eppstein, M. T. Goodrick, S. G. Kobourov, M. Nollenburg, "Lombardi drawings of graphs." *Graph Drawing*, 195–207 (2010)
31. A. Rusu, A. Crowell, B. Petzinger, A. Fabian, "PieVis: Interactive Graph Visualization Using a Rings-Based Tree Drawing Algorithm for Children and Crust Display for Parents." *Proc. IEEE Info Vis '11*, 465-470 (2011)
32. P. Eades, "A heuristic for graph drawing", Congressus Nutnerantiunt, **42**, 149–160 (1984)
33. T. M. J. Fruchterman, E. M. Reingold. "Graph drawing by force-directed placement." Softw., Pract. Exper., **21.11**, 1129-1164 (1991)
34. T. Dwyer, Y. Koren, K. Marriott, "Drawing directed graphs using quadratic programming." IEEE Transactions on Visualization and Computer Graphics, **12.4**, 536-548 (2006)
35. M. Bohanec, "DEXiTree: A program for pretty drawing of trees." *Proc. Information Society IS 2007*, 8-11(2007)
36. T. Dwyer, K. Marriott, P. Sbarski. "Hi-tree layout using quadratic programming." Diagrammatic Representation and Inference (Springer Berlin Heidelberg, 2010) 212-219
37. R. Chernobelskiy, K. Cunningham, M. T. Goodrich, S. G. Kobourov, L. Trott, "Force-directed lombardi-style graph drawing." *Graph Drawing*, 78–90 (2011)
38. M. L. Huang, J. Liang, Q. V. Nguyen, "A visualization approach for frauds detection in financial market." *Proc. IEEE Info Vis '09*, 197-202 (2009)
39. L. Jin, D. C. Banks, "Tennisviewer: A browser for competition trees." IEEE Computer Graphics and Applications, **17.4**, 63-65 (1997)
40. D. Turo, "Hierarchical visualization with treemaps: making sense of pro basketball data." *CHI '94* Conference Companion on Human Factors in Computing Systems, 441-442 (1994)
41. B. B. Bederson, "Quantum treemaps and bubblemaps for a zoomable image browser." *Proc. User Interface Systems and Technology*, 71-80 (2001)
42. A. Fiore, M. Smith, "Treemap Visualizations of Newsgroups." Technical Report, Microsoft Research, Microsoft Corporation, Redmond, WA (2001)
43. M. Smith, "Tools for Navigating Large Social Cyberspaces." Communications of the ACM, **45.4**, 51-55 (2002)
44. B. Shneiderman, "Tree visualization with tree-maps: 2-d space-filling approach." ACM Transactions on graphics, **11.1**, 92-99 (1992)
45. B. Johnson, B. Shneiderman. "Tree-maps: A space-filling approach to the visualization of hierarchical information structures." *Proc. IEEE Visualization '91*, 284-291 (1991)
46. M. Wattenberg, "Visualizing the stock market." *CHI'99 extended abstracts on Human factors in computing systems* (1999)
47. J. J. Van Wijk, H. Van de Wetering, "Cushion treemaps: Visualization of hierarchical information." *Proc. IEEE Info Vis '99*, 73-78 (1999)
48. M. Bruls, K. Huizing, J. J. Van Wijk. "Squarified treemaps." *Data Visualization 2000*, 33-42 (2000)
49. B. Shneiderman, M. Wattenberg, "Ordered treemap layouts." *Proc. IEEE Info Vis '01*, 73-78 (2001)
50. J. Heer, G. G. Robertson, "Animated transitions in statistical data graphics." IEEE Transactions on Visualization and Computer Graphics, **13.6**, 1240-1247 (2007)
51. B. B. Bederson, B. Shneiderman, M. Wattenberg, "Ordered and quantum treemaps: Making effective use of 2D space to display hierarchies." ACM Transactions on Graphics, **21.4**, 833-854 (2002)
52. J. Liang, M. L. Huang, "Highlighting in information visualization: A survey." *Proc. IEEE Info Vis '10*, 79- 85 (2010)
53. 15. J. Liang, Q. V. Nguyen, S. Simoff, M. L. Huang, "Angular Treemaps-A New Technique for Visualizing and Emphasizing Hierarchical Structures." *Proc. IEEE Info Vis '12*, 74-80 (2012)
54. M. Balzer, O. Deussen, "Voronoi Treemaps." *Proc. IEEE Info Vis '05*, 49-56 (2005)
55. M. Balzer, O. Deussen, C. Lewerentz, "Voronoi treemaps for the visualization of software metrics." *Proc. ACM SoftVis '05*, 165-172 (2005)
56. Q. Du, V. Faber, M. Gunzburger, "Centroidal Voronoi tessellations: applications and algorithms." SIAM review, **41.4**, 637-676 (1999)
57. A. Sud, D. Fisher, H. Lee, "Fast Dynamic Voronoi Treemaps." *Proc. IEEE International Symposium on Science and Engineering '10*, 85-94 (2010)
58. Y. Tu, H. Shen, "Visualizing changes of hierarchical data using treemaps." IEEE Transactions on Visualization and Computer Graphics, **13.6**, 1286-1293 (2007)
59. J. Liang, S. Simeon, Q. V. Nguyen, M. L. Huang, "Visualizing large trees with divide & conquer partition." *Proceedings of the 6th International Symposium on Visual Information Communication and Interaction*, 79-87 (2013)
60. J. Liang, Q. V. Nguyen, S. Simoff, M. L. Huang, "Divide and Conquer treemaps: Visualizing large trees with various shapes." Journal of Visual Languages & Computing, **31**, 104-127 (2015)
61. S. Tak, A. Cockburn, "Enhanced spatial stability with Hilbert and Moore treemaps." IEEE Transactions on Visualization and Computer Graphics, **19.1**, 141-148 (2013)
62. J. Hao, K. Zhang, M. L. Huang, "RELT–visualizing trees on mobile devices." Advances in Visual

63. Information Systems (Springer Berlin Heidelberg, 2007) 344- 357
64. J. Hao, C. A. Gabrysch, C. Zhao, Q. Zhu, K. Zhang, "Visualizing and Navigating Hierarchical Information on Mobile User Interfaces." International Journal of Advanced Intelligence, **2.1**, 81-103 (2010)
65. H. Shin, G. Park, J. Han, "Tablorer–An Interactive Tree Visualization System for Tablet PCs." *Proceedings of the 13th Eurographics*, **30.3** 1131-1140 (2011)
66. S. Zhao, M. J. McGuffin, M. H. Chignell, "Elastic hierarchies: Combining treemaps and node-link diagrams." *Proc. IEEE Info Vis '05*, 57-64 (2005)
67. K. Marriott, P. Sbarski, T. V. Gelder, D. Prager, A. Bulka, "Hi-trees and their layout." IEEE Transactions on Visualization and Computer Graphics, **17.3** 290-304 (2011)

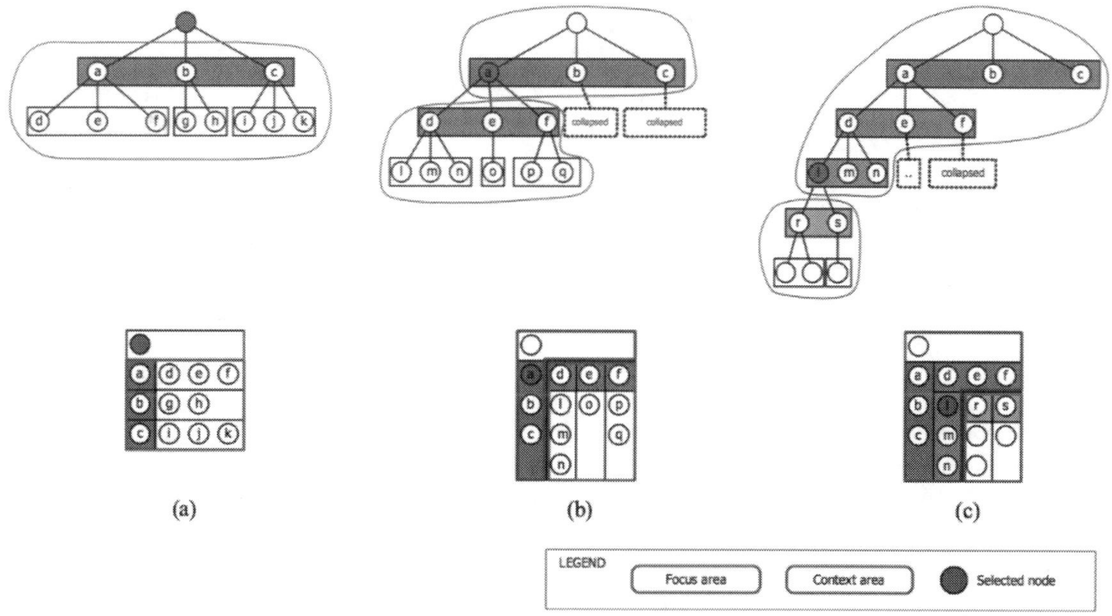

Figure 24. Three different examples of Tablorer

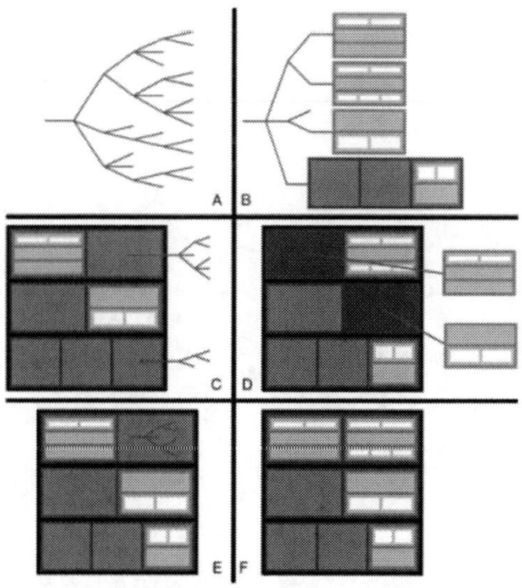

Figure 25. All possible combinations of treemap and node-link diagram

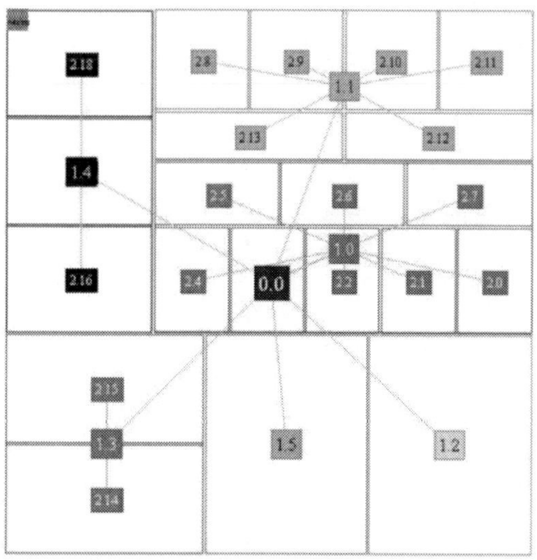

Figure 26. One example of EncCon approach

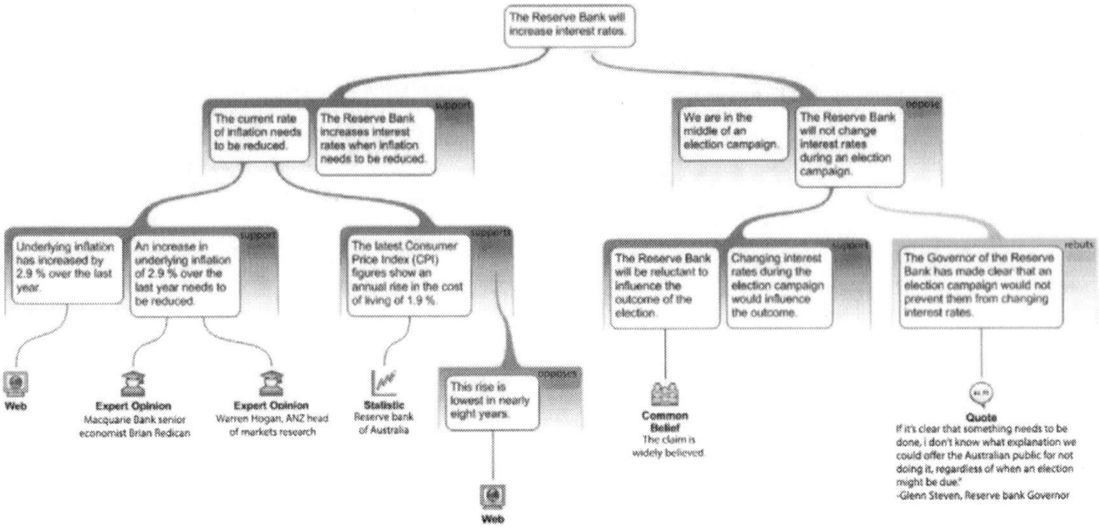

Figure 27. One example of Hi-Tree, an argument map

Improving the vector auto regression technique for time-series link prediction by using support vector machine

Jan Miles Co and Proceso Fernandez

Ateneo de Manila University, Department of Information Systems and Computer Science, Quezon City, Philippines

Abstract. Predicting links between the nodes of a graph has become an important Data Mining task because of its direct applications to biology, social networking, communication surveillance, and other domains. Recent literature in time-series link prediction has shown that the Vector Auto Regression (VAR) technique is one of the most accurate for this problem. In this study, we apply Support Vector Machine (SVM) to improve the VAR technique that uses an unweighted adjacency matrix along with 5 matrices: Common Neighbor (CN), Adamic-Adar (AA), Jaccard's Coefficient (JC), Preferential Attachment (PA), and Research Allocation Index (RA). A DBLP dataset covering the years from 2003 until 2013 was collected and transformed into time-sliced graph representations. The appropriate matrices were computed from these graphs, mapped to the feature space, and then used to build baseline VAR models with lag of 2 and some corresponding SVM classifiers. Using the Area Under the Receiver Operating Characteristic Curve (AUC-ROC) as the main fitness metric, the average result of 82.04% for the VAR was improved to 84.78% with SVM. Additional experiments to handle the highly imbalanced dataset by oversampling with SMOTE and undersampling with K-means clusters, however, did not improve the average AUC-ROC of the baseline SVM.

1 Introduction

One of the major problems in network analysis involves predicting the existence or emergence of links given a network. The link prediction task has many practical applications in various domains. In biology, link prediction has been used for identifying protein-protein interaction [1] and for investigating brain network connections [2]. In social media, link prediction has been used for identifying both future friends and possible enemies by analyzing positive and negative links [3]. Efforts have also been made to apply link prediction techniques in communication surveillance to identify terrorist networks by monitoring emails [4].

Most of the previous works on link prediction use a static network to predict hidden or future links. In the detection of hidden links, the network is based on a known partial snapshot, and the objective is to predict currently existing links [4]. In the prediction of future links, the network is based on a snapshot at time t, and the objective is to predict links at time t' ($t' > t$) [5]. In this framework, insight regarding the dynamics of the network is disregarded, and information on the occurrence and frequency of links across time is lost. Hence, recent works on link prediction use a dynamic network where the network is characterized by a series of snapshots that represent the network across time [4, 5].

Some of the recent works on link prediction have also explored different ways to semantically enrich the network representation. A study was done by [6] that uses a heterogeneous network where nodes represent more than one type of entity. In the prediction of co-authorship networks, a node may represent an author, a topic, a venue or a paper. In another study done by [7], the network was formed by combining multiple layers of network formed by different types of nodes. In this case, the network model is composed of the following layers: the first layer is a co-authorship network; the second layer is a co-venue network, while the third layer is a co-citing network. The final homogeneous network is an aggregation of the three networks where nodes represent authors.

In this study, we focus on link prediction for a dynamic and homogeneous network. Since Vector Auto Regression (VAR) has been shown to be one of the best techniques for time-series link prediction [8], we incorporate some ideas from this technique and explore ways of improving the prediction. In particular, because the VAR model assumes a linear dependence of the temporal links on multiple time-series, we propose the use of Support Vector Machine (SVM) in order to more robustly handle a non-linear type of dependency even while retaining the assumption that the dependency is on multiple time-series. This paper describes our experimentation on this proposed idea.

2 Related Literature

In this section, we briefly review the literature on VAR model and its application in link prediction, the SVM model and its problems on imbalanced datasets, and some of the techniques applied for handling such imbalanced datasets.

2.1 Vector Auto Regression for Link Prediction

The VAR econometric model is an extension of the univariate autoregressive model that is applied to multivariate data. It provides better forecast than univariate time-series models and is one of the most successful models for analyzing multivariate time-series [9]. In a recent work on dynamic link prediction, the VAR technique was applied in homogeneous networks represented by both unweighted and weighted adjacency matrices. For each of the unweighted and weighted adjacency matrices, five additional matrices were created according to different similarity metrics. These metrics are the Number of Common Neighbor (CN), Adamic-Adar Coefficient (AA), Jaccard's Coefficient (JC), Preferential Attachment (PA), and Resource Allocation Index (RA). Using a dataset created from DBLP, the performance of VAR was compared to static link prediction and to several dynamic link prediction techniques, which are Moving Average (MA), Random Walk (RW), and Autoregressive Integrated Moving Average (ARIMA). The VAR technique showed the best performance among the many link prediction techniques. For a more detailed discussion, refer to [8].

2.2 SVM and Imbalanced Datasets

Support Vector Machine (SVM) is a well-known learning model that is used mainly for classification and regression analysis. It computes for a maximum-margin hyperplane that separates the two classes of instances from a given dataset. The SVM does not necessarily assume that the dataset is linearly separable in the original feature space, and thus often uses the technique of projecting, via kernel trick, to higher-dimensional space where the instances are presumed to be more easily separable.

The SVM has been shown to be very successful in many applications including image retrieval, handwriting recognition, and text classification. However, the performance of SVM drops significantly when faced with a highly imbalanced dataset. A highly imbalanced dataset is characterized by instances from one class far outnumbering the instances from another class. This makes it difficult to classify instances correctly due to a small number of the sample size for one class [10]. This type of dataset is observed in our co-authorship network since there are significantly many potential co-authorship links and only few of these are realized.

2.3 SMOTE and SVM with Different Error Costs

In a work done by [10], three techniques were used to address the problem of highly skewed datasets. The first technique is to not undersample the majority instances, since doing otherwise leads to information loss. In the case of imbalanced datasets, the learned boundary of SVM tends to be too close to the positive instances. Hence, the second technique is to apply Different Error Costs (DEC) to different classes to force SVM to push the boundary away from the positive instances. The third technique is to apply Synthetic Minority Over-Sampling Technique (SMOTE) to make positive instances more densely distributed. Hence, making the boundary better defined. Ten (10) datasets where used to test the techniques. The performance of SVM with the original dataset was compared to the performance of the following techniques: SVM with undersampling, SVM with SMOTE, SVM with DEC, and SVM with a combination of SMOTE and DEC. In 7 out of 10 datasets, the best performance was achieved with a combination of SMOTE and DEC [10].

2.4 K-Means Clustering for Under-Sampling

K-means clustering is one of the simplest unsupervised learning algorithms and has been used by researchers to solve well-known clustering problems. In a work done by [11], K-means clustering was used as an undersampling technique for a highly imbalanced dataset. The training set was divided into two: the 1st set contains the minority instances, while the 2nd set contains the majority instances. The majority instances were partitioned into K clusters, for some $K > 1$. Each majority cluster was combined with the minority set to form a candidate training set, and the quality of the candidate training set was evaluated by using the Fuzzy Unordered Rule Induction Algorithm (FURIA). The best candidate training set was used for classification with C4.5 decision tree. This undersampling technique was applied to cardiovascular datasets from Hull and Dundee clinical sites. The proposed K-means undersampling method outperformed the use of the original dataset and the use of another K-means undersampling technique that was proposed by [12].

3 Methodology

3.1. Dataset

DBLP computer science bibliography contains bibliographic information on major computer science journals and proceedings. The co-authorship network in DBLP was used for the link prediction task. Following the dataset preparation in [8], we selected only the articles from 2003 to 2013, removing all items labelled as inproceedings, proceedings, book, incollection, phdthesis, masterthesis, and www. The reduced set was further trimmed by removing all authors who have 50 articles or less. The final dataset has 1,743 authors and 21,920 articles. The number of co-authorship links in this dataset is less than 1% of the total number of possible links.

To build the time-series models, the dataset was partitioned based on the year of publication. Each of the resulting 11 subsets, corresponding to years 2003 (t=0) up to 2013 (t=10), was processed to generate a snapshot of the dynamic co-authorship network graph. A snapshot is represented by an $n \times n$ unweighted adjacency matrix, where n = 1,743 (the number of authors), and an entry is 1 if the corresponding two authors have co-authored an article published on the given year, otherwise the matrix element is 0. Based from the unweighted adjacency matrix for each snapshot, five matrices were computed corresponding to the similarity metrics that are commonly used for link prediction:

• Number of Common Neighbor (CN)

$$CN(x,y) = |\Gamma(x) \cap \Gamma(y)| \tag{1}$$

• Adamic-Adar Coefficient (AA)

$$AA(x,y) = \sum_{z \in \Gamma(x) \cap \Gamma(y)} \frac{1}{\log(|\Gamma(z)|)} \tag{2}$$

• Jaccard's Coefficient (JC)

$$JC(x,y) = \frac{\Gamma(x) \cap \Gamma(y)}{\Gamma(x) \cup \Gamma(y)} \tag{3}$$

• Preferential Attachment (PA)

$$PA(x,y) = |\Gamma(x)| * |\Gamma(y)| \tag{4}$$

• Resource Allocation Index (RA)

$$RA(x,y) = \sum_{z \in \Gamma(x) \cap \Gamma(y)} \frac{1}{|\Gamma(z)|} \tag{5}$$

where $\Gamma(x)$ is the set of nodes adjacent to node (author) x.

3.2 Baseline VAR and SVM Models

We used the VAR technique with lag of 2, as proposed by [8], to form the baseline model. The predicted adjacency matrix at time t, denoted by $\hat{Y}_t^A$, is computed using the formula:

$$\begin{aligned}\hat{Y}_t^A = {} & C_{t-1} + \alpha_{t-1}^{A} Y_{t-1}^{A} + \alpha_{t-1}^{CN} Y_{t-1}^{CN} + \alpha_{t-1}^{AA} Y_{t-1}^{AA} + \alpha_{t-1}^{JC} Y_{t-1}^{JC} \\ & + \alpha_{t-1}^{PA} Y_{t-1}^{PA} + \alpha_{t-1}^{RA} Y_{t-1}^{RA} + \alpha_{t-2}^{A} Y_{t-2}^{A} + \alpha_{t-2}^{CN} Y_{t-2}^{CN} \\ & + \alpha_{t-2}^{AA} Y_{t-2}^{AA} + \alpha_{t-2}^{JC} Y_{t-2}^{JC} + \alpha_{t-2}^{PA} Y_{t-2}^{PA} + \alpha_{t-2}^{RA} Y_{t-2}^{RA}\end{aligned} \tag{6}$$

where each Y_j^i is an $n \times n$ matrix representing the actual values for the similarity metric i at time j, and the $n \times n$ matrix C_j and scalar coefficients α_j^i are time-based VAR model parameters. We applied linear regression, using the *lm*() function in R, to find the best fitting parameter set for each snapshot.

The VAR model described here assumes that the co-authorship link at time t is linearly dependent on 12 factors: the 6 metrics (including the adjacency relation) from 2 previous time periods, t-1 and t-2. However, a visualization of the dataset for the snapshot t=2, as shown in Figure 1, indicates that this may not necessarily be the case. In this figure, there is no clean linear model that fits the data. However, it should be noted that the two principal components accounted for just 58.3% of the variance in the dataset. We explore SVM to verify if it can improve the time-series prediction.

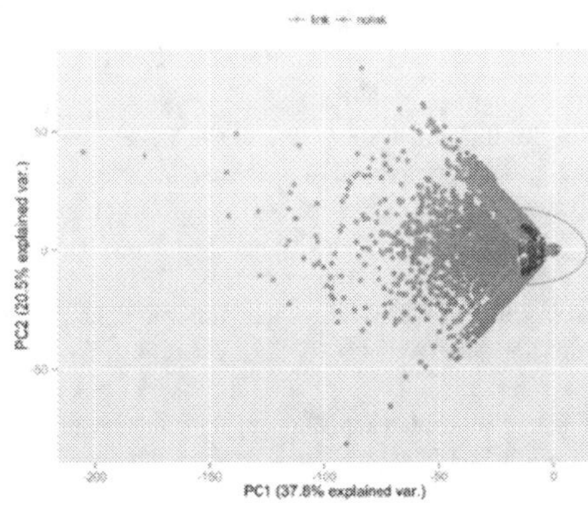

Figure 1. Data visualization of the instances from t=2, for the VAR and SVM models. Principal Component Analysis was used to project the 13-dimensional points to the 2-coordinate space by using the first two principal components of the instances.

For the SVM model, each instance representing the presence or absence of a co-authorship link is mapped to the 12-dimensional feature space, following the linear dependency assumed in the lag 2 VAR model of this study. These instances were further projected to higher dimension using an SVM linear kernel function. An SVM classifier is then used to predict the class for each instance, and these predictions are collected to construct the predicted adjacency matrix, $\hat{Y}_t^A$, of the network at time t.

3.3 Benchmarking

To compare the predictive powers of the VAR and SVM models, we applied backtesting. That is, we first built a VAR (or SVM) model for time t using the metric values for time t-1 and t-2, then used this model to predict the values at time t+1 using values from time t and t-1, and finally compared the prediction against the known values at time t+1. We were able to measure the performance of the VAR technique in 8 years, from Year 3 to Year 10.

The Area Under the Receiver Operating Characteristic Curve (AUC-ROC) was used to measure the performance of the predictive models. In the Receiver Operating Characteristic (ROC) Curve, the x-axis corresponds to the False Positive Rate (FPR) and the y-axis corresponds to the True Positive Rate (TPR), which are respectively computed as follows:

$$FPR = \frac{FalsePositive}{FalsePositive + TrueNegative} \tag{7}$$

$$TPR = \frac{TruePositive}{TruePositive + TrueNegative} \quad (8)$$

The AUC of a perfect model is 1, while a random model will have an expected AUC of 0.5. In our context, positives correspond to links while negatives correspond to no links.

3.4 Further Experimentation on SVM

3.4.1 SMOTE and Different Error Cost

We were unable to apply the first approach proposed by [10] of not undersampling, due to the large sample size of our negative instances. For our first attempt to improve our baseline SVM, we applied the Different Error Costs (DEC) technique that was suggested by [10], which is setting the cost ratio to the inverse of the imbalance ratio. For our training set, we created an imbalanced ratio of 1:2 and set the error cost to 1 for positive links and 2 for negative links.

For our second attempt to improve our baseline SVM, we applied SMOTE, which was also suggested by [10]. To create the training sets, we oversampled the positive instances by 100% and undersampled our negative instances by selecting negative instances until we have equal number of positive and negative instances. Hence, we doubled the number of both positive and negative instances from our baseline SVM experiment.

For our third attempt to improve our baseline experiment, we combined the use of DEC and SMOTE. We used SMOTE to create an imbalance ratio of 1:2 and set the appropriate error costs. We performed the three experiments, then computed the AUC-ROC and compared this to our baseline SVM result.

3.4.2 K-Means Clustering

For our last attempt to improve our baseline SVM, we used K-means clustering, as proposed by [11], to under-sample the majority instances. First, we separated the positive instances from the negative instances. Then we clustered the negative instances into $K = 2$ clusters and selected the largest cluster. To form the training set, we performed random sampling in the larger cluster until we have an equal number of positive and negative instances. We combined the sampled majority instances to the positive instances and used this dataset to train an SVM classifier. We evaluated the AUC-ROC and compared this to our baseline SVM result.

4 Results and Discussion

In this section, we present the result of our experiments in two parts: the result of comparing the performance of VAR technique and SVM classification algorithm and the result of our attempts to further improve our baseline SVM. The complete AUC-ROC results for all the experiments are shown in Table 1.

4.1. Baseline VAR and SVM models

SVM was able to outperform VAR with average AUC-ROC values of 84.78% and 82.04% respectively. A two-tailed paired t-test was conducted in order to determine if there is significant difference in the means, with at least 90% confidence level. The resulting p-value of 0.064 suggests that there is a statistically significant difference.

In 6 out of 8 years, SVM was able to achieve a better performance than VAR. Figure 2 shows a plot of the ROC of VAR and SVM for Year 10, the snapshot where the SVM was able to outperform VAR by the largest margin.

There were two snapshots where VAR was able to outperform SVM but these were only by small margins of 0.24 (in Year 3) and 0.18 (in Year 7). Figure 3 shows the ROC for Year 3.

The results suggest that the VAR model can be significantly improved by SVM classification algorithm when the VAR model multivariate time-series input data is used as the training set for linear SVM in the domain of time-series link prediction.

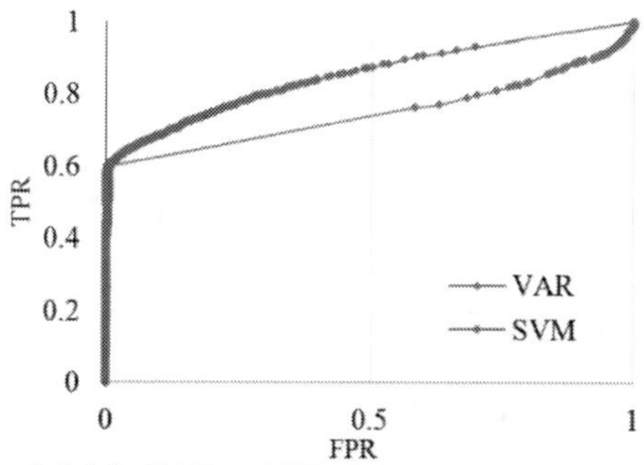

Figure 2. ROC of VAR and SVM in Year 10.

Table 1. AUC-ROC of Link Prediction Experiments.

	Year 3	Year 4	Year 5	Year 6	Year 7	Year 8	Year 9	Year 10	Average
VAR	**86.73**	85.51	85.21	81.54	84.82	81.15	76.46	74.90	82.04
SVM	86.49	86.02	85.71	**85.25**	84.64	83.21	82.33	**84.60**	**84.78**
DEC	81.31	84.08	73.27	85.24	**84.85**	**83.24**	82.38	84.31	82.34
SMOTE	86.56	86.00	**85.73**	85.13	84.60	83.20	82.35	84.46	84.75
DEC w/ SMOTE	70.94	**86.08**	81.68	74.49	81.75	83.21	**82.39**	80.34	80.11
K-Means	86.37	85.74	85.49	85.14	84.66	82.98	82.29	84.02	84.59

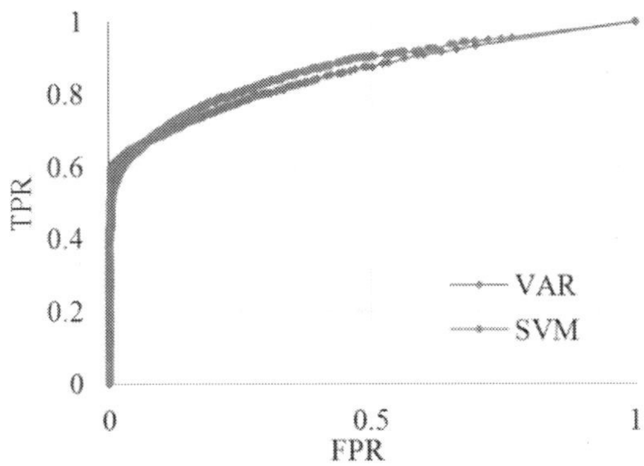

Figure 3. ROC of VAR and SVM in Year 3.

4.2. Experiments to Improve Baseline SVM

In our subsequent experiments, we were unable to improve the average AUC-ROC of our baseline SVM despite implementing multiple techniques for handling the highly imbalanced dataset. In Table 1, we highlight the AUC-ROC for each year where the highest AUC-ROC was achieved. However, we were able to improve the performance of our baseline SVM in 6 out of 8 years. Each of the three techniques, SVM with DEC, SMOTE, and DEC with SMOTE, were able to improve the results in two years each. DEC was able to improve SVM by 0.21% and 0.03% in Years 7 and 8, while SMOTE was able to improve SVM by 0.07% and 0.02% in Years 3 and 5. Finally, DEC with SMOTE was able to improve SVM by 0.06% both in Years 4 and 9.

From our results, we hypothesize that the reason DEC, SMOTE, and DEC with SMOTE failed to improve the average performance of our baseline SVM is that the techniques were designed to perform well on not undersampled data as proposed by [10]. Although the techniques were able to improve our baseline SVM in a few individual years, the improvements were not consistent enough to beat the average performance of the baseline SVM.

SVM with undersampling by K-means clustering also failed to improve our baseline SVM. We infer that there is inherent information loss in random sampling from the larger majority instances cluster that leads to the poor performance of SVM with undersampling by K-means. However, in the case of datasets with noisy majority instances, undersampling by K-means clustering might improve classification as shown by [11].

5 Conclusion and Further Studies

In this study, we performed dynamic link prediction using various algorithms. Primarily, we were able to improve the performance of the VAR model by transforming its input multivariate time-series data as a feature set vector that was used as a training set to linear SVM. The VAR model was able to achieve an average AUC-ROC of 82.04% while SVM was able to achieve an AUC-ROC of 84.78%. This implies that the performance of the VAR model can be improved by using SVM classification algorithm for dynamic link prediction.

In our attempt to further improve the performance of the baseline SVM, we experimented on several techniques such as SVM with different error costs (DEC), SMOTE, and undersampling by K-means clustering. In the case of un-noisy, large, and highly imbalanced datasets, we were forced to under-sample the majority instances, which might explain the poor performance of SVM with DEC, SMOTE, the combination of both techniques, and K-means clustering.

Based on these findings, we aim to improve other existing link prediction techniques by using other classification algorithms. We also intend to enrich our network model by using a weighted network and by using sematic information, which might result to better link prediction as shown in some previous works. Finally, we recommend further studies to improve oversampling and under-sampling techniques for binary classification specifically for link prediction, where dataset is often un-noisy, large, and highly imbalanced.

References

1. M. Kurakar, S. Izudheen. IJCTT **9**, *Link Prediction in Protein-Protein Networks: Survey*, 164-168 (2014)
2. Y. Yang, H. Guo, T. Tian, H. Li. TST **20**, *Link Prediction in Brain Network Based on Hierarchical Random Graph Model*, 306-315 (2015)
3. J. Tang, S. Chang, C. Aggarwal, H. Liu. WSDM, *Negative Link Prediction in Social Media* (2015)
4. Z. Huang, D.K.J. Lin, INFORMS JOC **21**, *The Time-Series Link Prediction Problem with Application in Communication Surveillance*, 286-303 (2009)
5. P.R. Soares, R.B. Prudencio, IJCNN, *Time Series Based Link Prediction* (2012)
6. J.B. Lee, H. Adorna, ASONAM, *Link Prediction in a Modified Heterogeneous Bibliographic Network* (2015)
7. M. Pujari, R. Kanawati. NHM **10**, *Link Prediction in Multiplex Networks* (2015)
8. A. Ozacan, S.G. Oguducu. ICIS, *Multivariate Temporal Link Prediction in Evolving Social Networks* (2015)
9. P.D. Gilbert, IJF **14**, *Combining VAR Estimation and State Space Mode Reduction for Simple Good Predictions*. 229-250 (1995)
10. R. Akbani, S. Kwek, N. Japkowicz. ECML **3201**, *Applying Support Vector Machine to Imbalanced Datasets*, 39-50 (2004)
11. M.M. Rahman, D.N. Davis. WCR **3**, *Cluster Based Under-Sampling for Imbalanced Cardiovascular Data* (2013)
12. S.J. Yen, Y.S. Lee. ESWA **36**, *Cluster-Based Undersampling approaches for imbalanced data distributions*, 5718-5727 (2009)

MATEC Web of Conferences 56, 01009 (2016) DOI: 10.1051/matecconf/20165601009
ICCAE 2016

The Fusion of the Visual and Thermal Images on the Basis of Determining the Image Fragments which Contain Essential Details

Zygmunt Kuś

Faculty of Automatic Control, Electronics and Computer Science Silesian University of Technology, Akademicka 16, Gliwice, Poland

Abstract. The aim of the following study was to develop a procedure which guarantees the data fusion of thermal and visual images. The first stage of the proposed algorithm consisted of images acquisition which guaranteed that the same parts of images represented the same parts of the observed terrain. The second stage depended on previous information about the searched object features. Two different situations were considered herein. In the case when we had the searched object's feature vector for both representations of a searched object, we could conduct the pattern recognition for each image. It was conducted separately for visual and thermal images. In this way, we obtained the important parts of the images which should be represented in a fused image. The other case examined in the paper, considered the situation in which we did not have the formalised information about the object. In this case, it was necessary to analyse whole images in order to define the potential parts of the images where the object could be found. This analysis should be helpful for an operator to indicate the parts of the images where there are some artefacts which can be the elements of the searched object. Therefore, in this case, the second stage of the algorithm consisted in calculating the local features of the images. These features constituted grey scale gradient computed for the pixels inside the aperture. This study presented the examples of the fused images obtained by means of the developed method.

1 Introduction

The fusion of various nature images is particularly valuable when it is used for a support of terrain observations or surveying the objects by the video terminal operator. The images fusion may be conducted in order to emphasise some essential details which are difficult to find in two images simultaneously. The other goal may be to observe some relationship between information contained in these images.

One has to bear in mind that we need to solve the following initial problems during designing the system of images fusion. Firstly, we need to answer what is the goal of the fusion; secondly, what parts of the image or what kind of the objects are interesting.

We may use the algorithms of pattern recognition to emphasise the parts of the images which contain the important objects when we know some features of these objects (e.g. in the form of a pattern vector).

Otherwise we can use the methods of an edge detection or gradient computation assuming that greater homogenous areas do not contain any important objects.

However, even in this case, it is necessary to define what we mean by a greater or smaller area. Therefore we have to know something about an object which we want to observe, for example in order to set the size of the aperture used for image processing during images fusion.

There are many publications which consider the problem of visual and thermal images fusion [1,2,3,4,5]. The comparison of seven well known image fusion methods is presented in [6]. The security area can be also supported by the images fusion. The examples of such systems are presented in [7,8] for detection of unattended packages and for helping drivers with driving at night or in bad weather conditions [7]. Moreover, the detection and recognition of clouds at night [9] may be improved by using the fusion of visual and thermal images.

The images fusion task constitute one of the problems which appear when we use image processing systems in such applications as: image recognition [10,11,12] an unmanned aerial vehicle [13,14] used for image acquisition [15,16,17,18], object tracking [19,20,21] or images fusion [22].

In the following paper, we will consider the problem of the visual and thermal images fusion. The proposed methods will base on two different approaches. Firstly, we will use information about observed objects, and secondly, we will analyse the images without any previous information.

2 The algorithm for visual and infrared images fusion

The first stage of the proposed algorithm consists of images acquisition which guarantee that the same parts of images represents the same parts of the observed terrain.

indicate the parts of the images where there are some artefacts which can be the elements of the searched object. Therefore, in this case, *the second stage* of the algorithm

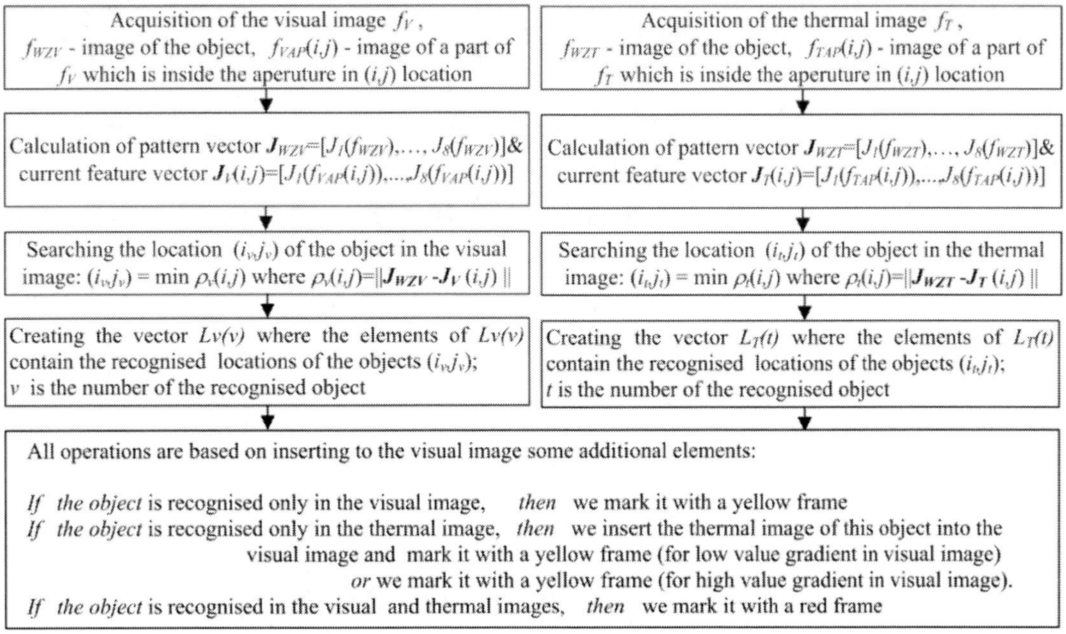

Figure 1. The block scheme of the algorithm of the visual and thermal images fusion for the case when we use pattern recognition.

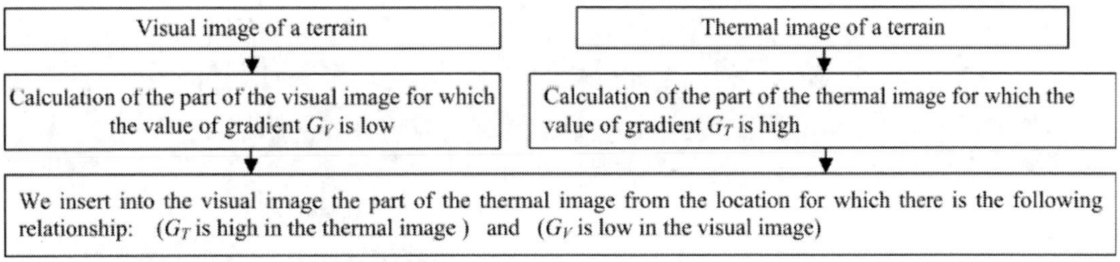

Figure 2. The block scheme of the algorithm of the visual and thermal images fusion for the case when we use the gradient function.

The second stage depends on the previous information about the searched object features. In the case when we have the searched object's feature vector for both representations of a searched object, we can conduct the pattern recognition for each image. It is conducted separately for visual and thermal images. The pattern vectors consist of moment invariants J_1 to J_8 defined in [23]. In order to assess the distance between vectors we will use Euclidian metric.

On the other hand, in the case when we do not have the formalised information about the object, it is necessary to analyse whole images in order to define the potential parts of the images where the object can be found. This analysis should be helpful for an operator to consist in calculating the local features of the images. These features can be, for example, grey scale gradient, the edges, mean value for the pixels inside the aperture, the local value of temperature or temperature gradient.

In the case when we may use pattern recognition, *the last stage* of the proposed method is to mark in the visual image the objects recognised on the basis of visual and thermal images.

In the case when we have no precise information about some objects, *the last stage* consists of inserting into a visual image these parts of a thermal image for which we observed high value of gradient.

The Fig. 1 presents the block scheme of the algorithm of the visual and thermal images fusion for the case when we use pattern recognition. The Fig. 2 presents the block scheme of the algorithm of the visual and thermal images fusion for the case when we use the gradient function.

The proposed algorithms guarantee that the video terminal operator will obtain the fused image which connects the most crucial parts of the visual and thermal images in a clear and comprehensive way.

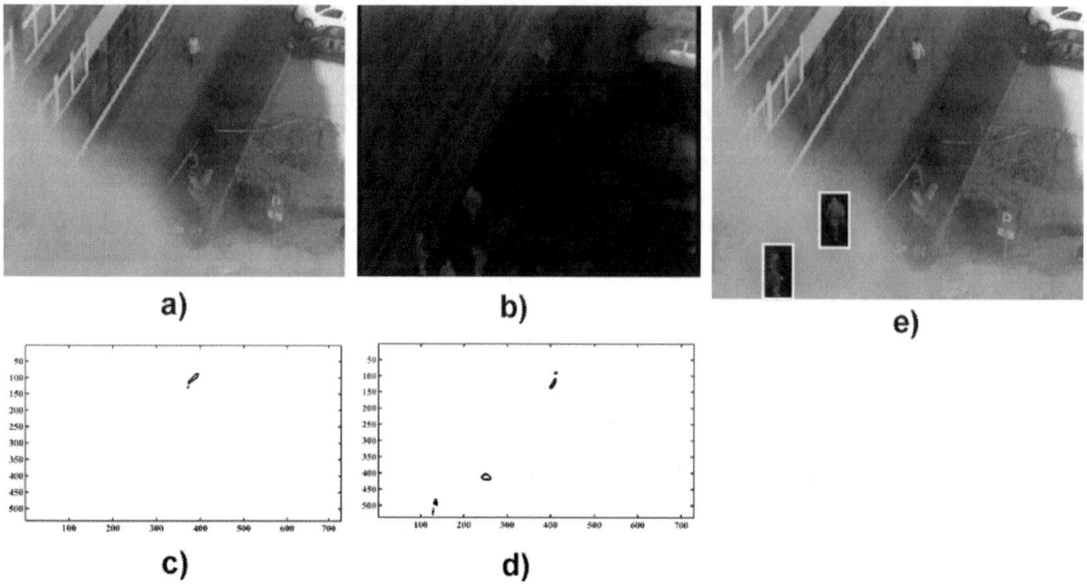

Figure 3. Images for the case 1: a) the visual image, b) the thermal image, c) the locations of the people recognised in the visual image, d) the locations of the people recognised in the thermal image, e) the fused image.

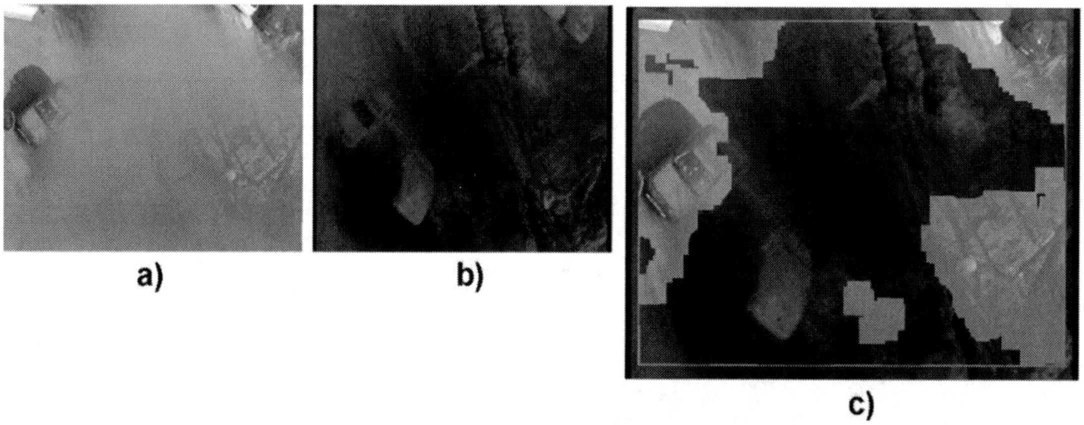

Figure 4. Images for the case 2: a) the visual image, b) the thermal image and c) the fused image.

3 Examples

In order to exemplify the presented above algorithms of the fusion, we will examine two pairs of visual and thermal images. The first experiment will be conducted with the help of a pattern vector calculated for people treated as observed objects and the second experiment will be conducted with the use of gradient calculation.

The following cases correspond to the following situations:

1) In this case, the part of the visual image is disturbed by smoke. The thermal image is not disturbed. According to

Fig. 3., we can observe that thanks to using the pattern recognition for the thermal image we can see where are the people from the area which is covered by smoke.

2) In this case, the part of the visual image is disturbed by smoke similarly as in the first case. However, we will not define any pattern vector in this case, but we will calculate the value of the gradient for both images. According to the Fig. 4., we can see that the visual image is filled in by the thermal image in these places where the value of gradient is low – in places disturbed by smoke.

The presented examples illustrate the usefulness of the proposed algorithms for supporting a person who is observing or supervising a terrain. Undoubtedly, more precise information about the observed objects results in better results of such an inspection.

4 Conclusions

We have developed two procedures which allow to conduct the data fusion of thermal and visual images. These methods concern two different tasks of terrain observation and supervision. The first method was developed assuming that we can formalise information about the observed objects, for example in the form of pattern vector. In the second one there were not such a limitation. For both methods, the first stage was the same and consisted of images acquisition which guaranteed that the same parts of images represented the same parts of the observed terrain.

The second stage was different for the abovementioned methods. It depended on previous information about the searched object features. In the first method, when we had the searched object's feature vector for visual and thermal representations of a searched object, it was possible to conduct the pattern recognition for each image. The visual and thermal pattern vectors were used for object recognition respectively for visual and thermal images. In this way, we obtained the significant parts of the images which should be represented in a fused image. The second method considered the situation in which we did not have the formalised information about the object. In this case, we had to analyse whole images in order to define the potential parts of the images where the objects important for the supervision could be found. Therefore the second stage of this algorithm consisted in calculating the local features of the images. We used the gradient as an examined feature.

Both methods of the image analysis resulted in the fusion images which may be helpful for an operator to indicate the important parts of the images.

This study presented the examples of the fused images obtained by means of the developed method.

Acknowledgements

This work was supported by PBS I ID 178438 path A from the Polish National Centre for Research and Development.

References

1. Xia Yu, Qu Shi-Ru and Cai Lei, Proc. of CCC, 3685-3690, (2013)
2. M. K. Bhowmik, D. Bhattacharjee, M. Nasipuri, D. K. Basu and M. Kundu, Proc. of IAS, 311-316, (2010)
3. Xiaojing Gu, Shaoyuan Sun and Jian'an Fang, Proc. of IS IITAW, 612-615, (2008)
4. Moulay A. Akhloufi, Celia Porcher, Abdelhakim Bendada, Proc. SPIE 9076, Airborne (ISR) SA XI, 90760O, (2014)
5. Bieszczad G., Sosnowski T., Madura H., Kastek M., Bareła J., Proc. of SPIE **7660**, 76603Z, (2010)
6. V. Aslantas, E. Bendes, A. N. Toprak and R. Kurban, Proc. of ICDP, 1-6, (2011)
7. A. Yigit and A. Temizel, Proc. of IEEE SPCAC, 617-620, (2010)
8. Jędrasiak K., Daniec K., Nawrat A., Koteras R., El.Rev., ISSN0033-2097, R.88 NR 10a/2012, 90-97, (2012)
9. Kai Liu, Zheng Kou and Limin Luo, Proc. of ICCET **5**, 420-423, (2010)
10. K. Jedrasiak and A. Nawrat, Computer Vision and Graphics, LNCS **5337**, 391-399, (2009)
11. T. Grzejszczak, El.Rev. **87(2)**, 201-204, (2011)
12. T. Grzejszczak, J. Nalepa and M. Kawulok, Adv. in ISC **226**, 439-449, (2013)
13. R. Zawiski and M. Błachuta, Proc. of 17th IC on MMAR, 475-480,(2012)
14. R. Zawiski and M. Błachuta, BPASTS, **62(3)**, 535-550, (2014)
15. P. Iwaneczko, K. Jedrasiak, K. Daniec and A. Nawrat, CVG, Lect.Not.CS **7594**, 87-94, (2013)
16. M. Błachuta, R. Czyba, W. Janusz and G. Szafrański, J of IRSTA **74(1-2)**, 413-420, (2014)
17. R. Czyba and G. Szafrański, Int. j. adv. robot. syst. **10**, (2013).
18. Bieszczad G., Sosnowski T., Madura H., Kastek M., Bareła J., Proc. of SPIE **7660**, 76603Z, (2010)
19. A. Nawrat and K. Jedrasiak, Proc.of 9th WSEAS ICRCMT, 67-72, (2009)
20. R. Gade, T. B. Moeslund, Sensors, **14**, issue 8, doi:10.3390/s140813679, 13679-13691, (2014)
21. Bieszczad G., Sosnowski T., Madura H., Proc. SPIE 8193, 81933R, (2011)
22. K. Jedrasiak, A. Nawrat, K. Daniec, R. Koteras, M. Mikulski and T. Grzejszczak, CVG LN in CS **7594**, 423-432, (2013)
23. Z. Kuś, Proc. of ICCAE, MATEC Web of Conferences (ISSN: 2261-236X), (2016).

The Evaluation of Study Success between Online Study and Classroom Study Environment

Phudinan Singkhamfu

College of Arts, Media, and Technology. Chiang Mai University, Chiangmai, 50200, Thailand

Abstract. Online study has increasingly become more attractive to students at university level due to convenience access to their instructors and to study resources. This study has developed online social network for study. It proposes to provide lesson content availability, past lecture, by sending online study lesson media to students' mobile phone or tablet. Approximately 85 undergraduate software engineering students participated for 1.5 semesters. In comparing the use of the study toll, and without the tool, the alterations were found between traditional classroom learning style and online study. Also, the study's aim was to attest the online study tool's efficiency. However, these results were not obvious when the achievement factor was controlled by the limitation of time. The primary purpose of this study is to evaluate these two groups of students with extended experiment time for a noticeable result by used questionnaires course examination, and inventory of ILP learning process. The observed, shows that students with online study tools scored higher on course examinations after measures by the mentioned methodology.

1 Introduction

In many ways, when accessed, computer technology greatly benefits students. Also the application allows for students to have convenient access to access to their instructors, as well as present course content. What was examined is the possibility that computer technology may assist students study skills more effectively. While the previous part of this research made use of study tools developed especially for this project, many introductory textbooks now provide students the option of accessing web sites and online study content. Resources such as these intend to help students master text book context and may be viewed as alternatives to the more traditional student study guide. Using computing tools, instructors can assist in preparing study experiences for students. They are made available and are sometimes offered at no extra cost from textbook publishing companies.

We have taken note a percentage of Thai students tend to ignore long texts found in books or long paragraphs which correspond to the content of theirs. The trend of social network communication is significantly popular among university students. We observe that short texts were viewed more frequently long texts. Facebook status is a god example; users regularly update what are they doing or have done. The message comes as a short message and every sender's friends are happy to read the status or post shared.

The concept of the software is that students would be motived to read abridged lesson content. An exploratory observation shows that students always use their mobile phones to frequently access their Facebook page when free. Post classroom learning processes can be adopted by simulating the Facebook status and replaced by shot lesson messages, then simultaneously sending out student peers the post to be shared.

Classroom interaction is reported to create benefits in relation to the rise of more active learning environments, the learning communities can base on application interface that represents a social network application, and the provision of greater feedback for lecturers. Also, it accelerates student motivation. However, using this application to conduct an interactive class is not an easy exercise in perspective. Lecturers may feel that classes are more interactive because they ask questions or accept questions, but they frequently fail to inspect the quality, content, frequency or length of each interaction, in addition to the number of participating students. Thoughts project aims to reduce the gap between in classroom and after class room activity, the lecturer can initiate interactivity after class by using short messages as is done on Facebook status to remind students of the in class lesson content. Students who already have seen the posted message will be able to pass "Impressed" and "Comment" on each single message, while the other students will receive notifications which other students have shared.

2 Methodology

Various designations of interactivity exist in the literature, which focus on the in classroom participation and technology Moore defines key interactions as learner-content, learner-instructor, and learner-learner [1]. Also there are different ways in defining interactivity aimed to the structure regarding loops rationality and originator. Yacci define the interactivity as a message loop, which is commenced and concluded by the student also the message content must be "mutually coherent" [2]. Earlier definitions developed from communication and educational theories are technology independent.

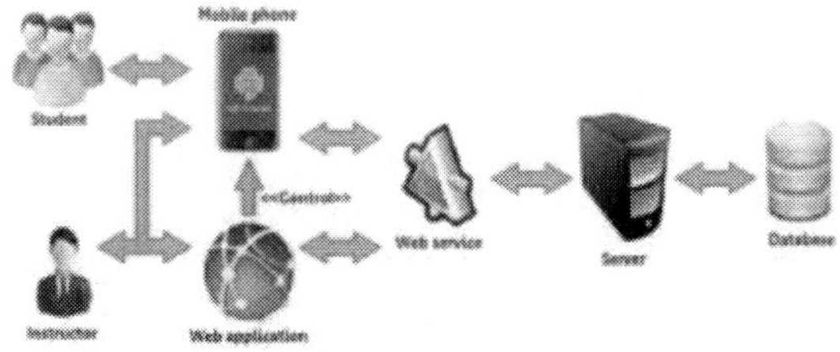

Figure 1. The system architecture of the application.

The applications were designed to enhance student enthusiasm and study success: By accessing instructor posts. With the available feature to criticize and leave comments, the students can interact with their peers or with the class instructor without interrupting as would be the case in class activities, [3]-[4] Fig 1. Students and Instructor are concern as the main focus of this design; the mobile application is ICT interface for students to make any in class and after class interaction, as well as for instructor use. The web application is a tool for instructor the application's aim is to make lesson plans and post them via the mobile application which provides students accessibility to this information. The application enables any instructor to create lesson submitting schedules as automatic submission or real time posting. Both sections provide the notification feature. Real time availability via one's mobile phone aims to increase the student's motivation to study [5]. One feature of the web application is that students are able to receive a join request from class peers. The web application creates a new class by instructors. Concerning classes, instructors can post their lesson content into the mobile application [6].

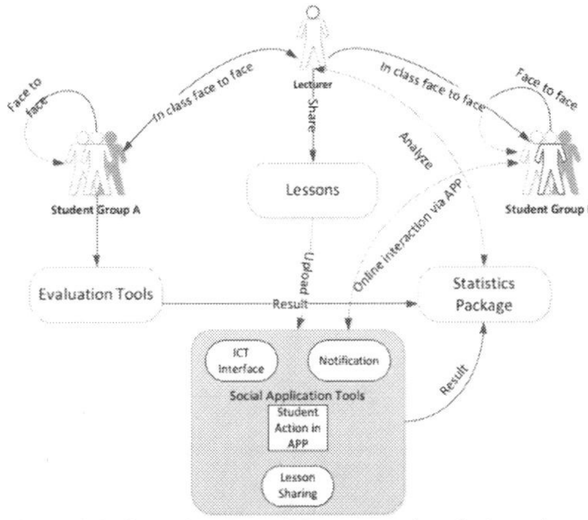

Figure 2. Information flow of the proposed analysis tools, as related to the lecturer, student group A and student group B

The Fig 2. Shows work flow of this research and relation of lecturer, student group A, and student group B. According to the figure only student group B in online access to the online application. In the other hand, student group A is maintaining the ordinary teaching style by face to face study with lecturer in class. Firstly, lecturer uploads scheduled short message lesson for student group B by use the social application tools [6]-[7]. There are similarities between both study group are both them also attend the lecture class and meet lecturer face to face, and they also have change to interaction with each other face to face in class room.

Fig. 3 Illustrated action flow of application, the first step begins as the lecturer utilizes the web application to create a short lesson message by posting, in real time or schedule to post at the pre-set time. The post sent to the mobile application while notifying students via their phones that the latest lesson detail has been sent. Students may only view newer all old lesson content through use of the mobile application. As such, students can click "Impressed" which works similarly to "LIKE" on Facebook, the "Impressed" function will also generate notifications to be sent to other student-peers, drawing attention to specific unread messages.

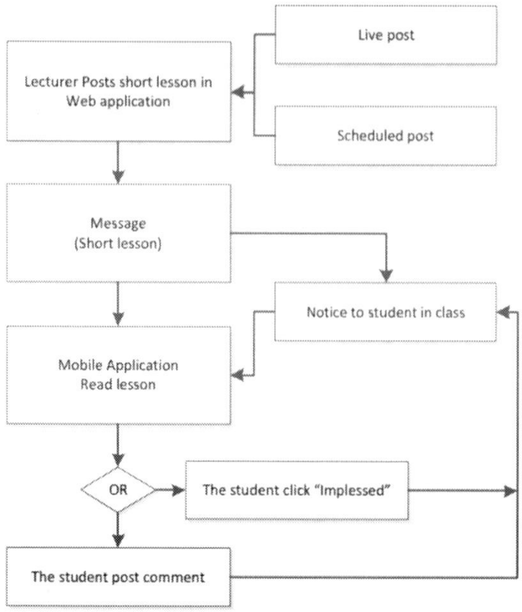

Figure 3. Flow of the social application tools [6].

Figures and tables, as originals of good quality and well contrasted, are to be in their final form, ready for reproduction, pasted in the appropriate place in the text. Try to ensure that the size of the text in your figures is approximately the same size as the main text (10 point). Try to ensure that lines are no thinner than 0.25 point.

On top of this, students are able to post comments regarding the message at hand, either asking for additional information or leaving a comment of their own, where by all other individual, having access to the application, can view all interactions, and are ever notified that a new thread has just been added to the post.[8]-[9]

2.1 Evaluation Instruments

2.1.1 Mid-term and Final examinations

Student activity data were collected from two multiple-choice examinations each covering half of the course lesson. Examination items were both prepared by the lecturer and approved by course committee.

2.1.2 Questionnaire

After completed each lecture examination, students were asked to complete five point rating questions. Two some of them were included as probable predictors of the online study tool what has been used. The first question, asked students ''the complicated lesson is hard to read in short message?'' and was after with the choices of ''I have no difficulty to read'' (rating=5), ''I have some difficulty to read'' (rating=3), and ''It is very difficult'' (rating=1). The second, asked students ''Do you response to any of your friend's activity regarding to the sent message?'' and was after with the choices of ''Yes every times'' (rating=5), ''Sometimes'' (rating=3) and ''I never look at it'' (rating=1).

2.1.3 Inventory of learning Processes (ILP)

Students were asked to complete the ILP [7] to provide an indication of the quality of their study approaches. The ILP provides scores on four subscales: (1) Deep Processing, (2) Methodical Study, (3) Fact Retention, and (4) Elaborative Processing.

3 Experimental

In the first semester of 2013, one of Chiang Mai University computer network classes, which consisted of 45 undergraduate software engineering students, took part in this research. During class, students used the application via their mobile phone. They pasted relative messages and enquiries; from their actions alone, the students were interacting in class. With their phone in hand, these students were involved in the lesson content.

At the end of class, in class interactivity using the application ends. The after class use of the application is largely implemented by the lecturer who send short message repeat the recent in class lesson to the web application, and set time of 4 hours after class to post via the mobile application. Every day before 9am.-8pm., the short message lessons were divided into 5 messages for each 50 minutes in class lesson, having 4 hours intervals between each post. Moreover, students individually attended pre and post sub exams. Our finding from related questionnaires and interviews conducted yelled positive result from both students' lecturers.

The ILP is the evaluation tool which helps to indicate the progressive result of study group who use the application tools, and compare with the other group.

Table 1 shows the ILP separated into four subscales: Deep Processing, Elaborative Processing, Fact Retention, and Methodical Study. The Deep Processing subscale measures the extent to which students focus on higher-level concepts and critical evaluation of the information. The Elaborative Processing subscale measures a student's leaning to individually action the information, though such as self-participation. The Fact Retention subscale observes a student's tendency to learn facts and details. Lastly, the Methodical Study subscale measures how often a student studies. Also the ILP assess the study styles of students from an information processing perception. Using true or false, students specify if each of 62 statements represents the way they normally study and learn.

Table 1. The ILP subscale and example of question [7].

Subscale	Example Question (Ture / Fault)
The Deep Processing	"I find it difficult to handle questions requiring comparison of different concepts"
Elaborative Processing	"I learn new words or ideas by visualizing a situation in which they occur"
Fact Retention	"I am good at learning formulas, names, and dates"
Methodical Study	"For the exam, I prepare a set of notes integrating the information from all sources in the course"

4 Conclusion

The result from student who done the ILP indicated that they were dramatically increased more attention the strategies related with Deep Processing and Elaborative Processing following the classroom teaching on the topic. Including processes slow by the Deep Processing and Elaborative Processing subscales is also more significant certain that these two subscales have presented important relations with academic achievement. Obviously, proves that education students about their specific online learning tools and new after-class study methods can help them gain more effective lesson understanding. And get better score in class, and use improved study abilities.

Table 2. Questionnaire shown result after semester.

Question	Agree	Disagree
The complicated lesson is hard to read in short message	36.7%	63.3%
Sending message too often bothering student	21.1%	78.9%
Response to every new message when the notification has alarm	30%	70%
Response to any of friend's activity regarding to the sent message	80.2%	19.8%
Acquiring more understanding of the lesson, what already have studied in class	76.4%	23.6%
When student read message because of incoming of new message	52.4%	47.6%
Student read message following friend click Impress	68.7%	31.3%

The mid-term examination shows benefit of using online tools, by increase class's score average, when comparing with previous semester. Main concern of the analysis was to define how the tools could influence student to read the message, and also inspiration other student to read it by the first student action. Table 2 Shown results of questionnaire from student after finished the research. Overall the positive result proved the origin concept of research in positive way. Student will have consideration on friend's activity more than incoming of the message, and their feels more comfortable to read short lesson message more often comparing with read very long sentence in study book.

References

1. Moore, M. G. *Editorial: three types of interaction. The American Journal of Distance Education*, **3(2)**, pp. 1–6, (1989)
2. Yacci, M. *Interactivity demystified: a structural definition for distance education and intelligent computer-based instruction. Educational Technology*, **40(4)**, 5–16,(2000)
3. Anderson, *An updated and theoretical rationale for interaction. IT Forum. Retrieved* 25 April, (2004)
4. Liu, T. -C., Wang, H. -Y., Liang, J. -K., Chan, T. -W., & Yang, J. -C. *Applying wireless technologies to a build highly interactive learning environment. In IEEE international workshop on wireless and mobile technologies in education*, Va¨xjo¨ , Sweden, (2002)
5. C.Markett, I. Arnedillo Sanchez, S. Weber, B. Tangney. *Using short message service to encourage interactivity in the classroom. Science Direct Computer & Education*, (2006)

6. Phudinan Singkhamfu, Patiphat Inkong, Chaithawat Suraphruk. *Study Assistance by Social methodology application for mobile device*, June 2014, *conference on Global Trends in Academic Research*, (2014)
7. Schmeck, R.Dillon, & R. Schmeck (Eds.),. *Learning styles of college students, Individual differences in cognition*. (pp. 233–279). New York: Academic Press, (1983)
8. Haewoon Kwak, Changhyun Lee, Hosung Park, and Sue Moon. *What is Twitter, a Social Network or a News Media? What is Twitter, a Social Network or a News Media*, (2010)
9. Moira Burke, Cameron Marlow and Thomas Lento. *Social Network Activity and Social Well-Being*. CHI 2010: Social Media Users, (2010)
10. Bell, J. Doing your research project: *A guide to first-time researchers in education and social science (3rd ed.). Buckingham: Open University Press*, (1999)
11. Ananny, M., Strohecker, C., & Biddick, K. *Shifting scales on common ground: developing personal expressions and public opinions. International Journal of Continuing Engineering Education and Life-Long Learning,* **14(6)**, pp. 484–505, (2004)
12. Sharinaz Ismail, Noor Shahira Md Noor, Mohd Sharifuddin Ahmad. *Social Network Analysis on Personal Knowledge Management. National Graduate Conference, University Tenaga Nasional*, Malaysia, (2014)

Underlying Structure of E-learning Readiness among Palestinian Secondary School Teachers

Fuad A.A.Trayek[1], Tunku Badariah Tunku Ahmad[1], Mohamad Sahari Nordin[1] , Mohammed AM Dwikat[2] , Enas Said Ali Abulibdeh[3] , Muath Asmar[4] and Siti Salwa Md. Sawari[5]

[1]*Kulliyyah of Education, International Islamic University Malaysia, 53100 KL, Malaysia*
[2]*Faculty of Engineering & Information Technology, An Najah National University Nablus, Palestine*
[3]*Education Department, Al Ain University of Science and Technology, United Arab Emirates*
[4]*Faculty of Economics & Social Sciences, An Najah National University Nablus, Palestine*
[5] *Faculty of Islamic Civilization, Universiti Teknologi Malaysia, Jalan Semarak, 54100 Kuala Lumpur Malaysia*

Abstract. This paper reports on the results of an exploratory factor analysis procedure applied on the e-learning readiness data obtained from a survey of four hundred and seventy-five (N = 475) teachers from secondary schools in Nablus, Palestine. The data were collected using a 23-item, self-developed Likert questionnaire measuring e-learning readiness based on Chapnick's conception of the construct. Principal axis factoring (PAF) with Promax rotation applied on the data extracted four distinct factors supporting four of Chapnick's e-learning readiness dimensions, namely technological, psychological, infrastructure, and equipment readiness. Together these four dimensions explained 56% of the variance. A reliability analysis produced high internal consistency estimates ranging between .81 (equipment readiness) and .91 (technological readiness) for the extracted factor structure. These findings provide sound empirical support for the construct validity of the items and for the existence of these four factors that measure e-learning readiness.

1 Introduction

E-learning is defined as content, materials and instruction delivered via all electronic and digital media, such as the Internet, intranet, satellite broadcasts, audio/videotapes, interactive TV and CD/DVD-ROM [1,2]. In recent years, e-learning has emerged as an alternative platform to provide access to education to varied student populations, including deprived and disadvantaged groups. In some educational settings, e-learning may be the only viable solution to impart educational content to the masses, such as the case in Palestinian cities where teacher and student mobility is made difficult by Israeli occupation.

Caught in the midst of the long-standing Israeli-Palestinian conflict, Palestinian teachers face many difficulties in imparting education to their students. In many towns and cities in the country, Palestinian teachers and students experience restricted movement and mobility, created by checkpoints, roadblocks, and gated passages manned by Israeli soldiers and tanks [3]. Going to school daily is a challenging, sometimes dangerous, endeavour for teachers and students. This debilitating situation has paved the way for ICTs to be widely used among Palestinians to learn, communicate and exchange information. Given the political and physical scenario, e-learning constitutes an alternative instructional delivery method to be used in Palestine, which enables a larger population of her citizens to receive education.

2 Literature Review

E-learning refers to the use of the Internet, multiple forms of electronic media, and information and communication technologies (ICTs) to support and enhance learning in educational institutions. According to Soliman [4], e-learning covers all ICT- and Internet-based teaching and learning methods used by educational institutions to enhance instructional quality and improve learning performance. An effective e-learning implementation should galvanize students greater towards independent modes of learning. But in educational settings, achieving a significant usage of e-learning is influenced by many factors, and each institution has to study its ability and preparedness to adopt e-learning in the teaching and learning system. This ability or preparedness is called *e-learning readiness,* defined by Peters [5] as the degree to which an institution or organization is prepared on several key aspects to integrate ICT and e-learning into its system. Considering its role in determining the success of an e-learning initiative, the readiness factor must be thoroughly assessed before an institution decides to introduce e-learning.

In this study, e-learning readiness was defined as the ability and preparedness of Palestinian secondary school teachers to implement e-learning at their respective schools. Their readiness was measured in terms of their e-learning knowledge, computer skills, and positive beliefs about the benefits of e-learning. Tubaishat and Azzedine assert that e-learning success can be achieved by understanding the needs and readiness of all stakeholders in a particular e-learning environment [6]. Towards this end, Chapnick outlined eight factors that can be considered in assessing e-learning readiness, namely psychological, sociological, environmental, human resource, financial, technological, equipment, and content readiness [7]. Darab and Montazer, on the other hand, suggested eleven key factors, i.e. policy, network, equipment, security, cultural, human resource, financial, technical infrastructure, content, standards, and management [8]. Five of their factors overlapped with those of Chapnick's.

Most key theorists maintain that it is very important to have relevant ICT training for teachers and instructors to enhance their preparedness and ability to take up e-learning [9]. Indeed, ICT-related factors are instrumental in shaping the success of e-learning, as highlighted by Bukaliya and Mubika [10], who found ICT knowledge and skills (in software and hardware), computer training, and infrastructure to be the main factors affecting teachers' readiness for the e-learning uptake in Zimbabwean schools. Quite recently three readiness dimensions, i.e. technological, psychological and equipment, were established as significant correlates of e-learning readiness, as they exercised quite a strong influence on teachers' ability to implement the initiative [11].

In contrast, other studies found no significant relationship between teachers' general readiness to embrace e-learning and their psychological predisposition towards it [12], suggesting that psychological readiness may not be quite as important as technical, content, or infrastructure readiness. In essence, however, the literature does suggest that several key factors come into place when considering e-learning readiness. Although we are well-informed by the current research literature on readiness factors, there is still a need to understand the underlying structure that explains the nature of the construct, especially based on reports of teachers in war-torn places such as Palestine. Apart from obtaining an understanding of its underlying structure, this exercise also serves as a means to validate items that can be used to measure teachers' readiness to use e-learning.

3 Objectives of the Study

The objectives of this study were twofold. The first was to identify the underlying factor structure that could explain the e-learning readiness of Palestinian secondary school teachers. The second was to validate the e-learning readiness measurement scale based on the data.

4 Methods

4.1 Respondents

The study was a cross-sectional survey involving 475 teachers from the secondary schools in Nablus, Palestine. Female teachers made up 50.3% of the sample (n = 239), while male teachers, 49.7% (n=236). The sample's mean age was 28.9 years.

4.2 Instrument

The readiness data were collected using a self-developed questionnaire that contained two parts. The first part requested demographic details of the respondents, while the second required the respondents to indicate their level of agreement to 23 items that measured e-learning preparedness in three aspects: psychological (7 items), technological (9 items), and equipment readiness (7 items). The items were primarily drawn from Chapnick's dimensions, as well as previous studies on e-learning, and measured the constructs on a 5-point Likert scale ranging from Strongly Agree (5) to Strongly Disagree (1). After content validation by five experts in the e-learning and educational research fields, the items were refined and pilot tested with 35 secondary teachers in Nablus.

4.3 Data Collection and Analysis

With the help of school principals, 554 copies of the questionnaire were distributed to all secondary school teachers in the city. Out of the 554 distributed, 475 were collected, constituting a response rate of 85.7%. To extract the underlying structure of e-learning readiness, the data were subjected to a robust exploratory factor analysis procedure, i.e. Principal Axis Factoring (PAF). Stevens [13] explained that "the purpose of exploratory factor analysis is to identify the factor structure or model for a set of variables". In this analysis, PAF was deemed suitable to address the research objectives. Promax rotation, rather than Varimax, was used because it was anticipated that theoretically, the e-learning factors would be correlated. In addition, Matsunaga [14] suggests the use of Promax, rather than the more popular Varimax, as it is the most suitable and robust technique for data obtained in social science research.

5 Results

5.1 Underlying Structure of e-Learning Readiness

The PAF procedures applied on the data produced acceptable results in terms of sampling adequacy and inter-item correlations. The Kaiser–Meyer–Olkin (KMO) measure of sampling adequacy, which ranges from 0 to 1 with the recommended cut-score of 0.7, was 0.90, indicating an adequate sampling adequacy for the application of PAF on the data. Bartlett's test of sphericity, which tests whether the correlation matrix is an identity matrix, was significant ($\chi 2$= 5972.1, df= 253, p = .001), indicating that the overall correlations within the correlation matrix were adequate. In summary, these results show factorability of the data, hence justifying the use of PAF in the analysis.

The scree plot drawn from the data (Figure 1) suggests that the items represented four underlying readiness factors. The eigenvalues of the factors and the total variance explained supported this four-factor structure.

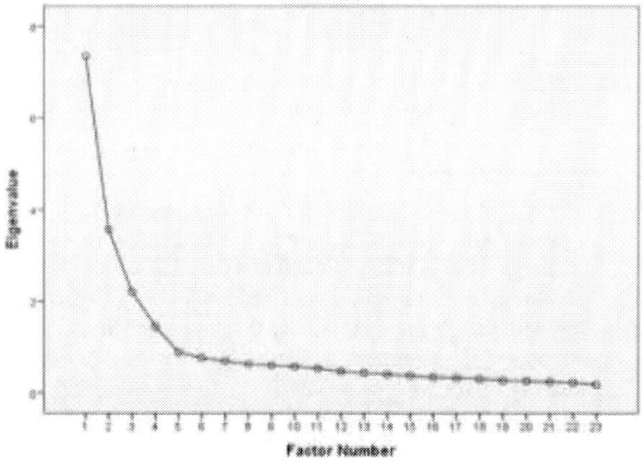

Figure 1.Scree Plot

An evaluation of the eigenvalue criteria with greater than 1.0 indicated the existence of four factors, which cumulatively explained 56% of the variance, with no cross-loading. The correlations among items were significant with communalities ranging between .234 and .725. Only four items (i.e. "I have the ICT facilities to implement e-learning in my classroom", "The school conducts ICT workshops to equip teachers with e-learning skills", "My school can provide the technical staff to support e-learning," and "The school's Internet connection is good") had a communality of less than 0.4.

A total of 23 items were assigned to factors based on their highest level of loading. The first factor reflected teachers' *technological readiness*, and comprised eight items with loadings greater than 0.49. The second factor reflected *psychological readiness*, comprising five items with loadings higher than 0.67. The third factor consisted of seven items with loadings of more than 0.67; the factor suggested *infrastructure readiness*.

The fourth factor reflected *equipment readiness*, and contained three items with loadings of more than 0.58. None of the items loaded highly on more than one factor. These findings provide empirical support for the validity of the readiness scale. The four factors are shown in Table 1 along with their representative items, factor loadings, eigenvalues, individual variance explained and communalities.

Table 1. The Four-Factor Structure with Factor Loadings, Eigenvalues, Variance Explained and Communalities

Factor and Items	Factor Loading	Eigen-value	Variance Explained (%)	Communalities
F1: Technological Readiness		**7.348**	**30.2**	
1. I have enough ICT skills to use e-learning	.858			.70
2. Accessing the Internet is not a problem to me	.650			.51
3. I have the basic technical knowledge to fix technical e-learning problems	.795			.64
4. I have enough ICT competencies to prepare e-learning materials	.881			.70
5. I know what e-learning is	.602			.46
6. I clearly understand the components of e-learning to use it for teaching	.822			.66
7. It is easy to acquire the skills to use e-learning	.606			.61
8. I am ready to integrate e-learning in my teaching	.497			.57
F2: Psychological Readiness		**3.560**	**13.5**	
1. Using e-learning for teaching is a good idea	.688			.60
2. I think e-learning can help students to be more active	.849			.69
3. I think e-learning is helpful to improve teaching and learning	.898			.78
4. I think e-learning is easy to use for teaching	.743			.51
5. It is easy to gain knowledge from the use of e-learning to teach	.674			.47
F3: Infrastructure Readiness		**2.203**	**7.9**	
1. I have the ICT facilities to conduct e-learning in my classroom	.553			.36

Table 1. *Continued*

Factor and Items	Factor Loading	Eigen-value	Variance Explained (%)	Communalities
2. The school conducts ICT workshops to equip teachers with e-learning skills	.605			.34
3. My school can provide the technical staff to support e-learning	.458			.33
4. The school's Internet connection is good	.534			.25
5. My school's ICT infrastructure can support e-learning	.755			.59
6. My school has enough computer labs to support e-learning	.690			.60
7. The school's computer labs have up-to-date facilities to implement e-learning	.658			.65
F4: Equipment Readiness		**1.446**	**4.3**	
1. The school has CD-ROM drives that work well	.587			.55
2. The school has printers that work well	.837			.63
3. The school has projectors that work well	.823			.68

5.2 Reliability of the e-Learning Readiness Data

To generate the data's reliability estimates, Cronbach's alpha coefficients were calculated for the extracted structure. The results are shown in Table 2.

Table 2. Reliability of the Extracted Factor Structure

Factor	No of Items	Cronbach's Alpha
1. Technological Readiness	8	.91
2. Psychological Readiness	5	.88
3. Infrastructure Readiness	7	.82
4. Equipment Readiness	3	.81
Total	23	

All of the factors exhibited good internal consistency with coefficients greater than .80, where an index of .70 or higher indicates a high internal consistency of the measurement scale [15]. Thus, it may be concluded that the e-learning readiness scale used in this study showed high internal consistencies as its extracted four-factor structured yielded estimates ranging between .81 and .91. Therefore, the items may be relied on to produce reliable estimates of Palestinian teachers' e-learning readiness, and may be replicated in future studies measuring the same constructs.

6 Conclusion

In sum, the results produced a reliable four-factor structure of teachers' e-learning readiness, and supported three of Chapnick's dimensions, namely technological, psychological and equipment readiness, and two of Darab and Montazer's factors, namely infrastructure and equipment readiness. The results also provide empirical evidence in support of Akaslan and Law [9] and Bukaliya and Mubika [10], who established the importance of ICT skills and training in influencing readiness. This study found that technological capabilities explained slightly more than 30% of teachers' e-learning readiness. This finding is an important one as it highlights the crucial role of ICT skills and efficacy in determining teachers' ability to implement e-learning. Second in importance is teachers' own beliefs that they can implement e-learning in their classroom, which is this study was shown to account for 13.5% of the variance explained.

In regard to validity of the readiness scale, the PAF results showed that all items had high factor loadings and could be categorized meaningfully into four logical factors, which cumulatively explained 56% of teachers' e-learning readiness. The total variance explained and the factor loadings suggest that the scale was successful in capturing the underlying dimensions of teachers' e-learning readiness. Thus, it can be reasonably argued that the scale provides a valid measure of the construct.

In addition, the Cronbach's alpha values for the factor structure were high (ranging from .81 to .91), suggesting that the scale has a high internal consistency. In conclusion, the reliability and validity results showed that the instrument developed for the study is a valid and reliable measurement tool that can be used to assess e-learning readiness in future studies. However, it would be more insightful for future researchers to consider other dimensions of Chapnick's and Darab and Montazer's models, such as human resource, content and financial readiness, which may be very instrumental in influencing teachers' uptake and implementation of e-learning in their classroom.

References

1. Clark, Ruth C., and Richard E. Mayer. "E-learning and the science of instruction: Proven guidelines for consumers and designers of multimedia learning." John Wiley & Sons, 2011.
2. Govindasamy, T. "Successful implementation of e-learning: pedagogical considerations." *The Internet and Higher Education,* 287-299, 2001.

3. Trayek, F. A. A., TB Tunku Ahmad, and M. S. Nordin. "E-learning readiness and its correlates among secondary school teachers in Nablus, Palestine." *Recent Trends in Social and Behaviour Sciences*. Lumban Gaol et al. (Eds). London: Taylor & Francis Group, 2014.
4. Soliman, Nagwa A. "Using E-Learning to Develop EFL Students' Language Skills and Activate Their Independent Learning." *Creative Education*, 752-757, 2014.
5. Peters, T. "E-Ready for What? E-readiness in developing countries: current status and prospects toward the millennium development goals." infoDev, bridges. Org, 2005.
6. Tubaishat, Abdallah, and Azzedine Lansari. "Are students ready to Adopt E-Learning? A Preliminary E-readiness study of a University in the Gulf Region." *International Journal of Information*, 1 (5), 2011.
7. Chapnick, S (2000). "Are you ready for elearning?." Retrieved from http://www.gc21.de/ibt/en/site/ gc21 /ibt/permnent/publicforum/dok/are_you_ready_for_e learning.pdf on October 18, 2014.
8. Darab, B., and Gh A. Montazer. "An eclectic model for assessing e-learning readiness in the Iranian universities." *Computers & Education*, 56 (3), 900-910, 2011.
9. Akaslan, Dursun, and E. L. Law. "Measuring Teachers' Readiness for E-learning in Higher Education Institutions associated with the Subject of Electricity." Proceedings of 2011 IEEE Global Engineering Education Conference: Learning Environments and Ecosystems in Engineering Education, EDUCON-2011, Amman, Jordan, 2011.
10. Bukaliya, Richard, and Augustine Kudakwashe Mubika. "Teacher competence in ICT: implications for computer education in Zimbabwean secondary schools." *International Journal of Social Sciences and Education*, 1(4), 414-425, 2011.
11. Trayek, F. A. A., TB Tunku Ahmad, and M. S. Nordin. "E-learning readiness and its correlates among secondary school teachers in Nablus, Palestine." *Recent Trends in Social and Behaviour Sciences*. Lumban Gaol et al. (Eds). London: Taylor & Francis Group, 2014.
12. Mulwa, Angeline Sabina, and Dorothy Ndunge Kyalo. "The influence of principals', teachers'and students'attitude on readiness to adopt e-learning in secondary schools in kitui district, kenya." European Scientific Journal, 9 (5), 2013.
13. Stevens, J. "Applied multivariate statistics for the social sciences (3rd ed.)." Mahwah, NJ: Lawrence Erlbaum Associates, 1996.
14. Matsunaga, Masaki. "How to Factor-Analyze Your Data Right: Do's, Don'ts, and How-To's." *International Journal of Psychological Research*, 3 (1), 97-110, 2010.
15. Iacobucci, Dawn, and Adam Duhachek. "Advancing alpha: Measuring reliability with confidence." *Journal of Consumer Psychology*, 13 (4), 478-487, 2003.

A Decision Frameworkon Small and Medium Enterprises to Adopt Cloud Software as a Service

Hsi-Peng Lu[1]and Yao-Yuan Liu[2,a]

[1]*Department of Information Management, National Taiwan University of Science and Technology, Taipei,Taiwan,ROC.*
[2]*Graduate school of Management, National Taiwan University of Science and Technology, Taipei, Taiwan, ROC*

Abstract. Software as a service (SaaS)is the future trend in software services. However, only a few successful SaaS providerscurrently exist. This study investigated how Taiwanese small and medium enterprises (SMEs) can utilize SaaS, giveSaaSprovidersbusiness strategy references, and supply a basis for information system operators to transfer service mode in order to innovate the provision of SaaS. The study is based on survey results obtained from 474 SMEs for a telecom company in Taiwan to determine the reasons SMEs utilizeSaaS. Taking the technology acceptance model (TAM) as the fundamental theoretical foundation, we explored the perceived usefulness, perceived ease of use, and SaaS system features as external variables. In addition, we analyzed each system characteristic on the Likert scale and found that, in addition to the necessary consideration for cost and convenience, SaaS also provides information and communication technology applications in system characteristics,which are including online sharing services, customer relationship management services, and business operation and orders purchasing services. The fact that SMEs wish to utilize SaaS has a positive and direct impact on business.

1 Introduction

1.1Research rationale

Traditionally, small and medium enterprises (SMEs)have utilized software packages to manage information because their resources are limited, and they cannot afford the high cost involved in the development of information systems. However, the data contained in software packages are prone to degradation, which leadsto situations in which risk estimation and operational growth are compromised. Software as a service (SaaS) has cross-platform, cross-network technical capabilities that are not subject to time and space constraints, and is consequently flexible, stable, scalable, and price efficient. Depending on the specific usage desiredby SMEs, SaaS can be utilized anytime and anywhere, via mobile technology on emerging broadband and 4G networks. SMEs can thus expand their future operation, efficiency and strengthen their competitive advantage by utilizing SaaS.

1.2 Purpose

Previous studies conducted on SaaS focused on the reasons why SMEs are using cloud services and determined that SMEs benefit from the framework and effectiveness of improved enterprise terminal IT services, as well as improved cost benefits, information security, and network privacy and stability. However, the relationship between

SaaS system characteristics and their use by SMEs has not been adequately investigated. In this study, we employed the technology acceptance model (TAM)and exploredthe factors and variables that affect the behavior and intention of SMEs that use cloud services (in particular, SaaS). The results of this study will supply SaaS providers with a reference for business strategy and will help IT operators to develop SaaS that conform to the needs of SMEs and facilitate innovation in the provision of SaaS.

2 Literaturereview

2.1Software as a service (SaaS)

Cloud resources can be dynamically extended and configured and applications modularizedto accord withenterpriseneeds. Further, SMEs can purchase authorization from suppliers to use services such as office, back-enddataanalysis,customerrelationship management, business process management, and human resources management applications according to their own needs. Users can experience software and services simply by connecting to the Internet, thereby eliminating information systems construction and software licensing costs. Further, resources can be distributed in accordance with the needs of users and payment remitted according to actual usage. Examples of such software services include Amazon's AWS, Salesforce's CRM, andMicrosoft's Office 365.

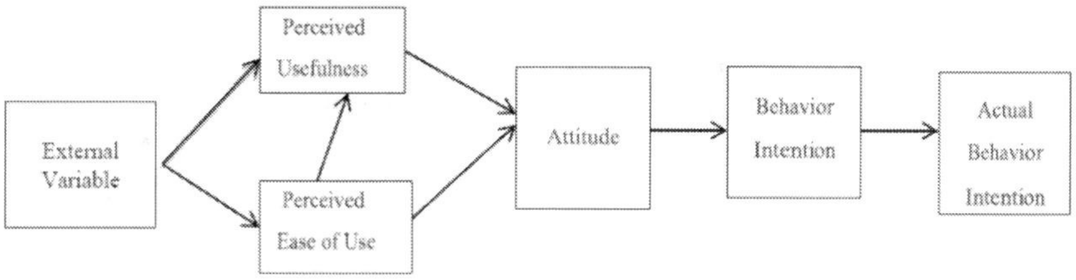

Figure 1. Technology acceptance model (Davis, 1986) [1]

2.2The technology acceptance model (TAM)

The technology acceptance model (TAM) as shown in Figure 1 was proposed byDavis [1]. It was developed based on rational action theory [2], self-efficiency, the cost-benefit model, and innovation. Similar to rational action theory, TAM holds that belief affects attitude, attitude further influences behavioral intention, and behavioral intention is the best variable with which to predict behavior [3]. Behavioral intention refers to the subjective probability of a person performing certain actions, theorizing that the stronger the intention, the higher the likelihood that such actions are actually performed [4]. TAM is employed primarily to explain the crucial factors of users accepting science and technology systems. Davis et al.[5] abandoned the subjective norm in TRA by means of experimentation, explaining that the acceptance intention for users of a specific system is affected by perceptions and beliefs, such as perceived usefulness and perceived ease of use. TAM, by definition, measures usefulness as a result of performance improvement users expect in particular application systems, and perceives ease of use as a result of expected effort dispensation by users.

TAM classifies system features, users' characteristics, organization factors, and training as external variables that influence acceptance and use by users. However, the effect is relatively small compared to that of perceived usefulness and perceived ease of use,andTAM does not indicate which variables affect perceived usefulness and perceived ease of use. TAM has been verified and is highly recognized. It is applied in various fields, including technology products and services, and in the continued development of TAM [6]. The core architecture of TAM is highly reliable and can be introduced in all categories [7]. From the theory and empirical results of TAM, it can be seen that the actual usage will be affected by behavior intention, and the behavior intention is further influenced by the user's attitude and cognizance. Furthermore, user attitude is decided by the cognition of usefulness and ease of use. Venkatesh and Bala[8] summarized the research conducted on TAM, and found that they focused on three areas: psychological viewpoint focusing on TAM architecture, verification and support of TAM structure, and increasing facets to expand TAM architecture. Subsequently, they developed a representative theoretical framework indicating four different determinants (i.e., external variables, which affect the perceived usefulness and perceived ease of use): individual differences, system characteristics, social influence and facilitating conditions (Figure 2).

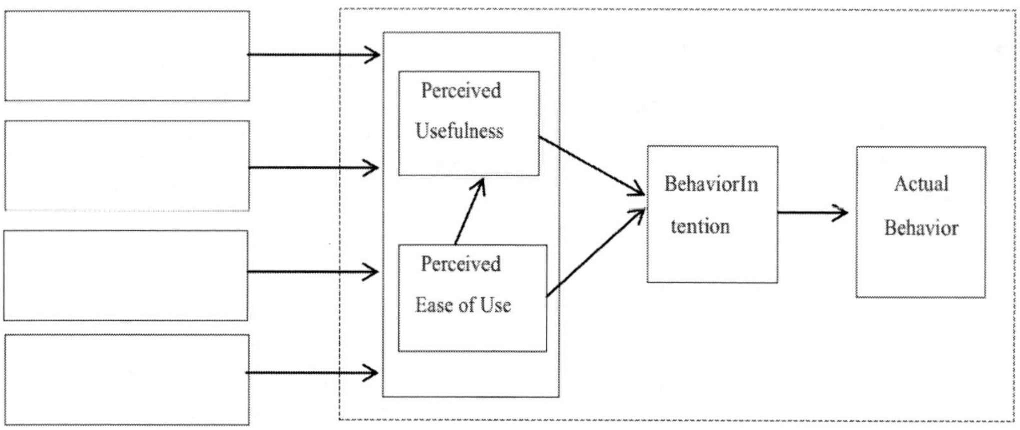

Figure 2. Theoretical architecture: determinants of perceived usefulness and perceived ease of use (Venkatesh andBala, 2008)[8].

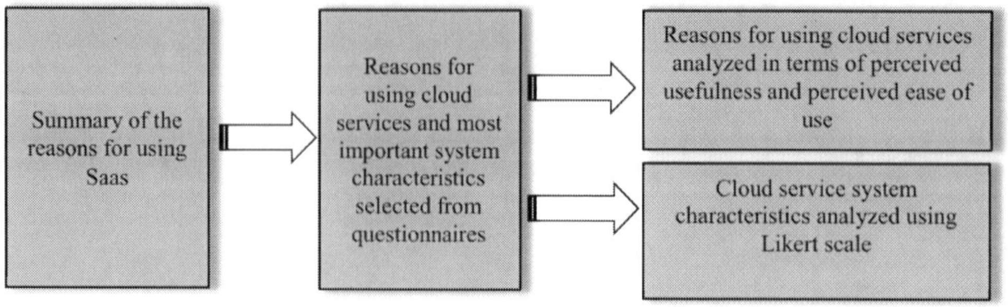

Figure 3. Rationale for using cloud service and system characteristics analysis procedure

2.3 Extension of TAM

TAM is widely applied in qualitative research, and many scholars have expanded TAM to improve the interpretations of user acceptance behaviors. King and He [9] reviewed 88 publications and proposed a more general architecture based on TAM. They divided research on TAM into four categories: (1) Research on external variables, including situation involvement, prior use or experience [10] and computer self-efficacy [11]; (2) improvement of the prediction of TAM for behavior intention by combining with other theories, including subjective standard and expectation [12], task-science and technology [13], and risk and trust [14]; (3) situational factors with moderating effect, such asscience and technology characteristics; and (4) subsequent measurements, such as attitude, perceived use, and actual use[15].

3 Research design

3.1Research method

The study consisted of survey questionnaires, and the responses from Taiwanese companies were used to determine the main factors to consider when importing cloud application systems for such users. A flowchart of the research method is presented in Figure 3.

3.2Establishment of research model and hypothesis

3.2.1 Survey design

The survey questionnairestargeted technology managers. The contents of the questionnaires were divided into three sections: basic information for enterprises, reasons for using cloud services, and, most importantly, cloud system characteristics. Thus, the first category(the enterprise category) dealt with number of computers and number of employees. The second category was the main consideration for using SaaS. The third category was concerned with the most crucial characteristics of SaaS. The main purpose of the questionnaireswas to determine the reasons for using SaaS and the characteristics of SaaS systems.

This study found that enterprises use cloud services for 12 main reasons (reduce build and maintenance costs, pay fees according to the usage, improve employee communication efficiency, attempt to apply technology in response to trends, can be used anywhere, centrally manage data, fast and easy to use, can be used in mobile terminal, high system expansion elasticity, access to the latest applications and services, fast computing speed, and high security) collected from various journals, publications, books, and vendor information, and seven service system characteristics (information and communication application and online sharing service, office administration service, CRM services, operation order management service, operation financial management service, human resource management service, business operation and orders production and manufacturing management services).These contents were all designed into the questionnaire.

3.2.2 Sampling method and survey object

The survey targeted 474 SMEs. Questionnaires were mailed to information managers in the SMEs, who were requested to respond.

3.3 The information analysis method

In this study, perceived usefulness and perceived ease of use were analyzed on a simple scale, and the system characteristics analyzed using the Likert-type total formula scale. The subjects were asked to respond usinga specific score to reflect the extent of their favor, which can help to identify the major factors in cloud computing services that enterprises consider among the system characteristics. In order to conform to the Likert construction and testing rules, the subjects were required to select one offive favor grades. Numberswere assigned to the values of the individual options, with two representing the least agreeable, and 10 the most agreeable.

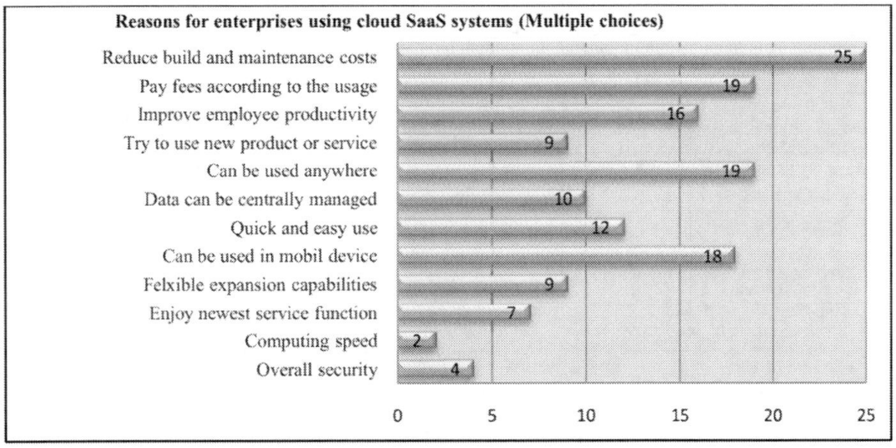

Figure 4. Questionnaire statistics showing crucial reasons enterprises use cloud applications.

4 Analysis of results

4.1 Analysis of basic enterprise information

In order to understand the sample structure of the returnedquestionnaires, the basic information pertaining to the objectives in the study was analyzed using the descriptive statistical method. Statistical variables of research objectives included industry category, the number of employees (Table 1), and the number of computers (Table 2). In terms of industry category responses, the majority sampled were information service corporations, with 19 units in total accounting for 35.8%, followed by members of the electric and electronic machinery industry, with 11 units accounting for 20.8%.

4.2Survey results and analysis

In this study, 474 questionnaires were sent by e-mail, and 53 responses were received—a return rate of 11.2%. From the response results, we found that among enterprises that had the intention of using commercial operation application services, the top five factors for employing SaaS were reduced build and maintenance costs, pay fees according to the usage structure, improved employee work efficiency, can be used anywhere, and mobile device applications (Figure 4) according to the summarization and categorization of perceived usefulness, perceived ease of use, and external variables (Table 3). The main consideration as to whether to purchase a cloud application was cost. Convenience was also an important determinant, followed by the function of system applications, by which they expect to improve employee work efficiency. In terms of system applications, analyzing demand strength on the Likert scale, the enterprises using commercial operation services are in maximum demand of ICT applicationsand online sharing services, followed by business orders and purchasing operation services, and CRM services (Table 4).

Table 1. Quantitative statistics for the number of employees in corporations that replied to the questionnaire.

Item	Category	Number of returned questionnaires	Proportion
Number of employees	1-5	11	20.7%
	6-10	5	9.4%
	11-20	6	11.3%
	21-50	9	16.9%
	51-100	9	16.9%
	101-200	7	13.2%
	201-500	2	3.7%
	Over 501	4	7.5%

Table 2. Quantitative statistics for the number of computers in corporations that replied to the questionnaire.

Item	Category	Number of returned questionnaires	Proportion
Number of computers	1-5	7	13.2%
	6-10	7	13.2%
	11-20	8	15.1%
	21-50	12	22.6%
	51-100	12	22.6%
	101-200	4	7.5%
	Over 201	3	5.6%

Table 3. Considerations analysis between PU,PEOU,EV

Items	Crucial considerations	Votes
Perceived usefulness (PU)	Can be used anywhere	19
	Can be used in mobile devices	18
Perceived ease of use (PEOU)	Reduce build and maintenance costs	25
	Pay fees according to the usage	19
External variables (EV)	Improve employee work efficiency	16

Table 4. Statistics of important factors of external variables

Service items	Scores
(1) Information and communication technology application characteristics, and online sharing services	7.4
(2) Office administration services	6.4
(3) Customer relationship management (CRM)	6.6
(4) Business operation and orders purchasing services	6.7
(5) Operation financial management services	6.4
(6) Human resource management services	6.3
(7) Business production and manufacturing management services	6.0

5 Conclusions and recommendations

5.1 Conclusion

The relationship between perceived usefulness and perceived ease of use and behavior intention can be obtained based on the analysis conducted in this study. Perceived usefulness and perceived ease of use positively influence behavior intention, and also can verify and support the core structure of TAM [3,6,8]. Schepers and Wetzels[7] summarized a common conclusion that perceived usefulness as having a stronger impact on behavior intention than perceived ease of use. However, in this study,cloud servicesare network applications, and the conclusion that the influence of perceived ease of use is superior to perceived usefulness corresponds to the results obtained by King and He[9]. They posited that the perceived ease of use played a more important role in internet application. Although cloud systems possess advantages such as elasticity, multiplicity, the ability to extend, and availability on demand, the cost of implementing SaaS is still the foremost consideration for SMEs, meanwhile, basic advantages of SaaS systems, such as improving the work efficiency of employees, being used anywhere, and being used in mobile device, are also very vital.

5.2 Contribution

This study utilized TAM theory and questionnaires to understand the considerations of SMEs that use SaaS.Further, external characteristicvariables were introduced. Two behavior intentions of SMEs that employ SaaS were found to be most prominent in the study results.

5.2.1Usefulness after ease

In the study, perceived ease of use not only had a greater impact on behavior intention than perceived usefulness, it also influences perceived usefulness positively. From these two points, a multiplier method in the cognitive stage is letting SMEs realize that SaaS systemscan have friendly interfaces, are easy to learn and use, and are not a waste of time because they have succinct interfaces with plain and brief illustrations. It can enhance their behavior intention and improve the possibility of actual actions.

5.2.2 Intensifying system characteristics to enhance the intention of SMEs

The results also indicate that system characteristics play a vital role in both perceived usefulness and perceived ease of use. Further, perceived usefulness can strengthen the information and communication technology applications in system characteristics, which are including online sharing service, business operation and orders purchasing services, and customer relationship management services, to satisfy the present and future demands of enterprises.

5.2.3 Obtaining crucial value for future product development

The results collected for cloud service providers, whether for product development in the future, marketing, or customer services, will be crucial references for vendors who can understand SaaS system functions of SMEs associated with this research.

5.3Research limitations and recommendations

This study was ultimately restricted by time and resources, and there are three insufficient aspects that need further investigation: (1)Questionnaires should be designed according to business size and specific industry (such as information services, chain industry, and manufacturing), and external variables built. (2) Reliability and validity should be considered in the analysis to improve the accuracy of the results. (3) The effect of using SaaS for internet externalities should be discussed.

References

1. F. D. Davis, A technology acceptance model for empirically testing new end-user information systems: Theory and results. Massachusetts Institute of Technology,Sloan School of Management, (1986).
2. M. Fishbein,I. Ajzen, Belief, attitude, intention and behaviour:An introduction to theory and research,Boston,Massachusetts:Addison-Wesley (1975).
3. F. D. Davis, Perceived usefulness, perceived ease of use, and user acceptance of information technology. MIS Quart.,**13**(3), 319-340 (1989).
4. I. Ajzen, The theory of planned behavior. Organ.Behav.Hum.Decis.Process.,**50**(2), 179-211 (1991).
5. F. D. Davis, R. P. Bagozzi, P. R. Warshaw, User acceptance of computer technology: A comparison of two theoretical models. Manag. Sci.'982-1003 (1989).
6. V. Venkatesh, F. D. Davis, A theoretical extension of the technology acceptance model: Four longitudinal field studies. Manag.Sci., **46**(2), 186-204 (2000).
7. J. Schepers, M. Wetzels, A meta-analysis of the technology acceptance model: Investigating subjective norm and moderation effects. Info. &Manag., **44**(1), 90-103 (2007).
8. V. Venkatesh, H.Bala, Technology acceptance model 3 and a research agenda on interventions. Decis. Sci., **39**(2), 273-315 (2008).
9. W. R. King,J. He, A meta-analysis of the technology acceptance model. Info.&Manag.,**43**(6), 740-755 (2006).
10. V. Venkatesh, M. G. Morris, Why don't men ever stop to ask for directions? Gender,social influence, and their role in technology acceptance and usage behavior. MIS Quart., **24**(1), 115-139 (2000).
11. V. Venkatesh, Determinants of perceived ease of use: Integrating control, intrinsic motivation, and emotion in the technology acceptance model. Info.Syst.Res., **11**(4), 342-365 (2000).
12. V. Venkatesh, M. G. Morris, G.B. Davis, F. D. Davis, User acceptance of information technology: Toward a unified view. MIS Quart., **27**(3), 425-478 (2003).
13. T. J. Larsen, A. M. Sorebo, O. Sorebo, The role of task-technology fit as usersmotivation to continue information system use. Comput.Hum.Behav.**25**(3), 778-784(2009).
14. J. M. O. Egea, M. V. R. Gonzales, Explaining physicians'acceptance of EHCR systems: An extension of TAM with trust and risk factors. Comput.Hum.Behav.,**27**(1),319-332 (2011).
15. T. S. Behrend, E. N. Wiebe, J. E. London, E. C. Johnson, Cloud computing adoption and usage in community colleges. Behav.&Info.Tech., **30**(2), 231-240 (2011).

Research on the Integration Method of Dynamic Heterogeneous Spatial Information Plotting Data

Xiao-lan Xie[1,2], Zhen Long[3], Wen-qi Liao[4] and Xiu-juan Guo[1]

[1] College of Information Science and Engineering, Guilin University Of Technology, 541006, Guilin ,Guangxi, China
[2] Guilin University of Technology, Guangxi Key Laboratory of Spatial Information and Geomatics, 541004, Guilin, China
[3] College of Mechanical and Control Engineering, Guilin University Of Technology, 541006, Guilin ,Guangxi, China
[4] College of Civil Engineering and Architecture, Guilin University Of Technology, 541004, Guilin ,Guangxi, China

Abstract. According to the dynamic heterogeneous spatial information plotting data, this paper presented a new integration method about dynamic heterogeneous spatial data, studied the spatial information plotting object and established the semantic ontology database. It declared the mapping rules in ontology, solved the semantic differences of spatial information plotting objects and finished the automatic or semi-automatic integration of plotting content to integrated automatic or semi-automatic of the content for different plotting situations(same name of different place, different name of same place etc) and different plotting data types(text, images, sound, video etc). So it has great significance to enhance the efficiency of the plotting data integration.

1 Introduction

Since 1990s, with the development of information and communication technology, as well as the wide use of remote sensing, geographic information system and satellite positioning technology, the acquisition mode of spatial information tends to be diversified, geographic spatial information is growing exponentially. Spatial information data show a distinct heterogeneity, that is, the characteristics of multi-source, multi-scale, multi temporal phase. However, due to the limitations of the data itself: a large number of spatial information data cannot be effectively used and the information needed by the user can not be satisfied in time, so the integration and update of spatial information data, information integration and fusion have become the cutting-edge tasks and key technology research in academic field and the key technology research. Due to the different application purposes of geographic data from different departments, it caused multi semantics, multi temporal and spatial, different storage format and the differences between data model and storage structure of spatial data in the same area and the same scale, it brings great difficulties to the data sharing and data integration between the GIS Department[1, 2, 3].

Especially in the different data production departments, in order to satisfy the different needs of users, the data standard and specification used are different. It exists great difference in the division of geographical elements, the classification of geographical elements, the definition of geographical elements, the definition of geographical element attribute, and the coding method of geographical feature, resulting the diversity of multi-source heterogeneous geographic information in semantic, it is the main reason that the data could not be used directly with each other and share. Spatial information plotting data comes from different databases and complex relationship, so the data is not unified and not standardized, the relationships are confused.

Therefore, in view of the above mentioned problems about spatial information plotting data, it needs a new efficient integration method of heterogeneous spatial information[4, 5, 6].

2 Data Integration Method

In this paper, we study an effective integration method of spatial information plotting data, which is helpful for fully understanding the spatial information plotting data to solve the defects of the existing technology. The method mainly includes the following steps:

(1) The semantic ontology base construction of spatial information plotting object

It can study the problem of semantic dissension, determine the professional field covered by the semantic ontology of the plotting object and the purpose of applying ontology, list the important terms and concepts in the field, in order to build the ontology framework and design meta-ontology. It also encodes the domain ontology and verifies whether the ontology meets the requirements. All these have done to complete the construction of the semantic ontology base of spatial information plotting object. The local ontology is constructed according to the specific application, and then the semantic mapping rules between heterogeneous spatial information are declared according to the relationship between ontology classes.

(2) The collection and extraction of the plotting data

①If there are business data of the front-end-processor, we can use the way of preposition exchange of the plotting data information to deal with the data. The department involved in the information exchange adopts the front-end-processor to achieve data exchange, which is composed of the operating system, the pre exchange plotting information database, the information exchange communication interface and the information exchange bridge. The relevant departments need to send the data exchanged to the front-end-processor through the Department of LAN. The front-end-processor transmits the data to the information exchange platform. The information exchange platform processes the data, and then sends the data to the front-end-processor in the network according to the business process. The business department can access the data needed through front-end-processor and the real-time and dynamic submission of the spatial information plotting data is realized.

②The collection and extraction tool setting of the plotting data could realize the secure, reliable, stable and efficient the plotting data collection and extraction system between business data in different business database. The system takes some measures like collection, extraction and loading to convert the business data provided by some unit into the original database of the plotting data through the data transmission, information encryption, format conversion and system management.

(3) The analysis and processing of plotting data

The data filtering conversion system is to filter and transform the data that has been collected in the intermediate database. It combines with the semantic ontology database of plotting object and mapping rules in ontology, realizes the processing of different data and suspicious data according to a certain filtering and transformation rule. The system includes fault-tolerant management and spatial topological relation integration, saves the correct data filtered into the final database of plotting data and provides support for the construction of each data subsystem.

(4) Form the final database of plotting data

The final database is the final plotting database which confirmed after the former collection, extraction, analysis and processing of plotting data[7]. It provides support for other business system. Finally it realizes the classification management of data, authority management, security authentication, security audit by setting the safety management mechanism. It encrypts data in different security levels and realizes data backup.

3 THE EXPERIMENTAL METHOD

As is shown in Figure 1, according to the integration method of dynamic heterogeneous spatial data, the whole technical scheme diagram can be divided into four parts: establishing the plotting object semantic ontology database, collecting the plotting data, processing plotting data, building the final database, it mainly includes the following steps.

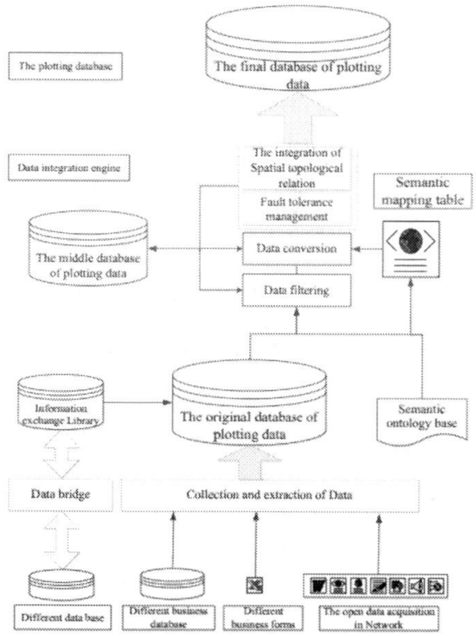

Figure 1. A flow chart of Integration method of dynamic heterogeneous spatial information plotting data.

3.1 Establishing the plotting object semantic ontology database

We use the open source Protégé ontology development tool developed by Stanford medical Infomatics from Stanford University to express spatial information plotting object semantic ontology. Protégé ontology development structure displays in a hierarchy tree structure, it can easily add or edit classes, subclasses, attributes, instances, etc. It provides basic functions of ontology construction and achieves visual expression of ontology through the OWL Viz Tab. It describes geographic knowledge according to the form of class and attribute. Using ontology expression language based on OWL to clearly express the semantics and mutual relations of geographic concepts. It achieved by the use of open source Jena development kit which used to develop the semantic ontology application program, support the simple reasoning based on rules and implement mapping rules in ontology.

3.2 Collecting the plotting data

The system takes some measures like collection, extraction and loading to convert the business data provided by some unit into the original database of the plotting data through the data transmission, information encryption, format conversion and system management.

3.3 Processing the plotting data

3.3.1 Data processing method

For these plotting data from various sources, we can plot the object semantic ontology database by combining with the space information, then eliminate duplication, remove ambiguity, number, classify, nominate and so on, finally form the standardized data. According to the underlying demand of plotting data, after the data processing and filtering, we should combine with the declared semantic mapping rules, convert the original database data into the form that target data required. When analyze and process the data, we convert these data at the same time. Moreover, we make logical operation to the filtered data according to the requirements of the application, convert into the form that target data required, so that ensure the data validity. Then develop the fault tolerance rules in the case of ensuring the integrity of the data. Meanwhile, arrange the topological relation existed in the data, eliminate the errors in the data, ensure the quality of the data. According to the features of the object with different plotting data types (text, images, audio, video, etc.), we set up related data filtering rule, transformation rule, fault-tolerant rule and the topological relation collation rule, and take some related settings on the processing system, so as to improve the automation and efficiency of plotting data conversion and filtering.

The data processing method has the following characteristics:

①Use the platform architecture and design based on J2EE, use the current advanced struts, hibernate and other technologies to realize, and support Tomcat, Weblogic, WebSphere, Apusic and other kinds of application environments;

② The database based on various domestic and international mainstream are provided, such as the interface definition and information processing of Oracle, SQL server, Access, Sybase and other data systems. The information source of data processing could be defined freely, and support various operating system platforms such as UNIX, Windows and Linux;

③Support a variety of information pretreatment and management strategies of data processing, provide an unified, efficient and multi-tasking processing engine on the data processing, it can define and configure specific data processing and processing mode flexibly and dynamically;

④ Use the multi-thread processing mechanism, support a set of multi-tasking management functions to process systems, a variety of tasks can be defined freely and centralized management, and the operation interface and the use is very simple and clear;

⑤ Support the information processing based on a variety of file type database systems, such as SHP, DBF, Excel, XML, GML, etc. The processing strategies and configuration of information use the unified and simple management platform and tools;

⑥ Not only support automatic processing, manual processing and other various processing methods to the resource data, but also realize output, statistic, analyze and other further service functions to the processed data;

⑦Provide the service of information processing and processing log in details, it can define the requisite log category and detailed grade or degree by itself;

⑧ Provide fully independent security access and control system, each operation and resource use can be accomplished by defining roles and users;

⑨Use the B/S architecture design and development, using and managing the interface will be intuitive and simple, it does not need special training to users, the products provide the function of on-line use help at the same time.

3.3.2 Data processing flow

①First we should finish the data preprocessing on the imputation plotting data, mainly include taking the standardized treatment of the content of plotting data, for example, standardize some key fields whether include "-", standardize full width, half width and spacing of Chinese and English name;

②The extract of information center system need to take some automatic processing to the processed data. According to the certain rules and standards, we can plot the data by the data collection tools, after data collection and extract, form the original library of plotting data. Then filter the data through the data integration engine in accordance with a certain data filtering rule, so that form a filtered plotting data repository[8, 9, 10]. Because of the complex relationship between data sources, it is necessary to use a data transformation rule to the filtered plotting data repository for data conversion. In order to ensure the integrity of the data, we need to set the fault-tolerant rule, take some fault-tolerant managements (save, eliminate, etc) to the incorrect data. Meanwhile, integrate existed topological relations in the plotting data, eliminate topology errors, then form the final library of plotting data. At last, the final library will be transformed into the subject library of plotting data which provide the application to other business systems;

③In the data processing system, we could configure data processing methods according to a certain rule, then implement the data processing, so that these processed data can be collected into the final library of plotting data .

The flow chart of data processing is shown in Figure 2.

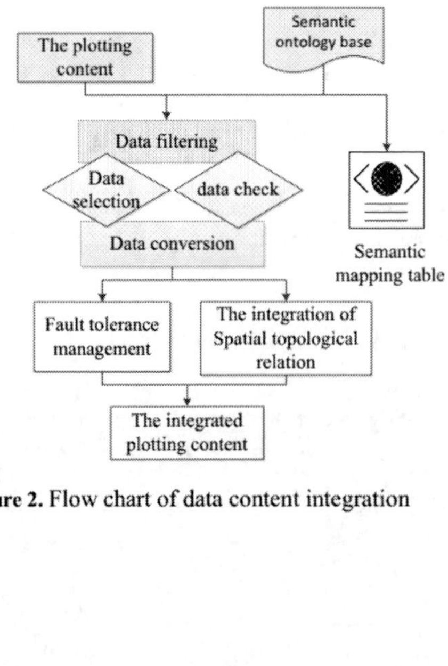

Figure 2. Flow chart of data content integration

3.4 Building the final database

The plotting data final database is the ultimate resource database which is formed by the data collection, extraction and processing analysis. It provides data support for other business system, and the classification of data management, authority management, security authentication, and security audit have been realized by setting the security management mechanism. Finally it encrypts the data in different security levels and realizes the secure backup of the data.

It mainly has the following advantages by taking the above overall frame structure:

①It uses the exchange front-end-processor and pre exchange of the plotting data resource information database, it can effectively separate switch network from department business network to ensure the safety performance of department business system and the plotting database;

②It uses the information exchange bridge to realize the on-line real-time exchange between the department business information database and the pre exchange plotting data information database, and meet the goal of online real-time plotting data information exchange between departments;

③ Establishing the information exchange platform with central exchange management function, which is convenient for system management and system expansion;

④ The system which is based on the information exchange platform can support all kinds of heterogeneous platforms or data sources, and facilitate the rapid expansion of the platform, in the premise of ensuring reliable, efficient and safe transmission of information;

⑤The whole system has good expansibility, which is in line with the standard of national government information resource exchange system and other related standards;

⑥By establishing the data integration mechanism of standard drawing object semantic ontology library, to constraint ontology through the terminology established, relying on the artificial intervention, establishing plotting object semantic mapping, so the spatial information semantic points difference can effectively be solved, and better data integration effect on painting has been realized.

4 Conclusion

As the different sources of the spatial information plotting data and the complex relationship, it leads to some features of the data, for example, the data is not unified, not standardized and the correlation between data is chaos. Therefore, aiming at the problem of the spatial information plotting data, this paper proposes a new method for the integration of dynamic heterogeneous spatial information data. According to the experiment results, the integrated method can increase the efficiency of data integration, so it has certain practical significance.

Acknowledgement

This research work was supported by the National Natural Science Foundation of China (Grant No.61540054), National High Technology Research and Development Program 863 under Grant No. 2013AA12A402, Natural Science Foundation of Guangxi Provincial under Grant No. 2013GXNSFAA019349.

References

1. Lin Fengcheng, Chung Lankun, Ku Wenyuan, Chu Linru, Chou Tienyin, Service component architecture for geographic information system in cloud computing infrastructure, Proceedings - International Conference on Advanced Information Networking and Applications, 368-373 (2013)
2. Yan Zhihui, Sun Qiang, Discussion on cloud GIS, WIT Transactions on Information and Communication Technologies, **57**, 387-394 (2014)
3. Diao Zhijian, Guo Song, A research of GIS software application based on cloud computing, Applied Mechanics and Materials, **513-517**, 2107-2110 (2014)
4. Schockaert Steven, Smart Philip D, Twaroch Florian A, Generating approximate region boundaries from heterogeneous spatial information: An evolutionary approach, Information Sciences, **181**, 257-283 (2011)
5. Li Weidong, Peng Shumin, Zhou Jinyan, Integration framework of multi-source heterogeneous spatial information and remote computing based on web services, 2010 2nd International Conference on Industrial and Information Systems, **2**, 457-460 (2010)
6. Wu Hao, He Chaoying, Liao Anping, Peng Shu, A framework for integrating heterogeneous spatial information and applications adaptively based on multi-agent and web service, Proceedings - 3rd International Conference on Multimedia Information Networking and Security, 253-257 (2011)
7. Condit Richard, Lao Suzanne, Singh Anudeep, Esufali Shameema; Dolins Steven, Data and database standards for permanent forest plots in a global network, Forest Ecology and Management, **316**, 21-31 (2014)
8. Bergamaschi Sonia, Guerra Francesco, Orsini Mirko, Sartori Claudio, A new type of metadata for querying data integration systems, SEBD 2007 - Proceedings of the 15th Italian Symposium on Advanced Database Systems, 266-273 (2007)
9. Wang Xin, Huang Linpeng, Xu Xiaohui, Zhang Y.I., Chen Junqing, Journal of Information Science and Engineering, **27,** 681-695 (2011)
10. Wang Xiaobin, Wang Wangx, Bao Mingyu, Research into framework of real-time data integration system, Applied Mechanics and Materials, **310**, 605-608 (2013)

An Imbalanced Data Classification Algorithm of De-noising Auto-Encoder Neural Network Based on SMOTE

Chenggang Zhang[1], Jiazhi Song[2], Zhili Pei[3] and Jingqing Jiang [3]

[1]Inner Mongolia University for the Nationalities, College of Mathematics, 028000 Tongliao, China
[2]Northeast Normal University, College of Computer Science and Information Technology, 130000 Changchun, China
[3]Inner Mongolia University for the Nationalities, College of Computer Science and Technology, 028000 Tongliao, China

Abstract. Imbalanced data classification problem has always been one of the hot issues in the field of machine learning. Synthetic minority over-sampling technique (SMOTE) is a classical approach to balance datasets, but it may give rise to such problem as noise. Stacked De-noising Auto-Encoder neural network (SDAE), can effectively reduce data redundancy and noise through unsupervised layer-wise greedy learning. Aiming at the shortcomings of SMOTE algorithm when synthesizing new minority class samples, the paper proposed a Stacked De-noising Auto-Encoder neural network algorithm based on SMOTE, SMOTE-SDAE, which is aimed to deal with imbalanced data classification. The proposed algorithm is not only able to synthesize new minority class samples, but it also can de-noise and classify the sampled data. Experimental results show that compared with traditional algorithms, SMOTE-SDAE significantly improves the minority class classification accuracy of the imbalanced datasets.

1 Introduction

Classification problem is one of the important research contents in the field of machine learning. Some of the existing classification methods can generally perform better when classifying balanced data. However, there are large amounts of imbalanced datasets in the field of practical application, such as network intrusion, text classification, credit card cheat-detection, medical diagnosis, etc., for which minority class recognition rate is more important [1-2]. To remedy the deficiency that minority class samples have on distributed information, SMOTE algorithm put forward by Chawla etc. [2]. Not only effectively synthesizes minority class samples, but, to a large extent, it also avoids over-fitting problem. The algorithm has already achieved a favourable effect in imbalanced datasets classification, but it brings a new problem such as noise.

Auto-encoder neural network based on thought of deep learning has already obtained huge success in the field of machine learning [3]. It initializes the network weights through unsupervised layer-wise greedy learning, and learn about data features and reduces the irrelevant and redundant data through constantly adjusting the network parameters, and then it fine-tunes the network parameter using back propagation (BP) algorithm. Stacked De-noising Auto-Encoder neural network, SDAE, can train more robust expressions of input data through adding noise into original data. Thereby it can improve the generalization ability of auto-encoder neural network to input data [4-6]. Imbalanced data classification algorithm of de-noising auto-encoder neural network based on SMOTE proposed in the paper can reduce the noise problem from SMOTE, and it de-noises and classifies the sampled data, which improves minority class classification quality.

2 Related works

2.1 SMOTE Algorithm

Synthetic minority over-sampling technique (SMOTE) algorithm is a kind of typical sampling method proposed by Chawla etc. in 2002 [5]. Compared to traditional over sampling technology, it can effectively avoid over fitting phenomenon of the classifier. The main gist is to insert synthesized minority class samples at the nearest neighbours of it, thus, to increase the number of minority class samples to balance the dataset. To be specific:

Suppose oversampling ratio is N. First, randomly choose K samples from P nearest minority class neighbours of each minority class sample. Then according to (1), synthesize each minority class sample and chosen K samples respectively to generate N new minority class samples; finally, add the new samples to original sample set to form a new training sample set.

$$x_{new} = x + rand(0,1) \times (y[i] - x) \tag{1}$$

Among which, $i = 1,2,\ldots N$, $rand(0,1)$ is a random number in range $(0,1)$; x_{new} is the new synthesized sample, x is minority class sample, $y[i]$ is the $i-th$ nearest neighbour of x.

2.2 Auto-encoder Neural Network (AE)

Auto-encoder neural network is an unsupervised learning neural network which reconstructs input data as much as possible. It initializes the network weights using the greedy layer-wise training method, and fines tune network parameter using the back propagation (BP) algorithm to optimize the overall performance. Using hidden layer outputs as the new input features, SAE (stacked auto-encoder) deep structure is formed through multiplication AE.

2.3 Stacked De-noising Auto-Encoder neural networks (SDAE)

On the basis of traditional de-noising auto-encoder neural network (AE), after adding noise with a certain probability distribution to the input data, DAE proposed by Vincent etc. [6], makes auto-encoder neural network (AE) learn how to remove the noise, and reconstruct undisturbed input as much as possible. Therefore, the features generated from the learning of input corrupted with noise are more robust, which improved the data generalization ability of auto-encoder neural network model to input data.

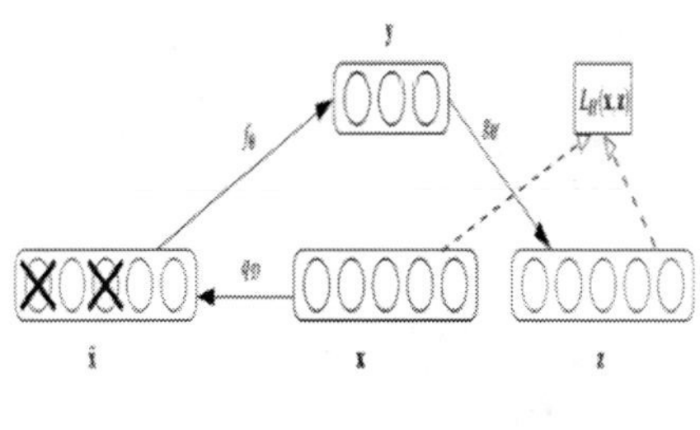

Figure 1. De-noising Auto-encoder neural network [6]

In Fig.1, original data x is added noise in a certain probability q_D to form disturbed data $\tilde{x}$ as auto-encoder input. And f is an activation function used to compute the activation values of each neuron of the hidden layer, as defined in (2).

$$h_{w,b}(\tilde{x}) = f(\sum w\tilde{x} + b) \tag{2}$$

Therefore, the cost function of de-noising auto-encoder neural network is defined according to (3).

$$J_{DAE}(w,b) = \frac{1}{m}\sum(\frac{1}{2}\|h_{w,b}(\tilde{x}) - z\|^2) \tag{3}$$

Here, w is weights between neurons, b is bias, m is number of samples among which *sigmoid* activation function $f(x) = 1/(1+e^{-x})$ with the range [0, 1] is used in the study. Using hidden layer outputs as the new input features, Stacked Auto-Encoder neural networks (SDAE) deep structure is formed through multiplicating DAE.

3 SMOTE-SDAE Algorithm

Combining the strengths of SMOTE and SDAE, the paper puts forward an imbalanced data classification algorithm of de-noising auto-encoder neural network based on SMOTE. First of all, use SMOTE to balance data set. Then aiming at the new data noise problem brought by SMOTE, more robust features are obtained after unsupervised layer-wise greedy training of SDAE. SDAE improves input data generalization ability of auto-encoder neural network, which improves the classification accuracy of minority class and overall samples. Detailed descriptions of the algorithm as follows:

1) Training phase:

a) Set parameters: $\theta = \{W, b\}$ among which W is network weight, b is bias; neuron number of the visual layer is *v*, neuron numbers of the hidden layer is *h1, h2*; q_D is Gaussian noise, *T* is total number of positive samples in data set, *N* is sample synthesis rate, k is number of chosen nearest neighbors, default is 5.

b) Load dataset: Let *Dadaset* is the original training set.

c) Oversample dataset: newDataset = SMOTE(*T, N, k*).

d) Add noise: Generate *D-newDataset* through adding Gaussian noise according to q_D into *newDataset.*

e) Unsupervised training: Study network parameters $\theta = \{W, b\}$ with layer-wise greedy learning.

f) Supervised training: Use *newDataset* to fine-tune network parameter with optimization algorithm L-BFGS.

2) Test phase: Test *N* with test set, and return *AUC,F-value* and *G-mean*.

4 Experiment and analysis

Table 1. AUC contrasts of different algorithms.

Dataset	Total Number of Samples	Number of Attributes	Negative Sample Size	Positive Sample Size	Imbalance Rate
Breast	277	9	196	81	2.42
German	1000	24	700	300	2.33
Heart	303	13	164	139	1.18
Ionosphere	351	34	225	126	1.79
Pima	768	8	500	268	1.87
Sonar	208	60	111	97	1.14
Spam base	4601	57	2788	1813	1.54
Wpbc	198	33	151	47	3.21

4.1. Dataset description

All the experimental data in the paper are eight binary classification datasets commonly used in the study of unbalanced data classification, which is obtained from UCI machine leaning databases, detailed descriptions are shown in table 1.

4.2. Evaluation index based on confusion matrix

In the study, minority class in classification learning is defined as positive, and majority class negative. Confusion matrix to evaluate two-class problem is shown in Table 2

Table 2. Confusion matrix for two-class problem.

Classification	Actual Positive Sample	Actual Negative Sample
Predict as Positive	TP	FP
Predict as Negative	FN	TN

In Table 2, TP stands for the number of minority class being classified as minority class. TN is the number of majority samples being assessed as majority class.

In learning imbalanced data, the effect of minority class on classification accuracy is far less than that of majority class. So classification learning taking classification accuracy as a criterion usually leads to low minority class recognition rate. The classifier tends to forecast a sample as majority class sample. The classic classification accuracy evaluation criteria do not apply to the classifier performance assessment of imbalanced data. Therefore, for imbalanced data classification, there have been new evaluation criteria, such as *AUC*, F-*value* and *G-mean* [7], etc. Their definitions are as follows:

AUC (Area Under roc Curve) provides a method to measure the classifier performance when it is hard to be judged because of ROC (Receiver Operating Characteristic) curve intersection. *AUC* value of a classifier is the area under the corresponding ROC curve. The larger the area, the better the performance of the classifier will be.

F-value is a classification evaluation index comprehensively incorporating *recall* and *precision*, as defined in (4).

$$F-value=\frac{(1+\beta^2)\times recall\times precision}{\beta^2\times recall+precision} \quad (4)$$

In (4), $precision=TP/(TP+FP)$, $recall=TP/(TP+FN)$, β ranges from $[0,+\infty]$.When $\beta=1$, in the experiment, *F-value* can be used to balance the equally important relation between *recall* and *precision*.

G-mean is the geometric mean of classification accuracy of minority class and majority class. It is defined according to (5).

$$G-mean=\sqrt{\frac{TN}{TN+FP}\times\frac{TP}{TP+FN}} \quad (5)$$

G-mean is the maximized accuracy of two classes under the condition of maintaining the classification accuracy balance of minority class ad majority class. *G-mean* is the maximum only in the case of simultaneous high classification accuracy of minority class and majority class.

4.3. Experimental results and analysis

Experiments are conducted on the proposed SMOTE-SDAE algorithm and SVM、SMOTE-SVM、SDAE、SAE algorithms, the results of which are compared using three evaluation methods mentioned above. The following three tables show the compared experimental results of AUC, F-value and G-means of each algorithm. Libsvm toolbox is used in all the SVM algorithms in the experiment, and RBF is adopted as Kernel function with parameter. The average of 10 times 10 fold cross-validations is used as a result. The experimental environment is win7 64bit , Matlab2012b , CPU 3.4GHz, RAM 4G.

Table 3.AUC contrasts of different algorithms

AUC	Breast	German	Heart	Ionosphere	Pima	Sonar	Spambase	Wpbc
SVM	0.7277	—	0.8295	0.9391	0.6571	0.8555	0.7077	0.6618
SMOTE-SVM	0.7479	0.7179	0.8736	0.9679	0.7045	0.9585	**0.8878**	0.7900
SAE	0.6253	0.7133	0.7412	0.9267	0.6765	0.8073	0.8318	0.5784
SDAE	0.6536	0.7313	0.7809	0.9136	0.6890	0.8318	0.8346	0.6386
SMOTE-SDAE	**0.7845**	**0.7412**	**0.9113**	**0.9792**	**0.7371**	**0.9767**	0.8871	**0.8949**

Table 4.F-value contrasts of different algorithms

F-value	Breast	German	Heart	Ionosphere	Pima	Sonar	Spambase	Wpbc
SVM	**0.8391**	**0.8230**	0.8432	0.9538	0.7856	0.8655	0.7835	0.8750
SMOTE-SVM	0.5844	0.4681	0.7428	0.9489	0.3908	0.8414	0.4892	0.7737
SAE	0.4322	0.4946	0.7205	0.8648	0.5300	0.7612	0.8028	0.2857
SDAE	0.4191	0.8084	0.7618	0.8836	0.5236	0.7997	0.8085	0.4255
SMOTE-SDAE	0.8319	0.8144	**0.9492**	**0.9842**	**0.8367**	**0.9787**	**0.9186**	**0.9021**

Table 5.G-mean contrasts of different algorithms

G-mean	Breast	German	Heart	Ionosphere	Pima	Sonar	Spambase	Wpbc
SVM	0.5071	—	0.8128	0.9322	0.4235	0.8462	0.5726	0.4830
SMOTE-SVM	0.6582	0.5656	0.8000	0.9596	0.5082	0.8574	0.5701	0.8008
SAE	0.5725	0.6041	0.7465	0.8812	0.6287	0.7722	0.8380	0.4677
SDAE	0.5494	0.7002	0.7795	0.9098	0.6196	0.8258	**0.8426**	0.5872
SMOTE-SDAE	**0.7297**	**0.7082**	**0.8450**	**0.9689**	**0.6440**	**0.9177**	0.7747	**0.8824**

5 Conclusions

Imbalanced data classification problem has always been a research focus in the field of data mining and machine learning. Directing at the SMOTE algorithm's problems such as noise when synthesizing new minority class samples, the current paper put forward a De-noising Auto-Encoder neural network algorithm based on SMOTE, which can balance data sets through generating new minority class samples, and de-noise and classify the sampled data sets. In comparison with traditional de-noising auto-encoder neural network and algorithms like SVM, the suggested algorithm improved minority class classification accuracy. Such problems as how to better synthesize minority class samples, neural network node number selection and high-dimensional data processing remain to be studied further.

Acknowledgement

This work was supported by The National Natural Science Foundation of China (Project No. 61163034, 61373067), and The Natural Science Foundation of the Inner Mongolia autonomous region of China (Project No. 2013MS0910, 2013MS0911).

References

1. X. Zhang, MATEC Web of Conference **22**,05003 (2015)
2. Ramentol E, Caballero Y, Bello R, et al, K.A.I.S.**33**,245(2012)
3. Chan P.K., Fan W., Prodromidis A. L, et al,IEEE Intel. Syst, **14**,67(1999)
4. G. E. Hinton, and R. Salakhutdinov, Scinence,**313**,504(2006)
5. Y. Bengio, Foundatons and trends in Machine Learning,(2009)
6. P. Vincent, H. Larochelle, Y. Bengio, PA.Manzagol, J Mach Learn Res,**11**,3371(2010)
7. Buckland M, Gey F, JASIST, **45,** 12(1994).

The Evaluation of Feed-in Tariff Models for Photovoltaic System in Thailand

Kangsadan Sagulpongmalee[1], and Apichit Therdyothin[2]

[1]Division of Energy Technology, School of Energy, Environment and Materials, King Mongkut's University of Technology Thonburi, 126 Pracha Uthit Road, Bang mod, Thung khru, Bangkok 10140, Thailand.
[2]Division of Energy Management Technology, School of Energy, Environment and Materials, King Mongkut's University of Technology Thonburi, 126 Pracha Uthit Road, Bang mod, Thung khru, Bangkok 10140, Thailand.

Abstract. Thailand is targeted to reach 6,000 MW of total installed PV capacity by 2036. Feed-in tariff (FIT) was one of the most successful PV mechanisms of Thailand for promoting PV generated electricity. The evaluation of the FIT models for PV in Thailand which was designed 3 models such as premium price FIT model (Adder) in the first FIT policy to motivate attention on investment in PV power plant. After that used fixed price FIT model for PV ground-mounted and front-end loaded FIT model for solar rooftop. In addition to, Thailand has project-specific tariff design which FIT rates are differentiated tariff payment levels by technology, capacity size, and quality of the resource. As result of FIT policies, the PV installation is 1,287 MW of cumulative capacity in 2014. Furthermore, the financial evaluation of FIT for PV project in Thailand found that Net Present Value (NPV) 32.97 million Baht, Internal Rate of Return (IRR) 13.22%, payback period 8.86 years and B/C ratio was 1.66 which must be implemented in conjunction with other financial support measures such as low interest loans, tax benefits, etc. The several incentives to promote PV in Thailand especially FIT shown as PV projects are to be profitable and incentives to investors.

1 Introduction

Renewable energy (RE) become important issue because of the conventional energy source is limited. At the end of 2013, the rate of depletion of world's fossil energy in term of reserves-to-production ratio (R/P) of oil, natural gas, and coal were 53.3, 55.1, and 113 years of global production, respectively [1]. This limitation and depletion of fossil energy reserves included with the increasing energy consumption, development of RE resources. The policy planners have tried to fill out how to promote RE to generate the energy through barriers such as the economics, institutional, political and legislative. Feed-in Tariff (FIT) was one of the most successful renewable mechanisms for promoting large-scale energy markets that offered necessary to attract inventor [2]. FIT program is an incentive plan which provides investors with a set electricity price from RE sources that is fed into the power grid [3]. As of early 2015, FIT policies have been adopted by 73 countries at the national level and in 35 states/provinces [4]. Many countries in Europe (France, Germany, Greece, Italy, Spain and the U.K.) have implemented of FIT support polices for different sized PV systems and the different ways of implementing FIT support policies in the various countries can lead to significantly different results [5]. Especially, the Italian PV market has experienced a progress in the PV market after the first mechanism of feed-in tariffs was promoted in 2005 and Italy has become one of the leading markets for PV power plants currently which the amount of PV installations in Italy was driven by the FIT explicitly [6]. In Malaysia has been introduced FIT in 2011 which has greatly improved the prospect of the RE resources, especially for PV, the solar PV panel instalments in Malaysia is also contributed by the efforts from the business and banking sector [7].

In Thailand, the final energy consumption in 2014 was about 75,804 ktoe and the energy demand of national had to depend on energy imports, was 69,248 ktoe [8] which effect to strengthen energy security in domestic. Currently, Photovoltaic (PV) is one of the fastest growing renewable energy technologies which Thailand has potential extremely. Thai government has introduced various incentives measures especially FITs were considered to promote PV to success the target since 2006. The National Energy Policy Council (NEPC) resolutions ratified Alternative Energy Development Plan (AEDP 2015) to establish renewable energy target (6,000 MW for solar PV target) which be the challenge target [9]. Thailand's feed-in tariff program has already been an attractive RE support mechanism but could be awareness in planning and regulatory framework for FIT [2]. Therefore cost of FIT project will include in electricity bill of ratepayer so FIT pricing is extremely important. The purpose of this paper is to evaluate and analysis the FIT models for photovoltaic system of Thailand after implementation of the FIT policies several years ago.

2 Remuneration of feed-in tariffs policy Page layout

The first feed-in tariff (FIT) was introduced in 1978 which signed on the National Energy Act of the US. The primary objective of FIT policies is to expand of RE generation. In generally, purchase agreements are typically offered within contracts ranging from 10-25 years and the payment levels offered for each kilowatt-hour which there are several designs and can be differentiated by technology type or fuel type, project size, resource quality, and project location to reflect actual project costs [10]. In addition to, Couture and Gagnon [11] have suggested the different ways to structure the remuneration of FIT tariff policy that varying on seven models around the world which have been classified into two broad categories comprised of dependent or independent from the actual electricity market price. The market-independent FIT policies (Figure 1. (a)) have four options. The first and most basic option is the fixed price FIT model which price fixed for the duration of the contract. The second FIT policy is the fixed price model with full or partial inflation adjustment. The third is the front-end loaded FIT model which higher payments are offered in the early years than the later years. The fourth is the spot market gap model which actual FIT payment composed of the gap between the spot market price and the required FIT price. The other category consists of the three different types of the market-dependent FIT policies (Figure 1. (b)). The first is premium price FIT model which the premium price remuneration paid over the retail or market electricity price at the time. The second is the variable premium FIT model which includes both caps and floor into the FIT structure which the total remuneration is premium rate plus with the current market price. The third is the percentage of the retail price FIT model which establishes a fixed percentage of the retail electricity price.

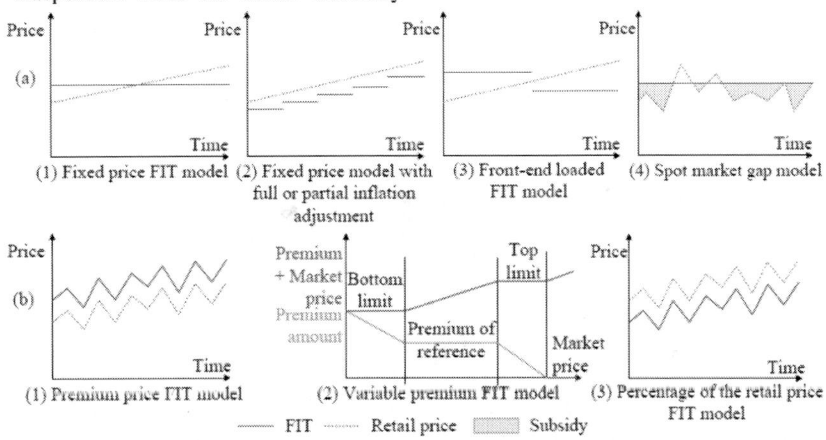

Figure 1. Models of feed-in tariff policy [11]

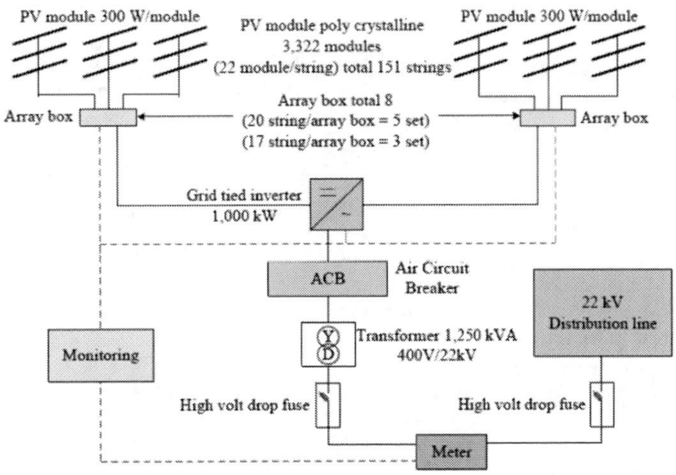

Figure 2. Layout of PV power plant system in Thailand capacity 1 MW.

3 Situation of PV in Thailand

Thailand as the country locates near the equator therefore; the majority of areas in Thailand have high sunlight intensity. The maximum solar radiation is 20-24 MJ/m^2-day during April and May and average daily solar radiation about 18.2 MJ/m2-day or 5.6-6.7 kWh/m^2-day. Thailand's electricity sector has evolved from a government monopoly to a semi-unbundled structure by enhanced single buyer. The Installation of PV power plant in Thailand, also called solar farm, has become the main PV installation. In 2014, Thailand PV installation is 1,287 MW of cumulative capacity which increased 823 MW in 2013. The most of PV power plants were built with imported PV components, mainly from Taiwan (45%), Japan (27%) and China (13%) and domestic has 5% of total PV system installation. The inverters of PV power plants were imported from United States (53%), Germany (26%) and Japan (10%) and the local inverter company has 8% of total PV system installation [12]. The Layout of solar farm system in Thailand capacity 1 MW as shown in Figure 2.

The PV growth in Thailand is driven and managed by the National Energy Policy Council (NEPC) [13]. The Renewable Energy Development Program (REDP) (2008-2022) was approved on 16 January 2009 by NEPC which set the PV target in 2022 of 500 MW. On 30 November 2011, the NEPC resolution approved the new plan, the 10-Year Alternative Energy Development Plan (AEDP) (2012-2021) to replace REDP which aims to increase the share of RE to be 25% of total final energy consumption in 10 years and PV's target was 2,000 MW in 2021 [14]. Furthermore, the NEPC also has been revised AEDP and increased PV target several times, including on 16 July 2013, 3,000 MW [15] and on 22 October 2014, 3,800 MW which comprised 2,800 MW solar farms, 800 MW solar farms on government land or managed by agricultural cooperatives, and 200 MW of PV rooftop [16] and the latest, on 17 September 2015, NEPC approved the Alternative Energy Development Plan 2015-2036 (AEDP 2015) comprised PV 6,000 MW [17].

4 The evaluation of feed-in Tariff models for PV in Thailand

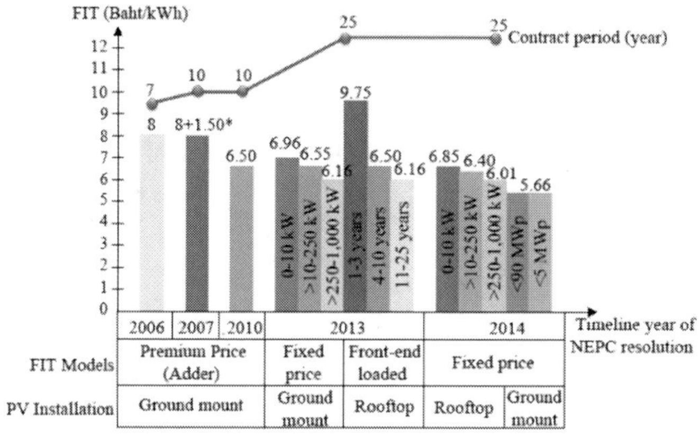

Figure 3. Evaluation of FIT rates and models in Thailand [2, 9, 14-16]

4.1. Evaluation of feed-in Tariff rates of Thailand

Thailand is one of the first countries in the ASEAN to implement FIT mechanism which adopted by the Thai Cabinet in 2002 and implemented premium FIT (Adder) in 2006 [2]. The Adder paid on top of the retail electricity price to guarantee attractive purchasing rates. In 2010, higher Adder was paid in the three southernmost provinces which have experienced political unrest. Through the PV world market prices have decreased [18] and Ministry of Energy representatives has raised concerns regarding transparency and consistency [2], and significantly reduced approval rates of new applications as result the FIT of Thailand has cut the rate down in 2011.

The NEPC resolution in 2013 approved to switch from a premium-price FIT payment to a fixed-price FIT payment for ground mount plant and front-end loaded FIT payment for solar rooftop (Figure 3.). Currently, The NEPC progressed on 14 May 2015, a total PV of 1,558 MW are already installed on grid and total in the pipeline amounts to 2,819 MW [19]. By the first RE plan, premium price FIT model can create confidence and profitable for investors are extremely FIT rate of 8 Baht/kWh added with retail electricity price about 3.50 Baht/kWh so the investors will receive total price of 11.50 Baht/kWh which is the highest rate but contract period of 7 and 10 years. After that, FIT rate adjusted reduction to 6.50 Baht/kWh followed to global PV market price. The next, FIT model changed to fixed price FIT model to PV ground mount and front-end loaded FIT model for solar rooftop but increase contract period to 25 years.

4.2. Evaluation of feed-in Tariff designs of Thailand

FITs in Thailand has several project-specific tariff designs which are differentiated tariff payment levels by technology, capacity size of the installation, and the project location.

4.2.1 Evaluation of Differentiation by Technology or Fuel Type

FITs Thailand were designed to offer different tariffs by fuel type and technology. In 2006, premium FIT were biomass, wind, hydro, and PV of 0.3, 2.5, 0.4-0.8, and 8 Baht/kWh respectively [2]. In 2013, FIT's Thailand was separated PV supporting into two technologies were ground mount and rooftop [15].

4.2.2 Evaluation of Tariff Differentiation by Project Size

Thailand differentiates by project size in its FIT policy for PV. In 2013, FIT has divided size for PV ground mount into three groups were 0-10 kW (6.96 Baht/kWh), 10-250 kW (6.55 Baht/kWh), and 250-1000 kW (6.16 Baht/kWh) [15] which provided the highest tariff for the smallest capacity PV installations.

4.2.3 Evaluation of Tariff Differentiation by Project Location

Differentiated by project location, Thailand offers a higher payment of installed PV rooftop for the residential or building more than PV ground mount. Furthermore, PV installation in southernmost provinces and off-grid areas were advanced higher FIT 1.50 baht/kWh [20].

4.3. Evaluation of PV project of Thailand's feed-in Tariff policy

The evaluation of Feed-in Tariff for PV system in Thailand for installation capacity plant 1 MW which have financial assumption such as initial investment 49.9 million Baht (Table 1.), operating and maintenance cost increased 12% per year from initial O&M cost, plant factor 16%, deterioration of income 0.8% per year of income on first year, interest rate 6%, paying loan back in 10 years, Debt-to-Equity Ratio (70:30) and contract period 25 years. According to those parameters, FIT rate for PV ground-mounted is 5.66 Baht/kWh, found that the solar PV power plant will receive the Net Present Value (NPV) 32.97 million Baht, the Internal Rate of Return (IRR) 13.22%, the payback period 8.86 years and the Benefit/Cost (B/C) ratio of 1.66 excluding income from carbon credits and including implemented in conjunction with other promotion measures such as low interest loans, tax benefits, etc.

Table 1. The initial investment for PV power plant in Thailand of capacity 1 MW in 2014.

Description	Material cost (Baht)	Labour cost (Baht)	Total (Baht)
1. Electrical work	29,190,725	1,583,661	30,774,386
2. Civil and architectural work	5,638,000	558,200	7,896,200
3. Sanitary work	590,000	69,150	659,150
4. Operation construction	-	3,615,500	3,615,500
5. Import tax and Value Added Tax (Vat 7%)	-	-	4,891,294
6. Others	-	-	2,100,000
Total			49,936,530

5 Conclusions

Thailand has successful in promoting PV which important attractive mechanism is FIT policy. The PV's growth in Thailand is driven by the energy policy of the government which is managed by the National Energy Policy Council (NEPC). In 2006, Thailand began to adopt this approach in FIT policy and in 2014, Thailand PV installation is 1,287 MW of cumulative capacity which higher than other countries in ASEAN. Thailand's FIT policy has been designed to differentiated tariff payment levels depending to project-specific comprise differentiated by technology, capacity size of the installation, and location of the project. In addition, the remuneration of FIT policy has been applied that varying on three models such as premium price FIT model, fixed price FIT model, and front-end loaded FIT model. According to several FIT designed which are appropriate to the geographical conditions of Thailand and as a result currently, Thailand's PV installed capacity increased significantly. The evaluation of PV power plant in Thailand using FIT policy found that the projects are capable of profitable but be implemented in conjunction with other financial support measures such as low interest loans, tax benefits, etc. Although PV power plant project will not make extremely profits for investors which FIT rate is expected to reduce in the future but solar power is a clean and environmentally friendly and energy security of the country therefore should be considered for promoting the PV project with reasonable FIT rate of each country.

Acknowledgements

The authors would like to thank to the thesis committees including Warunee Tia (Assoc. Prof.) and Janthana Kunchornrat (Ph.D.), School of Energy, Environment and Materials, King Mongkut's University of Technology Thonburi for any kinds of advice and cooperation.

References

1. BP Statistical Review of World Energy. *BP Statistical Review of World Energy June 2014*, Accessed May 1, 2015 at: <https://www.bp.com/content/dam/bp/pdf/Energy-economics/statistical-review-2014/.pdf>
2. S. Tongsopit, C. Greacen, Renew Energ. **60**, 439 – 445 (2013)
3. K. Kyoung-Kuk, L.Chi-Guhn.Energ Pol. **49**, 192-203 (2012)
4. REN21. *Renewables 2015 Global Status Report*, Accessed August 3, 2015 at: <http://www.ren21.net/wp-content/uploads/2015 /07/REN12-GSR2015_Onlinebook_low1.pdf>
5. A.Campoccia, L.Dusonchet, E.Telaretti, and G.Zizzo. Sol Energ. **107**, 530-542 (2014)

6. A.Marco, D.Umberto. Appl Energ, **135**, 721-729 (2014)
7. S. L. Wong, N. Norzita, A. Tuan, I. M. Inuwa, Renew Sustain Energ Rev. **41**, 42-52 (2015)
8. Department of Alternative Energy Development and Efficiency, Ministry of Energy, *Energy Balance of Thailand*, Accessed October 1, 2015 at: < http://www4.dede.go.th/dede/images/stories/stat_dede/sit_58.pdf>
9. National Energy Policy Council Resolutions 1/2014 on 15 August 2014 (in Thai). Accessed August 3, 2015 at: < http://www.eppo.go.th/nepc/kpc/kpc-NCPO-1.html>
10. T. D. Couture, K. C. C. Kreycik, E. Williams. *A Policymaker's Guide to Feed-in Tariff Policy Design*: National Renewable Energy Laboratory, Colorado, (2010)
11. T. Couture, Y. Gagnon, Energ Pol. **38**, 955–965 (2010)
12. Department of Alternative Energy Development and Efficiency: Ministry of Energy. *Thailand PV status Report 2012-2013*. Accessed August 3, 2015 at: < http://www4.dede.go.th/dede/images/stories/jum_circular_55/PV_Status_Report_eng_final.pdf>
13. Energy Policy & Planning Office: Ministry of Energy. The National Energy Policy Council (NEPC). Accessed August 3, 2015 at: < http://www.eppo.go.th/doc/doc-manage.html>
14. National Energy Policy Council Resolutions 6/2011 on 30 November 2011 (in Thai). Accessed August 3, 2015 at: < http://www.eppo.go.th/nepc/kpc/kpc-139.htm>
15. National Energy Policy Council Resolutions 2/2013 on 16 July 2013 (in Thai). Accessed August 4, 2015 at: < http://www.eppo.go.th/nepc/kpc/kpc-145.html>
16. National Energy Policy Council Resolutions 1/2014 on 22 October 2014 (in Thai). Accessed August 5, 2015 at: < http://www.eppo.go.th/nepc/kpc/kpc-146.html>
17. Department of Alternative Energy Development and Efficiency: Ministry of Energy. Alternative Energy Development Plan: AEDP 2015. Accessed October 6, 2015 at: < www.dede.go.th/download/files/PPT_AEDP2015.pdf>
18. National Energy Policy Council Resolutions 2/2010 on 28 June 2010 (in Thai). Accessed August 1, 2015 at: < http://www.eppo.go.th/nepc/kpc/kpc-N2.html>
19. National Energy Policy Council Resolutions 2/2015 on 14 May 2015 (in Thai). Accessed September 16, 2015 at: < http://www.eppo.go.th/nepc/kpc/kpc-N2.html>
20. National Energy Policy Council Resolutions 4/2007 on 4 June 2007 (in Thai). Accessed July 10, 2015 at: < http://www.eppo.go.th/nepc/kpc/kpc-113.htm>

MATEC Web of Conferences **56**, 02001 (2016) DOI: 10.1051/matecconf/20165602001
ICCAE 2016

Near-Field Cross Section Imaging of Wideband Millimeter Wave

Yingzhi Kan , Yongfeng Zhu and Qiang Fu

School of Electronic Science and Engineering, National University of Defense Technology, Hunan, China

Abstract. Near-field millimeter wave imaging has been a hot topic recent years for its importance applications in the area of anti-terrorism. The penetrating characteristic of millimeter wave is of significant importance to security, such as the concealed weapons detection, ground-penetrating radar imaging, through-barrier imaging and so on. Cross section imaging is a basic aspect for near-field millimeter wave imaging, which includes antenna array distribution and wideband signal processing. This paper utilizes back projection method in space area to realize ultra-band near-field cross section imaging. We induce two dimensional direction integral formulas to obtain the reconstruction image of the near-field imaging area, and the simulation results validate the effectiveness of this imaging algorithm.

1 Introduction

The demand for millimeter wave imaging techniques increases in the field of nondestructive testing (NDT) or in security applications since the threat of terrorism attacks is increasing these years[1, 2]. The security detection in public is to detect the threatened weapons under person's clothes or dangerous objects in the luggage. The X-ray detection system and the metal detectors are common methods for security check. These systems have their own advantages and disadvantages. Metal detector can only detect metal objects, but not be useful for some modern weapons such as plastic or chinaware gun, and liquid bombshell. On the other hand, some metal commodity may be detected as metal weapons to result in false alarms. To decrease false alarms, the imaging systems like the X-ray imaging system and the millimeter wave imaging system can both be utilized to overcome the defects of metal detectors. Comparing with X-ray which is dangerous when illuminate person, the millimeter wave system transmits low power with no harm to person. Besides, the millimeter wave system can penetrate clothes and luggage, which make it a prospective application in security area.

Near-field millimeter wave imaging system, could realize coronal plane imaging, cross section imaging and three dimensional imaging [3, 4]. Coronal plane imaging and cross section imaging are basic problems for studying three dimensional imaging. Coronal plane imaging achieves two dimensional resolutions by scan a rectangular planar aperture. Cross section imaging achieves two dimensional resolutions by a linear antenna array in one dimension and a wideband signal in another dimension. In this paper, we mainly discuss cross section imaging.

Near-field cross section imaging mode transmits wideband millimeter wave along a linear array [5]. To improve resolution, we can expand the linear antenna array to enhance the cross resolution, and increase the wideband of transmitted signal to enhance the range resolution. Here, we adopt stepped frequency wideband signal. As for near-field millimeter imaging system, the carrier frequency is high which can achieve ultra-band transmitted signal to reach millimeter resolution. Common cross section imaging algorithms, based on spectral decomposition theory, utilize space filtering and interpolation to obtain effective two dimensional space spectrum data to implement Fourier two dimensional imaging. In the imaging process, the main imaging error is introduced by interpolation which directly affects the accuracy and efficiency of the images. In this paper, we consider the ultra-band signal and small imaging scene, so that, it is not available to utilize interpolation algorithm since the wideband is not sufficient used in the means of interpolation in that case. To avoid interpolation, we introduce the back projection algorithm [6] to directly calculate two dimensional integral of the recorded echo signal in the form of frequency to reconstruct near-field image. The paper derives the formula of direct two dimensional integral and simulation verifies the effectiveness of the algorithm. This work is important to extend to study the three dimensional near-field millimeter wave imaging.

2 Algorithm description

Cross section image reflects the scattering distribution along the cross direction and the range direction. We set cross direction as x direction and range direction as z direction, and the returned scattered data is recorded by an two dimensional matrix.

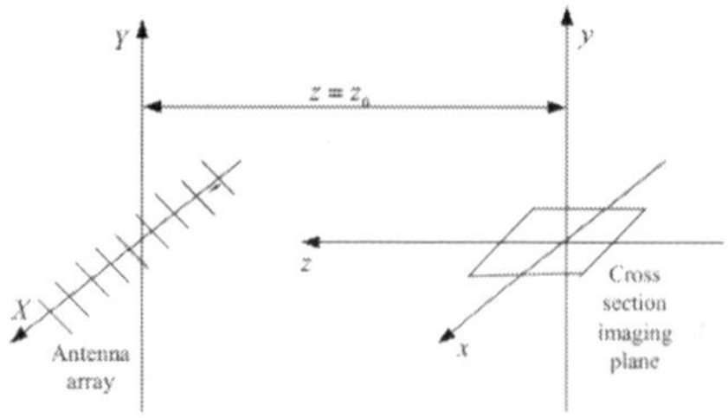

Figure 1. Geometry of cross section imaging

The basic imaging configuration is shown in Figure 1. The imaging transceiver is scanned, mechanically or electronically along a linear array aperture. At each sampling point in the aperture, a wide beam-width transmitting antenna is used to emit the millimeter wave signal, which is generally swept over a wide frequency bandwidth. The transmitting and receiving antennas are generally identical in construction, placed in close proximity to each other, and scanned together in a simultaneous source-receiver scanning configuration. The antenna placement is referred to as a quasi-monostatic configuration. The antennas are distinct, but can be considered to be a single antenna located at the midpoint between the two antenna phase centers. This approximation is valid for target ranges significantly larger than the antenna spacing.

The transceiver moves along X direction step by step, which transmits stepped frequency signal and records the scattered echo data in frequency domain. This measurement process is repeated at each sampling point. After ergodic all the sampling points in X direction, the system records a measurement matrix, of which each element represents a scattered data at a certain space position and a certain frequency. The destination of cross section imaging is to reconstruct scattering distribution of target in the $x-z$ plane according to the recorded data.

Generally, cross section imaging algorithms are based on spectral decomposition theory. The electric field distributed in x-z plane can be obtained according to the wave spectrum transferring relationship, which is equal to the scattering distribution function of the targets.

The classical algorithm has been detailed described in [5].The imaging algorithm mainly concludes a one-dimensional FFT, a space filtering, a two-dimensional interpolation and a two-dimensional IFFT. The imaging precision is affected by the interpolation process.

2.1 Direct integral algorithm

For near-field wideband millimeter wave imaging, when the imaging scene is small and the range between the target and the antenna array is short, the interpolation process become difficult to achieve to assure interpolation accuracy. To overcome this disadvantage we introduce two dimensional direct imaging methods.

Suppose targets located in $x-z$ plane and measuring antenna located in X direction. The X axis parallels to x axis and the perpendicular range between x and X is z_0. The length of the antenna array is L_x.

The recorded scattered data $E(X,k)$ is a two-dimensional matrix, which can be denoted as

$$E(X,k)=\iint \sigma(x,z)\exp(-jk\cdot r)dxdz \quad (1)$$

Where

$$\begin{aligned} r &= \sqrt{(x-X)^2+(z-z_0)^2} \\ k &= 4\pi f/c \\ k^2 &= k_x^2+k_z^2 \end{aligned} \quad (2)$$

Here, c is the speed of light, f is temporal frequency, k is the wavenumber frequency, and k_x, k_z are the three components of k along x,z direction respectively. The aim of imaging process is to reconstruct the 2-D reflectivity function $\sigma(x,z)$ from the recorded scattered data $E(X,k)$.

Then, we can get

$$\begin{aligned}\sigma(x,z)=&\iint E(k_x,k_z) \\ &\cdot\exp[j(k_x\cdot(x-X)+k_z\cdot(z-z_0))]dk_xdk_z\end{aligned} \quad (3)$$

Let

$$\begin{aligned} k_x &= k\cdot\cos\theta = k\cdot\frac{x-X}{r} \\ &= k\cdot\frac{x-X}{\sqrt{(z-z_0)^2+(x-X)^2}} \end{aligned} \quad (4)$$

$$\begin{aligned} k_z &= k\cdot\sin\theta = k\cdot\frac{z-z_0}{r} \\ &= k\cdot\frac{z-z_0}{\sqrt{(z-z_0)^2+(x-X)^2}} \end{aligned} \quad (5)$$

We obtain

$$\begin{aligned}\sigma(x,z)=&\iint E(k\cdot\frac{x-X}{r},k\cdot\frac{z-z_0}{r}) \\ &\cdot\exp(jk\cdot r)dk_xdk_z \\ =&\iint F(k\cdot\frac{x-X}{r},k\cdot\frac{z_0-Z}{r}) \\ &\exp(jk\cdot r)\cdot\left|\frac{\partial(k_x,k_z)}{\partial(k,x)}\right|dkdx\end{aligned} \quad (6)$$

Let

$$J=\left|\frac{\partial(k_x,k_z)}{\partial(k,x)}\right|=\begin{vmatrix}\frac{\partial k_x}{\partial k} & \frac{\partial k_x}{\partial x}\\ \frac{\partial k_z}{\partial k} & \frac{\partial k_z}{\partial x}\end{vmatrix}=k\cdot\frac{|z-z_0|}{r^2} \quad (7)$$

Then

$$\begin{aligned}\sigma(x,z)=&\iint E(k\cdot\frac{x-X}{r},k\cdot\frac{z-z_0}{r})\exp(jk\cdot r) \\ &\cdot k\cdot\frac{|z-z_0|}{(z-z_0)^2+(x-X)^2}dkdx\end{aligned} \quad (8)$$

Or

$$\begin{aligned}\sigma(x,z)=&\iint E(X,k)\exp(jk\cdot r) \\ &\cdot k\cdot\frac{|z-z_0|}{(z-z_0)^2+(x-X)^2}dkdx\end{aligned} \quad (9)$$

DOI: 10.1051/matecconf/20165602001

That is the direct two-dimensional integral reconstruction algorithm.

2.2 2-D resolution analysis

Image resolution reflects the details of reconstructed image, which is a first criterion and shows the accuracy of the imaging system. The ability of high resolution is quite important to detect concealed weapons. Image resolution can be considered as the ability to distinguish two scatters located very close to each other. So the resolution along cross direction is in [5].

$$\delta_x = \frac{\lambda}{2L_x}\sqrt{z_0^2 + L_x^2/4} \tag{10}$$

Generally, in far-field the range resolution is decided by the signal bandwidth. However in the near-field situation, since the range between the target and antenna cannot approximate, the wavefront is spherical, and the cross range resolution is decided by the occupied k_z .That is

$$\delta_z = \pi / (K_{z\max} - K_{z\min}) \tag{11}$$

The expression shows that the cross resolution of the near-field millimetre wave system is decided by the signal frequency, the perpendicular distance between the scanning plane and the scatters, and the scan size. The system parameters can be chosen by considering the image resolution requirement.

3 Simulation experiment

The transceiver antenna scans along a linear array of $(-15cm, 15cm)$, at each sampling point, the antenna transmits stepped frequency signal with $10GHz \sim 40GHz$ bandwidth. The frequency interval is $\Delta f = 100MHz$, the antenna sampling interval is $\Delta x = 3.75mm$. The area to be imaged is $20cm \times 20cm$, and the perpendicular distance to the antenna plane is $30cm$. There are three scatters located in the imaging area, as Figure 2.

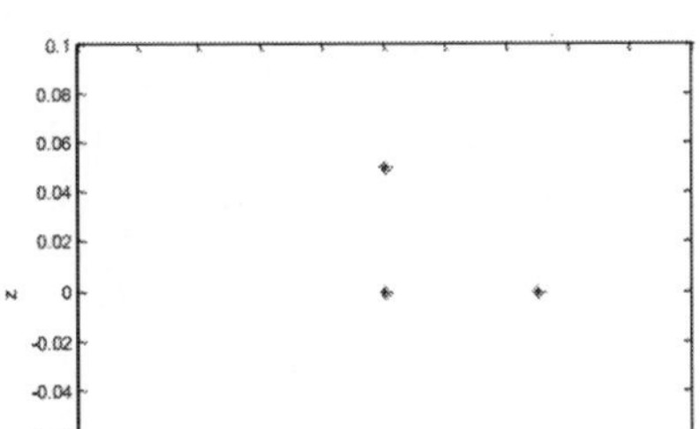

Figure 2. Simulated targets

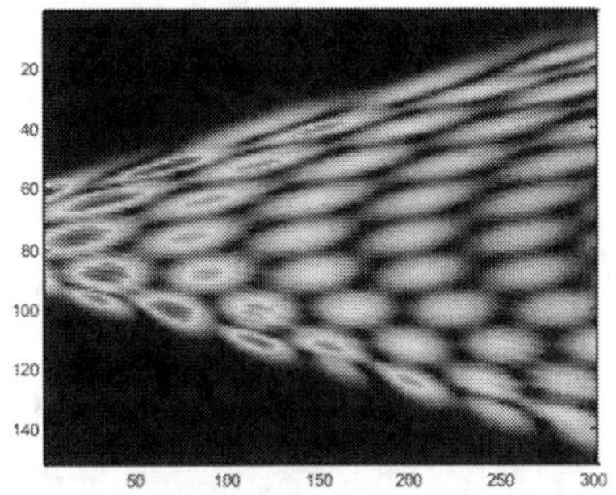

Figure 3. 2-D space spectrum

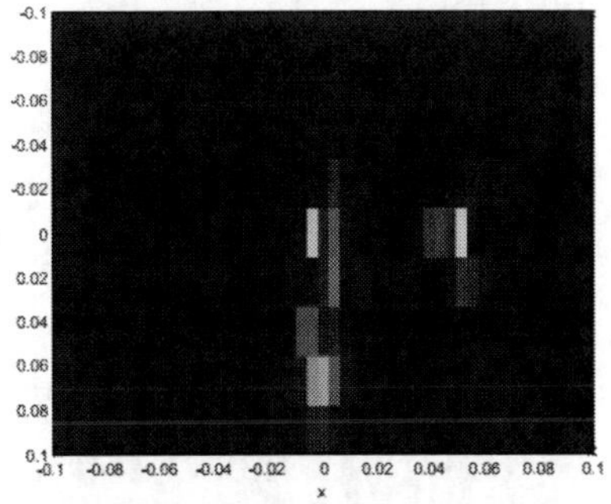

Figure 4. Interpolation based imaging result

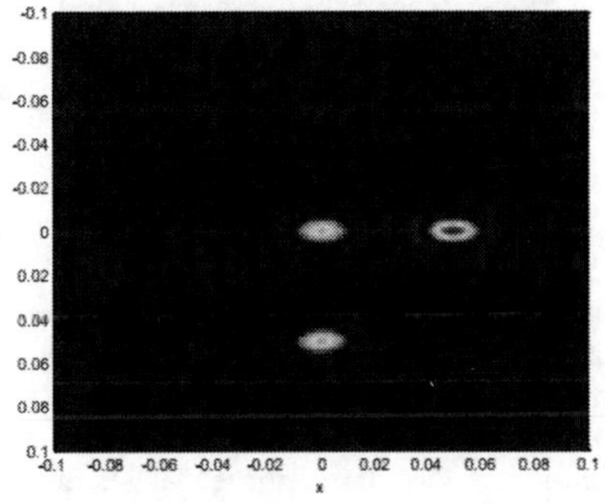

Figure 5. Direct two-dimensional integral results

The imaging results are shown from Figure 3 to Figure 5. Figure 3 is the 2-D space spectrum of the scattered data and it is clear that the spectrum is not uniform. Figure 4 is the imaging result of the conventional method. The cross section imaging result of the proposed method is as Figure 5. In Figure 5, the three point scatters are focused well and are located at the exact position. It is very obvious in the image and no calibration is required comparing to the original scene. Compared with the two-dimensional interpolation method, the proposed direct integral method is simple and direct. The imaging accuracy is decided directly by the sampling returned signal, not affected by the signal processing method. And because of the small imaging scene, the direct two-dimensional integral method is of high calculation efficiency. The experiment demonstrates the direct two-dimensional method is available for the near-field millimeter wave imaging.

As for the interpolation method, in the situation of near-field and small imaging scene, is not appropriate. The linear antenna array is short resulting in the two-dimensional space spectrum in k-space is curving. These make the two-dimensional interpolation difficult to accurately implement. Also, the range resolution is affected by the interpolation result since the resolution is decided by the span of K_z. The span of K_z is limited when the curvature of space spectrum is severe which directly restrict the range resolution and lower the bandwidth lower utilization rate. The direct two-dimensional integral method avoids interpolation and can make the full use of the wide bandwidth to realize high resolution imaging.

4 Conclusions

The paper discusses the wideband millimeter wave near-field cross section imaging. Since the two-dimensional Fourier method needs to resolve the interpolation problem, and in the certain case of near-field small imaging scene the interpolation problem become more difficult to deal with. To avoid interpolation and simplify the imaging process, we induce the back projection formula to achieve direct two-dimensional integral imaging of cross section, and analyze the resolution of the two dimensions. We set an imaging scene and related parameters to simulate the imaging progress of near-field cross section imaging. The simulation result is focused well to demonstrate the effectiveness of the proposed algorithm. And since the imaging scene is small the calculation is simple. It is an efficient imaging algorithm for near-field cross section imaging.

References

1. Frank Gumbmann, Phat Tran, Lorenz-Peter Schmidt. Sparse Linear Array Design for a Short Range Imaging Radar. 2009 EuMA
2. M. Ravan, Reza K. Amimeh, and Natalia K. Nikolova. Two-dimensional near-field microwave holography. INVERSE PROBLEMS,26, 2010.
3. David Sheen, Douglas McMakin, and Thomas Hall. Near-field Three-dimensional Radar Imaging Techniques and Applications. APPLIED OPTICS, 49(19), 2010.
4. ZHANG Yansheng, YANG Jungang, AN Wei. Single frequency two-dimensional synthetic aperture imaging system and imaging algorithm. Journal of National University of Defense Technology, 36(4), 2014.
5. Zhang Xilin, li Nanjing, Hu Chufeng, Li Ping. Radar Target Cross Section Characteristics Test and Imaging Diagnosis. China Astronautic Publishing House, 2009.
6. Huang Peikang, Yin Hongcheng, Xu Xiaojian. Radar Targets Characteristics. Publishing House of Electronics Industry, 2010.

Possibilistic Clustering Algorithm Incorporating Grey-Level Histogram and Spatial Information for Image Segmentation

Jing Yu, Yujun Kuang, and Xinchuan Fu

Mobilelink, UESTC, Chengdu, China

Abstract. Image segmentation is a process of segmenting an image into non-intersecting regions containing homogeneous pixels that are inhomogeneous with those in other adjacent regions. In this paper, a possibilistic clustering algorithm incorporating grey-level histogram and spatial information (PCA_HS) for image segmentation is proposed. The grey-level histogram speeds up the algorithm and the spatial information enhances its robustness to noise and outliers. To assess the proposed algorithm, four widely used validity indexes are computed and discussed. As the experimental quantitative and qualitative results on real images with and without noise show, PCA_HS can preserve the homogeneity and integrality of the regions and hence is more effective and efficient than traditional PCA.

1 Introduction

Image segmentation plays an important role in a variety of applications such as object recognition, image understanding and computer vision. The task of image segmentation is to partition an image into a number of non-overlapping regions with homogeneous characteristics while the union of any two spatially adjacent regions is inhomogeneous. To date, many researchers have proposed various segmentation algorithms [1-4]. Among these methods, fuzzy clustering is one of the most popular types of methods for image segmentation, which introduces fuzziness for the belonging-ness of each image pixel and thus can retain more information from the original image [5].

Fuzzy C-Means (FCM) algorithm was first introduced by Dunn and later extended by Bezdek, which now is one of the most widely used fuzzy clustering algorithms in image segmentation [6]. However, since the memberships of FCM are relative values depending on each other, it is sensitive to noise, outliers and other imaging artifacts. To overcome this weakness and produce memberships that have a more reasonable explanation of the degree of belonging for the data, Krishnapuram and Keller [7, 8] proposed a possibilistic approach to clustering (PCM1) and later a parallel one (PCM2) using a possibilistic type of membership functions. These algorithms with possibilistic memberships were proved to be more robust to noise and outliers than FCM. However, M. Barni, et al. [9] has proved that PCM1 to be sensitive to the initial prototypes since no link exists among clusters, which tends to produce coincident clusters. Furthermore, the performance of PCM1 heavily depends on the parameter η_i and the fuzzifier m [8, 9]. PCM2, on the other hand, is also sensitive to parameter η_i. In addition, it omits the fuzzifier m and thus cannot flexibly adjust any parameter to get reasonable clustering results when coincident clusters occur. Based on the FCM objective function, the partition coefficient (PC) and partition entropy (PE) validity indexes, Yang and Wu [10] proposed another possibilistic clustering algorithm (PCA). In fact, PCA is similar to PCM2, except for the inclusion of the fuzzifier m and the cluster number c into membership functions, and the replacement of η_i with a relatively constant parameter β. However, as the experimental results show, the direct application of PCA on an image will segment out many isolated pixels, which is undesirable.

An image has two important features that are useful for segmentation. One is the grey-level histogram, which can extremely decrease the spatial and computational complexity. To utilize this feature, Xu [11] proposed the Genetic Fuzzy C-Means Algorithm. Despite its time efficiency, this algorithm is sensitive to noise and outliers. Another important feature is the spatial information, which can enhance the robustness to noise and outliers. Based on FCM or PCM, Many improved clustering methods have been proposed by incorporating spatial information into the membership or objective functions [5, 6, 12, 13]. Chuang et al. [12] proposed sFCMp,q, which defined the spatial functions and incorporated them into the membership functions of FCM. The drawback of sFCMp,q is that the membership functions are computed in each iteration step, which is very time-consuming. Zuo [13] improved PCM1 by updating the memberships to the geometric mean of its neighbouring memberships in each iteration step, which is also time-consuming. Hence, in this study, our motivation is to take advantage of the grey-level histogram to speed up the algorithm and make use of the spatial information to enhance its robustness to noise and outliers as well.

The remainder of this paper is organized as follows. In the next section, PCA is simply reviewed, and then the proposed algorithm is described in detail, followed by a brief introduction of the four validity indexes generalized in [10]. In Section III, we use some natural images to compare the efficiency and robustness between PCA and PCA_HS. The validity indexes are computed and discussed after each segmentation. Finally, conclusions are drawn in section IV.

2 Methodologies

2.1 PCA

Based on the FCM objective function, the partition coefficient (PC) and partition entropy (PE) validity indexes, Yang and Wu [10] proposed another possibilistic clustering algorithm (PCA). The objective function of PCA is defined as below

$$J(U,C)=\sum_{i=1}^{c}\sum_{j=1}^{n}u_{ij}^{m}\|x_j-c_i\|^2 + \frac{\beta}{m^2\sqrt{c}}\sum_{i=1}^{c}\sum_{j=1}^{n}(u_{ij}^{m}\ln u_{ij}^{m}-u_{ij}^{m}). \quad (1)$$

where x_j is the grey value of the jth pixel, c_i represents the center value of the ith cluster, n is the total number of pixels, u_{ij} stands for the possibilistic membership of the jth pixel with respect to cluster i, the weighting exponent $m\in(1,+\infty)$ is called the fuzzifier, c denotes the cluster number and the sample co-variance β is expressed as below

$$\beta=\frac{1}{n}\sum_{j=1}^{n}\|x_j-\bar{x}\|^2, \quad \text{with} \quad \bar{x}=\frac{1}{n}\sum_{j=1}^{n}x_j. \quad (2)$$

It has been proved that the necessary conditions for a minimizer of the objective function in (1) are the following update equations

$$u_{ij}=\exp(-\frac{m\sqrt{c}}{\beta}\|x_j-c_i\|^2), \quad \begin{matrix} i=1,\dots,c, \\ j=1,\dots,n. \end{matrix} \quad (3)$$

and

$$c_i=\frac{\sum_{j=1}^{n}u_{ij}^{m}x_j}{\sum_{j=1}^{n}u_{ij}^{m}}, \quad i=1,\dots,c. \quad (4)$$

2.2 PCA_HS

A straightforward description of PCA_HS is this: In PCA_HS, the histogram of the image is first obtained. Afterwards, PCA is applied on it. Finally, spatial information is used to revise isolated pixels. The following parts depict these steps in detail.

To explain the process of employing the grey-level histogram, let k represent a certain grey value of the image and $h(k)$ denote the number of pixels with grey value k. For later use, let L stand for the number of different grey values in the image. Now the distance of a pixel with grey value k to the ith cluster is below

$$d_{ik}=\|k-c_i\|, \quad k\in\{k' \mid h(k')\neq 0\}. \quad (5)$$

The membership function in (3) turns to be

$$t_{ik}=\exp(-\frac{m\sqrt{c}}{\beta}\|k-c_i\|^2), \quad \begin{matrix} i=1,\dots,c, \\ k\in\{k' \mid h(k')\neq 0\}. \end{matrix} \quad (6)$$

Correspondingly, the center of a cluster in (4) becomes

$$c_i=\frac{\sum_{k\in\{k' \mid h(k')\neq 0\}}t_{ik}^{m}h(k)k}{\sum_{k\in\{k' \mid h(k')\neq 0\}}t_{ik}^{m}h(k)}. \quad (7)$$

The relationship between u_{ij} and t_{ik} is as follow

$$u_{ij}=t_{ik}, \quad \text{if} \quad x_j=k. \quad (8)$$

Finally, the objective function is expressed as below

$$J(T,C)=\sum_{i=1}^{c}\sum_{k\in\{k' \mid h(k')\neq 0\}}t_{ik}^{m}d_{ik}^{2}h(k) + \frac{\beta}{m^2\sqrt{c}}\sum_{i=1}^{c}\sum_{k\in\{k' \mid h(k')\neq 0\}}(t_{ik}^{m}\ln t_{ik}^{m}-t_{ik}^{m})h(k). \quad (9)$$

The membership matrix U of size $c\times n$ turns into T of size $c\times L$. For a grey scale image of 8 bits, L is no more than 256, while n is usually much bigger than 256, and thus the grey-level histogram can help speed up the algorithm a lot.

Now that PCA has been applied on the histogram, each pixel in the image was divided into a cluster with a cluster number decided by the following equation

$$m_j=i, \quad \max_i u_{ij}. \quad (10)$$

However, PCA with grey-level histogram is still sensitive to noise and outliers. For this reason, the spatial information is applied, and its main idea is that if hardly any neighbouring pixels are divided into the same cluster with the center pixel, it is reasonable to believe that this pixel is contaminated by noise and is wrongly divided. Therefore, we recursively reset its maximum membership function value to 0 by (11), and modify its cluster number by (12), until the new maximum value satisfies the second condition in (11). Finally adjust its grey value to the mean of its neighbouring pixels which are clustered to the same cluster with the center pixel.

Let N represent the edge length of the pixels in the window. A 3×3 window except for the central pixel itself is used throughout this work. NP denotes the number of the neighbouring pixels that didn't divide into the same cluster with the center pixel, and thus the membership function can be defined as follow, and the initial value of u'_{ij} is u_{ij} in (8).

$$u'_{ij} = \begin{cases} 0, & NP/(N^2-1) > \varepsilon_0, i = m'_{ij}, \\ u'_{ij}, & NP/(N^2-1) \le \varepsilon_0, i = m'_{ij}. \end{cases} \quad (11)$$

The selection of ε_0 is analysed below. In Figure 1, it can be assumed that the center pixel in (a) and (b) are wrongly divided but correctly divided in (c). Values of $NP/(N^2-1)$ in Figure 1 are 1, 0.875 and 0.75, respectively. Hence, it is rational to select $\varepsilon_0 \in [0.75, 0.875)$.

(a)

2	2	2
3	1	3
3	3	3

(b)

1	2	2
3	1	2
3	3	3

(c)

1	?	?
1	1	?
?	?	?

Figure 1. a 3×3 window in the clustering domain. (a) Central pixel and its neighbors in different clusters. (b) One neighboring pixel in the same cluster with the central pixel. (c) More than one neighboring pixels in the same cluster with the central pixel.

In fact, this is a kind of filter technique in the cluster domain rather than image domain. Through this, we can effectively revise the wrongly divided pixels, while protecting the edge pixels between regions from being affected. Pixels in the image are finally divided by the revised cluster numbers defined as below

$$m'_{ij} = i, \quad \max_i u'_{ij}. \quad (12)$$

The outline of PCA_HS is as below

Step 1: Decide the number of clusters; fix the fuzzifier m; set the iteration counter l=1; apply histogram for the grey-level image; estimate β by (2);

Step 2: Find the initial fuzzy partition matrix T by the standard FCM algorithm;

Step 3: **Repeat**

Compute $T^{(l+1)}$ by (6); Increment l;

Until $\| T^{(l-1)} - T^{(l)} \| < \varepsilon$;

Step 4: Compute u'_{ij} by (8) and (11);

Step 5: Output the clustering result by (12) and then compute the validity indexes.

2.3 Validity Indexes for PCA_HS

The normalized possibilistic c-membership is defined as below

$$\bar{u}_{ij} = \frac{u'_{ij}}{\sum_{k=1}^{c} u'_{kj}}, \quad \begin{matrix} i = 1,\dots,c, \\ j = 1,\dots,n. \end{matrix} \quad (13)$$

The generalized PC, PE, FS and XB defined in [10] can be sorted into two types, namely, the fuzzy partition type and the feature structure type [12]. GPC and GPE are the fuzzy partition type while GFS and GXB are the feature structure type. GPC and GPE are defined as below

$$GPC = \frac{1}{n}\sum_{i=1}^{c}\sum_{j=1}^{n} \bar{u}_{ij}^2. \quad (14)$$

and

$$GPE = -\frac{1}{n}\sum_{i=1}^{c}\sum_{j=1}^{n} \bar{u}_{ij} \ln \bar{u}_{ij}. \quad (15)$$

The idea of GPC and GPE is that the partition with less fuzziness means better performance. As a result, the best clustering is achieved when GPC is maximized and GPE is minimized.

The above indexes use only the membership functions and lack a direct connection to the featuring properties. GFS and GXB are defined as below

$$GFS = \sum_{i=1}^{c}\sum_{j=1}^{n} \bar{u}_{ij}^{m} \| x_j - c_i \|^2 - \sum_{i=1}^{c}\sum_{j=1}^{n} \bar{u}_{ij}^{m} \| c_i - \bar{c} \|^2. \quad (16)$$

and

$$GXB = \frac{\sum_{i=i}^{c}\sum_{j=1}^{n} \bar{u}_{ij}^{m} \| x_j - c_i \|^2}{n \min_{i \ne j} \| c_i - c_j \|^2}. \quad (17)$$

where $\bar{c}$ is defined as below

$$\bar{c} = \sum_{i=1}^{c} c_i / c. \quad (18)$$

The idea of GFS and GXB is that the partition should be compact in clusters and separated from other clusters. The best clustering is achieved when GFS and GXB are minimized.

3 Experiments and Discussions

In this section, we conduct the experiment with and without Gaussian or Salt & Pepper Noise on real images that are widely used by other researchers. The performance of PCA and PCA_HS are also compared. Throughout the experiment, the fuzzifier m equals 1.5, which is suggested in [8]. Two groups of the experimental results are presented as representatives.

3.1. Results on Noise-Free Images

Our first experiment applies PCA and PCA_HS on two noise-free images. The results are shown in Figure 2.

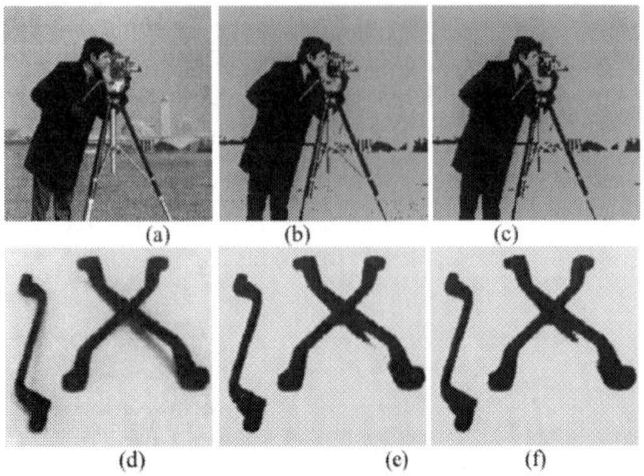

Figure 2. Segmentation results on noise-free images. First column: the original images; second column: PCA results of the corresponding images; third column: PCA_HS results of the corresponding images.

3.2 Results on Gaussian Noised Images

In this experiment, the original images are corrupted by zero-mean Gaussian Noise with its variance equal to 300 and the segmentation results are shown in Figure 3.

Figure 3. Segmentation results on zero-mean Gaussian noised images with variance equal to 300. First column: the noised images; second column: PCA results of the corresponding images; third column: PCA_HS results of the corresponding images.

3.3 Results on Salt & Pepper Noised Images

In this experiment, the original images are corrupted by Salt & Pepper Noise and the segmentation results are shown in Figure 4.

Figure 4. Segmentation results on Salt & Pepper Noised images (3%). First column: the noised images; second column: PCA results of the corresponding images; third column: PCA_HS results of the corresponding images.

3.4 Discussions

As can be seen from Figure 2 to Figure 4, for both the cameraman and the spanner with and without noise, PCA retains many isolated pixels, while PCA_HS can accurately identify them and yield more homogeneous and integrated regions.

For more objective assessment, the validity indexes of the cameraman and the spanner are computed and the results are listed in Table I. As shown, for the cameraman and the spanner with and without noise, GPC of PCA_HS is bigger than PCA, and meanwhile GPE, GFS and GXB are all smaller than PCA. Other experiments reach similar results. This indicates that PCA_HS is more effective than traditional PCA on real images with and without Gaussian or Salt & Pepper Noise.

TABLE I. VALIDITY INDEXES FOR IMAGES WITH AND WITHOUT GAUSSIAN AND SALT & PEPPER NOISE

Images	Algorithms	GPC	GPE	GXB	GFS(× 10^8)
Cameraman	PCA	0.9821	0.0322	0.02877	-2.1120
	PCA_HS	0.9830	0.0307	0.02858	-2.1163
Spanner	PCA	0.9955	0.0074	0.00620	-5.7159
	PCA_HS	0.9957	0.0073	0.00618	-5.7169
Gaussian Cameraman	PCA	0.9725	0.0473	0.03482	-1.8675
	PCA_HS	0.9762	0.0418	0.03200	-1.8854
Gaussian Spanner	PCA	0.9950	0.0086	0.00717	-5.4754
	PCA_HS	0.9952	0.0082	0.00681	-5.4784
Salt & Pepper Cameraman	PCA	0.9806	0.0350	0.03482	-2.0585
	PCA_HS	0.9815	0.0337	0.03200	-2.0785
Salt & Pepper Spanner	PCA	0.9954	0.0077	0.00717	-5.8128
	PCA_HS	0.9955	0.0076	0.00681	-5.8220

4 Conclusions

In this paper we proposed a possibilistic clustering algorithm incorporating the grey-level histogram and spatial information for image segmentation. Usually, the application of the grey-level histogram shrinks the membership matrix and thus can reduce the spatial and computational complexity. The spatial information is used for filtering in the cluster domain rather than image domain, and hence can enhance the robustness and meanwhile protect the edge pixels between regions from being affected. The efficiency analysis and the experiment results on real images with and without Gaussian or Salt & Pepper Noise indicate that PCA_HS is faster and more robust than traditional PCA for image segmentation.

Acknowledgment

The authors would like to thank the anonymous reviewers for their valuable comments and helpful suggestions in improving the presentation of the paper. This work was supported in part by the Fundamental Research Funds for the Central Universities.

References

1. M.O. Baradez, C.P. McGuckinb, N. Forrazb, R. Pettengell, and A. Hoppe, "Robust and automated unimodal histogram thresholding and potential applications," *Pattern Recognit.*, vol. 37, no. 6, pp. 1131-1148, 2004.
2. S.A. Hojjatoleslami, and J. Kittler, "Region growing: a new approach," *IEEE Trans. Image Process.*, vol. 7, no. 7, pp. 1079-1084, 1998.
3. B. Caldairoua, N. Passata, P.A. Habas, C. Studholme, and F. Rousseau, "A non-local fuzzy segmentation method: application to brain MRI," *Pattern Recognit.*, vol. 44 no.9, pp. 1916-1927, 2010.
4. B. Sowmya, and B.S. Rani, "Color image segmentation using fuzzy clustering techniques and competitive neural network," *Appl. Soft Comput.*, vol.11 no.3, pp. 3170-3178, 2011.
5. W.L. Cai, S.C. Chen, and D.Q. Zhang, "Fast and robust fuzzy c-means clustering algorithms incorporating local information for image segmentation," *Pattern Recognit.*, vol.40 no.3, pp. 825-838, 2007.
6. M. G. Gong, Y. Liang, J. Shi, W.P. Ma, and J.J. Ma, "Fuzzy C-Means Clustering With Local Information and Kernel Metric for Image Segmentation," *IEEE Trans. Image Process.*, vol. 22, no. 2, pp. 573-584, Feb. 2013.
7. R. Krishnapuram, J.M. Keller, "A possibilistic approach to clustering," *IEEE Trans. Fuzzy Syst.*, vol.1, no. 2, pp. 98-110, May 1993.
8. R. Krishnapuram, J.M. Keller, "The possibilistic C-means algorithm: insights and recommendations," *IEEE Trans. Fuzzy Syst.*, vol.4, no.3, pp. 385-393, Aug 1996.
9. M. Barni, V. Cappellini, and A. Mecocci, "Comments on 'A Possibilistic Approach to Clustering,'" *IEEE Trans. Fuzzy Syst.*, vol. 4, pp. 393-396, Aug. 1996.
10. M. S. Yang, K. L. Wu, "Unsupervised Possibilistic Clustering," *Pattern Recognit., vol.39, no.1, pp.* 5-21, 2006.
11. Y.F. Xu, "Image Segmentation Based on the Genetic Fuzzy C-Mean Algorithm," *Journal of Northwestern Polytechnical University*, vol.20, no.4, pp. 549-553, November, 2002.
12. K.S. Chuang, H.L. Tzeng, S. Chen, J. Wu, and T.J. Chen, "Fuzzy c-means clustering with spatial information for image segmentation," *Computerized Medical Imaging and Graphics*, vol. 30, no. 1, pp. 9-15, 2006.
13. H. Zuo, and W. Li, "Improved PCM Clustering Algorithm and Its Application in Image Segmentation," *Computer & Digital Engineering*, vol.38, no.11, pp.148-151, June 2010.

MATEC Web of Conferences 56, 02003 (2016) DOI: 10.1051/matecconf/20165602003
ICCAE 2016

An Optimized Structure on FPGA of Key Point Detection in SIFT Algorithm

Chenyu Xu[1], En Zhu[1], Tiange Su[1] and Xiaoran Li[2]

[1]School of Information Science and Engineering, Southeast University, Nanjing, China
[2]School of Electronic Science and Engineering, Southeast University, Nanjing, China

Abstract. SIFT algorithm is the most efficient and powerful algorithm to describe the features of images and it has been applied in many fields. In this paper, we propose an optimized method to realize the hardware implementation of the SIFT algorithm. We mainly discuss the structure of Data Generation here. A pipeline architecture is introduced to accelerate this optimized system. Parameters' setting and approximation's controlling in different image qualities and hardware resources are the focus of this paper. The results of experiments fully prove that this structure is real-time and effective, and provide consultative opinion to meet the different situations.
Keywords. SIFT; pipeline architecture; Gaussian window function; approximating errors

1 Introduction

Image matching is the process to identify the features of pictures to confirm the similarity and establish relevance between these similar pictures. The most essential parts of image matching is feature obtaining. The definition of features and the detection of them is the key to the whole process. There are many features in a picture, including edges and corners, and many algorithms are proposed to detect a specific type of features, like Canny operator and Harris corner. In 1999, David G. Lowe proposed a new algorithm [1], which is known as Scale-Invariant Feature Transform (SIFT), and revised it latter in 2004. This algorithm is established on the invariance of features no matter how environment changes. These changes can be described as multi-scales, and only those points which can show their invariance in each scale could be defined as key points. When all key points have been detected, corresponding descriptors will be generated and matching is the final step. With its distinctive process, SIFT algorithm has shown its advantages in many aspects, and been adopted widely.

Although SIFT algorithm has been realized on software platform consummately and proficiently nowadays, including a program in OpenCV patented by Lowe himself, it is another matter when SIFT algorithm is implemented to hardware platform. Time and resources consumption is the prime consideration; thus huge computations need to be transformed and simplified. Consequently, many scholars propose their designs for hardware implementation of SIFT algorithm. All these designs require the introduction of pipeline structure. This structure can utilize time most profitably. Those particular designs mainly differs in several specific steps of SIFT algorithm. The first step is the Generation of Gaussian Pyramid. ROI is introduced to generate suitable initial images in [2] and initial images are divided into smaller partitions in [3]. The size of Gaussian window function varies in [3]-[6], even the number of scales and octaves [7]. The value of σ also differs in [3]. Some IP kernels are introduced to calculate the filtering such as MAC and MUX in [2]. The second step is the Detection of Key Points. ADI and DER are introduced to obtain key points more accurately [7]. There are also some scholars using other methods like Harris corner to detect features [4]. The third step is the Calculation of Gradient and there is no difference in this part. Other steps include Generation of Descriptor and Matching. Some scholars complete these parts in combination with DSP [7], or other method like BRIEF descriptor [8], and we do not discuss these steps in details here.

After referring to these multiple designs, we propose our own optimized SIFT algorithm and implement it to hardware platform with Verilog. We mainly emphasize and discuss the setting of parameters and controlling of approximating errors in different conditions. The whole system is verified to be real-time and effective.

The following content of this paper is divided into 5 sections. Section II briefly discusses the algorithm proposed by David G. Lowe and its optimization to suit the hardware environment. Section III is the design of whole structure on hardware. Section IV is the results of experiments. Section V is the final conclusion of this paper.

2 Theory of SIFT Algorithm and Its Optimization

To be frank, when we implement SIFT algorithm on hardware platform, there are a couple of problems we need to handle, especially the time and resources consumption. Therefore, we need to figure out its principle first, and consider how to adjust and optimize SIFT algorithm to meet the hardware conditions without the introduction of DSP. This significant algorithm can be divided into two main stages: Key Point Detection and Key Point Description. In this paper, we mainly discuss the first stage, and we will briefly introduce the theory of this stage and its optimization in this section.

2.1 Theory of SIFT Algorithm

It has been verified that the only possible scale-space kernel is the Gaussian function [1]. The scale space of an image is defined in (1):

$$L(x,y,\sigma_i) = G(x,y,\sigma_i) * I(x,y) \tag{1}$$

The Gaussian filters are defined in (2):

$$G(x,y,\sigma_i) = \frac{1}{2\pi\sigma_i^2} e^{-(x^2+y^2)/2\sigma_i^2} \tag{2}$$

The complete realization of Gaussian scale space has a couple of steps: first, each scale is generated from the previous scale, and σ_i are different in each scales; second, after generating some scales of image, we can assemble them and call them an octave, then the last third scale is down-sampled by a factor of 2 to generate the first scale of next octave; third, the generating processes are the same in different octaves and the σ_i series are the same, too; fourth, the initial image should be up-sampled by a factor of 2 to generate the zero octave, and smoothed latterly by a Gaussian filter to create the first scale of zero octave.

In this process, σ_i series are crucial to keep continuity in scale space. In Lowe's work, the initial image is supposed to be incrementally convolved with Gaussians by a constant factor k [1], and the σ_i is defined as $k^{i-1}\sigma_0$. Here Lowe recommends using $\sigma_0 = 1.6$ [1]. If there are $s+3$ scales in one octave, k is defined as $2^{1/s}$.

After generating Gaussian Pyramid, we need to generate difference-of-Gaussian function, which is defined in (3):

$$D(x,y,\sigma) = L(x,y,k\sigma) - L(x,y,\sigma) \tag{3}$$

Then we can detect the local maxima and minima of the middle scales, by comparing each sample point in these scales to its eight neighbors in the current image and nine neighbors in the scale above and below [1]. Only the point which is larger than all of its neighbors or smaller than all of them is the extremum.

However, not all of the extrema are key points. Some of them have low contrast or are poorly localized along an edge [1]. The most effective way to remove or revise these points is to use Hessian matrix, which is defined in (4):

$$H = \begin{bmatrix} D_{xx} & D_{xy} \\ D_{xy} & D_{yy} \end{bmatrix} \tag{4}$$

All valid and reasonable key points should meet the restriction of (5) and here $r = 10$:

$$\frac{Tr(H)^2}{Det(H)} < \frac{(r+1)^2}{r} \tag{5}$$

The final step is to compute the gradient magnitude and orientation. When the calculation is finished, we can store the results into RAMs and they can be easily called by Generation of Descriptor.

2.2 Optimization of SIFT Algorithm

The main optimization is in the step of Generation of Gaussian Pyramid. There are a couple of consensus in all implementations: first, the generation of all scale of an octave should be completed at one time, the new σ_i is defined in (6); second, the size of Gaussian filter should be fixed and medium-sized; third, approximating errors exist in any computation; fourth, the generation of the zero octave should be abandoned.

$$\sigma_{i+1}' = \sqrt{\sigma_i'^2 + \sigma_{i+1}^2} \tag{6}$$

In this paper, we mainly discuss the realization of second and third point of consensus.

To the second point, on one hand, if the size of Gaussian filter is limited, differences compared with the standard values will be created. The differences will become more severe if the size of template is much smaller than the standard of $(6\sigma_i + 1) \times (6\sigma_i + 1)$. Besides, these differences will be accumulated when (6) is used repeatedly. On the other hand, the size of template should not be too enormous, considering the restriction of hardware resources. However, if we assume $\sigma_0 = 1.6$, the shortest width of window to avoid the differences is 41, which is too difficult to achieve. Only [7] and [9] emphasize this issue and announce that they manage to adopt this size in their work. Therefore, it is criticized to set a medium-sized template. Yet some scholars prefer to choose much smaller size of template. It seems that much smaller size of template could also obtain satisfying results. Whether the restriction of setting size of template is meaningful will be examined and elaborated in Section IV.

Related experiments show that the larger of the template size, the more accurate of the final results. If we could not enlarge the size of template, we could reduce the value of σ_0, while the results may not as effective as the size broadening.

To the third point, approximating errors will introduce some redundant key points. If the number of these extra key points is too large, the speed of the whole system will be degraded. The only method to solve these problems is to scale up the coefficients first, rather than remove the redundant points in Key Point Detection. One is the enlargement factor of each pixel; the other is the enlargement factor of Gaussian template.

The calculation of gradient magnitude and orientation also need to be revised. The formula for gradient magnitude is defined in (7):

$$m(x,y)=|L(x+1,y)-L(x-1,y)|+|L(x,y+1)-L(x,y-1)| \quad (7)$$

The orientation of gradient can be divided into some equal parts. In Lowe's work, it is 36 parts, and here we consider that 16 parts is enough.

3 FPGA Implementation of Optimized Algorithm

When the preparation has been finished, it is the time to implement this optimized algorithm. In this section, we briefly introduce the design of our pipeline architecture, and the main structure of each step of Data Generation.

3.1 Pipeline Architecture

The pipeline architecture is a kind of structure analogous to assembly line. There are some basic aspects: first, the data are sent in one by one in each data line; second, when the first datum is being carried on the second step of operation, the second datum will be carried on its first step of operation; third, the first octave of Gaussian Pyramid is the first part of the pipeline; fourth, Extrema Detection, Hessian Matrix Rejection, and Gradient Calculation could be operated at the same time.

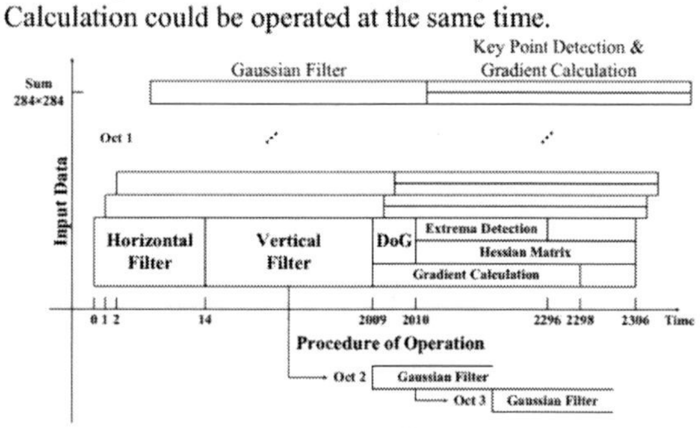

Figure 1. Structure of Pipeline

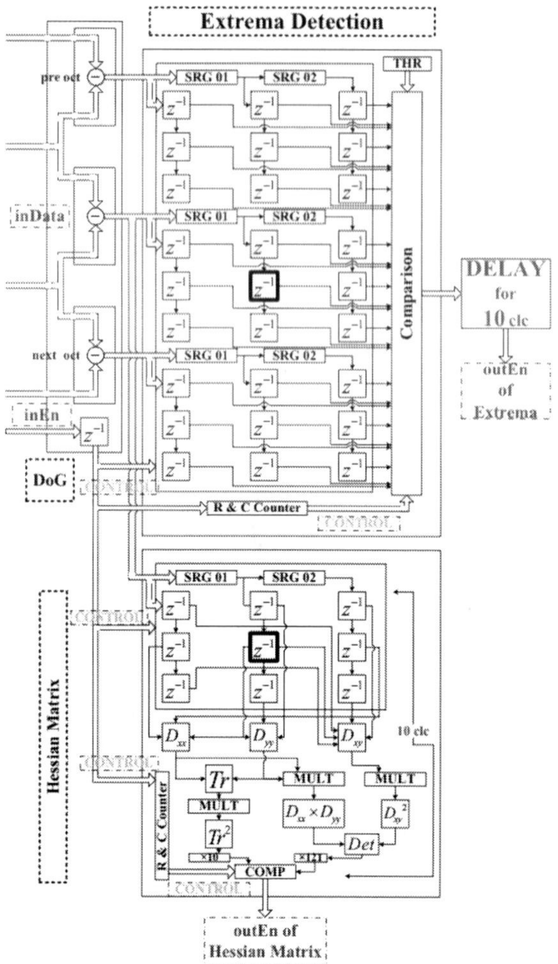

Figure 2. Structure of Key Point Detection

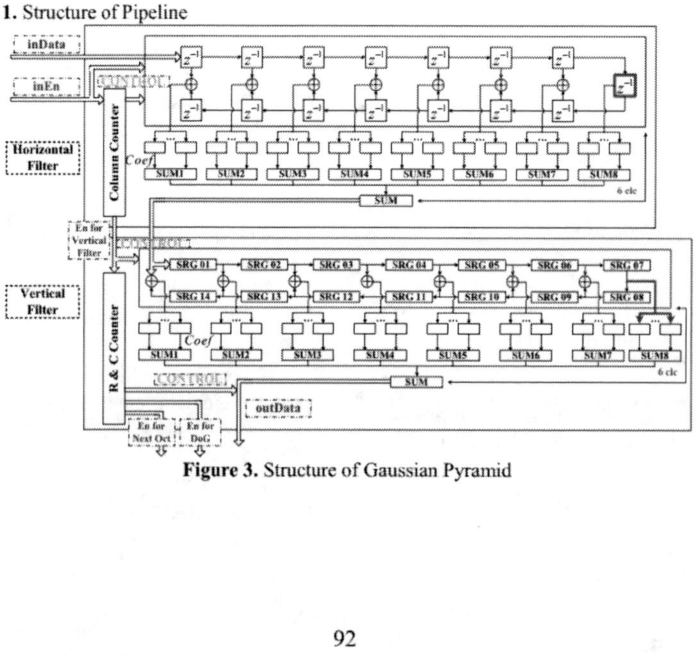

Figure 3. Structure of Gaussian Pyramid

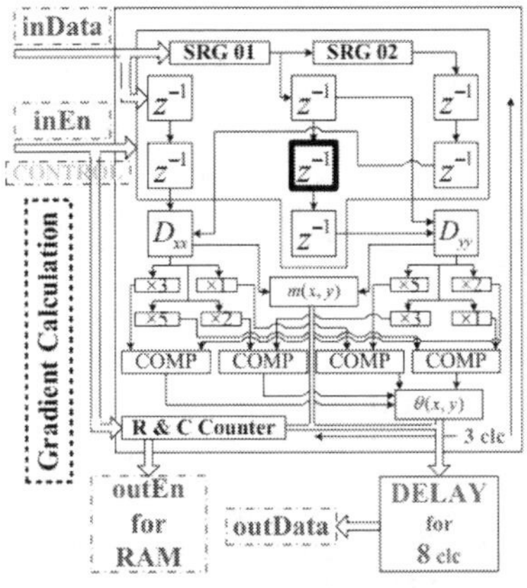

Figure 4. Structure of Gradient Calculation

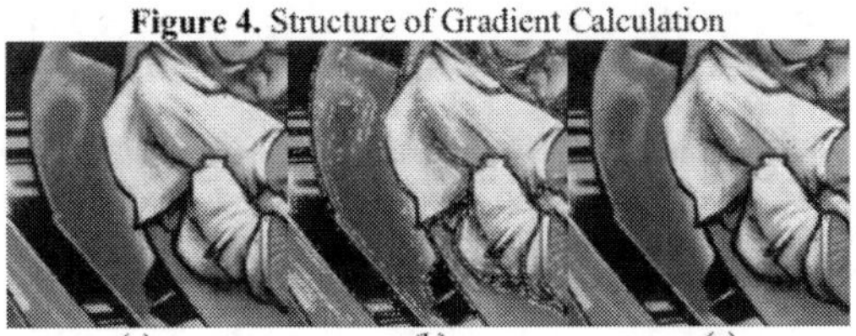

(a) (b) (c)

Figure 5. Results of Key Point Detection

Table 1. Numbers of Detected Key Point

Sca & Oct	o1 s2	o1 s3	o2 s2	o2 s3	o3 s2	o3 s3
OpenCV	61	55	32	13	7	3
No Approx	123	1	51	1	12	0
Miss No.	37	55	13	13	1	3
Mistake No.	99	1	32	1	6	0
Approx	1916	944	417	128	75	18
Miss No.	28	53	14	13	4	3
Mistake No.	1883	942	399	128	72	18

Apart from the common views above, we make some adjustment in our system: first, the size of the initial image is 284×284 ; second, Gaussian filtering is separated into horizontal filtering and vertical filtering, with advantages analyzed in 2; third, the output of Extrema Detection, Hessian Matrix Rejection, and Gradient Calculation should be sent out at the same time.

A crude description of the relations of each steps and direction of data flows is shown in Figure 1. The operating process of the first datum costs 2306 clocks. And from this moment, the computing results of the points in first octave will be sent out continuously. Whereas the outputs of the second and third octave will not be continuous.

3.2 Structure of Optimized System

Due to the constraints of resources, we set the number of s as 2, the size of Gaussian filter as 15×15 , the enlargement factor of each pixel as 1 and that of Gaussian template as 1024. The whole diagram of Gaussian Pyramid, Key Point Detection and Gradient Calculation is shown in Figure 2-4 respectively.

The most important parts in these structures are the shift registers and the enable signals. A shift register can store a whole row of image data, and the outputs of these registers are continuous in a line. These output data are for the following computations. The enable signals are the key to control the identification of the valid data.

Buffers could be introduced to reduce the consumption of resources and the order of filters should be reversed.

4 Experimental Results

In order to evaluate the performances of the proposed architecture and examine the effects of setting parameters in different conditions, we divide the experiment into two parts:

The first part has three steps: first, we test the processing time and observe the final results of Key Point Detection; second, we adjust the system in a further optimization and observe the final matching result; third, we change the size of window and observe the matching performances.

In the second part, noise will be introduced and the effects of parameters will be analyzed systematically.

The whole hardware implementation is realized on Xilinx ZC702, and clock frequency is 100MHz. The reference standards are the results in OpenCV. The results of tests are shown through MATLAB.

4.1 Experimental Results without Noise

When $\sigma_0 = 1.6$, and other parameters are the same as those in Part B, Section III, the result of the algorithm in OpenCV is shown in Figure 5(a), and the result of our hardware implementation is shown in Figure 5(b). Only the results of second scale in first octave are shown here. We can discover that too many key points detected because of the unreasonable setting of enlargement factors. However, if we examine the ideal result without approximating errors, shown in Figure 5(c), we could conclude that to set window width as 15 seems inappropriate. Related data is presented in Table 1.

Yet the time of the whole process is 966990 clocks, less than 1ms. Suppose that Descriptor Generation and Matching cost 3ms per 1000 key points. Whole SIFT algorithm along with matching costs no more than 5ms. It is a considerable processing speed and is enough for continuous real-time matching. However, the consumption of resources is much higher, for those shift registers employ too much LUTs. In this case, buffers are necessary.

Therefore, we adjust and revise the structure of our system. The enlargement factor of pixel and template are switched into 64 and 128 respectively. The result of Key Point Detection is presented in Table 2. We could find out that the result is much better though still unsatisfying. However, when we observe the final matching results, shown in Figure 6, we could find out that the matching result is accurate. Here the size of larger image is 800×640 and that of smaller one is 284×284, with an angle of rotation existing between two images.

Furthermore, if we set the width of window as 3 and enlargement factors are adaptive, the results are shown in Figure 7. We would discover that this extreme case could also achieve the same accurate result while this can never be required on software platform. Besides, all key points detected here are generated because of the approximating errors introduced by simplification of calculations.

Therefore, we could conclude that on one hand, SIFT algorithm has the best ability of robustness. On the other hand, the setting of parameters does not need to be so strict when we do hardware implementation. In addition, matching results are what we should pay attention to, rather than intermediate results.

4.2 Experimental Results with Noise

4.1 is based on the condition without noise, the results could be different if noise is introduced. Thus, we introduce four kinds of noise respectively. The maximum extents of noises when matching results are still correct are presented in Table 3. We can reveal that the smaller value of window width, the poorer ability of noise immunity. In this case, the restriction of window width is significant. Various factors into consideration, we could conclude that a medium size of Gaussian filter is the best choice for hardware realization and 15×15 is such one.

Table 2. Numbers of Key Point after Adjustment

Sca & Oct	o1 s2	o1 s3	o2 s2	o2 s3	o3 s2	o3 s3
Approx	408	125	88	39	16	7
Miss No.	42	62	10	20	2	3
Mistake No.	372	121	67	38	11	6

(a) (b)

Figure 6. Matching Results with $w = 15$

(a) (b)

Figure 7. Matching Results with $w = 3$

Table 3. Limited Extent of Noise

w	3	15	31
average	[4,4]	[6,6]	[7,7]
disk	2	4	4
Gaussian ($\sigma = 5$)	[4,4]	[7,7]	[8,8]
Motion ($\theta = 45^\circ$)	5	9	11

5 Conclusion

In this paper, we have proposed an optimized structure of Data Generation in SIFT algorithm, and its implementation on hardware platform. The whole system is based on a pipeline structure and verified to be real-time and effective. The results of experiments fully prove that the setting of parameters and controlling of approximations are crucial in hardware implementation, and this process will be affected by different conditions. Here we recommend using a medium-sized Gaussian window function.

References

1. D. G. Lowe, "Dinstinctive Image Features from Scale-Invariant Keypoints," Int. J. Comput. Vision, vol. 60, no. 2, pp. 91 - 110, 2004.
2. K. Mizuno, H. Noguchi, Guangji He, Y. Terachi, T. Kamino, H. Kawaguchi and M. Yoshimoto "Fast and Low-Memory-Bandwidth Architecture of SIFT Descriptor Generation with Scalability on Speed and Accuracy for VGA Video," in Proc. Int. Conf. Field Programmable Logic and Applications (FPL), 2010, pp. 608 - 611.
3. L. Yao, H. Feng, Y. Zhu, Z. Jiang, D. Zhao, and W. Feng, "An Architecture of Optimised SIFT Feature Detection for an FPGA Implementation of an Image Matcher," in Proc. Int. Conf. Field Program. Technol., 2009, pp. 30 - 37.
4. Takahiro Suzuki and Takeshi Ikenaga, "SIFT-Based Low Complexity Keypoint Extraction and Its Real-Time Hardware Implementation for Full-HD Video", in Proc. Signal & Information Processing Association Annual Summit and Conference (APSIPA ASC), 2012 Asia-Pacific.
5. Jie Jiang, Xiaoyang Li, Guangjun Zhang, "SIFT Hardware Implementation for Real-Time Image Feature Extraction," IEEE Trans. Circuits Syst. Video Technol., vol. 24, no. 7, pp. 1209 - 1220, Jul. 2014.
6. S. Zhong, J. Wang, L. Yan, L. Kang, and Z. Cao, "A real-time embedded architecture for SIFT," J. Syst. Arch., vol. 59, no. 1, pp. 16 - 29, Jan. 2013.
7. Han Xiao, Wenhao He, Kui Yuan, Feng Wen, "Real-time Scene Recognition on Embedded System with SIFT Keypoints and a New Descriptor," in Proc. 2013 IEEE int. Conf. Mechatronics and

Automation (ICMA), Aug. 4-7 Takamatsu, Japan, pp. 1317 – 1324.

8. J. Wang, S. Zhong, L. Yan, and Z. Cao, "An Embedded System-on-Chip Architecture for Real-time Visual Detection and Matching," IEEE Trans. Circuits Syst. Video Technol., vol. 24, no. 3, pp. 525 - 538, Mar. 2014.
9. M.Qasaimeh, A.Sagahyroon, T.Shanableh, "A Parallel Hardware Architecture for Scale Invariant Feature Transform (SIFT)," in Proc. Int. Conf. Multimedia Computing and Systems (ICMCS), 2014, pp. 295 - 300.

A Novel Image Fusion Algorithm for Visible and PMMW Images based on Clustering and NSCT

Jintao Xiong , Weichao Xie, Jianyu Yang, Yanlong Fu, Kuan Hu and Zhibin Zhong
University of Electronic Science and Technology of China, Chengdu 611731, Sichuan, China

Abstract. Aiming at the fusion of visible and Passive Millimeter Wave (PMMW) images, a novel algorithm based on clustering and NSCT (Nonsubsampled Contourlet Transform) is proposed. It takes advantages of the particular ability of PMMW image in presenting metal target and uses the clustering algorithm for PMMW image to extract the potential target regions. In the process of fusion, NSCT is applied to both input images, and then the decomposition coefficients on different scale are combined using different rules. At last, the fusion image is obtained by taking the inverse NSCT of the fusion coefficients. Some methodologies are used to evaluate the fusion results. Experiments demonstrate the superiority of the proposed algorithm for metal target detection compared to wavelet transform and Laplace transform.

1 Introduction

A passive millimeter wave (PMMW) imaging system achieves images by detecting the difference of millimeter band energy that objects radiated, which has great performance in penetrating and excellence ability of recognizing a metal target from its surrounding environment [1]. Passive millimeter wave imaging system has large potential in the field of hidden objects detection because it is not harmful to humans [2]. However, the main difficulty in the PMMW imaging systems is low image quality due to wavelength, which cannot meet the demand of practical application. Considering the fact that even PMMW images with low pixel resolution still provide valuable information, it is deserved to combine it with a visible image to enhance the characteristic.

Image fusion is a process of generating a single fused image using a set of input images which are assumed to be registered. The image for fusion may be either the same scene at the same time different sensors acquired, the same sensor at different time period [3]. The resulting image will be more informative than any of the input images.

The classical image fusion methods have weighted average, Wavelet fusion, Laplace fusion and NSCT fusion. NSCT has the characteristic of shift-invariant; therefore, this paper presents the application of it for image fusion. Before the image fusion, an image pre-processing which specifics to PMMW image is adopted. In this process, an image segmentation based on clustering is adopted. Figure 1 illustrates the diagram of image fusion scheme.

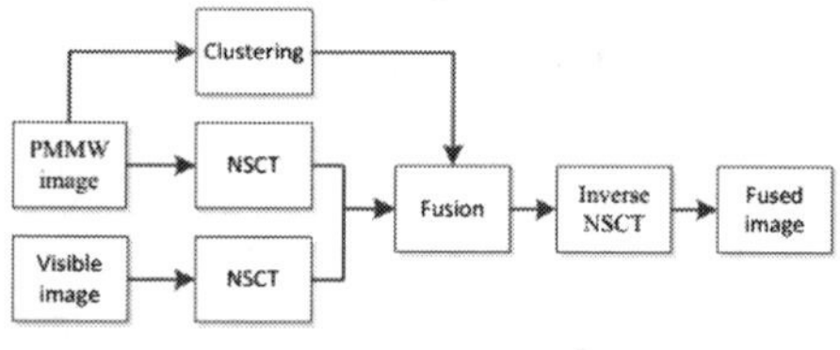

Figure 1. The diagram of fusion scheme.

2 Image segmentation based on clustering

Image segmentation is the process of partitioning a digital image into multiple segments. Several general-purpose algorithms and techniques have been developed for image segmentation, such as the method based on threshold segmentation [4].

Image segmentation based on clustering is relatively new image segmentation algorithm, which segmentation accuracy is higher than usual threshold method. Several clustering methods such as k-means clustering, fuzzy c-means (FCM) clustering and GMM (Gaussian Mixture Model) can be used to image segmentation [5, 6].

GMM is one of the most widely used model for statistical segmentation of PMMW images. In the GMM, the probability function is view as the combination of two or more normal Gaussian distribution [7]. GMM is a probabilistic model that assumes all the data points are generated from a mixture of a finite number of Gaussian distributions with unknown parameters; its expression can be defined as:

$$p(x)=\sum_{k=1}^{K}\pi_k N(x\mid\mu_k,\Sigma_k) \tag{1}$$

Where m is the number of models we used, μ is D-dimensional mean vector, $|\Sigma|$ means the determinant of the covariance matrix, and π_k is the mixing coefficients.

If sample category is not given, the concept of implicit variables is introduced to simplify the solution of model parameters. Suppose there are N data points, subject to certain distribution $P_r(x;\theta)$, trying to search a set of parameters θ to make the probability of generating these data points maximum, the expression of this probability is called Likelihood Function, can be defined as:

$$P=\prod_{i=1}^{N}P_r(x_i;\theta) \tag{2}$$

$$\log p(X;\pi,\mu,\Sigma)=\sum_{i=1}^{N}\log\{\sum_{k=1}^{K}\pi_k N(x_i;\mu_k,\Sigma_k)\} \tag{3}$$

In the formula (3), $X=\{x_1,x_2,...,x_N\}$ represents the vector of observation, category z_k each sample x_i belongs to is unknown. Next, a method called EM (Expectation Maximization) algorithm is applied to search the best model parameters, which used to maximum the expectation showed in formula (3).

The process of EM algorithm can be viewed as two steps: E-step and M-step. As to E-step, it is under the assumption that the model parameters are given, and seeking expectation with implicit variable z.

$$\gamma(i,k)=P_r(z_k\mid x_i;\pi,\mu,\Sigma) \tag{4}$$

$$\gamma(i,k)=\frac{\pi_k N(x_i\mid\mu_k,\Sigma_k)}{\sum_{j=1}^{K}\pi_j N(x_i\mid\mu_j,\Sigma_j)} \tag{5}$$

For M-step, the maximum likelihood function derivative separately to mean, covariance, mixing coefficients, and put the result to zero, new expression of three parameters can be obtained:

$$N_k=\sum_{i=1}^{N}\gamma(i,k) \tag{6}$$

$$\mu_k=\frac{1}{N_k}\sum_{i=1}^{N}\gamma(i,k)x_i \tag{7}$$

$$\mu_k=\frac{1}{N_k}\sum_{i=1}^{N}\gamma(i,k)(x_i-\mu_k)(x_i-\mu_k)^T \tag{8}$$

$$\pi_k=\frac{N_k}{N} \tag{9}$$

To verify the clustering algorithm based on GMM can be used in millimeter wave image segmentation, a series of scene detection experiments is carried out; some PMMW image samples are obtained for the verification and analysis to the algorithm.

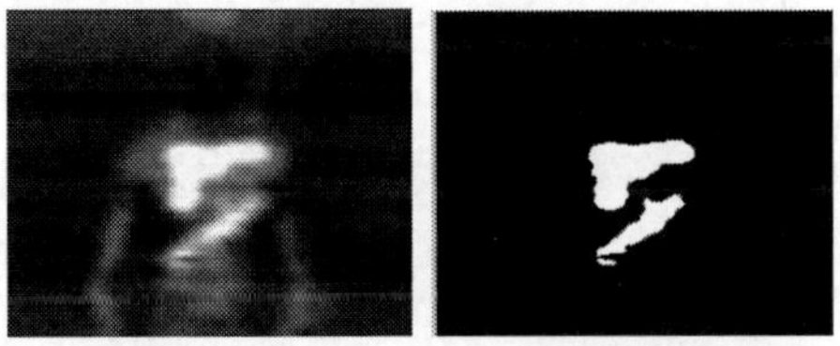

(a) PMMW image (b) Segmentation result

Figure 2. Person with two concealed targets

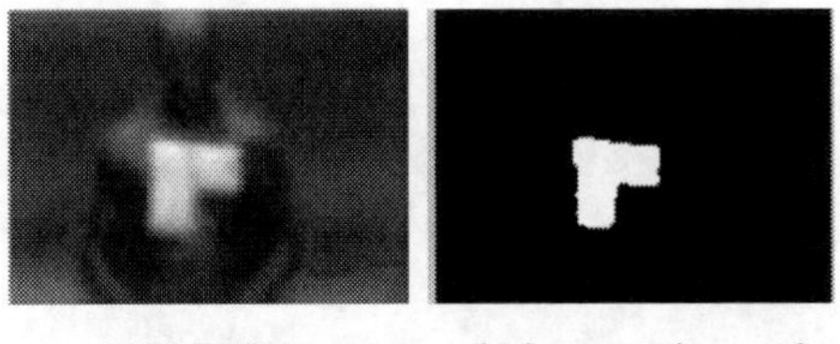

(a) PMMW image (b) Segmentation result

Figure 3. Person with one concealed target

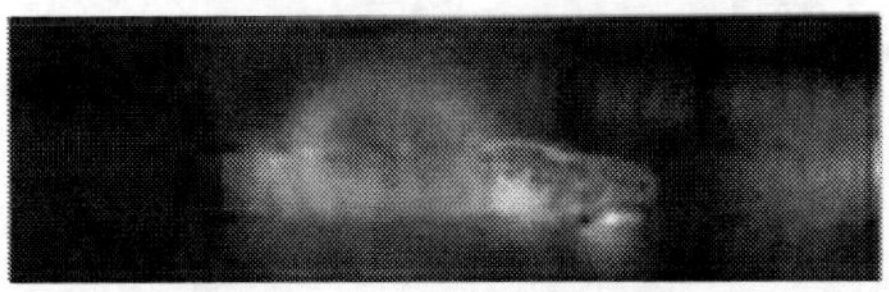

(a)PMMW image with concealed car

(b) Segmentation result

Figure 4. The concealed car

As shown in Figure 2-Figure 4, image segmentation through GMM has a well segmentation result.

3 Image fusion based on NSCT

NSCT is a method of multi-scale decomposition which has the characteristic of shift-invariance. NSCT contains NSP (NonSubsampled Pyramid Decomposition) and DFB (Nonsubsampled Directional Filter Banks). In the multi-scale transform of pyramid transform, the high frequency image is acquired through upsampling or downsampling the filtered image and then subtract the higher level image, however, in that case, image will lost the characteristic of shift-invariant. Unlike pyramid transform, NSCT will not sample the image, but sample the filter. The filter of each level is obtained through sampling higher level filter. Each level's low-frequency sub-band image and high-frequency sub-band image are acquired by filtering the image with the resulting filter [8]. In the process of reconstruction, filters must satisfy to Bezout equation:

$$H_0(z)G_0(z)+H_1(z)G_1(z)=1 \qquad (10)$$

$\{H_0(z),H_1(z)\}$ is the analysis filter, $\{G_0(z),G_1(z)\}$ is the synthesis filter [9] . The NSP is shown in Figure 5.

DFB is a two-channel filter, it is the improvement of fan filter bank proposed by Bamberger and Smith, its analysis filters and synthesis filters also need to meet Bezout equation.

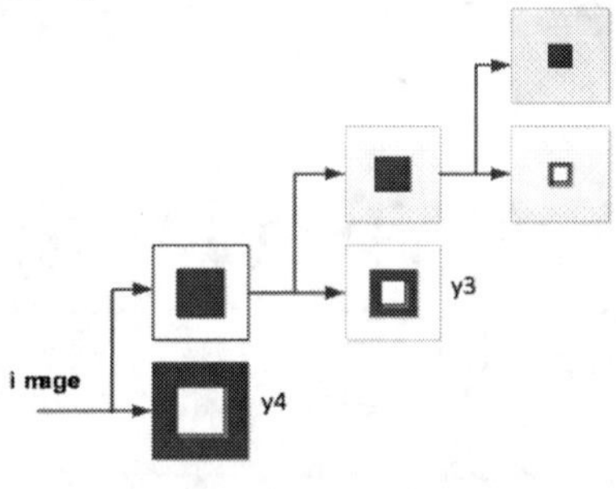

Figure 5. NSP with three stages

The NSCT is constructed by combining the NSP and DFB as shown as Figure 6.

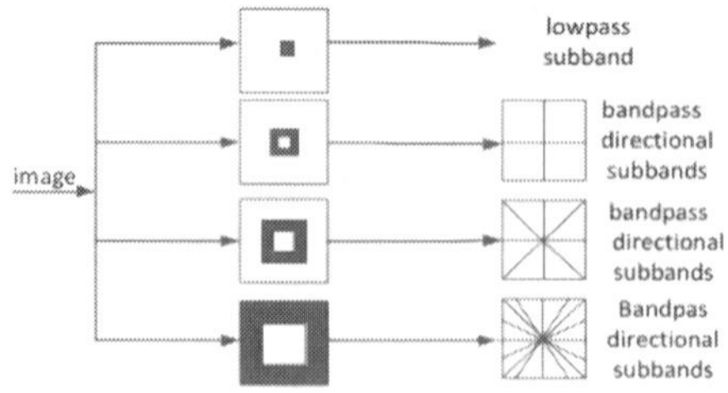

Figure 6. Segmentation map in frequency domain

This paper will take three times NSCT decomposition both to visible image and PMMW image, then separately fusion the sub-image with different fusion rule. The rule of absolute value is adopted for third layer's high frequency sub-images. Regional average gradient is used to fusion both first and second layer's high frequency sub-images. As for the low frequency images of third layer, a fusion criterion called regional variance is adopted in Article [10], the definition of regional variance is:

$$V^s(i,j)=\frac{1}{(M-1)(N-1)}\sum_{m=-(M-1)/2}^{(M-1)/2}\sum_{n=-(N-1)/2}^{(N-1)/2}\left|C^s(i+m,j+n)-\bar{C}\right|^2 \qquad (11)$$

Where m and n represent the length and width of region respectively, $C(i,j)$ is the coefficient of decomposition, superscript s is the sign of visible image or PMMW image. As is shown in the regional variance expression, regional variance is treated separately for the two images, and do not link the two images. In this paper, a new fusion criterion which used to distinguish the regional differences between two images is proposed.

$$Diff(i,j)=\frac{1}{(M-1)(N-1)}\sum_{m=-(M-1)/2}^{(M-1)/2}\sum_{n=-(N-1)/2}^{(N-1)/2} f(i,j,m,n)) \qquad (12)$$

$$f(i,j,m,n)=(V(i+m,j+n)-M(i+m,j+n))^2/(V(i,j)^2+M(i,j)^2)$$

Where $V(i,j)$ stands for visible image, $M(i,j)$ represents PMMW image. Fused image by compared $Diff(i,j)$ with a threshold ∂

$$F(i,j)=\begin{cases}\max(V(i,j),M(i,j)) & Diff(i,j)>\partial \\ k1*V(i,j)+k2*M(i,j) & Diff(i,j)\le\partial\end{cases} \qquad (13)$$

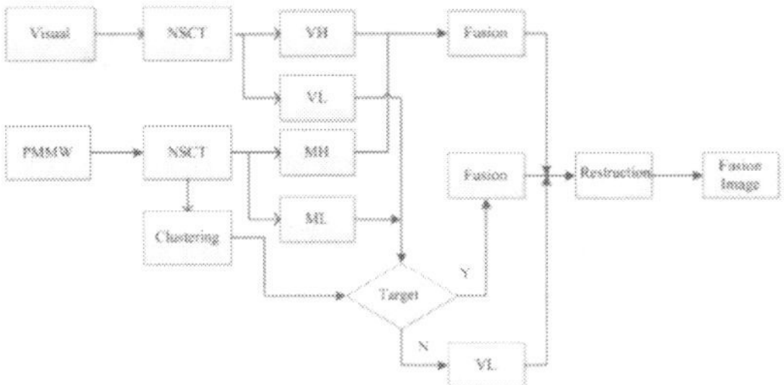

Figure 7. The diagram of fusion scheme.

In the process of image fusion, a novel image fusion algorithm is proposed. In classical fusion methods, visible image fusion with PMMW image or the segmentation result of PMMW directly, however, the former will cause the problem of blur non-target area, and the latter trends to lead the matter of poor visual effect. In this paper, we not fusion the visible image with PMMW or the segmentation result, but using the segmentation result as a label to mark one point of image whether belongs to target, if so, fusing images with above fusion rules, otherwise, treating visible image as the fusion result directly. In this way, the fusion image not only has the complex optical background, but also with nice visual effect. The diagram of fusion scheme is show as Figure 7, where VH/VL and MH/ML represent to decomposed high-frequency/low-frequency image of visible image and PMMW image respectively.

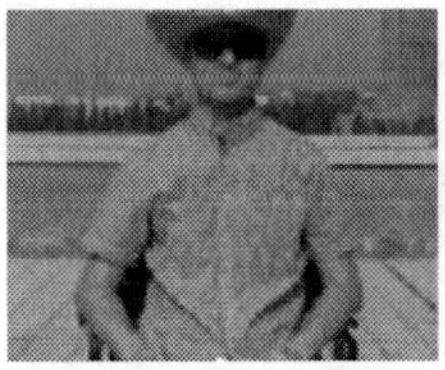
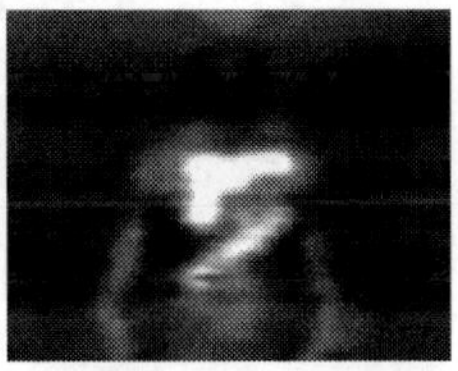

(a) Visible image of person with concealed targets (b) PMMW image of person with concealed targets

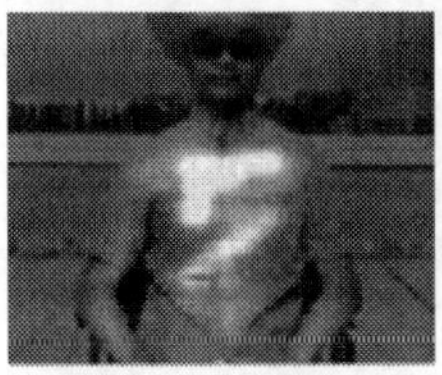

(c) Direct fusion result (d) LT fusion result

(e) WT fusion result. (f) The result of proposed method

Figure 8. Person with two concealed targets

(a) Visible image of concealed car

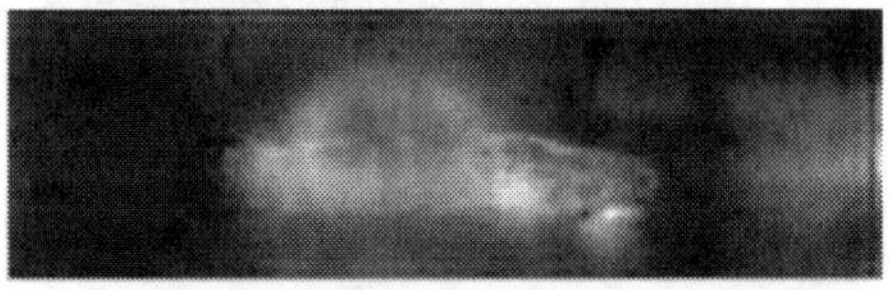

(b) PMMW image of concealed car

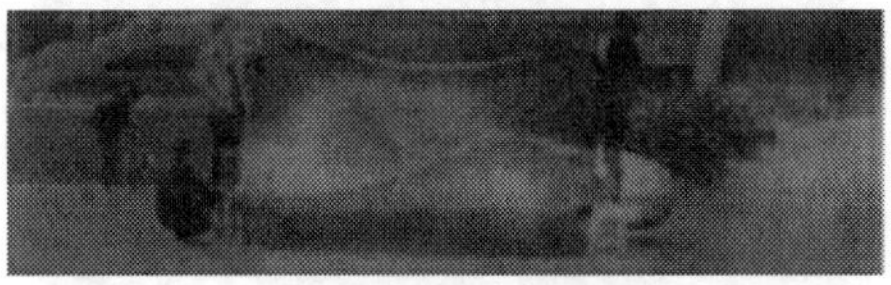

(c) Direct fusion result

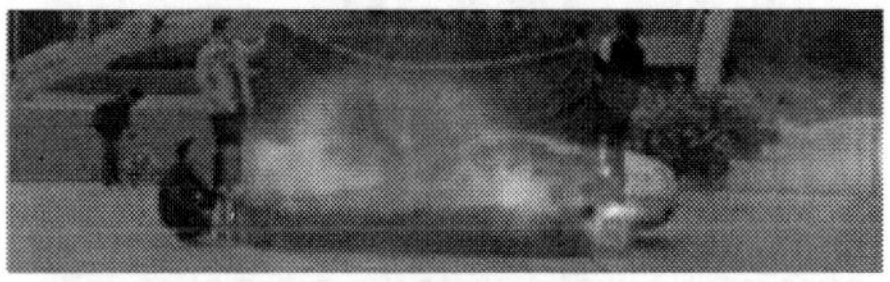

(d) LT fusion result

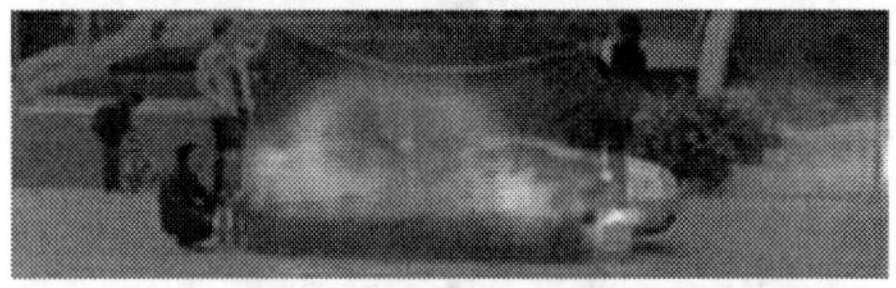

(e) WT fusion result

(f) The result of proposed method

Figure 9. The concealed car

As shown in figure 8-figure 9, the images fusion based on weighted average makes entire image becomes blurred, particularly non-target background area, and the optical image comparison shows that the entire image dimmed. Though, the fusion result based on Laplace and Wavelet is better than weighted average, still causing blurred background [11]. However the fusion algorithm based on clustering and NSCT not only well maintained complex optical image background, but also the target's contour information has been well maintained.

To evaluate the effects of fusion in better fashion, three objective evaluation indicators are introduced to evaluate the effect of fusion. Information entropy is a measure of image information richness index, a direct reflection to information image contained. Average gradient reflects the degree of detail of the image contrast of the image. Spatial frequency reflects the overall level of activity in the spatial domain of an image [12].

Table 1. Objective evaluation indicators corresponding to Figure 8

	EN	AG	SF
Figure 8. (a)	6.7779	5.8767	4.2737e+005
Figure 8. (b)	6.6533	2.2263	1.6907e+005
Figure 8. (d)	6.8367	3.9731	2.8977e+005
Figure 8. (e)	6.8273	4.0299	2.9443e+005
Figure 8. (f)	7.0015	6.2941	4.5341e+005

Table 2. Objective evaluation indicators corresponding to Figure 9

	EN	AG	SF
Figure 9. (a)	7.0697	6.3155	2.5980e+005
Figure 9. (b)	7.1125	3.0172	1.2640e+005
Figure 9. (d)	6.8197	4.3755	1.7771e+005
Figure 9. (e)	6.7773	4.3446	1.7728e+005
Figure 9. (f)	7.2728	7.4150	3.0520e+005

Where EN denotes entropy, AG represents average gradient, and SF stands for spatial frequency. Seen from the Table 1 and Table 2, the fusion results generated from our algorithm, comparing with weighted average, Wavelet and Laplace, contain more information, and the fusion image are clearer and overall activity are also higher.

4 Conclusion

EM (Expectation Maximization) algorithm can be used to estimate the parameters of Gaussian Mixture Model. Though Parametric Learning algorithm takes a relatively long time, its segmentation accuracy is higher than almost threshold method. Image fusion technique can overcome the shortcoming of single image information; a complementary form can be obtained through fusing image acquired from different sensors. According to the characteristics of millimeter wave image, a novel image fusion algorithm for visible and PMMW images based on clustering and NSCT is proposed. The proposed fusion algorithm, comparing to Laplace fusion and Wavelet fusion, not only on the image information contents and the image sharpness, but also the overall level of activity, have great improvement.

References

1. Dill S, Peichl M, Süß H. Study of passive MMW personnel imaging with respect to suspicious and common concealed objects for security applications[C]. SPIE Europe Security and Defence. International Society for Optics and Photonics, (2008): 71170C-71170C-8.
2. L. Yujiri, M. Shoucri, P. Moffa. Passive millimeter wave imaging. Microwave Magazine, IEEE, (2003), 4(3): 39-50.
3. Garg, Rishu, Preeti Gupta, and Harvinder Kaur. "Survey on multi-focus image fusion algorithms." Engineering and Computational Sciences (RAECS), 2014 Recent Advances in. IEEE, (2014).
4. Rajeshwar Dass, Priyanka, Swapna Devi, "Image Segmentation Techniques", International Journal on Electronics & Communication Technology (IJECT), **3**, Issue 1, pp. 66-70, Jan. - March (2012).
5. Lee Y, Choi S. Minimum entropy, k-means, spectral clustering[C]. Neural Networks, 2004. Proceedings. 2004 IEEE International Joint Conference on. IEEE, (2004), 1.
6. Choong M Y, Khong W L, Chin R K Y, et al. Clustering Algorithm in Normalised Cuts Based Image Segmentation[C]. Modelling Symposium (AMS), (2013) 7th Asia. IEEE, 2013: 166-171.
7. Li Y, Li L. A novel split and merge EM algorithm for Gaussian mixture model[C]. Natural Computation, 2009. ICNC'09. Fifth International Conference on. IEEE, (2009), 6: 479-483.
8. Da Cunha A L, Zhou J, Do M N. The nonsubsampled contourlet transform: theory, design, and applications[J]. Image Processing, IEEE Transactions on, (2006), 15(**10**): 3089-3101.
9. Liu Y, Liu Y. A novel fusion algorithm for infrared image and visible light image based on non-subsampled contourlet transform[J]. Scientific Research and Essays, (2014), 9(9): 374-379.
10. Zhang Q, Guo B L. Research on image fusion based on the nonsubsampled contourlet transform[C]. Control and Automation, 2007. ICCA 2007. IEEE International Conference on. IEEE, (2007): 3239-3243.

11. Burt P J, Adelson E H. The Laplacian pyramid as a compact image code[J]. Communications, IEEE Transactions on, (1983), 31(4): 532-540.
12. Yang B, Li S, Sun F. Image fusion using nonsubsampled contourlet transform[C]. Image and Graphics, 2007. ICIG 2007. Fourth International Conference on. IEEE, (2007): 719-72.

DOI: 10.1051/matecconf/20165602005

Rotation Invariant Automatic Infrared Target Recognition using G-Radon

Jin-Ju Won, Sungho Kim

Department of Electronic Engineering Yeungnam University, Gyeonsan, Gyeongbuk, 38541, Korea

Abstract. In recent research, automatic target recognition (ATR) of infrared targets has been taking a lot of interest to the researchers. A rotation invariant method is useful in target recognition, classification and image analysis to reduce the number of training data. In particular, rotation invariant method, Radon transform, is an effective technique that is used for medical care such as computerized tomography (CT) image. This paper proposes a new rotation invariant algorithm for target recognition. The proposed method combines the gradient information and radon transform. The propose method, called gradient Radon (G-Radon), is applied to synthesized infrared images and compared with traditional radon transform and Zernike moments for validation.

1 Introduction

In recent research, ATR of infrared targets has been taking a lot of interest to the researchers [1]. Conventional ATR algorithms [2] are based on the model-based method [3, 4] and learning-based method [5, 6]. Example of model-based research is image classification using hidden markov trees [3]. Learning-based researches are wavelet based target recognition and target recognition using support vector machines (SVM) [5, 6]. Histogram of oriented gradients (HOG) [7] is a famous algorithm with excellent performance when combined with SVM. But these algorithms are sensitive to rotation. Therefore rotation invariant method is necessary in target recognition and classification. The rotation invariant method is useful to reduce the number of training images. The simplest form of rotation invariant feature is the Fourier transformation of the boundary curve which is invariant with regard to translation and rotation [8]. A popular invariant feature is based on the moment including non-orthogonal moments [9]-[11] and orthogonal moments [12]-[14]. These methods have property such as less sensitive to noise and very accurate in image reconstruction. Scale invariant feature transform (SIFT) also has the property that for size invariant and rotation invariant. SIFT selects the local maxima and generates a descriptor as feature points by using a histogram of the surrounding area in an image. This method serves to strength the distortion or partial change of the image, but a significant amount of calculation occurs to generate the feature vector of 128 dimensions [15, 16].

This paper uses image gradient to extract target contour and Radon transform to extract rotation invariant feature. Radon transform is an effective technique that is used for medical care such as CT images. We propose a gradient Radon (G-Radon) with small amount of calculations and the high recognition rate as combination of the two methods. The conventional Radon transform is sensitive to translation and noise. Therefore we need to extract more robust feature such as gradient magnitude feature.

Section 2 presents the proposed gradient feature and Radon transform. Conventional Radon and G-Radon is compared for the rotated infrared targets images. After performing the G-radon, we compare features of upright targets and those of rotated targets. Section 3 validates the effectiveness of G-Radon for the synthetic infrared data set by comparing with the results using traditional Radon transform. Only an upright target image per target was used for the model learning and rotated 72 different target images were used for testing. The proposed G-radon transform, conventional Radon transform and Zernike moments are compared using confusion matrices. Through the experiments, we can validate the superior performance of the proposed G-radon feature. Section 4 concludes this paper and discusses about future works. The ultimate goal of this research is to achieve an improved ground target recognition and classification through the fusion of SAR/IR sensor information. This paper validates the target recognition for IR dataset only. In the future, we will verify the effects of SAR/IR fusion in ATR.

2 Proposed G-radon based ATR

The proposed feature extraction method is called G-radon. We use image gradients to extract targets' edge information for effective target recognition and compute gradient magnitude values. Next we perform the Radon transform and extract rotation invariant feature.

2.1 Gradient feature extraction

The directional gradients, G_x and G_y, are extracted from an image to obtain edge information. Edge magnitude is calculated eq. (1).

$$f(x,y) = \sqrt{(G_x^2(x,y) + G_y^2(x,y))} \quad (1)$$

Fig. 1 shows the edge extraction results for a test image (Fig. 1(a)). Fig. 1(b) presents the gradient image in the x-axis direction. The value on the left side of the image is positive value because the intensity of background is lower value and that of the target is higher value. Fig. 1(c) shows the gradient result in the y-axis direction. Fig. 1(d) represents the computed edge magnitude value using eq. (1).

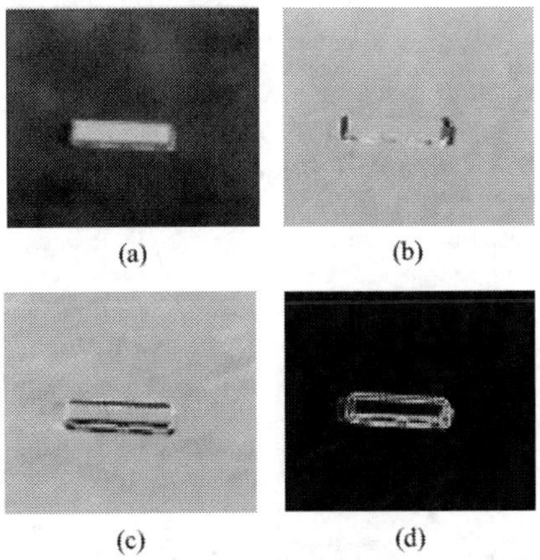

Figure 1. (a) Original input image, (b) computed row directional gradient result, (c) computed column directional gradient result, (d) gradient magnitude.

2.2 Rotation invariant Radon transform

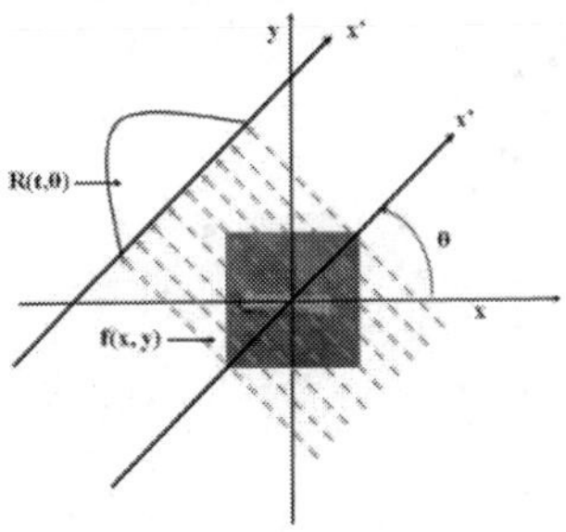

Figure 2. Radon transform of a two-dimensional function *f(x,y)*.

Radon transform [17] accumulates image in Radon transform space as shown Fig. 2. The extraction process is simple and the feature is robust to image noise due to the integral property. The radon transform defined as

$$R(t,\theta) = \int_{-\infty}^{\infty}\int_{-\infty}^{\infty} f(x,y)\delta(t - x\cos - y\sin\theta)dxdy \quad (2)$$

where t is the perpendicular distance for a straight line from the origin, θ is the angle between the distance vector and the x-axis. Range of θ is 0 to π. $R(\rho,\theta)$ is the projection data obtained by the image curvilinear integral.

δ means dirac delta function. In each θ, gradient image $f(x,y)$ can be expressed as follows

$$R(\rho,\theta) = \int_{-\infty}^{\infty}\int_{-\infty}^{\infty} f(\rho\cos - \rho\sin\theta,\ \rho\sin\theta + u\sin\theta)\delta(0)dxdy \quad (3)$$

when $x=1$, the delta function is 0. So it is converted to the following equation

$$R(\rho,\theta) = \int_{-\infty}^{\infty}\int_{-\infty}^{\infty} f(\rho\cos\theta - u\sin\theta, \rho\sin\theta + u\sin\theta)dud\rho \quad (4)$$

when the image is rotated by θ_0, Radon transform's projection data is also rotated by θ_0.

$$R(\rho, \theta + \theta_0) \quad (5)$$

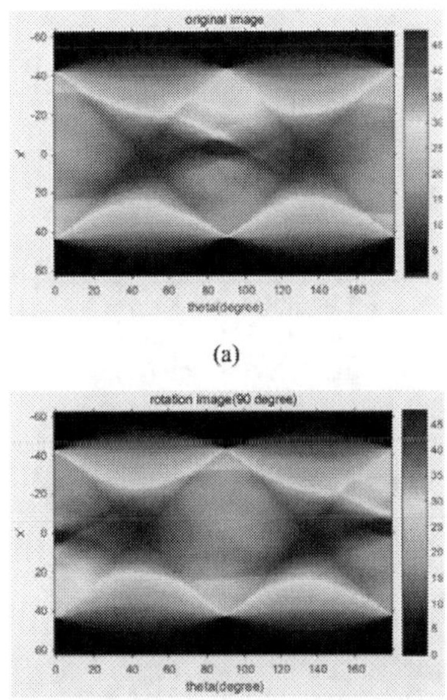

Figure 3. (a) Non-rotated image(0°), (b)Rotated image (90°).

Therefore radon transform is rotation invariant. Fig. 3 shows the application of Radon transform. Fig. 3(a) is a non-rotated image (0°). High value is distributed from 30° to 150°. These values are the values accumulated in radon space. The value is distributed differently according to the shape of target. Fig. 3(b) is a rotated image (90°). High value is distributed from 120° to 180° and from 0° to 60. Radon space is moved as much as the image is rotated.

The Fig. 4 shows the results of G-radon for the Bus (one is upright position, the other is rotated by 30°) and a fire truck (upright position). Blue solid line indicates the result of the Bus (0), orange broken line represents the result of the Bus (30) and yellow line with a circular mark indicates the fire truck (0). Bus (0) denotes upright position and Bus (30) represents an image rotated by 30°. The blue solid line show almost the same feature profiles as the orange broken line, but the yellow line with a circular mark show a different feature profiles. We can see that the result of G-radon is dependent on shape of target regardless of angle.

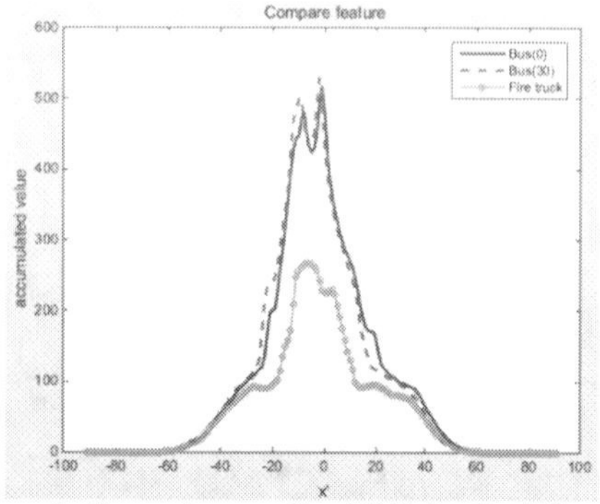

Figure 4. Comparison results of G-Radon for different targets with different rotations.

3 Experimental results

This section presents experimental results tested on synthesized IR image dataset. We show results of G-radon according to the number of targets. Finally, we show improved results by comparing G-radon with traditional methods such as Radon transform and Zernike moments.

3.1 Infrared target dataset

Infrared target dataset was prepared using a synthesis tool, OKTAL-SE [18, 19]. OKTAL-SE can synthesize a desired image using user's desired parameters such as season, weather, time, temperature, atmosphere environment, and thermal noise. Fig. 5 shows the processing flow of the scenario generator using OKTAL-SE. First, we insert 3D information such as building, terrain information and environmental information such as atmospheric radiation, temperature and special effects information such as cloud, fog, and image noise on the scenario generator. We create a Radiative Atmospheric file with the input parameter by using the SE-ATMOSPHERE. The resulting file is a scenario environmental information. It is used to create the scenario in the SE-SCENARIO. If you enter the desired targets and trajectories to build a test environment, it can provide an image of the simulated environment. Then, we create a synthetic image with a desire spectral band and sensor information using SE-RAY-IR. The resulting synthesized image can be found and added to the sensor noise in SE-SIGNAL-VIEWER.

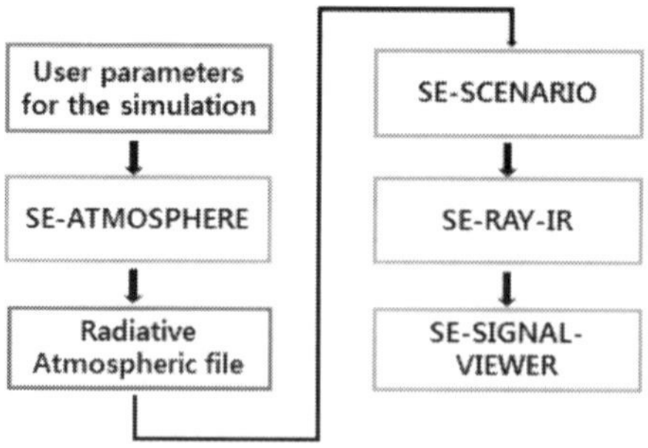

Figure 5. The synthetic infrared image generation flow using OKTAL-SE.

Target images are synthesized in the MWIR spectral band. Synthesis environment was created for the summer, fine, 10am. Other atmospheric environment, trajectory and special effect were not added. The pose of the IR camera was set at view point 0°, altitude 4km. Figure 6 shows the results of synthesized images. Dataset consists of 20 types of military targets and normal vehicles. The dimension of target data is 30 by 30 pixels. Each target is rotated by 72 different orientation in azimuth (0°, 5°, ..., 350°, 355°) from a fixed range. In the experiments, we separate dataset into train data (upright image) and test data (rotated images). Train data is learned and randomly pick out a target in test data.

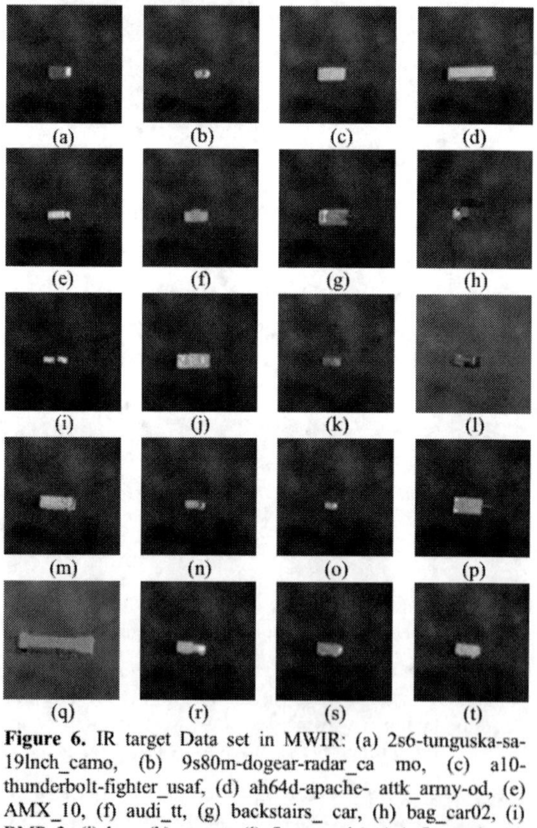

Figure 6. IR target Data set in MWIR: (a) 2s6-tunguska-sa-19lnch_camo, (b) 9s80m-dogear-radar_ca mo, (c) a10-thunderbolt-fighter_usaf, (d) ah64d-apache- attk_army-od, (e) AMX_10, (f) audi_tt, (g) backstairs_ car, (h) bag_car02, (i) BMP_3, (j) bus, (k) expert, (l) fire_ truck1, (m) firetruck, (n) ford_transit, (o) jeep, (p) leclerc, (q) naut_gangway, (r) su25, (s) T_72, (t) TMM.

3.2 Recognition performance

We compare the recognition rate of the traditional radon in order to confirm the performance of the proposed method for the different number of targets. We use a k-nearest neighbour classifier (k-NN, k=1) as a simple classifier. As shown in Fig. 7, the conventional Radon transform shows lower recognition rate than the proposed G-radon according to the number of test targets. The conventional Radon showed the recognition rate, 46%, for the 20 kinds of targets. The proposed method's recognition rate is also decreased depending on the number of targets. However, the proposed method's performance is better than the traditional method. The G-Radon showed the recognition rate, 68.2%, for the 20 kinds of targets. This result is 20% higher than the traditional method. The recognition rate curve of the G-radon decreases slowly than that of the Radon transform. The recognition result of each method on the number of target from 20 to 12 is shown in Table I.

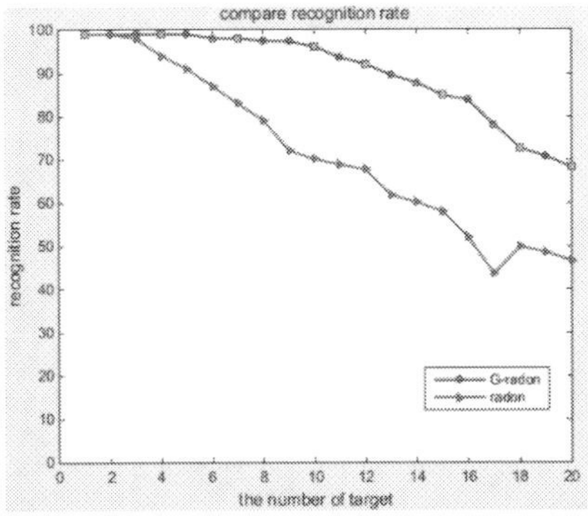

Figure 7. Recognition rate for G-radon and Radon transform.

Table 1. Comparisons of recognition rate

# of target	Proposed G-radon	Radon transform	Zernike transform
20	68.2	46.6	36.3
18	72.6	49.8	38
16	83.8	52	41.2
14	87.8	60	48.2
12	92.1	67.6	50.9

Fig. 8 show the recognition rate when the target is 10 as a confusion matrix for G-radon. The number of y-axis represents the true IDs of test targets and that of x-axis represents presented (recognized) IDs. The red color represents 100% of recognition rate and the blue color represents 0% of recognition rate. In figure 8, predicted values of the G-radon represents the red color. This result means that the predicted IDs are almost same true IDs. In Fig. 9 show the recognition rate for Radon transform. Only a few of the predicted values show a high recognition rate of 80% or more. Other predicted values represent recognition rate of 50% or less. In Fig. 10, the predicted values of Zernike transform are scattered. Most of the targets are not recognized.

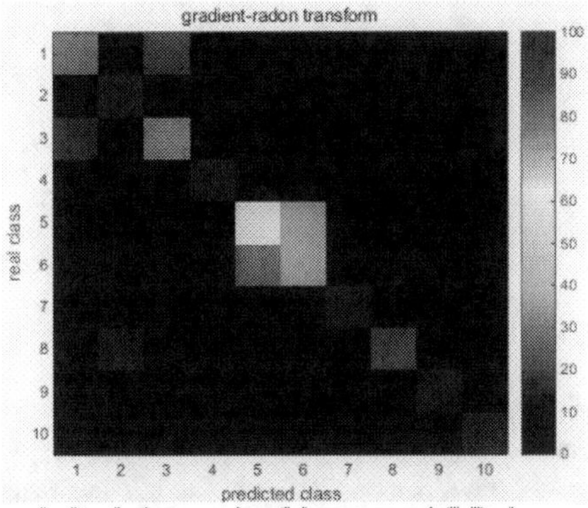

Figure 8. Confusion matrix of the proposed G-Radon.

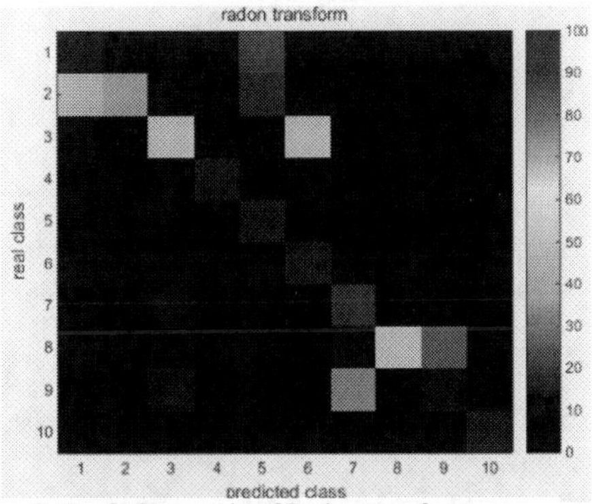

Figure 9. Confusion matrix of Radon transform

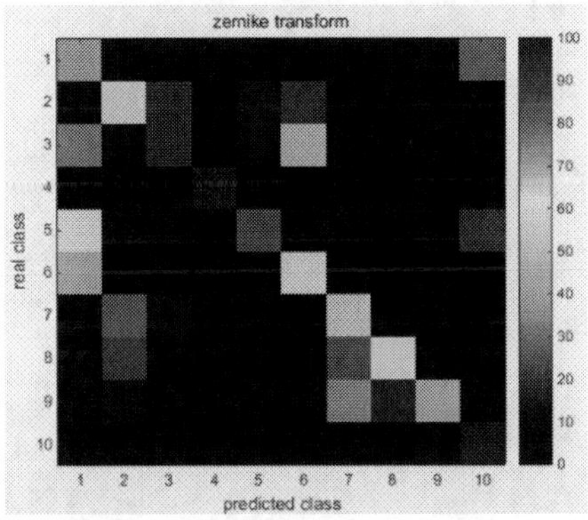

Figure 10. Confusion matrix of Zernike moments.

4 Conclusions and future works

This paper proposes a new rotation invariant algorithm, G-radon, to achieve effective recognition. G-radon extracts gradient magnitude value and then computes Radon transform. We compare the recognition rate of the traditional Radon for the different number of targets to confirm the performance of the proposed method. According to the recognition test on the synthesized infrared dataset, the proposed G-Radon shows upgrade rotation invariance compared with the conventional Radon transform and Zernike moments. In future, we will combine the advantages of HOG and rotation invariant descriptor and consider the development of robust algorithms for noise. The ultimate goal of the research is to achieve the improved ground target recognition and classification through the fusion of SAR/IR sensor information. This paper validate the target recognition for IR dataset only. In the future, we will verify the effectiveness SAR/IR sensor fusion for ATR problem.

Acknowledgement

This work was supported by the STRL (Sensor Target Recognition Laboratory) program of Defense Acquisition Program Administration and Agency for Defense Development. This research was supported by Basic Science Research Program through the National Research Foundation of Korea(NRF) funded by the Ministry of Science, ICT & Future Planning (NRF-2015R1A2A2A01002299).

References

1. F. Zhang, S. Liu, D. Wang, W. Guan, Image and Vision Computing, **27**, 4 (2009).
2. M. N. A. Khan, G. Fan, D. R. Heisterkamp, L. Yu, Computer Vision and Pattern Recognition Workshops (CVPRW), (2014).
3. P. Bharadwaj, L. Carin, IEEE Transactions on,Pattern Analysis and Machine Intelligence, **24**, 10 (2002)
4. Y. Lamdan, H. J. Wolfson, 2nd International Conference on Computer Vision (1988).
5. L. A. Chan, N. M. Nasrabadi,. International Journal on Artificial Intelligence Tools, **6**, 2 (1997).
6. Q. Zhao, J. C. Principe, IEEE Transactions on Aerospace and Electronic Systems, **37**, 2 (2001).
7. N. Dalal, B. Triggs, IEEE Computer Society Conference on Computer Vision and Pattern Recognition, (2005).
8. X. Wang, B. Xiao, J. F. Ma, X. L. Bi, Pattern Recognition, **40**, 12 (2007).
9. M. K. Hu, IRE Transactions on Information Theory, 8, 2 (1962).
10. J. Flusser, Pattern recognition, **33**, 9 (2000).
11. J. Flusser, Pattern Recognition, **35**, 12 (2002).
12. C. Kan, M. D. Srinath, Pattern recognition, **35**,1 (2002).
13. C. H. Teh, R. T. Chin, IEEE Transactions on Pattern Analysis and Machine Intelligence,**10**, 4 (1988).
14. R. Mukundan, S. H. Ong, P. A. Lee, IEEE Transactions on Image Processing, **10**, 9 (2001).
15. S. H. Weon, G. Y. Kim, H. I. Choi, Digital Contents Society, **12**, 3 (2011).
16. D. G. Lowe, International journal of computer vision, **60**, 2 (2004).
17. HØILUND, Carsten. The radon transform. Aalborg University, Vision, Graphics and Interactive Systems (VGIS), November, 12, (2007).
18. J. Latger, T. Cathala, N. Douchin, A. L. Goff, Defense and Security Symposium (pp. 654302-654302). International Society for Optics and Photonics, (2007).
19. J. J. Won, S. Kim, Synthetic SAR/IR Database Generation for sensor fusion-based A.T.R. (2014).

Action Recognition Based on Sub-action Motion History Image and Static History Image

Shichao Zhang ,Enqing Chen ,Lin Qi , Chengwu Liang

School of Information Engineering, Zhengzhou University, Zhengzhou 450000, China

Abstract. In this paper, we propose a robust and effective framework to largely improve the performance of human action recognition using depth maps. The key contribution is the proposition of the Sub-action Motion History Image (SMHI) and Static History Image (SHI) in a depth sequence. We evenly subdivide the normalized motion energy into a set of segments which corresponding frame indices are used to partition a video into different sub-actions segments. The Local Binary Patterns (LBP) descriptor is then computed from the SMHI and SHI for the representation of an action. We evaluate the proposed framework on MSR Action3D dataset. Experimental results indicate that the proposed approach outperforms the most of the art methods and demonstrate the effectiveness of the proposed approaches.

1 INTRODUCTIONS

Human action recognition has been an active field in computer vision, due to its extensive application in real-word, such as human computer interaction, medical health care, and video retrieval. In the past few decades, research has been mainly focused on recognizing actions from videos taken by ordinary RGB cameras. While significant efforts, recognizing actions accurately still remain a challenging task.

As the imaging techniques advance, the release of the Microsoft Kinect provides a new possibility to address these issues. Using Kinect, depth information can be captured simultaneously with RGB videos. Depth maps have several advantages with regard to RGB images in activity recognition. First, depth cameras provide an efficient and powerful human motion capturing technology which can accurately estimate the 3D skeleton joint positions from a single depth map [1]. Second, depth cameras are robust to the illumination changes, which bring great benefits to the activity recognition.

In this paper, we propose an effective and robust approach to recognize actions by extracting Local Binary Patterns (LBP) descriptors from Sub-action Motion History Images (SMHI) and Static History Images (SHI). The SMHI are generated by stacking sub-action motion energy of depth maps projected onto three orthogonal Cartesian planes. We propose an adaptive sub-action segment based on motion energy. The stacked motion energy of each sub-action represents particular appearances and shapes, which can be used to recognize corresponding actions. The SHI are obtained by accumulating static information simultaneously, which presents the static posture history by retaining the static part of body. It also contains the motion information which helps accurately identify the action.

The rest of the paper is organized as follows. Section II discusses the related work on depth-based action recognition. Section III, we describe the detailed procedures of the computing the feature descriptor. Section Ⅳ is the experiment and comparison. Finally, the conclusion of this paper is presented in Section Ⅴ.

2 RELATED WORK

With the introduction of the low-cost RGB-D cameras (Kinect), action recognition in depth videos has become a very active topic. In this field, different approaches have been proposed. In this section, we give a review of the research efforts for depth-based action recognition.

Li et al. [2] sample a bag of 3D points from the depth maps to describe a set of salient postures that correspond to the nodes in the action graph. And the action graph was used to model the dynamics of actions. What's more, the sampling scheme is view dependent and more accurate than using 2D silhouettes. Xia et al. [3] suggest a compact posture representation through a histogram of 3D joint locations （HOJ3D）. They aligned the spherical coordinates with the person ' s specific direction. Then, the Linear Discriminate Analysis (LDA) was used to extract the dominant features. The K-means clustering was performed to represent each posture as a visual word, and then a Bag of Words (BOW) model translates each action into a series of symbols. After that, a discrete HMM classifier used for action recognition. Yang et al. [4] employ a HOG [5] feature extraction after projecting the depth maps into three orthogonal planes and accumulating the depth maps throughout each posture into a motion image. A linear SVM classifier used to recognize actions.

As the spatial-temporal interest points can provide a compact representation of the image content by describing local parts of the scene, so it can be robust to clutter, occlusions, and intra-class variations. Recently, many methods are proposed for action recognition based on spatial-temporal interest points. Zhang et al. [6] proposed a 4D local spatial-temporal feature which combines both intensity and depth information. They first used separate filters along the 3D spatial dimensions and the temporal dimension to detect interest point. After that, they computed the intensity and depth gradients with a 4D hyper cuboids to obtain features for action sequence, and the Latent Dirichlet Allocation with Gibbs sampling was used to classify the action.

The Sparse Representations is the classic representation scheme and been applied in the field of action recognition. There have been several approaches for action recognition that use sparse representations. Zheng et al. [7] proposed a dictionary learning framework for cross-view action recognition with the assumption that sparse representations of videos from different views of the same action should be strictly equal. However, this assumption is too strong to flexibly model the relationship between different views.

3 PROPOSED METHOD

This section offers a detailed description for human action recognition from depth maps. The framework is demonstrated in Figure.1, there are two main segments in this part: sub-action motion history image (SMHI) and static history image (SHI) for action representation, and feature descriptor.

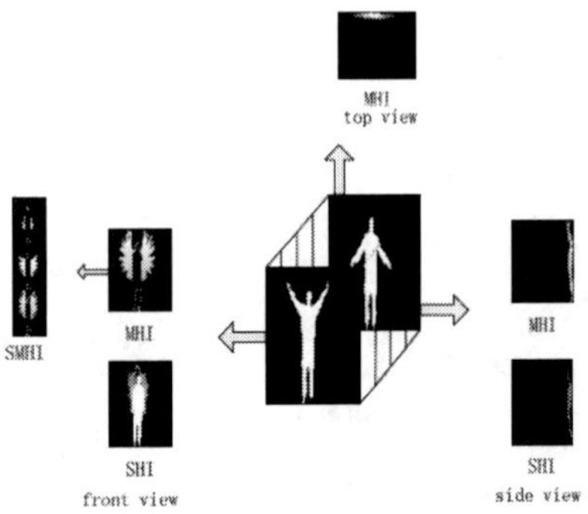

Figure 1. The framework of the proposed method.

3.1. Sub-action Motion History Images and Static History Images

Motion history image (MHI) and motion energy image (MEI) templates proposed by Bobick and Davis [8] describe where the motion happens and how the object moves, which presents the motion history from stacking the action sequence into a single gray scale image and preserving dominant motion information. However, the traditional MHI method has the limitation of scalability because only lateral motion of the action is analysed. Human activities are performed in 3D space, which means MHI performed in 2D space may miss some motion information of the action performed in the real world. In order to make full use of body shapes and motion information from depth maps, each depth frame is projected onto three orthogonal Cartesian planes, just like Liang [9]. So each depth image generates three 2D maps, which is front, side and top view, respectively. On each projected map, we obtain its motion history image by computing the absolute difference between two consecutive maps and calculating its sum. The motion energy is obtained by accumulating summations of non-zero elements of MHI.

Static history image (SHI) presents the static posture history by retaining the static part of body. As an action performs, the body has both parts of the movement and the static. When computing the differences, the stationary parts and the moving parts are preserved simultaneously. It also contains the motion information which helps accurately identify the action. Figure 2 shows the MHI and SHI generated from the front view of the action Two Hand Wave.

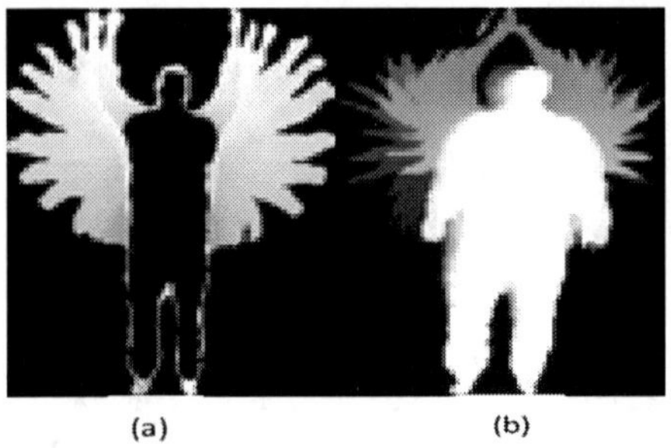

Figure 2. (a) MHI and (b) SHI from front side of one sample action

The motion update function $\varphi_m(x,y,t)$ and the static update function $\varphi_s(x,y,t)$ are defined to represent the regions of motion information and static posture with action performing [9]:

$$\varphi_M(x,y,t)=\begin{cases}1 & if\ D_t>\varepsilon_M\\ 0 & otherwise\end{cases} \tag{1}$$

$$\varphi_S(x,y,t)=\begin{cases}1 & if\ I_t-D_t>\varepsilon_S\\ 0 & otherwise\end{cases} \tag{2}$$

where x, y and t represent pixel position coordinates and time. $I_t\left(t\in(1,t)\right)$ is a depth sequence and $D_t\left(t\in(1,t)\right)$ is a absolute difference between two frames. ε_M is the motion threshold and ε_S is the static threshold, and we empirically set $\varepsilon_M=15$ and $\varepsilon_S=50$ in our experiments. T is the total number of frames in actions.

The motion history image (MHI) $H_M(x,y,t)$ and static history image (SHI) $H_S(x,y,t)$ can be generated by using motion update functio $\varphi_m(x,y,t)$ and $\varphi_s(x,y,t)$:

$$H_M(x,y,t)=\begin{cases} T & \varphi_m(x,y,t)=1 \\ H_M(x,y,t-1)-1 & otherwise \end{cases} \qquad (3)$$

$$H_S(x,y,t)=\begin{cases} T & \varphi_s(x,y,t)=1 \\ H_S(x,y,t-1)-1 & otherwise \end{cases} \qquad (4)$$

Additionally, to different people, they could have varied motion speed or frequency when they are required to perform the same activity. For the same action, different people to complete may cause intra-class variations increase; there is no help in classification. Just like the action Draw circle, the second image and the seventh image have the same posture (in fact, it has a little same action sequence). So, condensing the whole action sequence into a single image may lose some detailed information. It is not beneficial to recognize the actions. In order to handle this difficulty, we propose an adaptive sub-action segments based on the motion energy. To divide the whole action sequence into different sub-actions, the key information of each action fragment can be preserved, and the intra-class variations also can be reduced. For a depth sequence , we first obtain the projected maps $\left(I_1^i, I_2^i, I_3^i\right), i=(1,\cdots,T)$ according to front, side, and top views, i.e. I_1, I_2, I_3. After then, we compute the motion energy by accumulating summations of non-zero elements value of each MHI as:

$$E(p,t)=sum(H_M(x,y,t)) \qquad (5)$$

Where $E(p,t), p=(1,2,3)$ is the motion energy of the each MHI which obtained by three projected maps at t moment; $sum(\cdot)$ represents the summations of non-zero elements value of each MHI.

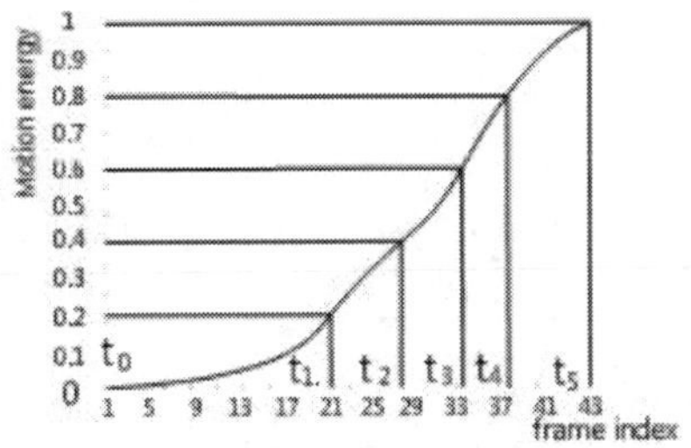

Figure 3. The sample of Sub-action segments.

Laptev et al. [10] use the temporal pyramid to take into account the temporal order of a video. Inspired by this, we propose adaptive sub-action segments approach as shown in Fig.3. We evenly subdivide the normalized motion energy into a set of segments. The different energy segments corresponding frame indices are used to partition a video into different sub-actions segments. The definition of the sub motion segmentation is shown as follows:

$$E(p,t)/E(p,T)=N/M \qquad (6)$$

$E(p,t), p=(1,2,3)$ is the motion energy of MHI at t moment and $E(p,T), p=(1,2,3)$ is the motion energy of the whole MHI; M represents to partition a video into M sub-actions segments; $N=(1,2,\cdots,M-1)$.

In this paper, we divide the front view into 3 sub-action segments to form respective motion history image. Which means $E(p,t_1)/E(p,T)=1/3$, $E(p,t_2)/E(p,T)=2/3$, and $E(p,t_3)/E(p,T)=3/3$. And we use the five different t moment to partition a video into 3 sub-actions, which is t_0t_1, t_0t_2, t_0t_3, respectively; t_0 represent the opening time and T is the total number of frames.

Considering the motion information is mainly concentrated on the front view. For the front view, we use a 3 sub-action MHI and SHI. For the side view, we use MHI and SHI, respectively. For the top view, we only use MHI. Thus, each action depth sequence can be modelled as a seven templates.

3.2 Feature Descriptor and Recognition

In order to make our feature has a better discriminative power; we further extract the LBP feature [11] for each template. The LBP operator labels the pixels of an image with decimal numbers that encode the local structure around each pixel. A subset of these patterns named uniform patterns is able to describe image texture. The histogram of LBP labels calculated over a region can be used as a texture descriptor. Since seven templates are used, the LBP operator is applied to overlapped blocks of the templates; this can make full use of texture information. The LBP histograms of the blocks for each template are concatenated to form the feature vector. The dimensionality of total feature is D=11151. It is relatively high and not conducive to classification. We employ the widely used principal component analysis (PCA) due to its effectiveness and simplicity.

After feature extraction, we use support vector machine (SVM) to classify the actions, due to its strong discriminative power.

Table 1. Evaluation of method on the on three subsets.

Method (%)	Test One				Test Two				Cross Subject Test			
	AS1	AS2	AS3	Overall	AS1	AS2	AS3	Overall	AS1	AS2	AS3	Overall
Bag of 3D Points [2]	89.5	89.0	96.3	91.6	93.4	92.9	96.3	94.2	72.9	71.9	79.2	74.7
HOJ3D [3]	98.5	96.7	93.5	96.2	98.6	97.9	94.9	97.2	87.9	85.5	63.5	79.0
3DMTM-PHOG [9]	97.3	97.4	98.7	97.8	100.0	100.0	100.0	100.0	93.4	82.3	96.4	90.7
EigenJoints [12]	94.7	95.6	97.3	97.3	97.3	98.7	97.3	97.8	74.5	76.1	96.4	82.3
Our method	97.3	96.7	98.7	97.6	100.0	100.0	100.0	100.0	95.3	90.3	95.5	93.7

4. EXPERIMENTS RESULTS

In this section, we evaluated the proposed method onthe MSR Action3D dataset [2]. In all experiments, we select the SMHI-HOG and SHI-HOG with 90% principal components for PCA. The empirical results show that the proposed method outperforms the most of the art methods. Especially in similarity actions, we can achieve a good recognition rate.

MSR-Action3D dataset [2] is a set of depth videos captured by a Kinect device. The dataset contains 567 depth sequences and 20 action types. Commonly, the dataset is divided into three actions subsets: AS1, AS2 and AS3. For each subset, there are three different tests model: Test One (T1), Test Two (T2), and Cross Subject Test (CST).

The performance of SMHI-HOG and SHI-HOG in terms of accuracies on all tests is shown in Table I. We also compare our proposed method with other methods on the Cross Subject Test in Table II. The proposed method achieves an accuracy of 93.7% which significantly outperforms the existing methods. The approach [2] uses a bag of 3D points to characterize a set of salient postures based on the original depth maps. The Histograms of 3D Joints [3] and EigenJoints [12] mainly depends on the accurate estimation of the joints positions, so it cannot achieve a good recognition rates. The DSTIP [13] proposed a spatiotemporal interest point detector and a depth cuboids similarity descriptor to recognize actions. It can effectively eliminate the noise in the depth maps but it is very complex, and the accuracy of which is 89.3%.

The proposed method outperforms 3DMTM-PHOG [9] by 3%, though both methods are based upon motion history images and static history images. Especially in Cross Subject Test AS2, we can achieve a recognition rate of 90.3%, which can outperform the method by 8%. This may use the sub-action motion history images which can capture more key information. Using the motion history images directly, this key information may be covered by other coarse information. Thus, the proposed method is quite robust.

Table 2. Evaluation of method on the Cross Subject Test.

Method	Accuracy (%)
Bag of 3D Points [2]	74.70
HOJ3D [3]	79.00
EigenJoints Depth Motion Maps [4]	88.73
3DMTM-PHOG [9]	90.70
EigenJoints [12]	82.30
DSTIP [13]	89.30
Random Occupancy Pattern [14]	86.50
STOP [15]	84.80
Action Let Ensemble [16]	88.20
HON4D [17]	88.89
Pose Set [18]	90.00
Our method	93. 70

5 CONCLUSIONS

In this paper puts forward an effective and robust method to perform human action recognition on depth sequences. We propose the adaptive sub-action segments which can effectively preserve detailed action information. The SMHI and SHI can represent human actions in a discriminative way. It is able to capture the key motion information and static posture information. The Local Binary Patterns descriptor is used to encode the feature. To divide the whole action sequence into different sub-actions, the intra-class variations can be reduced; it is conducive to classification. The experimental results on MSR Action3D dataset demonstrate the effectiveness and robustness. Especially for the similarity actions, our method achieves a better recognition rate. And the recognition rate is better than most of methods. It is proves that our sub-action segmentation method is effective.

ACKNOWLEDGMENT

This work was supported partly by the National Natural Science Foundation of China under Grants No. 61331021, No.61210005 and No. 61201251.

References

1. J. Shotton, A. Fitzgibbon, M. Cook, T. Sharp, M. Finocchio, R. Moore, A. Kipman, A. Blake, 2011. In: Proc. of the 2011 IEEE Computer Society Conference on Computer Vision and Pattern Recognition, 1297–1304.
2. W.Q. Li, Z.Y. Zhang, Z.C. Liu. IEEE Conf. on Computer Vision and Pattern Recognition Workshops, 2010, pp. 9–14.
3. L. Xia, C. Chen, J. Aggarwal. IEEE Conf. on Computer Vision and Pattern Recognition Workshops, 2012, pp. 20–27.
4. X.D. Yang, C.Y. Zhang, Y.L. Tian. Proc. of the 20th ACM Int'l Conf. on Multimedia,2012, pp. 1057–1060, (MM '12).
5. N. Dalaland, and B. Triggs. Histograms of oriented gradients for human detection. In CVPR, 2005.
6. Zhang, L.E. Parker. 4-dimensional local spatial-temporal features for human activity recognition. IEEERSJ Int'l Conf. on Intelligent Robots and Systems, 2011, pp.2044–2049.
7. Zheng, Z. Jiang, J. Phillips, and R. Chellappa. Cross-view action recognition via a transferable dictionary pair. In BMVC, 2012.
8. A. F. Bobick and J. W. Davis. Pattern Analysis and Machine Intelligence, IEEE Transactions on, 23(3):257–267, 2001.
9. B. Liang and L.H. Zhang, 3D Motion Trail Model based Pyramid Histograms of Oriented Gradient for Action Recognition. 22nd International Conference on Pattern Recognition 2014.
10. J. Laptev, M. Marszalek, C. Schmid, and B. Rozenfeld. Learning Realistic Human Actions from Movies. In CVPR, 2008.
11. T. Ojala, M. Pietikaimen, and T. Maenpaa. Multi-resolution gray-scale and rotation invariant texture classifications on Pattern Analysis and Machine Intelligence, 24(7):971-987,(2002).1,3,5.

12. X.D. Yang, Y.L. Tian. Effective 3d action recognition using EigenJoints. J. Vis. Commun. Image Represent. 25 (2014) 2–11.
13. L. Xia, J. Aggarwal. Spatial-temporal depth cuboids similarity feature for activity recognition using depth camera. IEEE Conf. On Computer Vision and Pattern Recognition, 2013, pp, 2834-2841.
14. J. Wang, Z. Liu, J. Chorowski, Z. Chen, and Y. Wu. Robust 3D Action Recognition with Random Occupancy Patterns. In ECCV, 2012.
15. A. Vieira, E. Nascimento, G. Oliveira, Z. Liu, and M. Campos. STOP: Space-Time Occupancy Patterns for 3D Action Recognition from Depth Map Sequences. In CIARP, 2012.
16. J. Wang, Z.C. Liu, Y. Wu, J.S. Yuan. Mining action-let ensemble for action recognition with depth cameras. IEEE Conf. on Computer Vision and, Pattern Recognition, 2012, pp. 1290–1297.
17. O. Oreifej and Z. Liu. HON4D: Histogram of Oriented 4DNormals for Activity Recognition from Depth Sequences. In CVPR, 2013.
18. C. Wang, Y. Wang and A. Yuille. An Approach to Pose based Action Recognition. In CVPR, 2013.

k-d Tree-Segmented Block Truncation Coding for Image Compression

Ryan Rey M. Daga[1,2] and Proceso L. Fernandez[1]

[1]Ateneo de Manila University, Loyola Heights, Katipunan Ave., Quezon City, Philippines
[2]University of the Philippines Visayas Tacloban College, Tacloban City, Leyte, Philippines

Abstract. Block truncation coding (BTC) is a class of image compression algorithms whose main technique is the partitioning of an image into pixel blocks that are then each encoded using a representative set of pixel values. It is commonly used because of its simplicity and low computational complexity. The Quadtree-segmented BTC (QTS-BTC), which utilizes a dynamic hierarchical segmentation technique, is among the most efficient in the BTC class. In this study, we propose a new BTC variant that introduces two ideas: (1) the use of a *k*-d tree for segmentation and (2) the use of a Mean Squared Error (MSE) threshold for dynamically determining the granularity of the blocks. We refer to this new BTC variant as the *k*-d Tree Segmented BTC (KTS-BTC), and we test this against some of the existing BTC variants by running the algorithms on a standard image compression dataset. The results show that the proposed variant yields low bit rates of the compressed images, even outperforming the state-of-the-art QTS-BTC, without a significant reduction in image quality as measured using the Peak Signal-to-Noise Ratio (PSNR). The utilization of *k*-d tree for image segmentation is further shown to have more impact than that of employing the MSE thresholding scheme as a block activity classifier.

1 Introduction

Nowadays, digital images are being used in many different fields. This has resulted to a continuous accumulation of stored and transmitted digital images, as well as an increase in the quality, and hence file size, of a non-trivial fraction of existing digital images. Compression techniques are thus becoming increasingly important because these help reduce the storage requirements and bandwidth costs.

There are several classes of image encoding schemes that have been proposed. These include the vector quantization [1], fractal image compression [2], wavelet image compression [3], and block truncation coding (BTC) [4]. In this study, we focus on the last one.

Block truncation coding was originally proposed by Mitchel et al., in 1979 for grayscale image compression [4]. This technique divides an image into non-overlapping blocks, with each block encoded using a representative set of pixel values. BTC has paved the way for other pixel-based moment-preserving compression techniques. Different methods have been introduced to suit both grayscale and colored images. Absolute moment block truncation coding (AMBTC) [5], generalized moment preserving quantization [6], multiple block partitioning [7] and even the integration of a heuristic such as harmony search [8] have been introduced.

BTC (and its variants) is a popular choice for compression because of its low computational cost, making it suitable for real time multimedia applications. It is generally able to quickly compress an image without much loss of image quality. However, BTC has a higher bit rate for image coding compared to other classes of compression algorithms. To address this issue, some proposed image coding schemes have employed representative patterns to encode bitmaps [9] [10].

Recently, the quadtree segmentation BTC [11] has been proposed to cut down the bit rate requirements of an image while maintaining the image quality. This segmentation technique exploits the spatial similarity among neighboring pixels in the image, and dynamically determines the size of the image segments (i.e., non-overlapping blocks) based on a threshold of difference between two means.

In this study, we propose a new BTC variant that extends the ideas from the quadtree segmentation BTC. In particular, we propose the use of a *k*-d tree (instead of a quadtree) for a more flexible segmentation, and the use of Mean Square Error (MSE), instead of difference of two means, for thresholding. These concepts are discussed in greater detail in later sections.

2 Preliminaries

In this section, we cover the important algorithms that can help in better understanding the quadtree segmentation BTC, from which several concepts in our proposed new algorithm are based.

2.1 Block Truncation Coding

In Block Truncation Coding (BTC), an image is first divided into non-overlapping $m \times m$ blocks. Then for each block, the mean $\bar{x}$ of the pixel values is computed in the standard way,

$$\bar{x} = \frac{1}{n}\sum_{i=1}^{n} x_i \quad (1)$$

where $n = m^2$.

A corresponding $m \times m$ bit plane1 is generated by encoding each pixel in the block by either 1 or 0 depending on its value relative to $\bar{x}$. Specifically, if the pixel value is greater than or equal to $\bar{x}$, it is coded as 1; otherwise, it is encoded as 0.

The bit values 1 and 0 are mapped to the statistical moments x_H and x_L, respectively, which are computed as follows:

$$x_H = \bar{x} + \sigma\sqrt{\frac{p}{q}} \quad (2)$$

$$x_L = \bar{x} - \sigma\sqrt{\frac{p}{q}} \quad (3)$$

Here, p is the number of 0's in the bit plane, q is the number of 1's, and

$$\sigma = \sqrt{\frac{1}{n}\sum_{i=1}^{n}(x_i - \bar{x})^2} \quad (4)$$

The bit rate of the resulting compressed greyscale image depends on the size of all blocks. For the typical 4×4 block size, each of the 16 pixels is encoded using 1 bit on the bit plane, and an overhead of two 8 bits are needed for the x_H and x_L representative values. Thus, 32 bits are used to encode a 16-pixel block, resulting in a bit rate of 2 bits per pixel (bpp).

2.2 Absolute Moment Block Truncation Coding

The Absolute Moment Block Truncation Coding (AMBTC) is a simple and fast variant of BTC that has been proposed for color images [12][5]. A given color image in RGB format is first decomposed into three greyscale images corresponding to the red, green and blue channels.

The encoding for each image channel then follows the BTC technique, except for the computation of the two statistical moments. The formula used in AMBTC are given in Equations 5 and 6.

$$\bar{x}_H = \frac{1}{q}\sum_{x_i \geq \bar{x}} x_i \quad (5)$$

$$\bar{x}_L = \frac{1}{p}\sum_{x_i < \bar{x}} x_i \quad (6)$$

where p and q are still the number of 0's and 1's, respectively, in the generated bit plane.

The resulting bit rate for images compressed using AMBTC is similar to that of the BTC. Using 4×4 blocks, the AMBTC bit requirements is 2 bpp for greyscale images and 6bpp for three-channel color images. Moreover, the AMBTC produces better image quality than BTC, where quality is measured in terms of Peak Signal-to-Noise Ratio (PSNR):

$$PSNR = 10\log_{10}\left(\frac{255^2}{MSE}\right) \quad (7)$$

In the above formula, MSE refers to the Mean Squared Error, which compares each pixel value $x_{i,j}$ at position (i, j) in the original image with the corresponding pixel value $y_{i,j}$ in the reconstructed image. In this study, since all input data images have dimension 512×512, the MSE is computed using the formula

$$MSE = \frac{\sum_{i=0}^{511}\sum_{j=0}^{511}(x_{i,j} - y_{i,j})^2}{(512)(512)} \quad (8)$$

2.3 Quadtree Segmented Block Truncation Coding

The quadtree segmented BTC is a hierarchical segmentation technique that partitions an image into variable-sized square blocks. During the quadtree segmentation process, a series of binary decisions with respect to different threshold values is made. A block type classifier is employed to determine the block activity, and it is usually based on a statistical information [11]. If a block is inactive (e.g., low variance) the segmentation of the given block is terminated, otherwise the block is further divided into four sub-blocks.

Chen et al. recently proposed the quadtree-segmented AMBTC [11]. In this coding scheme, a maximum of three-level quadtree structure, an example of which is shown in Figure 1, is imposed on the segmentation. The procedure of quadtree-segmented AMBTC is as follows:

Step 1: Decompose the color image into three grayscale images using the red, blue and green channels of the image.

Step 2: Partition each grayscale image into image blocks of size 16 ×16.

Step 3: Select an unprocessed 16 ×16 image block, and calculate two quantization levels $\bar{x}_H$ and $\bar{x}_L$ using the formula in AMBTC.

Step 4: If $|\bar{x}_H - \bar{x}_L| \leq THRESHOLD$, then encode this 16 ×16 block using its block mean and proceed to Step 8. Otherwise, subdivide this block into four 8×8 blocks.

Step 5: While there is still an unprocessed 8×8 block, select one such block, and calculate the mean and calculate two quantization levels $\bar{x}_H$ and $\bar{x}_L$. Otherwise, proceed to Step 8.

Step 6: If $|\bar{x}_H - \bar{x}_L| \leq THRESHOLD$, then encode this 8 ×8 block using its block mean and proceed to Step 5. Otherwise, subdivide this block into four 4×4 blocks.

Step 7: For each 4×4 image block, calculate the mean and calculate two quantization levels $\bar{x}_H$ and $\bar{x}_L$. If $|\bar{x}_H - \bar{x}_L| \leq THRESHOLD$, then encode this 8 ×8 block using its block mean. Otherwise, encode this block using a 4×4 bit plane. Following the technique in AMBTC. Return to Step 5 after processing all 4×4 blocks.

Step 8: If there is still any 16×16 blocks to be processed, return to Step 3.

We refer to this BTC variants as QTS-BTC in this paper. An example of a block segmentation in QTS-BTC is shown in Fig 1. Here, a 16×16 block is segmented into four 8×8 blocks, two of which are each further subdivided into four 4×4 blocks.

The bit rate in QTS-BTC is typically better than in AMBTC (or BTC) since it is possible to represent an entire block with a single 8-bit number instead of an $m \times m$ bit plane plus two 8-bit quantization values. However, the PSNR values are typically a little bit better in AMBT (BTC).

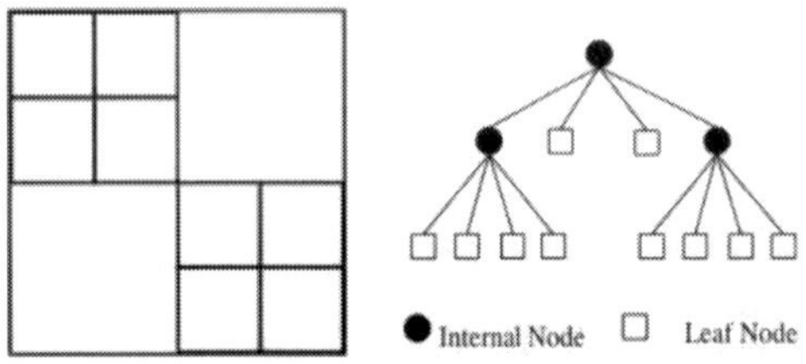

(b) A segmentation of a 16×16 blocks (a) Corresponding three-level quadtree

Figure 1. Example of a three-level quadtree segmentation

3 Proposed Compression Technique

In this study, we improve the QTS-BTC performance by applying two modifications:

1. The use of a *k*-d tree for the segmentation of a block
2. The use of the inequality MSE ≤ THRESHOLD t decide whether or not to subdivide a block further.

We refer to this new BTC variant as *k*-d tree-segmented BTC, or KTS-BTC. Figure 2 shows the overall process of the proposed image coding scheme.

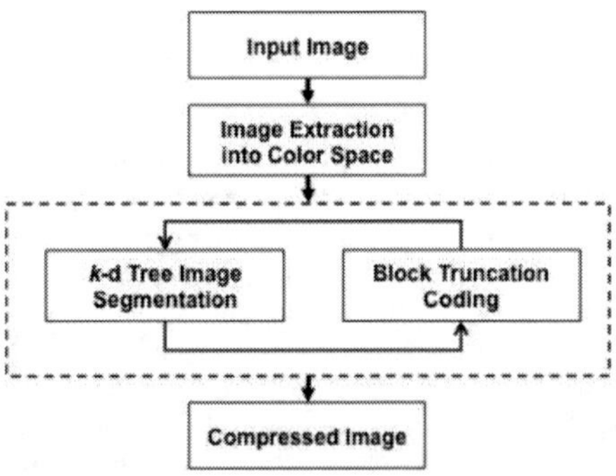

Figure 2. General flow of the proposed compression technique

An input color image is initially decomposed into three greyscale images corresponding to the three color channels, and then each greyscale image is partitioned into non-overlapping 16×16 blocks. For each block, the block mean is computed and the block activity is determined by calculating the MSE of replacing all pixel values by the mean. If a block is inactive (i.e., MSE ≤ THRESHOLD), then the block is encoded using its block mean. Otherwise, it is partitioned into two equally-sized sub-blocks.

The partitioning of a block follows the *k*-d tree partitioning of a 2-dimensional space. That is, the partitioning is alternately done on the x and y axes: a 16×16 block may be divided vertically (producing rectangular 8×16 blocks), then horizontally (producing 8×8 blocks, then vertically), then vertically (rectangular 4×8 blocks), and then finally horizontally (square 4×4 blocks).

The smallest blocks are 4×4 blocks. Each of theses 4×4 blocks is encoded either using the block mean if the block is inactive, or using a 4×4 bit plane following the AMBTC technique if the block is active. This process terminates when all blocks have been encoded.

Figure 3 shows an example of a *k*-d tree segmentation of a block. It should be noted that any block segmentation using the quadtree can be achieved using *k*-d tree segmentation (albeit using more levels in the tree, and thus more bits in the encoding), but the reverse is not true. Since there is possibility to encode an $m \times 2m$ vertical blocks with a single pixel value, which is not allowed in the QTS-BTC, then there is opportunity to have a lower overall bit rate. This, however, comes at the risk of reducing the PSNR. This risk is managed by using an MSE threshold, which may be set to some low value based on the desired PSNR.

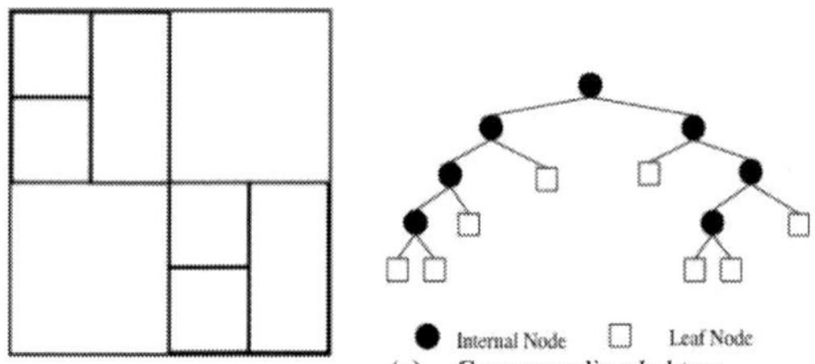

(a) Corresponding *k*-d tree

(b) A segmentation of 16×16 block

Figure 3. Example of a *k*-d tree segmentation of a 16×16

To encode the resulting variable-sized blocks properly, we use a bit flag 1 to indicate that a block is to be divided further, and 0 otherwise. Table 1 shows the prefixes used for the different block types, as well as the effective bit rates (in bits per pixel) for both greyscale and color images.

The number of bits used in the computation of the bit rates may be determined by careful calculations. For example, an inative 8×8 block uses the prefix 110. The first bit (1) indicates that the main block (16×16) is divided further. This 1-bit overhead cost is, however, shared among four 8×8 blocks. The second bit (1) implies that the 8×16 block is subdivided further, and this 1-bit overhead cost is shared by two. Finally, the last prefix bit (0) indicates that the 8×8 block is inactive, and this 1-bit overhead is shouldered entirely by the 8×8 block. Together with the 8-bit mean, the number of bits required to represent an inactive 8×8 block is this $\frac{1}{4}+\frac{1}{2}+1+8=$

9.75 bits. The values for the other block types can be similarly analyzed,

Table 1. Effective bit rates (in bpp) of the different block types on 1-channel and on 3-channel images

Block Type	Prefix	Num of Bits	Eff. Bit rate	
			1-ch	3-ch
16×16	0	9	0.035	0.105
8×16	10	9.5	0.074	0.223
8×8	110	9.75	0.152	0.457
4×8	1110	9.875	0.309	0.926
4×4, inactive	11110	9.9375	0.621	1.863
4×4, active	11111	33.937	2.123	6.369

4 Results and Discussion

To test the performance of the proposed KTS-BTC, we run it on a set of standard test images and compared the bit rates and PSNR values of the generated images against those produced by the AMBTC and QTS-BTC. These images, shown in Figure 4, were used in a previous study by Che et al. [11] and were taken from [13].

(c) Airplane (b) House (a) Lena

(f) Peppers (e) Splash (d) Tiffany

Figure 4. Color test images (512×512)

4.1 Compressed Image Results

Figure 5 shows some compressed images using AMBTC, QTS-BTC and the proposed KTS-BTC. There is no noticeable difference among the images, and this is validated by the small differences in the computed PSNR values.

Table 2 details the bit rates of the compressed images produced using these 3 techniques. The threshold values for the KTS-BTC and QTS-BTC used here are both 5. The proposed KTS-BTC yielded the lowest bit rate requirement. It was able to reduce the bit rates of the compressed images produced by AMBTC and QTS-BTC by an average of 22.62% and 14.83%, respectively.

As may be expected, the lower bit rate in KTS-BTC is achieved at the cost of slightly inferior PSNR (see Table 3). However, the difference is quite minimal, e.g., the average PSNR value from the AMBTC was reduced by about 0.53% only.

(a) AMBTC: 32.413 (b) Quadtree: 32.279 (c) *k*-d Tree: 32.253

Figure 5. Example of compressed images, using different compression techniques, with their corresponding PSNR values

Table 2. Bit rate requirements of color images using different image coding schemes

Image	AMBTC	Quadtree	*k*-d Tree
Airplane	6	4.175	3.71
House	6	5.103	4.844
Lena	6	5.989	5.293
Peppers	6	6.02	5.696
Splash	6	5.555	3.305
Tiffany	6	5.865	5.009
Average	**6**	**5.451**	**4.643**

Table 3. PSNR (dB) of color images using different image coding schemes

Image	AMBTC	Quadtree	*k*-d Tree
Airplane	32.413	32.279	32.253
House	30.498	30.480	30.461
Lena	33.109	33.112	32.995
Peppers	32.701	32.709	32.654
Splash	36.158	36.094	35.625
Tiffany	35.308	35.310	35.145
Average	**33.365**	**33.331**	**33.189**

These results indicate that the proposed KTS-BTC compression technique is generally better than QTS-BTC and AMBTC. We investigate further the effect of the two introduced ideas in KTS-BTC.

4.2 Effect of *k*-d Tree Segmentation

To determine the impact of the use of the *k*-d tree for segmentation, we implemented the KTS-BTC with a block activity classifier that uses $|\bar{x}_H - \bar{x}_L| \leq THRESHOLD$ to determine when to perform further segmentation.

Table 4 shows that across varying threshold values, the average bit rate is better in KTS-BTC, while the average PSNR value is slightly better in QTS-BTC. Computing the ratio of PSNR to bit rate reveals the superiority of the KTSBTC. Figure 6 shows a graph of this ratio across the possible threshold values.

Table 4. PSNR and BPP at different threshold values

Threshold	QTS-BTC		KTS-BTC	
	PSNR	BPP	PSNR	BPP
5	33.208	4.745	33.189	4.643
10	32.312	2.772	32.228	2.664
15	31.335	1.975	31.191	1.879
20	30.588	1.551	30.217	1.463
25	29.526	1.28	29.307	1.199

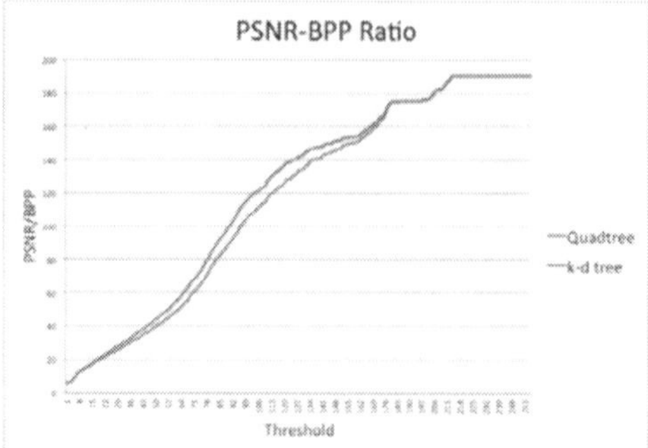

Figure 6. PSNR to BPP ratios at different threshold values using quadtree and *k*-d tree segmentation schemes

4.3 Effect of MSE Thresholding

To determine the impact of using the MSE thresholding, we implemented an MSE-thresholded QTS-BTC. To compare the results of the two QTS-BTC variants properly, it is inappropriate to look at the PSNR to BPP ratio across various threshold values, since the thresholding techniques are different. Instead, we plot the PSNR vs BPP curves. Figure 7 shows that the MSE thresholding is only minimally better.

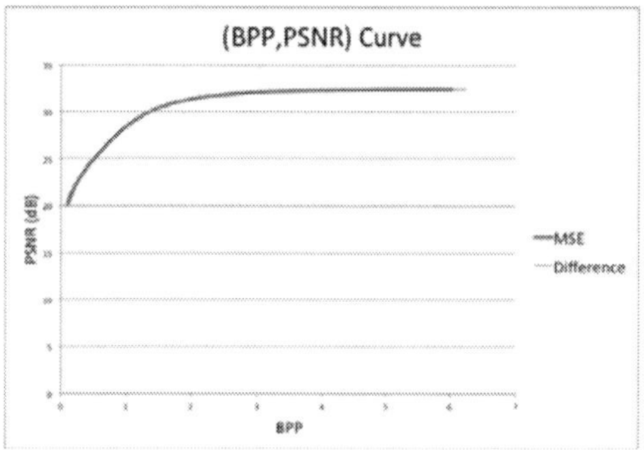

Figure 7: PSNR vs Bit Rate curves of QTS-BTC using the two different thresholding schemes

5 Conclusion

In this paper, we proposed a new BTC compression technique which we refer to as the *k*-d Tree Segmented BTC (KTS-BTC). This variant introduces two ideas: (1) the use of a *k*-d tree for segmentation and (2) the use of MSE for block activity thresholding.

The proposed algorithm was compared against the AMBTC and the state-of-the-art QTS-BTC. The superiority of KTS-BTC was established through actual implementation on a standard dataset. The results showed a significant improvement in bit rates, at the cost of a minimal reduction in PSNR values. A further investigation of the effects of the two introduced ideas showed that the improvement is mainly due to the use of *k*-d trees for segmentation.

Future studies can explore dynamically selecting a better default *k*-d tree-based initial partitioning (horizontal versus vertical), the use of bigger blocks, and the use of other block activity thresholding schemes.

References

1. C.C. Chang, Y.C. Hu, Consumer Electronics, IEEE Transactions on **44**, 1201 (1998)
2. Y. Fisher, Fractals **2**, 347 (1994)
3. B.E. Usevitch, Signal Processing Magazine, IEEE **18**, 22 (2001)
4. E.J. Delp, O.R. Mitchell, Communications, IEEE Transactions on **27**, 1335 (1979)
5. M.D. Lema, O.R. Mitchell, Communications, IEEE Transactions on **32**, 1148 (1984)
6. D. Halverson, N. Griswold, G. Wise, Acoustics, Speech and Signal Processing, IEEE Transactions on **32**, 664 (1984)
7. R.R.M. Daga, P. Fernandez, *Multi-partition Block Truncation Coding for Image Compression*, in *NCITE 2015 Proceedings of the National Conference of Information Technology Education* (PSITE, 2015), pp. 35 – 40
8. R.R.M. Daga, J.P.T. Yusiong, IJCSI International Journal of Computer Science Issues **9** (2012)
9. P. Nasiopoulos, R.K. Ward, D.J. Morse, Communications, IEEE Transactions on **39**, 1245 (1991)
10. B.C. Dhara, B. Chanda, Pattern Recognition **37**, 2131 (2004)
11. W.L. Chen, Y.C. Hu, K.Y. Liu, C.C. Lo, C.H. Wen, International Journal of Signal Processing, Image Processing and Pattern Recognition **7**, 65 (2014)
12. S. Vimala, M. Sathya, K.K. Devi, International Journal of Computer Applications **51**, 48 (2012)
13. *University of southern california - signal and image processing institue image database*, http://sipi.usc.edu/database/, Accessed: 2015-10-20

A surveillance video compression algorithm based on regional dictionary

Suyu Wang[1, 2], Yujie Chen[1, 2] and Yuheng Bai[1, 2]

[1]Beijing Engineering Research Center for IoT Software and Systems, 100124Beijing, China
[2]School of Software Engineering, Beijing University of Technology, 100124 Beijing, China

Abstract. Surveillance video system as an effective means of public safety widely appeared in people's life in recent years. Traditional video coding technology, the encoding process is complex, computation, and conflicts with hardware limitations of the monitoring system. This paper presents a kind method used to design and train redundant dictionary based on image region. Each frame is divided into same-sized area with equal blocks, then the image areas of the same position are put together as the training sample library of regional redundant dictionary. In this way, redundant dictionary compression results in the occupied volume can be reduce. On top of this dictionary, surveillance video compression coding algorithm based on the regional dictionary is designed and implemented. Experimental results show that the proposed algorithm can effectively improve the compression ratio of the algorithm.

With the need for Social and public security and big data analysis, video surveillance system is increasingly applied to people's life [1]. Surveillance systems made of various cameras are lying over all positions of the society and provide services for people's safe life. According to IDC (Data Corporation International) research report estimates, the total amount of global data will reach 40ZB in 2020, among which the data of surveillance video is 5.8ZB. And China will account for 21%, which means that in 2020, China will have 1.2ZB (One billion and two hundred million TB) surveillance video data needed to be stored, transmitted and analyzed. Therefore, facing with massive surveillance video, considering the transmission and storage costs, we need to carry out new research and achieve a major breakthrough in surveillance video encoding and analysis technology, it is very necessary to research higher compression efficiency of surveillance video encoding technologies. This paper proposes a method for studying of balancing the redundant dictionary size and the overall algorithm of encoding effect. The way to reduce the volume of redundant dictionary can be found out by analysing the compression result. The redundancy of each image block in the way of training the dictionary itself can be reduced by improving the original training method. A redundant dictionary trained by a region which is as a simple formed by multiple unit image blocks is applied to all of the image block in the area. This method use of redundant information can reduce the number of redundant dictionary, improve the training efficiency further, and do not cause significant losses for the compression algorithm of encoding efficiency. This paper also proposes a algorithm that combined with surveillance video compression and coding algorithm based on key frame to build the public background of surveillance and use the redundant dictionary to sparse decomposition the background image block or the image block with the approximate background, then quantize and compress correspondingly, while the foreground image block is reserved separately. Decoding the foreground background combined decoding, so as to improve the quality of encoding video in the premise of a small amount of loss compression ratio.

1 Algorithms' outline

The surveillance video compression coding algorithm based on the regional dictionary which is similar to that based on sub block dictionary, and it is divided into two stages: the training of the regional dictionary and the video compression and coding. The compression results of the surveillance video compression coding algorithm are generally made up of encoding's results of surveillance video frames and redundant dictionaries. The training process of the regional dictionary is the same as that of the sub block mostly, but there are also differences in details. Firstly, each frame of the video captured from the whole surveillance video was divided into a number of image regions with the same size in the same way, each of the image regions has the same shape and the same number of image blocks. After that, the same region in the picture are collected together, as the training sample library of the region's redundant dictionary. The image block took out from the training sample library randomly is trained redundant dictionary using K-SVD algorithm to get redundant dictionary of the corresponding area. Then, uniform quantization, subtraction Background, JPEG-LS encoding, after those get the encoding results.

2 Sparse representation theory based on redundant dictionary

The sparse representation theory of signal points out that [2], when the signal is projected to a certain transformation base, the obtained transform coefficients are mostly zero or very small absolute value, that is, the transform coefficient vector is sparse or approximate sparse, then we can transform the transformation vector as a simple expression of the original signal in the transform domain. To obtain a sparse representation of the signal, it is usually necessary to select the appropriate transformation basis for the characteristics of the signal. At this point, the transformation base is also called sparse group, the coefficient of sparse decomposition and sparse matrix can be combined to reconstruct the original signal [3].

In the model of signals sparse representation, a linear weighted combination of a few atoms in a redundant dictionary can be used to represent the signal [4]. Sparse decomposition is the process of decomposing signals based on the redundant dictionary and getting the sparse representation coefficients. It means that redundant dictionary is very important in sparse representation, and it will affect the effect of sparse representation and the efficiency of sparse decomposition, whether the redundant dictionary is appropriate.

The construction of redundant dictionary based on sparse representation has two ways[5]: (1) the method based on generating functions, such as Contour lets transformation; (2) the method based on sample training, such as the Optimum-Direction method[6], GPCA[7], K-SVD algorithm (K-singular value decomposition) [8]. K-SVD is the algorithm for redundant dictionary training in this paper.

The basic idea of the algorithm is based on the given training sample, and the dictionary is updated by searching the best representation of the training sample under the strict sparsity degree. To sum up ,the K-SVD algorithm can be said to be the iteration of two steps:

• Sparse decomposition procedure. Samples are sparsely decomposed with any tracking algorithms.

• The step of dictionary updating. If the coefficient matrix of the sparse decomposition is given, the singular value decomposition method is used to decompose the residual matrix of the dictionary, and update the corresponding atom in the dictionary and the corresponding sparse decomposition coefficient simultaneously. Then back to the sparse decomposition step, multiple iterations are completed by the loop.

The best dictionary of the sparse representation of data samples $\{y_i\}_{i=1}^N$ will be found through solving the formula. The formula is as follows:

$$\min_{D,X}\{\| Y-DX \|_F^2\} \quad subject\ to \quad \forall i, \| x_i \|_0 \leq T_0 \quad (1)$$

3 The surveillance video compression coding algorithm based on the regional dictionary

3.1 Description of training process of regional dictionary

In this paper, the compression result of the surveillance video compression coding algorithm is made up of encoding's results and the redundant dictionary. The redundant dictionary is trained for the surveillance video encoded specially. It is not adaptive to other surveillance video or signal, and the algorithm must use the redundant dictionary to reconstruct the original surveillance video, so redundant dictionary also needs to be counted in the compression results. The atomics of redundant dictionary are dense and the redundant dictionary, the key to the decomposition and reconstruction of the video in the process of sparse representation, should not be compressed. So in the compression results, it has a large proportion that reduces the overall compression ratio. If the volume of redundant dictionary can be controlled, better compression effect should be got, but at the same time, we should pay attention to the reconstructed quality of the surveillance video.

Although the volume of redundant dictionary is fixed. With the increase of the length of encoded surveillance video, the proportion of redundant dictionary in the compression results will fall, because the volume of each frame encoded residual is too small, the reduction is slow and slow. Therefore, it is very important to reduce the overall size of the redundant dictionary, which is the key to improve the compression ratio of the algorithm. In this paper, each frame of the video intercepted from the whole surveillance video is segmented in the same way. Each frame was divided into a number of regions with same size, and each of the image regions has the same shape and the same number of image blocks. After that, the same region in the picture are collected together as the training sample library of the region's redundant dictionary. The image block took out from the training sample library randomly is trained redundant dictionary using K-SVD algorithm to get redundant dictionary of the corresponding area. Since each region's dictionary training is independent of each other without no overlap in the image block, it can be executed in parallel to speed up the training efficiency.

3.2 Compressed encoding

In the video compression algorithm based on the regional dictionary, during the process of sparse decomposition, every frame of the surveillance video is read in order, and the image block is processed one by one. Firstly, each image block is decided which region it belongs to, and then according to corresponding region dictionary the coefficients of sparse decomposition of the image block are obtained, finally the coefficients of sparse decomposition of the video are obtained. After the uniform quantization, differential background [9], JPEG-LS encoding, encoding results can be got.

In the process of sparse decomposition, each image block of each frame is decomposed by the corresponding position regional dictionary trained by the K-SVD algorithm, the coefficients of all the image blocks are the coefficients of this frame. The coefficients are between 1 and 0, and there are a large number of decimal numbers. Therefore, the coefficients are quantized to 0-255 with details remained utmost, which is convenient for the following processing and preservation. As the background of each frame is similar, the sparse decomposition coefficients of the background are also similar. In order to remove the redundant information, the first step is establishing a public background of the surveillance video, the public background is also sparsely decomposed by the same set of redundant dictionary, the coefficients obtained are quantified uniformly, while the other frames are different from the coefficients of public background, and retain the residual error for the final entropy encoding only. In front, the sparse decomposition and uniform quantization have has caused loss to the video quality. In order to avoid losing more video information, the last of the encoding uses JPEG_LS which is Near-lossless Coding. Complete compression encoding process is as follows:

• Sparse decomposition: Each frame of the public background and the surveillance video is decomposed, and the image block of the video frame is decomposed by the regional dictionary of the corresponding region and then the sparse decomposition coefficients of each frame are obtained;

• The final result S_i of each frame is obtained through uniform quantization, differential background and JPEG-LS encoding.

4 Experimental result

4.1 The description of experimental material

The material used in this section is the 432nd to 562nd frames of S5-T1-G, a group of PETS2006 video sequences. In the 130 frames, the foreground has a large change, pure background without the foreground can be obtained when public background is established. In order to have a reference, the algorithm will be compared to the MJPEG Motion JPEG (Motion, Joint, Photographic, Experts, Group). MJPEG is a video compression format without considering the difference of the different frames of the video basically, and it uses JPEG encoding to compress each frame image independently. MJPEG is the main product of video surveillance equipment manufacturers, which has a large market, and its characteristics are similar to the algorithm in this paper.

On the quality evaluation of the recovered after compression, the paper uses two methods. one is comparing the objective value with the test sample, the other one is the subjective evaluation. The common objective evaluation criteria are MSE, SNR, PSNR and so on currently, which can be transformed with the formula. This paper chooses PSNR (the peak signal to noise ratio) as the standard. Supposing the original image is $f(x,y) \in R^{m \times n}$,the restored image is $\hat{f}(x,y)$, then the PSNR is as follows:

$$PSNR = 10\log_{10}\left\{\frac{255^2 mn}{\sum_{x=0}^{m-1}\sum_{y=0}^{n-1}(f(x,y)-\hat{f}(x,y))^2}\right\} \quad (2)$$

4.2 Analysis of experimental results

This paper selects 432nd to 562nd frames. In the 130 frames, the foreground has a large change, pure background without the foreground can be obtained when public background is established. While the unit image block is selected as 8 * 8, then the region composed by 11 * 4 image blocks is trained redundancy dictionary based on the region. As a result, the number of redundant dictionaries is greatly reduced, from 6336 to 144. During the sparse decomposition, first the image block is judge which region it belongs to, and then it is sparsely decomposed by the corresponding region dictionary, as well as the reconstruction.

Table 1. Comparison between surveillance video compression coding algorithm based on the regional dictionary and the algorithm in the video compression algorithm MJPEG.

algorithm	compression ratio	PSNR(dB)
Motion JPEG algorithm	15.0207	31.8202
	9.7771	34.9453
	5.9689	36.6654
Algorithm based on the regional dictionary	20.2682	31.6745
	10.4001	34.6090
	5.4762	36.7784

The compression ratio and PSNR of several algorithms is compared showed in Table 1. The algorithm based on regional dictionary is compared with MJPEG algorithm, when the reconstruction quality is low, namely the PSNR value is low, the compression ratio is better than MJPEG algorithm; but when PSNR is higher, the compression ratio is similar with MJPEG algorithm.

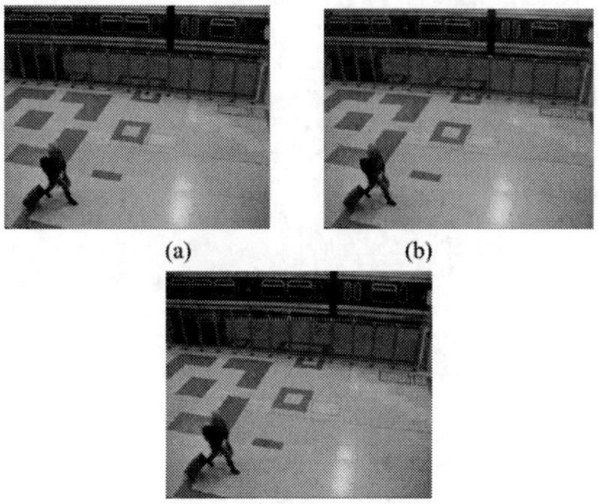

(a) (b) (c)

Figure 1. Overall subjective contrast.

Figure 2. Local subjective contrast(1).

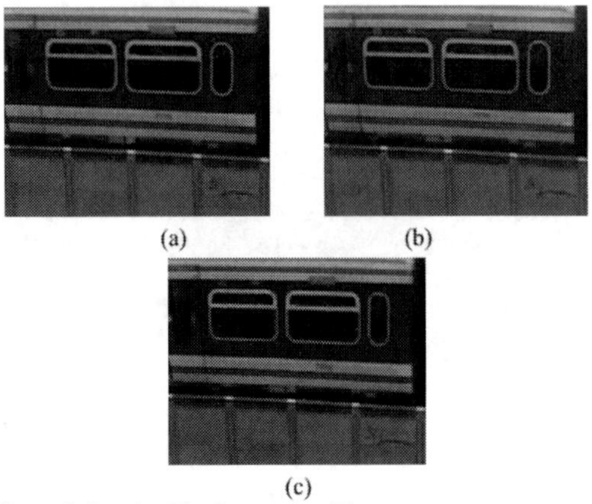

Figure 3. Local subjective contrast(2).

Figure 1 is the whole subjective comparison of a frame, figure a is the original label of the 477th frame material, figure b is the MJPEG encoding map of the frame, figure c is the reconstruction based on the regional dictionary algorithm of the frame. Figure 2 and figure 3 are a subjective comparison of the details. a, b, and c are the same meanings. The regional dictionary's performance is very good in the two algorithm through subjective comparison, the quality of reconstructed image and original is similar, and also no less than MJPEG algorithm. Although each image block is not corresponding to a redundant dictionary, the background area can still be clearly reconstructed. Moreover, with the regional dictionary displacing the block dictionary, reconstruction ability of foreground based on regional dictionary algorithm becomes stronger without separate encoding. Because samples of the regional dictionary samples are more diverse, the atomic structure is also more, the algorithm can find a result which is more similar to the original image.

Seen from the contrast of foreground details of Figure 2, the outline and details of reconstructed foreground based on the algorithm of the regional dictionary are still relatively vague.

5 Summary

The paper proposes a new algorithm for surveillance video, which is based on the regional dictionary. The key of the algorithm is to train a regional dictionary which is more conforms to surveillance video, and it can bring a good sparse decomposition and restoration effect. The experimental results show that the image compression coding algorithm based on the regional dictionary has a good effect on the background reconstruction, and has a general effect on the foreground reconstruction. There are still some shortcomings in this paper, the most obvious one is that, the process of the foreground is relatively simple. In the algorithm, the foreground is only judged by the average error. If it can be judged by the corresponding algorithm, it should be able to improve the quality of the picture further.

Acknowledgments

This work is supported by the National Natural Science Foundation of China (No. 61201361), Science Foundation of the Beijing Education Commission (KM20131 0005028), Training Programme Foundation for the Talents in Beijing City (2013D005015000008), and Training Programme Foundation for the Youth Talents by Beijing Education Commission.

References

1. Zhong Bin. Video surveillance system and intelligent video surveillance technology analysis [J]. electronic technology and software engineering, **23**,187-188(2014)
2. Deng Chengzhi. Image sparse representation theory and its application [D]. Huazhong University of Science and Technology (2008)
3. JIANG Ping. Research on construction method of based on compressed video sensing dictionary [D]. Xidian University (2008)
4. Lian Qiusheng, Shi Baoshun, Chen Anzhen. Research Advances on Dictionary Learning Models ,Algorithms and Applications[J]. Acta Automatica Sinica, **41(2)**:240-260 (2015)
5. Rubinsteinr, Bruckstelna, Eladm. Dictionaries for Sparse Representation Modeling[J] .Proceedings of the IEEE, **98(6)**:1045-1057(2010)
6. KreutzDelgado K, Murray J F, Rao B D, et al. Dictionary learning algorithms for sparse representation[J]. Neural computation, **15(2)**:349-396(2003)
7. Aharon M, Elad M, Bruckstein A. K-SVD: An Algorithm for Designing Overcomplete Dictionaries for Sparse Representation[J]. Signal Processing, IEEE Transactions on, **54(11)**:4311-4322 (2006)
8. Sun Yubao. Image sparse representation model and its application in image processing inverse problem [D]. Nanjing University of Science and Technology (2010)
9. Zhang Xianguo. Research on surveillance video based on background model[D]. Peking University (2013)

Applications of hand feature points detection and localization algorithms

Tomasz Grzejszczak and Michał Niezabitowski

Silesian University of Technology, Institute of Automatic Control, 16 Akademicka Street, 44-100 Gliwice, Poland

Abstract. Growing interest in hand gesture recognition system, bring new possibilities in developing new ways of intuitive control. Controlling the everyday surrounding can be much easier with use of natural body language. This article presents the review of hand gestures systems basing on color images, focusing on hand feature points detection. Additionally, the ways of hand feature points control possibilities are discussed, along with possible examples of hand gestures control. Hand gestures are introduced with discussion about their usage in everyday equipment control.

1 Introduction

It can be observed that the development of new technologies tends to focus on the intuitiveness of control [1-6]. Not so long ago, touch screens were the cutting edge technology, now they are commonly used in nearly all devices. In the future, the touchless technology can become the new standard of intuitive control. With growing hardware and software possibilities focusing on gesture recognition, it will not be long until gesture based control would be as common as touch panels [7].

Gestures was always the most intuitive ways of communication between humans. We use gestures every day, sometimes unconditionally. Moreover, for deaf people, gestures and sign language is the only way of communication.

This article proposes the ways of controlling with use of hand gestures, along with a review of color image based, hand feature recognizing algorithms [8]. The aim of this paper is to show the present research in hand feature points detection and propose the interesting ways of usage of their results.

2 Hand feature points

Many literature solutions, concerning hand gesture recognition, focus on hand feature points detection. Feature points are characteristic regions of hand, that are easy to detect. The proposed hand model consists of 25 characteristic points, that guarantee the possibility of full hand reconstruction with use of those points coordinates. Human hand, as well as whole human skeleton, consist of constant length rigid bones, that are connected in joints, that in case of fingers are called knuckles [9-11]. Moreover, human hand consists of 27 bones, with 14 phalanges of fingers, 5 metacarpal bones and 8 wrist bones.

Each finger, except thumb, consist of 3 phalanges, that can be visually observed. Those bones positions can be approximated without the use of x-ray image. Having in mind that their length is constant, it is enough to represent their position with use of two points line segment starting and ending in knuckles or finger tips.

The palm region and wrist region consists of multiple bones, that cannot be visually separated. Thus, the wrist region is represented as a line connecting two contour points, while the palm is represented as a circle or parabola, connecting wrist points and binger base points.

With addition of 4 concave points between fingers, the total of 25 feature points is a whole set of feature points that describes the hand arrangement. The visualization of hand feature points positions are presented in Fig. 1.

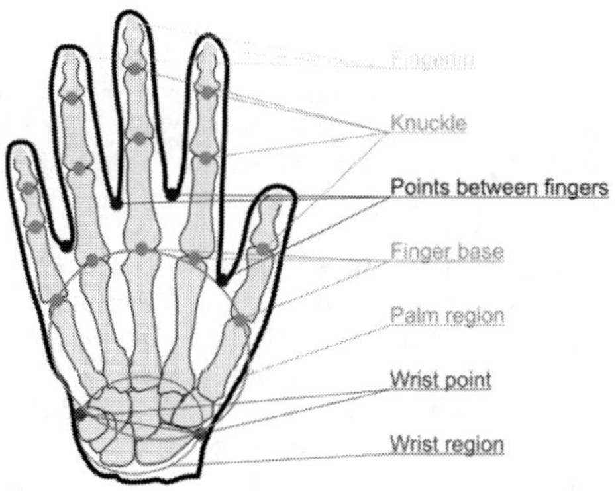

Fig. 1 Human hand with marked bones and proposed feature points.

However, some points cannot be seen in particular hand orientation, other are limited by algorithm capabilities. The missing points should be concluded from incomplete information set, in case where the whole hand representation is needed.

Detecting and tracking hand feature points can become the easy and intuitive way of control. Human hand complexity allows to make various movements, while algorithms tracking mutual distances between feature points can bring huge possibilities in intuitive control.

3 Related literature

With growing interest of intuitive control, there are multiple solutions on gesture recognition, each aiming in different aspect of hand gesture. The main approaches uses the hardware sensors called data gloves, special markers that allow easy tracking, 3D depth cameras or traditional color images. A detailed review on types of gesture recognition technologies, can be found in [12]. All approaches differ. It is important to choose one approach and compare the results with similar solutions. In the related work, a solution based on color images and computer vision algorithms has been chosen. Usually, the mask of hand is found with use of skin detection algorithm [13], and then, this mask is subject to various transformations.

It is important to notice that there are two main research fields in hand gesture recognition domain. Namely, the methods can be divided as appearance based methods and feature based methods.

The capabilities of the algorithms described in the literature differ from the simple rise fingers counting [14], through rise finger position evaluation [15], to full fingertip and finger base description, with detecting the palm region [16, 17], or even the concave points between fingers [18].

In the previous research, described in authors other publications, an algorithm of wrist detection algorithm has been described [19]. This algorithm has become the base of the hand feature points detection and localization with use of directional image, capable of accurate fingertip and finger base detection. Moreover, the literature solutions has been discussed and tested, and the mutual comparison can be found in [20].

The detection algorithm must have a purpose. The most common application is a Sign Language translator. The part of Sign Language for deaf people covers each letter of alphabet by the different hand arrangement. The aim of some solutions is to translate the gestures shown by user [14]. Unfortunately, the task is difficult, and many solutions bound the number of recognizable gestures [21]. The typical gesture set of a Sign Language varies in about 25-30 gestures, dependent on language.

Other common application is to use the gestures as a augmented way of intuitive control. The commonly available smart TVs and game consoles includes the possibility of simple gesture control [22, 23]. However, the recognition is limited to simple body tracking. Other solutions uses the hand position and gestures to replace the computer mouse or create the advanced controller [24, 25].

Finally, the most of hand recognition algorithms, does not have any application at all. The solutions focus on accuracy of rise finger counting or finger identification, without giving any practical usage.

4 Feature points detection

There are many hand feature points extraction algorithms described in the literature, however one can divide the approaches by used image processing method. Nearly all feature detection algorithms assumes a hand mask as an algorithm input. The most common image processing methods used in hand feature points detection are: contour analysis, distance transform and template matching. The division is presented in Fig. 2.

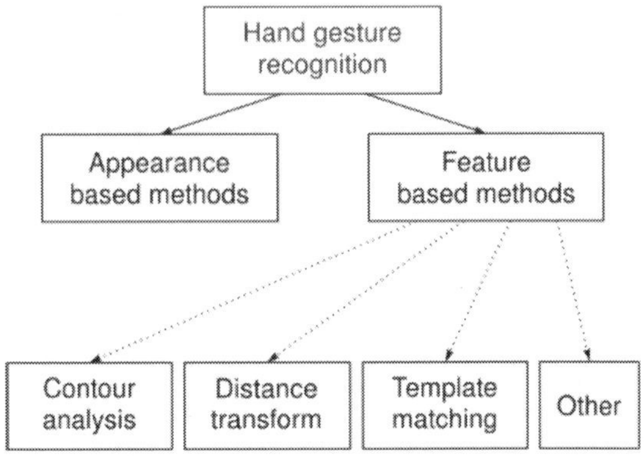

Figure 2 The division of hand gesture recognition methods with presentation of most common algorithms and solutions.

The mask contour based algorithms uses the hand mask silhouette, that is a set of points that lie on the border of hand region and background region. Usually, the algorithms are build basing on the mutual distances or angles of contour points, wrist point and palm center. The function of distance or angle between wrist or palm point and consecutive contour points is obtained and analyzed. Usually the local maxima of this function are considered to be the finger tips.

Distance transform of the hand mask brings useful information about palm region and fingers. The position of global maximum of a distance transform is treated as a palm center and the value is treated as palm radius. The mask points that lie outside of the palm region are considered as fingers. All points that can be used in finger representation can be found by analyzing the local maxima of distance transform.

Template matching algorithms are basing on the observation that fingers in hand mask are similar to round objects. Usually a circle image with radius similar to finger width is fitted into mask image with use of cross correlation or other template matching technique. After obtaining the resulting probability map, the global maximum is searched. The position of maximum is saved and the values in vicinity of this point are set to zero, in order to find the next global maximum. The positions of maxima are treated as fingertip point. Additionally, the algorithms are equipped with false positive detection method.

There are also other approaches in hand feature points detection, however the presented above are the most common.

5 Possible applications

Human hand is very complex, yet it is easy to perform a various and quick movements. Tracking only fingertips and their mutual distance can lead to multiple applications. It can be observed, that touch platforms such as smartphones of computer touchpad can introduce multiple, yet simple gestures that can carry various meanings, such as rotate, zoom, navigate or perform a specified macro command. Few years ago, many types of button and switch control were replaced by the touch panels. In the future, new algorithms would focus on creating the touchless technology.

In case of a computer control support, hand gestures would be the good replacement for mouse. The single finger tracking algorithms are easy to implement, leaving place for several button actions represented as gestures. Moreover, the combination of gestures and movements can cover complicated actions as drag and drop, select all, copy or paste.

Human hand is capable of preforming complex tasks. A hand feature tracking algorithm can record those tasks and reproduce them with use of specified robot manipulators. The path of each finger can be followed with high precision. This can lead to creation of a universal robot manipulator capable of human imitation.

Moreover, gestures can be used to precisely control the complicated machines. Nowadays, difficult tasks are performed with use of an operator equipped with a joystick, many times working in different reference. It would be much easier and more intuitive to operate those machines with use of simple hand movements. We can imagine a construction worker lifting a heavy load, just by moving a hand in air.

However, there exist a problem with fast movements. An operator can have a cramp, or become tired, or a vision system can detect an impulsive error. This could lead to damaging fast position changing during the complex precision tasks. In order to overcome this issue, it should be kept in mind that a security filter should be implemented in those types of solutions. The filter should detect the drastic changes, preventing the fast movement or approximating the impulsive errors.

6 Exemplary gestures

Most solutions focus on simple, one finger tracking (Fig. 3a). However, there are many possibilities of hand position arrangement, leading to intuitive control of many appliances. Fig. 3 shows few examples of feature tracking possibilities. Analyzing the following images, few applications are introduced:

a) simple one finger tracking can be used to control nearly everything position, from mouse pointer to robot arm position;
b) detecting the mutual position of two fingers can introduce the pinch gesture, widely used in touch panels to enlarge or rotate objects. Detecting two fingers in 3D can be used to control the position and clamping of a manipulator gripper;
c) introducing the third finger can create the slider control, with middle finder indicating the volume value;
d) grabbing gesture can be used to grab and move objects;
e) performing a "C" shaped gesture, and rotating it can be used to control all types of knobs;
f) performing a "L" shaped gesture, and rotating it can be used to rotate, for example pictures on computer;

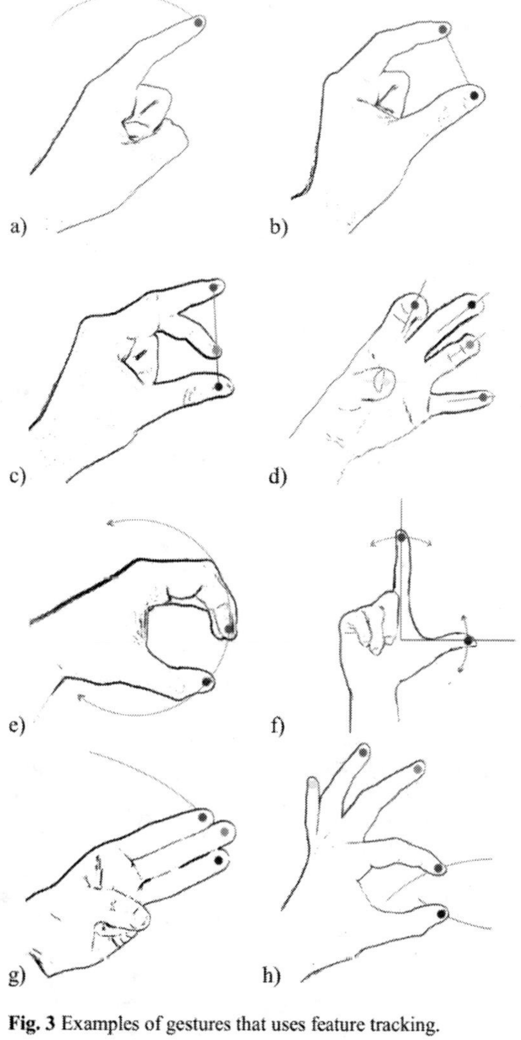

Fig. 3 Examples of gestures that uses feature tracking.

g) tracking multiple fingers is similar to one finger tracking, but an additional feature can be introduced. For example, tracking one finger moves mouse pointer. When two fingers are present, a selection region is drawn. When three fingers are present, drag and drop command is performed. Rising thumb can open a context window;

h) multiple actions can also be introduced while pinching gesture is performed. Position of two fingers can be correlated with position of manipulator gripper. The number of rise fingers can for example indicate the speed of manipulator.

Basically each everyday action can be performed, as it would be used on the traditional equipment. Those are only some examples of hand positions. There are many more, depending on the actual application and need of a user.

7 Conclusions

It is not the question whether the gesture control will become popular. The question is "when?". Simple tasks concerning TV or game control can already be managed with use of gestures. More advanced applications are already developing. Human hand is capable of multiple position arrangement without much effort. This article discussed many possibilities of everyday equipment control with use of gestures.

Unfortunately, the precision is still the problem. In conclusion, simple tasks, such as hand tracking are commonly used in wide applications. The more advanced solutions still needs to be improved.

Acknowledgment

The research presented here were done by authors as parts of the research projects BKM/524/RAU1/2013/t.2 and BK/265/Rau1/2014, respectively. The calculations were performed with the use of IT infrastructure of GeCONiI Upper Silesian Centre for Computational Science and Engineering (National Centre for Research and Development grant no POIG.02.03.01-24-099/13).

References

1. M. Alberdi, M. Amundarain, A. Garrido and I. Garrido: Renewable Energy Vol. **48** (2012), 16-26.
2. M. Goretti Sevillano, I. Garrido and A.J. Garrido: International Journal of Innovative Computing, Information and Control Vol. **8**(9) (2012), 6473-6489.
3. A. Babiarz: AIP Conference Proceedings, Vol.**1637**, (2014), 47-54.
4. J. Klamka, A. Czornik, M. Niezabitowski and A. Babiarz: Proceedings of the 11th IEEE International Conference on Control and Automation, Taichung, Taiwan (2014), 1210-1214.
5. T. Szkodny and A. Łęgowski: Intelligent Robotics and Applications, Springer (2015), 309-320.
6. A. Babiarz: Man-Machine Interactions 4, Advances in Intelligent Systems and Computing, Springer Vol.**391** (2016), 151-160.
7. K. Palenta and A. Babiarz: Man-Machine Interactions 3, Advances in Intelligent Systems and Computing, Springer Vol.**242** (2014), 115-122.
8. T. Grzejszczak and A. Łęgowski and M. Niezabitowski: Proceedings of the 20th International Conference on Control Systems and Computer Science (CSCS 2015), (2015), 71-77.
9. A. Babiarz, R. Bieda, K. Jaskot and J. Klamka: Bulletin of the Polish Academy of Sciences: Technical Sciences Vol. **61**(4) (2013), 955-971.
10. A. Babiarz, A. Czornik, J. Klamka, M. Niezabitowski and R. Zawiski: Proceedings of the 19th International Conference on Methods and Models in Automation and Robotics (MMAR 2014) (2014), 508-513.
11. A. Babiarz, J. Klamka, R. Zawiski and M. Niezabitowski, Proceedings of the 11th IEEE International Conference on Control and Automation (2014), 947-952.
12. P. Halarnkar, S. Shah, H. Shah, H. Shah and J. Shah: International Journal of Engineering Science and Technology Vol. **4**(11) (2012), 4648-4654.
13. M. Kawulok, J. Kawulok and J. Nalepa: Pattern Recognition Letters Vol. **41**(0) (2014), 3-13.
14. N. Tanibata, N. Shimada and Y. Shirai: Proceedings of the International Conference on Vision Interface (2002), 391-398.
15. Y. Sato, Y. Kobayashi and H. Koike: Proceedings of the fourth IEEE International Conference on Automatic Face and Gesture Recognition (2000), 462-467.
16. W. Hürst and C. van Wezel: English Multimedia Tools and Applications Vol. **62**(1), 233-258.
17. E. Stergiopoulou and N. Papamarkos: Engineering Applications of Artificial Intelligence Vol. **22**(8) (2009), 1141-1158.
18. L. Dung and M. Mizukawa: Journal of Robotics and Mechatronics Vol. 21(6) (2009), 726-738.
19. Z. Feng, B. Yang, Y. Chen, Y. Zheng, T. Xu, Y. Li, T. Xu and D. Zhu: Pattern Recognition Vol. **44**(5) (2011), 1089-1105.
20. T. Grzejszczak, J. Nalepa and M. Kawulok: Proceedings of the 8th International Conference on Computer Recognition Systems CORES 2013 Vol. **226**, 439-449.
21. T. Grzejszczak, A. Gałuszka, M. Niezabitowski and K. Radlak: Technological Innovation for Collective Awareness Systems (2014), 167–174.
22. Y. Li: Proceedings of the 3rd IEEE International Conference on Software Engineering and Service Science (ICSESS) (2012), 196–199.
23. W.-P. Lee, C. Kaoli and J.-Y. Huang: Knowledge-Based Systems Vol. **56**(0) (2014), 167-178.
24. D.-H. Shin, Y. Hwang and H. Choo: Behaviour & Information Technology Vol. **32**(2) (2013), 156-172.
25. T. Osunkoya and J.-C. Chern: Gesture-based human-computer-interaction using kinect for windows mouse control and power point presentation (Department of Mathematics and Computer Science, Chicago State University, Chicago, IL, vol. 60628, 2013).

Dynamical Pattern Vector in Pattern Recognition with the Use of Thermal Images

Zygmunt Kuś

Faculty of Automatic Control, Electronics and Computer Science Silesian University of Technology, Akademicka 16, Gliwice, Poland

Abstract. The goal of the following paper was to develop the methodology of object tracking in adverse conditions. Suddenly appearing clouds, fog or smoke could be the examples of atmospheric conditions. We used thermal and visible images in each moment during object tracking. We computed the pattern vectors of the tracked object on the basis of the visual and thermal images separately. The pattern vector and current feature vector for an image of a given type are used to compute the distance between the object pattern vector and feature vector calculated for a given location of the aperture. It is calculated for both: the visual and thermal image. The crux of the proposed method was the algorithm of selection which distance (for visual or thermal image) was used for object tracking. It was obtained by multiplying the values of the distances between a pattern vector and current feature vector by some coefficients (different for thermal and visual images). The values of these coefficients depended on the usefulness of a given type of an image for pattern recognition. This usefulness was defined by the variability of the particular pixels in the image which is represented by calculating gradient in the image. On top of that, this study presented the examples of the object recognition by means of the developed method.

1 Introduction

Thermal imagery, which is obtained from an infrared sensor and can display invisible thermal energy emitted from objects, is a crucial source of information in low visibility conditions. This information can be very useful for image processing system which is used for surveillance, object tracking or security applications.

We can obtain complementary information of the objects which are located in an inspected scene when we use both: thermal and visible images.

In recent years, thermal imaging technology has been widely used in industrial, commercial and military applications. These applications include detection, tracking and identification of people [1,2,3], tracking vehicles on the road [4], or animals which can appear on the road [5].

Thermal imagery is also useful for a diagnosis of equipment [6] or for inspecting the production process [7,8].

The images fusion task appears when one has to solve problems which one encounters when we use image processing systems in such applications as: image recognition [9,10,11,12], an unmanned aerial vehicle [13,14] used for image acquisition [15,16,17,18], object tracking [19,20,21] or images fusion [22,23].

In the following paper, we will consider the problem of object tracking and pattern recognition in the conditions of distorted visibility of the object. We assume that both visual and thermal images can be sometimes distorted. The proposed method is based on the separation of two parts of pattern vector in a pattern recognition task. One part depends on a visual image and the other one depends on a thermal image. The process of pattern recognition is based on this part of the pattern vector for which the image is not disturbed.

2 The Basic Assumptions of Pattern Recognition

The proposed method assumes that we will acquire the visual and thermal images ($f_V(i,j)$ and $f_T(i,j)$) of the size of MxN in each moment of object tracking. All images were transformed to grey scale images. The mutual position of the conventional camera and thermal camera was parallel positioned. This way, we obtained visual image which corresponded to the same area in thermal camera.

We will compute the pattern vectors ($\boldsymbol{J}_{WZV}$ and $\boldsymbol{J}_{WZT}$) of the tracked object on the basis of the visual and thermal object's images ($f_{WZV}(i,j)$ and $f_{WZT}(i,j)$ of the size of mxn) separately.

The pattern vectors consist of moment invariants J_1 to J_8 defined in Eq.1.

The majority of the abovementioned definitions of the moment invariants were presented in [24]. In [25] the use of these IF in image recognition was discussed. The

$$J_1 = \frac{I_1}{m_{00}^2};\quad J_2 = \frac{I_2}{m_{00}^4};\quad J_3 = \frac{I_3}{m_{00}^5};\quad J_4 = \frac{I_4}{m_{00}^5};\quad J_5 = \frac{I_5}{m_{00}^{10}};\quad J_6 = \frac{I_6}{m_{00}^7};\quad J_7 = \frac{I_7}{m_{00}^4};\quad J_8 = \frac{I_8}{m_{00}^5};$$
$$I_1 = M_{20} + M_{02};\qquad I_2 = (M_{20} - M_{02})^2 + 4M_{11}^2;\qquad I_3 = (M_{30} - 3M_{12})^2 + (3M_{21} - M_{03});$$
$$I_4 = (M_{30} + M_{12})^2 + (M_{21} + M_{03})^2;$$
$$I_5 = (M_{30} - 3M_{12})(M_{30} + M_{12})[(M_{30} + M_{12})^2 - 3(M_{21} + M_{03})^2] + (3M_{21} - M_{03})(M_{21} + M_{03})[3(M_{30} + M_{12})^2 - (M_{21} + M_{03})^2];$$
$$I_6 = (M_{20} - M_{02})[(M_{30} + M_{12})^2 - (M_{21} + M_{03})^2] + 4M_{11}(M_{30} + M_{12})(M_{21} + M_{03});\tag{1}$$
$$I_7 = M_{20}M_{02} - M_{11}^2;\qquad I_8 = M_{30}M_{12} + M_{21}M_{03} - M_{12}^2 - M_{21}^2;$$
$$M_{pq} = \sum_{i=0}^{N-1}\sum_{j=0}^{M-1}(i - x_s)^p (j - y_s)^q f(i,j) \text{ where } x_s = \frac{m_{10}}{m_{00}};\quad y_s = \frac{m_{01}}{m_{00}};$$
$$m_{pq} = \sum_{i=0}^{N-1}\sum_{j=0}^{M-1} i^p j^q f(i,j) \text{ - a moment of order } (p+q);\qquad f(i,j) \text{ - image pixel;}$$

pattern vectors ($\boldsymbol{J}_{WZV}$ and $\boldsymbol{J}_{WZT}$) and current feature vectors ($\boldsymbol{J}_V$ and $\boldsymbol{J}_T$) are defined in Eq.2.

$$\begin{aligned}\boldsymbol{J}_{WZV} &= [J_1(f_{WZV}),\dots,J_8(f_{WZV})];\\ \boldsymbol{J}_{WZT} &= [J_1(f_{WZT}),\dots,J_8(f_{WZT})];\\ \boldsymbol{J}_V(i,j) &= [J_1(f_{VAP}(i,j)),\dots,J_8(f_{VAP}(i,j))];\\ \boldsymbol{J}_T(i,j) &= [J_1(f_{TAP}(i,j)),\dots,J_8(f_{TAP}(i,j))];\end{aligned}\tag{2}$$

where $f_{VAP}(i,j)$ / $f_{TAP}(i,j)$ denotes a part of the image $f_V(i,j)$/ $f_T(i,j)$ which is inside the aperture in the position (i,j). The pattern vector and current feature vector for an image of a given type (visual or thermal) are used to compute the distances $\rho_V(i,j)$ and $\rho_T(i,j)$. These distances are calculated for a given location of the aperture as it is presented in Eq.3. In order to assess this distance between vectors we will use Euclidian metric.

$$\begin{aligned}&\rho_V(i,j) = a^{\rho_V^*(i,j)};\quad \rho_T(i,j) = a^{\rho_T^*(i,j)};\\ &\text{where}\quad a \in (0,1);\\ &\rho_V^*(i,j) = \|\boldsymbol{J}_{WZV} - \boldsymbol{J}_V(i,j)\|;\\ &\rho_T^*(i,j) = \|\boldsymbol{J}_{WZT} - \boldsymbol{J}_T(i,j)\|;\end{aligned}\tag{3}$$

We assume that $a \in (0,1)$ and then we obtain $\rho_V(i,j)$ and $\rho_T(i,j)$ with values close to 0 for the object located beyond (i,j) position and 1 otherwise. We propose such a modification of $\rho^*(i,j)$ in order to obtain normalised values of an error function in the range (0, 1). The crux of the proposed method is the algorithm of deciding which $\rho(i,j)$ (for visual or thermal image) is used for object tracking.

We assume that the usefulness of visual or thermal images may be defined by the variability of the particular pixels in the image which is represented by the function of gradient in the image. The low value of gradient indicates that the image or its part is not useful for our goal. Hence we obtain the values of the coefficients α_C, α_L, β_C and β_L as it is presented in Eq.4.

$$\begin{aligned}&grad(f_V) \ge g_0 \Rightarrow \alpha_C = 1;\\ &grad(f_V) < g_0 \Rightarrow \alpha_C = 0;\\ &grad(f_T) \ge g_0 \Rightarrow \beta_C = 1;\\ &grad(f_T) < g_0 \Rightarrow \beta_C = 0;\\ &grad(f_{VAP}(i,j)) \ge g_0 \Rightarrow \alpha_L(i,j) = 1;\\ &grad(f_{VAP}(i,j)) < g_0 \Rightarrow \alpha_L(i,j) = 0;\\ &grad(f_{TAP}(i,j)) \ge g_0 \Rightarrow \beta_L(i,j) = 1;\\ &grad(f_{TAP}(i,j)) < g_0 \Rightarrow \beta_L(i,j) = 0;\end{aligned}\tag{4}$$

g_0 is the threshold value of the gradient.

The coefficients calculated in Eq.4 allow us to create the function $\rho(i,j)$ which consists of two parts which depend on visual and thermal pattern vectors separately. The function $\rho(i,j)$, presented in Eq.5, is based on the correlation between the values of coefficients α_C, α_L, β_C, β_L and values of $\rho_V(i,j)$ and $\rho_T(i,j)$.

$$\begin{aligned}\rho(i,j) = {}&\alpha_C \cdot \alpha_L(i,j) \cdot \rho_V(i,j) +\\ &+ \beta_C \cdot \beta_L(i,j) \cdot \rho_T(i,j)\end{aligned}\tag{5}$$

The coefficients close to 1 mean that the quality of the image is correct and at the same time $\rho_V(i,j)$ and $\rho_T(i,j)$ close to 1 mean that the object is recognized in (i,j) position. In this case, both parts, for visual and thermal pattern vectors, are used for pattern recognition. The coefficients close to 0 mean that the quality of the image is low (fog, smoke) and the corresponding part (visual or thermal) of the pattern vector is not used.

3 The Algorithm of the Use of the Dynamical Features Vector

The presented method is based on the simultaneous use of a visual and thermal image.

The first stage of the proposed algorithm consists of visual and thermal images acquisition. This process should provide two images in which the same parts of images represent the same parts of the observed terrain.

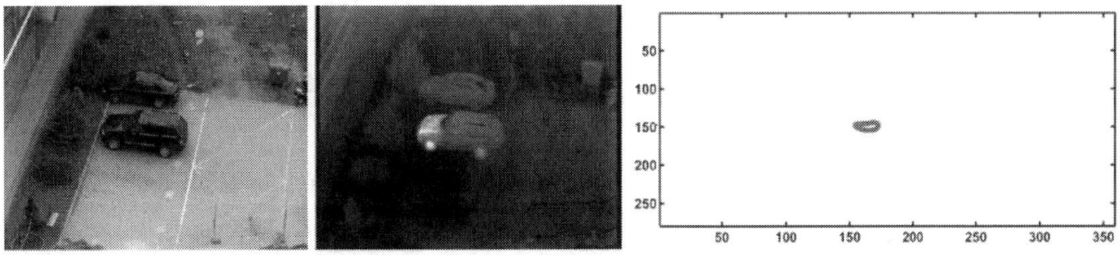

a) b) c)

Figure 1. The visual image a), thermal image b) and calculated location c) of the searched object for the case 1.

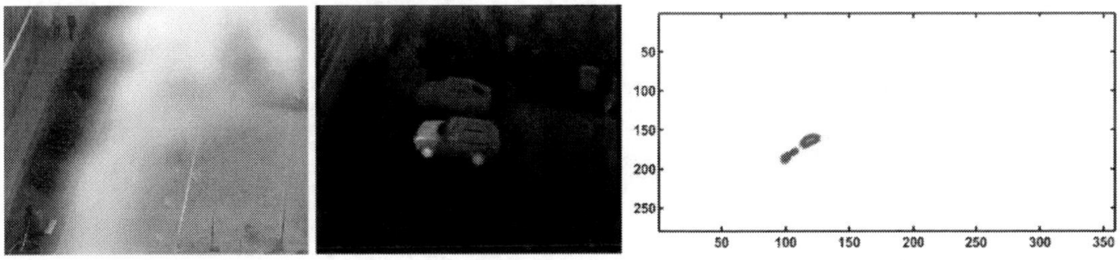

a) b) c)

Figure 2. The visual image a), thermal image b) and calculated location c) of the searched object for the case 2.

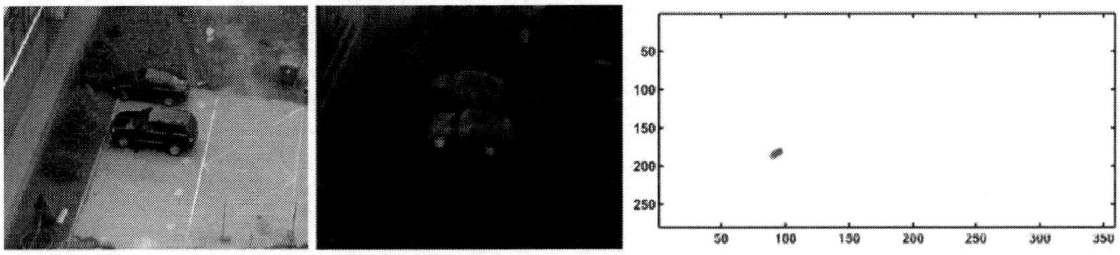

a) b) c)

Figure 3. The visual image a), thermal image b) and calculated location c) of the searched object for the case 3.

The second stage is conducted with the assumption that we have good quality visual and thermal images which allow to obtain correct pattern vectors of the object for both images. This stage consists of calculating the searched object's pattern vectors for both representations of a searched object, separately for visual and thermal images according to Eq.2.

In the third stage, it is necessary to analyse whole images in order to define the potential parts of the images where the object may be found. This analysis is conducted for whole images (α_C, β_C denote the usefulness of the whole images for pattern recognition) and for every location of the aperture in the image ($\alpha_L(i,j)$, $\beta_L(i,j)$ denote the usefulness of a given part of an image inside the aperture for pattern recognition). This procedure will be helpful for an image processing system to take into account during calculations only these parts of the images in which there are some artefacts which may constitute the elements of the searched object. Therefore, in this case, we will calculate grey scale gradient.

The last stage is to calculate distance between the pattern vector and current feature vector. In this stage, we use the parts of the current feature vectors according to the analysis carried out in the previous stage. It is realised by calculating the abovementioned distance (Eq.5) with the proper coefficients. The values of these coefficients depend on whether the whole image or its part is disturbed by smoke or other disturbances. The object is located in position (i_m, j_m) in the image if and only if $\rho(i_m, j_m) = \min \rho(i, j)$.

4 Examples

The proposed examples will present the problem of vehicle tracking in various visibility conditions. The figures present the visible and thermal images pairs and the location of the searched object. The following cases correspond to the following situations:

1) An off-road vehicle was the tracked object. According to Fig. 1, both visual and thermal images are of a good quality (α_C, = 1; β_C =1); the parts of images containing object can be used to obtain pattern vectors for visual and thermal object images. The recognition will be

conducted on the basis of both images. Fig. 1c presents correctly calculated location of the object.

2) According to Fig. 2, the visual image is partially distorted by smoke whereas the thermal image is correct (α_C, = 1; β_C =1). In this case, searching the object will be conducted on the basis of both images; however, $\alpha_L(i,j)$ in the smoky regions will assume zero values and in these places only thermal image will be used. Fig. 2 presents object location which is slightly shifted with respect to Fig.1c; however, we can assume that it is correctly calculated location of the object.

3) According to Fig. 3, the visual image is correct whereas thermal image is distorted because the temperature of the part of the searched object has changed. In this case, object searching will be conducted on the basis of both images; nevertheless, the value of the thermal image is minor. However, the thermal image cannot be eliminated from calculations because without additional information about the changes of the object temperature we are not able to interpret the thermal image correctly. Fig. 3c presents correctly calculated location of the object.

5 Conclusions

The paper has developed the methodology of object tracking in adverse conditions. In order to conduct the tests, we used the example of the disturbance in the form of smoke and changes of the searched object temperature. The obtained results could be used for other disturbances such as suddenly appearing clouds, fog or adverse atmospheric conditions. We used thermal and visible images, in each moment during object tracking. We described searched object features by computing the pattern vectors containing moment invariants. These vectors were calculated on the basis of the visual and thermal images separately. The distance between the object pattern vector and current feature vector was calculated for a given location of the aperture separately for visual and thermal images. The crux of the proposed method was the algorithm of selection which distance (for visual or thermal image) was used for object tracking. This algorithm was based on multiplying the values of the distances between a pattern vector and current feature vector by some coefficients (different for thermal and visual images). The algorithm of the calculation of these coefficients, based on a gradient function which corresponds to the usefulness of the image for pattern recognition, was presented in the paper. There were proposed coefficients which were calculated for the whole image (global) and for a current aperture location (local). Finally, this study presented the examples of the object recognition by means of the developed method.

Acknowledgements

This work was supported by PBS I ID 178438 path A from the Polish National Centre for Research and Development.

References

1. A. Treptow, G. Cielniak and T. Duckett, IROS, , 2103-2108, (2005)
2. C. N. Padole and L.A. Alexandre, Proc. of IEEE CSC CVPR, 9-14, (2010)
3. I. Ciric, Z. Cojbasic, V. Nikolic and D. Antic, Proc. of 11th TMSCBS **2,** 587-590, (2013)
4. W. Woeber, D. Szuegyi, W. Kubinger and L. Mehnen, Proc. of 55th IS ELMAR, 357-360, (2013)
5. Debao Zhou and Jingzhou Wang, Proc. of EO, **3**, 342-345, (2011)
6. Shih-Yuan Huang, Chi-Wu Mao and Kuo-Sheng Cheng, IEEE IM, **54(6)**, 2381-2388, (2005)
7. G. Ginesu, D. D. Giusto, V. Margner and P. Meinlschmidt, IEEE IE **51(2)**, 480-490, (2004)
8. Jędrasiak K., Daniec K., Nawrat A., Koteras R., El.Rev., ISSN0033-2097, R.88 NR 10a/2012, 90-97, (2012)
9. K. Jedrasiak and A. Nawrat, Computer Vision and Graphics, LNCS **5337**, 391-399, (2009)
10. T. Grzejszczak, El.Rev. **87(2)**, 201-204, (2011)
11. T. Grzejszczak, J. Nalepa and M. Kawulok, Adv. in ISC **226**, 439-449, (2013)
12. Bieszczad G., Sosnowski T., Madura H., Kastek M., Bareła J., Proc. of SPIE **7660**, 76603Z, (2010)
13. R. Zawiski and M. Błachuta, Proc. of 17th IC on MMAR, 475-480,(2012)
14. R. Zawiski and M. Błachuta, BPASTS, **62(3)**, 535-550, (2014)
15. P. Iwaneczko, K. Jedrasiak, K. Daniec and A. Nawrat, CVG, Lect.Not.CS **7594**, 87-94, (2013)
16. M. Błachuta, R. Czyba, W. Janusz and G. Szafrański, J of IRSTA **74(1-2)**, 413-420, (2014)
17. R. Czyba and G. Szafranski, Int. j. adv. robot. syst. **10**, (2013).
18. Bieszczad G., Sosnowski T., Madura H., Kastek M., Bareła J., Proc. SPIE 8012, 80120L, (2011)
19. A. Nawrat and K. Jedrasiak, Proc.of 9th WSEAS ICRCMT, 67-72, (2009)
20. R. Gade, T. B. Moeslund, Sensors, **14**, issue 8, doi:10.3390/s140813679, 13679-13691, (2014)
21. Bieszczad G., Sosnowski T., Madura H., Proc. SPIE 8193, 81933R, (2011)
22. K. Jedrasiak, A. Nawrat, K. Daniec, R. Koteras, M. Mikulski and T. Grzejszczak, CVG LN in CS **7594**, 423-432, (2013)
23. Moulay A. Akhloufi, Celia Porcher, Abdelhakim Bendada, Proc. SPIE 9076, Airborne (ISR) SA XI, 90760O, (2014).
24. Papoulis A., PRVSP McGraw-Hill, New York, 3rd edition, (1991)
25. Milan Sonka, Vaclav Hlavac, Roger Boyle, IP, AMV, Thompson, (2008).

MATEC Web of Conferences 56, 02011 (2016) DOI: 10.1051/matecconf/20165602011
ICCAE 2016

NUI framework based on real-time head pose estimation and hand gesture recognition

Hyunduk Kim, Sang-Heon Lee and Myoung-Kyu Sohn

Department of IoT and Robotics Convergence Research, DGIST, Daegu, Korea

Abstract. The natural user interface (NUI) is used for the natural motion interface without using device or tool such as mice, keyboards, pens and markers. In this paper, we develop natural user interface framework based on two recognition module. First module is real-time head pose estimation module using random forests and second module is hand gesture recognition module, named Hand gesture Key Emulation Toolkit (HandGKET). Using the head pose estimation module, we can know where the user is looking and what the user's focus of attention is. Moreover, using the hand gesture recognition module, we can also control the computer using the user's hand gesture without mouse and keyboard. In proposed framework, the user's head direction and hand gesture are mapped into mouse and keyboard event, respectively.

1 Introduction

The history of interaction and interface design is a flow and step from complex interaction to simple interaction between human and computer [1]. The word natural interaction came from Natural User Interface (NUI) that use human body interaction and voice interaction, verbal and non-verbal communication, becoming a one of Human Computer Interaction (HCI) area. It is an evolution from Graphical User Interface (GUI). GUI is the translation from command to graphic for easier purpose for users. Before GUI era, Command Line Interface (CLI) was the starting computer interaction generation which just used codified and very strict command. NUI is a human computer interaction which targeting on some of human abilities such as body movement, touch, motion, voice, vision and using cognitive functions to interact with computer or machine [2-4].

For many human computer interaction applications it would be helpful to know where a user is looking at and what user focus of attention is. Such information can be obtained from tracking the orientation of a human head, or gaze [5]. A person's gaze direction is determined by two factors: the orientation of the head, and the orientation of the eyes. In this paper, we limit our discussion to the head pose orientation. Then the gaze estimation can be formulated as a pose estimation problem. Determining head pose is one of the most important topics in the field of computer vision. There are many applications with accurate and robust head pose estimation algorithms, such as human-computing interfaces, driver surveillance systems, entertainment systems, and so on. For this reason, many applications would benefit from automatic and robust head pose estimation systems. Accurately localizing the head and its orientation is either the explicit goal of systems like human computer interfaces or a necessary pre-processing step for further analysis, such as identification or facial expression recognition. Due to its relevance and to the challenges posed by the problem, there has been considerable effort in the computer vision community to develop fast and reliable algorithms for head pose estimation [6-8].

Recently, hand gesture recognition systems based on vision have undergone an increasingly popularity due to their wide range of potential applications in the field of human computer interaction, such as multimedia application control, video-games, and medical systems. These interfaces are considered more natural, intuitive, friendly, and less intrusive for the user than traditional HCI devices (mouse, keyboard, remote control, etc.). Although the use of keyboard and mouse can be still very useful for some applications, there are situations and applications, where hand-based interfaces can be a key advantage [9].

In general, approaches relying solely on 2D images are sensitive to illumination changes and lack of distinctive features. Since 3D sensing devices have become available, computer vision researchers have started to leverage the additional depth information for solving some of the inherent limitations of image-based method. In this paper, we develop interaction modules based on 3D depth, head pose estimation and hand gesture recognition, and we propose NUI framework based these interaction modules.

2 Head pose estimation module

In this paper, proposed system consist of two process, input system and main system. In the input system, color image and depth map are initially acquired from RGB-D camera, such as ASUS Xtion Pro and Kinect. And then, depth information is converted to real world coordinate system due to estimate the position of head in the real world coordinate system. The main system consists of three processes. In the head pose estimation process, position and orientation of head are estimated using random forests classifier. Due to this classifier, system can be operated in real time and deal with the large set of training data. In the user's gaze tracking process, calculate the position of users gaze on the display device using proposed geometric model. In the Correction process, estimated user's gaze position is corrected using Kalman filter. Due to this process, it is possible to perform stable gaze tracking. Figure 1 shows the structure of head pose estimation module [8].

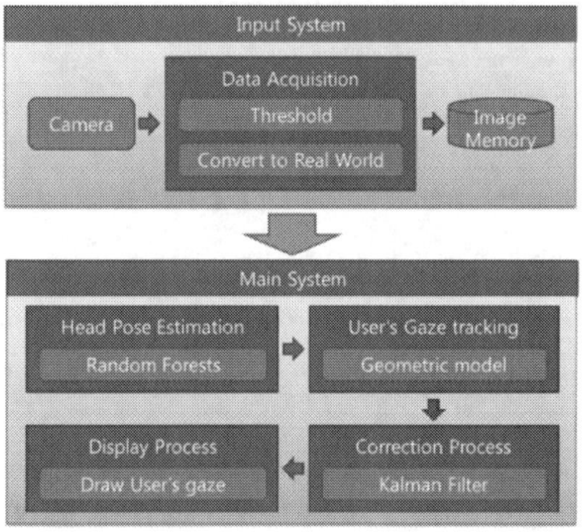

Figure 1. The structure of head pose estimation module

2.1. Head pose estimation

In this process, we estimate the position and orientation of head using random forests classifier with both hue and depth information. Decision trees can map complex input spaces into simpler, discrete or continuous output spaces, depending on whether they are used for classification of regression purposes. A tree splits the original problem into smaller ones, solvable with simple predictors, thus achieving complex, highly non-linear mappings in a very simple manner. A non-leaf node in the tree contains a binary test, guiding a data sample towards the left or right child node. The tests are chosen in a supervised-learning framework, and training a tree boils down to selecting the tests which cluster the training such as to allow good predictions using simpler models.

Random forests are collections of decision trees, each trained on a randomly sampled subset of the available data; this reduces over-fitting in comparison to trees trained on the whole dataset, as shown by Breiman [10]. Randomness is introduced by the subset of training examples provided to each tree, but also by a random subset of tests available for optimization at each node. Figure 2 shows the example of head pose estimation using random forests [8].

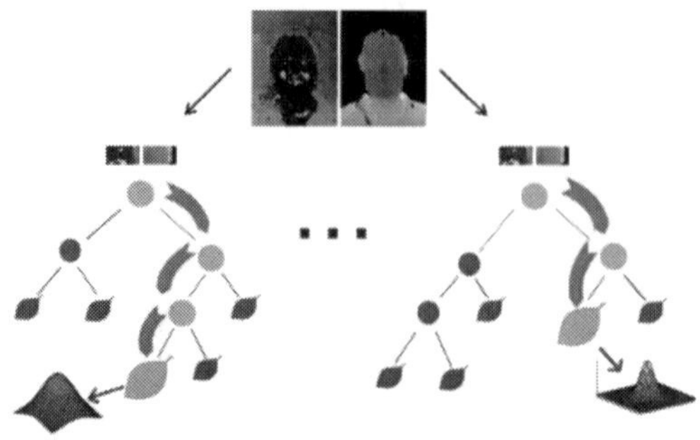

Figure 2. Head pose estimation using random forests

2.2. Gaze tracking

In this process, we calculate the position of user's gaze on the display device from estimated head position and orientation using geometric model, which consists three steps. Let (x, y, z) and (θ_x, θ_y) be the estimated head position and orientation respectively. In real world coordinate system, we assume that camera is located at (0, 0, 0). And then user's gaze position in the real world coordinate system is estimated in the first step. Let $w_1 \times h_1$ be the resolution of RGB-D camera and let f and p be the focal length and pixel size of RGB-D camera, respectively. Then, user's gaze position in pixel coordinate system is converted in the second step. Let $w_2 \times h_2$ be the resolution of diplay device. Then, user's gaze on the display device can be determined in the third stpe. Figure 3 shows the overview of the geometric model for gaze tracking.

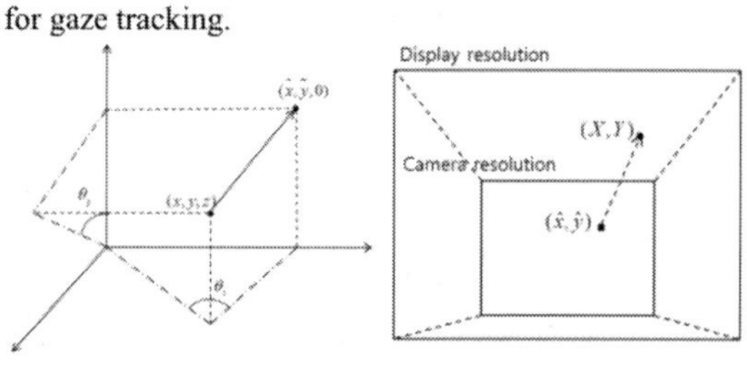

(a) (b)

Figure 3. Overview of geometric models: (a) Real-world coordinate system, (b) Pixel coordinate system

In general, resolution of RGB-D camera is 640×480 and that of display device is Full-HD or 4K. The estimated user's gaze contains the noise due to the difference between camera resolution and display resolution. In order to solve this problem, we perform correction process using Kalman filter. The Kalman filter has numerous applications in technology because of its convenient form for real time processing and easy to implement given a basic understanding [8].

3 Hand gesture recognition module

The Hand Gesture Key Emulation Toolkit (HandGKET) facilitates integration of hand gesture control with PC applications such as games and media center [11]. This middleware recognizes user's hand gestures and generates keyboard or mouse events to control applications in your computer. By editing the gesture-key mapping script, user can control existing off-the-shelf applications which have different control keys. HandGKET consists of three parts. First part is recognition module. This module analysis hand movements and recognize hand gesture. Second part is gesture-key mapping script. When HandGKET runs, first load the gesture key mapping script, and stores it in inter-memory. Third part is event generator, which create keyboard or mouse event according to gesture-key mapping information stored in inter-memory. Figure 4 shows the structure of hand gesture recognition module named HandGKET.

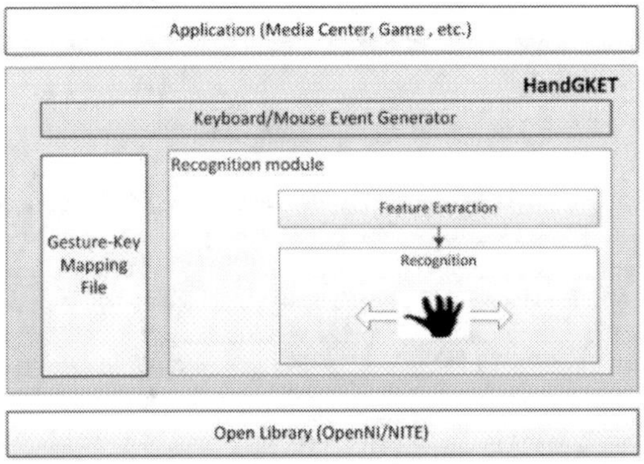

Figure 4. The structure of hand gesture recognition module

3.1. Recognition module

In the recognition module, hand gesture is recognized using hand trajectory and threshold. The type of hand gesture is defined by round trip gesture. For example, in the case of hand-moving operation to the right direction, hand must is moved to right at the starting point and is returned to the starting point in order to complete the operation.

In the recognition module, hand gesture is recognized using hand trajectory and threshold. The type of hand gesture is defined by round trip gesture. For example, in the case of hand-moving operation to the right direction, hand must is moved to right at the starting point and is returned to the starting point in order to complete the operation.

By setting the starting point, recognition accuracy can be increased in the continuous gesture sequences. The starting point is updated by motion energy in real time and is determined by hand position when the movement of hand is not detected for a certain amount of time after hand is detected. Another advantage of HandGKET is that it can set the threshold value based on real world distance. Despite the same gesture, it can be recognized as different gesture by setting a different threshold value. This makes it possible to increase the number of gestures and to extend the functionality of the toolkit. Figure 5 shows the example of the round trip gesture.

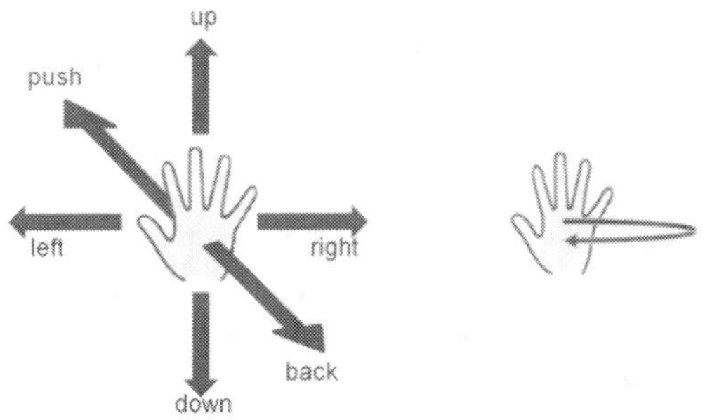

Figure 5. The example of round trip gesture.

3.2. Gesture-Key mapping script

Figure 6 shows the example of gesture-key mapping script. In the Figure 6, first column means the hand gesture and the threshold value is set in the second column. The hand gesture is recognition if the size of gesture is bigger than threshold value. Third column is type of event, such as key click, double key click, key hold, key toggle, and so on. Fourth and fifth columns mean the generated and released event. Sixth column means additional information, for example, the interval of event time

```
# for one-hand gestures
HAND_LEFT   5  KEY_CLICK  VK_LEFT    0 0
HAND_RIGHT  5  KEY_CLICK  VK_RIGHT   0 0
HAND_RIGHT 30  KEY_CLICK  M          U U
HAND_UP     5  KEY_CLICK  VK_UP      0 0
HAND_DOWN   5  KEY_CLICK  VK_DOWN    0 0
HAND_PUSH  10  KEY_CLICK  VK_RETURN  0 0
HAND_BACK   7  KEY_CLICK  VK_ESCAPE  0 0
```

Figure 6. The example of a gesture-key mapping script.

4 Proposed NUI framework

In this paper, we design a NUI framework based on real time head pose estimation and hand gesture recognition. The user's head direction and hand gesture are mapped into mouse and keyboard event, respectively. Using proposed NUI framework, we can control our computer without mouse and keyboard. Figure 7 shows the structure of proposed NUI framework.

The proposed NUI framework consists of two modules, head pose estimation module and hand gesture recognition module. The head pose estimation module first detects face region, and then estimation head position and orientation. The user's gaze is estimated using head position and orientation. By mapping user's gaze into mouse event, we can control contents without mouse device.

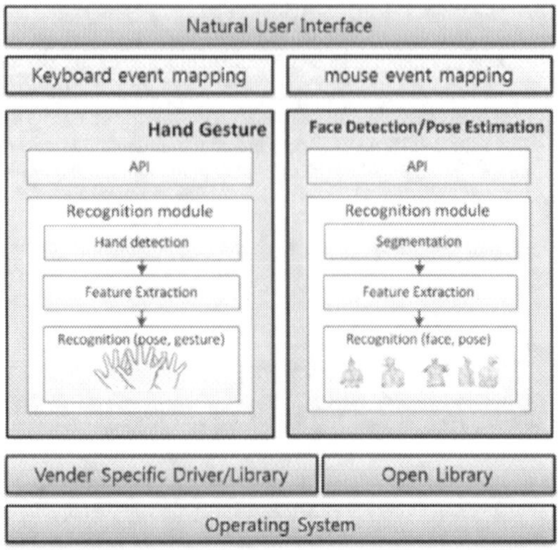

Figure 7. The structure of propose NUI framework.

The hand recognition module first detects hand region, and then recognize hand gesture. The hand gesture is mapped into keyboard event when the user takes predefined hand gesture. Using this, we can control contents without keyboard device. Figure 8 shows the application of proposed NUI framework.

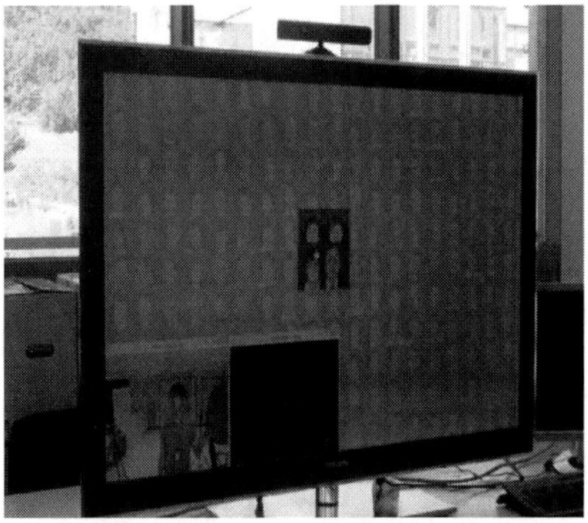

Figure 8. The application of proposed NUI framework.

5 Conclusion

In this paper, we proposed natural user interface (NUI) framework without traditional devices, such as mouse and keyboard. Proposed NUI frame work consists of two modules, head pose estimation module and hand gesture recognition module. Each module is mapped into mouse and keyboard event. For this reason, user can naturally interact with content and feel the immersive. The future work is development of more natural interface based on head pose and hand gesture as well as voice and emotion.

Acknowledge

This work was supported by the DGIST R&D Program of the Ministry of Education, Science and Technology of Korea(14-IT-03). It was also supported by Ministry of Culture, Sports and Tourism (MCST) and Korea Creative Content Agency (KOCCA) in the Culture Technology (CT) Research & Development Program (Immersive Game Contents CT Co-Research Center).

References

1. A. Valli, The Design of Natural Interaction," *Multimedia Tool Applications*, **38**(3), pp. 295-305 (2008)
2. J. Calle, P. Martínez, D. Del Valle, and D. Cuadra, Towards the achievement of natural interaction, *Engineering the User Interface*, Springer, London, pp. 1-9 (2009)
3. A. Del Bimbo, Special issue on natural interaction, *Multimedia Tools and Applications*, **38**(3), pp. 293-294 (2008)
4. W. Xu and E. J. Lee, Human-Computer Natural User Interface Based on Hand Motion Detection and Tracking, *J. of Korea Multimedia Society*, **15**(4), pp. 501-507 (2012)
5. R. Stiefelhagen, J. Yang, and A. Waibel, A model-based gaze tracking system, *Int. J. on Artifical Intelligence Tools*, **6**(2), pp. 193–209 (1997)
6. E. Murphy-Chutorian and M. M. Trivedi, Head pose estimation in computer vision: A survey, *IEEE Trans. on Pattern Analysis and Machine Intelligence*, **31**(4), pp. 607–626 (2009)
7. H. Kim, S. H. Lee, M. K. Sohn, and D. J. Kim, Illumination invariant head pose estimation using random forests classifier and binary pattern run length matrix, *Human-centric Computing and Information Science*, **4**(9), pp. 1-12 (2014)
8. H. Kim, M. K. Sohn, D. J. Kim, and N. Ryu, User's Gaze Tracking System and Its Application Using Head Pose Estimation, *IEEE Int. Conf. on Artificial Intelligence, Modelling and Simulation*, pp. 166-171 (2014)
9. A. I. Maqueda, C. R. del-Blanco, F. Jaureguizar, and N. García, Human–computer interaction based on visual hand-gesture recognition using volumetric spatiograms of local binary patterns. *Computer Vision and Image Understanding*, **141**, pp. 126-137 (2015)
10. L. Breiman, Random forests, *Machine learning*, **45**(1), pp.5–32 (2001.)
11. http://openni.ru/files/handgket/index.html

MATEC Web of Conferences 56, 02012 (2016) DOI: 10.1051/matecconf/20165602012
ICCAE 2016

An Evolutionary Approach to Driving Tendency Recognition for Advanced Driver Assistance Systems

Jong-Hyun Lee and Chang Wook Ahn

Department of Computer Engineering, Sungkyunkwan University (SKKU), Suwon 440-746, Republic of Korea

Abstract. Driving tendency recognition is important for constructing Advanced Driver Assistance Systems (ADAS). However, it had not been a lot of research using vehicle sensing data, due to the high difficulty to define it. In this paper, we attempt to improve the learning capability of a machine learning method using evolutionary computation. We propose a driving tendency recognition method, with consideration of data characteristics. Comparison of our classification system with conventional methods demonstrated the effectiveness and accuracy over 92% in our system. Our proposed evolutionary approach is confirmed that improve the classification accuracy of the learning method through evolution in the experiment.

1 Introduction

Automobile safety driving support apparatus is one of the major issues in the world automobile industries. The technology has been actively developed and commercialized. Some technologies such as pre-crash safety [1], forward vehicle collision warning [2], brake assist [3], blind spot warning [4], and lane keeping assist [5] are being commercialized and applied to luxury cars. The products are expected to be extended to all cars as well as luxury cars.

Safety of cars driving is significantly affected by driving tendencies of the driver as well as the safety support systems because the support of driver assistance systems is limited and human is still driving manly. However, the driver's inclination is difficult to define as objective information. The existing research has been carried out to quantify subjective driving through the driver's inclination Experience Index rating [6]. However, the researches rarely used the combined information with the sensor and driving data. Also, there are additional barriers to develop the safety driving support system that can recognize aggressive driving behavior of the driving tendency. The problems are that people have too many type of driving patterns, additional equipment such as camera and bio-signal sensors increase the cost, the driving tendency may change as situation and time, and so on.

In this paper, we design a driving pattern recognition algorithm for the driver assistance systems that use only basic signal data of electronic control units (ECU) in the vehicle collected by controller area network (CAN). The main idea is that the signal combinations in the many factors such as vehicle acceleration, engine speed, torque, and steering wheel angle reflect the driver's inclination.

The algorithm requires a high accuracy and flexibility due to the complex and changing the objective function. Also, there are many dozens of data sources (i.e., features) that can be influence by the driver's inclination among thousands of the sources. The data consisting of many features are collected dozens of times every second in the real driving environment. It is difficult to define the optimal pre-processing function that is able to preserve the important information of the driver, but the raw data have too huge size to learn the pattern using existing methods in this environment. To solve this problem, we design high performance machine learning algorithm based on evolutionary computation and random forest algorithm for the big and complex dataset.

The method we propose in this research is novel in terms of technology convergence of evolution and machine learning technique. Also, it is meaningful that the challenge issue about the driving tendency is attempted to solve by applying advanced information technology. In this work, random forest which is widely used for pattern learning is tried to improve by evolutionary computation for adapting to the tendency changes and obtaining the high accuracy of driving pattern classification. For the validation of the proposed algorithm, we collect real driving data for ten expert drivers and perform the comparison experiment with existing algorithms.

The rest of the paper is organized as follows: Section 2 describes the proposed evolutionary approach for the recognition system in detail and in Section 3 the experimental results are depicted. Conclusion is placed in Sections 4.

2 Proposed Approach

Operation tendency of a vehicle reflects the emotion states of the driver. The tendency recognition is the main point of the driving assistance systems, particularly for the intelligent active safety alert systems. Also, it is significantly important in the improvement of effectiveness and stability of intelligent safety systems that alert a collision to driver.

2.1 Data Acquisition

The tendency unpredictability appears in the operation. Sometimes the signals may be clearly distinct and be too minuteness to distinguish. For this reason, the recognition system uses the sensor information as possible that associate with operations of the driver.

In this study, vehicle data were collected in Korean K9 model of Hyundai Motors by CAN on-board diagnostics (OBD) device with universal serial bus. About a hundred sensor data such as height, velocity, steering, gear, lane keeping, and fuel consumption were selected among a thousand data in ECU based on the relationship with the driver. The frequency for data collection was 1000~2500Hz. The raw data was used to prevent loss of information. In order to collect a variety of drivers' patterns, ten expert drivers travelled the same course of approximately 15km at the similar time of day as shown in Figure 1. In order to ease tensions on the unfamiliar environment, the course driving for each driver was performed twice; the second data has been used in the experiments. As the result, the huge size of data was several gigabytes were collected. The new method to handle and learn the data was required for the big and complex dataset.

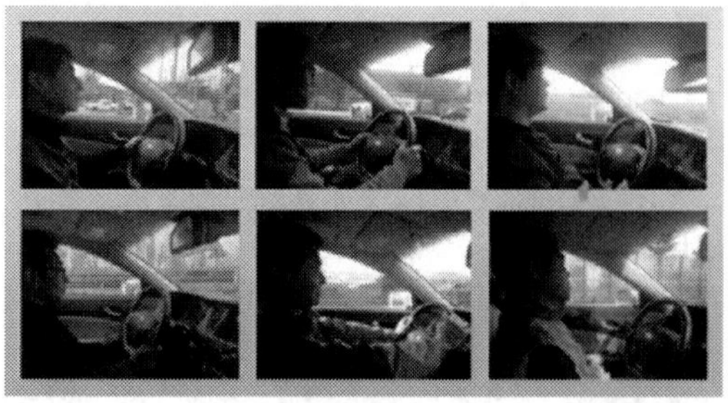

Figure 1. Ten expert drivers travelled the same course.

2.2 Driving Tendency Recognition

Random forest learning algorithm [7] is widely used for handling big data, pattern classification, and recognition. The ensemble principle is the key point of random forest. The algorithm constructs many weak-classifiers consisting of decision tree for randomly selected feature or sample subset in a training dataset. The each decision tree is learned in the given data subset to produce a decision making model. In other words, the feature points in their dataset are examined to divide the conditions for deriving an effective decision. In the classification, the algorithm collects the predicted decision results that are obtained by the models, respectively. The final decision is made by a majority rule of the results.

The main point of this research is that the subset sampling process of features and samples can be improved by evolutionary computation [8] because it is perform by the random rule. Although the random rule is an advantage to relieve the overfitting, it is bring up the decrease the learning capability. In this research, it is important to obtain the search capability of the learning algorithm due to the characteristic of the complex objective function. Thus, the random forest algorithm is designed as evolutionary approach.

2.2.1 Overall process of the proposed method

The proposed algorithm is performed as shown in the Figure 2. Selected data subsets by the rules that are randomly generated at the initial state are assigned to each weak classifier. The classifiers are constructed based on the subsets, and then they predict the class (i.e., each driver) of training samples and make a decision result by the majority rule. The results are evaluated based on the classification accuracy to give the fitness for evolutionary individuals. The population consisted of the individuals pass though the genetic process such as selection, crossover, mutation, and again evaluation iteratively. When the individuals are converged, the algorithm stops repeating and then derives one of the best in individuals as the result that means the optimal subsets for the classification. In the next step of using the trained model, it performs ensemble classification using the random forest that generated from the data of the optimum individual.

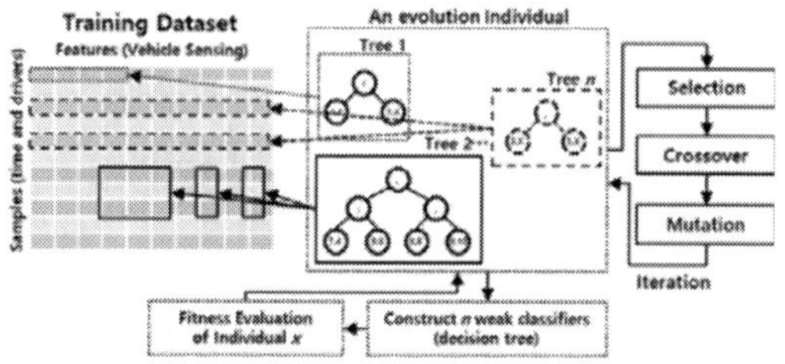

Figure 2. The evolutionary process of the proposed learning algorithm.

2.2.2 Structure of an evolution individual

An individual is designed for data subset selection in the tree structure. The tree has two types of nodes branch and leaf. The leaf nodes are assigned the coordinates of the subset in the full dataset. The m-th column n-th row is represented as {n, m}. Also, it can have an operator that selects all in a row or column. In the initial, the node is assigned a random value within the data size. Then, that can be changed by the genetic operators such as crossover and mutation. The branch nodes have the range selection operator of the comma and colon. The comma is union operator, which combines multiple references into one reference. The colon is range operator, which produces one reference to all the cells between two references, including the two references. The operators may constitute a region with reference to the value of the child node. Sometimes the subset region can cause incomplete problems for the features and samples. Then, the classifier uses the dataset in complete area.

2.2.3 Genetic operators for the evolution

The crossover operator consisting of internal and external crossover as shown in Figure 3 leads the interactions between the individuals or the trees in an individual. These crossover methods are designed in consideration of the characteristics of a number of tree-based individual. In the external crossover, the two subtrees are selected in different two individuals and are swapped. The internal crossover is performed in the same individual.

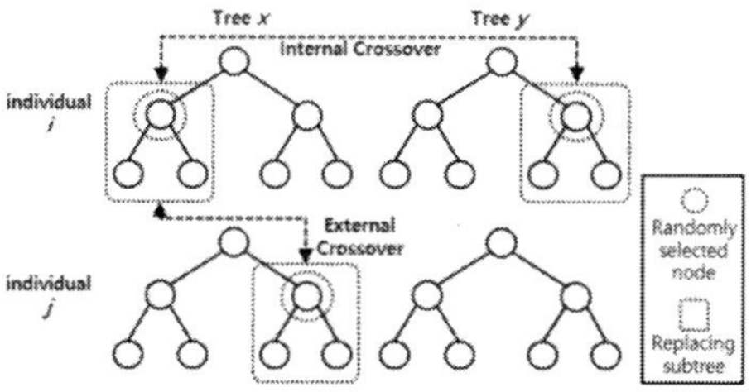

Figure 3. A genetic operator: internal and external crossover.

Selection and mutation process follow the rule of existing tree based evolutionary algorithms. The selection replicates the higher fitness object to the next generation through a competitive. The mutation performs such as changing the value of any node of an individual for increasing the search capability.

2.3 Use of Driving Data

Each driver has a specific driving tendency. The tendency may be evaluated by the statistical values that appear in the operation such as involvement lane, over speeding, and sudden starts and stops. In this research, we gave the risk ratings to the experts based on the safety driving instructions of traffic agency for the full driving data of them. In other words, some drivers in a lot of the violation had a high-risk rating. Other drivers keeping regulations had a good safety rating. It means that can generate a learning model for the risk driving based on the tendency. They can be used for evaluation of diver's inclination in the driving. For this, the accuracy of the learning model is more important than anything. Thus, the experiments in the following section perform with respect to the learning accuracy.

3 Experimental Results

The simulation environment of the proposed method was constructed in C++ with parameters including the number of evolution individuals, depth limitation of tree in the individuals, iteration, probabilities of crossover, and mutation set at 100 individuals, 4 depths, 100 generation, 0.7, and 0.1, respectively. The maximum depth of decision trees is unlimited. The number of decision trees is 100.

Figure 4 shows an accuracy variation of the proposed Random Forest in the learning process of the generations. Early generations seem low accuracy due to the incompleteness of the data sets. However, it is improved by the evolution of tree subset selectors to improve the classification accuracy through evolution. also, it seems the average, the maximum, and minimum accuracy of the one hundred individuals increase depending on the progress of the generation. The proposed approach was confirmed that improve the classification accuracy of the learning method through evolution in this experiment.

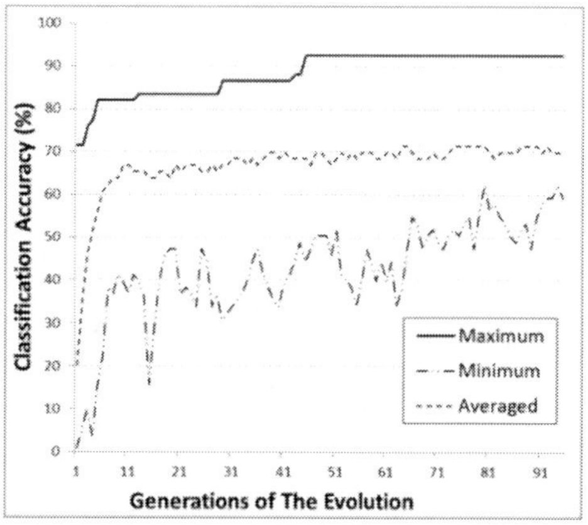

Figure 4. The experiment result of the proposed of the generations progress.

For comparing accuracy, we adopted the classification experiment to the typical classifiers such as support vector machine (SVM), multilayer perceptron (MLP), AdaBoost.M1, classification and regression trees (CART) of decision tree, and random forest. The parameters of the conventional classifiers were set at a typical value of WEKA [9]. The accuracy result in this paper was obtained by 10-fold cross validation.

Table 1 displays the classification results for conventional classifiers (implemented in WEKA) and our proposed method. By referring to the results of Table 1, we see that our classification system resulted in accuracy 7.03% higher than average of the conventions, and also 3.91% higher than SVM that is the best conventional classifier.

Table 1. The comparison experiment results with the conventional algorithms and the proposed evolutionary approach in the classification accuracy (%).

SVM	MLP	Adaboost M1	CART	Random Forest	Evolutionary Approach
88.57	86.71	83.72	82.82	85.43	92.48

4 Conclusions

The recognition of driving tendencies based on basic vehicle sensor data is a challenging task in the safety driving assistance system. This paper targets the improvement of driver recognition accuracy, and proposes a new learning method based on the evolutionary approach and random forest. Our proposed method was promoted the search capability by the evolution. Experimentally, we validated that the proposed mechanism is working well and the superiority in comparison to conventional classifiers in the selected dozens of sensor data in ECU by the relationship with the driver.

In future work, the improvement of the structure of select area operators is required. Adaptive system design has been considered to mitigate the sensitivity of the parameter settings. Also, it has devised a way to verify the operation model based on the risk in the application step. It plans to develop the other applications.

Acknowledge

This research was supported by X-Project Program through the National Research Foundation of Korea(NRF) funded by the Ministry of Science, ICT & Future Planning (NRF-2015R1A2A1A16074929).

References

1. E. Infantes, M. E. Caspar, S. Kramer, S. Schaub, T. Langner, A. Eggers, and P. Lemmen, *The Effect of Pre-Crash Safety Systems to Occupant Protection in Offset Frontal Impacts*. SAE Technical Paper **2015-26-0164**, (2015)
2. K. D. Kusano, and H. C. Gabler, *Safety benefits of forward collision warning, brake assist, and autonomous braking systems in rear-end collisions*. IEEE Transactions on Intelligent Transportation Systems, **13(4)**, 1546-1555, (2012)
3. A. Badea-Romero, F. J. Páez, A. Furones, J. M. Barrios, and J. L. de-Miguel, *Assessing the benefit of the brake assist system for pedestrian injury mitigation through real-world accident investigations*. Safety science **53**, 193-201, (2013)
4. B. F. Wu, H. Y. Huang, C. J. Chen, Y. H. Chen, C. W. Chang, and Y. L. Chen, *The blind spot warning system for daytime and nighttime.*, 2012 IEEE International Conference on In Industrial Technology (ICIT), pp. 591-597, (2012, March)
5. K. Mineta, K. Unoura, and T. Ikeda, *Development of a lane mark recognition system for a lane keeping assist system*. SAE Technical Paper **2003-01-0281**. (2003).
6. D. Yagil, *Interpersonal antecedents of drivers' aggression,* Transportation research part F: traffic psychology and behaviour **4(2)**, 119-131, (2001)
7. A. Liaw, and M. Wiener, *Classification and regression by randomForest*. R news **2(3)**, 18-22, (2002)
8. T. Bäck, and H. P. Schwefel, *An overview of evolutionary algorithms for parameter optimization*. Evolutionary computation **1(1)**, 1-23, (1993)
9. M. Hall, et al. *The WEKA data mining software: an update*. ACM SIGKDD explorations newsletter **11.1**, 10-18 (2009)

Based on Gaussian Mixture Model to Explore the General Characteristic and Make Recognition of XinTianYou

Juan Li[1], Kai Wu[2], Yinrui Wang[2] and Xinyu Yang[2]

[1]Centre for Music Education, Xi'an Jiaotong University, 710049 Xi'an, China
[2]Department of Computer Science & Technology, Xi'an Jiaotong University, 710049 Xi'an, China

Abstract. *XinTianYou*, a folk song style from Shannxi province in China, is considered to be a precious traditional culture heritage. Research about *XinTianYou* is important to the overall Chinese folk music theory and is potentially quite useful for the culture preservation and applications. In this paper, we analyze the general characteristics of *XinTianYou* by using the pitch, rhythm features and the combination of these two features. First, we use the Gaussian Mixture Model (GMM) to cluster the *XinTianYou* audio based on pitch and rhythm respectively, and analyze the general characteristics of *XinTianYou* based on the clustering result. Second, we propose an improved Features Relative Contribution Algorithm (CFRCA) to com-pare the contributions of pitch and rhythm. Third, the probability of a song being *XinTianYou* can be estimated based on the GMM and the cosine similarity distance. The experimental results show that *XinTianYou* has large pitch span and large proportion of high pitch value (about 22%). Regarding the rhythm, we find that moderato is dominated while lento-moderato keep a similar ratio as moderato-allegro. The similarity between pitch features of all *XinTianYou* songs is more significant than rhythm features. Additionally, the average accuracy of *XinTianYou* recognition reaches 92.4% based on our method

1 Introduction

XinTianYou is an improvisational folk song style which stems from Northern Shannxi, China. It is a unique music form and a treasure of traditional Chinese culture. Exploring the general characteristics of *XinTianYou* can improve the understanding of the aesthetic characteristics and the artistic style of this genre of music, and potentially benefit the cultural preservation.

In this paper we aim to find and understand the general characteristics of *XinTianYou*. We use the Gaussian Mixture Model (GMM) to analyze the *XinTianYou* audio features including pitch and rhythm and do the clustering. Then analyze the general characteristics of *XinTianYou* based on the clustering results and music theory. A new method named Complete Features Relative Contribution Algorithm (CFRCA) is introduced to compare the contribution of pitch and rhythm. Lastly, we use GMM and the cosine similarity distance to recognize *XinTianYou*. Potentially our method can also be applied to other types of folk songs, which enlighten the study of finding the "general characteristics" of different kinds of folk songs and further find the general characteristics or gene of Chinese culture.

There is a variety of related work in Chinese folk music area that has looked at finding general characteristics of *XinTianYou*. In [1] the authors point out that the tune of *XinTianYou* is generally divided into two types: one has slow speed, long-drawn-out melody and broad range, while the other has a more rigorous structure, graceful and smooth melody. Chen. et al [2] think that *XinTianYou* has free rhythm and unchained melody styles. Zhang. et al [3] points out that the basic structure of *XinTianYou* is stable and clear. The basic structure contains two continuous 4th melodies. *XinTianYou* also features sonorous drawn-out strains and flexible tempo. Liu [4] states that *XinTianYou* has simple tunes but rich mode scale, and "double 4th interval structure" melody is an important feature of *XinTianYou*. These previous studies focus on the music theory. They cannot support their conclusion with precise data evidence and recognize *XinTianYou* automatically because of the lack of quantitative analysis and scientific experiment.

In the area of music information retrieval, quantitative analysis and identification method have been used. But lots of research work focuses on the classification of music genres [5-7], which mainly cares about the accuracy of classification rather than the general characteristics.

In this paper, we apply mathematic methods and modern data analysis techniques to analyze the general characteristics and make recognition of *XinTianYou*

. We first apply GMM to the pitch and rhythm features of the music data, then explore the general characteristics, and use GMM combined with cosine similarity to identify *XinTianYou*. The experiment shows that our results are consistent with music theory and our method can identify *XinTianYou* effectively.

The paper is organized as follows: we describe the use of GMM for audio features in Section 2 then analyze the general characteristics of *XinTianYou* based on pitch and rhythm in Section 3 next we demonstrate the experimental results in Section 4 and finally the conclusions and future work in Section 5.

2 The processing of GMM based on features of audio data

Gaussian Mixture model (GMM) [8-9] has been widely used in the acoustic area. According to the statistics theory, using a linear combination of Gaussian probability density functions can approximate arbitrary distribution, thus theoretically GMM could describe the statistical distribution of music elements. We apply GMM to cluster the features of *XinTianYou* audio data. The clustering result can represent overall information of the audio data, and is useful for analyzing its general characteristics and further the reorganization of *XinTianYou*.

2.1 Formatting the title, authors and affiliations

The database we use in this paper contains audio data of 109 *XinTianYou* songs. They are selected from Integration of Chinese folk songs [1], which is one of the most authoritative collections of Chinese folk songs. This collection contains the most rich, comprehensive and authentic Chinese folk song music [10]. After getting audio data for each song, we decompose each audio data to short segments in order to get locally stable signal. To make sure the adjacent segments transfer smoothly, we allow overlap between them. Assume that a piece of audio data has been sampled with frequency F, we can get NL samples. The number of frames Nframes of this data can be computed as follows:

$$N_{frames} = \left\lfloor \frac{N_L - N_M}{N_f - N_M} \right\rfloor \tag{1}$$

In which Nf is frame length - the number of samples in each frame, and NM, is overlap length - the number of samples in the overlapping region. Practically F, Nf and NM are set to 22050Hz, 1024 and 896 respectively.

After the above preprocessing, we compute two types of features: the pitch-domain features LSP (Linear Spectrum Pairs) [11], PF (Pitch Frequency) [12], and time-domain features Tempo [13], FP(Fluctuation Pattern) [14]. We use the software MARSYAS [15] (Music Analysis, Retrieval and Synthesis for Audio Signals) and MIR-TOOLBOX[16] to extract these features.

2.2 The process of GMM

GMM Gaussian mixture model uses the weighted linear combination of Gaussian probability density functions to representation the statistical distribution of data. Here we build a GMM for each of the features extracted from our audio data.

Each audio piece includes Nframes frames and for each frame we compute features. Each feature is N-dimensional (N may change for different features). Then each feature vector for each frame can be described as: $x_j = [x_{j,1}, x_{j,2}, \ldots, x_{j,N}], j = 1, 2, \ldots, N_{frames}$.

2.2.1 Dynamic K-means to initialize GMM

When building GMM, we need to find a proper cluster number to balance the accuracy and computational efficiency. Here we apply the dynamic K-Means algorithm to determine the cluster number automatically.

The idea of dynamic K-Means is finding a cluster number which can minimize the following error function:

$$E_K = \sum_{i=1}^{K} \sum_{x \in L(k)} | x - \mu_k |^2 \tag{2}$$

Where K is clustering number, L(k) is the k-th cluster, x is feature vector, μk is the mean of cluster L(k). Through dynamic K-Means algorithm, we can determine the cluster number automatically, and calculate the mean, variance and weights of each cluster. Then we initialize the Gaussian mixture model with these parameters. In our experiment, the K for feature LSP, PF is 30 and for feature Tempo, FP is 6, 10 respectively.

2.2.2 Estimate parameters of GMM

Assume each feature of each frame data obeys the Gaussian mixture distribution:

$$x_i \sim \sum_{k=1}^{K} \pi_k N(x_i; \mu_k, \Sigma_k) \tag{3}$$

We use the Expectation Maximization (EM) method to find the Maximum Likelihood Estimation (MLE) in order to estimate the model parameters πk, μk and Σk. More specifically we find the parameters which make the expectation of the following logarithmic likelihood function to be maximal:

$$\sum_{i=1}^{N_{Frames}} \log\{ \sum_{k=1}^{K} \pi_k N(x_i; \mu_k, \Sigma_k)\} \tag{4}$$

The EM algorithm is an iterative algorithm that contains two main steps. First, in the E-step, it tries to "guess" the probability that sample xi is produced by each Gaussian component; second, in the M-step, it updates the parameters of the model based on the guesses.

For each feature we can compute a GMM by using above method. All the GMMs can form a database. The process of building such a database is illustrated in Fig. 1. In the following sections we show how to use this database for the analysis and recognition of *XinTianYou*.

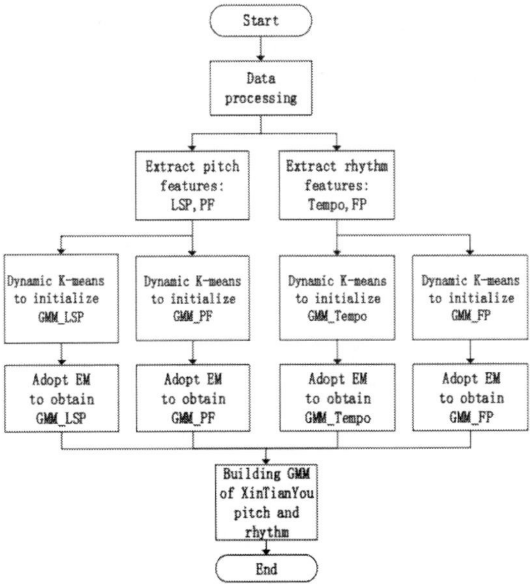

Figure 1. The process of creating the database of GMM parameters

3 General characteristics analysis based on GMM

The basic elements of music include pitch level, rhythm, strength and timbre. The pitch level and length elements are the dominant features music. This is because the artistic effects of a piece of music do not change radically if the strength and timbre changed, and we can still recognize the basic melody. But if the pitch level or length are changed, the music might become unrecognizable. Therefore, the most important basic elements of music are the pitch level and length. So we choose pitch, rhythm, and the combination of them to explore the general characteristics of our data.

3.1 General characteristics on pitch

Pitch, indicates the subjective feeling of human when hearing sound. The value of pitch depends on the fundamental frequency of sound. Higher fundamental frequency results in higher pitch value. We use two typical features to represent pitch: Linear Spectrum Pairs (LSP) and Pitch Frequency (PF).

The parameters of LSP are recognition coefficients $\{a_i\}(i=0,1,2,\ldots,p)$, where

$$A(z)=1+\sum_{i=1}^{p}a_i z^{-i} \quad (5)$$

where A(z) is the amplitude. From (5) can be seen, lower-order coefficients describe the overall shape of the signal, and higher-order coefficients denote more detailed information. The change of pitch value results in the corresponding change of A(z) This feature is mainly reflected by the lower-order coefficients, which located in the first few dimensions in LSP.

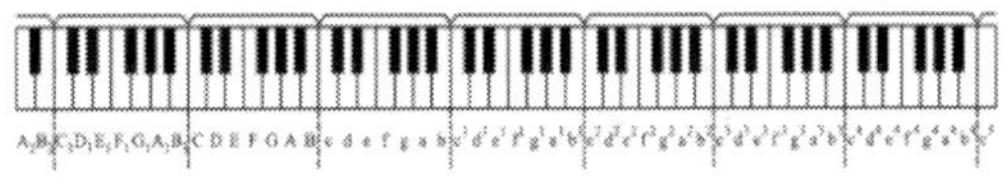

Figure. 2. The distribution of music pitch

Fig. 2 shows different musical labels which represent different pitches. The first few dimensions of LSP change according to pitches significantly. The larger the pitch values, the higher the LSP values. In Table 1, we show the some of the pitches in Fig. 2 and the first two dimensions of the corresponding LSP.

We use the GMM to cluster the values of LSP parameter extracted from our data. Table 2 shows the GMM parameters.

Table 1. Pitch and the corresponding LSP.

Pitch	LSP	Pitch	LSP
c	(0.021590, 0.041759)	a^1	(0.065057, 0.078752)
d	(0.023863, 0.043392)	b^1	(0.072334, 0.083915)
e	(0.026762, 0.047466)	c^2	(0.075741, 0.085291)
f	(0.028137, 0.048744)	d^2	(0.085412, 0.096189)
g	(0.031064, 0.050118)	e^2	(0.095214, 0.103180)
a	(0.034433, 0.050691)	f^2	(0.100769, 0.107648)
b	(0.038145, 0.053089)	g^2	(0.112784, 0.117654)
c^1	(0.040163, 0.053993)	a^2	(0.126203, 0.129951)
d^1	(0.044813, 0.060637)	b^2	(0.141350, 0.144610)
e^1	(0.049778, 0.065306)	c^3	(0.149566, 0.152258)
f^1	(0.052413, 0.066771)	d^3	(0.167541, 0.170068)
g^1	(0.058259, 0.070933)	e^3	(0.187734, 0.190275)

We can find the centre of different clusters by GMM_LSP from Table 2. The different mean of clusters is mapped into ranges of pitch values. For example, the cluster 1 corresponding to the mean value (0.0478, 0.0599) falls into the pitch ranges d1-e1, then the weight of cluster is accumulated into pitch ranges d1-e1. We get the following histogram (Fig.3) by summing all weights in the same pitch ranges.

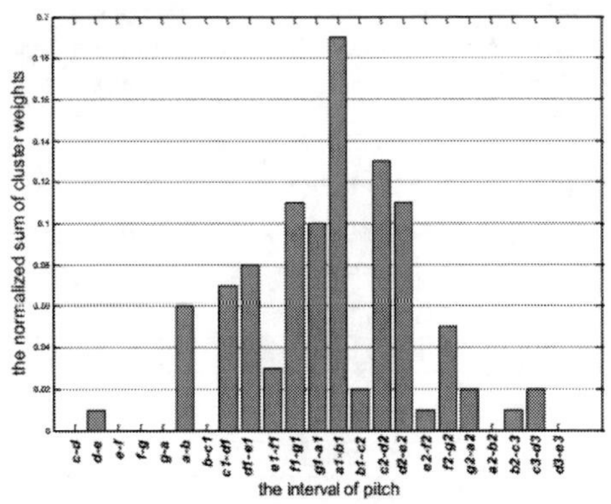

Figure 3. The histogram of pitch feature

Table 2. GMM parameters of LSP (GMM_LSP).

cluster	μ	Σ	π	*cluster*	μ	Σ	π
1	(0.0478,0.0599)	(0.0008,0.0018)	0.08	16	(0.0819,0.0886)	(0.0011,0.0010)	0.02
2	(0.0725,0.0811)	(0.0015,0.0010)	0.01	17	(0.0879,0.0934)	(0.0019,0.0016)	0.04
3	(0.0953,0.1057)	(0.0001,0.0001)	0.01	18	(0.0791,0.0860)	(0.0007,0.0009)	0.03
4	(0.0676,0.0779)	(0.0044,0.0045)	0.06	19	(0.0662,0.0753)	(0.0010,0.0012)	0.07
5	(0.0839,0.0898)	(0.0001,0.0007)	0.02	20	(0.0514,0.0629)	(0.0016,0.0016)	0.01
6	(0.0936,0.0978)	(0.0012,0.0010)	0.03	21	(0.0582,0.1019)	(0.0145,0.0246)	0.02
7	(0.0561,0.0664)	(0.0010,0.0011)	0.06	22	(0.0342,0.0460)	(0.0017,0.0022)	0.03
8	(0.1011,0.1045)	(0.0033,0.0026)	0.04	23	(0.0595,0.0704)	(0.0011,0.0014)	0.01
9	(0.0749,0.0813)	(0.0001,0.0003)	0.01	24	(0.0291,0.0460)	(0.0031,0.0045)	0.05
10	(0.0877,0.1028)	(0.0115,0.0275)	0.01	25	(0.0243,0.0463)	(0.0019,0.0034)	0.05
11	(0.1115,0.1090)	(0.0100,0.0238)	0.01	26	(0.1382,0.1395)	(0.0102,0.0099)	0.01
12	(0.0426,0.0525)	(0.0003,0.0011)	0.03	27	(0.0631,0.0738)	(0.0009,0.0016)	0.05
13	(0.0443,0.0568)	(0.0014,0.0019)	0.02	28	(0.0707,0.0785)	(0.0003,0.0006)	0.07
14	(0.0380,0.0509)	(0.0014,0.0023)	0.07	29	(0.1439,0.3094)	(0.0122,0.0223)	0.01
15	(0.0760,0.0837)	(0.0011,0.0010)	0.03	30	(0.1142,0.1161)	(0.0054,0.0050)	0.02

Fig. 3 shows that the pitch of *XinTianYou* spans a large range and mainly gathers in high tone area. Here we analyze the histogram in details. First, we can see that the overall pitch range is between d-e and c3-d3. Generally speaking, the high pitch of soprano is about c2-c3, and tenor is c1-c2. The clusters fall into pitch range d2-e2 reach nearly 22%, which shows the pitch is gathered in a higher value area. Also we can see that the proportion of altos is the majority part of the histogram. This is because of the limitation of human voice and the need of altos for the preparation of climax. Moreover, the center of cluster ranges from one-lined octave a-b to three-line octave c3-d3, which is across three octaves. This shows that the pitch spans a relatively large range. In addition, the histogram shows some very small values, such as those for slots b2-c3, b-c1 and e2-f2. These values look like 0 because of the visualization of the histogram. The reason for these small values is that the Chinese Five-tone modes do not include semitone relations except that the six and seven tones do have semitone, but the six and seven tones rarely happen in Chinese folk songs. Lastly, we can see that there is even a recognizable value in pitch range c3-d3, which seems too high for human to sing. In fact such high notes are rarely happen in *XinTianYou*. The reason is that LSP may confuse overtones with normal notes and causes this unusual high pitch.

Table 3. Pitch and corresponding PF.

Pitch	PF	Pitch	PF
c	8452.703682	a^1	10540.322318
d	8637.746753	b^1	10735.355463
e	8841.169540	c^2	10841.908323
f	8941.068626	d^2	11043.199109
g	9137.720976	e^2	11249.289466
a	9338.307723	f^2	11340.560287
b	9539.031137	g^2	11544.899454
c^1	9640.591765	a^2	11742.506746
d^1	9839.225916	b^2	11939.553027
e^1	10036.915030	c^3	12042.834835
f^1	10141.457661	d^3	12227.168391
g^1	10343.886322	e^3	12428.597918

To analyze the pitch, we also compute the feature PF. The correspondence between PF and pitch is shown in Table 3.

It can be seen from Table 3 that there are a simple linear relationship between pitch and PF. The difference of PF value is 100 corresponding a semitone, such as e->f, b->c, and 200 corresponding a whole tone such as c->d, d->e, f->g. Based on PF, the GMM clusters results are showed as Table 4.

We also compute a histogram for the PF cluster by using the cluster mean. The weights for clusters are accumulated into 23 different pitch ranges. The result is showed as Fig. 4.

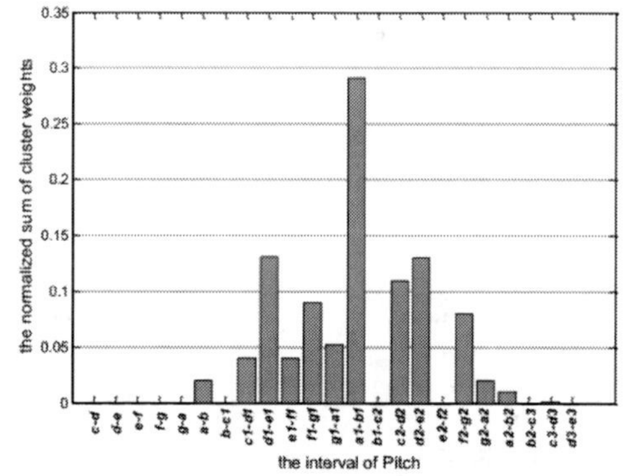

Figure 4. The histogram of pitch feature

Based on the distribution of PF, it can be further concluded that the pitch of *XinTianYou* spans in a large range and mainly gathers in high pitch area. The pitch range above d2-e2 takes nearly 23%. The proportion of altos takes a majority part of the data. The pitch range is from a-b to c3-d3, which covers three octaves. This demonstrates again that *XinTianYou* spans a relatively large range. The proportion of the pitch above c^3 is different from the result of LSP. The histogram value for pitch c^3-d^3 is near to 0. This is because PF feature only consider frequency up to 2400Hz and filters the overtone. In addition, the histogram values for b-c^1, b^1-c^2, e^2-f^2, b^2-c^3 are the similar to the results by LSP.

Table 4. GMM parameters of PF

cluster	μ	∑	π	*cluster*	μ	∑	π	*cluster*	μ	∑	π
1	10241.17	5.8891	0.03	11	11044.82	6.2639	0.07	21	9984.156	2.7769	0.001
2	10439.45	2.2246	0.001	12	12066.48	42.7593	0.001	22	9743.09	4.0011	0.01
3	10038.77	1.4168	0.01	13	11421.28	114.4496	0.08	23	10533.91	13.9345	0.001
4	10140.27	1.6062	0.01	14	9839.809	0.9368	0.04	24	10578.82	6.6313	0.001
5	9494.753	80.377	0.02	15	10140.85	0.7101	0.04	25	10730.61	3.2575	0.06
6	10937.21	4.61	0.03	16	10645.72	1.5175	0.05	26	9641.595	1.1343	0.03
7	11218.53	61.4448	0.06	17	10596.56	751.5881	0.10	27	10540.76	0.8913	0.06
8	9936.723	1.6054	0.04	18	9839.517	2.0984	0.01	28	11741.93	0.7763	0.02
9	10332.38	9.3718	0.01	19	10440.42	0.6573	0.03	29	10342.73	1.2871	0.05
10	10036.36	1.1352	0.04	20	11937.94	7.8618	0.01	30	10841.98	0.8463	0.08

3.2 General characteristics on rhythm

Rhythm denotes the organization of the length and strength of sound. It includes time-related elements such as tempo and beat. Here we use Tempo and FP to represent the rhythm information.

Tempo is computed by first detecting the peak of the signal curve, then counting the number of beats per time unit. It reflects the rhythm information to some extent. Table 5 shows the tempo values and corresponding speed.

Table 5. Speed and corresponding Tempo

Speed	Tempo	Speed	Tempo
Lento	50.6668	Mf	90.6661
Rubato	60.0928	Allegretto	99.3841
Mp	69.8375	Allogro	109.9577
Moderato	80.7498	Molto mosso	120.1858

We compute the GMM for tempo feature. The parameters of clusters are shown in Table 6 below.

Table 6. The parameters of GMM_Tempo clusters

cluster	μ	∑	π
1	60.3513	1.929	0.2
2	100.2964	3.3397	0.09
3	70.109	1.2834	0.29
4	50.6639	8.1217	0.07
5	80.0148	1.8544	0.27
6	90.002	2.1225	0.08

We build a histogram by using the centroid of each cluster. The vertical axis values of the histogram are the proportion of the clusters falls into each tempo range. The histogram is showed in the Fig. 5.

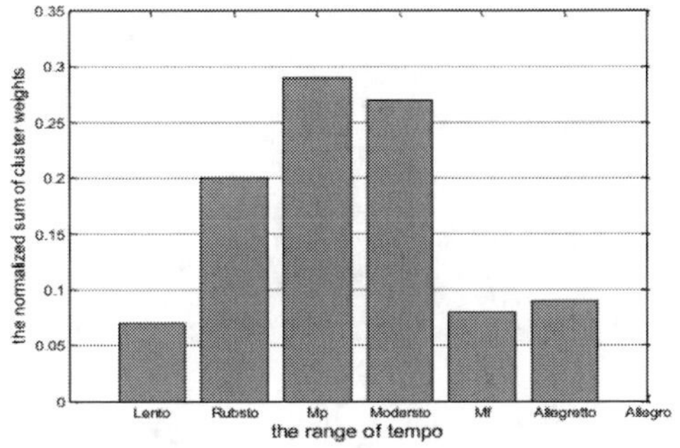

Figure 5. The histogram of rhythm feature

The general characteristics of *XinTianYou* rhythm is that the distribution of tempo spans from Lento to Allegro. The central part of the whole range is dominant while the two endings keep a similar ratio. The values from Mp to Mf are dominated, which is consistent with the frequency of walking. In addition, the figure shows that the value for Allegro is zero. This is because the Allegro is clustered into Allegretto because there are only few instances in *XinTianYou*.

Fluctuation Pattern (FP) is also used to analyze the distribution of speed. The FP values and their corresponding rhythm types are shown in Table 7.

Table 7. Speed and corresponding FP

Speed	FP	Speed	FP
Lento	1.704	Mf	3.124
Rubato	2.062	Allegretto	3.395
Mp	2.426	Allegro	3.645
Moderato	2.832	Molto mosso	3.931

The GMM parameters of Tempo clusters are shown in Table 8.

Table 8. The parameters of GMM_FP clusters

$Model_i$	μ	Σ	π
1	2.7270	0.1901	0.12
2	3.0051	0.2483	0.17
3	3.4658	0.5831	0.08
4	2.7594	0.4926	0.07
5	2.4353	0.2724	0.08
6	2.9121	0.2015	0.09
7	3.25	0.0902	0.10
8	1.712	0.1655	0.09
9	2.2024	0.1643	0.17
10	3.830	0.8542	0.02

Again we compute the histogram as shown in Fig. 6, from which we can see a similar distribution as Tempo. Compared to Fig. 5, the Allegro to Molto mosso has a larger value. This denotes that there is few Molto mosso in *XinTianYou*.

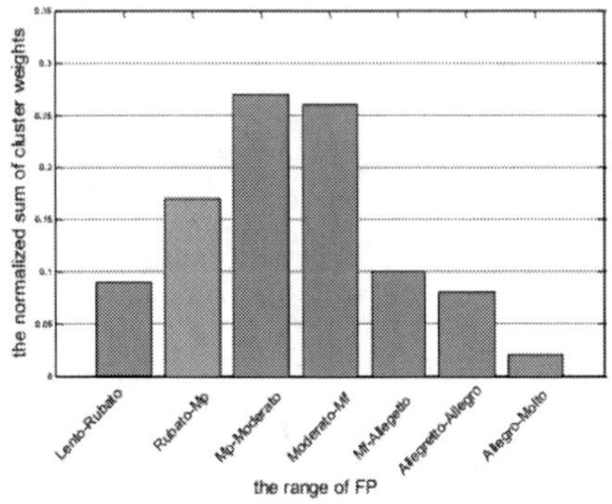

Figure 6. The proportion of *XinTianYou* speed

3.3 Analysis on the general characteristics of pitch and rhythm's combination

As introduced in Section 3.1 and 3.2, the general characteristics of pitch of *XinTianYou* is the overall high value and the large span, and the general characteristics of rhythm is that moderato is the dominant part and lento-moderato has a similar ratio as moderato-allegro. In this section, we aim to analyze the combination of these two factors. We propose to use a method named CFRCA (Complete Features Relative Contribution Algorithm) for this purpose. This method is inspired by FRCA (Features Relative Contribution Algorithm) [17] which can be used to compute all features' relative contributions. In the rest of this section we will provide more details of both FRCA and CFRCA.

First of all, we normalize the extracted features to map the feature values to the same range. More specifically we use a Standardized deviation based method which is commonly used. The calculation formula is as follows:

$$x_i' = \frac{x_i - \bar{x}}{s} \qquad (6)$$

where $s = \sqrt{\frac{1}{n}\sum (x_i - \bar{x})^2}$.

FRCA(Features Relative Contribution Algorithm)

For the features sequence $(x_1, x_2, \ldots, x_N)$,the relative contribution of feature x_i can be calculated as follows:

$$C_1(i) = \sum_{j>i}[J(i,j) - J(i+1,j)] + \sum_{j<i}[J(i,j) - J(i,j-1)] \quad (7)$$

Where $J(i, j)$ represents the cosine similarity between the *i-th* and *j-th* features, which can be computed by $J(i,j) = \prod_{1\le g<h\le M} \cos_{i,j}(g,h)$; M represents the number of songs.

CFRCA (Complete Features Relative Contribution Algorithm)

The result of FRCA, the relative contribution $C_1(i)$, just represents the relative importance of feature x_i in a specific feature arrangement. The relative contribution $C_2(i)$ of feature x_i in other sequences can be different form $C_1(i)$. To make the contribution measurement independent to the feature arrangement, we put propose to use CFRCA, which considers all permutations of N features: $P = \{p_1, p_2, \ldots, p_{A_N^N}\}$, where p_i represents the *i-th* permutation. Then the average of the contribution values computed by all permutations is used for the final contribution:

$$C(i) = \frac{1}{A_N^N}\sum_{k=1}^{A_N^N} C_k(i) \qquad (8)$$

where $C_k(i)$ represents the relative contribution of *i-th* feature in *k-th* permutation.

By doing this we can get the contribution of each feature which fully consider the permutations between all features. Fig. 7 shows the pseudo code of CFRCA.

```
Algorithm : Complete Features Relative Contribution Algorithm
Input  : Feature set : X={x1, x2, ..., xN}
Output : The complete relative contribution of feature:
         C={C(1),C(2),...,C(N)}

for (i =1;i <= N; i++)
    Sum = 0;
    for (j = 1; j<= A_N^N ;j++)
        Cj(i) = relative_ contri(i,j);
        //compute the ith feature's relative contribution
        //in jth permution
        Sum = Sum + Cj(i) ;
    end
    C(i) = Sum / A_N^N ;
end

Procedure relative_contri(i,j)
P = {p1, p2, ..., p_{A_N^N}}
   // Set of full permutation by feature sequence
l = find_location(pj , i) ;
   // find ith feature's location in jth permutation
Contri = 0;
for (m =1; m<=N ; m++)
if(m > l)
   Contri = Contri+∏cos(pj(l,m))- ∏cos(pj(l+1,m));
else if (m < l)
   Contri =Contri+∏cos(pj(m,l))-∏cos(pj(m,l-1));
end
return Contri;
```

Figure 7. Complete Features Relative Contribution Algorithm

The experimental result of CFRCA is shown in Fig. 8. We can see that the contribution of pitch is relatively higher than rhythm. The contribution of LSP is the highest in all four features, which shows that LSP is the most representative feature of pitch. Although lower than LSP, the contribution of PF is still outstanding. The lowest contribution is from FP. The reason is that FP represents more information of rhythm compare to Tempo and the contribution of rhythm is low [18].

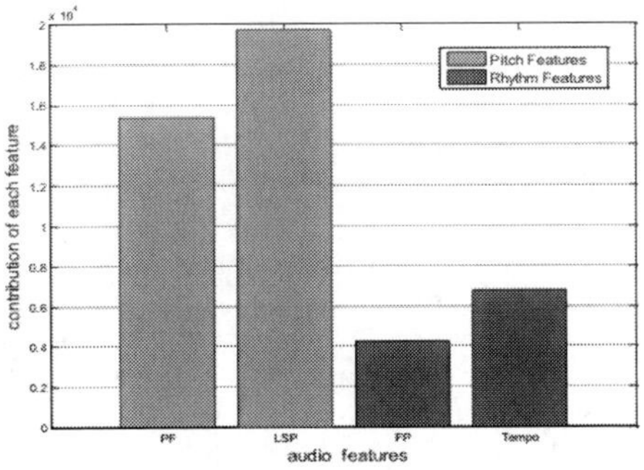

Fig. 8. The contribution of each feature by using CFRCA

In conclusion, from this section we find that *XinTianYou* has the following general characters: first, the pitch of spans a large range and mainly gathers in the high tone area. Second, the distribution of tempo spans from Lento to Allegro. The central part of the whole range is dominant while the two endings keep a similar ratio. Third, the pitch feature is more representative than the rhythm feature.

4 Recognition of XinTianYou based on GMM

In this section we use the GMM computed in Section 3.1 and Section 3.2 to recognize *XinTianYou*.

We randomly select 80 from the overall 109 songs for the training set and the remaining 29 songs and another 30 non-*XinTianYou* folk songs as test set. To measure the similarity between the test data and the GMM models, we use the following likelihood equation:

$$P(n) = \frac{1}{N_{frames}(n)} \sum_{i=1}^{N_{frames}(n)} \sum_{k=1}^{K} \pi_k \cos(x_i, \mu_k)$$

$$(n=1, 2, 3, \ldots, 80) \qquad (9)$$

Given a test song, we compare each frame of it to each GMM center and compute the cosine similarity. Then we compute the mean of all these similarities and use it for the final measurement.

To make our results take the form of probability, we normalize the value of the likelihood value for all 80 training data. More specifically each similarity value is divided by the maximum value. As a result the probability interval of *XinTianYou* is $[\min(P(x))/\max(P(y)),1]$, x=1,2,3,…80; y=1,2,3,…80. The similarity distribution can be got by sorting the P(n) values of the training set.

This make the P(n) value for test data is higher. We then compute P(n) for test data. We consider the test song is more likely to be *XinTianYou* if P(n) is higher.

In the experiment, we use the cross-validation that randomly selected five combinations of training set (80 pieces) and test set (29 pieces of *XinTianYou* and 30 pieces of non-*XinTianYou*). Table 9 shows the similarity range and probability interval of training sets. Table 10 shows the true positives (the number of 29 pieces of *XinTianYou* correctly classified as *XinTianYou*), and Table 11 shows the false positives (the number of 30 pieces of non-*XinTianYou* incorrectly classified as *XinTianYou*).

Table 9. The similarity range and probability interval of training sets

Experiment No.	Similarity range	Probability interval
i	(0.999381,0.999774)	(99.9602%,1)
ii	(0.999377,0.999791)	(99.9585%,1)
iii	(0.999574,0.999777)	(99.9796%,1)
iv	(0.999483,0.999897)	(99.9586%,1)
v	(0.999466,0.999768)	(99.9698%,1)

Probability Interval is small here because there exists general characteristics in pitch and rhythm of *XinTianYou*, which make different folk songs of test sets substituted into the model can also get high probability.

Table 10. The recognition of 29 pieces of *XinTianYou* in test sets

Experiment No.	Test sets Probability interval	True positives	Accuracy
i	(99.9725%, 100.0004%)	27	93.1034%
ii	(99.9766%, 100.0005%)	28	96.5517%
iii	(99.9601%, 99.99994%)	26	89.6551%
iv	(99.9703%, 100.0002%)	27	93.1034%
v	(99.9583%, 100.0000%)	26	89.6551%
Average		26.8	92.4138%

Table 11. The recognition of 30 pieces of non-*XinTianYou* in test sets

Experiment No.	Test sets probability interval	False positives	False identification rate
i	(79.9855%, 99.9607%)	2	6.67%
ii	(81.0654%, 88.3482%)	0	0%
iii	(81.9811%, 99.9810%)	3	10.00%
iv	(83.9703%, 99.9601%)	4	13.33%
v	(78.8953%, 99.9712%)	2	6.67%
Average		11	7.33%

Probability Interval is small in Table 9. This again proves that general characteristics in pitch and rhythm of *XinTianYou*. From Table 10 we can see that there are 26.8 true positives on average and the average accuracy is 92.4138%, and Table 11 shows the False identification rate is only 7.33% on average. The high accuracy and the low False identification rate shows that our method can recognize *XinTianYou* quite effectively. Besides, we can see some the testing interval boundaries are greater than 100% in the Table 10. This is because we use the training sets' probability interval as references, and the resulting probability interval of test sets can be out of this range.

5 Conclusions

In this paper we use pitch and rhythm features to analyze *XinTianYou*, a traditional Chinese folk song style. The Gaussian Mixture Model is used for the clustering analysis and finding the general characters of *XinTianYou* data. We analyze the results with consideration of music theory, and also propose a method to recognize *XinTianYou*.

The experimental results show that the pitch of *XinTianYou* spans a large range, which is over 3 octaves. High pitch values take a proportion about 22%. The experiment of rhythm of *XinTianYou* shows that moderato is the dominant part while lento-moderato keeps a similar ratio as moderato-allegro. The pitch features of *XinTianYou* are more representative than rhythm features. Moreover, by using our method the average accuracy of recognition *XinTianYou* reaches 92.4%.

"General characteristics" provides a new way of looking into the commonality or gene of Chinese folk songs. The method we use in this paper can also be applied to other styles of music other than *XinTianYou*. In the future we plan to explore other types of Chinese folk song styles, and look for the general characteristics between them.

References

1. Integration of Chinese folk songs · Shaanxi Juan. The Editorial Committee, Integration of Chinese folk songs: Shaanxi Juan [M]. Chinese ISBN center, (1994).
2. Chen Haoyue. Characteristics and style of Shaanxi folk songs [J]. Stage (journal of Peking university), (2014), 3:238-239.
3. Zhang Zhentao. Into the modern northern shaanxi folk songs [J]. Chinese musicology, 2012, (4):75-86.
4. Liu Dajian. Three elements of constituting the style of XinTianYou [J]. Art research, (2008), (1):42-44.
5. Fu Z, Lu G, Ting K M, et al. A Survey of Audio-Based Music Classification and Annotation [J]. Multimedia IEEE Transactions on, (2011), **13**(2):303 - 319.
6. Li T, Ogihara M, Li Q. A comparative study on content-based music genre classification[C]//Proceedings of the 26th annual international ACM SIGIR conference on Research and development in information retrieval. ACM, 2003: 282-289.
7. Sturm B L. Classification accuracy is not enough [J]. Journal of Intelligent Information Systems, (2013), 41(3): 371-406.
8. Jouvet, D., Vincent, E., Dung Tran. Investigating stranded GMM for improving automatic speech recognition [J]. //Hands-free Speech Communication and Microphone Arrays (HSCMA), 2014 4th Joint Workshop on IEEE, (2014): 192-196.
9. Dileep A, D D A, Dileep A D, et al. Class-specific GMM based intermediate matching kernel for classification of varying length patterns of long duration speech using support vector machines[J]. Speech Communication, (2014):126–143.
10. Qiao Jianzhong. It should be emphasized that the literature value of "integration" "Integration of Chinese folk songs" [J]. People's music, (1986), (2):38-41.
11. MIN Gang; ZHANG Xiongwei; YANG Jibin; CHEN Yanpu. Sparse representation and performance analysis for LSP parameters via dictionary learning [J]. Journal of PLA University of Science and Technology (Natural Science Edition), (2014):121-126.
12. Tolonen, Karjalainen. A Computationally Efficient Multipitch Analysis Model, IEEE Transactions on Speech and Audio Processing, 2000, 8(6):708-716.
13. Lartillot O. Mirtempo: tempo estimation through advanced frame-by-frame peaks tracking [J]. Proceedings of the Music Information Retrieval Evaluation eXchange (MIREX), Utrecht, Netherlands, (2010):1-2.
14. Bosteels K, Kerre E E. Fuzzy audio similarity measures based on spectrum histograms and fluctuation patterns [M]//Computational Intelligence in Multimedia Processing: Recent Advances. Springer Berlin Heidelberg, (2008): 213-231.
15. Tzanetakis G, Cook P. Marsyas: A framework for audio analysis [J]. Organised sound, (2000), 4(03): 169-175.
16. Olivier Lartillot, Petri Toiviainen. A MATLAB TOOLBOX FOR MUSICAL FEATURE EXTRACTION FROM AUDIO [M]. Proc. of the 10th Int. Conference on Digital Audio Effects (DAFx-07), Bordeaux, France, September 10-15, (2007):237-244
17. Zhen B, Wu X, Liu Z, et al. On the Importance of Components of the MFCC in Speech and Speaker Recognition [J]. ACTA SCIENTIARUM NATURALIUM-UNIVERSITATIS PEKINENSIS, (2001), 37(3): 371-378.
18. GOTO M. A chorus-section detecting method for musical audio signals [A]. Hongkong: IEEE Press, (2003).437-440.

Signal Frequency Estimation Based on Kalman Filtering Method

Yandu LIU , Di YAN and Haixin ZHENG

Equipment Academy, 101416 Beijing, China

Abstract. In order to further improve the frequency estimation precision under high dynamic, the frequency estimation method based on Kalman filter is studied. Two methods of Kalman filter directly using for frequency estimation are studied, taking the phase of echo signal and the echo signal for measurements, which adopt nonlinear and linear Kalman filtering algorithm to estimate the Doppler frequency. Based on the study of performance of these two methods, the Doppler adaptability of the two methods is studied. The simulation results show that the speed precision is improved. But, the two methods of this research are limited by the fixed frequency conversion, and when Doppler changed a large range, the filtering is easy to diverge.

1 Introduction

LEO satellite flight dynamics complicated, speed and all derivative values are large, so it belongs to frequency estimation problem under high dynamic condition. Phase-locked loop is the most commonly used in satellite signal processing frequency estimation method[1-4], but under high dynamic conditions it requires using higher-order loop, which cause stability problems. Kalman filter is an optimal estimation method, with excellent performance and noise filtering characteristics predict [5, 6]. In order to further improve the high frequency dynamic state estimation accuracy, this paper focus on the frequency estimation method based on Kalman Filter.

1.1 Kalman Filter

Kalman filter is a recursive process, the current estimate of the time and past time-related and affects an estimated value of future time. Discrete linear system model is [7]:

$$\mathbf{X}_k=\mathbf{\Phi}_{k,k-1}\mathbf{X}_{k-1}+\mathbf{\Gamma}_{k,k-1}\mathbf{W}_k \quad (1)$$

$$\mathbf{Z}_k=\mathbf{H}_k\mathbf{X}_k+\mathbf{V}_k \quad (2)$$

$\mathbf{\Phi}_{k,k-1}$ is a state deviation transition matrix, $\mathbf{H}_k$ is measurement matrix, which represents the relationship between the amount of measurement and status. $\mathbf{\Gamma}_{k,k-1}$ is System noise driving matrix, and the statistical characteristics of the system noise $\mathbf{W}_k$ and measurement noise $\mathbf{V}_k$ is:

$$E[\mathbf{W}_k]=0,\mathbf{R}_{WW}(k,j)=\mathbf{Q}_k\delta_{kj} \quad (3)$$

$$E[\mathbf{V}_k]=0,\mathbf{R}_{VV}(k,j)=\mathbf{R}_k\delta_{kj} \quad (4)$$

$$\mathbf{R}_{WV}(k,j)=0 \quad (5)$$

The statistical characteristics of initial state $\mathbf{X}_0$ is :

$$E[\mathbf{X}_0]=\bar{\mathbf{X}}_0,Var[\mathbf{X}_0]=\mathbf{P}_0 \quad (6)$$

So the basic process of Kalman filter is as follows:

Step1: Calculation step prediction of mean square error matrix.

$$\mathbf{P}_{k,k-1}=\mathbf{\Phi}_{k,k-1}\mathbf{P}_{k-1}\mathbf{\Phi}_{k,k-1}^T+\mathbf{\Gamma}_{k,k-1}\mathbf{Q}_{k-1}\mathbf{\Gamma}_{k,k-1}^T \quad (7)$$

Step 2: Calculation state step prediction.

$$\hat{\mathbf{X}}_{k,k-1}=\mathbf{\Phi}_{k,k-1}\mathbf{X}_{k-1} \quad (8)$$

Step3: Calculation Kalman filter gain.

$$\mathbf{K}_k=\mathbf{P}_{k,k-1}\mathbf{H}_k^T(\mathbf{H}_k\mathbf{P}_{k,k-1}\mathbf{H}_k^T+\mathbf{R}_k)^{-1} \quad (9)$$

Step4: Calculation mean square error matrix of filter.

$$\mathbf{P}_k=(\mathbf{I}-\mathbf{K}_k\mathbf{H}_k)\mathbf{P}_{k,k-1} \quad (10)$$

Step5: Calculation Status filter value.

$$\mathbf{X}_k=\mathbf{\Phi}_{k,k-1}\mathbf{X}_{k-1}+\mathbf{K}_k[\mathbf{r}_k-\mathbf{h}(\hat{\mathbf{X}}_{k,k-1})] \quad (11)$$

Step1 and step 2 push the time from k-1 moment to k moment, so it is called time update process; step 3,4,5 using the observed quantity on time to update the value, so it is called measurement update process.

1.2 Phase Locked LOOP(PLL) and Kalman FILter(KF)

Kalman filter's time update equation and measurement update equations are:

$$\hat{\mathbf{X}}_{k+1,k}=\mathbf{\Phi}_{k+1,k}\mathbf{X}_k \quad (12)$$

$$\mathbf{X}_k=\hat{\mathbf{X}}_{k,k-1}+\mathbf{K}_k[\mathbf{y}_k-h(\hat{\mathbf{X}}_{k,k-1})] \quad (13)$$

Combined type two, then:

$$\hat{\mathbf{X}}_{k+1,k}=\mathbf{\Phi}_{k+1,k}\hat{\mathbf{X}}_{k,k-1}+\mathbf{\Phi}_{k+1,k}\mathbf{K}_k[\mathbf{y}_k-h(\hat{\mathbf{X}}_{k,k-1})] \quad (14)$$

In order to facilitate the derivation of transfer function, neglect the system noise and the measurement noise matrix. Fig.1 is the recursive process of the formula (14).

 DOI: 10.1051/matecconf/20165603002

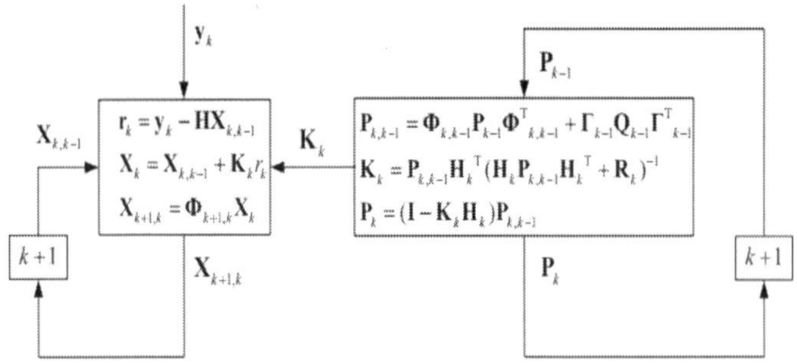

Fig. 1. Kalman filtering recursive process.

The recursive process is converted into the structure diagram of the algorithm shown in Fig.2, and the delay module z^{-1} in the loop is the estimate of the current time prediction, and the module k+1 in Fig.1 is corresponding to the step estimate[8, 9].

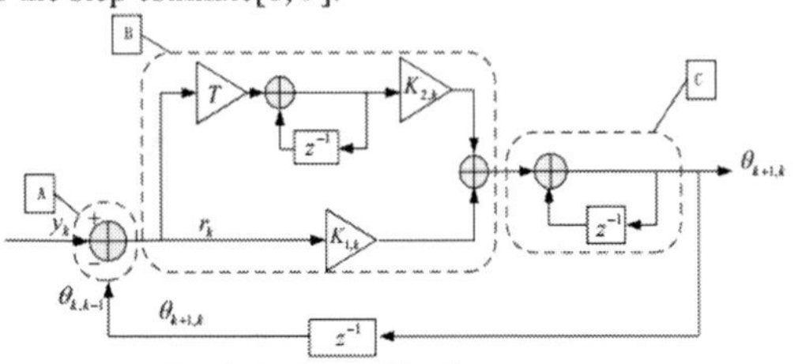

Fig. 2. Kalman filtering structure.

The overall structure of the Fig.2 and PLL is very similar, compared with the transfer function of PLL and KF, Part A is equivalent to the phase detector, Part B is in similar to the PLL loop filter and Part C is equivalent to the NCO in PLL.

2 Kalman frequency measurement

Through the study of the last section, the estimation accuracy of the Kalman filter and the SNR threshold are better than the PLL. In this section, the method of estimating the frequency of Kalman filter is studied directly. According to the difference of measurement, it is divided into two methods. First method is a EKF filter without phase detection, and the measurement is the signal itself. Second method is a KF filtering with the phase detector, and measurement is the phase of the phase detector output.

2.1 EKF method without phase detector

Principle block diagram as shown in Fig. 3, intermediate frequency (IF) signal is mixed and shifted to the baseband. Then feed into the Kalman filter after resample. Because of the nonlinear relationship between the measurement and the phase, EKF algorithm is used.

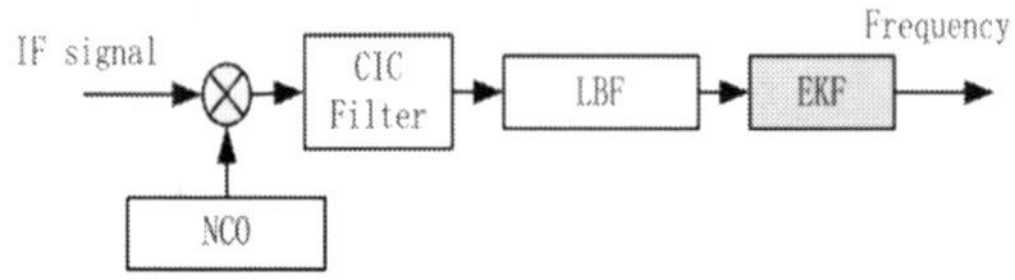

Fig. 3. EKF filter without phase detection method.

The CIC extractor and the low-pass filter generate different amplitude gain for different frequency offset signal, and the difference is large. So the amplitude of the state vector is increased, and the five order EKF is formed.
State vector is:

$$\mathbf{X}=[A \quad \theta \quad \Delta\omega_0 \quad \Delta\omega_1 \quad \Delta\omega_2]^T \tag{15}$$

A is signal amplitude, θ is the phase, and the status transfer matrix is[10]:

$$\mathbf{\Phi}_{k,k-1}=\begin{bmatrix} 1 & 0 & 0 & 0 & 0 \\ 0 & 1 & 2\pi T & \pi T^2 & \frac{1}{3}\pi T^3 \\ 0 & 0 & 1 & 2\pi T & \pi T^2 \\ 0 & 0 & 0 & 1 & 2\pi T \\ 0 & 0 & 0 & 0 & 1 \end{bmatrix} \tag{16}$$

And the Measurement matrix is:

$$\mathbf{H}=[\sin(x(2)) \quad x(1)\times\cos(x(2)) \quad 0 \quad 0 \quad 0] \tag{17}$$

In order to validate the algorithm, the signal of 70dBHz and 50dBHz is simulated, as shown in Fig.4. The results are compared with the theoretical values. The simulation results show that the new five order EKF model is correct. The theoretical amplitude A_{theory} is calculated as (18) [11]

$$A_{theory} = \sqrt{2\frac{\Sigma O^2/N}{1+\frac{1}{10^{SNR/10}}}} \tag{18}$$

N is the signal points, SNR is the known SNR. The comparison between the amplitude and the theoretical value is shown in Fig.4 and Fig.5.

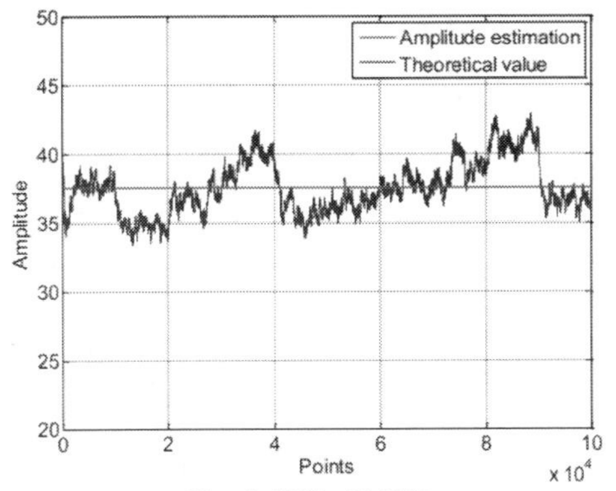

Fig. 4. SNR=50dBHz

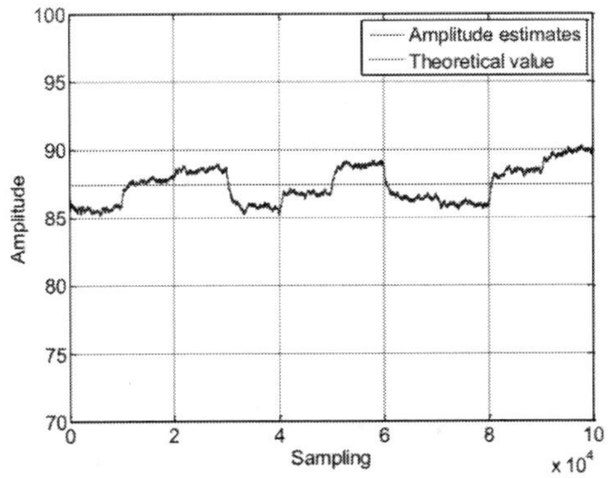

Fig. 5. SNR=70dBHz.

2.2 KF method with phase detector

In the last section, using EKF algorithm, the nonlinear error is introduced in the process of model linearization, and the total estimation error of the nonlinear error in high SNR can be large. In order to avoid nonlinear error introduced by the linearization of the nonlinear model, with the linear variation of the phase measurements, using a linear KF algorithm, as shown in Fig.6, with the aid of phase detecting is studied.

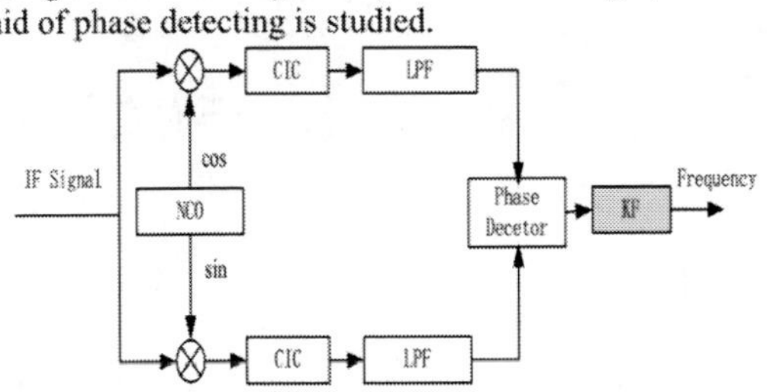

Fig. 6. KF method with phase detector

Actual signal phase is continuously with time increasing, but the arctangent discriminator phase shifting to range $[-\frac{\pi}{2},\frac{\pi}{2}]$, the phase value is periodically from $\frac{\pi}{2}$ mutations to $-\frac{\pi}{2}$, resulting in new information changes caused by the estimated value of the shock. Using the method of identifying the phase mutation and improving the new information, the method can effectively inhibit the effect of the phase mutation.

The comparison of the results of the new interest calculation method before and after the improvement of the new information is made by simulation.

The method shown in Fig.7 can better solve the filtering instability caused by the measurement of the amount of the measurement, and not only recover the carrier phase accurately and estimate the residual Doppler frequency accurately.

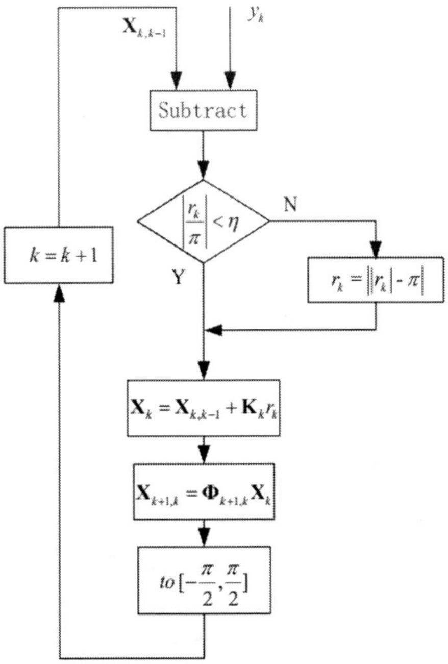

Fig. 7. Improvement of the new interest calculation method process.

When SNR=45dBHz,f_d=20Hz , The results of the phase and frequency estimation are shown in Fig. 8.

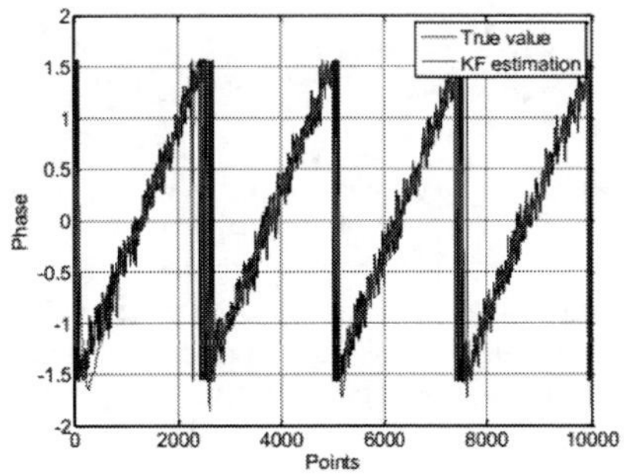

Fig.8(a). Former phase estimator compared with true value before the improvement

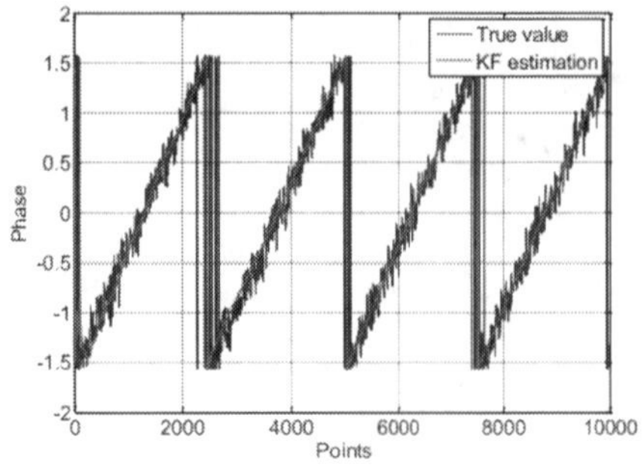

Fig. 8(b). Improved phase estimator compared with true value

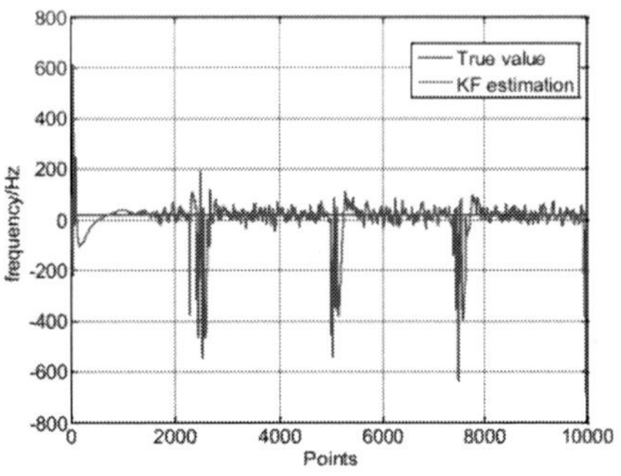

Fig.8(c). Former frequency estimator compared with true value before the improvement

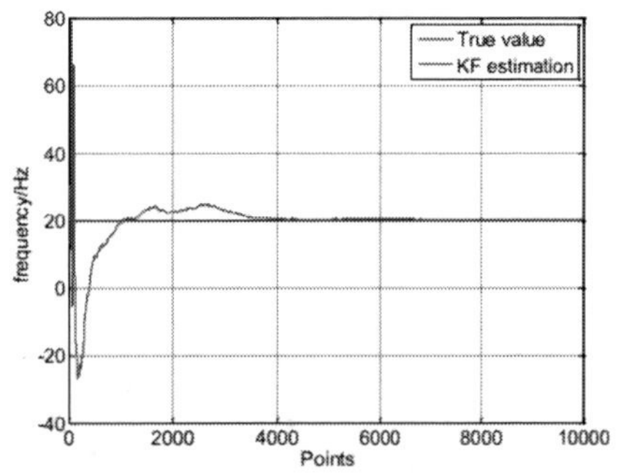

Fig.8(d). Improved frequency estimator compared with true value

Fig.8 shows that point mutations in the phase, improved phase estimation results in the presence of multiple zero crossing frequency estimation results big mutation; the results of improved phase estimation is more smooth, and no mutation in frequency estimation results.

When $SNR = 45\text{dBHz}, f_d = 500 + 1000t$ Hz, The results of the phase and frequency estimation are shown in Fig.9.

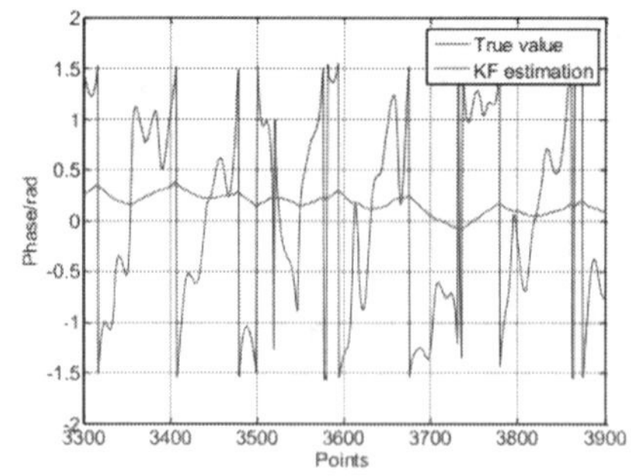

Fig.9(a). Former phase estimator compared with true value before the improvement

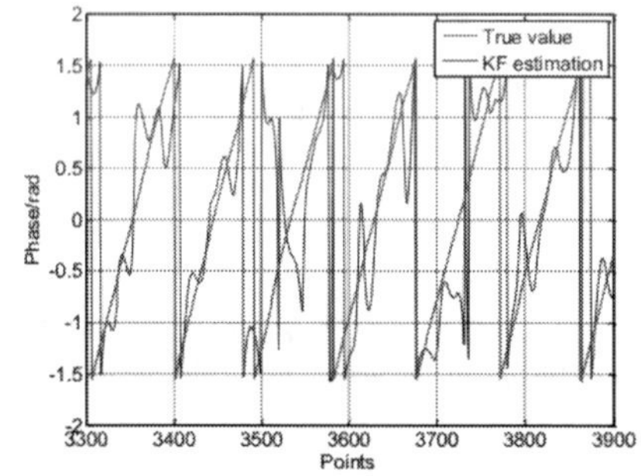

Fig.9(b). Improved phase estimator compared with true value

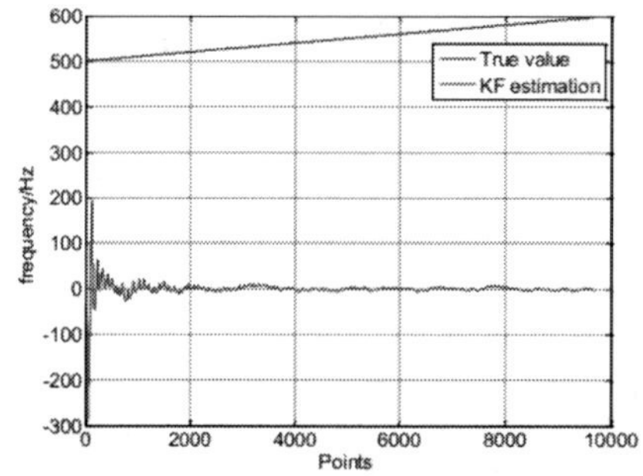

Fig.9(c). Former frequency estimator compared with true value before the improvement

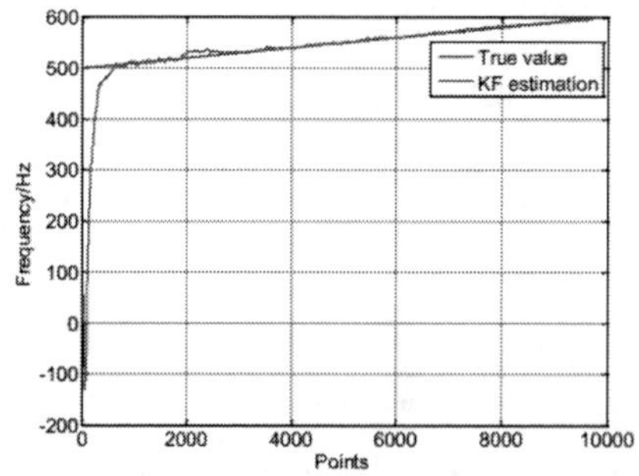

Fig.9(d). Improved frequency estimator compared with true value

Fig.9 shows that the Kalman filter appeared divergent phenomenon when the Doppler frequency is large, and the improved Kalman filter can recover the phase and frequency information accurately.

3 Frequency estimation results correction

The two method is pre - processing of FIR filter before the processing of Kalman filter, which brings the time delay of fixed point number. Set the order of filters $M=50$, signal doppler first rate of change is $\dot{f}_d=10000\text{Hz/s}$, doppler error caused by delay is Δf_d, so

$$\Delta f_d=\frac{M}{2}T_s\dot{f}_d=2.5\text{Hz}$$

T_s is the signal sampling period, $\frac{M}{2}T_s$ is the delay of the filter. As shown in Fig.10, the Doppler error caused by the delay.

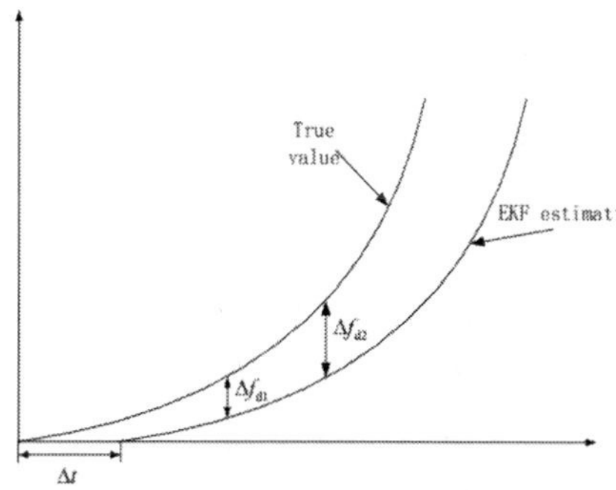

Fig.10. Improvement of the new interest calculation method process.

Fig.10 shows that with the simulation time, the Doppler error caused by the delay is becoming bigger and bigger. The error of the delay must be corrected. Let the state vector $\mathbf{X}_k$ be obtained after setting the moment k, and the modified state quantity $\mathbf{X}_{kcorr}$ is

$$\mathbf{X}_{kcorr}=\begin{bmatrix}1&0&0&0&0\\0&1&2\pi T&\pi T^2&\frac{1}{3}\pi T^3\\0&0&1&T&\frac{1}{2}T^2\\0&0&0&1&T\\0&0&0&0&1\end{bmatrix}\mathbf{X}_k \quad (19)$$

$T=\frac{M}{2}T_s$ is the delay of the filter is introduced.

4 Frequency Estimation performanc analysis

There are two main error sources of EFK method, error σ_{11} caused by measurement error and σ_{12} caused by linearization of the EFK. FK Method estimates of the error source is only one: the error due to measurement noise σ_2.

Set input intermediate frequency signal is:

$$s(t)=\sin[2\pi(f_c+f_d)t+\varphi]+n(t) \quad (20)$$

f_c is signal frequency, f_d is Doppler frequency, φ is the early phase, $n(t)$ is white noise sequence and $E(n^2)=\sigma^2$.

NCO output two orthogonal signals are:

$$N_I(t)=\sin(2\pi f_c t) \quad (21)$$

$$N_Q(t)=\cos(2\pi f_c t) \quad (22)$$

After mixing and filter the double frequency:

$$I(t)=A\sin(2\pi f_d t+\varphi)+n_I(t) \quad (23)$$

$$Q(t)=A\cos(2\pi f_d t+\varphi)+n_Q(t) \quad (24)$$

The size of A is caused by CIC extraction filter gain

$$E(n_I^2)=E(n_Q^2)=A^2\sigma^2 \quad (25)$$

After the arctangent :

$$y(t)=\tan^{-1}\left(\frac{I(t)}{Q(t)}\right) \quad (26)$$

the noise sequencey translated by $I(t)$、$Q(t)$ to $y(t)$ is $n_y(t)$, the variance is σ_y^2, when:

$$y(t)=2\pi f_d t+\varphi+n_y(t) \quad (27)$$

According to the principle of error transfer [11]

$$\sigma_y^2=\left(\frac{\partial y}{\partial I}\right)^2\sigma_I^2+\left(\frac{\partial y}{\partial Q}\right)^2\sigma_Q^2$$

$$=\left(\frac{Q}{I^2+Q^2}\right)^2A^2\sigma^2+\left(\frac{I}{I^2+Q^2}\right)^2A^2\sigma^2$$

$$=\frac{A^2\sigma^2}{I^2+Q^2}$$

$$=\sigma^2$$

To Solving $\frac{\partial y}{\partial I}$、$\frac{\partial y}{\partial Q}$ theoretical expression of $I(t)$、$Q(t)$ is used when:

$$I(t)=A\sin(2\pi f_d t+\varphi) \quad (28)$$

$$Q(t)=A\cos(2\pi f_d t+\varphi) \quad (29)$$

EKF method use $I(t)$ as measurement, and KF method use $y(t)$ as measurement. In order to facilitate the input of the two methods, $I(t)$ will be normalized as:

$$I'(t)=\sin(2\pi f_d t+\varphi)+n'(t) \quad (30)$$

$$E(n'^2)=\sigma^2 \quad (31)$$

Type(30) and type(27) of noise power are the same, but the signal form. The measurement in type(27) is the linear function of state vector, but the measurement in type(30) is the nonlinear function of state vector, which makes the type(27) can use KF algorithm. Meanwhile the type(30) can only use EKF algorithm which just introduce nonlinear error σ_{12}. Only measurement matrix calculation methods of EKF and KF algorithms are different, so the main recursive process almost unanimously, also using the same parameters in practical application, so the estimated error caused by the measurement noise is the same. Therefore, under normal circumstances, the KF method is better than the EKF method , when $\sigma_{11}+\sigma_{12}>\sigma_2$.

Compared the accuracy of the estimation of the two methods, the simulation results are shown in Fig.11.

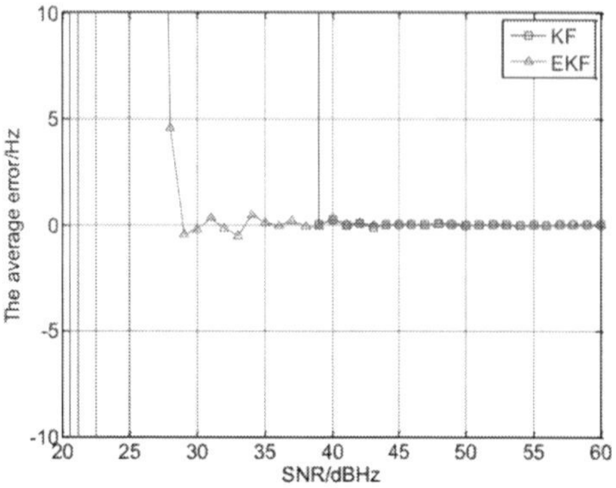

Fig.11(a). Average error of frequency estimation under different SNR.

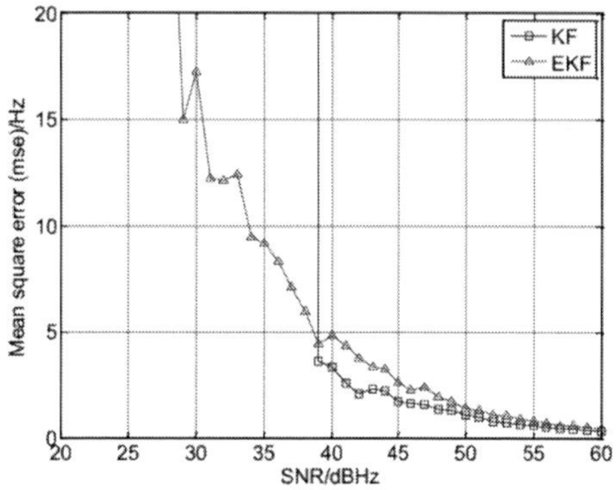

Fig.11(b). The root mean square error of frequency estimation under different SNR

5 Conclusions

Fig.11 shows that the EKF method of a SNR threshold is lower than that of the KF method, which is because the periodic mutation of the measurement has more influence on the algorithm under the low SNR, resulting in the divergence of the algorithm. However, when the SNR threshold is above, KF method, the accuracy is higher than that of the EKF method.

Therefore, the two methods of this research are limited by the fixed frequency conversion, and when Doppler changed a large range, the filtering is easy to diverge. If the closed-loop structure of PLL is introduced, the residual Doppler frequency after the frequency conversion is maintained at the appropriate interval, and the above divergence can be avoided.

References

1. Wagdy M F, Vaishnava S. A Fast-Locking Digital Phase-Locked Loop[C]//Proceedings of the Third International Conference on Information Technology: New Generations. IEEE Computer Society, 2006: 742-746.
2. El-Laithy D M, Zekry A, Abouelatta M. Speeding-up Phase-Locked Loops based on Adaptive Loop Bandwidth [J]. International Journal of Computer Applications, 2013, 61(3): 1-6.
3. Rao M, Lo Presti L, Fantino M, et al. A software receiver adaptive phase lock loop method [J]. 2009.
4. Aung M, Hurd W J, Buu C M, et al. The Block V Receiver fast acquisition algorithm for the Galileo S-band mission [J]. 1994.
5. Vilnrotter V A, Hinedi S, Kumar R. Frequency estimation techniques for high dynamic trajectories[J]. Aerospace and Electronic Systems, IEEE Transactions on, 1989, 25(4): 559-577.
6. Natali F D. AFC tracking algorithms [J]. Communications, IEEE Transactions on, 1984, 32(8): 935-947.
7. Anguirre S, Hinedi S. Two novel automatic frequency tracking loops [J]. Aerospace and Electronic Systems, IEEE Transactions on, 1989, 25(5): 749-760.
8. Hurd W J, Statman J I, Vilnrotter V A. High dynamic GPS receiver using maximum likelihood estimationand frequency tracking [J]. Aerospace and Electronic Systems, IEEE Transactions on, 1987 (4): 425-437.
9. Aguirre S. An automatic frequency control loop using overlapping DFTs [J]. Telecommunications and Data Acquisition Progress Report, 1988, 94: 222-231.
10. Hinedi S. An extended Kalman filter based automatic frequency control loop[C]//The Telecommunications and Data Acquisition Report. 1988, 1: 219-228.
11. Wang J, Liang Q, Liang K, et al. A new extended Kalman filter based carrier tracking loop[C]//2009 3rd IEEE International Symposium on Microwave, Antenna, Propagation and EMC Technologies for Wireless Communications. 2009: 1181-1184.

 DOI: 10.1051/matecconf/20165603003

The Sorting Methods of Support Vector Clustering Based on Boundary Extraction and Category Utility

Weigao Chen, Xin Jia, Weigang Zhu and Xiaojing Tang

Equipment Academy, Beijing, China

Abstract. According to the problems of low accuracy and high computational complexity in the classification of unknown radar signals, a method of unsupervised Support Vector Clustering (SVC) based on boundary extraction and Category Utility (CU) of unknown radar signals is studied. By analyzing the principle of SVC, only the boundary data of data sets contribute to the support vector extracted. Thus firstly, for reducing the data set, at the same time reducing the computational complexity, the algorithm is designed to extract the boundary data through local normal vector. Then using CU select the optimal parameters. At last distinguish different categories and get the sorting results by Cone Cluster Labelling (CCL) and Depth-First Search (DFS). Through comparing the simulation results, the proposed method which is based on boundary extraction and CU is proved to have turned out quite good time effectiveness, which not only improves the accuracy of classification, but also reduces the computational complexity greatly.

1 Introduction

With the wide application of electronic information technology in the military field, it is significant to improve the whole operational efficiency of the electronic warfare equipment, and also presents a great challenge to each core components of the electronic countermeasure. As radar countermeasure's reconnaissance front, radar signal sorting bears the brunt of this challenge. The purpose of radar signal sorting is to extract the pulse sequence belonging to the same radar from the disorderly interleaved pulse stream. It is the premise of signal analysis and processing [1]. For the classification of unknown radar signals, prior knowledge is very scarce, so with the help of unsupervised clustering in data mining, the inner characteristic and distribution rule of data set can be obtained without prior knowledge. SVC algorithm uses the kernel function nonlinear mapping method to make the features which is not obvious in the data space enlarged, extracted and cleared in the high dimensional feature space to benefit the clustering. Nature of SVC is a convex quadratic optimization problem, the results have a strong global optimization and generalization ability, in addition, the algorithm can also adapt to the clusters of arbitrary boundary shape, so it is very suitable for unknown disorderly radar signal sorting. However, under the condition of high density and large data quantity, the calculation of optimization and cluster label process are huge, which greatly inhibits the performance of SVC. So in order to solve the problem, we propose a support vector clustering algorithm (BECU SVC) based on boundary extraction and CU. The algorithm firstly uses local normal vector to extract the boundary of de-noised data, which reduces the data set [2]; secondly, extracts the support vectors data; then under the guidance of CU [3], using improved CCL [4] to cluster label; at last accomplishes the category classification of surplus data by DFS.

2 Boundary Extraction Based on Local Normal Vector

Because of the influence of false alarm detection, clutter, jamming, etc. the radar signals transmitted to signal processor carry a lot of noise data, so the noise reduction processing is the precondition of radar signal sorting. Because SVC algorithm needs to select the appropriate data from the data set to construct the support vector function, and these 'suitable data' are only distributed in the contour region of the data set, and the remaining data inside the contour have no contribution to SVC, so these data are redundant. Therefore, in order to abandon redundant data and reduce the computation of SVC, it's necessary to extract the de-noised data set. Consider a universal two-dimensional projection data distribution, which contains three typical boundary data points, as shown in Figure 1.

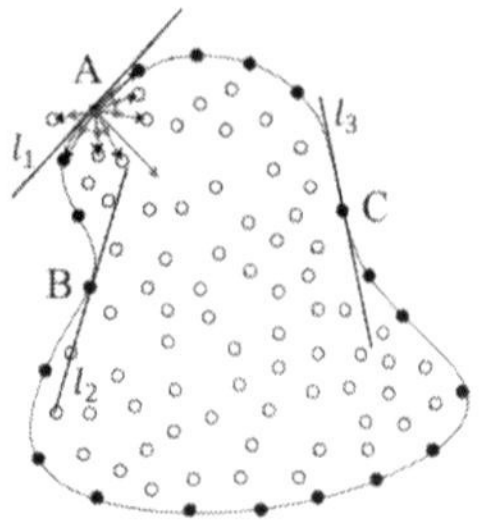

Figure 1. Two-dimensional projection of the data set

The solid dots in figure represent the contour data points, hollow dots are internal or external data points. A, B, C respectively represent three typical boundary distribution points, A point is located in the bulge of boundary, the hyper tangent plane l_1 divides internal data points into the same side; B, C points locate in the sunken boundary, the most points are divided into one side by the hyper tangent plane l_2, l_3, and the left points locate on the other side. In Figure 1, for an example of A point, the local normal vector and the closest 7 points (Euclidean Distance) of A are indicated. Observe all the vectors, we can find for the points of convex (including high-dimensional hyper convex) profile data set, if the angle $\theta < \frac{\pi}{2}$ between local normal vectors of these profile points and unit vectors of other points, these points belong to the same cluster.

In general, we use the k nearest neighbors (kNN) method to calculate the local normal vector of $\boldsymbol{x}_i (i = 1,2,\cdots,n)$. At first Euclidean Distance is used to find the nearest k points $\boldsymbol{x}_j$ ($j = 1,2,\cdots,k$) of $\boldsymbol{x}_i$ in the high dimensional data space, calculate the unit vector $\boldsymbol{v}_{ij}$

$$\boldsymbol{v}_{ij} = \frac{x_j - x_i}{|x_j - x_i|},\ j = 1,2,\cdots,k \tag{1}$$

As shown in Fig.2, the approximate local normal vector can be expressed as sum of k unit vectors, i.e.:

$$\boldsymbol{n}_i = \sum_{j=1}^{k} \boldsymbol{v}_{ij} \tag{2}$$

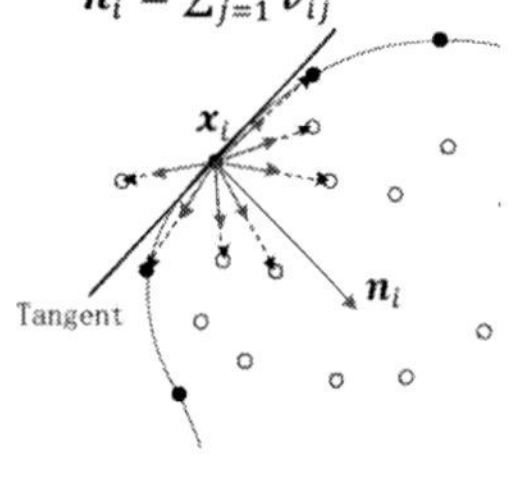

Figure 2. The schematic of local normal vector generated

Transformed the local normal vector $\boldsymbol{n}_i$ into the local unit normal vector $\boldsymbol{n}_i^T$, calculated the angle θ_{ij} between $\boldsymbol{n}_i^T$ and each unit vector $\boldsymbol{v}_{ij}$:

$$\theta_{ij} = \cos^{-1}(\boldsymbol{n}_i^T \cdot \boldsymbol{v}_{ij}),\quad 0 \le \theta_{ij} \le \pi \tag{3}$$

Point x_j locate in the inner side of the hyper tangent plane, which side has a large amount of data, if the angle $\theta_{ij} < \frac{\pi}{2}$; point x_j locate on the hyper tangent plane, if the angle $\theta_{ij} = \frac{\pi}{2}$; and if the angle $\theta_{ij} > \frac{\pi}{2}$, x_j locate in the outside of the hyper tangent plane. All or most of the values of the data points x_i on the boundary are satisfy $\theta_{ij} \le \frac{\pi}{2}$. Thus, the following method can be used to determine whether point x_i is a boundary data point:

$$p_i = \frac{1}{k}\sum_{j=1}^{k} l_j,\ l_j = \begin{cases} 1, & if\ \theta_{ij} \le \frac{\pi}{2}; \\ 0, & \text{else.} \end{cases} \tag{4}$$

For the boundary point $\boldsymbol{x}_i$ in Fig.2, the $p_i = 1$ if without the effect of interference data which locate in the outside. In Fig.1, because points B and C locate in sunken boundary region, there are few data points on the outside of the hyper tangent plane, thus for the Eq.4, $p_i = 1 - \gamma$ (in general the value of γ is very small). By the above analysis, it can be set that the corresponding point $\boldsymbol{x}_i$ is the boundary data of data set, if $p_i \ge 1 - \gamma$. Through detecting the de-noised data using above method, we can get a simple boundary data set to serve the follow-up SVC progress.

3 Cluster Labelling Method of CCL Based on CU

With the technology theory of Support Vector Machine, SVC can solve the problem of sort learning under the condition of no supervision. The SVC algorithm mainly includes two stages of optimization and cluster label. In the optimization phase, data are mapped to a high dimensional space by kernel function, and then seek the minimum hypersphere which obtains the all data set in the high dimensional space, and then can get support vector and the minimum hypersphere. In the cluster label phase, Complete Graph (CG) [5] method is used to determine the label of support vector data by detecting the points between two support vectors whether locate in the minimum hypersphere.

For given dataset $\{x_1, x_2, \cdots, x_n\}$, $x_i \in R^d$, a total of n samples of d dimensional, a minimum hypersphere of radius is R which covers all data points can be found in the feature space, we can get a soft-margin target function of SVC:

$$\min\ R^2 + C\sum_{i=1}^{n} \xi_i$$
$$\text{s.t.}\ \ \|\phi(x_i) - a\|^2 \le R^2 + \xi_i,\quad \xi_i \ge 0 \tag{5}$$

Where constant C is the penalty factor, which balances the relationship between radius and slack; $\xi = (\xi_1, \xi_2, \cdots, \xi_n)^T$ is the slack which permits few samples exist outside of the hypersphere; which ϕ is the mapping from data space to feature space; $a = (a_1, a_2, \cdots, a_d)^T$ is the center of the hypersphere; $\|\cdot\|$ refers to the Euclidean norm. Convert Eq. (5) to Lagrangian function and introduce Gaussian kernel function $K(x, y) = \phi(x)^T \cdot \phi(y) = \exp(-q\|x - y\|^2)$, which q is the width of the kernel function. Then Eq. (5) can be simplified to:

$$\min\ \sum_{i,j=1}^{n} \beta_i \beta_j K(x_i, x_j)$$
$$\text{s.t.}\ \ \sum_{i=1}^{n} \beta_i = 1,\quad 0 \le \beta_i \le C \tag{6}$$

Which $\beta_i, \beta_j \ge 0$ is the Lagrangian Multiplier. At the same time, the distance between the projection of sample data in feature space and the center of hypersphere can be obtained:

$$R^2(x) = \|\phi(x_i) - a\|^2$$
$$= 1 - 2\sum_{j=1}^{n} \beta_i K(x_i, x) + \sum_{i,j=1}^{n} \beta_i \beta_j K(x_i, x_j) \tag{7}$$

Because of Eq. (6) need to meet the condition of Karush-Kuhn-Tucker, thus: 1) When $\beta_i = 0$, the corresponding $\phi(x_i)$ locate in the minimum hypersphere in feature space, and x_i locate in the cluster in the data space; 2) When $\beta_i = C$, the corresponding $\phi(x_i)$ locate outside of minimum hypersphere in feature space, and x_i locate outside of cluster in the data space, these points are defined as the bounded Support Vectors (bSVs); 3) When $0 < \beta_i < C$, the point $\phi(x_i)$ in the feature space locate on the surface of minimum hypersphere, the corresponding x_i locate on the edge of the cluster, these points are defined as the Support Vectors (SVs).

Through the above analysis, we can find that Eq. (7) is just related to the SVs and the bSVs, which is called the Function Support. We can get the radius of minimal hypersphere $R(v_i) = r$ by substitution some SVs v_i into Eq. (7).After the SVs are obtained, the incidence matrix A is established by CG cluster label method. Using incidence matrix traversal of all support vectors, label the support vectors interconnected as the same cluster. Finally, complete the remaining data category based on the cluster label of support vectors.

Because of random sampling test and construction of the correlation matrix, CG cluster labelling algorithm has a greater randomness and computation. Thus in order to avoid randomness and improve the efficiency of cluster labelling, Sei-Hyung Lee proposed the CCL cluster labelling algorithm, with the help of support vector structure-distribution relationship in high dimensional feature space, it constructs high dimensional cone for each support vector, and these support vector cones approximately covering the minimum hypersphere. Transformed to the data space, the contour of data set can be covered by hypersphere accordingly, and then label support vectors by judging whether each hyper sphere intersects or not.

3.1 CCL cluster calibration principle

Set V is the produced support vector data set after optimization phase, $v_i, v_j \in V$, $1 \le i, j \le N_{sv}$, due to the introduction of the Gaussian kernel function, $K(v_i, v_i) = 1$, $\|\phi(v_i)\| = 1$, projecting all data to the surface of unit hypersphere B whose origin is O. The minimal hypersphere S center is b, then:

$$\|b\| \le \|\beta_1\phi(v_1)\| + \cdots + \|\beta_n\phi(v_n)\| = 1 \quad (8)$$

Eq.(8) shows that the minimal hypersphere S center b locates in the unit hypersphere's surface or inside. Because the support vectors are distributed on the surface of minimal hypersphere, and all the data are projected onto the surface of unit hypersphere, so the support vectors locate on the dotted circle with the minimal hypersphere S and the unit hypersphere B, and the data within the cluster locate on the region P dotted with S and B. As shown in Fig.3.

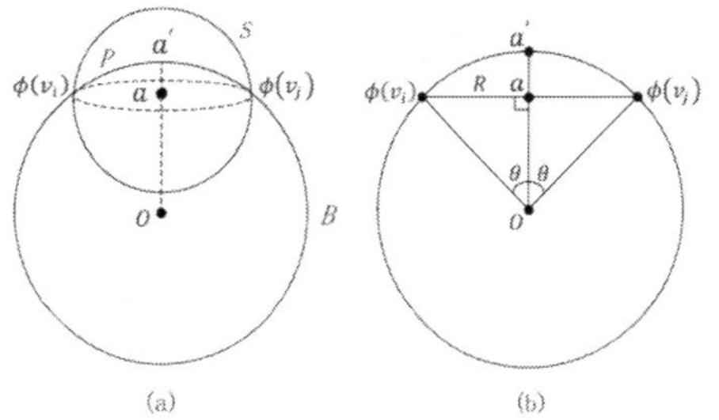

Figure 3. The space distribution of support vectors (a) three-dimensional; (b) two-dimensional cross section.

Where a is center of the support vector dotted circle, a′ is the intersection point of Oa's extension line and the unit ball B, Fig.3 (b) is the cross section of Fig.3 (a). It can be seen from the above analysis, for any support vector $\phi(v_i)$, the angle between $\overrightarrow{O\phi(v_i)}$ and $\overrightarrow{Oa}$ is θ, i.e. $\angle\phi(v_i)Oa = \theta$. Set ε_{v_i} is the cone formed by $\overrightarrow{Oa}$ rotate along $\overrightarrow{\phi(v_i)}$, and a′ is the intersection point for all the cones, which can be understood as the union of all cones covering cap region P approximately. Correspondingly in the data space, ε_{v_i} corresponding to the hypersphere ρ_{v_i} which center is v_i, radius is $\|v_i - g_i\|(g_i \in R^d$, $\angle\phi(v_i)O\phi(g_i) = \theta)$, $U_i(\rho_{v_i})$ cover the contour of the data set P′ approximately, and $\phi(P') \subseteq P$.

The CCL algorithm label the data that is not larger than the angle θ as the same type in feature space, transformed into data space, we can label by judging the value of radius $\|v_i - g_i\|$. Because $\|\phi(v_i)\| = \|\phi(g_i)\| = \|a'\| = 1$, thus $\cos\theta = \angle\phi(v_i)O\phi(g_i) = \angle\phi(v_i)Oa' = \langle\phi(v_i)\cdot a'\rangle = \langle\phi(v_i)\cdot\phi(g_i)\rangle = K(v_i, g_i)$, and then:

$$\|v_i - g_i\| = \sqrt{-\frac{\ln(\cos\theta)}{q}} \quad (9)$$

For any $v_i \in V$, $\angle\phi(v_i)Oa' = \theta$, the radius of hypersphere in data space are equal correspondingly, so let $\|v_i - g_i\| = Z$. b and a are coincide, and a is minimal hypersphere's center, the radius of the hyper sphere is R. Due to $\|\phi(v_i)\| = 1$, from Fig.3 (b), it can be seen that $\cos\theta = \|a\|$, $\|a\|^2 = 1 - R^2$, $\|a\| = \sqrt{1 - R^2}$, introduce to Eq. (9), and then:

$$\|v_i - g_i\| = Z = \sqrt{-\frac{\ln(\sqrt{1-R^2})}{q}} \quad (10)$$

Next we can determine two support vector v_i and v_j whether or not for the same category by judging the two hypersphere ρ_{v_i}, ρ_{v_j} which are formatted by v_i、v_j whether intersect or not in data space, i.e. if $\|v_i - v_j\| \le 2Z$, labelling the support vector v_i and v_j to the same category. After getting the value of Z, the Depth-First Search (DFS) can be used to complete the labelling of all the SVs, then, according to the Euclidean distance, the remaining data are classified into the nearest SVs' label.

3.2 Parameter optimization based on CU

From Eq. (10), it can be seen that the value of q will directly determine the value of Z, thereby affecting the number of generated clusters. If $q \to 0$, $Z \to \infty$, all hypersphere ρ_{v_i} are formed by SVs intersect each other, and every ρ_{v_i} cover the entire data set, thus all the data can be labelled to the same category. If $q \to \infty$, $Z \to 0$, the entire data set is covered by $\mathrm{U}_i(\rho_{v_i})$, and all the hyperspheres are disjoint, so as to generate categories, whose number is the same as the data samples. In summary, how to select the kernel function width is very important to the whole clustering results.

To solve this problem, this paper proposes a method of parameter optimization based on CU, which guides the optimization of the kernel function width q by measuring the overall cluster quality with calculating the CU of clustering result. The CU is defined as:

$$\mathrm{CU} = \frac{\sum_l \Pr[C_l] \sum_i \sum_j \left(\Pr[a_i = v_{ij} | C_l]^2 - \Pr[a_i = v_{ij}]^2 \right)}{k} \tag{11}$$

Where C_l ($l = 1,2,\cdots,k$) is the l th category of clustering result, a_i is the i th attribute, $v_{ij} = v_{i1}, v_{i2}, \cdots, v_{ij}$ is the attribute value. The upper formula can be understood as the probability $\Pr[a_i = v_{ij} | C_l]$ of a_i, whose attribute's value is v_{ij} for a certain category C_l is a good estimate relative to the probability $\Pr[a_i = v_{ij}]$ of a_i, whose attribute's value is v_{ij} for all categories. Extend Eq. (14), make it suit the radar parameters:

$$\mathrm{CU} = \frac{1}{k} \sum_l \Pr[C_l] \frac{1}{2\sqrt{\pi}} \sum_i \left(\frac{1}{\sigma_{il}} - \frac{1}{\sigma_i} \right) \tag{12}$$

Where σ_i is the standard deviation of a_i attribute, and σ_{il} is the standard deviation of a_i attribute in the l category. Because the essence of clustering is to estimate the value of each category attribute accurately, thus the CU can better reflect the clustering results.

Through analyzing the experimental results, we can get that with the increase of q, the number of support vectors are increased, and the clustering contour is shrinking gradually, while the number of bSVs is controlled by penalty factor C. For parameter optimization, firstly set $C = 1$, $q_1 = \frac{max\|v_i - v_j\|^2 + min\|v_i - v_j\|^2}{2max\|v_i - v_j\|^2 \cdot min\|v_i - v_j\|^2}$ in the range of $\left[q_{min} = \frac{1}{max\|v_i - v_j\|^2}, q_{max} = \frac{1}{min\|v_i - v_j\|^2} \right]$, calculate the CU of clustering results; and then let q_2 is the middle point of $[q_{min}, q_1]$, and q_3 is the middle point of $[q_1, q_{max}]$, calculate CU_2 and CU_3; if there are certain categories with one point, decrease the value of C, getting the CU by taking middle point method; at last compare these gotten CUs and select the q^* corresponding to CU_{max} as the appropriate kernel function width.

4 Improved Signal Sorting Process

The proposed support vector clustering method based on boundary extraction and CU is divided into three parts: boundary extraction, support vector extraction and parameter optimization based on CU, the process is as shown in the figure 4:

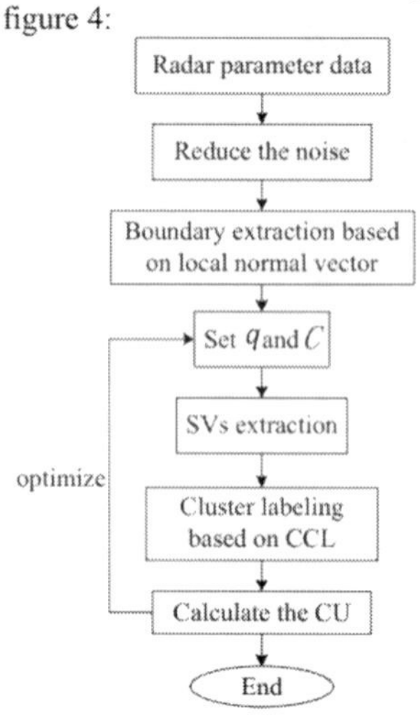

Figure 4. The process of support vector clustering based on boundary and CU.

Step 1: the parameter data extracted from radar signal buffer, after the noise reduction processing, boundary information is extracted by local normal vector;

Step 2: parameter setting, the SVs data is extracted by SVC;

Step 3: first label the SVs through CCL method, and then use DFS classifying the remaining samples; then calculate the CU of clustering results and guidance parameters optimization until get the last clustering results.

5 Simulation Test Analysis

To verify the effectiveness of support vector clustering method based on boundary extraction and CU (BECU-SVC), four groups of radar pulse data with different characteristic parameters are being simulated, totally the number of samples are 7617. Select RF, DOA and PW as the characteristic parameters of radar signal sorting, the parameter information of each group is shown in Table 1. At the same time, for simulating the real environment of reconnaissance, 5% additional jitter error are mixed to each parameters.

The original data are normalized, and the local normal vector is used to extract the contour of data set. In order to achieve the purpose of reducing the data samples, and try not to lose the contour information. Set the parameters $k = 30$, $\gamma = 0.04$ of the Sect.2 to extract the boundary data through multiple simulation experiments. The data distributions of the boundary extraction fore-and-aft are shown in Fig. 5.

Table 1. The Feature Information of Samples.

Signal type	DOA/(°)	PW/(μs)	RF/(MHz)		Number of pulses
			Type	Range	
Pulse sequence 1	42-50	1.2-1.4	Frequency agility	2080-2250	1824
Pulse sequence 2	55-78	1.0-1.2	Frequency agility	2750-2850	1853

Pulse sequence 3	72-80	1.2-1.25	Frequency hopping	2150-2350	2049
Pulse sequence 4	56-64	1.3-1.45	Frequency agility	2650-2810	1891

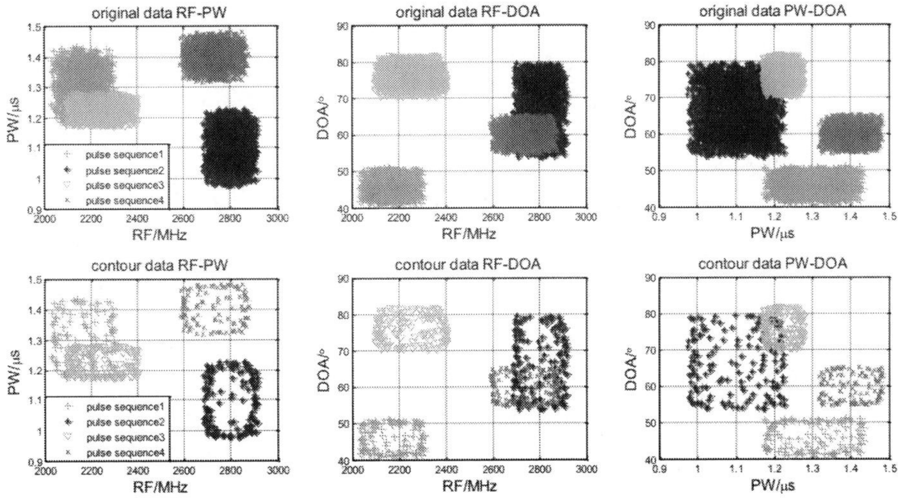

Figure 5. Comparison of data distribution before and after boundary extraction.

From the data distribution in Fig. 5, it is clear that the contour of each pulse sequence is clear after the boundary extraction. The samples for SVC are reduced from 7617 to 503 by using these contour data, which reduces 93.4% approximately, and reduce the computational cost greatly.

As mixed pulse sequences in Table 1 for the sample data, and process these samples by CG, CCL and BECU-SVC respectively. Under the guidance of CU, set the optimal parameter, and the results obtained by 100 Carlo Monte are shown in Table 2.

Table 2. Sorting Result Statistics.

	CG	CCL	BECU-SVC
Total pulses	7617	7617	7617
Missing pulses	34	0	0
Error sorting pulses	197	93	59
Correct sorting pulses	7420	7524	7558
CU	2.838	3.092	3.262
Sorting accuracy (%)	97.4	98.8	99.2
Time-consuming (s)	326.762	63.672	8.723

By comparing the results of Table 2, we can find the results of CG algorithm have certain randomness, the results with more missing and error pulses, at the same time the incidence matrix needs great computation, thus the time-consuming of CG is the longest. CCL algorithm only needs to calculate the value of Z, through judging the relationship between SVs and Z to complete labelling in data space after extracting the SVs samples, thus avoiding the randomness and the construction of large incidence matrix, but because there is no reduction of the sample data, the calculation is still very large in solving the quadratic optimization problem. In this paper, the BECU-SVC method uses the boundary extraction to reduce the sample data which construct kernel matrix and participate in the quadratic optimization programming. At the same time, improved CCL method is used to reduce the computational complexity of cluster labelling, thus it can improve the efficiency of sorting significantly, and can get better accuracy, the time-consuming of BECU-SVC method is only 2.67% of the traditional CG algorithm.

6 Conclusion

With the rapid development of information technology, the production of new technology system for radar is promoted, and it has brought more complex and changeable signal patterns. Therefore in the future battlefield, we will be threatening by emerging in endlessly unknown radar signals. In order to meet the challenge, aiming at the problem of high time-consuming for unknown radar signal sorting with less prior knowledge, a new method of support vector clustering based on boundary extraction and CU for radar signal sorting is studied. Through comparing the simulation results, it is proved that the proposed BECU-SVC algorithm can not only improve the accuracy of signal sorting, but also greatly reduce the computational complexity, and has good timeliness. However, in the process of experiment, there is still room for parameter optimization of q, which will be the focus of the next step.

References

1 He Ming-Hao, *Radar countermeasure signal processing,* (2010) 6-9.
2 Yuan Ping, YunFengChang, Yajian Zhou, et al. Knowl Inf Syst, **43**, 2 (2015) 281-310.

3 Ian H. Witten, Eibe Frank and Mark A. Hall. *Data Mining: Pratical Machine Learning Tools and Techniques, Third Edition* (2014) 203-205.
4 Sei-Hyung Lee, Karen M.Daniels. Pattern Anal Applic, **15**, 3 (2012) 327-344.
5 Ben-Hur A, Horn D, Siegelmann H. T., et al. JMLR, (2012) 125-137.

A software radio platform based on ARM and FPGA

Xin.Yang [1], Zhengxiang.Li[2]

[1]Embedded system design, Communication University of China, Beijing, China
[2]Embedded system design, Communication University of China, Beijing, China

Abstract. The rapid rise in computational performance offered by computer systems has greatly increased the number of practical software radio applications. A scheme presented in this paper is a software radio platform based on ARM and FPGA. FPGA works as the coprocessor together with the ARM, which serves as the core processor. ARM is used for digital signal processing and real-time data transmission, and FPGA is used for synchronous timing control and serial-parallel conversion. A SPI driver for real-time data transmission between ARM and FPGA under ARM-Linux system is provided. By adopting modular design, the software radio platform is capable of implementing wireless communication functions and satisfies the requirements of real-time signal processing platform for high security and broad applicability.

1 Introduction

With the rapid development of Software Radio technology, Software Radio technology has become a hotspot of research. Software Radio is a new architecture for wireless communication based on digital signal processing and microelectronics technology. In this paper, an ARM and FPGA based software radio platform is designed according to the needs of actual system in digital signal processing and wireless communication, which take full advantage of the flexibility of ARM and the parallelism of FPGA.[1] The choice of ARM is Freescale' s IMX6, which has a rich instruction set and programming flexibility; FPGA has a great advantage in the speed and parallel computing, suitable for requirements for real-time data transmission.

This paper begins with an overview of the software radio platform, including hardware components and data-stream transmission on the platform. It then moves on to describe the design of data transmission interface between ARM and FPGA. An explanation of how the SPI driver can implement real-time data transmission between ARM and FPGA under ARM-Linux system will be presented after that. The sequential logic and the transmission of multi-data in FPGA is explained at the end of this paper.

2 Platform Overview

The core architecture of this platform is based on ARM system and FPGA system. ARM system uses a Freescale' s IMX6Q microprocessor which encompasses a quad-core platform running up to 1.2 GHz. FPGA system uses a chip of Altera's Cyclone III series chips which contains six groups of phase-locked loop CLK management and can run up to220MHz. A/D converter module uses ADS7886 which is a 12-bit and 1MHz Sample Rate Serial ADC. The serial interface in ADS7886 is controlled by the CS and SCLK signals for connections with microprocessors.

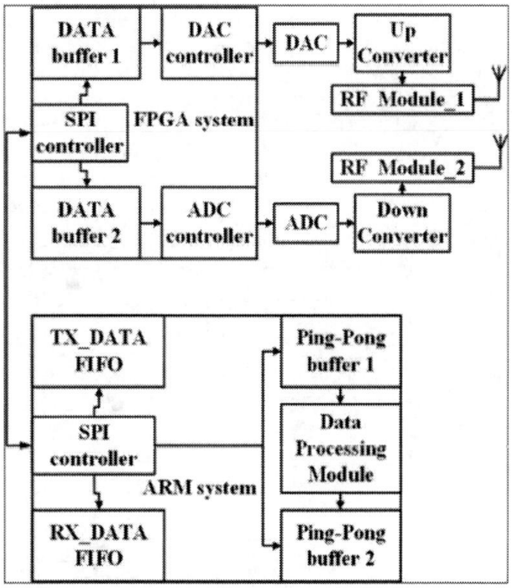

Figure 1. Diagram of embedded software radio platform.

Figure.1 shows the diagram of embedded software radio platform. The platform is mainly composed of ARM system, FPGA system, ADC&DAC module and

RF module. The function of ARM system is storing the real-time data and processing the real-data. The ARM chip IMX6Q complete instructions and data transmission by the address of read and write bus.

Embedded Linux kernel is responsible for mission management and integration of enhanced SPI serial data transmission, guaranteeing real-time data transmission. The role of FPGA is synchronous timing controlling and serial-parallel converter. FPGA is in charge of data acquisition from the ADC, digital-to-analogue conversion by controlling the DAC and controlling the speed of data transmitting to ARM.[2] ADC&DAC module consists of ADC, DAC and peripheral circuit. RF module is used to receive and process radio-frequency signal.

2.1 Data Transmission Design

2.1.1 Real-time Data Transmission Design

The difficulty of platform design lies in the design of real-time data transmission interface, because it is the data transmission channel, and is also a bridge to connect the system front-end and back-end.[3] Compared with general processing platform, the real-time platform requires demands on faster response time, and deviation of synchronous sequential logic will cause serious consequences. It means that the hardware of platform must have the ability to process and transmit data with high speed.[4]

What is innovative about real-time data transmission interface of this platform is that FPGA and ARM use master-slave mode control structure by Enhanced serial peripheral interface of ARM. ARM SPI bus controller works on slave mode, and FPGA SPI bus controller works on mater mode to provide chip select signal and CLK signal for ARM. FPGA can read valid data processed by ARM from data cache of ARM through SPI bus, and then FPGA convert serial data read out from ARM into parallel data. After that, FPGA will send parallelized data to DAC which will handle the digital to analog conversion. Meanwhile, FPGA can write avail data sampled by ADC into data cache of ARM through SPI bus, and then ARM will store valid data and process valid data.

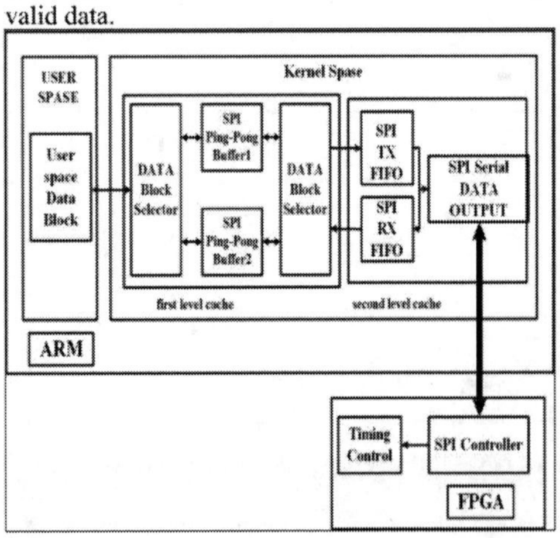

Figure 2. Data streams between ARM and FPGA

Figure.2 shows data streams between ARM and FPGA. By reason that the Enhanced serial peripheral interface is a full-duplex, synchronous, four-wire communication block. Once ARM-Linux user space has completed their data handling, ARM-Linux kernel space can receive processed data in blocks from ARM-Linux user space (the maximum data block size is 120Kbytes), and then FPGA can read out serial processed data from ARM and convert serial data into parallel data. After that, FPGA will send parallelized data to DAC, which will handle the digital to analog conversion. Meanwhile, ARM-Linux kernel space can transfer data sampled by ADC in blocks to ARM-Linux user space (the maximum data block size is 120Kbytes), ARM can process data in ARM-Linux user space. Real-time system must have ability to process and transmit data with high speed. In order to ensure constant and continuous data transmission, ARM-Linux system uses ping-pong operation of two pieces of data buffer (the maximum data buffer size is 120K bytes) to achieve real-time data in series. When the valid data block of user space is writing into Ping-pong Buffer0 in kernel space, SPI TX_FIFO can read valid data from Ping-pong Buffer1 which has completed writing into the valid data block. Similarly, when the valid data block of user space is writing into Ping-pong Buffer1 in kernel space, SPI TX_FIFO can read valid data from Ping-pong Buffer0 which has completed writing into the valid data block. In this way, SPI TX_FIFO can read valid data from data buffer continuously. (first level cache of ARM internal data)

FPGA SPI bus controller provide chip select signal and clk signal for ARM to control the read or write speed of RAM. Therefore, FPGA can read processed data from data buffer of ARM at a constant rate, FPGA also can write real-time data sampled by ADC into data buffer of ARM at a constant rate.

As ARM SPI bus controller works on slave mode, and FPGA provides chip select signal and CLK signal for ARM. In order to ensure ARM can output data continuously at a constant rate, half- interruption method is used to collect and transfer data.[5] When the number of valid data is less than 32, ARM-Linux will generate an interrupt and use interrupt service routine to fill SPI TX_FIFO. (second level cache of ARM internal data)

In the design of data transmission interface, ARM is responsible for the data caching and processing, FPGA is responsible for timing control for ARM.

2.1.2 FPGA Data Transformation Design

In the platform, FPGA is responsible for completing serial-parallel conversion and synchronous timing control of ARM as well as ADC & DAC. FPGA works as the coprocessor together with the ARM, which serves as the core processor. ARM is used for digital signal processing and real-time data transmission, and FPGA is used for synchronous timing control and serial-parallel conversion.

FPGA works as the coprocessor together with the ARM, which serves as the core processor. ARM is used for digital signal processing and real-time data transmission, and FPGA is used for synchronous timing control and serial-parallel conversion. A SPI driver for real-time data transmission between ARM and FPGA under ARM-Linux system is provided.[6] The design of real-time data transmission interface takes full advantage of ARM and FPGA. ARM contains various peripheral interfaces and enough storing resource. Besides, it can run an embedded operating system Linux to manage all the tasks. FPGA has excellent reconfiguration flexibility and parallel computational ability, which enhance the performance of both control and computation of the platform. The CLK of ADC & DAC and timing control of ARM is provided by FPGA. ARM is used to process data and store data.

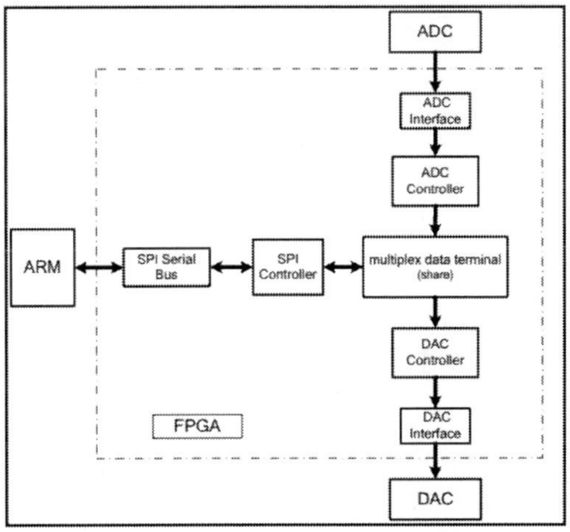

Figure 3. FPGA SPI bus controller

Figure.3 shows SPI bus data streams and SPI bus controller in FPGA.FPGA SPI bus controller provide chip select signal and clk signal for ARM to control the read or write speed of RAM.A SPI driver for real-time data transmission between ARM and FPGA under ARM-Linux system is provided.data streams between ARM and FPGA. By reason that the Enhanced serial peripheral interface is a full-duplex, synchronous, four-wire communication block. Once ARM-Linux user space has completed their data handling, ARM-Linux kernel space can receive processed data in blocks from ARM-Linux user space (the maximum data block size is 120Kbytes), and then FPGA can read out serial processed data from ARM and convert serial data into parallel data. After that, FPGA will send parallelized data to DAC.

Therefore, FPGA can read processed data from data buffer of ARM at a constant rate, FPGA also can write real-time data sampled by ADC into data buffer of ARM at a constant rate.A/D converter module uses ADS7886 which is a 12-bit and 1MHz Sample Rate Serial ADC. The serial interface in ADS7886 is controlled by the CS and SCLK signals for connections with microprocessors.

Multiplex data terminal is a module inside FPGA, which can implement sharing data channel between ARM and ADC&DAC. FPGA can read processed data from data buffer of ARM through the SPI bus. Meanwhile, FPGA can write real-time data sampled by ADC into data buffer of ARM through the SPI bus.The data transmission between ARM and FPGA is a full-duplex transmission.

2.2 ARM-Linux SPI bus driver design

Driver is the interface between the operating system and hardware and it shielded the hardware details of the application. Only the hardware system combined with the highly efficient and reliable driver could it work properly in the operating system.

In order to realize the communication between ARM and FPGA in ARM-Linux system for master-slave mode to transmit data. A specialized SPI bus driver and software interface is designed to provide data transmission channel between the operating system and hardware.

Once ARM-Linux user space has completed their data handling, ARM-Linux kernel space can receive processed data in blocks from ARM-Linux user space (the maximum data block size is 120Kbytes), and then FPGA can read out serial processed data from ARM and convert serial data into parallel data. After that, FPGA will send parallelized data to DAC, which will handle the digital to analog conversion. Meanwhile, ARM-Linux kernel space can transfer data sampled by ADC in blocks to ARM-Linux user space (the maximum data block size is 120Kbytes), ARM can process data in ARM-Linux user space.[7] Real-time system must have ability to process and transmit data with high speed. In order to ensure constant and continuous data transmission, ARM-Linux system uses ping-pong operation of two pieces of data buffer (the maximum data buffer size is 120K bytes) to achieve real-time data in series. When the valid data block of user space is writing into Ping-pong Buffer0 in kernel space, SPI TX_FIFO can read valid data from Ping-pong Buffer1 which has completed writing into the valid data block. Similarly, when the valid data block of user space is writing into Ping-pong Buffer1 in kernel space, SPI TX_FIFO can read valid data from Ping-pong Buffer0 which has completed writing into the valid data block. In this way, SPI TX_FIFO can read valid data from data buffer continuously.

Figure.4 shows SPI bus data streams and FIFO in ARM.SPI TX_FIFO can read valid data from Ping-pong Buffer1 which has completed writing into the valid data block. Similarly, when the valid data block of user space is writing into Ping-pong Buffer1 in kernel space, SPI TX_FIFO can read valid data from Ping-pong Buffer0 which has completed writing into the valid data block. In this way, SPI TX_FIFO can read valid data from data buffer continuously.

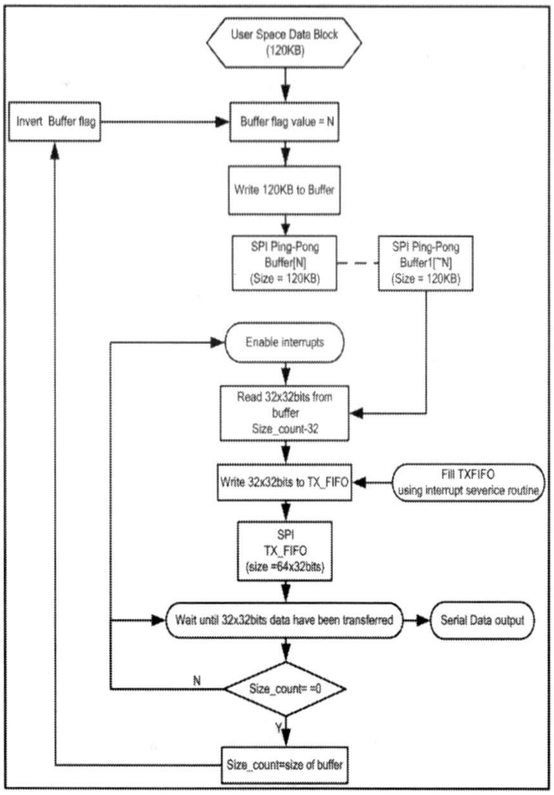

Figure 4. Software flow pattern of SPI bus driver

The processed data in ARM-Linux user space can be written into ARM-Linux kernel space in blocks, and then FPGA can read out serial processed data from ARM through the SPI bus. Meanwhile, the real-time data sampled by ADC can be transmitted to ARM-Linux kernel space through the SPI bus, and then the real-time data in ARM-Linux kernel space can be written into ARM-Linux user space in blocks.

In order to ensure constant and continuous data transmission, ARM-Linux system uses ping-pong operation of two pieces of data buffer (the maximum data buffer size is 120K bytes) to guarantee the real-time data processing. When the valid data block of user space is writing into Ping-pong Buffer0 in kernel space, SPI TX_FIFO can read valid data from Ping-pong Buffer1 which has completed writing into the valid data block. Similarly, When the valid data block of user space is writing into Ping-pong Buffer1 in kernel space, SPI TX_FIFO can read valid data from Ping-pong Buffer0 which has completed writing into the valid data block. The time that FPGA finish read data from a data buffer is about , this time is longer than the time that Linux controls worst response time (50ms).[8] In this way, FPGA can transmit data continuously.

3 Conclusion

This software radio platform takes full advantage of the flexibility of ARM and the parallelism of FPGA, realizes the design of high-speed data transmission and processing.

With the rapid development of Software Radio technology, Software Radio technology has become a hotspot of research. Software Radio is a new architecture for wireless communication based on digital signal processing and microelectronics technology.[9] In this paper, an ARM and FPGA based software radio platform is designed according to the needs of actual system in digital signal processing and wireless communication, which take full advantage of the flexibility of ARM and the parallelism of FPGA. The choice of ARM is Freescale' s IMX6, which has a rich instruction set and programming flexibility; FPGA has a great advantage in the speed and parallel computing, suitable for requirements for real-time data transmission.

This work presented an embedded scalable cluster platform with software defined radio applications. FPGA serves as the coprocessor and works together with the ARM which serves as the core processor in this platform. ARM is used to complete digital signal processing and real-time data transmission, And FPGA is used to complete synchronous timing control and serial-parallel conversion.[10] A SPI driver for real-time data transmission between ARM and FPGA under ARM-Linux system is provided.

Experimental results show that the real-time and high speed performance of the software radio platform can meet the design requirements, and it is capable of implementing wireless communication functions and satisfies the requirements of real-time signal processing platform for high security and broad applicability.

References

1. Wang Ya-ting, *High-speed data acquisition technology research based on ARM and FPGA*, 2007.
2. Long Zai-chuan, Feng Lin. "*Embedded CCD acquisition system based on ARM and FPGA*", Optoelectronics • Laser, vol. 18, no.11, pp.1296-1298, 2007.
3. ZHU Xiao-peng, XIAO Tie-jun, Zhao Hui. "*ARM+FPGA for realtime data acquisition system design*", Computer Engineering and Design, vol. 30, no.13, pp.3088-3090, 2009.
4. XU Xin-shun, JIA Zhi-ping. Embedded Linux Application Programming, People's Posts & Telecom Press, 2008.
5. J. van de Belt, P. D. Sutton, and L. E. Doyle, "*Accelerating software radio: Iris on the Zynq SoC*" in Very Large Scale Integration (VLSISoC), 2013 IFIP/IEEE 21st International Conference, pp. 294-295, 2013.
6. Zhao Xiao-pu, Li Li, Song Hua "*Gigabit Ethernet interface in S3C2440A System*", Electronic Design and Application, vol. 8, pp.101-103, 2007.
7. A. Love, W. Zha, and P. Athanas. "*In pursuit of instant gratification for FPGA design*" in Field

Programmable Logic and Applications (FPL), 2013 23rd International Conference , pp. 1-8, 2013.

8. J. van de Belt, P. D. Sutton, and L. E. Doyle, "*Accelerating software radio: Iris on the Zynq SoC*" in Very Large Scale Integration (VLSISoC), 2013 IFIP/IEEE 21st International Conference, pp. 294-295, 2013.
9. J. Delorme, J. Martin, A. Nafkha, C. Moy, F. Clermidy, P. Leray, and J. Palicot. "*A FPGA partial reconfiguration design approach for cognitive radio based on NoC architecture*" in Circuits and Systems and TAISA Conference, 2008. NEWCAS-TAISA 2008.
10. Wu Wei, Hao Jian-Xin, Lu Li. "*High-speed data acquisition system based on ARM and FPGA*", Modern electronic technology, no.1, pp.126-128, 2009.

MATEC Web of Conferences **56**, 04001 (2016) DOI: 10.1051/matecconf/20165604001
ICCAE 2016

Application of Artificial Neural Network (ANN): Development of Central-based ANN (CebaANN)

Su Yeon Jeong[1], Tae Seon Yoon[2] and Chae Yoon Jeong[1]

[1]*Hankuk Academy of Foreign Studies, Student, 17035 Gyeonggi-do Yongin, Korea*
[2]*Hankuk Academy of Foreign Studies, Science and Information Department, 17035 Gyeonggi-do Yongin, Korea*

Abstract. Nowaday, the number of known protein structures is significantly less than the number of known amino acid sequences. It is because the regularity of amino acid depend on structure is not clear and the number of thermodynamic conditions are too many. There are some cases that discovering protein structure by experiment. However, It needs much time and cost for increasing the number of amino acid sequences, thus, there is less efficiency. So the empirical method which predict theoretically the structure of protein has been developed. We suggest Central-Based Artificial Neural Network as prediction method of protein structure. CebaANN can analyze similarity more detail by making part of center that affect outcome bigger. In experiment we got 85% of prediction probability at E structure, but we got 34% of probability at total.

1 Introduction

Nowaday, the number of known protein structures is significantly less than the number of known amino acid sequences. It is because the regularity of amino acid depend on structure is not clear and the number of thermodynamic conditions are too many. There are some case that discovering protein structure by experiment. However, It needs much time and cost for increasing the number of amino acid sequences, thus, there is less efficiency.

Empirical methods which theoretically predict the structure of protein has been proposed to overcome this limitation. First is ab initio method based on the thermodynamic hypothesis of Anfinsen. Second is method that using protein which is already discovered as template. First method is possible in a brief time, however, accuracy of the structure is unable to give us a satisfactory result. Therefore, second method which means two protein which have similar structure is usually used.

Because it is unable to compare existing protein sequences and structure, one by one, algorithm such as artificial neural network and Support Vector Machine is usually used at analyzing link between sequence and structure and deriving similarity by comparing result of general structure and result of input structure.

Artificial Neural Network is algorithm which was invented from neural network and consisting of input, hidden, output nodes. In this paper, we developed cebaANN which is application of ANN, and propose it is algorithm for protein structure prediction. CebaANN has existing structure [input - hidden - output]. However, it can analyze similarity more detail by making part of center that affect outcome bigger. In experiment we got 85% of prediction probability at E structure, but we got 34% of probability at total.

2 Relative research

2.1 Artificial neural network

An artificial neural network is an interconnected group of nodes, akin to the vast network of neurons in a brain. Here, each circular node represents an artificial neuron and an arrow represents a connection from the output of one neuron to the input of another. G and H structure is numbering by P(a), In G and H structure is numbering by P(a), machine learning and cognitive science, artificial neural networks (ANNs) are a family of statistical learning models inspired by G and H structure is numbering by P(a), biological neural networks (the central nervous systems of animals, in particular the brain) and are used to estimate or approximate functions that can depend on a large number of input and are generally unknown. Artificial neural networks are generally presented as systems of interconnected "neurons" which exchange messages between each other. The connections have numeric weights that can be tuned based on experience, making neural nets adaptive to inputs and capable of learning.

For example, a neural network for handwriting recognition is defined by a set of input neurons which may be activated by the pixels of an input image. After being weighted and transformed by a function (determined by the network's designer), the activations of these neurons are then passed on to other neurons. This process is repeated until finally, an output neuron is activated. This determines which character was read.

Like other machine learning methods - systems that learn from data - neural networks have been used to solve a wide variety of tasks that are hard to solve using ordinary rule-based programming, including computer vision and speech recognition.

2.2 DSSP classification

The Protein structure is the biomolecular structure of a protein molecule. Proteins are polymers — specifically polypeptides — formed from sequences of amino acids. Each unit of a protein is called an amino acid residue because it is the residue of every amino acid that forms the protein by losing a water molecule. By convention, a chain under 40 residues is often identified as a peptide, rather than a protein.[1]To be able to perform their biological function, proteins fold into one or more specific spatial conformations, driven by a number of non-covalent interactions such as hydrogen bonding,ionic interactions, Van der Waals forces, andhydrophobic packing. To understand the functions of proteins at a molecular level, it is often necessary to determine their three-dimensional structure.

The Dictionary of Protein Secondary Structure, in short DSSP, is commonly used to describe the protein secondary structure with single letter codes. The secondary structure is assigned based on hydrogen bonding patterns as those initially proposed by Pauling et al. in 1951 (before any protein structure had ever been experimentally determined). There are eight types of secondary structure that DSSP defines:

- G = 3-turn helix (310 helix). Min length 3 residues
- H = 4-turn helix (α helix). Min length 4 residues
- I = 5-turn helix (π helix). Min length 5 residues
- T = hydrogen bonded turn (3, 4 or 5 turn)
- E = extended strand in parallel and/or anti-parallel β-sheet conformation. Min length 2 residues
- B = residue in isolated β-bridge (single pair β-sheet hydrogen bond formation)
- S = bend (the only non-hydrogen-bond based assignment)
- C = coil (residues which are not in any of the above conformations)

In this paper, we only use 6 structures (G, H, I, T, E, B)

2.3 Chou-fasman parameters

The Chou-Fasman method of secondary structure prediction depends on assigning a set of prediction values to a residue and then applying a simple algorithm to those numbers. We were in the process of converting a short for each amino acid to a number, using this table.

Table 1. Chou-fasman parameters.

Protein Name	Chou-fasman parameters						
	P (a)	P (b)	P(turn)	f(i)	f(i+1)	f(i+2)	f(i+3)
Alanine	142	83	66	0.06	0.076	0.035	0.058
Arginine	98	93	95	0.07	0.106	0.099	0.085
Aspartic Acid	101	54	146	0.147	0.11	0.179	0.081
Asparagine	67	89	156	0.161	0.083	0.191	0.091
Cysteine	70	119	119	0.149	0.05	0.117	0.128
Glutamic Acid	151	37	74	0.056	0.06	0.077	0.064
Glutamine	111	110	98	0.074	0.098	0.037	0.098
Glycine	57	75	156	0.102	0.085	0.19	0.152
Histidine	100	87	95	0.14	0.047	0.093	0.054
Isoleucine	108	160	47	0.043	0.034	0.013	0.056
Leucine	121	130	59	0.061	0.025	0.036	0.07
Lysine	114	74	101	0.055	0.115	0.072	0.095
Methionine	145	105	60	0.068	0.082	0.014	0.055
Phenylalanine	113	138	60	0.059	0.041	0.065	0.065
Proline	57	55	152	0.102	0.301	0.034	0.068
Serine	77	75	143	0.12	0.139	0.125	0.106
Threonine	83	119	96	0.086	0.108	0.065	0.079
Tryptophan	108	137	96	0.077	0.013	0.064	0.167
Tyrosine	69	147	114	0.082	0.065	0.114	0.125
Valine	106	170	50	0.062	0.048	0.028	0.053

3 Central-based artificial neural network (CebaANN)

Artificial neural networks are learning algorithm obtained an idea from the structure of the biology of neural networks. Artificial neural network is a model with problem solving that artificial neurons which form a network by binding synapse change weights in progress of learning. In here, it can compare binds of synapse to connection between nodes, and compare strength of bind to weight.

CebaANN is the application of artificial neural network that the structure is changed for making input value's affect more powerful as close to center input node. Unlike typical neural network that each input node is connected each hidden node, CebaANN has network that center input node is connected all of hidden nodes and as going outside, the number of hidden nodes which is connected input node decreases. For the structure, network must be constituted with the specific condition:

- output node should be one, and (n = the number of input nodes, m = the number of hidden nodes) m should be [(n+1)/2]

- threshold is (average of input node values / m) * sum of weights between input and hidden * sum of weights between hidden and output.
- threshold should be constant. so while in the progress, the sum of weight between input and hidden and between hidden and output should be constant.
- CebaANN is teacher learning method (if the target really have the result, base value is 1, and if not, base value is 0). So it have two case that apply the algorithm. Case 1 is output>threshold and base value is 0, Case 2 is output<threshold and base value is 1.

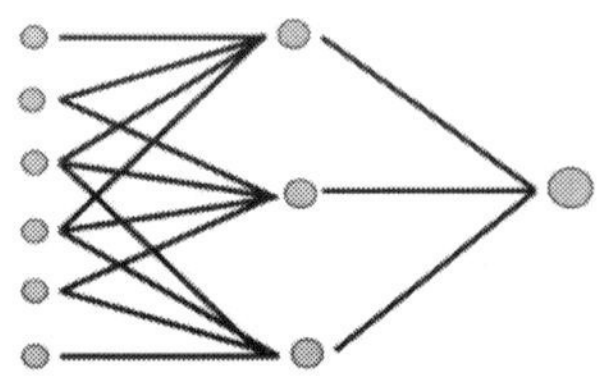

Figure 1. The structure of CebaANN_type 1.

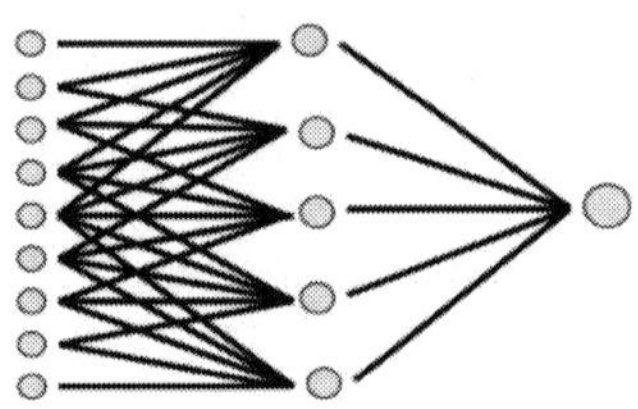

Figure 2. The structure of CebaANN_type 2

3.1. The method how change weights (Balancing method)

Balancing method :

If Case 1, follow 3.1.1 and 3.1.2 and make new output that applying changed weights. Repeat these three steps until output < threshold.

If Case 2, increase weights on the route of maximum input value (opposite of 3.1.1) and decrease weights on the route of minimum input value (opposite of 3.1.2) and make new output that applying changed weights. Repeat these three steps until output > threshold.

During the progress, if the number of Maximum or Minimum aren't one, decide the maximum for using as maximum which is close to center. if minimum is in same situation, decide the minimum like same, too. Because as go center, it's affect is more big.

3.1.1 Decreasing weights on the route of maximum input value

Decreasing input-hidden weight and hidden-output weight of the route whose input-hidden weight * hidden-output weight is minimum among the routes' that maximum input value passes. And increasing other input-hidden weights which is connected same hidden node as much as 1/(2*[m/2]) of change amount. As result, the sum of input-hidden weights is constant, and output is decreased.

3.1.2 Increasing weights on the route of minimum value

Increasing input-hidden weight and hidden-output weight of the route whose input-hidden weight * hidden-output weight is maximum among the routes' that minimum input value passes as much as change amount of a). And decreasing other input-hidden weights which is connected same hidden node as much as 1/(2*[m/2]) of change amount. As result, the sum of input-hidden weights and the sum of hidden-output weights are constant, and output is decreased.

4. Experiment (Using cebaANN)

We did experiment that getting weights which can predict protein structure when protein sequence pass the CebaANN.

4.1. Making material of experiment

Target protein sequence dataset : RS126. We use 2/3 of dataset as making special weights, and use others as measuring correct percent of prediction.

4.1.1 Separating target protein sequences

Seperating target protein sequence as amino acid sequence whose length is the number of input node (In this experiment, nine) from first amino acid. and increase the position of start amino acid as one. Now, we say this as input sequence.

4.1.2 Numbering each elements

Each element of input sequnces is numbering by chou-fasman parameters. G and H structure is numbering by P(a), and E and B structure is numbering by P(b), and T and S structure is numbering by P(turn). It's because G and H are helixs, and E and B are β-sheet or bridge, and T and S are turn or bend. And then, we can get three numbered sequence at one input sequence. Now, we say these three sequences as input group.

4.1.3 Matching input group with structure.

Matching input group with input sequence's disclosed structure. only one structure among six or 'none' is matching with input group. We say this matching as material. After, at the progress of running algorithm, elements' of input groups become input, and matched structure become base value.

4.2. Running CebaANN

Classifying materials into six groups of structure. And then, run CebaANN at each six group. During this progress, if material has targeted structure, base value become 1, and if not, base value become 0. After running is finished, special weights of each six group were made. First element of input group is runned at G, H group, second element of input group is runned at T, S group, Third element of input group is runned at E, B group.

4.3 Predicting structure.

At first, make protein sequence (which want to predict structure) as input group (4.1.1, 4.1.2). And then, passing input group at special weights of six groups.

If all of output is smaller than threshold, it has no structure. But if some of output is bigger, structure is decided by list of the highest-to-lowest parameter value of middle amino acid among structure whose output is bigger. If the number of highest parameter value isn't one, it would be decided as structure which has bigger output value. total.

5. Result

5.1. Special weights

5.1.1 Hidden-Output weights

special weights on the route between Hidden and Output.

Table 2. Hidden-Output weights.

(hidden)	1	2	3	4	5
G	9	236	260	200	295
H	328	180	188	172	132
E	67	109	209	272	343
B	244	152	144	164	296
T	648	337	148	-35	-98
S	396	-795	276	220	903

5.1.2 Input-Hidden weights

special weights on the route between Input and Hidden.

Table 3. Input-Hidden weights.

	1	2		(Input node) 3		
	(hid)1	1	2	1	2	3
G	-11078	3864	148	4200	241	193
H	353	105	187	438	449	359
E	-586	-250	-177	365	200	28
B	395	205	319	115	133	139
T	7966964	1966195	3835343	-3240436	-343323	457980
S	1.1E+180	-3E+179	1.7E+182	-3E+179	-2E+180	1E+179

(Input node) 4				5	
(hidden)1	2	3	4	1	2
2372	151	118	172	1637	112
44	78	75	79	43	137
990	352	198	120	474	214
205	224	224	188	83	69
-2957871	-4678738	-6698963	33440169	-3733851	461165
-3E+179	3.6E+179	1E+179	8.8E+178	-3E+179	4.4E+178

(Input node) 5			6			
3	4	5	hidden)2	3	4	5
119	27	63	351	333	427	257
92	18	48	152	145	158	204
153	59	98	409	213	125	107
50	43	1	254	279	238	135
2907195	-4.6E+07	46248755	726552	823860	27577772	-4.3E+07
1E+179	8.8E+178	3.7E+178	-2E+182	-4E+179	-4E+179	-1E+179

(Input node) 7			8		9
(hidden)3	4	5	4	5	5
238	247	157	127	82	442
330	308	421	438	605	-266
412	274	222	423	338	239
309	283	159	248	219	483
2510930	-2.1E+07	1888390	5975314	21967080	-2.8E+07
1E+179	8.8E+178	3.7E+178	8.8E+178	3.7E+178	3.7E+178

5.2. Correct percent of prediction

Correct percent : (the number of sequence which predicted structure equals the target structure) / (the number of sequence which has the target structure)

Table 4. percent of prediction

structure	Total	G	H	E	B	T	S	None
(%)	34.58	0.0	0.0	85.36	0.0	0.0	0.0	8.69

6. Conclusion

Nowaday, the prediction method of protein structure is developed so much. However the necessity of exact prediction method of protein is still exist. There are many amino acid sequence which doesn't discovered its structure. And because protein structure is clue to find protein's function, prediction method can help many people as diverse way. CebaANN's prediction is only act at E structure, but it's percent is high. It can prove that CebaANN has possibility to be accurate prediction method. We need to find why it only act at E structure and why it doesn't work at other structure. So we would expand experiment ahead. The limitation was it only do work at E structure. And We will make improvements at that part. This experiment was meaningful that we do new challenge.

References

1. Machin, David. "Biomathematics." (1976)
2. Taylor, William Ramsay. "The Classification of Amino Acid Conservation." *Journal of Theoretical Biology* 119.2 (1986): 205-18

DOI: 10.1051/matecconf/20165604001

3. P. Prevelige, Jr. and G. D. Fasman, Chou-Fasman Prediction of Secondary Structure, in Prediction of Protein Structure and the Principles of Protein Conformation, ed. G. B. Fasman (ISBN 0-306-43131-9, Plenum, New York)

Analysis of Human Papillomavirus Using Datamining - Apriori, Decision Tree, and Support Vector Machine (SVM) and its Application Field

Younghoon Cho[1], Seungwon Burm[1], Nayoung Choi[1] and Taeseon Yoon[2]

[1]*Department of Natural Science, Hankuk Academy of Foreign Studes, 50, Oedae-ro 54beon-gil, Mohyeon-myeon, Cheoin-gu, Yongin-si, Gyeonggi-do, Republic of Korea*
[2]*Department of Computer Science and Engineering, Korea University, Seoul, Repuulic of Korea*

Abstract. Human Papillomavirus(HPV) has various types (compared to other viruses) and plays a key role in evoking diverse diseases, especially cervical cancer. In this study, we aim to distinguish the features of HPV of different degree of fatality by analyzing their DNA sequences. We used Decision Tree Algorithm, Apriori Algorithm, and Support Vector Machine in our experiment. By analyzing their DNA sequences, we discovered some relationships between certain types of HPV, especially on the most fatal types, 16 and 18. Moreover, we concluded that it would be possible for scientists to develop more potent HPV cures by applying these relationships and features that HPV virus exhibit.

1 Introduction

Human Papillomavirus, or HPV, is a DNA virus which is involved in the papillomavirus family. It can induce a number of infections in keratinocytes of the skin or mucous membranes. HPV is known worldwide to induce various genital cancers, such as cervical cancer, vaginal cancer, and anal cancer. Most viruses pass through cell membranes and progress self-proliferation to survive. During this process, HPV produces several kinds of proteins that make host cells infinitely progress the cell-division. E6 protein inhibits p53 which induce an apoptosis. And E7 protein inhibits Rb which restricts cell cycle. Therefore, host cells for HPV keep cell division permanently and eventually turn into cancer cells.

According to numerous studies concerned, HPV has been proved its importance in revealing the mechanism of cancer development. It is the necessary cause of human cancer, especially cervical cancer. In other words, without the infection of HPV and the presence of HPV DNA, several types of cancer will not develop. This fact implies that if we can curb the HPV infection and gene expression, it is possible to prevent and cure cervical cancer and other types of genital cancer. Since these cancers are highly fatal to human, scientists are trying to reveal the overall process of human papillomavirus's expression in human body in order to develop the medication for cervical cancer. Recently, HPV vaccines were invented, and are being offered for girls aged under 15. However, to minimize the adverse effects and maximize the efficiency of HPV vaccines, it is crucial to reveal the molecular structure of Human Papillomavirus to create a best-suited medication for HPV only. The molecular structure of HPV is determined by the sequence of amino acids. And the sequence of amino acids is encoded in mRNA using 4 types of bases: A, G, C, and T. In this paper, we analyzed RNA sequences of human papillomaviruses classified as 'highest risk' and 'probably high risk' in order to find out similarities and differences of those two categories.

2 Relative Research

2.1 Human Papillomavirus (HPV)

Human papillomavirus(HPV) causes many different types of diseases and it is classified by the risk levels into different types. By analyzing various cases, HPV types were classified as 15 high-risk types(16, 18, 31, 33, 35, 39, 45, 51, 52, 56, 58, 59, 68, 73, 82), 3 probable high-risk types(26, 53, 66), and 12 low-risk types(6, 11, 40, 42, 43, 44, 54, 61, 70, 72, 81, CP6108)[1]. It has been found that DNA of human papillomavirus (HPV) types 16 and 18 is closely associated with human genital cancer. It supports the concept that HPV type 16 and 18 are key factors in the aetiology of genital cancer. Furthermore, searching about the vaccine [2] HPV 16/18 AS04-adjuvanted vaccine was immunogenic, generally well tolerated, and effective against HPV-16 or HPV-18 infections and the research analyzed efficacy in the final event-driven analysis of the women who were vaccinated at months 0, 1, and 6. The HPV 16/18 AS04-adjuvanted vaccine showed high efficacy against CIN2+ associated with HPV-16/18 and non-vaccine oncogenic HPV types [3]. In the field of cervical cancer, genital HPV has the role as the central etiologic factor in cervical cancer worldwide [5]. Also the presence of HPV in virtually all cervical cancers implies the highest worldwide attributable fraction so far reported for a specific cause of any major human cancer [6]. Furthermore, ingration of HPV-16 DNA, which occurs in cervices, can result in the increased expression of the viral E6 and E7 oncogenes through altered mRNA stability and occur cervical cancers. Also the demonstration that more than 20 different genital HPV types are associated with cervical cancer has important implications for cervical cancer-prevention strategies that include the development of vaccines targeted to genital HPVs [4].

2.2 Datamining Algorithms

In this study, we used three different datamining algorithms: Decision Tree Algorithm, Apriori Algorithm, and Support Vector Machine Algorithm.

2.1.1 Support Vector Machine (SVM)

Support Vector Machines are supervised learning models with associated learning algorithms which analyze data and recognize patterns, utilized for classification and regression analysis.

2.1.2 Apriori Algorithm

Apriori is an algorithm for frequent item set mining and association rule learning over transactional databases. It operates by recognizing the frequent individual items in the database and stretching them out to bigger and bigger item sets as long as those item sets show up sufficiently often in the database.

2.1.3 Decision Tree Algorithm

A decision tree is a decision support tool which uses a tree-like graph or model of decisions and their possible results. It is one way to display an algorithm. They are generally used in operations research, especially in decision analysis in order to help identify a strategy which is most likely to reach a goal.

3 Method

In this study, we used 11 base sequences of different types of HPV (HPV 16, HPV 18, HPV 26, HPV 31, HPV 33, HPV 35, HPV 53, HPV 66, HPV 68a, HPV 68b, HPV 82). As we mentioned in the relative research, three of them (HPV 26, HPV 53, HPV 66) are probable high-risk group, and the others are high-risk group. If we discover several similarities and differences in base sequences between these two groups, it means that we could find amino acids that play a dominant role in evoking cervical cancer. So, we extracted full RNA sequences of these viruses from NCBI (National Center for Biotechnology Information). And we applied several algorithms to figure out similarities and differences between these viruses.

In this paper, we used 3 types of datamining algorithms: Decision Tree Algorithm, Apriori Algorithm, and Support Vector Machine (SVM). Decision Tree Algorithm has its strength to clarify distinct differences between amino acid sequences, while Apriori Algorithm has its strength to declare similarities. And SVM can provide more accurate results since it can use higher dimensions than dimensions other algorithms use.

We conducted three experiments for each algorithm. The base sequence of each virus is too vast to analyze at once. So, we have to divide it into several parts. So, when we analyze HPV base sequences, we conducted three experiments: 9-windows, 13-windows, and 17-windows.

Especially, for SVM algorithm, we carried out the experiments with 10 - fold cross validation, and we applied four types of functions: normal, polynomial, polynoima2, and RBF.

4 Results

4.1 Decision Tree Algorithm Results

4.1.1 9 window results

Table 1 shows the results of 9-windows Decision Tree algorithm. It is remarkable that every virus possesses its unique rule, and that every extracted rule has the probability of at least 0.75. This value is high enough to affirm that every HPV virus has its own distinguishable trait. Also, we can see frequent repeatance of Threonine, Leucine, and Valine in Table 1. It may indicate that these three amino acids play a key role in HPV virus. Moreover, we can find that the amino acids extracted from position 2 and position 9 represent the distinguishable rule of each virus. So, we can find that posiion 2 and position 9 is an important factor that makes HPV viruses different.

Table 1. 9 window rule (POS=Position)

HPV	Rule Content
HPV 16	pos2=D pos3=T pos6=L
HPV 18	pos6=I pos7=L pos9=F
HPV 26	pos3=K pos4=T pos8=E
HPV 31	pos1=L pos4=S pos9=T
HPV 33	pos2=F pos4=V pos9=T
HPV 35	pos1=L pos2=P pos9=S
HPV 53	pos2=V pos7=Y pos8=L
HPV 66	pos1=I pos2=Y pos3=R
HPV 68a	pos1=I pos2=H pos9=K
HPV 68b	pos2=P pos4=T pos9=V
HPV 82	pos1=T pos5=G pos6=F

4.1.2 13 window results

Table 2 shows the results of 13-windows Decision Tree algorithm. All extracted rules have the probability of 0.75, which is quite high as we mentioned before. Table 2 is similar to Table 1 in that all rules are composed of 3 amino acids. What is different is first, in Table 2, position 7 are more frequent than position 2 or 9, and second, relatively various types of amino acids appeared. For instance, HPV 26 has Glutamine, which doesn't appear in Table 1. Also, HPV 31, HPV 68a, and HPV 82 have their unique figure of Lysine.

Table 2. 13 window rule (POS=Position)

HPV	Rule Content
HPV 16	pos2=F pos7=V pos10=Y
HPV 18	pos7=P pos9=Y pos12=I
HPV 26	pos3=E pos7=Q pos13=T
HPV 31	pos4=S pos7=K pos9=L
HPV 33	pos5=V pos7=P pos11=S
HPV 35	pos1=S pos7=I pos11=C
HPV 53	pos2=L pos7=N pos9=A
HPV 66	pos1=L pos7=Y pos8=Y
HPV 68a	pos3=P pos7=E pos13=K
HPV 68b	pos2=T pos5=A pos7=P
HPV 82	pos7=L pos8=K pos13=C

4.1.3 17 window results

Table 3 shows the results of 17-windows Decision Tree algorithm. It is noticeable that many rules of various HPV are constituted of 2 amino acids. Also, in 17-windows results, position 6 plays a key role in extracting rules. Furthermore, HPV 16 and HPV 33 have Cysteine, and this finding is unique to these two viruses.
So, it may implicate the key difference between HPV viruses.

Table 3. 17 window rule (POS=Position)

HPV	Rule Content
HPV 16	pos6=C pos7=L pos16=T
HPV 18	pos6=L pos13=V pos14=L
HPV 26	pos2=P pos9=E
HPV 31	pos5=Q pos6=S
HPV 33	pos6=C pos7=Q
HPV 35	pos5=F pos6=K pos15=T
HPV 53	pos6=G pos17=A
HPV 66	pos4=L pos5=V pos17=L
HPV 68a	pos6=E pos10=D
HPV 68b	pos1=S pos6=V pos12=I
HPV 82	pos6=A pos17=A

4.2 Apriori Algorithm

4.2.1 9 window results

Table 4 shows the results of 9-windows Apriori Algorithm. According to Table 4, it is clear that Leucine plays a dominant rule in all HPV virus types. All HPV viruses have a considerable rate of Leucine, and this fact indicates that we can utilize Leucine Synthesis Inhibitor to treat cervical cancer. Also, HPV 33 and HPV 68b have Isoleucine as a key component. Since Isoleucine is an isoform of Leucine, we may find the relationship between these two amio acids. Moreover, we can find Valine in HPV 16, HPV 18, HAV 35, and HPV 68a and Threonine in HPV 26, HPV 31, HPV 68b, and HPV 82. It is predictable that two amino acids, Valine and Threonine, are exclusive. In other words, when one HPV virus has valine, it doesn't have thereonine. This rule may be able to discover the difference between lethal and non-lethal Human Papillomaviruses. Also, this table implies the differences between high-risk group and probable high-risk group. High-risk group has both Leucine and Valine, while probable high-risk group doesn't have. This fact implies that Leucine and Valine plays a key role in evoking high-risk cervical cancer.

Table 4. 9 window rule (Amino Acid(Frequency))

HPV	Rule Content
HPV 16	L(44) V(30)
HPV 18	L(34) V(30)
HPV 26	L(32) T(31)
HPV 31	L(30) S(29) T(29)
HPV 33	I(29) L(31)
HPV 35	L(30) V(30)
HPV 53	L(35)
HPV 66	L(43)
HPV 68a	L(40) V(33)
HPV 68b	I(30) T(29)
HPV 82	L(36) T(32)

4.2.2 13 window results

Table 5 shows the results of 13-windows Apriori Algorithm. According to Table 5, it is also clear that Leucine plays a dominant rule in all HPV virus types.

Also, Isoleucine, Valine and Threonine: these three amino acids play a significant role in extracting rules. Especially, HPV 68b has 6 amino acids, and those 6 amino acids have almost equal importance. Moreover, in 13-windows results, unlike 9-windows results, Threonine and Valine co-exist in many HPV viruses.

Table 5. 13 window rule (Amin Acid(Frequency))

HPV	Rule Content
HPV 16	L(44) V(30)
HPV 18	L(34) V(30)
HPV 26	L(32) T(31)
HPV 31	L(30) S(29) T(29)
HPV 33	I(29) L(31)
HPV 35	L(30) V(30)
HPV 53	L(35)
HPV 66	L(43)
HPV 68a	L(40) V(33)
HPV 68b	I(30) T(29)
HPV 82	L(36) T(32)

4.2.3 17 window results

Table 6 shows the results of 17-windows Apriori Algorithm. According to Table 6, it is also clear that Leucine plays a dominant rule in all HPV virus types. Also, unlike Table 4 and Table 5, extracted rules include more amino acids. Also, Threonine and Valine, and Serine take part in extracting distinguishable rules.

Table 6. 17 window rule (Amino Acid(Frequency))

HPV	Rule Content
HPV 16	I(22) L(35) V(22)
HPV 18	I(21) L(26) T(21) V(22)
HPV 26	L(25) S(22) T(20)
HPV 31	L(20) S(20) T(21)
HPV 33	I(22) L(22) V(22)
HPV 35	I(20) L(28) T(21) V(22)
HPV 53	I(21) L(26)
HPV 66	L(30) V(24)
HPV 68a	L(29) T(20) V(21)
HPV 68b	I(25) L(21) R(20) S(20) T(21) V(23)
HPV 82	L(27) V(21)

4.3 Support Vector Machine Results

Considering the fact that results of Apriori and Decision tree don't have a lot in common, we conducted the third experiment using SVM. However, because of the excessive data, infinite loop was created and we were only able to use the normal types. We used 10 fold cross validation, and the phrase "Average loss on test set 90.0000 Zero/one-error on test set 90.00% (35 correct, 315 incorrect, 350 total)" was repeated 10 times. The correct ratio was low due to the large amount of data.

After experiencing the difficulty of experiment using such big data, we decided to compare a few specific types considered to have big differences. From the results made by Apriori and Decision tree, we found that among the high-risk types, HPV 68b has the least similarity with HPV 16, 18 which is most frequently in cases of cancer. As a result, we repeated the experiment HPV 18 and HPV 68b in depth.

4.3.1 9 window results

During experiment, we made data types of <HPV 18 : +1, HPV 68B : -1>. In 9-windows, +1 has 292 data units and -1 has 291 data units. For training, we used 240 data units, while for test, we used 60 data units.

In Table 7, it represents 9-windows results of SVM algorithms. Average of Accuracy rate is highest in RBF, and lowest in Polynomial2 and Normal. Since the accuracy rate is quite low (about half), it is clear that HPV 18 and HPV 68b is not clearly divided into two parts. However, this result is still meaningful since the result of RBF has 76.416%, which means that HPV 18 and HPV 68b have different properties in amino acid sequences.

Table 7. 9 window rule (Accuracy on Test(%))

	1st	2nd	3rd	4th	5th	average
Normal	57.5	51.6	51.6	49.1	46.6	51.834
	50.8	56.6	50.0	50.8	53.3	
Poly-1	54.1	50.8	60.0	51.6	45.8	52.25
	52.5	48.3	59.1	49.1	50.8	
Poly-2	58.3	55.8	50.0	45.8	49.1	51.749
	43.3	50.8	59.1	55.0	50.0	
RBF	79.1	80.0	73.3	73.3	78.3	76.416
	70.8	79.1	75.0	75.8	79.1	

4.3.2 13 window results

In 13-windows, +1 has 202 data units and -1 has 201 data units. For training, we used 160 data units for one time, while for test, we used 40 data units.

In Table 8, it represents 13-windows results of SVM algorithms. The accuracy rate of the experiment is highest in RBF, and lowest in polynomial2. The overall accuracy rate of 13-windows is generally higher than that of 9-windows. Also, this result indicates that HPV 18 and HPV 68b have some different traits, but it is not clearly divided.

Table 8. 13 window rule (Accuracy on Test(%))

	1st	2nd	3rd	4th	5th	average
Normal	58.7	56.2	66.2	51.2	63.7	56.7508
	65.0	50.0	50.0	55.0	51.2	
Polynomial	55.0	57.5	63.7	57.5	57.5	57.75
	63.7	62.5	57.5	52.5	50.0	
Poly-2	55.0	57.5	57.5	55.0	55.0	54.875
	53.7	52.5	48.7	55.0	58.7	
RBF	75.0	72.5	77.5	76.2	80.0	78.00
	76.2	75.0	82.5	77.5	87.5	

4.3.3 17 window results

In 17-windows, +1 has 155 data units and -1 has 154 units. For training, we used 120 data units for one time, while for test, we used 30 data units.
In Table 9, it represents the results of 17-windows SVM algorithms. Unlike 9 and 13 windows, the average accuracy of experiment is highest in polynomial2, and lowest in normal. Since polynomial2 function can use higher dimensions than normal function, it seems that they can classify different data sets into two groups. Also, polynomial2 and RBF function represents relatively high accuracy rate (76.833 and 76.499). So, we can conclude that HPV 18 and HPV 68b have several stark differences in their base sequences.

Table 9. 17 window rule (Accuracy on Test(%))

	1st	2nd	3rd	4th	5th	average
Normal	63.33	58.33	60.00	60.00	56.67	58.16
	58.33	53.33	66.67	45.00	60.00	
Polynomial	65.00	76.67	66.67	58.33	70.00	61.5
	60.00	65.00	56.67	38.33	58.33	
Polynomial2	90.00	78.33	73.33	76.67	75.00	76.833
	80.00	73.33	73.33	76.67	71.67	
RBF	78.33	81.67	73.33	75.00	83.33	76.499
	81.67	75.00	70.00	73.33	73.33	

5 Conclusion

In this study, we have found out that leucine, isoleucine, threonine, valine are the most dominant amino acids in Human Papillomavirus. Leucine acts on building muscles and regulating blood sugar. One of the major functions of isoleucine is proteinogenesis in the body. Threonine helps to maintain the proper protein balance in the body. Valine has a stimulant effect, so it is needed for muscle metabolism, tissue repair, and the maintenance of a proper nitrogen balance in the body. Most of the factors that induce cancer other than cervical cancer are mostly composed of leucine. As a result, inhibitors of leucine's synthesis are widely used as anti-cancer medicine, but it also distracts the synthesis of leucine that is essential to human body. Unlike other viruses, HPV is created with diverse amino acids. Malaria for instance, is mostly made of Leucine, which is a main amino acid which creates human muscle tissue. However HPV consists of Leucine, Valine, Threonine, Isoleucine, Cysteine, and many others in equal manner. This implicates that HPV can operate in various types of tissues composed of different amino acids. Furthermore, since the sequence and the types of amino acid is very different in every types of HPV, it will produce diverse kinds of proteins, which will cause different symptoms in human body. Since the extracting rules and components are distinct between all types of HPV, in further research, finding out the functions of proteins produced from different sequence of amino acids that attacks human body is necessary.
Through this experiment, we figured out that there are differences and similarities between viruses that we analyzed. From the initial part of this study, we have classified Human Papillomaviruses into three groups: 15 high-risk types, 3 probable high-risk types, and 12 low-risk types. Especially, HPV16 and HPV18, which are the key factors in inducing cervical cancers, show high percent of accordance in its amino acid composition, especially considering the percentage of Leucine and Valine. Also, HPV 18 and other Human Papillomavirus show some remarkable differences. This result indicates that we can make medications that have less adverse effect when we consider these structural differences between them. Regarding this fact, we concluded that we can treat cervical cancers by inhibiting the synthesis of certain kinds of amino acids which are prevalent in HPV16 and HPV18.

References

1. N. Muñoz, F. X. Bosch, S. Sanjosé, R. Herrero, X. Castellsagué, K. V. Shah, J.F. Snijders, Chris J.L.M. Meijer, N Engl J Med **348 :6** 518-27 (2003)
2. E. Schwarz, U. K. Freese, L. Gissmann, W. Mayer, B. Roggenbuck, A Stremlau, H. Hausen. Nature **314(6006)**, 111-4 (1985)
3. J. Paavonen, P. Naud, J. Salmeron, C.M. Wheeler, S.N. Chow, D. Apter, H. Kitchener, X. Castellague, J.C. Teixeira, S.R, Skinner, J. Hendrick, U. Jaisamrarn, G. Limson, S. Garland, A. Szarewski, B.Romanowski, F.Y. Aoki, T.F. Schwarz, W.A. Poppe, F.X. Bosch, D. Jenkis, K.Hardt, T. Zahaf, D. Descamps, F. Struyf, M. Kehtinen, G. Dubin, Lancet, **374(9686)**, 301-14 (2009)
4. S. Jeon, P.F. Lambert, Proc Natl Acad Sci U.S.A, **92(5)**, 1654-8 (1995)
5. J.M. Walboomers, M.V. Jacobs, M.M. Manos, F.X. Bosch, J.A. Kummer, K.V. Shah, P.J. Snijders, J. Peto, C.J. Meijer, N. Munoz, J Pathol, **189(1)**, 12-9 (1999)
6. F.X. Bosch, M.M. Manos, N. Munoz, M. Sherman, A.M. Jansen, J. Peto, M.H. Schiffman, V. Moreno, R. Kurman, K.V. Shah, J Natl Cancer Inst., **87(11)**, 796-802 (1995)

MATEC Web of Conferences 56, 05001 (2016) DOI: 10.1051/matecconf/20165605001
ICCAE 2016

Combined Resource Allocation System for Device-to-Device Communication towards LTE Networks

Fakhar Abbas[1], Ye Fang [1], Muhammad Irshad Zahoor[1], and Kashif Sultan[2]

[1]College of information and Communication Engineering, Harbin Engineering University, Harbin, PR, China.
[2]School of Computer and Communication Engineering, University of Science and Technology Beijing, Beijing, PR, China.

Abstract. The LTE networks are being developed to grant mobile broadband services in the fourth generation (4G) systems and allow operators to use spectrum more efficiently.D2D communication is a promising technique to provide wireless services and enhance spectrum exploitation in the LTE Heterogeneous Networks (HetNets).D2D communication in HetNets allows users to communicate with each other directly by reusing the resources when communicating via the base stations. But during the downlink period, both the D2D receiver and the Heterogeneous Users equipment's (HUE) experience interference caused by resource allocation. In this article, we identify and analyze the interference problem of HetNets caused by D2D transmitter during download. We propose a combined resource allocation and resource reuse method for LTE HetNets, where resource allocation to HUEs is employed on the basis of comparative fair algorithm and resource reuse to D2D users is employed on acquisitive empirical algorithm. This approach evaluates whether D2D mode is suitable or not by path loss evaluation, after that decreases the interference to HUE by selection of the minimum channel gain between HUE and D2D transmitter each time to mitigate interference. Our simulation results show that the efficiency and throughput of HetNets is improved by using the proposed method.

1 Introduction

As demands for high data rates increasing, the deficiency of radio resources impedes further growth of HetNets services. D2D communication tends to solve the problem by improving spectrum efficiency significantly. D2D communication enables mobile devices to communicate directly with each other, through reusing resources in HetNets [1] [2]. D2D communication scheme is incorporated which brings benefits in three ways. Firstly, D2D communication can increase the system throughput and the spectral efficiency by reusing the user resource. Secondly, the D2D communication decreases the transmission power of D2D transmitter. Thirdly, D2D communication has the advantage that licensed spectrum is allocated to local communication thus ensuring a secure and robust (against interference) environment [2]. D2D communication is also increasing necessity for local communication, such as games between mobile devices, video sharing, data flooding, local advertisement and information broadcast in hot point area like market. Machine to Machine communications in small distance in the Internet things also becomes a focus, as a local communication technology [2].

A major challenge in D2D communication is interference between different devices. Several attempts have been made to deal with the interference issues in D2D communication for LTE Networks. The D2D communication described in [3] can increase system throughput by reusing the resource of the cellular user. The interference cellular communication and D2D communication can be coordinated through proper power control and resource allocation. The D2D communication reduces the transmit power of terminal, which increases the working time of terminal and improves the energy efficiency. D2D communication can also decrease the load of base station (BS) via direct transportation. Besides, the D2D communication has the advantage that licensed spectrum is allocated to local communication. Licensed spectrum can guarantee a planned situation instead of an uncoordinated one. In [2], the method for D2D communication session association and organization involving procedures in the LTE System Architecture Evolution are proposed. The BS is capable of coordinating the interference between cellular communication and D2D communication by proper power control and resource allocation with the channel state information (CSI) of all involved links. In [4], a resource allocation method has been proposed to reduce the system interference. It, however, consisted of limited number of HUE and D2D pairs, and one D2D pair reused the resource of only one HUE in the system model. In [5] authors described three D2D resource allocation methods and the power optimization for only one cellular user and one D2D pair. In [6] the authors described mode selection method on the basis of approached presented in [5]. The authors in [7] proposed two methods to coordinate the mutual interference between cellular communication and D2D systems. The authors in [8] proposed time hopping (TH) based radio resource allocation schemes to improve the robustness of the hybrid network. These above mentioned approaches are based on the use of stand-alone schemes. This is, however, an interesting task to investigate the use of combined schemes.

Our aim is to reduce the interference which caused the interference to primary service in non-orthogonal method. The BS coordinates the transmit power for both links. We focus on the case in which HUE and D2D pairs are multiplexed on the same downlink OFDM physical resource block (PRB). Time Division Duplex (TDD) is used for the proposed scheme. Two UEs in a D2D pair share the given PRBs by time. Taking inspiration from the approach presented in [9] [10], this approach further seeks to improve resource allocation. First, there is a subcarrier and power combine allocation scheme to HUEs. We calculate the simulation result on both the comparative fair (CF) and Round Robin (RR) algorithm. Then, subcarrier reuse allocation scheme to the D2D pairs. We also present the comparison of path loss and the Signal to interference and noise ratio (SINR) to the threshold. Path loss is the main element used for approach selection. If the path loss of two D2D UEs is smaller than between any of them to the BS, the D2D mode is preferred. The resource reuse allocation is activated if the SINR of receiver in both of the HUE and the D2D pair is equal or bigger than the minimum allowed SINR.

The rest of the paper is organized as follows: In Section II, we present the system model. We formulate the problem in Section III. We then discuss the proposed approach in Section IV. Resource allocation and reuse algorithm are discussed in Section V. We present the simulations results in Section VI. Finally, we conclude the paper in Section VII.

2 System Model

In this part, we described network model and channel model with theoretical and mathematical analysis.

2.1 Network Model

In LTE system, the bandwidth is divided into equal size physical resource blocks (RBs). Each physical resource block (PRB) exists in 0.5 milli seconds time slots corresponding to 180 kHz in the frequency domain with subcarrier spacing of 15 kHz. In LTE, the multiple access method for downlink (DL) is Orthogonal Frequency Division Multiple access (OFDMA).Fig.1 shows the high level overview of LTE DL physical resource architecture. LTE adopts Orthogonal Frequency Division Multiplex (OFDM) based radio interface due to its higher spectral efficiency and flexibility against multi path delay spread. The physical properties of SC-FDMA require that RBs allocated to a single user to be contiguous in frequency. The minimum scheduling period in the frequency domain is one physical RB, therefore the smallest unit of resource that can be assigned comprises of two RBs. In this architecture, D2D UEs may communicate directly over the D2D links. However, BS needs to establish the D2D connection and also remains in control of resource allocation to limit the interference experienced by the HetNets receivers. The SINR is calculated from received signal and interference power level.

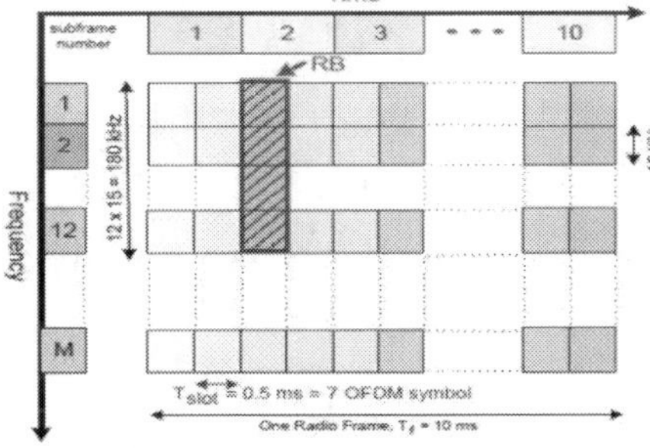

Figure 1. LTE Downlink physical resource

In order to guarantee the reuse of the HetNets resource, the interférence must be controlled well in a required range as provided by (1) and (2).

$$SINR_{HUE} <= SINR_{threshold} \quad (1)$$

$$SINR_{d2d(UE_1)} <= SINR_{thershold} \quad (2)$$

Equation (1) describes signal to interference and noise ratio (SINR) of Heterogeneous Users equipment's (HUE) which is less than the threshold level. "(2)," describes Signal to interference and noise ratio (SINR) of D2D user (UE1) is less the threshold level. The interference depends on the transmitter power and the distance between the two devices. Consequently intelligent selection of the physical resource block (PRB) for reuse would result in improved performance in terms of network throughput. If D2D UE's pair is close to the BS and the path loss between any of them and the BS is bigger than the path loss of the D2D pair, the HUE will suffer from significant interference which is troublemaking to the network throughput. To make sure that the D2D mode is ideal; the path loss of two D2D UEs must satisfy the following condition.

2.2 Channel Model

To evaluate important parameters SINR and channel gain, we consider both distant dependent path loss and shadow fading path loss. f_c Is the carrier frequency, d[m] is distance between base station and user while h_{BS} is the BSs antenna height. BS device path loss parameter for

Line-of-sight (LOS) and Non-Line-sight (NLOS) is as follows:

$$Pathloss = 22.\log 10(d[m] + 42.0 + 20\log 10(f_c[Ghz]/5)$$
$$Pathloss = 36.5.\log 10(d[m] + 40.9 + 26\log 10(f_c[Ghz]/5) \quad (3)$$

D2D communication path loss parameters for LOS and NLOS are as follows:

$$Pathloss = 15.9.\log 10(d[m] + 46.7 + 20\log 10(f_c[Ghz]/5)$$
$$Pathloss = 40.\log 10(R[km] + 30\log 10(f_c[Mhz]/5) + 49 \quad (4)$$

R is the D2D separation, f_c is carrier frequency. The slow fading or shadowing on a wireless channel is caused by obstacles in the propagation path of the radio waves and is location dependent [11]. When shadowing is modeled only by log normally distributed random variables, the variables do not meet the spatial correlation properties. Finding an accurate fading model is a parable the slow fading or shadowing on a wireless channel is caused by obstacles in the propagation path of the radio waveband obviously is location dependent. The linear power gain between the BS and a user is as follows

$$Gain = 10^{-(PL+shd)/10} \quad (5)$$

Power gain between two users for D2D communication is as follows.

$$Gain = 10^{-(PL)/10} \quad (6)$$

3 Problem Definitions

D2D commutation takes place in the primary HetNets. We identify the radio resources to be important and D2D communication shares the same resources with HetNets communication simultaneously rather than using dedicated resources. Therefore, the interference of D2D communications to the HetNets needs to be restricted to maintain a target performance level of the HetNets. If the distance between D2D is too long, the D2D will require high power. In that case, the primary HetNets would experience from interference.

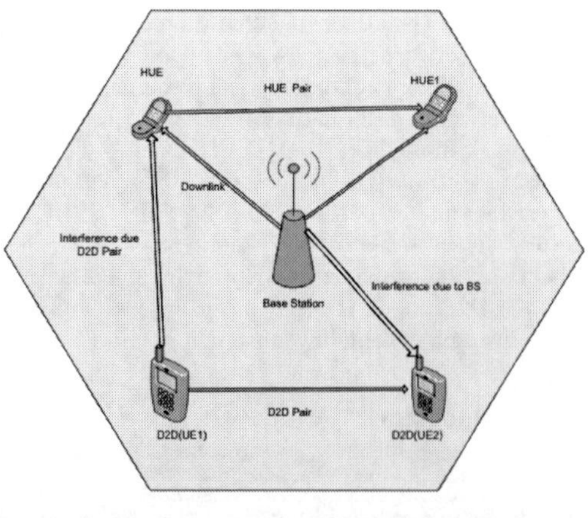

Figure 2. Interference problem in LTE Downlink when HetNets shares channel with the D2D network.

Fig.2 illustrates the interference problem when primary HetNets service shares channel with the D2D service in LTE Downlink network with HUE and D2D pair (D2D(UE1)–transmitter, D2D(UE2) –receiver).

During the downlink of HetNets, HUE is exposed to interference when D2D (UE1) transmits using the same allocated sub-band. Also, D2D receiver UE2 will suffer interference from the base station (BS), whichever shares the same sub-band with the HetNets service. It can be estimated that the amount of interference does not depend only on D2D transmit power but also on spatial distance between D2D transmitter and the HUE. According to analysis of power control scheme for D2D communication, we reject D2D connections if the maximum distance between D2D is higher. Further we assume that both UEs need to be in the same network for possible connection.

4 Proposed Approach

The base stations (BS) schedules the primary HetNets in normal mode, and we allocate one resource block from the RBs assigned to HUE to D2D link. The total number of available RBs for the Downlink is M. The BS serves a set H of HUEs and a set D of D2D pairs. We formulate the problem of assigning appropriate RBs for underlying D2D communication as an optimization problem that achieves higher throughput without impairing the existing HetNets. The underlying assumption is $H > D$.

4.1 UE Resource allocation

For primary HUE service, to optimize bandwidth allocation in the system such that the data rate requirements of a maximum number of users can be fulfilled. Aim is to find a bandwidth allocation that allows each user to satisfy its rate condition while maximizing the sum of user data rates subject to total power limit. The rate adaptive (RA) optimization problem can be state as below:

$$\max_{x_{hm}, p_{hm}} imize \sum_{h=1}^{H} \sum_{m=1}^{M} x_{hm} b_{hm} \quad (7)$$

Where H is the set of HUE and M is the total number of available resource block, h is the number of HUE, $x_{hm} \in \{0,1\}$ is subcarrier allocation indicator for all values of HUE, m. $p_{hm} >= 0$ is power for all value of h, m. And $\sum_{h=1}^{H} \sum_{m=1}^{M} p_{hm} \le P_T$, where P_T is the overall transmitting power constraint. $\sum_{h=1}^{H} x_{hm} <= 1$, for all values of $m \in \{1,2,3,4,....,M\}$, $R_i : R_j = \phi_i : \phi_i$ overall power and user proportionality constraints for all values of $i, j \in \{1,2,3,.....,H\}$. Normalized transmit rate is R_h / ϕ_h , where ϕ_h is rate influence of HUE. The throughput b_{hm} is determined by the Shannon capacity model in equation (8). p_{hm} Is received power, N_0 is the receiver's noise, I inter-cell interference and G_{bh} is the channel power gain between the base station (BS) and HetNets UE_h.

$$b_{hm} = \log 2(1+\frac{p_{hm}G_{bh}}{N_0+1}) \quad (8)$$

For user *h*, the total granted transmit throughput can be calculated as fallows, where B is the total bandwidth.

$$R_h = \frac{B}{M}\sum_{m=1}^{M} x_{hm}b_{hm} \quad (9)$$

During the download phase of HetNets, any UE is exposed to interference when any D2D user is allowed to transmit using the same allocated sub band. The amount of interference will depend not only on D2D transmit power but also on the channel gain between D2D transmitter and the HUE.

4.2 UE Resource allocation

The downlink signal to interference noise ratio of user Het (UE) is as follows:

$$SINR_h = \frac{P_B G_{Bh}}{N_0 + I + \sum_{m=1}^{M} x_{dh} P_{d2d(UE_1)} G_{d2d(UE_1)h}} \quad (10)$$

The signal to interference noise ratio for the device to device pair receiver $d2d(UE_1)$ is as follows:

$$SINR_{d2d(UE_2)} = \frac{P_{d2d(UE_1)} G_{d2d(UE_1,UE_2)}}{N_0 + I + \sum_{m=1}^{M} x_{dh} P_B G_{Bd2d(UE_2)}} \quad (11)$$

$G_{Bd2d(UE2)}$ Is the channel gain between BS and d2d, $P_{d2d(UE_1)}$ is D2D user received power. $G_{d2d(UE_1,UE_2)}$ Denotes the channel gain between two D2D UEs, P_B is BS power, N_0 is receiver's noise and thermal noise density and I is inter-cell interference.

For resource reuse method, we need to maximize the sum rates of the primary HUE and secondary D2D UEs by applying the Shannon capacity formula. To maximize the sum throughput of D2D and HUE when sharing DL resources of HUE, the problem is formulated as follow:

$$\max_{x_{dm}, P_{dm}} imize \sum_{d=1}^{D}\sum_{m=1}^{M} x_{dm}b_{dm} + \sum_{h=1}^{H}\sum_{d=1}^{D}\sum_{m=1}^{M} x_{hm}x_{dm}(b_{hm}^{c} - b_{hm}) \quad (12)$$

Where $x_{dm} \in \{0,1\}$ for all values of d , m and $p_{dm} \geq 0$ is power for all values of d , m . D is total number of D2D pairs; M is total number of available resource blocks during the downlink and H is the set of HUEs that BS serves. $\sum_{d=1}^{D} x_{dm} <= 1$ For all values of $m \in \{1,2,3,.....,M\}$, (b_{hm}^{c} - b_{hm}) is the throughput difference of HUE.

$$SINR_{HUE} <= SINR_{threshold} \quad (12.1)$$

$$SINR_{d2d(UE_1)} <= SINR_{thershold} \quad (12.2)$$

Signal to interference noise (SINR) condition to control interference. The amount of interference depends on transmitted power and distance between two users.

$$PL_{D2D} <= PL_{d2d(UE_1),BS} \quad (12.3)$$

$$PL_{D2D} <= PL_{BS,d2d(UE_2)} \quad (12.4)$$

The path loss of two D2D UEs must satisfy the above conditions. The speed of b_{hm} is determined by Shannon capacity formula with inter-cell interference. The throughput of b_{hm} is similar to b_{dm} .

$$b_{hm} = \log 2(1+SINR_h) \quad (13)$$

The sum throughput of whole scheme is the sum of two equations (7) and (12).

5 Combined resource allocation and resource reuse method

The allocation of resources between D2D and HetNets is determined by the Base station. If D2D users are assigned resources that are orthogonal to those occupied by the HUE, they cause no interference to each other and the analysis is simpler. On the other hand, the resource usage efficiency can be higher in non-orthogonal mode.

This work aims at reducing the interference in the non-orthogonal mode. Under non-orthogonal mode, the D2D and HUE reuse the same resources, causing interference to each other. The BS coordinates the transmit power for both links. The most constructive solution to the problem of rate adaptive optimization of equation (7) should iteratively combine both the subcarrier allocation and bit allocation. In the bit allocation step, the total power is assumed to be allocated to subcarriers according to the same SINR. According to $R_i : R_j = \phi_i : \phi_i$ for all values of $i, j \in \{1,2,3,.....,H\}$, the subcarrier allocation method initially determines the number of subcarriers allocated to each user. In order to assure user rate proportionality, the user with the minimum normalized transmit rate R_h / ϕ_h is permitted to choose a subcarrier at each iteration.

Below is the summarized proposed proportional subcarrier allocation algorithm. This is an Acquisitive policy, where the user with the minimum stabilize transmit rate is preferred to choose a best subcarrier each time.

Algorithm 1: HetNets Service RBs Distribution method

1. Initialization.
2. For $HUE = 1:H$, Sorted list of b_{hm} in ascending order, HUE is a user to find the optimal physical resource blocks $m* = \arg\max_{m \in s} b_{hm}$.
3. Set $x_{hm^*} = 1$, Removal of resources allocated from the set of subcarriers $S = S - \{m*\}$.
4. if $S \neq \varnothing$, determine the set of subcarriers is empty, then
5. Find out each iteration normalized transmission bit rate of the minimal user $h = \arg\min_h \{R_h / \phi_h\}$.
6. Find the optimal allocation of resource blocks to the user $m^* = \arg\max_{m \in S} b_{h*m*}$.
7. Set $x_{h*m*} = 1$, $S = S - \{m*\}$ until $S = \varnothing$.

Below is the D2D service reuse resource block distribution method. This is an alternative acquisitive empirical RBs selection algorithm. From Equation

$$\max imize_{x_{hm}, P_{hk}} \sum_{h=1}^{H} \sum_{m=1}^{M} x_{hm} b_{hm} \text{ , we find}$$

That lower channel gain will result in higher $SINR_h$. Therefore, we select primary UE with higher channel quality indicator split RB with the D2D pair with lower channel gain.

Algorithm 2: D2D Service RBs Distribution method

1. Initialization.
2. Set $HUE = 1:H$, $d = 1:\mathrm{D}$ While $D \neq \varnothing$ and $HUE \leq H$, HUE to select the optimal or suboptimal reusable resources $h* = \arg\max sum(x_{hm})$, the pairing of D2D users $d* = \arg\min G_{d2d(UE_1)h*}$.
3. If $PL_{D2D} <= PL_{d2d(UE_1)B}$ && $PL_{D2D} <= PL_{hd2d(UE_2)}$ and $SINR_h <= SINR_{threshold}$ && $SINR_{d2d(UE_2)} <= SINR_{threshold}$.
4. Set $x_{d^*h} = 1$, Removal of Resources taking Complete D2D Users $D = D - \{d^*\}$; otherwise go to step 2.

The sum throughput of whole scheme is the sum of two equations (7) and (12).

6 Evaluation and Simulation

6.1 Simulation Parameters

We perform general simulations to evaluate the efficiency and throughput of proposed methods. Each user calculates SINR from the received signal power and interference power level, and then feeds back the channel quality indicator (CQI) to base station (BS). Table I summarizes a list of simulation parameters and their default values. Link variation and various modulation and coding schemes are performed based on the CQI at the BS. The subcarrier and bit allocation signal are transmitted by control channel to user. For HUE, two different scheduling algorithms are employed: Round robin (RR), comparative Fair (CF). The transmitter power of HUE is controlled using the same fixed SNR target in a limitation (SNR=13dB). The transmitters of D2D pair's use the same fixed transmit power ($P_d = 0.05w$).

Table 1. Simulation Parameters and Values

Parameter	Values
Cell Layout	Hexagonal Network
Spectrum allocation	20 MHz
HUE distribution	uniformly
Carrier Frequency	2 GHz
Radius (Inter site distance)	500 m
RB Bandwidth	180 KHz
Base station Tx power	20 w
Base station height	30 m
Number of subcarrier per RB	100
Shadowing fading	8 dB
UE noise range	9 dB

6.2 Results Discussion

In this part we evaluate and discuss the simulation results.

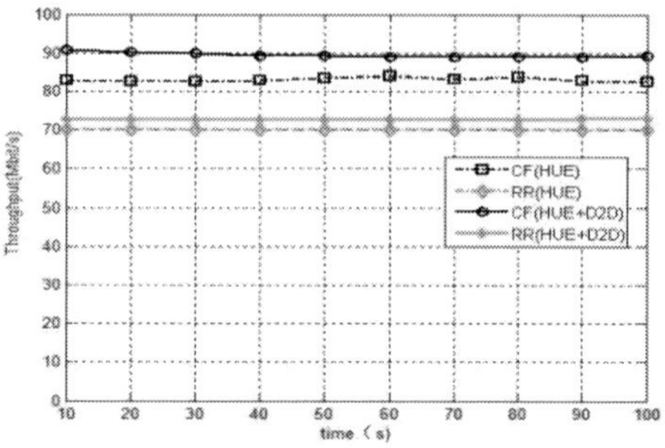

Figure 3.Scheme throughput for different scheduling algorithms with the proposed acquisitive resource block reuse algorithm.

Scheme throughput change with the time for different scheduling algorithms (RR, CF) in case of scenario 1 and 2 as shown in Fig. 3. Scenario 1 is a pure HetNets with H=40. Scenario 2 is a mixed network with H=26, D=4 four D2D pair. For both cases, the RR performance is poor because it never considers the current channel condition and always allocates the same resources to all users. The CF performance is good because CF not only does consider the channel quality indicator (CQI) but also allocates resources in a fair proportion. In addition, for both (RR, CF) scheduling algorithms, the scheme throughput is higher when D2D communication reuse the RBs of HetNets.

Scenario 1: HUE with set of H=40.

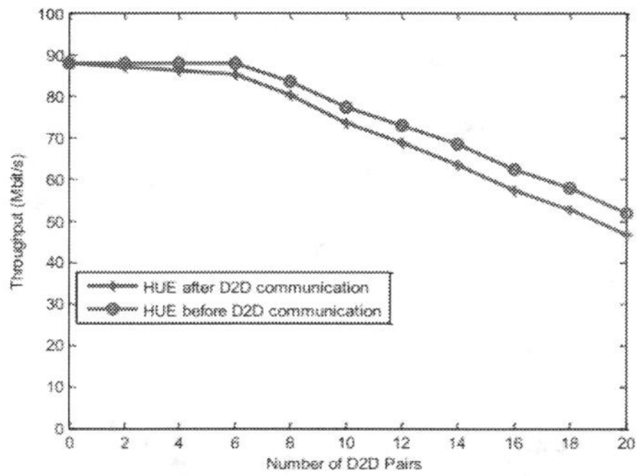

Figure 4.1 Scheme throughput comparison in HUE before and after D2D mode (Scenario D2D pair=0:2:20)

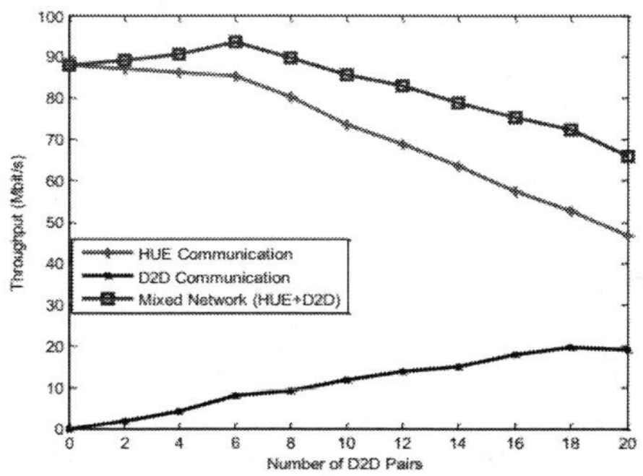

Figure 4.2 Throughput factor in HUE, D2D and mixed network mode (Scenario D2D pair=0:2:20)

Scheme throughput in both HetNets and the mixed network, when the total number of links in the HUE is fixed N =40. The evaluation of throughput in HUEs before and after D2D communication is shown in Fig. 4.1.The HetNets throughput is lower after D2D communication than before, because of the interference from the D2D connections. Scheme throughput factor in HUE, D2D and mixed network which clearly explains that mixed network throughput is higher as compared to HetNets and D2D mode shows in Fig. 4.2.

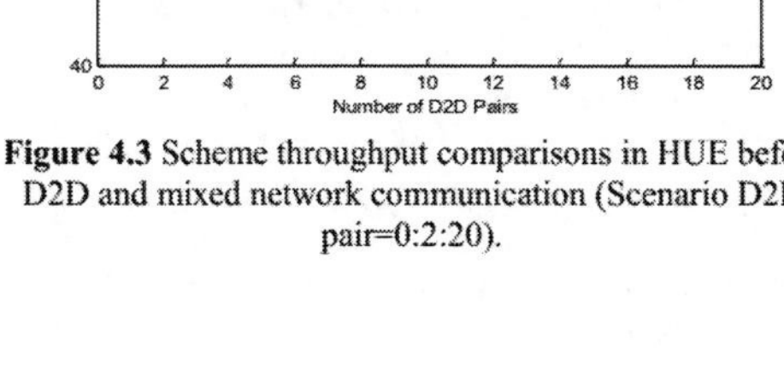

Figure 4.3 Scheme throughput comparisons in HUE before D2D and mixed network communication (Scenario D2D pair=0:2:20).

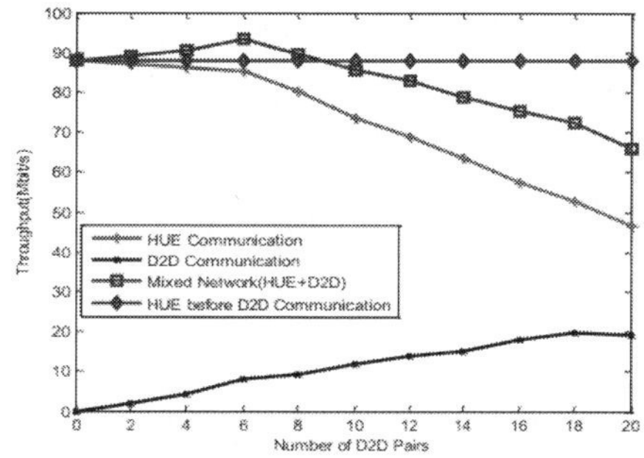

Figure 4.4 Scheme throughput comparison between HUE, D2D, mixed network and HUE before D2D communication (Scenario D2D pair=0:2:20)

Component of the scheme throughput in HUE before D2D and mixed network communication are shown in Fig. 4.3. Comparison of system throughput between pure HetNets and mixed network are shown in Fig. 4.4. When, the scheme is pure HetNets. After D2D links, the scheme throughput is increased. Due to the fixed total links in the HUE, the scheme throughput cannot increase considerably with the *D* increase as the H decrease. Results show clearly that when the primary service reaches the upper limit, the system throughput has increased.

Scenario 2: Mixed network HUE and D2D users with set of H=26, D=4.

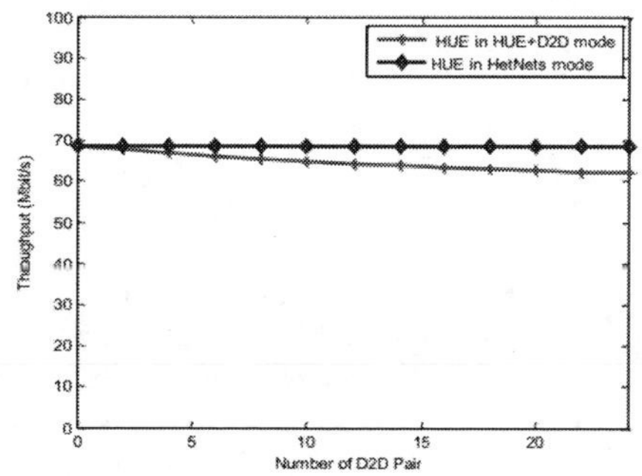

Figure 5.1 Comparison of throughput in HUE before and after D2D form

(Scenario D2D pair=0:2:24).

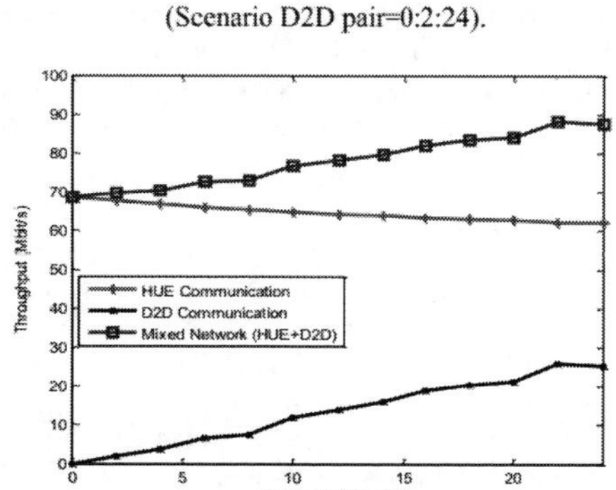

Figure 5.2 Throughput component in HUE, D2D and mixed network form(Scenario D2D pair=0:2:24).

System throughput both in the HetNets and the mixed network when the number of HUEs in the HetNets is fixed. The comparison of throughput from HUE between pure HetNets mode and mixed mode is shown in Fig 5.1. Because the number of HUE is fixed to 26, the throughput in pure HetNets is a straight line. The HetNets throughput is worse in mixed mode, which is due to the interference from the D2D network. Component of the scheme throughput in D2D, HetNets and mixed network which clearly describe that mixed network throughput is higher as compared to D2D and HetNets network shows in Fig 5.2.

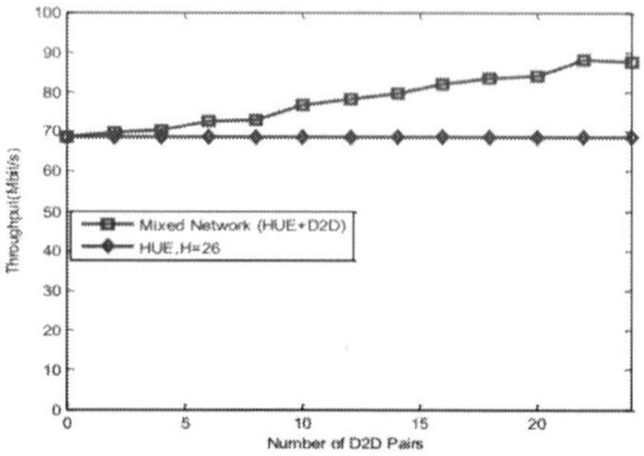

Figure 5.3 Scheme throughput comparisons in HUE and mixed network communication (Scenario pair=0:2:24).

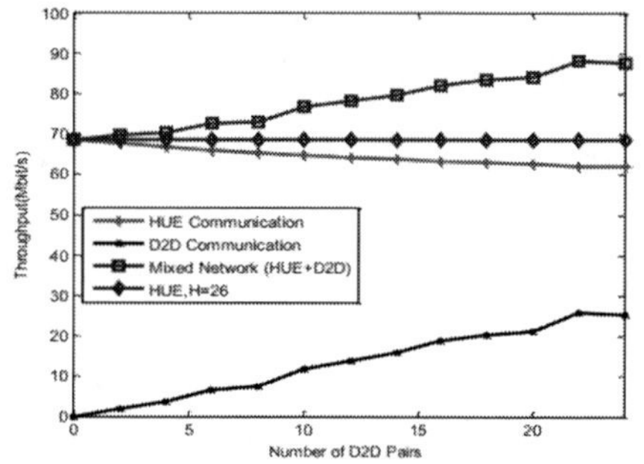

Figure5.4 Scheme throughput comparison between HUE, D2D, mixed network and HUE before D2D communication (Scenario pair=0:2:24).

Component of the scheme throughput in mixed network and HetNets shows in Fig. 5.3.The comparison of scheme throughput between pure HetNets, D2D and mixed network shows in Fig. 5.4, when the scheme is pure HetNets. After D2D communication towards the HetNets, the scheme throughput increase for a long range. Due to the increasing number of D2D links in the HUE, the scheme throughput increase extensively. We can find out that the D2D links are not up to an upper limit.

7 Conclusion

D2D is an advanced technology that can increase the spectral efficiency and throughput by reusing the resource of HUE. This technology offers a beneficial complement to organization approach. The network performance improves when D2D communications share resources with the primary network. But, D2D transmitter causes interference to the UE receiver during the D2D LTE downlink HetNets period. In this article we have investigated how to reduce the effects of interference between D2D and HUE, in order to improve the system sum throughput for D2D.We have proposed a combined resource allocation and resource reuse method with objective of mitigation the interference between D2D and HUE. Proposed method ensures the D2D mode selection is a realistic way and mitigates the interference. We formulated that the system performance improves through calculation of the sum throughput in the scheme before and after D2D communication in LTE HetNets. Simulation results show that efficiency and throughput have been improved for the proposed approach

Acknowledgment

This paper is funded by the International Exchange Program of Harbin Engineering University for innovation-oriented Talents Cultivation, the National Natural Science Foundation of China (Grant NO. 61301095, NO. 51374099), the Natural Science Foundation of Heilongjiang Province, China (Grant No. F201345) and the Fundamental Research Funds for the Central Universities of Ministry of Education of China (Grant No. HEUCF150812).

References

1] Z. Liu, T. Peng, S. Xiang, and W. Wang, "Mode selection for Device-To-Device (D2D) communication under LTE-Advanced network," Proc. 2012 IEEE International Conference on Communication (ICC), 5563–5567, (2012).

[2] K. Doppler, M. Rinne, C. Wijting, C. B. Ribeiro, and K. Hugl, "Device-to-Device communication as an underlay to LTE-advanced networks," IEEE Commun. Mag., 47(12), 42–49, (2009).

[3] C. H. Yu, O. Tirkkonen, K. Doppler, and C. Ribeiro, "On the performance of Device-to-Device underlay communication with simple power control," Proc. IEEE Vehicular Technology Conference, 1–5, (2009).

[4] P. Janis, V. Koivunen, C. Ribeiro, J. Korhonen, K. Doppler, and K. Hugl, "Interference-aware resource allocation for Device-to-Device radio underlaying cellular networks," Proc. IEEE 69th Vehicular Technologies Conference, 1–5, (2009).

[5] C. H. Yu, O. Tirkkonen, K. Doppler, and C. Ribeiro, "Power optimization of Device-to-Device communication underlaying cellular communication," Proc. IEEE International Conference on Communication, 1-5, (2009).

[6] K. Doppler, C. H. Yu, C. Ribeiro, and P. Janis, "Mode selection for Device-to-Device communication underlaying an LTE-Advanced network," Proc. IEEE Wireless Communications and Networking Conference (WCNC), 1–6, (2010).

[7] T. Peng, Q. Lu, H. Wang, S. Xu, and W. Wang, "Interference avoidance mechanisms in the hybrid cellular and Device-to-Device systems," Proc. IEEE 20th International Symp. on, Personal, Indoor and Mobile Radio Communications, 617-621, (2009).

[8] T. Chen, G. Charbit, and S. Hakola, "Time hopping for D2D communication in LTE cellular system," Proc. IEEE Wireless Communications and Networking Conference (WCNC), 1-6, (2010).

[9] G. Yu, Z. Zhang, Y. Chen, P. Cheng, and P. Qiu, "Subcarrier and Bit allocation for OFDMA systems with proportional fairness," Proc. IEEE Wireless Communications and Networking Conference (WCNC), 1717-1722, (2006).

[10] M. Zulhasnine, C. Huang, and A. Srinivasan, "Efficient resource allocation for Device-to-Device communication underlaying LTE network," Proc. IEEE 6th International Conference on Wireless and Mobile Computing, Networking and Communications, 368-375, (2010).

[11] F. Graziosi and F. Santucci, "A general correlation model for shadow fading in mobile radio systems," IEEE Commun. Lett. 6(3), 102–104, (2002).

Regret of Multi-Channel Bandit Game in Cognitive Radio Networks

Jun Ma and Yonghong Zhang

School of Electronic Engineering, University of Electronic Science and Technology of China (UESTC), Chengdu, China
Email: majunpaper@foxmail.com; zhangyhh@uestc.edu.cn

Abstract. The problem of how to evaluate the rate of convergence to Nash equilibrium solutions in the process of channel selection under incomplete information is studied. In this paper, the definition of regret is used to reflect the convergence rates of online algorithms. The process of selecting an idle channel for each secondary user is modeled as a multi-channel bandit game. The definition of the maximal averaged regret is given. Two existing online learning algorithms are used to obtain the Nash equilibrium for each SU. The maximal averaged regrets are used to evaluate the performances of online algorithms. When there is a pure strategy Nash equilibrium in the multi-channel bandit game, the maximal averaged regrets are finite. A cooperation mechanism is also needed in the process of calculating the maximal averaged regrets. Simulation results show the maximal averaged regrets are finite and the online algorithm with greater convergence rate has less maximal averaged regrets.

Keywords-cognitive radio networks; adversarial bandit problem; congestion game; online learning; dynamic spectrum access.

1 Introduction

With the emergences of new wireless services and applications, the demand for spectrum increases. However, some studies show that many licensed spectrum bands have not been utilized efficiently [1]. In order to alleviate this contradiction, cognitive radio users called secondary users (SUs) are proposed and allowed to access spectrum bands belong to primary users (PUs) as long as those bands are sensed to be idle. How to select a proper channel to sense and access will affect the idle spectrum usage. Therefore, the channel selection problem is important to each SU.

Firstly, some works about channel selection under incomplete information are introduced. In [2], the problem of how to select a channel set for a SU to access at a time has been investigated under the condition that the statistical information of PUs' traffic is assumed as a stationary and simple distribution which is unknown to SUs in advance. This work is mainly based on classical bandit models [3]. Due to the stochastic nature of cognitive radio networks (CRNs), the real primary traffic distribution is not always stationary. Under this scenario, the channel set selection problem is analyzed in [4]. Furthermore, a similar problem of multiple SUs who need to select and access a channel at a time is studied in [5] considering the effect of the interactions among SUs on individual benefits. Previous works have studied the channel selection problem with incomplete information by bandit models.

From the perspective of game theory, the channel selection problem can be also modeled as a congestion game. Congestion game is a kind of non-cooperative game with the utility of each player by using a certain resource depending on the total number of players who are using the same resource [6], [7]. Some works in the field of game theory have shown that congestion games are a kind of potential game in fact [8]. Potential games are used to study the channel selection problem in [9] and the network selection problem [10] while some online learning algorithms are applied to obtaining the Nash equilibrium (NE) solutions. Potential games always have potential functions which can guarantee at least a pure strategy NE [8]. Although these works claim that NE can be found under incomplete information, they need more information than our model because the potential functions will be given before searching the NE. These works do not consider the convergence rate of the online algorithm. The online algorithm with greater convergence rate will reduce the search time while increase the transmission data time. Therefore, the problem of how to evaluate the convergence rate of online algorithm under incomplete information should be studied.

In this paper, the channel selection problem with incomplete information is modeled as a multi-channel bandit game from the view of bandit models The *MAR* is used to evaluate the convergence rates of online algorithms.

2 System Model and Problem Formulation

Consider a CRN with N SUs as well as S SU base stations and M (M=S<N) primary channels which belong to PUs. All SUs and PUs operate in the slot transmission structure. Each PU has only one channel. If any one of SU nodes wants to communicate with a SU base station, it must use the idle slots of primary channels. A primary channel just serves a SU base station. For example, there are N=3, M=S=2 in a CRN. When SU n=1 communicates with SU base station A, SU node n=1 can only utilize the idle slots of channel m=1 which belongs to PU 1. Similarly, if SU n=2 communicates with SU base station B, SU n=1 can only utilize the idle slots of channel 2 which belongs to PU 2. Here, we assume connecting different SU base station do not affect the further communication with the destination of each SU. The spectrum sensing is assumed to be perfect for each SU and all primary channels are idle during the total simulation time for ease of researching. This assumption is rational. The idle time of some primary channels are much longer than the data transmission durations of secondary users who have little data to transmit (like the temperature sensors). The effect of primary traffics on the performances of convergence rates of online algorithms will be studied in our future works.

Each SU selects a channel from M channels at a time to sense and access if the channel is sensed to be idle and the number of using this channel is not too much. Each SU is rational and selfish with the goal of maximizing its own averaged transmission rate during the process of channel selection. At each slot, each SU selects a channel at a time and receives a reward if the channel is accessed, which is analogous to the process of a gambler selecting the arm of a slot machine and receiving a reward. If a SU transmits successfully, the reward is the channel transmission rate which is given by (1) and 0 for the transmission failure. The channel transmission rate of SU n at slot t in channel m can be written in (1), where W is the channel bandwidth, P is the transmit power and σ^2 is the thermal noise level, which are same to all SUs for the simplification of research. Note that $g^t_{n,m}$ denotes the channel gain at slot t and changes over the time but keeps unchanged during each slot.

$$r^t_{n,m} = W\log_2\left(1+\frac{Pg^t_{n,m}}{\sigma^2}\right) \tag{1}$$

When more than one SUs access the same channel simultaneously, the individual transmission success probability decreases because of the mutual interference. Therefore, the rewards obtained by SUs are affected by the number of SUs using the same channel at the same time. We assume $\boldsymbol{s}^t = (s^t_1, \dots, s^t_N)$ is a pure strategy profile at slot t for all SUs, where s^t_n denotes the selection strategy of SU n at slot t. $\mathbf{c}^t(\boldsymbol{s}^t) = (c^t_1, \dots, c^t_M)$ denotes the total number of SUs in each channel corresponding to the strategy profile $\boldsymbol{s}^t$ at slot t. The total number c^t_m affects the successful transmission probability $p(c^t_m)$ of SUs who use the channel m at slot t. When c^t_m increases, $p(c^t_m)$ of SUs decreases. In this paper, we adopt a common MAC protocol (the slotted Aloha) [7] and the specific expression of $p(c^t_m)$ is given in (2). When only one SU is in the channel m, c^t_m=1, it is certain to transmit its data successfully with probability 1. As mentioned before, we model the channel selection problem as a bandit problem. However, the classic bandit model is not fit for our situation because the rewards of each accessing a channel for each SU are not independently drawn from a fixed and unknown distribution.

$$p(c^t_m) = \begin{cases} 0 & c^t_m = 0 \\ 1 & c^t_m = 1 \\ \left(\frac{1}{c^t_m}\right)\left(1-\frac{1}{c^t_m}\right)^{c^t_m-1} & c^t_m > 1 \end{cases} \tag{2}$$

Here, we utilize a variant of the classic bandit model called adversarial bandit [11] which is a non-stochastic bandit problem to model our scenario. The adversarial bandit is also used in [4] to find the optimal channel. We assume that all SUs use the same online algorithm to find the equilibrium channel for themselves at the same time. In our model, we use T to denote the slot at which each user makes a selection using online learning algorithms. This structure is illustrated in Fig.1. At slot T, each SU will select a channel to access based on a specific distribution over M channels. This specific distribution is decided by the updating rule of an online learning algorithm. After all SUs have completed the process of channel selections at slot T, the pure strategy profile can be represented by $\boldsymbol{s}^T$.

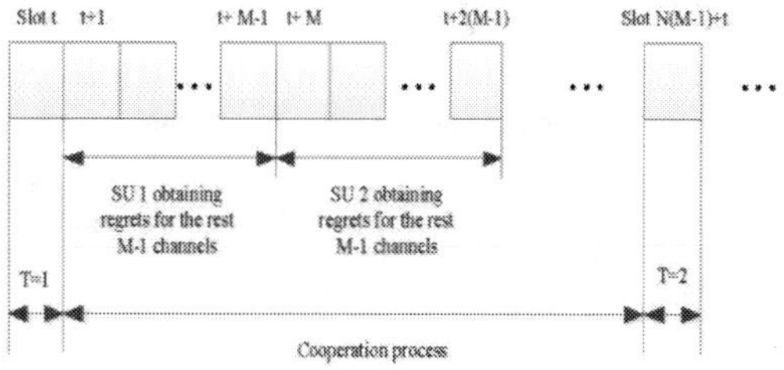

Figure 1. An illustration of the process of calculating regrets under incomplete information in our model

The following M-1slots will be used to calculate the regret for each channel which has not selected at slot T. In order to express our idea clearly, (j, s^T_{-n}) denotes $(s^T_1, \dots, s^T_{n-1}, j, s^T_{n+1}, \dots, s^T_N)$, which is $\boldsymbol{s}^T$. When SU n does not select a channel j at slot T, it may care about how much reward SU n has lost for not playing the pure strategy j at slot T under the condition that all the other SUs keep their selection strategies (s^T_{-n}) at slot T unchanged. If SU n knows its own payoff function and the strategy set $\boldsymbol{s}^T$ at slot T, SU n can calculate the regret of not selecting the channel j and the regret $R^T_n(j,T)$ is given as follows [11],

$$R^T_n(j,T) = U_n\left(j, s^T_{-n}\right) - U_n(\boldsymbol{s}^T) \tag{3}$$

Where U_n denotes the SU n's reward function, U_n= $r^t_{n,m}$ if SU n transmits successfully and otherwise U_n=0. However, in our model, each SU does not know its own payoff function. Hence, each SU needs the cooperation with other SUs. The cooperation among SUs will help each SU to find the equilibrium channel, which has been showed in Fig.1. Therefore, each SU has an incentive to cooperate with others. We use an example to explain the cooperation process. For example, if SU n does not select the channel j at slot T, it may select channel j at slot $t+1$ while other SUs should select the channels which have been chosen at slot T, namely, s_{-n}^{T}. In other words, the channel selection process will be repeated at slot $t+1$ and SU n will select channel j without doubt while other SUs keep s_{-n}^{T} unchanged. Therefore, SU n will repeat this process for M-1 slots because there are M-1 channels not be chosen at slot T. Hence, the next M-1slots are used to calculate the regrets which are produced by not selecting the channel j ($j \in M-1$) at slot T. The selecting order for M-1 channels is decided by each SU. All SUs know M and When SU n completes the regret calculations for M-1 channels, SU n+1 will be informed and continue to repeat the same process with SU n for M-1 slots until SU N completes this process. The next time for all SUs using the online algorithm to select a channel is at slot $t+N(M-1)$. For example, in the Fig.1, the secondary selection time decided by online algorithms is T=2. In our model, if the strategy set (j, s_{-n}^{T}) is the same with s^{T}, the strategy j is no need to try again. The total number of selecting the channel decided by online algorithms is Ta. If the online algorithm selects total T=Ta times, the averaged regret of SU n for channel j ($j \in M$) after Ta times is as follows,

$$\begin{aligned}\overline{R_n^j}(Ta) &= \frac{1}{Ta}\left(\sum_{T=1}^{Ta}[U_n(j,s_{-n}^T)-U_n(s^T)]\right)\\ &= \frac{1}{Ta}\left(\sum_{T=1}^{Ta}U_n(j,s_{-n}^T)-\sum_{T=1}^{Ta}U_n(s^T)\right)\end{aligned} \quad (4)$$

Here, we define the MAR as $R_n^{\max} = \max_j(\overline{R_n^j}(Ta))$ which is the maximal averaged regret and use the MAR to evaluate the performances of online learning algorithms.

3 Online Learning Scheme

In this section, we use two existing online learning algorithms to find the NE solution and calculate the MAR for all SUs using two existing online learning algorithms. The first one is based on the algorithm Exp3 [11]. The second one is based on the stochastic learning algorithm [12].

3.1 Online Learning Algorithm Based on Exp3

Each SU calculates the selection probability distribution $P_n(T)=(p_{n,1}(T),\dots,p_{n,M}(T))$ over M channels based on (5) according to the corresponding weight value for each channel. The weight value is updated by the normalized reward based on (8). Here, the normalized reward is the ratio between the actual reward and maximal probability reward which is constrained by the hardware of SUs. The maximal probability reward in our simulation part is produced by all random communication conditions.

Algorithm 1: Exp3 based online learning algorithm

1: Initialize total times Ta, parameter $0<b<1$, $R_n(0)=0$, $R_{n,m}(0)=0$, and the weight value $\omega_{n,m}^{T=1}=1$ of each channel $m \in M$ for each SU $n \in N$.

2: Each SU randomly selects a channel according to the probability $p_{n,m}^{T}$ and its expression is as follows,

$$p_{n,m}^{T} = (1-b)\frac{\omega_{n,m}^{T}}{\sum_{m=1}^{M}\omega_{n,m}^{T}} + \frac{b}{M} \quad (5)$$

3: Each SU receives a reward r_n^T, if $r_n^T \neq 0$ and then

$$R_n(T) = R_n(T-1) + r_n^T \quad (6)$$

4: Each SU updates the weight value $\omega_{n,m}^{T+1}$ of each channel based on the normalized reward $\bar{r}_n^T$. The updating rule for each SU is as follows,

$$\hat{\Omega}_{n,m}^{T} = \frac{\bar{r}_n^T}{p_{n,m}^T} \quad (7)$$

$$\omega_{n,m}^{T+1} = \omega_{n,m}^{T}\exp\left(b\frac{\hat{\Omega}_{n,m}^{T}}{M}\right) \quad (8)$$

5: Cooperation begin

From n=1 to N, each SU n selects channel m from the set of channels M according to the predetermined order and the selection strategy of other SUs s_{-n}^{T} is always unchanged. SU n can obtain the rewards for each channel from the set of channels M, If the channel m has been selected at slot T, the reward is $U_n(m,s_{-n}^T)=r_n^T$, else

$$R_{n,m}(T)=R_{n,m}(T-1)+U_n(m,s_{-n}^{T}) \quad (9)$$

When SU n completes its process of calculating the $R_{n,m}(t)$, the SU n+1 is informed and continues until SU N.

6: Cooperation end

7: T=T+1 and back to Step 2.

8: Calculate the MAR based on (4) and find the MAR

3.2 Stochastic Learning Algorithm

Each SU selects a channel at random according to a dynamic distribution $P_n(T)=(p_{n,1}(T),\dots,p_{n,M}(T))$ over M channels and uses the normalized reward to update the selection probability distribution for the next round based on (11). At the first slot (T=1), the probability distribution $\mathbf{P}_n(T)$ of selection channel is assumed as the uniform distribution over M channels.

Algorithm 2: stochastic learning algorithm

1: Initialize total times Ta, parameter $0<b<1$, $R_n(0)=0$, $R_{n,m}(0)=0$ for each channel $m \in M$ for each SU $n \in N$, and $P_n(T=1)$ is the uniform distribution over M channels.

2: Let each SU select a channel randomly according to the probability distribution $\mathbf{P}_n(T)$.

3: Each SU receives a reward r_n^T, if $r_n^T \neq 0$ and then

$$R_n(T) = R_n(T-1) + r_n^T \tag{10}$$

4: Each SU updates $P_n(T+1)$ as follows,

$$\mathbf{P}_n(T+1) = \mathbf{P}_n(T) + b\overline{r}_n^T(I_m - \mathbf{P}_n(T)) \tag{11}$$

where b is the learning rate, $\overline{r}_n^T$ is the normalized reward, I_m is the indicator function.

5: Cooperation begin

From n=1 to N, each SU n selects channel m from the set of channels M and according to the predetermined order and the selection strategy of other SUs s_{-n}^T is always unchanged. SU n can obtain the rewards for each channel from the set of channels M,

If the channel m has been selected at slot T, the reward is $U_n(m,s_{-n}^T) = r_n^T$, else

$$R_{n,m}(T)=R_{n,m}(T-1)+U_n(m, s_{-n}^T) \tag{12}$$

When SU n completes its process of calculating the $R_{n,m}$(t), the SU n+1 continues until SU N

6: Cooperation end

7: T=T+1 and back to Step 2.

8: Calculate the MAR based on (4) and find the MAR

During the cooperation, if SU n has selected channel m at slot T, SU n has no need to try again. Therefore, we think $U_n(m,s_{-n}^T) = r_n^T$.

3.3 The Relationship between NE and MAR

In this subsection, we use the definition of regret to evaluate the convergence rates of online algorithms. We can find based on the theorem 1 that the MAR will be reduced if the selection strategy converges to the equilibrium channel quickly.

Theorem 1: When the multi-channel bandit game has a pure strategy NE, the MAR of online algorithms for each SU is finite.

Proof: The total selection times which is decided by online algorithms is Ta. We assume channel J is the equilibrium strategy for SU n. At a special selection time T', each SU will select optimal channel J and does not change its selection after slot T'. In other words, when the total Ta is large enough, $\sum_{T=T'}^{Ta} R_n(T) = \frac{1}{Ta-T'}(U_n(J,s_{-n}^T) - U_n(s^T)) = 0$ after each SU passes slot T'. This is because the (J, s_{-n}^T) is the same with s^T at slot T= T' and the reward of each accessing the optimal channel J is random for SU n. When all SUs do not change their selections, the result of searching optimal channel will converge to an NE SU n. Therefore, $\sum_{T=1}^{Ta}[U_n(J,s_{-n}^T) - U_n(s^T)] < \infty$ and $R_n^{\max} = \max_j(\overline{R_n^j}(Ta)) < \infty$.

4 Simulation Results

The performances of online learning algorithms are evaluated in this section. There are N=4 SUs and M=2 primary channels in a CRN covering a 500m×500m area. S=2 SU base stations locate in (0,250), (500,250) which are accessible for all SUs. When the 4 SUs are selected randomly, the locations of SUs are fixed during the simulations. All SUs adopt the channel model of Flat/Light tree density proposed in [13]. The transmission power is 10^{-3}W and the noise power level for all SUs is assumed to be 10^{-12}W. The bandwidth for all SUs is assumed as 1Hz and the learning rate b=0.01 for the two algorithms. In figures, Ta is showed in abscissa axis, which is from 1000 to 20000. Each Ta runs 100 simulations.

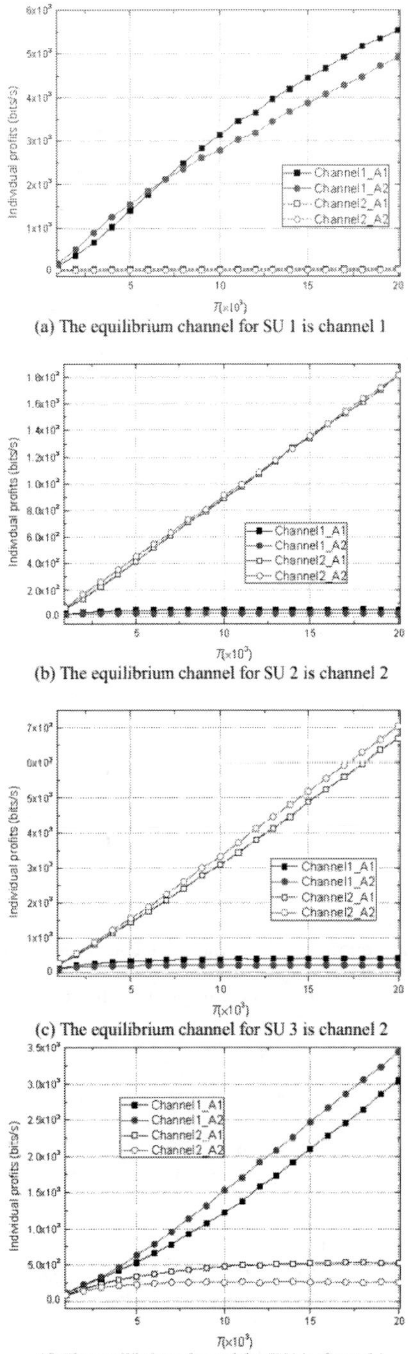

(a) The equilibrium channel for SU 1 is channel 1

(b) The equilibrium channel for SU 2 is channel 2

(c) The equilibrium channel for SU 3 is channel 2

(d) The equilibrium channel for SU4 is channel 1

Figure 2. Individual profits obtained from different channels using two online learning algorithms.

We can obtain NE based on (2) which is that SU1 and SU4 selects channel 1 as well as channel 2 for SU2 and SU3. The NE strategy can be written as (1, 2, 2, 1). First, we are interested in the behaviors of SUs in different channels. In Fig. 2, we show the individual accumulated transmission rates of each SU obtaining from channel 1 and channel 2 using algorithm 1 (A1) and algorithm 2 (A2). From the Fig. 2, we can see that each SU has a dominant strategy which is consistent with the pure strategy NE (1,2,2,1). From Fig. 2(a), we can find SU 1 has obtained more individual profits from the channel 1 regardless of A1 and A2. The same result is also showed in Fig. 2(d). Therefore, the channel 1 is the equilibrium solutions for both SU 1 and SU 4. In Fig. 2(b), SU 2 has received more individual profits from selecting the channel 2 and so dose SU 3 which is showed in Fig.2(c).

In Fig.3, we can note that all *MARs* are finite. In order to evaluate the convergence rates of two online algorithms, we compare the *MARs* of two online algorithms for all SUs in Fig. 3. From Fig.3, we can see the *MAR* of the equilibrium strategy for each SU by A1 is more than A2. Due to the lower convergence rate, each SU using A1 will select non-equilibrium channel more frequently, which will increase the *MAR*. The differences of *MAR*s between SUs using the same online algorithm are dependent on the different locations of SUs.

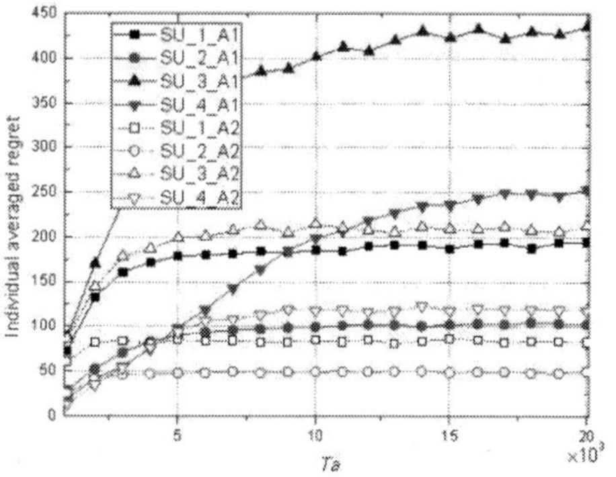

Figure 3. Comparison of 4 SUs' MAR of two selection strategies

5 Conclusion

In this paper, we have modeled the channel selection problem under incomplete information from the perspective of connection between then bandit model and the game model. When the channel selection game has a pure strategy NE, the *MAR* of selection strategies is finite. Two existing online learning algorithms are used to obtain the equilibrium channel for all SUs. The stochastic learning algorithm outperforms the Exp3 based online learning algorithm in terms of convergence rate to the Nash equilibrium solution, which is because the *MAR* of stochastic learning algorithm is less than Exp3. This work provides the connection between non-cooperative games and bandit problems, which will help us to illustrate the convergence rate of different online algorithm without knowing the payoff functions and other SUs' selection strategies.

References

1. B. Wang and K.J.R. Liu, "Advances in cognitive radio networks: A survey," IEEE J. Sel. Topics Signal Process., vol. 5, no. 1, pp. 5–23, Feb. 2011.
2. Z. Zhou, J. Hai, T. Peng, and J. Slevinsky, "Channel exploration and exploitation with imperfect spectrum sensing in cognitive radio networks," IEEE J. Sel. Areas Commun., vol. 31, no. 3, pp. 429–441, Mar. 2013.
3. Auer P, Cesa-Bianchi N, Fischer P. "Finite-time analysis of the multiarmed bandit problem," Mach. Learn., vol.47, no.2-3, pp. 235-256, 2002.
4. X. Fang, D. Yang, and G. Xue, "Taming wheel of fortune in the air: an algorithmic framework for channel selection strategy in cognitive radio networks," IEEE Trans. Veh. Technol., vol. 62, no. 2, pp. 783–796, Feb. 2013.
5. K. Liu and Q. Zhao, "Distributed learning in multi-armed bandit with multiple players," IEEE Trans. Signal Process., vol. 58, no. 11, pp. 5667–5681, Nov. 2010.
6. L. Blumrosen and S. Dobzinski, "Welfare maximization in congestion games," IEEE J. Sel. Areas Commun., vol. 25, no. 6, pp. 1224–1236, Aug. 2007.
7. L. M. Law, J. Huang, and M. Liu, "Price of anarchy for congestion games in cognitive radio networks," IEEE Trans. Wireless Commun., vol. 11, no. 10, pp. 3778–3787, Oct. 2012.
8. D.Monderer and L.S.Shapley, "Potential games," Games Econ. Behav., vol. 14, pp. 124–143, 1996.
9. Y. Xu, J. Wang, Q. Wu, A. Anpalagan, and Y.-D. Yao, "Opportunistic spectrum access in unknown dynamic environment: a game-theoretic stochastic learning solution," IEEE Trans. Wireless Commun., vol. 11, no. 4, pp. 1380–1391, Apr. 2012.
10. T. Li-Chuan, C. Feng-Tsun, Z. Daqiang, et al., "Network selection in cognitive heterogeneous networks using stochastic learning," IEEE Commun. Lett., vol. 17, no. 12, pp. 2304–2307, Dec. 2013.
11. P. Auer, N. Cesa-Bianchi, Y. Freund, and R. E. Schapire, "The non-stochastic multiarmed bandit problem," SIAM Journal on Computing,vol. 32, no. 1, pp. 48–77, Nov. 2002.
12. P. Sastry, V. Phansalkar, and M. Thathachar, "Decentralized learning of nash equilibria in multi-person stochastic games with incomplete information," IEEE Trans. Syst., Man, Cybernm., vol. 24, no. 5, pp. 769–777, May 1994.
13. V. Erceg, L. J. Greenstein, S. Y. Tjandra, et al., "An empirically based path loss model for wireless channels in suburban environments," IEEE J. Sel. Areas Commun., vol. 17, no. 7, pp. 1205–1211, Jul. 1999.

Channel Modelling and Performance Analysis of the Near Space based on the DS Signal System

Ke Sun , Xiaomin Hou and Qiang Zhao

Equipment Academy, Beijing 101416, China

Abstract. In order to describe the near space communication channel efficiently and analyse the the reasons that cause quality deterioration of ground-to-air communication link, design the Channel modelling program of the near space and establish the near space channel model based on the DS signal system which considers rain attenuation and multipath fading. Complete the analysis of rainfall impact on signal fading and estimate related parameters under different elevation. Set up the statistical model of multipath fading. Complete the simulation of Ka-band geostationary near space channel performance under different elevation. Draw bit error rate performance simulation curve under different weather conditions. The result of MATLAB simulation shows that the model can reflect the fading characteristic of the near space communication channel. This program improves the signal fading and interference tolerance. The findings can provide a better simulation Platform for the follow-up job according to the correctness and practicability of established model and be of great significance in exploration and research of near space communication.

Keywords-near space; rain attenuation; multipath fading; channel model

1 Introduction

In recent years, the near space communication system is reflected in the advantages which are the conventional satellite communication system difficult to achieve as a new means of communication. The near space communications platform has many advantages and high military value, such as effective communication, short communication time, small multipath fading, large communication coverage, and strong invulnerability. But the research on near space communications platform focused mainly on aircraft manufacturing, and the achievement of communication technologies is extremely limited, Signal transmission technology is even little suitable for the near space communications[1]. The fundamental reason is the lack of an effective description of the communication channel and construction.

Near space at an altitude of 20 km to 100 km in space area, between the satellite platform and aviation platform, which is the must go through place to outer space. At the same time, it is a strategic space for human to develop. As a new research field, many of the technique need to be innovative designed, concluding the measurement & control and information transmission system. Scientific research institutions at home and abroad had already started the study of tt&c system in near space because it has a series of advantages: Low cost, rapid deployment, less ground equipment, flexible use, easy to recycle.

There are many patents about spread spectrum in National Patent Office of America. These patents contain design, synchronization, sending, receiving and positioning of PN-code, etc. In the domestic, Tsinghua University, Shanghai Jiaotong University and Beijing University of aeronautics and astronautics and Aerospace science and technology group etc have made a series of innovative research results in the satellite signal capturing and tracking algorithm.

Because of the complexity of communication environment in near space, high-speed movement of communication carrier causes the influence of PN-code phase delay to wireless mobile communication been enhanced. Therefore, the communication system can work effectively and reliably or not, to a large extent depends on the presence of PN-code capture system. At the same time, the tt&c platforms must possess information security protection, such as anti-interference, resistance to intercept, secret transmission.

Rong-Wei Bao, et al [2] built the channel model to analyze the influence of rain attenuation on the channel at different elevation angles, but did not consider the effects of the ground. Zhong-Min Gan [3] discuss the effects of the elevation on the near space communications. Mainly rain attenuation will seriously affect the near-space communication at low elevation angles. But he did not discuss how to improve communication performance near space at moderate elevation and high angles. Yi-Dong Su [4] built the communication channel model near space Ka-band cell system coverage and analyzed round reflection, rain attenuation on the channel under different elevation angles. But he did not analyze the performance of carrier transmission signal.

With the further development of wireless communication, as one important branch of communication network, near space communication is attracting more and more attention. The characteristics and needs of the near space communication system are combined in this paper. Complete the analysis and simulation of construction of Ka-band geostationary near space channel communications platform. Exactly grasp near space channel characteristics. Complete the construction of the near space communication system transmission model.

2 The statistical model of near space channel

According to system components and platform height of the near space, near space communication channel is in line with all the characteristics of Ka-band satellite channel. Therefore, the channel model established is referred to the Ka-band geostationary satellite channel model. Near space platform communication channel is a complex time-varying channel. Not only the path loss and rain attenuation, but also multipath fading and ionosphere are the presence of the transmission signal. Among them, the influence of the ionosphere is negligible at frequencies above 1GHz.

Table 1. The classification and values of Ka-band atmospheric attenuation

Atmospheric attenuation of Ka-band	Atmospheric absorption attenuation	Tropospheric scintillation decay	Cloud attenua -tion	Rain attenuat -ion
Attenuation values/dB	≈1	<1	≈1	42

As can be seen from Table Ⅰ, it is just considered that atmospheric attenuation is related to the impact of rain attenuation and signal attenuation due to the surrounding environment [5]. Therefore, near space channel model built is mainly composed of two parts: the first is the impact of rain attenuation, the second is the impact of multipath fading.

2.1 The statistical model of rain attenuation

Rain attenuation is generated from the radio waves by the absorption of raindrops and the impact of scattering. It is mainly related to geometry of raindrops, rain intensity, rain range, the signal frequency, polarization and so on. Rain attenuation is a major factor affecting the quality of Ka-band communication link.

The effect of rain attenuation is expressed as a time-varying stochastic model [6]:

$$r_l(t) = a(t)exp(j\phi(t))s_l(t) \quad 0 \le t \le T \qquad (1)$$

In the formula, $s_l(t)$ and $r_l(t)$ is the modulated signal which is respectively used to send and receive. T is the width of modulated symbols. $a(t)$ and $\phi(t)$ is respectively envelope and phase of equivalently low communication channels. Both of them are real random variables. The distributions of probability can be approximated by the Gaussian distribution model [7]:

$$p(a) = \frac{1}{\sqrt{2\pi}\sigma_1} exp(-\frac{(a-m_1)^2}{2\sigma_1^2}) \qquad (2)$$

$$p(\phi) = \frac{1}{\sqrt{2\pi}\sigma_2} exp(-\frac{(\phi-m_2)^2}{2\sigma_2^2}) \qquad (3)$$

In the formula, $p(a)$ is probability density function of signal envelope, $p(\phi)$ is probability density function of signal phase, $p(\phi)$ and σ_2^2 is respectively variance of envelope and phase. m_1 and m_2 is respectively corresponding mean.

Rain attenuation would cause a very big impact on near space communication at low elevation. So this paper mainly considers the situation at low elevation [8].

Table II is composed of relevant parameters of Ka-band geostationary near space channel at the elevation angle of 14.2° under different weather conditions.

Table 2. The channel envelope and phase parameters of the near space

Weather	Mean m_1	Variance σ_1^2	Mean m_2	Variance σ_2^2
Biotite	0.346	0.00272	0.0154	0.00864
Thunderstorm	0.436	0.01386	0.0068	0.00414
Light rain	0.483	0.00003	0.0088	0.00546
Rain	0.662	0.02	-0.0089	0.03077

2.2 The statistical model of multipath fading

The signals which are reflected, scattering and diffracted in the communication process will encounter a variety of objects, such as buildings, trees, etc. The received signals are the synthesized signals which are reachable from each path. Envelopes and phases of components of each path are different. Therefore the amplitude fluctuation of composite signals would be more severe. In summary this is the multipath fading. When we analyze the probability distribution model of transmission characteristics of the channel, we usually use the Rice distribution to describe the multipath fading. It is the envelope probability density function of the signal [9]:

$$f_r(r) = 2rK\exp\left[-(1+r^2)K\right]I_0(2rK) \quad (4)$$

In formula, Rice factor is $K = \frac{1}{2R^2}$ which R^2 is the average scattering power caused by multipath.

If terrestrial user terminals are at different locations, the elevation will be different. With the elevation decreases, the value of factor K in line with the Rice distribution will be also reduced assigned in accordance with Table III [10].

Table 3. Rice distribution statistical properties of different elevation

Coverage area	Statistical parameters/dB	Ideal values /dB
High elevation	$10<K<20$	$K=15$
Moderate elevation	$8<K<10$	$K=10$
Low elevation	$2<K<6$	$K=5$

3 Channel modelling program of the near space

3.1 The signal system design

Since the spread spectrum system has more powerful ability of anti-human interference [11], narrow-band interference, anti-multipath interference than the conventional communication system, use the spread spectrum communication technology to improve the signal fading and interference tolerance in the near space communication links in this paper.

The spread spectrum sequences applied to spread spectrum system are commonly m sequence, gold sequence, M sequence and so on. Therefore, this paper decides to select the gold code sequence whose length is 1023.

4.2 The design of modem

According to the characteristics of the near space monitoring and control communication system and transmission characteristics of the spatial channel[12]-[14], the choice of digital modulation techniques in this paper is BPSK modulation of high power efficiency design due to sufficient bandwidth of Ka-band.

The block diagram of BPSK modulation and demodulation system is shown in Fig. 1.

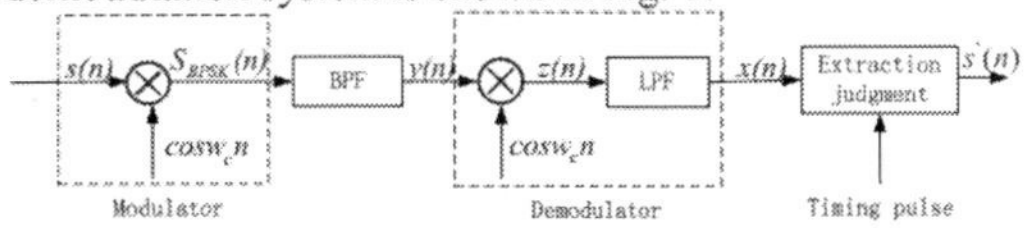

Figure 1. BPSK modulation and demodulation system block diagram.

4.3 The process design of simulation

According to the near space channel statistical model, establish the simulation flow diagram as shown in Fig. 2. The binary signal generator generates a continuous stream of binary bits. Pseudo-code sequence generator generates gold code sequence. Multiply them to form a spread spectrum code sequence. After BPSK modulation, multiplicative fading and additive noise pollution of AWGN channel, despread to form the carrier modulated signal and demodulate to recover the original signal. The verdict will be compared with the original symbol as statistical errors. This completes the simulation process design.

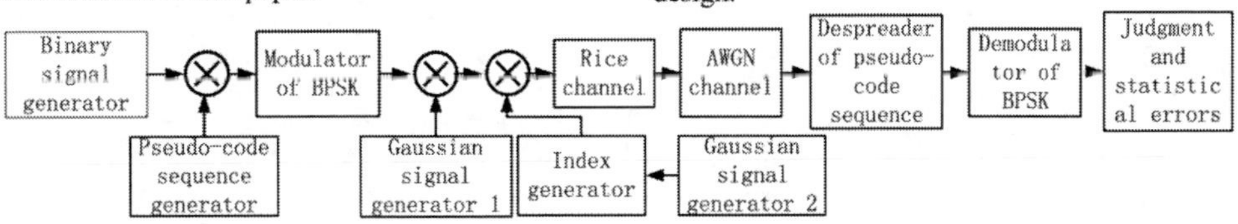

Figure 2. Simulation flow chart of the near space communications system.

The mean and variance generated by Gaussian signal generator 1 is respectively m_1 and σ_1^2. The Gaussian noise process $a(t)$ represents fading process of signal envelope. The Gaussian process $\phi(t)$ of the mean m_2 and variance σ_2^2 is generated by Gaussian signal generator 2. The phase fading process $exp(j\phi(t))$ is generated by index generator. It can produce a total attenuation factor $C(t) = a(t)exp(jf(t))$ that used to simulate the near space channel system with the rice channel and AWGN additive noise.

4 The performance analysis of near space channel model

4.1 The performance analysis of the rain attenuation model

Combining with data in Table II, simulate the BER performance of BPSK system in the four types of weather and pure AWGN channel at the elevation of 14.2°. Statistical results of BER are shown in Fig. 3.

Fig. 3 is shown that the bit error rate of the system is changed with SNR curve under different weather conditions. The figure shows that the fading caused by the rain is the least. Compared with pure AWGN channel, the loss of SNR at 10^{-2} of BER is about 4dB under rain weather conditions. The fading caused by the thunderstorm is relatively enlarged. The loss of SNR at 10^{-2} of BER is about 8dB under thunderstorm weather conditions. The fading caused by the biotite is the largest. The loss of SNR at 10^{-2} of BER is about 10dB under biotite weather conditions. The performance of the system cannot be significantly improved when the SNR is greater than 16dB.

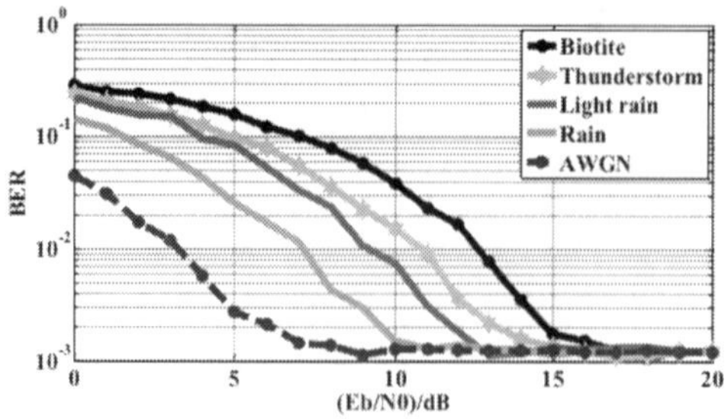

Figure 3. BER simulation of the near space rain attenuation channel.

4.4 The performance analysis of the multipath fading model

Combining with data in Table III and Rice channel formula, simulate the BER performance of BPSK system in the three types of elevation and pure AWGN channel. Statistical results of BER are shown in Fig. 4 to 6.

Fig. 4 to 6 are shown that the bit error rate of the system is changed with SNR curve under different weather conditions at the three types of elevation. At different elevation angles, the fading caused by the biotite is the largest. Compared with the pure rain attenuation channel at the elevation of 14.2°, the loss of SNR at 10^{-2} of BER is reduced by 3dB at low elevation, 2dB at moderate elevation, 1dB at high elevation. The fading caused by the thunderstorm and little rain is relatively enlarged. Compared with the pure AWGN channel, the loss of SNR at 10^{-2} of BER is about respectively 4dB and 3dB at low elevation, 5dB and 4dB at moderate elevation, 6dB and 5dB at high elevation. The fading caused by the rain is the least. Compared with the pure AWGN channel, the loss of SNR at 10^{-2} of BER is about 2dB at high elevation. The performance of the system cannot be significantly improved when the SNR is greater than 14dB.

In contrast with the situation of no fading, the lower the elevation, the lower the bit error rate under all the weather conditions. In the case of direct sequence spread spectrum modulation, increase the signal to noise ratio to be the lowest error rate 10^{-3}.The simulation results show that near space Ka-band channel quality has been significantly improved with lower elevation.

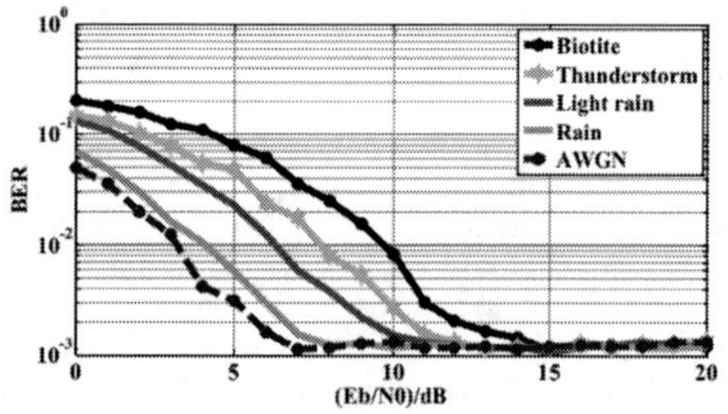

Figure 4. BER simulation of the near space low elevation attenuation channel.

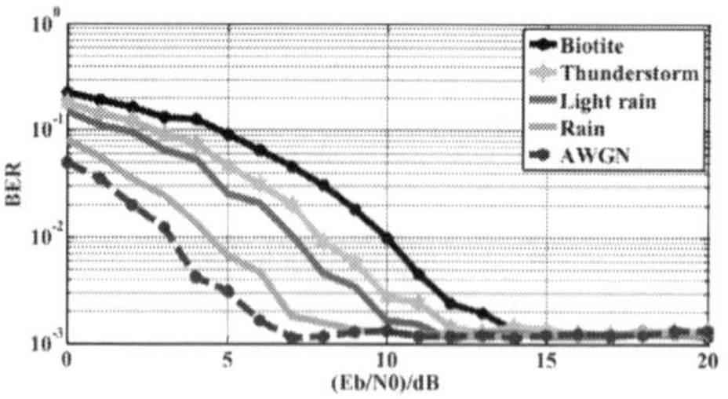

Figure 5. BER simulation of the near space moderate elevation attenuation channel.

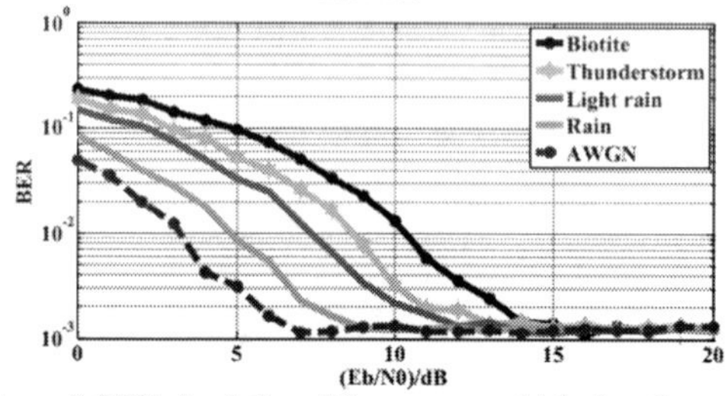

Figure 6. BER simulation of the near space high elevation attenuation channel.

5 Conclusions

By analyzing the characteristics of the near space platform communication system and the unique nature of Ka-band channel, analyze the probability distribution of the amplitude and phase of the communication signal on the basis of analyzing the compositions of the near space communication platform, characteristics of the rain attenuation and multipath fading. Show the Ka-band channel model in near space platform. Through further analysis of the physical meaning of the model parameters, estimate the channel model parameters based on the theory of derivation, moreover design of the program of DS signal system channel modelling. At last, simulate and compute the near space channel at different elevations and a variety of weather conditions. At the same time it proves the feasibility and rationality of the model parameters and reflects the significant anti-fading performance of DS signal system. If we can use the appropriate channel coding technology, the transmission performance is expected to have a greater increase. Therefore, this article provides a better environment of simulation for the follow-up job according to the correctness and practicability of established model.

References

1. M. C. Yang, and Y. C. Jiang, "Fading characteristics of HAPS communication channel in dynamic environment with low elevation angle," Journal of South China University of Technology, vol. 37, no. 1, pp. 27-33, Jan 2009.
2. R. W. Bao, and S. X. Chen, "Near space static channel model rearch," Journal of Air Force Engineering University: Natural Science Edition, vol. 11, no. 6, pp. 22-26, May 2010.

3. Z. M. Gan, "Near space platform: emergency communication effective means," Digital Communication World, vol. 4, no. 6, pp. 45-49, Sep 2008.
4. Y. D. Su, and S. X. Chen, "Near space Ka-band channel model," Journal of Air Force Engineering University: Natural Science Edition, vol. 12, no. 5, pp. 76-80, Oct 2011.
5. S. Karapantazis, and P. F. Niovi, "Broadband communications via high-Altitude platforms: A survey," IEEE Trans. Surveys&Tutorials, vol. 7, no. 1, pp. 2-31, Nov 2005.
6. W. Z. Li, C. L. Law, and V. K. Dubey, "Ka-band land mobile satellite channel model incorporating weather effects," IEEE Trans. Communication Letters, vol. 5, no. 5, pp. 194-195, Feb 2001.
7. C. Loo, "Statical models for land mobile and fixed satelite communication at Ka band". IEEE 46th Veh Technol Conf, 1996, Canada, pp. 1023-1027.
8. A. Z. Alejandro, C. R. Jose'Luis, D. P. Jose'Antonio. High-altitude platforms for wireless communications. United Kingdom: John, Wiley & Sons Ltd, 2008, pp. 123-135..
9. L. C. Jose, A. D. Jos, "Channel model based on semi-Markovian processes. An approach for HAPS systems," Proceedings of the 14th International Conference on Electronics, Communications and Computers, Veracruz, Mexico, 2004, pp. 52-56.
10. J. Han, H. Xu , and H. J. Wu. "HAPS Communication in the low elevation angle the performance analysis," Radio communication technology, vol. 29, no. 6, pp. 51-52, Mar 2003.
11. A. H. Wang, and W. X. Luo. "Ka frequency satellite communication channel building and system performance simulation," Journal of Communication, vol. 22, no. 9, pp. 61-69, July 2001.
12. E. Lutz, D. Cygan, and M. Dippold. "The land mobile satellite communication channel-recording, statistics and channel model," IEEE Trans. on vehicular technology, vol. 40, no. 2, pp. 375-386, July 1991.
13. C. X. Fan. "A kind of new development mobile communication method: research of stratosphere communication". Modern Electronic Technology, vol. 19, pp. 1-3, Jan 2005.
14. C. Lin. "Near space fading channel modeling and its performance analysis," Computer Engineering, vol. 38, no. 6, pp. 80-82, Mar 2012.

MATEC Web of Conferences 56, 05004 (2016) DOI: 10.1051/matecconf/20165605004
ICCAE 2016

A Method of Angle Error Extraction for BPSK Modulated Signals Based on Instantaneous Autocorrelation

SONG Xiao-rui [1], ZHENG Hai-xin [2] and ZENG Chuang-zhan [1]

[1] *Department of Graduate Management, Equipment Academy, 101416, Beijing, China*
[2] *Department of Optical and Electronic Equipment, Equipment Academy, 101416, Beijing, China*

Abstract. In this paper, a method of angle error extraction for binary phase-shift keying (BPSK) modulated signals based on instantaneous autocorrelation is proposed for the monopulse angle tracking problem of BPSK signal under a low bit signal to noise ratio (SNR). It reduces the effect of noise on the angle error extraction efficiently by using the characteristics of the instantaneous autocorrelation of noise. The effectiveness and reliability of the theoretical algorithm are proved by MATLAB simulation. As simulation results shown, this method has a good performance under ideal Gaussian noise background. When bit SNR is greater than 2dBHz, the relative error of the estimation results can be less than 5%.

1 Introduction

The angle error extraction of BPSK signal is a process of computing the ratio between the amplitude of sum signal and different signal by a series of signal processing [1, 2]. The sum and different signals are received by a Cassegrain antenna for target tracking. The angle error extraction is an important part of the monopulse technology, which is widely used in target tracking in aerospace, communications, modern search radar, radio astronomy, sonar, optical tracking and other fields.

There are two conventional methods of angle error extraction of BPSK signal so far. The first one is based on envelope detection [3, 4], that is, by intercepting a small portion of the wideband signal to make it into a narrowband signal which can be used to estimate the angle error by amplitude detection. This method is a non-coherent demodulation and ineffective under low SNR conditions. The second one is based on I/Q processor [5], that is, by using a PLL to recover the carrier as a local oscillation signal to finish the synchronous detection with which the amplitude modulated information can be recovered to estimate the angle error. This method is a coherent demodulation with a lower threshold than former one and has a widely use in digital monopulse receivers. However, it will be seriously affected by correlated noise in sum and different signals and the precision of angle error estimations will worsen quickly under a low SNR.

Aimed at solving the problems of angle error extraction under low SNR conditions and estimation precision seriously affected by correlated noise in sum and different signals, this paper carry out a method to implement the angle error extraction of BPSK signal under a low SNR based on instantaneous autocorrelation signal processing.

2 Signal instantaneous autocorrelation characteristics

The signal instantaneous autocorrelation is defined as a product of in-phase component and quadrature component [6, 7]:

$$ST(t)=s_I(t)\cdot s_Q(t) \tag{1}$$

From the properties of Hilbert transform, the phase difference between in-phase component and orthogonal component is 90°, that is, a quarter of a signal period delay, $\tau_0=T/4$, and T is a signal period. Thus, signal instantaneous autocorrelation is equivalent to signal autocorrelation with a fixed delay.

The received signal is given as:

$$r(t)=s(t)+n(t) \tag{2}$$

Then we have autocorrelation of received signal as follows,

$$\begin{aligned}YT_r(t,\tau)&=r(t)\cdot r^*(t-\tau)\\&=s(t)s^*(t-\tau)+s(t)n^*(t-\tau)\\&+n(t)s^*(t-\tau)+n(t)n^*(t-\tau)\end{aligned} \tag{3}$$

where noise $n(t)$ is incoherent with the signal. When $\tau\neq 0$, we have

$$E\left[n(t)\cdot n^*(t-\tau)\right]=0 \tag{4}$$

then (3) can be expressed as:

$$YT_r(t,\tau)=s(t)s^*(t-\tau) \tag{5}$$

From the above derivation, it is can be known that the autocorrelation of received signal can reduce noise effect efficiently. And signal instantaneous autocorrelation can be given as:

$$\begin{aligned} ST(t) &= s_I(t)\cdot s_Q(t) \\ &= s(t)\bullet s^*(t-\tau_0) \\ &= YT_r(t,\tau_0) \end{aligned} \tag{6}$$

Equation(6) shows that the substance of signal instantaneous autocorrelation is an autocorrelation with a fixed delay, which is a quarter of a signal period. Therefore, the noise effect can be reduced efficiently after the instantaneous autocorrelation signal processing to provide a better SNR condition to angle error estimation.

3 Angle error extraction based on instantaneous autocorrelation

Section II shows that signals can have a better SNR condition for follow processing after the instantaneous autocorrelation. With this feature, this paper provides a method of angle error extraction based on instantaneous autocorrelation. The process of angle error extraction is shown in Fig. 1.

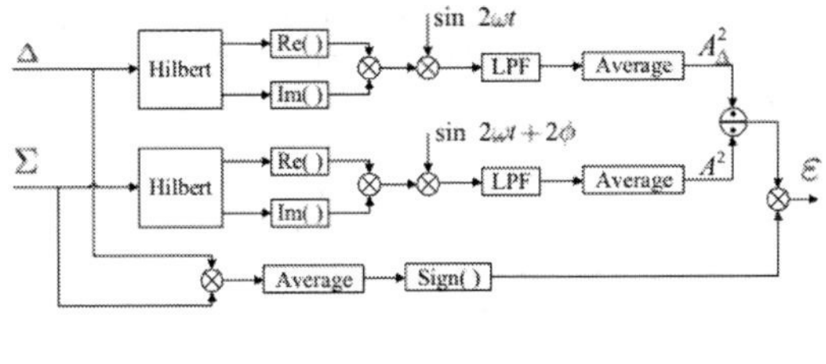

Figure 1. Block diagrams of the process of angle error extraction based on instantaneous autocorrelation.

The sum and different signals with ideal Gaussian noise are given as:

$$\Sigma = Am(t)\cos(\omega t) + n_s \tag{7}$$

$$\Delta_{az} = A_{az}m(t)\cos(\omega t + \phi_A) + n_{az} \tag{8}$$

$$\Delta_{az} = A_{az}m(t)\cos(\omega t + \phi_A) + n_{az} \tag{9}$$

where, $m(t)$ is the encoded data with values of +1 or -1; Σ is the sum signal; Δ_{az} is the azimuth difference signal; Δ_{el} is the elevation difference signal. And ϕ_A is the phase difference between the azimuth different and the sum signals; ϕ_E is the phase difference between the elevation different and the sum signals. In addition, A stands for the amplitude of sum signal, A_{az} for the amplitude of the azimuth different signal, and A_{el} for the amplitude of the elevation different signal. n_s, n_{az} and n_{el} are the noise in the sum, azimuth different and elevation different signal, separately. Therefore, according to the basic principle of monopulse angle tracking system, we can get the azimuth and the elevation angle error:

$$\varepsilon_{az} = A_{az} / A \tag{10}$$

$$\varepsilon_{el} = A_{el} / A \tag{11}$$

Computing the Hilbert transform of the sum and different signals, we can take the real parts as in-phase components and the imaginary parts as quadrature components:

$$\Sigma_{SI} = Am(t)\cos(\omega t) + n_s \tag{12}$$

$$\Delta_{az\_SI} = A_{az}m(t)\cos(\omega t + \phi_A) + n_{az} \tag{13}$$

$$\Delta_{el\_SI} = A_{el}m(t)\cos(\omega t + \phi_E) + n_{el} \tag{14}$$

$$\Sigma_{SQ} = Am(t)\sin(\omega t) + n_s^* \tag{15}$$

$$\Delta_{az\_SQ} = A_{az}m(t)\sin(\omega t + \phi_A) + n_{az}^* \tag{16}$$

$$\Delta_{el\_SQ} = A_{el}m(t)\sin(\omega t + \phi_E) + n_{el}^* \tag{17}$$

where, Σ_{SI}, Δ_{az_SI} and Δ_{el_SI} are the in-phase components of the sum, azimuth different and elevation different signals, separately. Similarly, Σ_{SQ}, Δ_{az_SQ} and Δ_{el_SQ} are the quadrature components of the sum, azimuth different and elevation different signal separately.

The signals after instantaneous autocorrelation are:

$$\begin{aligned} ST_\Sigma &= A^2\sin(2\omega t) \\ &+ n_s^* Am(t)\cos(\omega t) \\ &+ n_s Am(t)\sin(\omega t) + n_s^* n_s \end{aligned} \tag{18}$$

$$\begin{aligned} ST_{\Delta_{az}} &= A_{az}^{\ 2}\sin(2\omega t + 2\phi_A) \\ &+ n_{az}^* A_{az}m(t)\cos(\omega t + \phi_A) \\ &+ n_{az}A_{az}m(t)\sin(\omega t + \phi_A) + n_{az}^* n_{az} \end{aligned} \tag{19}$$

$$\begin{aligned} ST_{\Delta_{el}} &= A_{el}^{\ 2}\sin(2\omega t + 2\phi_E) \\ &+ n_{el}^* A_{el}m(t)\cos(\omega t + \phi_E) \\ &+ n_{el}A_{el}m(t)\sin(\omega t + \phi_E) + n_{el}^* n_{el} \end{aligned} \tag{20}$$

Since the results of noise autocorrelation are zero with non-zero delay, we can get

$$E(n^* n) = 0 \tag{21}$$

The square value of sum and different amplitude can be solved by down converting process of (18), (19) and (20). The solution functions are:

$$\begin{aligned} \Sigma_{AMP} &= E\left[ST_\Sigma \bullet \sin(2\omega t)\right] \\ &= A^2 \end{aligned} \tag{22}$$

$$\begin{aligned} \Delta_{az\_AMP} &= E\left[ST_{\Delta_{az}} \bullet \sin(2\omega t + \phi_A)\right] \\ &= A_{az}^{\ 2} \end{aligned} \tag{23}$$

$$\begin{aligned} \Delta_{el\_AMP} &= E\left[ST_{\Delta_{el}} \bullet \sin(2\omega t + \phi_E)\right] \\ &= A_{el}^{\ 2} \end{aligned} \tag{24}$$

To be here, the local oscillator signals used in down conversion are from numerically controlled oscillators(NCO) in sum, azimuth different and elevation different channels separately. The difference between the three lies in the initial phases.

The sign of the angle error can be differentiated by the functions as follows:

$$Symbol_{az} = sign(\Delta_{az} \bullet \Sigma) \tag{25}$$

$$Symbol_{el} = sign(\Delta_{el} \bullet \Sigma) \quad (26)$$

At last, the azimuth and elevation angle error are estimated as:

$$\varepsilon_{az} = Symbol_{az} \bullet \sqrt{\Delta_{az\_AMP} / \Sigma_{AMP}} \quad (27)$$

$$\varepsilon_{el} = Symbol_{el} \bullet \sqrt{\Delta_{el\_AMP} / \Sigma_{AMP}} \quad (28)$$

4 Simulation and Analysis

In this section, a simulation will be carried out with Matlab to proof the validity of the algorithm. Because the methods of azimuth and elevation angle error extraction are the same [8-10], only the azimuth angle error extraction will be simulated here. And the elevation angle error is estimated in the same way. The simulation parameter settings are as follows:

The code rate of BPSK signal: $R_b = 2\text{Mbps}$

The carrier frequency: $f_c = 70\text{MHz}$

The sampling frequency: $f_s = 60\text{MHz}$

The truth value of azimuth angle error: $\varepsilon_{az} = -0.5$

The baseband signal is a sequence of 0 and 1 generated at random, and the noise in signals is ideal Gaussian noise.

100 times Monte Carlo experiments are conducted to estimate the azimuth angle error under different SNR conditions. In the experiments, the bit SNR varies from -6dBHz to 16dBHz in a 0.5dBHz step. The result of the simulation is shown in Fig. 2.

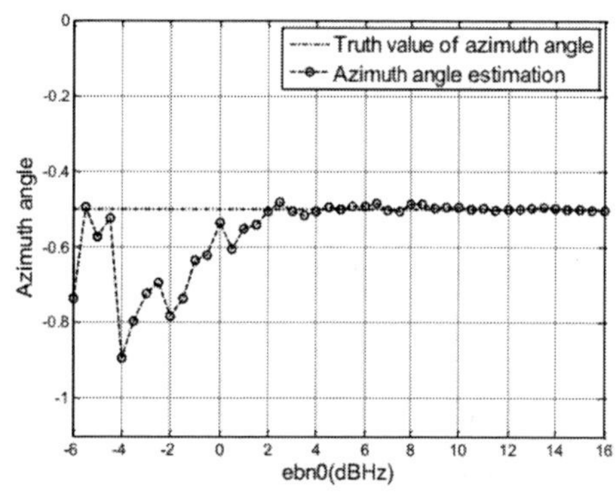

Figure 2. Azimuth angle error estimations under different bit SNR conditions.

It is shown in Fig. 2 that, all the relative errors between the azimuth angle error estimations and the truth values are less than 5% when $Eb/N_0 > 2\,\text{dBHz}$. The estimation precision can meet the needs of monopulse tracking system [11, 12]. Under the same simulation conditions, the comparison between different methods of angle error extraction is shown in Fig. 3 as follows.

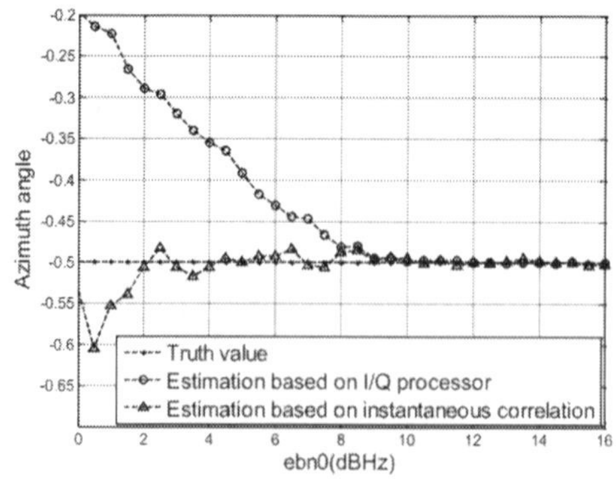

Figure 3. Angle error estimations of two methods.

Fig. 3 shows that, the threshold of the angle error extraction based on I/Q processor is about 8dBHz, but it of the angle error extraction based on instantaneous autocorrelation is about 2dBHz. In other words, the threshold has a 6dBHz decrease so that the performance of angle error extraction can improve significantly. In addition, unlike extraction method based on I/Q processor, the precision of angle error extraction based on instantaneous autocorrelation is not affected by noise independent or not.

The derivation in Section III shows that there are two error sources in the angle error extraction based on instantaneous autocorrelation. The first one is symbol estimation error, and the second one is the absolute value estimation error. The analysis of the two error sources is as follows.

Firstly, an analysis in symbol estimation error is carried out. Under different SNR conditions, the number of false symbols in 100 times Monte Carlo experiments is counted and the result is shown in Fig. 4.

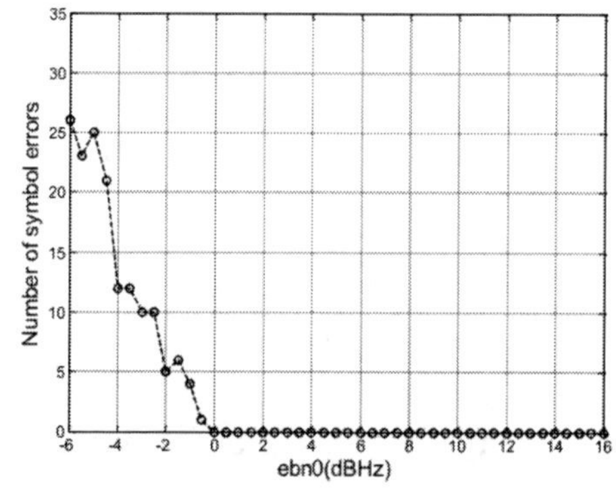

Figure 4. The statistic of false symbol estimations.

In Fig. 4, the number of false symbols is zero when $Eb/N_0 > 0\,\text{dBHz}$, which shows that there is no symbol estimation error under this condition. When $Eb/N_0 < 0\,\text{dBHz}$, the precision of symbol estimation will begin worsening and the number of false symbols will increase rapidly with the SNR decreasing.

Secondly, an analysis in absolute value estimation error is carried out. Fig. 5 shows the azimuth angle absolute value estimations in 100 times Monte Carlo experiments under different SNR conditions.

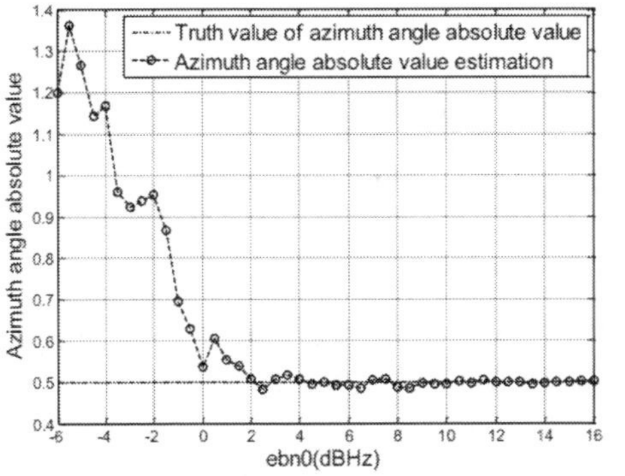

Figure 5. Azimuth angle absolute value estimations.

Fig. 5 shows that the relative errors between the azimuth angle absolute value estimations and the truth values are less than 5% when $Eb/N_0 > 2$ dBHz, meeting the requirements of angle tracking precision. And the precision cannot be meet with a bit SNR greater than 0dBHz less than 2dBHz.

To sum up, the threshold of the symbol estimation is about 0dBHz, but the threshold of the absolute value estimation is about 2dBHz. Therefore, the main restriction to the angle error extraction which is based on instantaneous autocorrelation, is the angle error absolute value estimation precision under a low SNR condition.

5 Conclusions

In this paper, a method of angle error extraction for BPSK modulated signals is provided. The main innovation point of the method is using the characteristics of the instantaneous autocorrelation of BPSK signals to reduce the effect of noise on the angle error extraction. As a result, the problems of angle error extraction under low SNR conditions and estimation precision seriously affected by correlated noise in the sum and different signals can be solved. However, this method is only apply to the three-channel monopulse angle tracking system, and how to realize the application in the dual-channel or the single channel monopulse angle tracking system remains to be solved.

Acknowledgment

The authors wish to thank Zhu Jiang, Yan Di, and Li Zi-bo. This work was supported in part by a grant from the Laboratory of TT&C, Department of Optical and Electronic Equipment, Equipment Academy.

References

1. Levanon, Nadav. *Radar principles*. New York: Wiley-Interscience, 1988, 320 p.1 (1988).
2. Nickel, Ulrich. "Overview of generalized monopulse estimation", *Aerospace and Electronic Systems Magazine*, IEEE, 21.6 (2006): 27-56.Luigi T.De Luca, *Propulsion physics* (EDP Sciences, Les Ulis, 2009)
3. Burgos-Garcia, M., et al. "Digital Monopulse Receivers For Phase Modulated Signals." in Radar Conference, 2003. Proceedings of the International, 2003:640-645.
4. Hofstetter, E. M., and D. Delong. "Detection and parameter estimation in an amplitude-comparison monopulse radar.", IEEE Transactions on Information Theory, 15.1(1969):22 - 30.
5. M. Sherman S, K. Barton D. Monopulse Principles and Techniques. Artech House, 2011.
6. ZHOU Min, FENG Quanyuan. "A new feature parameter for MFSK/MPSK recognition." in The 2011 International Conference on Intelligent Science and Information Engineerin (ISIE). Wuhan, IEEE, 2011:21-23.
7. FU Shiping, YANG Li, SHAO Wei. "Digital Signal Modulation Classification Algorithm Based on Instantaneous Parameter". Modern Defence Technology, 2013,41(6):71-75.
8. Conroy, J. T., Joseph L. LoCicero, and Donald R. Ucci. "Communication techniques using monopulse waveforms." In Military Communications Conference Proceedings, 1999. MILCOM 1999. IEEE. Vol. 2. IEEE, 1999.
9. Seifer, Arnold D. "Monopulse-radar angle tracking in noise or noise jamming.", IEEE Transactions on Aerospace & Electronic Systems, 28.3(1992):622-638.
10. Mosca, E. "Angle Estimation in Amplitude Comparison Monopulse Systems.", IEEE Transactions on Aerospace & Electronic Systems, aes-5.2(1969):205-212.
11. Armstrong, B. C., and H. D. Griffiths. "Use Of Difference Channel Information For Detection In Monopulse Radars.", Radar & Signal Processing IEE Proceedings-F, 38.3(1991):199 - 210.
12. Nickel, U. "Performance of corrected adaptive monopulse estimation.", Radar, Sonar and Navigation, IEE Proceedings –F, 146.1(1999):17-24.

MAX-SLNR Precoding Algorithm for Massive MIMO System

Jing Jiang, Honglin Wang, Yongbin Xie and Wujun Yang

Xi'an University of Posts & Telecommunications, School of Communication and Information Engineering, Xi'an, China

Abstract. Pilot Contamination obviously degrades the system performance of Massive MIMO systems. In this paper, a downlink precoding algorithm based on the Signal-to- Leakage-plus-Noise-Ratio (SLNR) criterion is put forward. First, the impact of Pilot Contamination on SLNR is analyzed, then the precoding matrix is calculated with the eigenvalues decomposition of SLNR, which not only maximize the array gains of the target user, but also minimize the impact of Pilot Contamination and the leak to the users of other cells. Further, a simplified solution is derived, in which the impact of Pilot Contamination can be suppressed only with the large-scale fading coefficients. Simulation results reveal that: in the scenario of the serious pilot contamination, the proposed algorithm can avoid the performance loss caused by the pilot contamination compared with the conventional Massive MIMO precoding algorithm. Thus the proposed algorithm can acquire the perfect performance gains of Massive MIMO system and has better practical value since the large-scale fading coefficients are easy to measure and feedback.

1 Introduction

At the end of 2010, Thomas l. Marzetta in Bell Laboratories puts forward the concept of large-scale multiple antennas systems, which is also known as Massive MIMO or Large-Scale MIMO [1], where a large number of antennas are deployed at each cell site. Thanks to large-scale antenna array, cell capacity and spectrum efficiency of users at cell's edge can be improved greatly [2-3] by using gain of array and gain of interference suppression .As antennas' number at base station tends to infinity, the channel capacity tends to infinity according to Information theory. While this theory is not feasible in the application of large-scale MIMO, because obtaining channel state information (CSI) gets harder and pilot contamination in uplink channel estimation becomes inevitable. Pilot contamination had been put forward and analyzed firstly in the literature [4]: large-scale MIMO system usually works in TDD which has channel reciprocity in order to avoid the feedback overhead of CSI, where base station does MIMO precoding in downlink by utilizing uplink channel state information. When users send uplink pilots to base station, pilots of different cells can't guarantee being orthogonal from each other. Non-orthogonal pilots in other cells will produce interference to uplink channel estimation, which makes channel estimation results include not only the CSI from local users to the local base station, but also the contaminated CSI from users in neighboring cells to the local base station. According to this channel estimation results, large-scale MIMO processing can't be accurate, which results in degraded system performance [4] [5]. How to reduce pilot contamination, at present, is becoming a hot spot in academia of large-scale MIMO. The following literatures provide the reference to precoding algorithm which eliminates pilot contamination. How pilot contamination affects the performance of the system is analyzed, and the multi-cell MMSE precoding algorithm had been put forward in literature [6]. In [6], the error between the desired signal and the detected signal in local cell and interference from other cells to local desired users can be minimized at the same time. One kind of multi-layer precoding about large-scale MIMO was put forward in literature [7], the first layer of precoding matrix is used to reduce inter-cell interference and the second precoding matrix is used in the beamforming of sending signals. The third layer of precoding matrix is used for multi-user interference elimination. In the algorithm, channel statistics information is used to reduce the inter-cell interference caused by pilot contamination. Moreover, the implementation of the layer 1 and layer 2 precoding matrix is set in the analog circuits of hardware, and implementation of the third layer is designed in the baseband, so that this kind of composite analog/digital structure reduces complexity of hardware. Similar idea also appears in the literature [8], precoding matrix is divided into inner precoding matrix and outer the precoding matrix, which means intra-cell interference elimination and eliminating inter-cell interference. The concept of pilot contamination precoding (PCP) was introduced in literature [9], it mainly says that the base station do the linear combination with slow fading coefficient from terminal to each pilot-reused cell to obtain the slow fading coefficient matrix, then the pilot contamination matrix can be obtained by the product of the slow fading coefficient matrix and conventional single cell linear precoding matrix. It has been proven that the algorithm can eliminate the influence of pilot contamination on precoding based on the SINR of the algorithm and capacity limit. Literature [10] proposed a downlink precoding scheme based on MAX-SINR criteria. Under the condition of keeping base station transmission power constant and making the user's SINR not lower than the setting threshold, it optimizes the weights of precoding to eliminate pilot contamination.

Among the precoding algorithms of MIMO system performance optimization, except the above laws, there is an algorithm based on maximizing Signal-to-Leakage plus-Noise-Ratio(MAX-SLNR). Sadek, Tarighat and Sayed made a perfect expression about classic SLNR precoding algorithm [11]-[12], SLNR precoding can effectively reduce the interference from other users, and maximize the desired signal power of each user, besides, MAX-SLNR allows us substituting generalized eigenvalue problem for the multi-user beamforming.

Based on SLNR and pilot contamination precoding, our algorithm not only optimizes the performance of target users, but also eliminates pilot contamination in the uplink channel estimation and downlink multi-user interference. Compared with multi-cell MMSE algorithm, the algorithm avoids the matrix inversion processing, which makes the complexity of the algorithm reduced greatly. Further, the algorithm takes into account the simplified solution based on slow fading coefficient, so that the pilot contamination can be effectively eliminated under the condition of without the small scale fading information. Due to the large-scale fading information is easy to measure, and feedback cycle is long, of which cost is far less than the small scale fading information feedback, algorithm has more practical application value. Finally, the proposed algorithm can effectively reduce the pilot contamination degradation on system performance through the simulation.

2 System Model

In this paper, we consider uplink training and transmit precoding in a multi-cell scenario with I cells in the Massive MIMO TDD system, where each cell consists of a base station with N_t antennas and K users with single antenna. Same pilots are reused in L cells among I cells. In uplink training, the signal received at the base station of j-th cell is:

$$\boldsymbol{y}_j = \sum_{l=1}^{L}\sum_{k=1}^{K}\sqrt{p_r\tau\beta_{lk}^j}\boldsymbol{h}_{lk}^j\boldsymbol{x}_j + \sum_{i=1,i\neq l}^{I}\sum_{k=1}^{K}\sqrt{p_r\tau\beta_{ik}^j}\boldsymbol{h}_{ik}^j\boldsymbol{x}_i + \boldsymbol{n}_j \quad (1)$$

where $\boldsymbol{x}_j$ is τ length pilots which is reused in L cells, $\boldsymbol{x}_i$ is the pilot used in i-th cell, which is orthogonal with $\boldsymbol{x}_j$. β_{lk}^j is large-scale fading coefficient between the k-th user of l-th cell and the base station of j-th cell. $\boldsymbol{h}_{lk}^j$ is small-scale fading matrix between the k-th user of l-th cell and the base station of j-th cell, elements in matrix are independent and identically distributed ($i.i.d.$) zero-mean, circularly-symmetric complex Gaussian $CN(0,1)$ random variables. Let the pilot power during tansmission at each user be p_r, $\boldsymbol{n}_j$ is additive noise. By utilizing training sequences ,take the channel estimation with LS algorithm in uplink, the channel state information between the k-th user of j-th cell and the base station of j-th cell can be expressed as:

$$\hat{\boldsymbol{h}}_{jk}^j = \boldsymbol{y}_j\cdot\boldsymbol{x}_j^H/\sqrt{p_r\tau} = \sqrt{\beta_{jk}^j}\boldsymbol{h}_{jk}^j + \sum_{l=1,l\neq j}^{L}\sum_{k=1}^{K}\sqrt{\beta_{lk}^j}\boldsymbol{h}_{lk}^j + \boldsymbol{n}_j' \quad (2)$$

In (2), the first part $\sqrt{\beta_{jk}^j}\boldsymbol{h}_{jk}^j$ is the real channel matrix, the second part $\sum_{l=1,l\neq j}^{L}\sum_{k=1}^{K}\sqrt{\beta_{lk}^j}\boldsymbol{h}_{lk}^j$ is the pilot contamination caused by the cells that use the same training sequences.

In Massive MIMO system, the signal received at the k-th user of j-th cell is:

$$y_{jk} = \sqrt{p_f}\boldsymbol{f}_{jk}^j\boldsymbol{a}_{jk}\boldsymbol{q}_{jk} + \sum_{i=1,i\neq j}^{I}\sum_{k=1}^{K}\sqrt{p_f}\boldsymbol{f}_{jk}^i\boldsymbol{a}_{ik}\boldsymbol{q}_{ik} + z_{jk} \quad (3)$$

In (3), let transmission power at base station be p_f; $\boldsymbol{q}_{ik}$ is signal transmitted in downlink, which satisfys $E[\boldsymbol{q}_{ik}\boldsymbol{q}_{ik}^H] = \boldsymbol{I}$; $\boldsymbol{a}_{jk}$ is downlink precoding vector, which satisfies $tr\{\boldsymbol{a}_{ik}^H\boldsymbol{a}_{ik}\} = 1$; z_{jk} is additive noise at the k-th user of j-th cell, which satisfies $z_{jk}\in CN(0,1)$; $\boldsymbol{f}_{jk}^j$ is downlink channel matrix between the k-th user of j-th cell and the base station of j-th cell, first part of (3) is the desired signal, second part is the interference signal from the base station at adjacent cells.

3 Precoding Algorithm

Signal-leakage-noise-ratio (SLNR) is defined as the ratio of the energy of the user's useful signal to noise power and the sum of the energy of user's leakage to other cells. According to the downlink system model (3), the SLNR at the k-th user of j-th cell can be expressed as:

$$SLNR_{jk}=\frac{\left|f_{jk}^{j}a_{jk}\right|^{2}}{\left|z_{jk}\right|^{2}+\left|\sum_{i=1,i\neq j}^{I}\sum_{k=1}^{K}f_{ik}^{j}a_{jk}\right|^{2}}=\frac{\left|f_{jk}^{j}a_{jk}\right|^{2}}{1+\left|\sum_{i=1,i\neq j}^{I}\sum_{k=1}^{K}f_{ik}^{j}a_{jk}\right|^{2}} \quad (4)$$

During downlink transmission, According to SLNR's eigenvalue decomposition of (4),precoding vector a_{jk} can be obtained at base station .According to channel reciprocity , downlink channel matrix is $f_{jk}^{j}=\left(\sqrt{\beta_{jk}^{j}}h_{jk}^{j}\right)^{T}$ .Instead of obtaining the real uplink channel state information directly, the sender get the uplink channel matrix $\hat{H}_{jk}^{j}$ estimated by launch pilot training sequence sent from user to base station through (2). Analysis of (2) shows that the estimated channel matrix contains the true channel state information, and pilot contamination caused by reusing pilot sequences in different cells. Power of the desired signal calculated by the channel estimation is:

$$\left|\left(\hat{h}_{jk}^{j}\right)^{T}a_{jk}\right|^{2}=\left|\left(\sqrt{\beta_{jk}^{j}}h_{jk}^{j}+\sum_{l=1,l\neq j}^{L}\sum_{k=1}^{K}\sqrt{\beta_{lk}^{j}}h_{lk}^{j}+n_{j}^{'}\right)^{T}a_{jk}\right|^{2}$$

$$=\left|\left(\sqrt{\beta_{jk}^{j}}h_{jk}^{j}\right)^{T}a_{jk}\right|^{2}+\left|\left(\sum_{l=1,l\neq j}^{L}\sum_{k=1}^{K}\sqrt{\beta_{lk}^{j}}h_{lk}^{j}\right)^{T}a_{jk}\right|^{2}+\frac{I}{\sqrt{p_{r}\tau}} \quad (5)$$

For the channel from different users to the base station as h_{jk}^{j} and h_{jk}^{j}, and the noise $n^{'}$ is independent, so as the antennas' number of base station tends to infinity, cross multiplication of independent vector is 0 according to the law of large numbers, then (5) can be simplified. In the end, the first term of (5) is power of the desired signal, which can be obtained at the sender according to the real channel.The second term is energy of interference caused by pilot contamination in channel estimation. Under the condition of obtaining uplink channel estimation information merely, energy of the pilot contamination must be minimum in order to make the energy gain of desired signal greater. Furthermore, energy of the desired signal calculated according to the channel estimation information can get as close as possible to energy of the useful signal obtained in the real channel. Therefore, when the sender gets channel estimation information merely, influence of pilot contamination in the uplink channel estimation must be considered, and (4) can be derived in further:

$$SLNR_{jk}=\frac{\left|\hat{h}_{jk}^{j\,T}a_{jk}\right|^{2}}{1+\left|\sum_{l=1,l\neq j}^{L}\sum_{k=1}^{K}\sqrt{\beta_{lk}^{j}}h_{lk}^{j\,T}a_{jk}\right|^{2}+\left|\sum_{i=1,i\neq j}^{I}\sum_{k=1}^{K}\sqrt{\beta_{ik}^{j}}h_{ik}^{j\,T}a_{jk}\right|^{2}} \quad (6)$$

In order to maximize SLNR, downlink precoding vector a_{jk} can be calculated as:

$$a_{jk}=\arg\max\frac{\left(\hat{h}_{jk}^{j\,T}\right)^{H}\hat{h}_{jk}^{j\,T}}{1+\sum_{l=1,l\neq j}^{L}\sum_{k=1}^{K}\beta_{lk}^{j}\left(h_{lk}^{j\,T}\right)^{H}h_{lk}^{j\,T}+\sum_{i=1,i\neq j}^{I}\sum_{k=1}^{K}\beta_{ik}^{j}\left(h_{ik}^{j\,T}\right)^{H}h_{ik}^{j\,T}} \quad (7)$$

(7) can be expressed as (8) .

according to (8), When the antenna number of base station tends to infinity, according to the law of large numbers, $\left(h_{lk}^{j\,T}\right)^{H}h_{lk}^{j\,T}=I$, then (8) can be expressed as (9).

Therefore precoding vector a_{jk} can be further simplified as the value in relation to large-scale fading in other cells.

$$a_{jk}\propto max.eigvector\left[\left(1+\sum_{l=1,l\neq j}^{L}\sum_{k=1}^{K}\beta_{lk}^{j}\left(h_{lk}^{j\,T}\right)^{H}h_{lk}^{j\,T}+\sum_{i=1,i\neq j}^{I}\sum_{k=1}^{K}\beta_{ik}^{j}\left(h_{ik}^{j\,T}\right)^{H}h_{ik}^{j\,T}\right)^{-1}\left(\left(\hat{h}_{jk}^{j\,T}\right)^{H}\hat{h}_{jk}^{j\,T}\right)\right] \quad (8)$$

$$\lim_{N_t\to\infty}a_{jk}=\text{argmax}\frac{\left(\hat{h}_{jk}^{j\,T}\right)^{H}\hat{h}_{jk}^{j\,T}}{1+N_t\sum_{l=1,l\neq j}^{L}\sum_{k=1}^{K}\beta_{lk}^{j}I+N_t\sum_{i=1,i\neq j}^{I}\sum_{k=1}^{K}\beta_{ik}^{j}I}$$

$$=\text{argmax}\frac{\left(\hat{h}_{jk}^{j\,T}\right)^{H}\hat{h}_{jk}^{j\,T}}{\frac{1}{N_t}+\sum_{l=1,l\neq j}^{L}\sum_{k=1}^{K}\beta_{lk}^{j}I+\sum_{i=1,i\neq j}^{I}\sum_{k=1}^{K}\beta_{ik}^{j}I}=\text{argmax}\frac{\left(\hat{h}_{jk}^{j\,T}\right)^{H}\hat{h}_{jk}^{j\,T}}{\sum_{l=1,l\neq j}^{L}\sum_{k=1}^{K}\beta_{lk}^{j}I+\sum_{i=1,i\neq j}^{I}\sum_{k=1}^{K}\beta_{ik}^{j}I} \quad (9)$$

4 Numerical Results

In this section, link-level simulation evaluation based on the proposed MAX-SLNR algorithm has been taken. The proposed algorithm is compared with the conventional Massive MIMO precoding algorithm-matched filter(MF) algorith[2] .The simulation parameters are shown in Table I .

Table 1. parameters of link level Massive MIMO simulation based on MAX-SLNR

Simulation Parameters of Massive MIMO	
Parameters	Set Value
Number of cells	2
Antenna number of base station	100
Number of users per cell	1
Antenna array distance	0.5λ
Modulation scheme	QPSK
Antenna number of user	1
Antenna array	uniform linear array
Base station /user transmission power	20dB/1dB
Large-scale fading coefficient (slight pilot contamination)	$\beta_{1k}^1=0.9$ $\beta_{2k}^1=0.1$
Large-scale fading coefficient (severe pilot contamination)	$\beta_{1k}^1=0.5$ $\beta_{2k}^1=0.5$

In the uplink, the uplink pilots are configured at terminals in two cells, cell 2 is defined as contamination cell, reusing the same pilot as the cell 1.Firstly, we use the hadamard matrix to generate orthogonal pilot sequences, after QPSK modulation, the original pilot sequences are sent by terminals in different cells, through the corresponding complex gaussian channel. Wireless channel is rapidly changing in reality, assuming that there is only a symbol transmission during a coherent time, satisfying channel reciprocity.

In the downlink, base station in cell1 send data symbols during 1024 different coherence time channel to k-th user in cell 1. After QPSK modulation, the original bitstream start precoding processing for different channel. At this time we compare our MAX-SLNR algorithm with matched filter (MF) precoding algorithms. Downlink MF precoding vector between the k-th user of j-th cell and the base station of j-th cell can be obtained by :

$$\boldsymbol{a}_{jk}=\left(\hat{\boldsymbol{h}}_{jk}^{j\,T}\right)^H \Big/ \left\|\left(\hat{\boldsymbol{h}}_{jk}^{j\,T}\right)^H\right\| \qquad (10).$$

Precoding vector based on MAX-SLNR can be calculated by (9) . After going through complex gaussian channel and adding gaussian noise signals, signal is received by MMSE detection. Then, bit error ratio is shown. We draw two kinds of algorithm performance curves with the complete channel information and performance curves only using the large-scale information. The results are as follows:

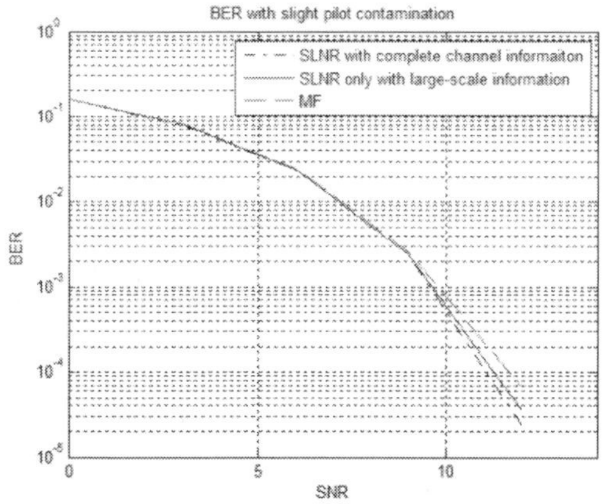

Figure 1 (a)

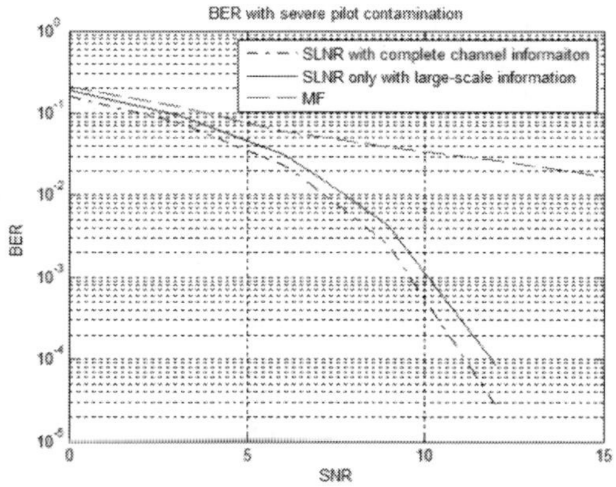

Figure 1 (b)

Figure 1. simulation results of Massive MIMO precoding based on MAX-SLNR

In figure 1 (a) , assuming that the large-scale fading coefficients as $\beta_{1k}^1=0.9$, $\beta_{2k}^1=0.1$ .under the condition of slight pilot contamination, as SNR increases, BER curve of the proposed algorithm based on MAX-SLNR is similar to curve of MF algorithm mentioned in literature[2]. In figure 1 (b) , considering scenario with severe pilot contamination, let the large-scale fading coefficients of two cells be $\beta_{1k}^1=0.5$, $\beta_{2k}^1=0.5$ .Under the condition that cell 1 suffers severe pilot contamination from cell 2, there is an obvious degradation of performance about MF algorithm of large-scale MIMO system. With the increase of SNR, affected by the pilot contamination, imprecise precoding seriously affects the performance of the system, so that BER curve cannot fall. And for MAX-SLNR algorithm based on pilot contamination term and inhibition of interference leakage brought by other cells' users, BER curve decreases obviously with the increase of SNR, the performance is much better than the MF algorithm. Compared with slight pilot contamination scenario, there is no obvious degradation of performance with serious pilot contamination. Moreover, the proposed MAX-SLNR algorithm only relying on large scale information can obviously inhibit the pilot contamination. The algorithm also has obvious advantage compared with MF algorithm performance.

5 Conclusions

The simulation results show that in the case of slight pilot contamination, performance curve of the proposed MAX-SLNR algorithm is approximated at that of the conventional Massive MIMO algorithms-MF algorithm. Affected by severe pilot contamination, MAX-SLNR algorithm can effectively reduce the system performance degradation brought by pilot contamination, and reduce the energy of interference leak to adjacent cells, thus gain of large-scale MIMO system performance could be maximized. This paper also puts forward the simplified solution based on the large-scale fading information, which can greatly reduce the channel of information feedback overhead between cells, and simplify the complexity of the channel measurement, which makes the algorithm more practical. In this paper, we assume that the Massive MIMO channel is independent of each other, but large-scale MIMO channel is quite complex in application. We will study the curves of MAX-SLNR performance under the condition of different Channel in the near future.

Acknowledgement

This paper is sponsored by National 863 Project(2014AA01A705) and the Natural Sience Foundation of ShanXi Province(2014JM8320).

References

1. T. L. Marzetta, IEEE Trans. Commun., vol. **9**, no. 11, pp. 3590-3600, (2010)
2. F. Rusek, D. Persson, B. K. Lau, E. G. Larsson, T. L. Marzetta, O. Edfors, F. Tufvesson, IEEE Signal Process. Mag., vol. **30**, no. 1, pp. 40-60, (2012)
3. H. Q. Ngo, E. G. Larsson, T. L. Marzetta, IEEE Trans. Commun., vol. **61**, no. 4, pp. 1436-1449, (2012)
4. J. Jose, A. Ashikhmin, T. L. Marzetta, *Proc. IEEE International Symposium on Information Theory*, pp. 2184-2188,(2009)
5. H. Q. Ngo, E. G. Larsson, T. L. Marzetta, *Proc. IEEE International Conference on Acoustics, Speech and Signal Processing,* pp. 3464–3467,(2011)
6. J. Jose, A. Ashikhmin, T. L,Marzetta, S. Vishwanath, IEEE Trans. Wireless Commun, vol. **10**, no. 8, pp. 2640-2651, (2011)
7. Alkhateeb Ahmed, Leus Geert, W. Heath Jr. Robert, *Proc. Asilomar Conference on Signals Systems and Computers, Pacific Grove,* CA, Nov. 2014.
8. A. Liu, V. K. N. Lau, IEEE Trans. Signal Process, vol. 62, no. 18, pp. 4786-4797,(2014)
9. Li Liangbin, A. Ashikhmin, T. L.Marzetta, *Proc of 51st Annual Allerton Conference on Communication, Control, and Computing,* pp. 226-232.(2013).
10. Jing Jiang， Zheng Xu, International Journal of Future Generation Communication and Networking， vol. **7**,no. 3, pp. 107-116,(2014)
11. A. Tarighat, M. Sadek, A. H. Sayed, *Proc of IEEE International Conference on Aoustics, Speech, and Signal Processing* , pp. 1129-1132.(2005)
12. M. Sadek, A. Tarighat, A. H.Sayed, IEEE Trans. Wireless Commun., vol. **6**, no. 5, pp. 1711-1721, (2007)

A Novel Reliability Enhanced Handoff Method in Future Wireless Heterogeneous Networks

YuPeng Wang , and KaiLi Li

Shenyang Aerospace University, Department of Electrical and Information Engineering, Shenyang, China

Abstract. As the demand increases, future networks will follow the trends of network variety and service flexibility, which requires heterogeneous type of network deployment and reliable communication method. In practice, most communication failure happens due to the bad radio link quality, i.e., high-speed users suffers a lot on the problem of radio link failure, which causes the problem of communication interrupt and radio link recovery. To make the communication more reliable, especially for the high mobility users, we propose a novel communication handoff mechanism to reduce the occurrence of service interrupt. Based on computer simulation, we find that the reliability on the service is greatly improved.

1 Introduction

To satisfy the increasing demand on the service rate, multiple types of base stations (BSs) are considered in the future network deployment, which consists of heterogeneous networks. However, various BS coverage sizes cause unreliable communication link when users moves from one big BS to a small BS or vice versa [1-5].

To solve the above problem and reduce the service interrupt rate, we propose a reliability enhanced handoff mechanism, which avoids the high-speed users to be handed over to the small BS due to its possible short stay, and muting the small BS signal transmission for some time to alleviate the interference when the high-speed use passing by. In this way, communication overhead can be greatly saved due to the avoidance on the unnecessary handoff and radio link recovery process.

The remaining part of this paper is organized as follows. In section 2, we formulate handoff problem for high-speed users in heterogeneous networks. The proposed enhanced handoff method is presented in section 3. The proposed method is evaluated in section 4. Conclusions are drawn in section 5.

2 Problem Formulation

In this paper, we consider the small BSs are deployed in the coverage of big BS for the service rate enhancement as in Figure1, where users may suffer the interference from both big and small BSs.

Radio link failure is detected by BS based on the user's physical channel signal transmission. If BS finds that the user's signal transmission is in out-of-synchronization state and does not recover within some fixed time interval, then BS will think the radio link is suffering some problems and all the signals from the user will be neglected. After that, a process called radio link recovery will be initiated to make the user's signal transmission be back to the in-synchronization state. Usually, the synchronization state is based on the received signal-to-noise and interference ratio (SINR) of the synchronization preamble or pilot, and the value SINR should be smoothed as the filter in (1).

$$F_n = (1-\alpha)\cdot F_{n-1} + \alpha\cdot M_n \quad (1)$$

where M_n is the simultaneous received SINR; F_n is the updated filtered SINR, which is used to decide the user's synchronization state and link quality; F_{n-1} is the previous SINR, where F_0 is set to M_1 when the first SINR is obtained; and $a = 1/2^{(k/4)}$ is typically used with k is the filter coefficient [6]-[7].

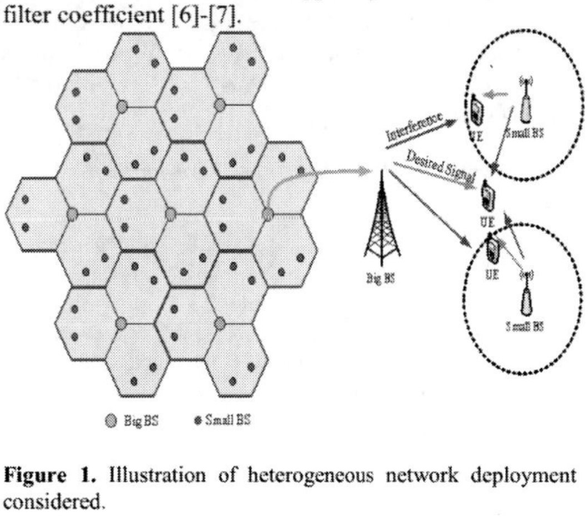

Figure 1. Illustration of heterogeneous network deployment considered.

3 Proposed Reliability Enhanced Handoff Method

3.1 Muted Subframe

To relieve the interference from other BSs when high-speed users passing the small BS, some subframes of small BS are muted, which are also known as almost blank subframes (ABS). In these muted subframes, there is little signal transmission except the necessary control signals. In this way, the interference limited user is protected from the strong interference. But the subframe utility is a little reduced by the existence of muted subframes. To achieve a better trade-off, the muted subframe configuration should be optimized, i.e., the density and pattern are the key factors of performance.

3.2 Proposed Reliability Enhanced Handoff Method

To avoid the unnecessary handoff of high-speed user into the small BS's coverage and protect them from interference when passing by the small BS, additional steps such as user speed classification and muted subframe related configuration are introduced into the existing handoff procedure as in [8].

Here, we take a typical scenario as an example to illustrate the implementation of our proposed idea for a better understanding. The detailed procedure of the proposed algorithm can be found in Fig. 2. The proposed additional steps are as follows:

Step-Measurement Reports with High-speed USER Indicator

When the high-speed user gets closer to the small BS, an event called A3 may be triggered due to the stronger receiving power from the small BS than the serving big BS. In addition to the conventional measurement report, the high-speed user may also send an additional high-speed user indictor (i.e., one bit information with value '1' indicating high-speed user and valuser'0' indicating other users) to the big BS to notify that the handoff to small BS should be avoided.

The user's mobility state estimation can be made by counting the number of handoffs within a specific time as usual or using the user handoff history information exchanged between BSs.

Step-Small BS Muted Subframe Turn-on Request

If the user is in high-speed state, big BS sends the "Muted Subframe Turn-on Request" message to the target small BS. When the big BS is providing VoIP service to the high-speed user now, the muted subframe turn-on Request also includes the next transmission timing information of the VoIP service.

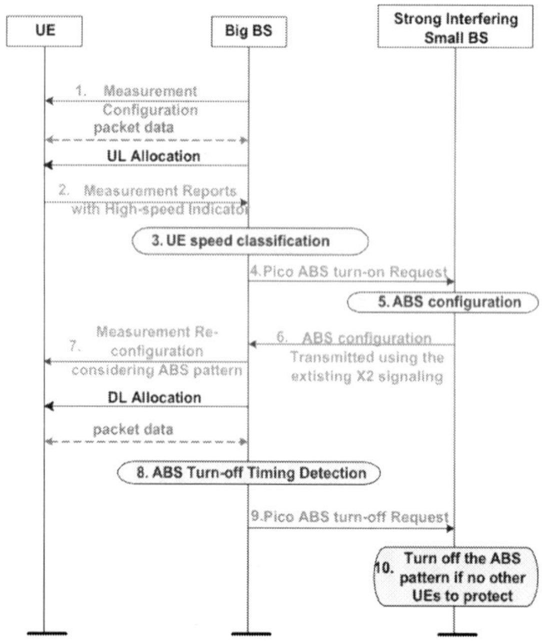

Figure 2 Detailed procedure for the proposed reliability enhanced handoff method

Step- Small BS Muted Subframe Configuration

When the small BS receives the "Muted Subframe Turn-on Request" from the big BS, some frames will be muted to protect high-speed user, but the muted subframe density configuration method is out of the scope of this paper. According to the small BS's current muted subframe status, the following actions can be taken by small BS.

a) If there is no muted subframe available at small BS now and the high-speed user is served with VOIP, then small BS will configure the muted subframe considering the high-speed user's next transmission timing information in the "Muted Subframe Turn-on Request" and VoIP data transmission interval to decide which subframe to be muted.

b) If there is no muted subframe at small BS now and the high-speed user is not requiring VoIP service, then Small BS will configure any subframe as the muted sub-subframe.

c) If the small BS has already configured the muted subframe and the incoming "Muted Subframe Turn-on Request" is not for the VoIP user, no additional configuration is required.

d) If the small BS already configured the muted subframe and the incoming "Muted Subframe Turn-on Request" is for the VoIP user, small BS should re-configure the muted subframe considering the next transmission information in the "Muted Subframe Turn-on Request".

Step-Small BS Muted Subframe Configuration Exchange

After small BS has configured or re-configured the muted subframe for the high-speed user, it will transmit the configuration information to the corresponding big BSs which have high-speed users to be protected by the muted subframe at this small BS, using the current existing interface between BSs.

Step- High-speed USER RRC Reconfiguration

After receiving the muted subframe configuration from small BS, the big BS will reset the measurement configuration through RRC signals and make the scheduling decisions considering the small BS muted subframe configuration.

Step-Muted Subframe Turn-off Timing Detection

The muted subframe adopted at small BS is used to protect the high-speed user, therefore the muted subframe should be turned off to avoid the resource waste when the high-speed user leaves the small BS's coverage.
There are two ways to detect the timing to turn-off muted subframe at small BS:

a) Explicit Detection

When the high-speed user detects the received RSRP or RSRQ from small BS below some predefined threshold, then it will send some specific message to its serving big BS to indicate the interference from the small BS is not serious now and the muted subframe at small BS can be turned off now.

b) Implicit Detection

When the serving Big BS receives another event A3 triggered measurement report with target BS ID is not the same small BS as that in the previous event A3 triggered measurement report from the high-speed user, the serving big BS knows that the high-speed user has already left the coverage of previous small BS and the muted subframe at that small BS can be turned off.

Step-Small BS Muted Subframe Turn Off

When big BS thinks there is no need to keep the muted subframe on for its serving high-speed user, it will send the "Muted Subframe Turn-off Request" to the corresponding small BS. After receiving this request, the small BS will decide whether to turn off the muted subframe or not.

Table 1 System simulation parameters

Parameter	Value
HTN scenario	3GPP downlink, outdoor Pico/Hotzone, configuration 1, model 1
Cellular layout	Hexagonal layout with wrap around, 7 eNodeBs, 3 cells per eNodeB
System frequency	2GHz carrier, 10 MHz bandwidth
ISD	500m (case 1)
Max MeNB Tx power	46 dBm
Max Pico Tx power	30 dBm
Number of Picos per cell	2, fixed at cell edge
Number of UE per cell	25, uniformly and randomly distributed
Traffic model	Full buffer
Scheduling algorithm	Round Robin
Scheduling delay	6ms
Scheduling granularity	5PRBs
Downlink HARQ	Asynchronous HARQ with CC, Maximum three retransmissions
Channel model	SCM urban macro high spread for 3GPP case 1
Number of MeNB BS antenna	1 Tx antenna
Number of Pico/Hotzone antenna	1 Tx antenna
Number of UE antenna	2 Rx antennas
Downlink receiver type	MRC
Channel estimation error	None
Pico Bias Configuration for Cell Extension	0, 6, 9 dB
UE Mobility Model	Bouncing Circle
High Mobility Classification	>60 km/h
Running time	60 s

4 Evaluation Results

4.1 Simulation Parameters

The detailed simulation parameters are outlined in Table 1 as same as that in [9]. Users are associated to a big or a small BS according to its best downlink reference signal receiving power. In addition to the data transmission, HARQ re-transmission is also considered in the performance evaluation.

4.2 Simulation Results

In this section, we present the benefits of the proposed method in the aspects of the system's throughput and radio link failure rate. To simplify the analysis and make a fair comparison, we assume the handoff is perfect without any failure. To avoid ping-pong and the short time of stay problem, the high speed user is not allowed to be hand off to the small BS for any cases.

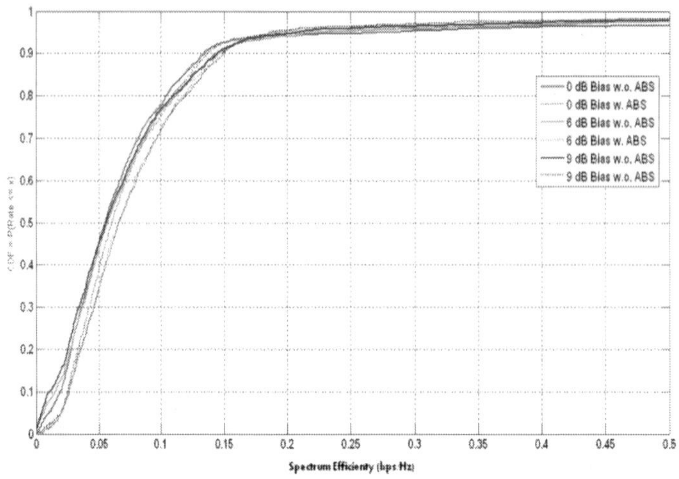

Figure 3. CDF of user spectrum efficiency in the cases of with and without muted subframe (ABS).

From the simulation results in Table 2, we find that the small BS's interference induced radio link failure is not negligible for the high-speed user if we avoid the high-speed user being handed off to the small BS even with 0dB bias for small BS's coverage extension. We also find that the small BS muted subframe (*a.k.a* ABS) induced gain on RLF reduction is over 60 %, and in the case of large bias configuration such as 6 dB and 9 dB, both the high-speed and low-speed users enjoy this gain.

Figure 3 shows the UE spectrum efficiency comparison between the cases of with and without ABS at small BS in the aspects of cumulative distribution function (CDF). A detailed comparison is shown in Table 3. From the results in Table 3, we find the ABS at small cell increases the user's received SINR, which covers the system resource loss due to ABS at small BS.

5 Conclusions

In this paper, we propose a reliability enhanced handoff method, which modifies the existing handoff procedure by adding some mobility related steps for system performance improvement. From the results, we observe that the performance of both radio link failure rate and system efficiency is greatly improved. For some system configuration such as high small BS bias case, even low speed users will benefit from the proposed method.

Table 2. RLF Rate with different small BS bias configuration

Small BS Bias	Speed (km/h)	w.o. ABS (RLFs/UE/S)	w. ABS (RLFs/UE/S)	Gain on RLF Reduction
0 dB	30	0	0	0
	120	0.030	0.006	78.46 %
6 dB	30	0.010	0.001	86.76 %
	120	0.056	0.018	66.48 %
9 dB	30	0.046	0.008	81.63 %
	120	0.071	0.026	63.50 %

Table 3. Spectrum efficiency comparison between the cases of with and without muted subframe (ABS) at small BS

Small BS Bias	Average SE for all UEs w.o. ABS (bps/Hz)	Average SE for all UEs w. ABS (bps/Hz)	ABS gain on Average SE for ALL UEs	Average SE for High-speed UEs w.o. ABS (bps/Hz)	Average SE for High-speed UEs w. ABS (bps/Hz)	ABS gain on Average SE for high speed UEs
0 dB	0.0905	0.0925	2.21 %	0.0486	0.0585	20 %
6 dB	0.0887	0.0933	5.19 %	0.0484	0.0611	26.24 %
9 dB	0.0804	0.0893	11.07 %	0.0472	0.0623	31.99 %

Acknowledgement

This research is supported by the research grant (with No. of L2014067) of Liaoning Provincial Education Department of China.

References

1. A. Racz, A. Te,evary, and N. Reider, "Handover performance in 3GPP long term evolution (LTE) systems," in Proc. MWCS, 2007, pp. 1-5.
2. A. Ulvan, R. Bestak, and M. Ulvan, "The study of handover procedure in LTE-based femtocell network," in Proc. IEEE WMNC, 2010, pp. 1-6.
3. D. Aziz, and R. Sigle, "Improvement of LTE handover performance through interference coordination," in Proc. of IEEE Veh. Tech. Conf. (VTC-spring), 2009, pp. 1-5.
4. K. Kitagawa, T. Komine, T. Yamamoto, and S. Konishi, "Performance evaluation of handover in LTE-Advanced systems with pico cell range

extension," in Proc. IEEE PIMRC, 2012, pp. 1071-1076.
5. R2-120652, On UE-speed-based methods for improving the mobility performance in HetNets, ALU, Feb. 2012.
6. Evolved Universal Terrestrial Radio Access (E-UTRA); Mobility enhancements in heterogeneous networks, 3GPP TS 36.839, V11.1.0, Jan. 2013.
7. Evolved Universal Terrestrial Radio Access (E-UTRA); Radio resource control (RRC); protocol specification, 3GPP TS 36.331, V11.3.0, Mar. 2013.
8. Evolved Universal Terrestrial Radio Access (E-UTRA) and Evolved Universal Terrestrial Radio Access Network (E-UTRAN); Overall description; Stage 2, 3GPP TS 36.300, V11.5.0, Mar 2013.
9. Evolved Universal Terrestrial Radio Access (E-UTRA); Further advancements for E-UTRA physical layer aspects, 3GPP TS 36.814, V9.0.0, Mar. 2010

MATEC Web of Conferences **56**, 05007 (2016) DOI: 10.1051/matecconf/20165605007
ICCAE 2016

A Quantitative Evaluation Method of Satellite Network Control Protocol based on Multi-attribute Utility Theory

Jun Zhu and Ruimin Lu [1]

Nanjing Telecommunication Technology Research Institute, Nanjing Jiangsu 210007, China

Abstract. Satellite network control protocol plays an important part in satellite communication system. Different design schemes of protocol have very different protocol attribute values. It is difficult to select a proper protocol solution from many solution alternatives with quite different protocol attributes, especially some mutual conflicted protocol attributes. Thereupon, this paper provides a multi-attribute utility theory based quantitative method to evaluate these protocol design schemes, in order to make a wise decision on how to select protocol solution. This method introduces several critical protocol attributes by utilizing a Petri net model, and aggregates them into a decision making matrix. On the basis of multi-attribute utility theory, quantitative evaluation values are achieved with the support of user attribute preference weight vectors. In the end, the paper presents experiments and discussions.

1 Introduction

Satellite Network Control Protocol (SNCP) is one of the most important parts in satellite communication system. It allows users to achieve high reliability and high performance communications. But to design and develop a satisfying SNCP is not an easily solved problem. Each SNCP should satisfy several important attributes. For example, SNCP should have high performance and reliability. So the attributes performance and reliability are two important attributes of SNCP. Besides, it is very comprehensive that the investment on designing and developing SNCP should be as little as possible. So are the attributes implemented difficulty, development period and maintainability. As a result, when we try to develop a SNCP, we should focus on the protocol attributes mentioned above.

However, different design schemes (protocol alternatives) of SNCP have very different protocol attributes. For example, one protocol design scheme has higher performance, but need much more investment than another scheme; Or, one scheme has lower reliability, but need more future maintainability. In such situation, it is difficult to select a proper protocol solution from so many solution alternatives with quite different protocol attributes, especially some mutual conflicted attributes. Thereupon, it is necessary to provide a quantitative method to evaluate these SNCP design schemes, so as to make a wise decision on how to select protocol solution.

To solve this problem, this paper proposes a Quantitative Evaluation Method of satellite network control protocol (SNCP-QEM) based on Multi-Attribute Utility Theory (MAUT) [1][2]. As Fig. 1 shows, this method introduces several critical protocol attributes, including performance, reliability, implement difficulty, development period, investment and maintainability. The attributes performance and reliability are predicted by employing a SNCP quantitative Generalized Stochastic Petri Nets (GSPN) model [3]. The other four attributes are estimated by analyzing the protocol design schemes. And then, these protocol attributes from different protocol solutions are aggregated into a decision matrix, which will be utilized as the input of quantitative evaluation algorithm. Based on MAUT, the algorithm employs user decision-making related attribute weight vector to achieve quantitative evaluation values and corresponding solution selection ranks. Consequently, it offers fundamental support to choose SNCP solution. At last, the paper presents effectiveness and performance evaluations of SNCP-QEM and some discussions.

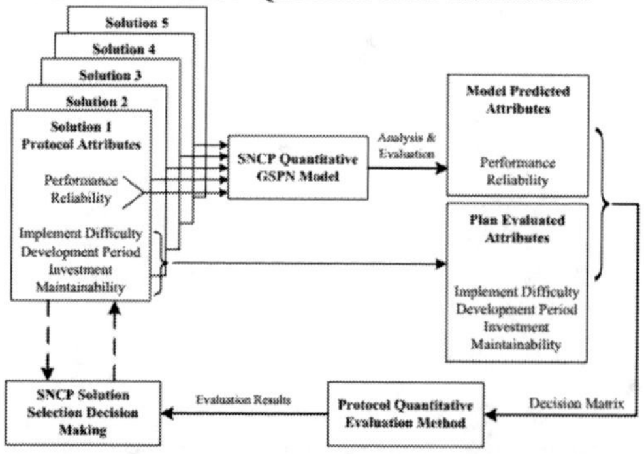

Figure 1. Principle of SNCP Quantitative Evaluation Method

2 Related work

Multi-attribute utility theory has been widely applied in the domain of communication. In this section, we try to present some typical approaches in related domain.

On one hand, MAUT is used for network selection, handoff selection, scheme selection and so on. For example, a network access selection algorithm based on multi-radio parallel transmission is proposed by Zhang etc., to make full use of radio resources in heterogeneous wireless networks. In this algorithm, the optimal scheme is determined based on cost function value of every network scheme by using TOPSIS method with user preferences [4]. Dimitris etc. propose a framework to address network selection problem by the employment of multi attribute decision making methods and by using certain Quality of Service (QoS) indicators acting as figures of merit which influence the decision process [5]. Multi attribute decision making algorithm is also applied to vertical handoff decision in a WiMAX-WLAN environment to facilitate user with better quality of service, by N. P. Singh etc [6]. Zhang etc. present a hybrid QoS-clustering web services composition algorithm based on multi-attribute decision making theory, to solve the web service selection problem with hybrid QoS. This algorithm takes all possible QoS expression types, such as real numbers, interval numbers, triangular fuzzy numbers, into consideration [7]. In addition, Guo etc. generalize and formulate multi-service oriented devices aggregation problem based on QoS-metric with multi-attribute decision making. A multi-service oriented devices aggregation mechanism, which concludes two processes dimensionless processing and device aggregation approach, is introduced [8].

On the other hand, MAUT is also employed for quantitative evaluations. A multi-attribute decision-making method is provided by Yu etc., to identify the key nodes in complex networks. Each node is regarded as a solution, and each importance evaluation criterion as one solution's attribute [9]. Li etc. introduce a dynamic trust quantization model with multiple decision factors based on information entropy. The multiple decision factors include direct trust, risk function, feedback trust and so on, which are incorporated to reflect trust relationship's complexity and uncertainty [10]. Liu etc. propose a multi-attribute ranking method based on the technique for order preference by similarity to ideal object, so as to evaluate the node importance comprehensively [11].

However, there is little detailed information in current literature that the MAUT has been applied to address the problem on how to quantitatively evaluate and select satellite network control protocol. As a result, it makes sense to provide a quantitative evaluation method of SNCP based on multi-attribute utility theory.

3 Satellite network control protocol attributes analysis and acquisition

This section devotes to introducing critical protocol attributes and the ways on how to acquire these attributes.

3.1 Protocol attributes

Satellite network control protocol attributes conclude many aspects. However, for the reason of evaluating the protocol, SNCP-QEM concerns on the protocol attributes listed in Table I, such as performance, reliability, implement difficulty, development period, investment, maintainability.

These protocol attributes are classified into two groups. One is 'qualitative attribute', and the other is 'quantitative attribute'. The 'qualitative' attributes cannot be precisely measured by values. It comprises implement difficulty, maintainability. On the other side, the 'quantitative attributes' can be assigned with precise values. Attributes performance and reliability belong to the category 'quantitative attribute'.

Table 1. Satellite Network Control Protocol Attributes

Attributes	Qualitative or Quantitative	From Where (Source)	Importance Extent
Performance	Quantitative	Model Prediction	★★★★★
Reliability	Quantitative	Model Prediction	★★★★★
Implement Difficulty	Qualitative	Scheme Evaluation	★★
Development Period	Quantitative	Scheme Evaluation	★★☆
Investment	Quantitative	Scheme Evaluation	★★★
Maintainability	Qualitative	Scheme Evaluation	★★★★

Where do these attributes come from? It is about how to get the values of these attributes. In our method, some protocol attributes values reside on scheme evaluation, which means that the protocol design scheme determines the attribute values. For example, implement difficulty, development period and investment are evaluated by protocol design scheme. Meanwhile, the other attributes lie on model prediction. We propose a quantitative model to predict some critical protocol attributes, such as performance, reliability.

Besides, Table I shows the importance extent for each attributes. Importance extent reveals how important the protocol attribute is for satellite network control protocol. It is thought that the attributes performance, reliability are the most important attributes, so they are assigned with five stars. And the implement difficulty and investment are less important. In the following part of this paper, the importance extent of protocol attributes determines the weight vectors in the quantitative evaluation method.

3.2 Model predicted protocol attributes

To obtain the predicted protocol attributes, we introduce a satellite network control protocol quantitative GSPN model.

DEFINITION 1. (Generalized Stochastic Petri Nets) A 8-tuple model $N=(P,T,\prod,I,O,H,M_0,W)$ is used to represent generalized stochastic Petri nets. The definition of the 8 elements can see the reference [12].

DEFINITION 2. (Satellite Network Control Protocol Quantitative GSPN Model) A kind of GSPN model $N_{CS}=(P_{CS},T,\prod,I_{CS},O_{CS},H,M_0,W)$ is used to describe satellite communication system network control protocol, where

1) $P_{CS}=P_I\cup P_S$ and $P_I\cap P_S=\varnothing$. P_I is a kind of signaling place, which is transmitted between different satellite signaling protocol nodes (between satellite and ground or between two satellites) through satellite physical channels to interchange network control information.

2) $I_{CS}=I_S\cup I_I$ and $I_S\cap I_I=\varnothing$. I_I is a kind of internal normal arcs in satellite protocol nodes $I_I\subseteq(P_I\times T)$. I_S is a kind of receive signaling arcs, which are connected with signaling place P_S ($I_S\subseteq(P_S\times T)$). In other side, $O_{CS}=O_S\cup O_I$ and $O_S\cap O_I=\varnothing$. O_I is a kind of internal normal arcs $O_I\subseteq(T\times P_I)$. O_S is a kind of send signaling arcs, which is connected with signaling place P_S ($O_S\subseteq(T\times P_S)$).

By utilizing the rules of this modeling method, a typical protocol example 'establishment procedure of communication connection' is mapped to GSPN model N_{CC} shown in Fig. 2. This model describes the whole control flow and protocol signaling interaction behaviors. Signaling places P_S in N_{CC} are marked with gray rounds. And sub-models which are indicated with dashed frames respectively correspond to ground user terminal 1, satellite 1, satellite 2 and ground user terminal 2.

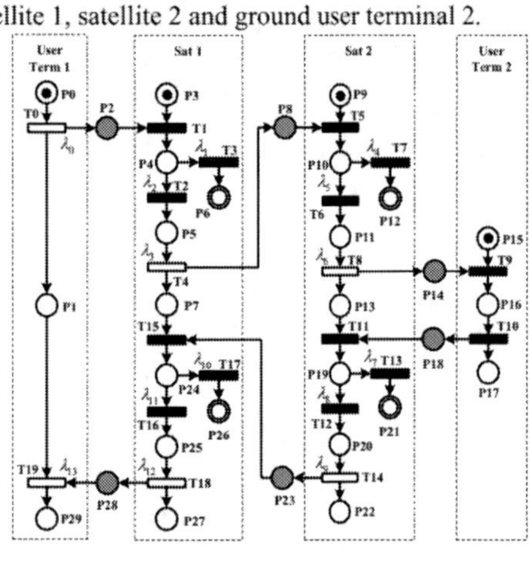

Figure 2. GSPN Model NCC of Protocol Example

We assign transition weights of this quantitative model N_{CC} with specific values according to the design scheme of SNCP, in order to get quantitative analysis results. The quantitative analysis procedure relies on GSPN tools PIPE2, which can give out the predictions of protocol performance and reliability. The detail of quantitative analysis is presented in section IV.

3.3 Scheme evaluated protocol attributes

Scheme evaluated protocol attributes primarily depend upon SNCP design scheme. For example, the protocol attribute investment indicates that how much money the design scheme will cost during the whole development phase. Attribute implement difficulty is estimated by analyzing the development scheme on the difficulty extent of protocol development. Development period shows how much time the design scheme needs to finish the protocol development task.

4 Protocol quantitative evaluation method based on multi-attribute utility theory

This section focuses on the MAUT based algorithm of protocol quantitative evaluation method and its computational complexity.

4.1 Introduction of protocol quantitative evaluation method

The SNCP quantitative evaluation algorithm employs the MAUT that the selected protocol solution shall have the least disparity with the best protocol solution A^+ (ideal solution) and the largest disparity with the worst protocol solution A^- (anti-ideal solution) [1]. The utility value of each protocol attribute mentioned in section III is monotonic. For example, if attribute $x_i\le x_j$, the utility function *f(x)* has the relation $f(x_i)\le f(x_j)$. Moreover, the best protocol solution is composed of all the best protocol attributes, and the worst protocol scheme is composed of all the worst protocol attributes. The essential of this algorithm is to regard the SNCP quantitative evaluation and solution selection method of m protocol solution alternatives and n protocol attributes as a geometry system constituting of m points in n-dimensional space. Meanwhile, the key point is how to select a protocol solution which owns the least Euclidean distance with the ideal solution.

4.2 Quantitative evaluation algorithm

Suppose that there are m ($m\ge 2$) protocol design solution alternatives. We have to carefully choose one of them as the final solution by evaluating protocol solution with multiple attributes utility. Define vectors $A_1,A_2,\cdots,A_i,\cdots,A_m$ ($1\le i\le m$) as theses m protocol solution alternatives. Thereupon, protocol alternative i is expressed with $A_i=(x_{i1},x_{i2},\cdots,x_{ij},\cdots,x_{in})$, where $x_{i1},x_{i2},\cdots,x_{ij},\cdots,x_{in}$ are the n different attributes in protocol solution alternative i. In addition, define these n ($n\ge 2$) protocol attributes as $X_1,X_2,\cdots,X_j,\cdots,X_n$ ($1\le j\le n$), then the attribute j vector is $X_j=(x_{1j},x_{2j},\cdots,x_{ij},\cdots,x_{mj})$, where $x_{1j},x_{2j},\cdots,x_{ij},\cdots,x_{mj}$ are the attribute values of one protocol attribute in different protocol solution alternatives.

In this protocol quantitative evaluation method, the protocol attributes are classified into two categories: efficiency attributes and cost attributes. The more the values of efficiency attributes are, the better the quantitative evaluation result is. For example, performance belongs to efficiency attributes. Whereas, the cost attributes are reversed. The more the values of cost attributes are, the worse the evaluation result is. Attributes implement difficulty; investment and development period are cost attributes.

The quantitative evaluation algorithm consists of the following four steps.

STEP 1. Build a weight normalized decision matrix **V**.

Build m protocol solution alternatives and n protocol attributes decision matrix **D**.

$$D = \begin{matrix} \\ A_1 \\ A_2 \\ \vdots \\ A_i \\ \vdots \\ A_m \end{matrix} \begin{matrix} X_1 & X_2 & \cdots & X_j & \cdots & X_n \\ \left[\begin{matrix} x_{11} \\ x_{21} \\ \vdots \\ x_{i1} \\ \vdots \\ x_{m1} \end{matrix}\right. & \begin{matrix} x_{12} \\ x_{22} \\ \vdots \\ x_{i2} \\ \vdots \\ x_{m2} \end{matrix} & \begin{matrix} \cdots \\ \cdots \\ \\ \cdots \\ \\ \cdots \end{matrix} & \begin{matrix} x_{1j} \\ x_{2j} \\ \vdots \\ x_{ij} \\ \vdots \\ x_{mj} \end{matrix} & \begin{matrix} \cdots \\ \cdots \\ \\ \cdots \\ \\ \cdots \end{matrix} & \left.\begin{matrix} x_{1n} \\ x_{2n} \\ \vdots \\ x_{in} \\ \vdots \\ x_{mn} \end{matrix}\right] \end{matrix} \quad (1)$$

Then, do the operation of attribute weight normalization to the decision matrix **D**. This step completes that the original attribute values are transformed to dimensionless attribute values, in order to lay the foundation for comparing protocol attribute values. Calculate new normalized decision matrix **R**.

$$R = \begin{bmatrix} r_{11} & r_{12} & \cdots & r_{1j} & \cdots & r_{1n} \\ r_{21} & r_{22} & \cdots & r_{2j} & \cdots & r_{2n} \\ \vdots & \vdots & & \vdots & & \vdots \\ r_{i1} & r_{i2} & \cdots & r_{ij} & \cdots & r_{in} \\ \vdots & \vdots & & \vdots & & \vdots \\ r_{m1} & r_{m2} & \cdots & r_{mj} & \cdots & r_{mn} \end{bmatrix} \quad (2)$$

Where $r_{ij} = x_{ij} \Big/ \sqrt{\sum_{i=1}^{m} x_{ij}^2}$. So each protocol attributes have the same vector length.

Furthermore, make matrix **R** weighted, and build weight normalized decision matrix **V**. Let $\vec{w} = (w_1, w_2, \cdots, w_j, \cdots, w_n)$ as attribute weight vector, where $\sum_{j=1}^{n} w_j = 1$. Matrix **V** is acquired by multiplying each column of matrix **R** with related weight vector w_j . Thereby, weight normalized decision matrix **V** equals as follows.

$$V = \begin{bmatrix} w_1r_{11} & w_2r_{12} & \cdots & w_jr_{1j} & \cdots & w_nr_{1n} \\ w_1r_{21} & w_2r_{22} & \cdots & w_jr_{2j} & \cdots & w_nr_{2n} \\ \vdots & \vdots & & \vdots & & \vdots \\ w_1r_{i1} & w_2r_{i2} & \cdots & w_jr_{ij} & \cdots & w_nr_{in} \\ \vdots & \vdots & & \vdots & & \vdots \\ w_1r_{m1} & w_2r_{m2} & \cdots & w_jr_{mj} & \cdots & w_nr_{mn} \end{bmatrix}$$

$$= \begin{bmatrix} v_{11} & v_{12} & \cdots & v_{1j} & \cdots & v_{1n} \\ v_{21} & v_{22} & \cdots & v_{2j} & \cdots & v_{2n} \\ \vdots & \vdots & & \vdots & & \vdots \\ v_{i1} & v_{i2} & \cdots & v_{ij} & \cdots & v_{in} \\ \vdots & \vdots & & \vdots & & \vdots \\ v_{m1} & v_{m2} & \cdots & v_{mj} & \cdots & v_{mn} \end{bmatrix} \quad (3)$$

STEP 2. Define protocol ideal solution and anti-ideal solution.

Suppose that collection J is efficiency attributes collection, and J' is cost attributes collection. Define satellite network control protocol ideal solution A^+ and anti-ideal solution A^-.

$$A^+ = \left\{ \left(\max_i v_{ij} \mid j \in J \right), \left(\min_i v_{ij} \mid j \in J' \right) \mid i \in M \right\} = \left\{ v_1^+, v_2^+, \cdots, v_j^+, \cdots, v_n^+ \right\} \quad (4)$$

$$A^- = \left\{ \left(\min_i v_{ij} \mid j \in J \right), \left(\max_i v_{ij} \mid j \in J' \right) \mid i \in M \right\} = \left\{ v_1^-, v_2^-, \cdots, v_j^-, \cdots, v_n^- \right\} \quad (5)$$

Thereby, protocol ideal solution A^+ and anti-ideal solution A^- stand for the most preferable and the least preferable protocol solution.

STEP 3. Compute the distances between protocol solution alternatives and the two protocol ideal solutions (ideal and anti-ideal solutions).

In this step, compute the Euclidean distances between each protocol solution and ideal solution. Euclidean distance means the real distance of two points in n-dimensional space. The distance of each solution A_i is measured by n-dimensional Euclidean distance. As a result, the distance between protocol alternative and ideal solution and the distance between protocol alternative and anti-ideal solution are presented as follows.

$$S_{i^+} = \sqrt{\sum_{j=1}^{n} (v_{ij} - v_j^+)^2} \text{ and } S_{i^-} = \sqrt{\sum_{j=1}^{n} (v_{ij} - v_j^-)^2} \quad (6)$$

Next, calculate the relative distance between each solution alternative A_i and ideal solution A^+. Let C_{i^+} be the distance between A_i and A^+.

$$C_{i^+} = S_{i^-} \big/ \left(S_{i^+} + S_{i^-} \right), 0 < C_{i^+} < 1 \quad (7)$$

According to the above definition, if $A_i = A^+$, then $C_{i^+} = 1$; if $A_i = A^-$, then $C_{i^+} = 0$. When C_{i^+} is approaching 1, the solution alternative A_i is approaching ideal solution A^+.

$$v(A_i)(\vec{w}_j) = C_{i^+} \quad (8)$$

In SNCP-QEM, C_{i^+} is regarded as the quantitative value of QEM, which is also symbolized with $v(A_i)(\vec{w}_j)$ in this paper. Depend upon these values, we can determine how to quantitatively evaluate satellite network control protocol in different situations.

STEP 4. Rank preference order according to the distances C_{i^+} of each solution alternative.

The protocol solution collection { A_i } is ranked in descending order according to the distances C_{i^+} . Then the solution preference order of protocol solutions is achieved. In accordance with the order result, it is helpful for users to select the most appropriate protocol solution from many alternatives to meet some specific requrements or conditions.

4.3 Computational complexity analysis

Formula (1) is an ergodic process of all the elements x_{ij} in decision matrix **D**, and the computational complexity is about $O(m \times n)$. In formula (2), the decision matrix **D** is normalized into decision matrix **R**. The computational complexity of computing $\sqrt{\sum_{i=1}^{m} x_{ij}^2}$ for each protocol attribute is $O(m)$. Then, for the n protocol attributes, the computational complexity of formula (2) becomes $O(m \times n) + O(m \times n)$. Formula (3) calculates the weight normalized decision matrix **V**, whose computational complexity is also $O(m \times n)$. So the computational complexity in STEP 1 is $O(m \times n)$. In SETP 2, calculating the maximum and minimum values ($\max_i v_{ij}$ and $\min_i v_{ij}$) for each protocol attribute costs around $O(m)$ computing time. This operation repeats for all of the n protocol attributes. The computational complexity of STEP 2 including formula (4) and (5) is $O(2 \times m \times n) = O(m \times n)$. In STEP 3, formula (6) generates the computational complexity $O(n)$. Formula (7), that computes the distance between each protocol solution alternative and ideal solution A^+, has the computational complexity of $O(m \times n)$. Thereby, the computational complexity in STEP 3 is $O(m \times n)$. In SETP 4, if we choose high efficient fast sort algorithm, the computational complexity is $O(n \log n)$.

Above all, the computational complexity of the whole protocol quantitative evaluation method is $O(3 \times m \times n) + O(n \log n) = O(m \times n)$. It means that the proposed method owns high execution efficiency.

5 Experiment and discussion

This section centers on protocol attributes evaluation and algorithm effectiveness evaluation. The typical protocol example establishment procedure of communication connection, mentioned in section III is taken for case study.

5.1 Protocol attributes evaluation

Assume that there are 6 protocol solution alternatives, which are represented by A_1, A_2, A_3, A_4, A_5, A_6. Table II and III show the values of model predicted protocol attributes and scheme evaluated protocol attributes. The predicted attribute values are achieved by use of protocol quantitative GSPN model N_{CC}. The model employs different protocol solutions as model inputs, and gets the predictions of protocol attributes performance and reliability. Meanwhile, the scheme evaluated attribute values mainly depends on protocol design schemes.

As a result, we can get the conclusions as follows: (1) By employing Single Attribute Decision Making (SADM) method, the solution rank result is $A_4>A_6>A_5>A_2>A_1>A_3$ according to the efficiency attribute performance; Meanwhile, the rank result is $A_1>A_3>A_2>A_5>A_6>A_4$ according to the cost attribute reliability. (2) Different solutions have contradictory protocol attribute values. For example, protocol solution A_4 has the highest performance but the worst reliability; Protocol solution A_1 has the best reliability, but worse performance. (3) It becomes troublesome, when we have to select one protocol solution from a large number of solutions that own great protocol attribute diversity. These protocol attributes conflict with each other, that increases the difficulty of evaluating protocol solution and even solution selection procedure, as Fig.3 shows. For example, if we need a protocol solution which has high performance and reliability, it is obvious that more investment and development period are needed. If the foundation is limited, the solution decreases the performance and reliability as a sacrifice. So it is difficult to find out an appropriate solution from these tanglesome protocol solutions.

Table 2. Model Predicted Protocol Attributes

Attributes / Solution	Model Predicted Attributes performance (X_1)	reliability (X_2)
1	1.326	1.313
2	1.481	1.884
3	1.294	1.604
4	1.654	2.201
5	1.578	2.066
6	1.584	2.076

Table 3. Scheme Evaluated Protocol Attributes

Attributes / Solution	Scheme Evaluation Attributes			
	Implement Difficulty (X_3)	Development Period (X_4)	Investment (X_5)	Maintainability (X_6)
1	4	12	65	5
2	6	10	46	9
3	5	8	52	2
4	7	7	70	6
5	8	5	87	3
6	2	9	75	4

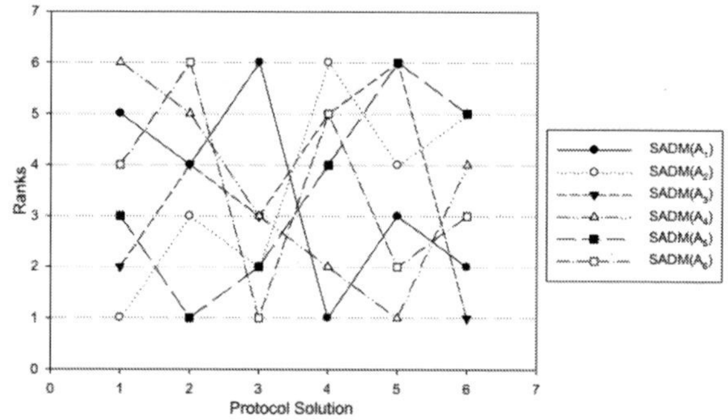

Figure 3. Solution Rank Result (SADM)

In one word, it is necessary to propose a quantitative evaluation method for several protocol solutions with several mutual-conflicted attribute values.

5.2 Algorithm effectiveness evaluation

Assume the following 6 weight vector $\vec{w}_i$ categories. (1) prefer **ONE** protocol attribute. Prefer performance (X_1): $\vec{w}_1 = (0.5, 0.1, 0.1, 0.1, 0.1, 0.1)$; prefer reliability (X_2): $\vec{w}_2 = (0.1, 0.5, 0.1, 0.1, 0.1, 0.1)$; prefer implement difficulty (X_3): $\vec{w}_3 = (0.1, 0.1, 0.5, 0.1, 0.1, 0.1)$; prefer development period (X_4): $\vec{w}_4 = (0.1, 0.1, 0.1, 0.5, 0.1, 0.1)$; prefer investment (X_5): $\vec{w}_5 = (0.1, 0.1, 0.1, 0.1, 0.5, 0.1)$; prefer maintainability (X_6): $\vec{w}_6 = (0.1, 0.1, 0.1, 0.1, 0.1, 0.5)$. (2) prefer **TWO** protocol attributes. Prefer X_1 & X_2: $\vec{w}_7 = (0.3, 0.3, 0.1, 0.1, 0.1, 0.1)$; prefer X_1 & X_3: $\vec{w}_8 = (0.3, 0.1, 0.3, 0.1, 0.1, 0.1)$; prefer X_2 & X_5: $\vec{w}_9 = (0.1, 0.3, 0.1, 0.1, 0.3, 0.1)$. (3) prefer **THREE** protocol attributes. Prefer X_1, X_2 & X_3: $\vec{w}_{10}=(0.233,0.233,0.233,0.1,0.1,0.1)$; prefer X_1, X_2 & X_4: $\vec{w}_{11}=(0.233,0.233,0.1,0.233,0.1,0.1)$; prefer X_3, X_4 & X_5: $\vec{w}_{12}=(0.1,0.1,0.233,0.233,0.233,0.1)$. (4) prefer **FOUR** protocol attributes. Prefer X_1, X_2, X_3 & X_4: $\vec{w}_{13}=(0.2,0.2,0.2,0.2,0.1,0.1)$; prefer X_1, X_2, X_3 & X_5: $\vec{w}_{14}=(0.2,0.2,0.2,0.1,0.2,0.1)$; prefer X_2, X_3, X_4 & X_6: $\vec{w}_{15}=(0.1,0.2,0.2,0.2,0.1,0.2)$. (5) prefer **FIVE** protocol attributes. Prefer X_1, X_2, X_3, X_4 & X_5: $\vec{w}_{16}=(0.18,0.18,0.18,0.18,0.18,0.1)$; prefer X_2, X_3, X_4, X_5 & X_6: $\vec{w}_{17}=(0.1,0.18,0.18,0.18,0.18,0.18)$. (6) prefer **ALL** protocol attributes. Prefer X_1, X_2, X_3, X_4, X_5 & X_6: $\vec{w}_{18}=(0.166,0.166,0.166,0.166,0.166,0.166)$. These six categories of weight vectors represent that the protocol designers lay different stress on protocol attributes, and pay attention to certain protocol design aspects.

On the basis of QEM method introduced in Section IV, we apply the above weight vectors to evaluation procedure, and get the evaluation results as Table IV shows. In Table IV, 12 weight vectors are adopted for quantitative evaluation. The larger the QEM value is, the better the protocol solution is under the same weight vector. Thereupon, the solution prefer rank results under corresponding weight vectors are: $A_6>A_5>A_3>A_4>A_1>A_2$, $A_1>A_3>A_6>A_5>A_2>A_4$, $A_6>A_1>A_3>A_2>A_4>A_5$, $A_3>A_1>A_6>A_5>A_4>A_2$, $A_3>A_1>A_2>A_6>A_5>A_4$, $A_6>A_1>A_3>A_5>A_2>A_4$, $A_6>A_3>A_1>A_5>A_2>A_4$, $A_6>A_3>A_1>A_5>A_4>A_2$, $A_3>A_6>A_5>A_1>A_4>A_2$, $A_3>A_6>A_1>A_5>A_4>A_2$, $A_3>A_6>A_1>A_5>A_4>A_2$, $A_6>A_5>A_3>A_4>A_1>A_2$.

1. Algorithm Effectiveness Evaluation

a) SADM vs. QEM

The evaluation contrasts the effectiveness of SADM algorithm and QEM algorithm. The prefer rank results are shown in Fig. 4. By contrast with SADM rank results according to single attribute performance or reliability (real line in Fig. 4), QEM rank results (dash line in Fig. 4) are quite different, because QEM method considers not only the single attribute performance or reliability but also the other five protocol attributes. However, SADM method focuses on only one attribute and does not put any other protocol attributes into consideration. For example, the solution A_4 rank of SADM(X_1) is 1, and its rank of QEM(X_1) is 4. Although solution A_4 has the best performance, the other protocol attributes such as reliability are not good. So the solution A_4 rank in QEM method decreased to 4 for the reason of considering other protocol attributes.

In conclusion, the QEM is much more balanced than SADM method, by putting all of the protocol attributes into consideration on the basis of emphasizing one or more attributes.

b) QEM vs. QEM (different categories of weight vectors)

Fig. 4 illustrates the effectiveness comparisons of SADM and QEM methods, and Fig. 5 shows the QEM evaluation effectiveness of different weight vector categories, such as Fig. 5(a) indicates the QEM value comparisons between prefer one attribute category and prefer two attribute category. We can draw the conclusions that: (1) The QEM values are quite different between different categories, because the weight vectors depend on these different preferences of users. And the user preferences represent different concern points on SNCP. (2) From Fig. 5(b)(c)(d), the QEM method not only focuses on the protocol attributes that user prefers, but also puts all the other attributes into all-around considerations. For example, in Fig. 5(c) the QEM value of solution A_2 in QEM(X_1, X_2 & X_3) under weight vector $\vec{w}_{10}$ is $v(A_2)(\vec{w}_{10}) = 0.3340$, which is different from the value ($v(A_2)(\vec{w}_{11}) = 0.4695$) of solution A_2 in QEM(X_1, X_2 & X_5) under weight vector $\vec{w}_{11}$. Since that the attribute X_5 value of A_2 is high and the attribute X_3 value is low, the QEM value of solution A_2 in QEM(X_1, X_2, X_3 & X_5) under weight vector $\vec{w}_{14}$ achieves a tradeoff ($v(A_2)(\vec{w}_{11}) < v(A_2)(\vec{w}_{14}) = 0.4135 < v(A_2)(\vec{w}_{10})$). (3) From Fig. 5(a) to Fig. 5(e), the line diagrams get into a balance. Fig. 5(a) presents mixed and disordered status, while Fig. 5(e) (prefer all of the attributes) presents a kind of unified status. When all of the protocol attributes are evenly put into consideration (QEM(X_1, X_2, X_3, X_4, X_5 & X_6) with $\vec{w}_{18}$), the QEM values reach a steady state. As a result, such steady QEM values can be adopted as the basis for the protocol solution selection.

Above all, this evaluation vividly exhibits the effectiveness of SNCP-QEM method. Different values reside on the different weight vectors, which reflect the concern points of users. And SNCP-QEM supplies a theoretical solution selection way.

2. Algorithm Performance Evaluation

The performance test environment is listed as follows: Intel Core i7-2600 CPU 3.4GHz, 8GB RAM, Windows 7 Ultimate 64bit SP1, and JDK 7u21.

Fig. 6 illustrates the QEM performance evaluation results by using a 3-dimention mesh diagram. This mesh diagram uses the numbers of protocol solution alternatives (m) and protocol attributes (n) as independent variables, and uses average execution time (t) as dependent variable. In the diagram, X axis stands for solution alternative number; Y axis stands for protocol attribute number; and Z axis stands for average execution time (ns). The average execution time of QEM is from 9.39×10^2 to 7.75×10^3 ns, when $3 \leq m \leq 10$ and $3 \leq n \leq 10$. It is concluded that the algorithm of SNCP-QEM owns high execution efficiency. Meanwhile, the average execution time grows up as the numbers of solution alternatives and protocol attributes increase. As a result, it is verified by this performance evaluation that the computational complexity of QEM satisfies $O(m\times n)$.

In one word, the QEM method, proposed in this paper has high enough performance to satisfy the quantitative evaluation requirements.

6 Conclusion

To sum up, this paper proposes a MAUT-based quantitative evaluation method of SNCP, in order to provide some theoretical and quantitative evidence for SNCP solution selection. The method involves several critical protocol attributes, such as performance,

 DOI: 10.1051/matecconf/20165605007

reliability, implement difficulty. Some protocol attributes performance and reliability are predicted by use of a satellite network control protocol quantitative GSPN model. Then, QEM aggregates all of the protocol attributes from different protocol solutions into a decision matrix. By utilizing the matrix as the input of quantitative evaluation algorithm, it employs user decision-making related attribute weight vectors to achieve quantitative evaluation values and corresponding solution selection ranks. Consequently, it offers fundamental supports and suggestions for selecting SNCP solution. In the end, the effectiveness and performance evaluation of SNCP-QEM give the conclusion that SNCP-QEM has a remarkable effect in quantitatively evaluating SNCP and a efficient execution procedure.

Table 4. Protocol Quantitative Evaluation Result Under 12 Group Weight Vector By SNCP-QEM

Weight Vector / Solution	Prefer 1 Attributes			Prefer 2 Attributes		Prefer 3 Attributes		Prefer 4 Attributes		Prefer 5 Attributes		Prefer All Attributes
	$\vec{w}_1$	$\vec{w}_2$	$\vec{w}_3$	$\vec{w}_7$	$\vec{w}_9$	$\vec{w}_{10}$	$\vec{w}_{12}$	$\vec{w}_{13}$	$\vec{w}_{14}$	$\vec{w}_{14}$	$\vec{w}_{15}$	$\vec{w}_{16}$
1	0.4346	0.7026	0.6406	0.5852	0.5963	0.6039	0.4757	0.4969	0.5206	0.5016	0.5211	0.5177
2	0.3635	0.3382	0.3311	0.342	0.5143	0.334	0.4095	0.3244	0.2525	0.3892	0.3194	0.3210
3	0.5391	0.6844	0.5319	0.6323	0.7324	0.5882	0.6176	0.6006	0.7001	0.6284	0.7064	0.6973
4	0.5289	0.2509	0.2157	0.3744	0.3491	0.3052	0.4	0.3922	0.3953	0.395	0.3970	0.4032
5	0.5592	0.3792	0.2071	0.4658	0.3661	0.3495	0.4312	0.4590	0.5387	0.4449	0.5162	0.5178
6	0.6694	0.4123	0.8653	0.5267	0.4565	0.6927	0.6407	0.6447	0.6611	0.6177	0.6372	0.6383

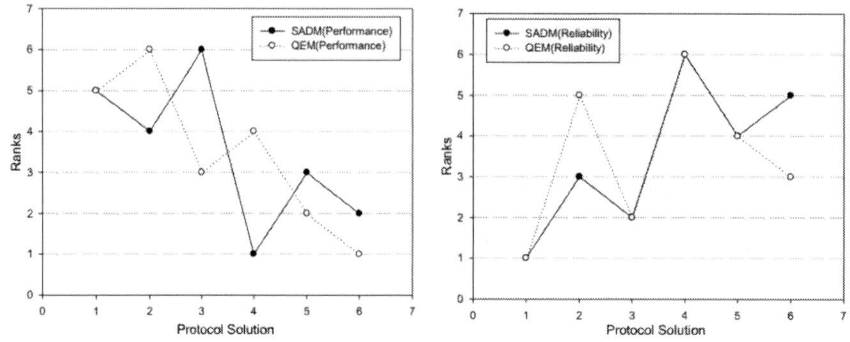

Figure 4. SADM vs. QEM Rank Result (Performance and Reliability)

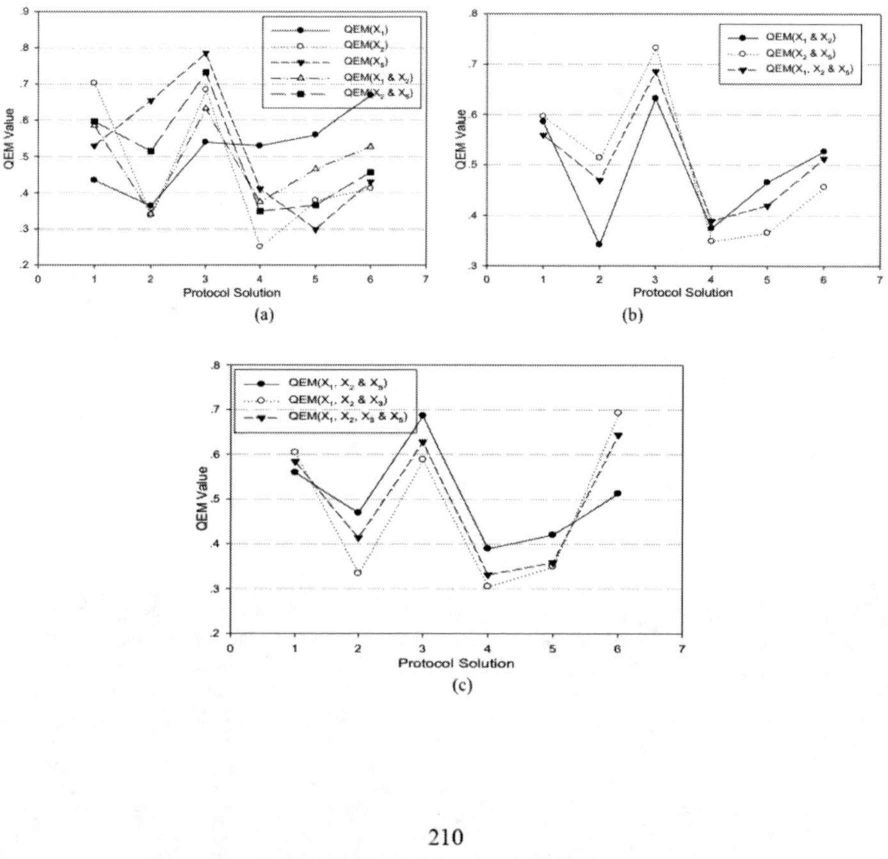

 DOI: 10.1051/matecconf/20165605007

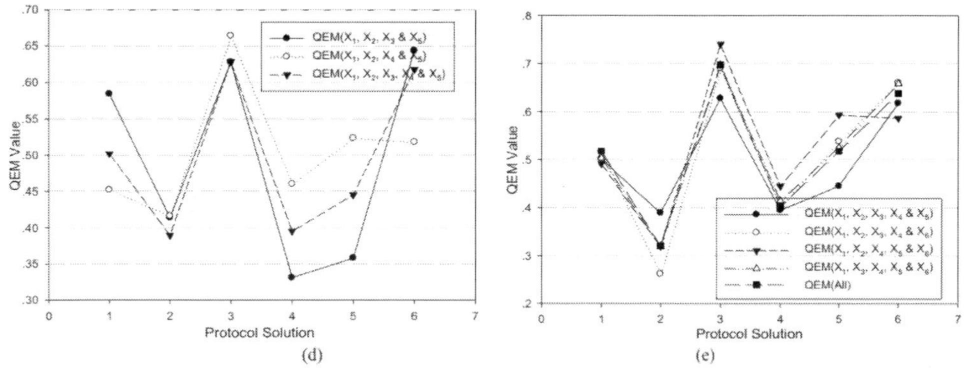

Figure 5. SNCP-QEM Results (different categories of weight vectors)

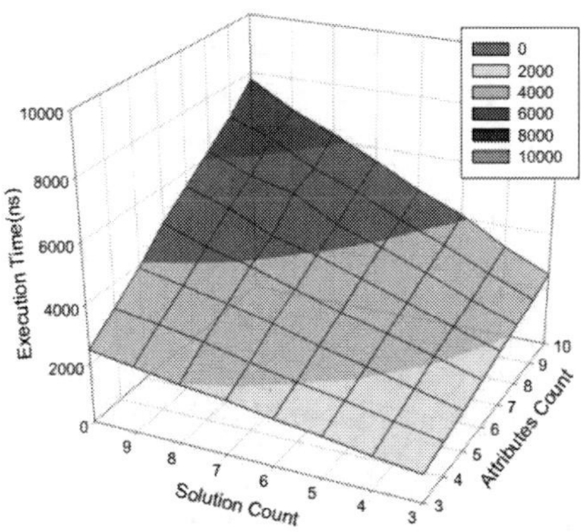

Figure 6. SNCP-QEM Performance Evaluation

Acknowledgment

The authors thank their current research sponsors, the National Natural Science Foundation of China (Grant 61301158). They are also very thankful to their past and present colleagues in Nanjing Telecommunication Technology Research Institute, who contributed to the research described in this paper.

References

1. C. L. Hwang, Y. J. Lai, T. Liu, Springer-Verlag (1981)
2. K. Yoon, Ph.D. dissertation, Kansas State University, Manhattan (1980)
3. C. Girault, R. Valk., Springer Verlag (2003)
4. L. Zhang, Q. Zhu, J. of Sig. Proc., **30**, 10 (2014)
5. Dimitris E. Charilas, Athanasios D. Panagopoulos, Ourania I. Markaki, Wireless Pers. Comm., 74, 147–165 (2014)
6. N. P. Singh, and Brahmjit Singh, Wireless Network, **20**, 1203–1211 (2014)
7. L. Zhang, C. Zhang, J. of Beijing Univ. of Posts and Telecom., **34**, 5 (2011)
8. S. Guo, C. Zhong, F. Liu, F. Zheng, X. Qiu, China Comm., 1 (2014)
9. H. Yu, Z. Liu, Y. Li, Acta Phys. Sin., **62**, 2 (2013)
10. X. Li, X. Gui, Chinese J. of Computers, **32**, 3 (2009)
11. Z. Liu, C. Jiang, J. Wang, H. Yu, Knowledge-Based Systems, **84**, 56–66 (2015)
12. M.A. Marsan, G. Balbo, John Wiley & Sons (1995)

Interference Alignment for the Uplink Heterogeneous Cellular Networks

Yingyuan Gao, Gang Wang, and Siqian Cui

Communication Research Center, Harbin Institute of Technology, Harbin, China

Abstract. Heterogeneous cellular networks, due to its multi-tier topological structure, provide significant improvements in terms of increased data rates and cell coverage. However, there are important intra-tier and inter-tier interference management problem to be solved. In this paper, we model a two-tier uplink heterogeneous network, which include a macrocell and K overlaid femtocells. Meanwhile we constitute a heterogeneous configuration that each femto BS is equipped with two antennas and the macro BS is equipped with $A(2 \le A \le 2K)$ antennas. We proved that the sum-DoF outer bound of the network is $4K^2/4K - A$, and the DoF outer bound for the femtocell and macrocell is $(2K-A)/(4K-A)$ and $A/(4K-A)$ respectively. Moreover, we present the achievable scheme, which is based on interference alignment. The simulation results show that the sum rate of the whole HetNet increases as the number of macro BS's antennas increases.

1 Introduction

As Heterogeneous Network (HetNet) [1] provides flexible and efficient topology to effectively improve spectral efficiency and avoid coverage holes, modern cellular network deployments are transitioning from largely homogeneous (one-tier) deployments to highly heterogeneous networks (two-tier) comprised of different low-cost low-power base station nodes (LPN) such as picocells, femtocells and distributed antenna systems [2-4]. Although the LPNs provide better quality of service (QoS) for mobile devices in comparison with a conventional one-tier cellular network, they also bring a serious cross-tier interference between tier-1 and tier-2 when they share the same licensed band [5]. Thus, to both fully exploit the possible gains through HetNet deployments and control the interference between tiers, we need to utilize the differences in base station types and change the conventional single-layer homogenous networks approach to include these differences.

Interference alignment, as an advanced beamforming technique has been utilized in a variety of networks, such as K user point-to-point interference channel [6], X channel [7] and conventional cellular downlink networks [8]. The key idea of interference alignment is to consolidate the interference signals from the undesired sources as much as possible while separate the desired signal from interference at the desired receiver. Interference alignment approach has also been applied to MIMO cellular networks in [8-11]. However, these works have focused on two-cell configuration with one data stream for each user [8], [9], and their extension to general MIMO cellular networks (with arbitrarily number of cells) and to the two-tier heterogeneous network is highly non-trivial.

In this paper, we construct a two-tier uplink heterogeneous network, which include a macrocell and K overlaid femtocells. We assume each femtocell has two single antenna mobile users. Meanwhile we constitute a heterogeneous configuration that each femto BS is equipped with two antennas and the macro BS is equipped with $A(2 \le A \le 2K)$ antennas. We analysed the sum-DoF outer bound of the network is $\frac{4K^2}{4K-A}$, in particular, the DoF outer bound of each message for the femto BS and macro BS is $\frac{2K-A}{4K-A}$ and $\frac{A}{4K-A}$ respectively. Moreover, we present the achievable scheme, which is based on interference alignment. Finally, the simulation results show that the sum rate of the whole HetNet increases as the number of macro BS's antennas increases.

The rest of this paper is organized as follows: In Section II, the system model is presented. Then the DoF outer bound is analyzed in Section III. The achievable scheme based on interference alignment is stated in detail in Section IV. Section V shows the simulation results. Finally, the conclusions are drawn in Section VI.

2 System Model

In this section, we describe the uplink HetNet as shown in Fig. 1, which is composed of one macrocell and K overlaid femtocells. All cells use the same frequency band. Each femto BS $i\left(i\in\{1,2,\cdots,K\}\triangleq\mathcal{K}\right)$ equipped with two antennas serves two femto users. The macro BS 0 which equipped with $A(2\le A<2K)$ antennas serves all the femto users simultaneously. Specifically, each femto user has independent message $W_u^{[i]}$ and $W_{iu}^{[0]}$ intended for the femto BS i in the femtocell and macro BS 0, respectively, where $u\in\{1,2\}$. This uplink HetNet is fully connected and is characterized by the following input-output relationship:

$$\mathbf{Y}^{[j]}=\sum_{i=1}^{K}\sum_{u=1}^{2}\mathbf{H}_u^{[ji]}X_u^{[i]}+\mathbf{Z}^{[j]}\quad j\in\{0,1,\cdots,K\}\qquad(1)$$

where $\mathbf{Y}^{[j]}\in\mathbb{C}^{2\times1}$, $X_u^{[i]}\in\mathbb{C}^{1\times1}$ and $\mathbf{Z}^{[j]}\in\mathbb{C}^{2\times1}$ are the output signal vector at the BS j, the symbol send from user u and the additive white Gaussian noise (AWGN) vector at BS j, respectively. We assume all noise terms are i.i.d zero mean complex Gaussian with unit variance. $\mathbf{H}_u^{[ji]}\in\mathbb{C}^{2\times1}(j\in\mathcal{K})$ is the channel coefficient matrix between femto BS j and user u in the femtocell i, and $\mathbf{H}_u^{[0i]}\in\mathbb{C}^{A\times1}$ is the channel coefficient matrix between macro BS 0 and user u in the femtocell i. We assume that all channel coefficient values are drawn i.i.d. from a continuous distribution and the absolute value of all the channel coefficients is bounded between a nonzero minimum value and a finite maximum value. Perfect knowledge of all channel coefficients is available to all BSs and users. The total power across all users is assumed to be equal to 2ρ. We indicate the size of the message set by $\left|W_u^{[i]}(\rho)\right|$ and $\left|W_{iu}^{[0]}(\rho)\right|$. For codewords spanning t_0 channel uses, the rates $R_u^{[i]}(\rho)=\frac{\log\left|W_u^{[i]}(\rho)\right|}{t_0}$ or $R_{iu}^{[0]}(\rho)=\frac{\log\left|W_{iu}^{[0]}(\rho)\right|}{t_0}$ are achievable if the probability of error for all messages can be simultaneously made arbitrarily small by choosing an appropriately large t_0. The capacity of the uplink HetNets is the set of all achievable rate tuples. We define the spatial degrees of freedom for message $W_u^{[i]}(\rho)$ and $W_{iu}^{[0]}(\rho)$ as:

$$\begin{cases}d_u^{[i]}=\lim\limits_{\rho\to\infty}R_u^{[i]}(\rho)/\log(\rho)\\ d_{iu}^{[0]}=\lim\limits_{\rho\to\infty}R_{iu}^{[0]}(\rho)/\log(\rho)\end{cases},i\in\mathcal{K},u=1,2.\quad(2)$$

Since we are interested in the symmetric DoF which is the largest value of DoF that is achievable by every message, we define all $d_u^{[i]}$ for the femto BS i as d_f and define all $d_{iu}^{[0]}$ for the macro BS 0 as d_m.

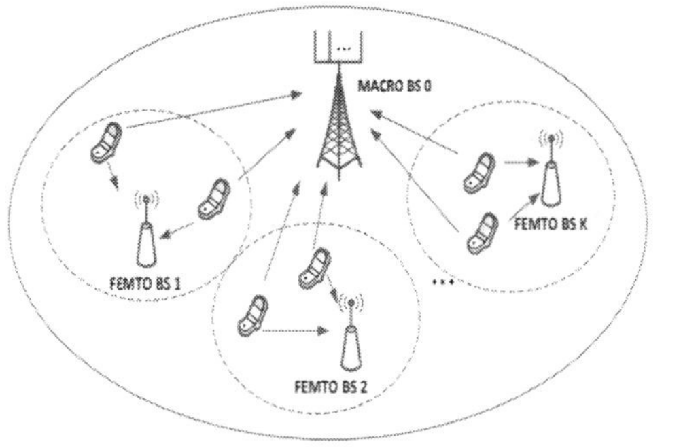

Figure 1. System model

3 DoF Outer Bound Analysis

In this section, we analyze the DoF outer bound for the uplink HetNet defined in Section II. Due to the reciprocity of linear beamforming-based alignment, which states that if interference alignment is feasible in the reciprocal network, then it is also feasible in the original network, and the achievable DoF are the same between the dual networks, we defer the DoF outer bound in the reciprocal downlink HetNet for convenience.

Firstly, consider user u in the femtocell i, $i\in\mathcal{K},u=1,2$, let us eliminate all messages except the desired messages $W_u^{[i]}$ from femto BS i and $W_{iu}^{[0]}$ from macro BS 0, and one undesired message $W_u^{[j]}$ from any other femto BS j, $(j\in\mathcal{K},j\neq i)$. Clearly, eliminating other messages cannot hurt the rates of the remaining messages. Now we argue that all three remaining messages are resolvable by user u in the femtocell i. Since the desired messages are decodable by design, user u in the femtocell i can reliably reconstruct and subtract $W_u^{[i]}$ and $W_{iu}^{[0]}$ from its received signal. This gives user u an invertible channel to femto BS j, from which it can reliably resolve message $W_u^{[j]}$. Thus, user u is able to resolve all $W_u^{[i]}$, $W_{iu}^{[0]}$ and $W_u^{[j]}$. Since the user u has only 1 antenna, the total DoF cannot be more than 1. This give us the outer bound:

$$d_u^{[i]}+d_u^{[j]}+d_{iu}^{[0]}\le1\quad i,j\in\mathcal{K},i\neq j\qquad(3)$$

which can be simplified as

$$2d_f+d_m\le1\qquad(4)$$

Secondly, when A is even, consider the first A single antenna users, the corresponding desired femto BSs of the users and the macro BS 0. If we allow full cooperation among A user and full cooperation among their corresponding femto BSs, then it is equivalent to the two users MIMO multiple-access channel (MAC) with A antennas at each node. Because the desired A messages are decodable by design, the new A antenna cooperative users can reliably reconstruct and subtract $W_u^{[1]}$ to $W_u^{[A/2]}(u=1,2)$ from its received signal. This gives the cooperative users an invertible channel to macro BS 0, since the new cooperative users has A antennas, it can reliably reconstruct all the messages from macro BS 0.

Due to allowing cooperation among transmitters does not hurt the capacity, the number of DoF achieved by all the messages is no more than A. This gives us another outer bound

$$\sum_{i=1}^{A/2}\sum_{u=1}^{2} d_u^{[i]} + \sum_{i=1}^{K}\sum_{u=1}^{2} d_{iu}^{[0]} \le A \tag{5}$$

which can be simplified as

$$Ad_f + 2Kd_m \le A \tag{6}$$

Considering (4) and (6) together, we get

$$\begin{cases} d_f = \dfrac{2K-A}{4K-A} \\ d_m = \dfrac{A}{4K-A} \end{cases} \tag{7}$$

and the sum-DoF of the network can be expressed as

$$DoF_{\text{sum}} \le \frac{4K^2}{4K-A} \tag{8}$$

When A is odd we can get the same outer bound.

4 Interference Alignment Scheme

In this section, we constitute the achievable scheme and present it in detail. The achievable scheme is based on interference alignment over multiple symbol extensions of the channel.

Consider a $M_n = (4K-A)n^N$ symbol extension of the original channel for any arbitrary $n \in N$. The value of N will be specified later. In the extended channel, we use the over-bar notation indicates quantities in the extended channel, we rewrite the input-output relationship of the network as

$$\bar{\mathbf{Y}}^{[j]} = \sum_{i=1}^{K}\sum_{u=1}^{2} \bar{\mathbf{H}}_u^{[ji]} \bar{\mathbf{X}}_u^{[i]} + \bar{\mathbf{Z}}^{[j]}, \mathcal{J} \in \{ \quad \cdots K\} \tag{9}$$

where $\bar{\mathbf{X}}_u^{[i]}$ is the $M_n \times 1$ column vector representing the M_n symbol extension of the transmitted symbol $X_u^{[i]}$, i.e.

$$\bar{\mathbf{X}}_u^{[i]} = \begin{bmatrix} X_u^{[i]}\left(M_n(t-1)+1\right) \\ X_u^{[i]}\left(M_n(t-1)+2\right) \\ \vdots \\ X_u^{[i]}\left(M_n t\right) \end{bmatrix} \tag{10}$$

Similarly, $\bar{\mathbf{Y}}^{[j]}$, $\bar{\mathbf{Z}}^{[j]}$ represent M_n symbol extensions of $\mathbf{Y}^{[j]}$ and $\mathbf{Z}^{[j]}$, respectively. $\bar{\mathbf{H}}_u^{[ji]}, \mathcal{J} \in \mathcal{K}$ represents the $2M_n \times M_n$ channel matrix from user u in the femtocell i to BS i, and $\bar{\mathbf{H}}_u^{[0i]}$ represents the $AM_n \times M_n$ channel matrix from user u in the femtocell i to macro BS 0.

The encoding strategy is as follows, which is also illustrated in Fig. 2. Suppose we use a linear scheme such that each user u in the femtocell i selects the beamforming matrix $\bar{\mathbf{V}}_u^{[i]}$ and $\bar{\mathbf{V}}_{iu}^{[0]}$ for precoding message $W_u^{[i]}$ and $W_{iu}^{[0]}$ intended for its corresponding femto BS i and macro BS 0, respectively, where $i \in \mathcal{K}, u = 1,2$. $\bar{\mathbf{V}}_u^{[i]}$ and $\bar{\mathbf{V}}_{iu}^{[0]}$ are $M_n \times \left|\bar{\mathbf{V}}_u^{[i]}\right|$ and $M_n \times \left|\bar{\mathbf{V}}_{iu}^{[0]}\right|$ matrix respectively, whose columns are beamforming directions. The key problem is how to align the interferences caused by all the undesired users as much as possible at the femto BSs and macro BS 0. That is, design $\bar{\mathbf{V}}_u^{[i]}$ and $\bar{\mathbf{V}}_{iu}^{[0]}$ to minimize the size of the interference space of each femto BS and the macro BS. In order to realize this goal, we first analyse the signal and interference space at all the femto BSs and macro BS.

The transmit signal sent by user u of the femtocell i is $\bar{\mathbf{X}}_u^{[i]} = \bar{\mathbf{V}}_u^{[i]}\mathbf{x}_u^{[i]} + \bar{\mathbf{V}}_{iu}^{[0]}\mathbf{x}_{iu}^{[0]}$, where $\mathbf{x}_u^{[i]}$ is $\left|\bar{\mathbf{V}}_u^{[i]}\right| \times 1$ data streams from user u to femto BS i, and $\mathbf{x}_{iu}^{[0]}$ is $\left|\bar{\mathbf{V}}_{iu}^{[0]}\right| \times 1$ data streams from user u to macro BS 0. Then we rewrite the input-output relationships as follows. The received signal at femto BS $j, j \in \mathcal{K}$ is

$$\bar{\mathbf{Y}}^{[j]} = \sum_{i=1}^{K}\sum_{u=1}^{2} \bar{\mathbf{H}}_u^{[ji]}\left(\bar{\mathbf{V}}_u^{[i]}\mathbf{x}_u^{[i]} + \bar{\mathbf{V}}_{iu}^{[0]}\mathbf{x}_{iu}^{[0]}\right) + \bar{\mathbf{Z}}^{[j]} \tag{11}$$

where the desired signal is

$$\bar{\mathbf{Y}}_D^{[j]} = \sum_{u=1}^{2} \begin{bmatrix} \bar{\mathbf{H}}_{u1}^{[jj]}\bar{\mathbf{V}}_u^{[j]} \\ \bar{\mathbf{H}}_{u2}^{[jj]}\bar{\mathbf{V}}_u^{[j]} \end{bmatrix} \mathbf{x}_u^{[j]} \tag{12}$$

and the interference signal is

$$\bar{\mathbf{Y}}_I^{[j]} = \sum_{i=1, i\neq j}^{K}\left(\sum_{u=1}^{2} \begin{bmatrix} \bar{\mathbf{H}}_{u1}^{[ji]}\bar{\mathbf{V}}_u^{[i]} \\ \bar{\mathbf{H}}_{u2}^{[ji]}\bar{\mathbf{V}}_u^{[i]} \end{bmatrix} \mathbf{x}_u^{[i]}\right) + \sum_{i=1}^{K}\left(\sum_{u=1}^{2} \begin{bmatrix} \bar{\mathbf{H}}_{u1}^{[ji]}\bar{\mathbf{V}}_{iu}^{[0]} \\ \bar{\mathbf{H}}_{u2}^{[ji]}\bar{\mathbf{V}}_{iu}^{[0]} \end{bmatrix} \mathbf{x}_{iu}^{[0]}\right) \tag{13}$$

The received signal at macro BS 0 is

$$\bar{\mathbf{Y}}^{[0]} = \sum_{i=1}^{K}\sum_{u=1}^{2} \begin{bmatrix} \bar{\mathbf{H}}_{u1}^{[0i]} \\ \vdots \\ \bar{\mathbf{H}}_{uA}^{[0i]} \end{bmatrix} \left(\mathbf{V}_u^{[i]}\mathbf{x}_u^{[i]} + \bar{\mathbf{V}}_{iu}^{[0]}\mathbf{x}_{iu}^{[0]}\right) + \bar{\mathbf{Z}}^{[j]} \tag{14}$$

where $\bar{\mathbf{H}}_{ua}^{[0i]}, a = 1,2,\cdots,A$ represents the channel matrix from user u of the femtocell i to the a-th receive antenna of macro BS 0. The desired signal is

$$\bar{\mathbf{Y}}_D^{[0]} = \sum_{i=1}^{K}\sum_{u=1}^{2} \begin{bmatrix} \bar{\mathbf{H}}_{u1}^{[0i]}\bar{\mathbf{V}}_{iu}^{[0]} \\ \vdots \\ \bar{\mathbf{H}}_{uA}^{[0i]}\bar{\mathbf{V}}_{iu}^{[0]} \end{bmatrix} \mathbf{x}_{iu}^{[0]} \tag{15}$$

and the interference signal is

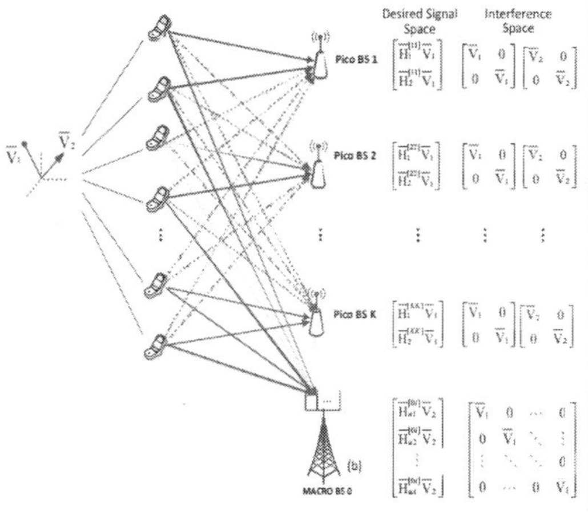

Figure 2. The encoding strategy for the HetNet.

$$\bar{\mathbf{Y}}_i^{[0]} = \sum_{i=1}^{K}\sum_{u=1}^{2}\begin{bmatrix}\bar{\mathbf{H}}_{u1}^{[0i]}\bar{\mathbf{V}}_u^{[i]}\\ \vdots \\ \bar{\mathbf{H}}_{uA}^{[0i]}\bar{\mathbf{V}}_u^{[i]}\end{bmatrix}\mathbf{x}_u^{[i]} \tag{16}$$

We wish to design all the precoding vectors so that signal space cast overlapping shadows at BSs where they constitute interference while they are distinguishable at the BSs where they are desired. As a result, each BS can decode its desired signal by zero-forcing the interference vectors. As each femto BS has two antennas while each user has only one antenna, the symbols sent from any two users cannot align with each other at the femto BS, i.e., at femto BS j,

$$\bar{\mathbf{H}}_u^{[ji]}\otimes\bar{\mathbf{V}}_u^{[i]}\cap\bar{\mathbf{H}}_u^{[jk]}\otimes\bar{\mathbf{V}}_u^{[k]}=\varnothing, i\neq j\neq k \tag{17}$$

where $\otimes$ denotes the Kronecker product operation. This is because the channel matrix from any two users to femto BS is invertible almost surely. Then the only way that alignment may take place is to let the space caused by interference from all the other users except user i and k to intersect with the union of the space spanned by interference from user i and k, i.e.

$$\left(\bar{\mathbf{H}}_u^{[ji]}\otimes\bar{\mathbf{V}}_u^{[i]}\bigcup\bar{\mathbf{H}}_u^{[jk]}\otimes\bar{\mathbf{V}}_u^{[k]}\right)\bigcap\bar{\mathbf{H}}_u^{[jq]}\otimes\bar{\mathbf{V}}_u^{[q]}\neq\varnothing \tag{18}$$

$$\Rightarrow\begin{bmatrix}\bar{\mathbf{H}}_{u1}^{[ji]}\bar{\mathbf{V}}_u^{[i]} & \bar{\mathbf{H}}_{u1}^{[jk]}\bar{\mathbf{V}}_u^{[k]}\\ \bar{\mathbf{H}}_{u2}^{[ji]}\bar{\mathbf{V}}_u^{[i]} & \bar{\mathbf{H}}_{u2}^{[jk]}\bar{\mathbf{V}}_u^{[k]}\end{bmatrix}\bigcap\begin{bmatrix}\bar{\mathbf{H}}_{u1}^{[jq]}\bar{\mathbf{V}}_u^{[q]}\\ \bar{\mathbf{H}}_{u1}^{[jq]}\bar{\mathbf{V}}_u^{[q]}\end{bmatrix}\neq\varnothing \tag{19}$$

where $q\neq i\neq j\neq k$. Based on this idea, in order to maximally align the interference, let us choose all the precoding vector $\bar{\mathbf{V}}_u^{[i]}$ for sending message $W_u^{[i]}$ to femto BS i as $\bar{\mathbf{V}}_1$ and choose all the precoding vector $\bar{\mathbf{V}}_{iu}^{[0]}$ for sending message $W_{iu}^{[0]}$ to macro BS 0 as $\bar{\mathbf{V}}_2$. For femto BS $j, j\in\mathcal{K}$, let $\bar{\mathbf{V}}_1$ and $\bar{\mathbf{V}}_2$ satisfy the following alignment conditions:

$$\begin{aligned}\bar{\mathbf{V}}_1 &\approx \bar{\mathbf{H}}_{u1}^{[j1]}\bar{\mathbf{V}}_1 \approx \bar{\mathbf{H}}_{u2}^{[j1]}\bar{\mathbf{V}}_1 \approx\cdots\approx \bar{\mathbf{H}}_{u1}^{[j(j-1)]} \approx \bar{\mathbf{H}}_{u2}^{[j(j-1)]}\bar{\mathbf{V}}_1\\ &\approx \bar{\mathbf{H}}_{u1}^{[j(j+1)]}\bar{\mathbf{V}}_1 \approx \bar{\mathbf{H}}_{u2}^{[j(j+1)]}\bar{\mathbf{V}}_1 \approx\cdots\approx \bar{\mathbf{H}}_{u1}^{[jK]}\bar{\mathbf{V}}_1 \approx \bar{\mathbf{H}}_{u2}^{[jK]}\bar{\mathbf{V}}_1,\\ &u=1,2.\end{aligned} \tag{20}$$

$$\bar{\mathbf{V}}_2 \approx \bar{\mathbf{H}}_{u1}^{[j1]}\bar{\mathbf{V}}_2 \approx \bar{\mathbf{H}}_{u2}^{[j1]}\bar{\mathbf{V}}_2 \approx\cdots\approx \bar{\mathbf{H}}_{u1}^{[jK]}\bar{\mathbf{V}}_2 \approx \bar{\mathbf{H}}_{u1}^{[jK]}\bar{\mathbf{V}}_2, u=1,2 \tag{21}$$

Then at femto BS $j, j\in\mathcal{K}$, all the interference signals sent along $\bar{\mathbf{V}}_1$ will be aligned into the vector space $\bar{\mathbf{V}}_1\times\bar{\mathbf{V}}_1$, i.e.

$$\text{span}\begin{pmatrix}\bar{\mathbf{H}}_{u1}^{[ji]}\bar{\mathbf{V}}_u^{[i]}\\ \bar{\mathbf{H}}_{u2}^{[ji]}\bar{\mathbf{V}}_u^{[i]}\end{pmatrix}\approx\text{span}\begin{bmatrix}\bar{\mathbf{V}}_1 & 0\\ 0 & \bar{\mathbf{V}}_1\end{bmatrix}, i\in\mathcal{K}, u=1,2. \tag{22}$$

and all the interference signals sent along $\bar{\mathbf{V}}_2$ will be aligned into the vector space $\bar{\mathbf{V}}_2\times\bar{\mathbf{V}}_2$, i.e.

$$\text{span}\begin{pmatrix}\bar{\mathbf{H}}_{u1}^{[ji]}\bar{\mathbf{V}}_{iu}^{[0]}\\ \bar{\mathbf{H}}_{u2}^{[ji]}\bar{\mathbf{V}}_{iu}^{[0]}\end{pmatrix}\approx\text{span}\begin{bmatrix}\bar{\mathbf{V}}_2 & 0\\ 0 & \bar{\mathbf{V}}_2\end{bmatrix}, i\in\mathcal{K}, u=1,2. \tag{23}$$

For macro BS 0, as macro BS has A antennas while each user has only one antenna, the symbols sent from any A users cannot align with each other at the macro BS, thus we use the same many-to-many alignment method. Especially, let $\bar{\mathbf{V}}_1$ satisfy the following alignment conditions:

$$\bar{\mathbf{V}}_1 \approx \bar{\mathbf{H}}_{ua}^{[01]}\bar{\mathbf{V}}_1 \approx\cdots\approx \bar{\mathbf{H}}_{ua}^{[0K]}\bar{\mathbf{V}}_1, u = \quad a = \quad \cdots A \tag{24}$$

Then at macro BS 0, all the interference signals sent along $\bar{\mathbf{V}}_1$ will be aligned into the vector space $\underbrace{\bar{\mathbf{V}}_1\times\cdots\times\bar{\mathbf{V}}_1}_{A\text{ times}}$, i.e.

$$\text{span}\begin{bmatrix}\bar{\mathbf{H}}_{u1}^{[0i]}\bar{\mathbf{V}}_1\\ \vdots\\ \bar{\mathbf{H}}_{uA}^{[0i]}\bar{\mathbf{V}}_1\end{bmatrix}\approx\text{span}\begin{bmatrix}\bar{\mathbf{V}}_1 & 0 & \cdots & 0\\ 0 & \bar{\mathbf{V}}_1 & \cdots & 0\\ \vdots & \vdots & \ddots & \vdots\\ 0 & 0 & \cdots & \bar{\mathbf{V}}_1\end{bmatrix}, \quad i\in\mathcal{K}, u=1,2. \tag{25}$$

From (20) and (24), for vector $\bar{\mathbf{V}}_1$ there are totally $N_1=4K(K-1)+2AK=4K^2+2AK-4K$ interference carrying channels. Relabel all these channels as $\boldsymbol{F}_1$ to $\boldsymbol{F}_{N_1}$. From (21), for vector $\bar{\mathbf{V}}_2$ there are totally $N_2=4K^2$ interference carrying channels. Also relabel all these N_2 channels as $\boldsymbol{F}_1^{'}$ to $\boldsymbol{F}_{N_2}^{'}$. Since $A\geq 2$ which means $N_1\geq N_2$ is always hold, we assume that there are another $2AK-4K$ imaginary interference carrying channels $\boldsymbol{F}_{N_2+1}^{'}, \boldsymbol{F}_{N_2+2}^{'}, ?\cdot\cdot\ \boldsymbol{F}_{N_1}^{'}$ for vector $\bar{\mathbf{V}}_2$. So both $\bar{\mathbf{V}}_1$ and $\bar{\mathbf{V}}_2$ need to satisfy N_1 conditions. Define $\mathcal{I}_1=\bigcup_k \boldsymbol{F}_k\bar{\mathbf{V}}_1$, which is the union of all interference terms due to signals transmitted along $\bar{\mathbf{V}}_1$, and define $\mathcal{I}_2 = \bigcup_k \boldsymbol{F}'_k \bar{\mathbf{V}}_2$, which is the union of all interference terms due to signals transmitted along $\bar{\mathbf{V}}_2$. The conditions (20), (24) for $\bar{\mathbf{V}}_1$ and (21) for $\bar{\mathbf{V}}_2$ can be expressed as $\bar{\mathbf{V}}_1 \approx \mathcal{I}_1$ and $\bar{\mathbf{V}}_2 \approx \mathcal{I}_2$. These conditions are satisfied simultaneously by the CJ scheme construction:

$$\bar{\mathbf{V}}_1 = \bigcup_{m=1}^{2K-A} \left\{ \left(\prod_{k=1}^{N} \boldsymbol{F}_k^{\alpha_k} \right) \mathbf{w}^{[m]}, \forall \alpha_k \le n-1, \alpha_k \in \mathbb{Z}_0^+ \right\} \quad (26)$$

$$\bar{\mathbf{V}}_2 = \bigcup_{m=2K-A+1}^{2K} \left\{ \left(\prod_{k=1}^{N} \boldsymbol{F}_k'^{\beta_k} \right) \mathbf{w}^{[m]}, \forall \beta_k \le n-1, \beta_k \in \mathbb{Z}_0^+ \right\} \quad (27)$$

$$\mathcal{I}_1 = \bigcup_{m=1}^{2K-A} \left\{ \left(\prod_{k=1}^{N} \boldsymbol{F}_k^{\alpha_k} \right) \mathbf{w}^{[m]}, \forall \alpha_k \le n, \alpha_k \in \mathbb{Z}_0^+ \right\} \quad (28)$$

$$\mathcal{I}_2 = \bigcup_{m=2K-A+1}^{2K} \left\{ \left(\prod_{k=1}^{N} \boldsymbol{F}_k'^{\beta_k} \right) \mathbf{w}^{[m]}, \forall \beta_k \le n-1, \beta_k \in \mathbb{Z}_0^+ \right\} \quad (29)$$

where $\mathbf{w}^{[m]}, m = 1, 2, \cdots, 2K$, is $M_n \times 1$ column vectors which is generated such that all the entries of the matrix $\left[\mathbf{w}^{[1]}\, \mathbf{w}^{[2]} \cdots \mathbf{w}^{[2K]}\right]$ are chosen i.i.d from some continuous distribution whose support lies between a finite minimum value and a finite maximum value. $\mathbb{Z}_0^+$ consists both positive integers and 0. $\bar{\mathbf{V}}_1$ and $\bar{\mathbf{V}}_2$ can be verified to have $(2K-A)n^N$ and An^N elements, $\mathcal{I}_1$ and $\mathcal{I}_2$ can be verified to have $(2K-A)(n+1)^N$ and $A(n+1)^N$ elements, respectively. Thus, when n goes to infinity, $\bar{\mathbf{V}}_1 \approx \mathcal{I}_1$ and $\bar{\mathbf{V}}_2 \approx \mathcal{I}_2$.

At femto BS $j, \forall j \in \mathcal{K}$, desired signal occupied $2\left|\bar{\mathbf{V}}_1\right|$ dimensions and aligned interferences occupy $2\left|\bar{\mathbf{V}}_1\right| + 2\left|\bar{\mathbf{V}}_2\right|$ dimensions. To avoid overlaps between desired signals and interferences the size of received signal space, $2M_n$, must be at least as big as the sum of the dimensions of desired signals and interferences, which is $2M_n \le 2(4K-A)n^N$. At macro BS 0, we can also get $AM_n \le A(4K-A)n^N$. Thus, that is the reason why we set $M_n \le (4K-A)n^N$.

Through interference alignment, we have guaranteed that the interference vectors span a small enough signal space. Further, through the similar proof of [6, 9] we can further prove that the linear independence of the desired signal from interferences at all femto BSs and macro BS. Therefore, all BSs can decode their desired streams using zero forcing.

This completes the whole achievable scheme.

4 Simulation Results

In this section, we analyze the performance of the proposed IA scheme under the difference antenna configurations. Fig. 3 shows the achievable average sum rate of the proposed IA scheme ($K = 8$) versus SNR under different macro BS's antenna configurations. It is shown that the sum rate of the whole HetNet increases as the number of macro BS's antennas increases. Meanwhile, from the figure we can also see that as the number of macro BS's antennas increases the proportion of the data streams assigned to the macrocell is increased while the proportion of the data streams assigned to the picocell is decreased. This indicates the best allocation strategy for this kind of heterogeneous network under difference macro BS's antenna configuration.

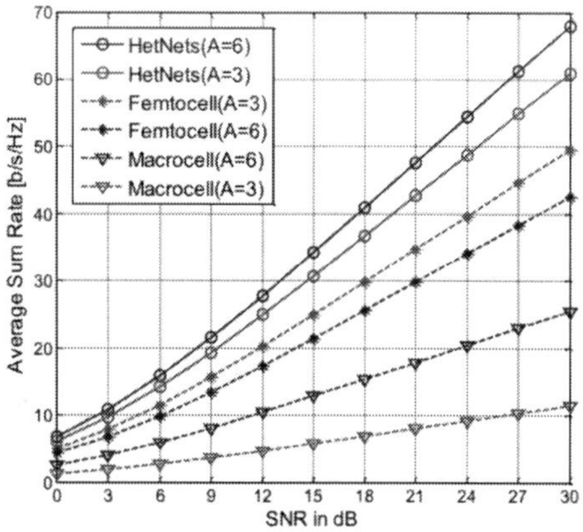

Figure 3. Simulation result

ACKNOWLEDGMENT

This work is supported by the National Basic Research Program of China (No. 2013CB329003) and National Nature Science Foundation of China (91438205).

References

1. Qualcomm, "LTE advanced: heterogeneous networks," white paper, Jan.2011. [Online]. Available: http://qualcomm.com/documents/files/lte-advanced-heteroge¬neous-networks.pdf.
2. Picochip, "The case for home base stations," white paper, Apr.2007.[Online].Available:http://www.femtoforum.org/femto/Files/File/picoChipFemtocellWhitePaper1.1.pdf .
3. J. Zhang and J. G. Andrews, "Distributed antenna systems with randomness," IEEE Trans. Wireless Commun., vol. 7, no. 9, pp. 3636-3646, Sep. 2008.
4. V. Chandrasekhar, J. G. Andrews, and A. Gatherer, "Femtocell networks: a survey," IEEE Commun. Magazine, vol. 46, no. 9, pp. 59-67, Sep. 2008.

5. A. Ghosh, N. Mangalvedhe, R. Ratasuk, B. Mondal, M. Cudak, E.Visotsky, T. A. Thomas, J. G. Andrews, P. Xia, H. S. Jo et al., "Heterogeneous cellular networks: From the¬ory to practice", IEEE Commun. Mag., vol. 50, no. 6, pp. 54-64, 2012.
6. V. Cadambe and S. Jafar, "Interference alignment and the degrees of freedom of the K user interference channel", IEEE Trans. Inf. Theory, vol. 54, no. 8, pp. 3425-3441, Aug. 2008.
7. V. Cadambe and S. Jafar, "Interference alignment and the degrees of freedom of wireless X networks", IEEE Trans. Inf. Theory, vol. 55, no. 9, pp. 3893-3908, Sep. 2009.
8. C. Suh, M. Ho, and D. Tse, "Downlink interference align¬ment", IEEE Trans. Commun., vol. 59, no. 9, pp. 2616-2626, Sep. 2011.
9. J. Sun, Y. Liu, and G. Zhu, "On the degrees of freedom of the cellular network," in Proc. IEEE ICC'10, May 2010, pp. 1-5.
10. Y. M. and O. T., "Interference alignment in the 2×(1+ N) MIMO model," in Proc. IEEE GLOBECOM'10, Dec. 2010, pp. 1963-1967.
11. W. Shin, N. Lee, J.-B. Lim, C. Shin, and K. Jang, "On the design of interference alignment scheme for two-cell MIMO interfering broadcast channels," IEEE Trans. Wireless Commun., vol. 10, pp. 437-442, Feb. 2010.

Jammer Suppression in DS-CDMA Communications using Parafac-based blind separation

Lingyun Xu[1,2] , Xianwei Zeng[2] and Xiaofei Zhang[1]

[1] Nanjing University of Aeronautics & Astronautics, Nanjing, China
[2] Nangjing Telecommunication Technology Institute, Nanjing, China

Abstract. In this paper we propose to apply parafac-based source separation techniques for jammer suppression in direct spread spectrum communication systems. The jammer excision is formulated as an optimization problem and a new algorithm is presented which is based on the parafac tri-iterative least square algorithm. By jointly diagonalizing the time delay correlation matrix of the observed signals and using the new proposed method, a better solution is achieved. The proposed algorithm can successfully separate communication signals and jamming signals. Simulation results reveal that our proposed algorithm has the better blind signal separation performance than joint diagonalization method. Our proposed algorithm doesn't require whitening processing. Moreover our proposed algorithm works well in the underdetermined condition, where the number of sources exceeds than the number of antennas.

1 Introduction

Jammer suppression techniques in spread spectrum (SS) communications have been under active research due to the inevitable gains in the overall system performance and capacity [1]. Although bandwidth expansion gives an inherent temporal mitigation capability for SS communications, usually called a processing gain, additional techniques are needed. This is because bandwidth expansion results in bandwidth dependent mitigation capability. Interference mitigation techniques, however, alleviate the need for wider spectrum for reliable communications.

In commercial cellular SS and DS-CDMA systems, many types of interference can appear, starting from multiuser interference inside each sector in a cell to inter-operator interference. Unintentional jamming can also be present due to co-existing systems at the same band, whereas intentional jamming arises mainly in military applications. Jamming can be mitigated by the use of multiple antenna sensors utilizing spatial diversity. However, when using conventional array receivers, directions of arrivals of signals must be estimated first. This in turn requires exact prior knowledge of the positions of the receiving antenna sensors. By using blind source separation (BSS) techniques as proposed in [2-3], DOA estimation can be relaxed, making it possible to achieve performance gains when applied to uncalibrated arrays in which the positions are known only roughly or not all. Belouchrain and Amin [2] were the first to use the blind source separation (BSS) techniques to aid conventional detection for jammer mitigation. They have used BSS based on second-order statistics to separate desired signaland jamming signals. However, their work suffers from some weakness. In general, there are two types of BSS methods for separating signal, which are namely second order statistic-based BSS [4-5] and high order statistic-based BSS [6]. Second-order statistic-based BSS methdos require assumption about the second-order statistics such as nonstationarity or nonwhiteness. Only one source with gaussain characteristic is required. Joint diagonalization of a set of covariance matrices in [4] can be used for blind signal separation, but this algorithm requires whitening processing. The process will influence the separate performance, as the statistical error of this stage can not be modified in the separation stage. However, conventional blind signal separation methods including joint diagonalization method cannot work in the underdetermined condition, where the number of source exceeds the number of sensors. On the other hand, high-order statistic methods require high computation.

This paper introduces a new source separation technique exploiting the time coherence of the source signals. The proposed approach is based on parafac blind separation. Parafac separation has been first introduced as a data analysis tool in psychometrics. Parafac decomposition is thus naturally related to linear algebra for multi-way arrays [7, 8]. Parafac decomposition was used widely in blind estimation of multi-input multi-output system, polarization sensitive array signal processing, array parameter estimation and so on [9-10].

Our work links the blind signal separation problem to the parafac model and derives a novel blind signal separation algorithm whose performance is better than joint diagonalization method [4]. Our proposed algorithm doesn't need whitening processing. Instead, our proposed algorithm relies on a fundamental result of Kruskal [8] regarding the uniqueness of low-rank three-way array decomposition. Our proposed algorithm works well in the underdetermined condition, where the number of sources exceeds than the number of sensor.

This paper is structured as follows. Section 2 develops data model. Section 3 discusses identifiability issues and deals with algorithmic issues. Section 4 presents simulation results, and section 5 summarizes our conclusions.

Denote: $[.]^H$, $[.]^T$ $[.]^*$, $[.]^+$ and $\| \|_F$ stand for the matrix conjugate transpose, matrix transpose, complex conjugate, pseudo-inverse and Frobenius norm, respectively.

2 Data model

A standard spread spectrum system with direct sequence spread is assumed. Consider the following basic linear mixture model:

$$\mathbf{x}(t) = \mathbf{A}\mathbf{s}(t) + \mathbf{n}(t) \quad (1)$$

where the stochastic vector $\mathbf{x}(t) = [\mathbf{x}_1(t), \mathbf{x}_2(t), \cdots, \mathbf{x}_M(t)]^T \in \mathbb{C}^{M\times 1}$ represents multi-antenna observations, the components of the stochastic vector $\mathbf{s}(t) = [s_1(t), s_2(t), \cdots, s_F(t)]^T \in \mathbb{R}^{F\times 1}$ correspond to source signals which include direct spread spectrum communication signal and jamming signals, $s_1(t)$ is the desired spread spectrum communication signal , the others are jamming signals. $\mathbf{n}(t) \in \mathbb{C}^{M\times 1}$ denotes additive noise. The a priori unknown mixing matrix $\mathbf{A} \in \mathbb{C}^{M\times F}$ characterizes the way the sources are combined in the observations. In this paper we assume that the sources are individually correlated in time. The covariance matrices of the observations then satisfy:

$$\begin{aligned}
\mathbf{X}_1 &\overset{def}{=} E\{\mathbf{x}_t\mathbf{x}_{t+\tau_1}^H\} = \mathbf{A}\mathbf{D}_1\mathbf{A}^H \\
\mathbf{X}_2 &\overset{def}{=} E\{\mathbf{x}_t\mathbf{x}_{t+\tau_2}^H\} = \mathbf{A}\mathbf{D}_2\mathbf{A}^H \\
&\vdots \\
\mathbf{X}_K &\overset{def}{=} E\{\mathbf{x}_t\mathbf{x}_{t+\tau_K}^H\} = \mathbf{A}\mathbf{D}_K\mathbf{A}^H
\end{aligned} \quad (2)$$

where $\mathbf{D}_k = E\{\mathbf{s}_t\mathbf{s}_{t+\tau_k}^H\}$ is diagonal, for $k = 1, 2, \cdots, \mathrm{K}$. One of the delays τ_k can be equal to zero. For simplicity, we have dropped the noise terms; they can be considered as a perturbation of (2). The problem we want to solve is the estimation of $\mathbf{A}$ from the set { $\mathbf{X}_k$ }.For (2), joint diagonalization method[4] can be used for blind signal separation. The mixing matrix A can be estimated by means of a joint diagonalizer of the matrices $\mathbf{X}_k, k = 1, 2, \cdots, \mathrm{K}$. In this paper, a novel blind signal separation method is investigated.

Define a matrix $\mathbf{H} \in \mathbb{C}^{K\times F}$ by $(\mathbf{H})_{kf} = (\mathbf{D}_k)_{ff}$, $k = 1, 2, \cdots, \mathrm{K}$, $\mathrm{f} = 1, 2, \cdots, \mathrm{F}$. (2) can be denoted as

$$\mathbf{X}_k = \mathbf{A}diag_k(\mathbf{H})\mathbf{A}^H, k = 1, 2, \cdots, \mathrm{K} \quad (3)$$

where $diag_m(.)$ is understood as an operator that extracts the mth row of its matrix argument and constructs a diagonal matrix out of it. In the presence of noise, the received signal model becomes

$$\tilde{\mathbf{X}}_k = \mathbf{A}diag_k(\mathbf{H})\mathbf{A}^H + \mathbf{W}_k, k = 1, 2, \cdots, \mathrm{K} \quad (4)$$

where $\mathbf{W}_k$ is the received noise. Stack the matrices $\mathbf{X}_k$ $(k = 1, 2, \cdots, \mathrm{K})$ in a trilinear model which is shown

$$x_{m,n,k} = \sum_{f=1}^{F} a_{m,f}a_{n,f}^* h_{k,f}, \quad (5)$$
$$m = 1,\ldots,\mathrm{M},\ n = 1,\ldots,\mathrm{M},\ k = 1,\ldots,K$$

where $a_{m,f}$ stands for the (m,f) element of $\mathbf{A}$ matrix, and similarly for the others. The parafac trilinear model displays the reflection for three kinds of diversity, which is shown in Fig.1. $\mathbf{X}_k = \mathbf{A}diag_k(\mathbf{H})\mathbf{A}^H$, $k = 1, 2, \cdots, K$, can be interpreted as slicing the 3-D data in a series of slices (2-D data) along the sensor direction. The symmetry of the trilinear model allows two more matrix system rearrangements,

$$\mathbf{Y}_m = \mathbf{A}^* diag_m(\mathbf{A})\mathbf{H}^T + \mathbf{W}_m \quad m = 1, \ldots, M \quad (6)$$

$$\mathbf{Z}_n = \mathbf{H}diag_n(\mathbf{A}^*)\mathbf{A}^T + \mathbf{W}_n \quad n = 1, \ldots, M \quad (7)$$

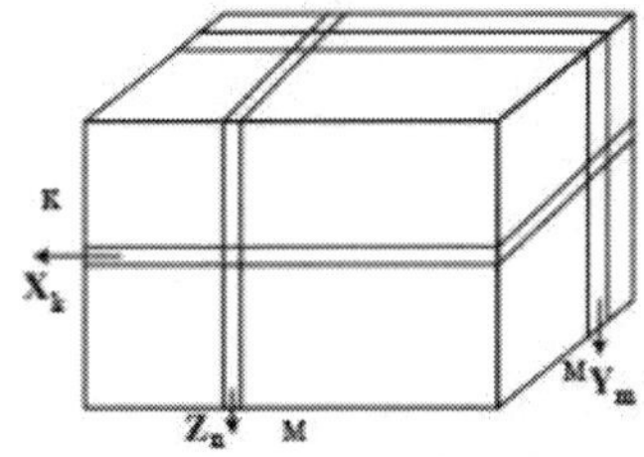

Fig.1 the parafac blind separation model

3 The proposed algorithm description

3.1 Parafac trilinear decomposition

TALS (Trilinear Alternating Least Square) algorithm is the common data detection method for parafac blind decomposition [11]. TALS algorithm is discussed in detail as follows. According to Eq.(3), stack the $\mathbf{X}_k$ $(k = 1, 2, \cdots, \mathrm{K})$, we can get

$$\begin{bmatrix} \mathbf{X}_1 \\ \mathbf{X}_2 \\ \vdots \\ \mathbf{X}_K \end{bmatrix} = \begin{bmatrix} \mathbf{A}diag_1(\mathbf{H}) \\ \mathbf{A}diag_2(\mathbf{H}) \\ \vdots \\ \mathbf{A}diag_K(\mathbf{H}) \end{bmatrix} \mathbf{A}^H \quad (8)$$

The cost function can be constructed via least squares and given by

$$\min_{A,H}\left\|\begin{bmatrix}\tilde{\mathbf{X}}_1\\ \tilde{\mathbf{X}}_2\\ \vdots\\ \tilde{\mathbf{X}}_K\end{bmatrix}-\begin{bmatrix}\mathbf{A}diag_1(\mathbf{H})\\ \mathbf{A}diag_2(\mathbf{H})\\ \vdots\\ \mathbf{A}diag_K(\mathbf{H})\end{bmatrix}\mathbf{A}^H\right\|_F \quad (9)$$

where $\tilde{\mathbf{X}}_k$, $k=1,2,\cdots,$ K are the noisy slices. Least squares update for $\mathbf{A}^*$ is

$$\hat{\mathbf{A}}^H=\begin{bmatrix}\hat{\mathbf{A}}diag_1(\hat{\mathbf{H}})\\ \hat{\mathbf{A}}diag_2(\hat{\mathbf{H}})\\ \vdots\\ \hat{\mathbf{A}}diag_K(\hat{\mathbf{H}})\end{bmatrix}^+\begin{bmatrix}\tilde{\mathbf{X}}_1\\ \tilde{\mathbf{X}}_2\\ \vdots\\ \tilde{\mathbf{X}}_K\end{bmatrix} \quad (10)$$

where $\hat{\mathbf{A}}$ and $\hat{\mathbf{H}}$ denote previously obtained estimates of $\mathbf{A}$ and $\mathbf{H}$ .Similarly, from the second way of slices: $\mathbf{Y}_m=\mathbf{A}^*diag_m(\mathbf{A})\mathbf{H}^T$, $m=1,\cdots,M$. The cost function is

$$\min_{A,H}\left\|\begin{bmatrix}\tilde{\mathbf{Y}}_1\\ \tilde{\mathbf{Y}}_2\\ \vdots\\ \tilde{\mathbf{Y}}_M\end{bmatrix}-\begin{bmatrix}\mathbf{A}^*diag_1(\mathbf{A})\\ \mathbf{A}^*diag_2(\mathbf{A})\\ \vdots\\ \mathbf{A}^*diag_M(\mathbf{A})\end{bmatrix}\mathbf{H}^T\right\|_F \quad (11)$$

where $\tilde{\mathbf{Y}}_m, m=1,\cdots,M$ are noisy slices. And the LS update for $\mathbf{H}$ is

$$\mathbf{H}^T=\begin{bmatrix}\hat{\mathbf{A}}^*diag_1(\hat{\mathbf{A}})\\ \hat{\mathbf{A}}^*diag_2(\hat{\mathbf{A}})\\ \vdots\\ \hat{\mathbf{A}}^*diag_M(\hat{\mathbf{A}})\end{bmatrix}^+\begin{bmatrix}\tilde{\mathbf{Y}}_1\\ \tilde{\mathbf{Y}}_2\\ \vdots\\ \tilde{\mathbf{Y}}_M\end{bmatrix} \quad (12)$$

Finally, from the third way of slices: $\mathbf{Z}_n=\mathbf{H}diag_n(\mathbf{A}^*)\mathbf{A}^T$, $n=1,2,\cdots,M$. And then LS update for $\mathbf{A}$ is Finally, from the third way of slices: $\mathbf{Z}_n=\mathbf{H}diag_n(\mathbf{A}^*)\mathbf{A}^T$, $n=1,2,\cdots,M$. And then LS update for $\mathbf{A}$ is

$$\hat{\mathbf{A}}^T=\begin{bmatrix}\hat{\mathbf{H}}diag_1(\hat{\mathbf{A}}^*)\\ \hat{\mathbf{H}}diag_2(\hat{\mathbf{A}}^*)\\ \vdots\\ \hat{\mathbf{H}}diag_M(\hat{\mathbf{A}}^*)\end{bmatrix}^+\begin{bmatrix}\tilde{\mathbf{Z}}_1\\ \tilde{\mathbf{Z}}_2\\ \vdots\\ \tilde{\mathbf{Z}}_M\end{bmatrix} \quad (13)$$

According to (10), (12) and (13), matrices $\mathbf{A}^*$, $\mathbf{H}$ and $\mathbf{A}$ are updated with conditioned least squares, respectively. The matrix update will stop until convergence. TALS algorithm can be initialized randomly.

TALS is optimal when noise is additive i.i.d. Gaussian. TALS algorithm has several advantages: it is easy to implement, guaranteed to converge and simple to extend to higher order data. But TALS algorithm has slow convergence. In this paper, we use the COMFAC algorithm [11] for trilinear decomposition. COMFAC algorithm is essentially a fast implementation of TALS, and can speeds up the least square fitting.

3.2 Uniqueness and Identifiability

In this subsection, we first will introduce a key concept, k-rank, and then discuss sufficient condition and necessary condition for uniqueness of parafac blind separation. The k-rank of a matrix is an important concept in the parafac blind source separation.

Definiton 1[8]: Consider the matrix $\mathbf{A}\in\mathbb{C}^{M\times F}$, denote rank($\mathbf{A}$)= r when $\mathbf{A}$ contains r columns which are linear independent. Moreover, if the linear independence in every column of $\mathbf{A}$ exists only for $l\le F$,which does not hold for more than l columns, note that the k-rank of $\mathbf{A}$ should have $k_A=l$, s.t. $k_A\le rank(\mathbf{A})$, $\forall\mathbf{A}$.

Theorem1[8] $\mathbf{X}_k=\mathbf{A}diag_k(\mathbf{H})\mathbf{A}^H$, for $k=1,2,\cdots,$ K

where $\mathbf{A}\in\mathbb{C}^{M\times F}$, $\mathbf{H}\in\mathbb{C}^{K\times F}$. Considering the matrice is full rank and full k-rank, if

$$2\min(M,F)+\min(F,K)\ge 2F+2. \quad (14)$$

then $\mathbf{A}$ and $\mathbf{H}$ are unique up to permutation and scaling of columns, that is to say, any other $\bar{\mathbf{A}}$ and $\bar{\mathbf{H}}$ that construct $\mathbf{X}_k$ ($k=1,2,\cdots,$ K) is related to $\mathbf{A}$ and $\mathbf{H}$ via

$$\bar{\mathbf{A}}=\mathbf{A}\mathbf{\Pi}\mathbf{\Delta}_1, \bar{\mathbf{H}}=\mathbf{H}\mathbf{\Pi}\mathbf{\Delta}_2. \quad (15)$$

where $\mathbf{\Pi}$ is a permutation matrix, and $\mathbf{\Delta}_1,\mathbf{\Delta}_2$ are diagonal scaling matrices satisfying $\mathbf{\Delta}_1\mathbf{\Delta}_1^*\mathbf{\Delta}_2=\mathbf{I}$.

When $M\ge F$ (M is number of sensors; F is the number of sources), $F\ge 2$, then identifiable condition is $K\ge 2$. When $M\le F$, $F\ge 2$, $K>F$, then identifiable condition is $F/2+1\le M\le F$. So our proposed algorithm works well in the underdetermined condition, where the number of sources is larger than the number of sensors. It should be pointed out that the joint diagonalization methods require stronger conditions in terms of the number of sensors as compared to the our proposed algorithm. Indeed, $M\ge F$ is required for the joint diagonalization algorithms [4], whereas this constraint is not needed for our proposed algorithm.

3.3 Blind signal Separation using parafac decomposition

Blind signal separation algorithm using parafac blind separate techniques is proposed in this paper. This algorithm firstly uses parafac decomposition to estimate the mixing matrix $\hat{\mathbf{A}}$, and the source signal is $\hat{\mathbf{S}}=\hat{\mathbf{A}}^+\mathbf{X}$. Based on the above analysis, the Parafac blind source separation for jammer suppression in DS-CDMA system is summarized as follows:

Step1: Random initialization for the matrices for $\mathbf{A}$ and $\mathbf{H}$;

Step 2: LS update for matrix $\mathbf{A}^*$ according to (10);

Step 3: LS update for matrix $\mathbf{H}$ according to (12);

4 Simulation and analysis

In this section, we present the simulation results to show the performance. A system with K=3 signals was considered. Signal-to-noise ratio (SNR) and the signal to jammer ratio (SJR) were defined with respect to the desired signal.

Simulation 1: There are 3 sources and 5 sensors to receive signal. The mixing matrix is $\mathbf{A}=[\mathbf{a}(\theta_1)\ \ \mathbf{a}(\theta_2)\ \ \mathbf{a}(\theta_3)]$, where $\mathbf{a}(\theta)=[1,\exp(-j\pi\cos\theta),\cdots,\exp(-j4\pi\cos\theta)]^T$. And $\theta_1=10^\circ$, $\theta_2=20^\circ$ and $\theta_3=30^\circ$. $s_1(t)$ is the desired direct sequence signal, the source rate is 1 kbit/s, using BPSK modulation mode, the length of PN code is 16, the amplitude is 1, the carrier frequency is 600kHz. $s_2(t)=5\sin(350\mathrm{t})$ is the tone jamming signal. $S_3(t)=4\sin(1500t^2)$ is the linear frequency modulation jamming signal. The source signals $s_1(t)$, $s_2(t)$, $s_3(t)$ are shown in Fig.2. The received signal of sensors is shown in Fig.3. Fig.4 shows the separated signal with our proposed algorithm at SNR=15dB and SJR=-16 dB. But there is a little change in the amplitude and order of the signals, it is common in the BSS techniques and does not affect the estimation of the signals. In order to qualify BSS performance, the similar coefficient between the separated signal and source signal is defined as

$$\rho_{ij}=\frac{\left|\sum_{t=0}^{T}a_i(t)b_j(t)\right|}{\sqrt{\sum_{t=0}^{T}a_i^2(t)\sum_t^T b_j^2(t)}}$$

where $a_i(t)$ and $b_j(t)$ are separated signal and source signal, respectively. The similar coefficients between the separated signal in Fig.4 and source signal in Fig.2 at SNR=15dB are 0.9987, 1.0000 and 0.9967.

Simulation 2: The proposed algorithm and joint diagonalization method [4] firstly estimate the mixing matrix, and then separate signal. The mixing matrix $\mathbf{A}$ estimation performance determines the signal separation performance. We present Monte Carlo simulations that are to assess the mixing matrix estimation performance of the proposed algorithm. The number of Monte Carlo trials is 500. Define $\mathrm{MSE}=\frac{1}{500\times M\times F}\sum_{m,f}\sum_{i=1}^{500}\left\|\hat{a}_{m,f,i}-a_{m,f}\right\|^2$, where $\hat{a}_{m,f,i}$ is (m,f) element of the estimated mixing matrix of the ith Monte Carlo simulation, and $a_{m,f}$ is (m,f) element of the perfect mixing matrix. M=5 is the number of sensors, and F=3 is the number of the sources. K=4 in this simulation. Their performance under different SNR is shown in Fig.5. From Fig.5 we find that our proposed algorithm has better the mixing matrix estimation performance than joint diagonalization method. So our proposed algorithm has better signal separation performance.

Simulation 3: The parameters of spread spectrum signal and interference signal remains as simulation 1. Fig. 6 shows the error rate of spread spectrum signal with the increase of SNR while keeping SJR fixed to -10dB. Fig.7 gives the error rate curve with the change of SJR at SNR equal to 6dB. In order to compare performance, no jamming, not separation jamming, joint diagonalization blind separation and the proposed algorithm based on parafac blind separation are given in Fig.6-7. From Fig.6-7 we can see that the performance can be improved after blind source separation. The performance of the proposed algorithm is better that other conditions.

Simulation 4: Conventional blind signal separation methods including joint diagonalization method cannot work in the underdetermined condition, where the number of sources exceeds the number of sensors. Our proposed algorithm has no this constrain. Blind signal separation in the underdetermined condition is investigated in this simulation. The number of sources is 5, and the number of sensors is 4. K is set 20 in this simulation. Fig.8 shows the mixing matrix estimation performance in the underdetermined condition. From Fig.8 we find that our proposed algorithm works well in the underdetermined condition. Our proposed algorithm overcomes the shortcoming of conventional BSS methods.

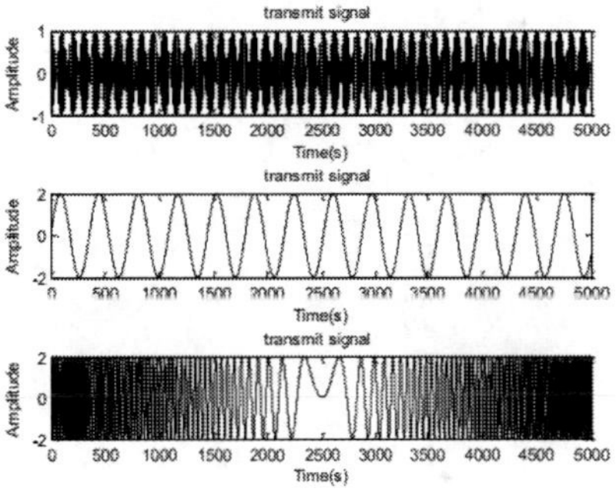

Fig.2 the source signals

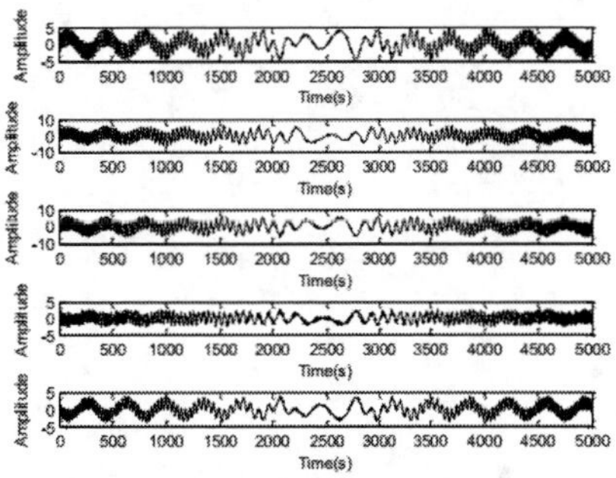

Fig.3 the received mixing signals

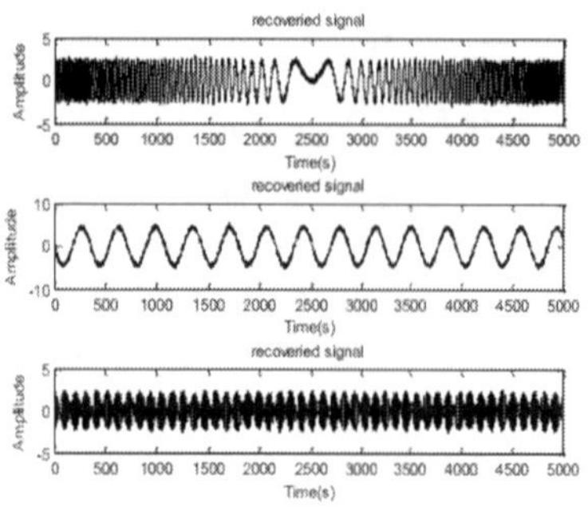

Fig.4 the separated signals based on parafac blind source separation

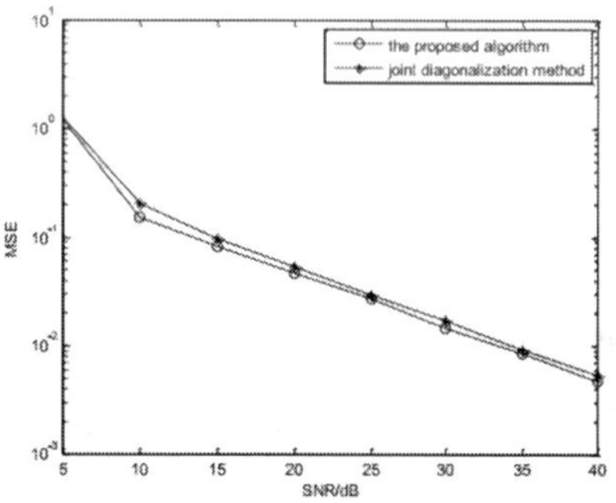

Fig.5 the algorithm performance comparison

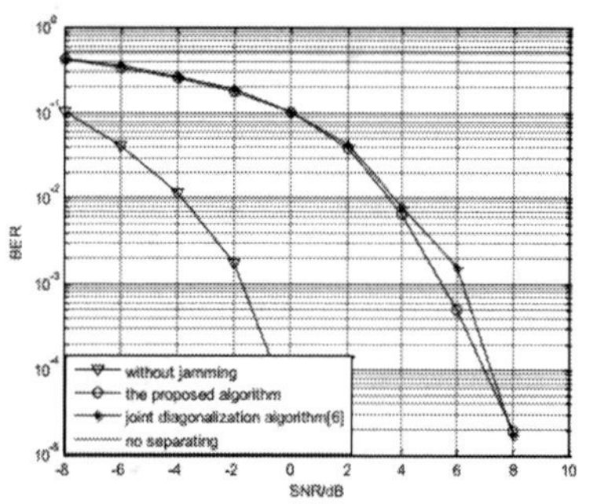

Fig.6 BER performance comparison at SJR=-10dB

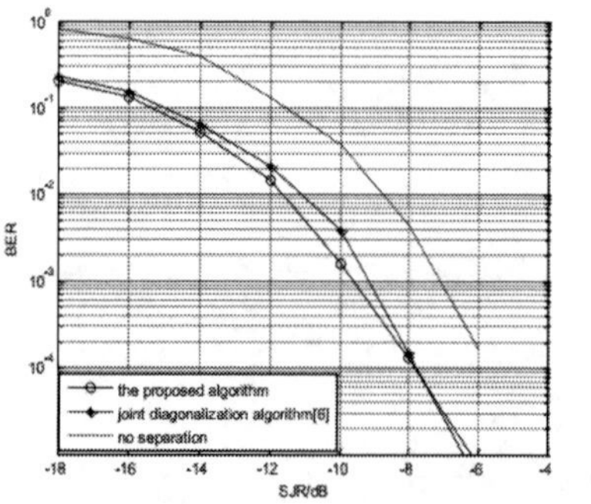

Fig.7 BER performance comparison at SNR=6dB

5 Conclusion

Jamming problems are important in DS-CDMA communications because both intentional and unintentional jamming can occur in several paractical circumstances. In this paper, we have considered suppression of jammer signal in the direct sequence spread spectrum communication system. Our work links the blind signal separation problem to the parafac model and derives a novel blind signal separation algorithm whose performance is better than joint diagonalization method. Our proposed algorithm doesn't need whitening processing. Instead, our proposed algorithm relies on the uniqueness of low-rank three-way array decomposition. Furthermore, our proposed algorithm works well in the underdetermined condition, where the number of sources exceeds than the number of sensors. Theory analysis and simulation results show the feasibility and effectiveness of the proposed algorithm and the research work is expected to provide a new thought for the development of communications anti-jamming.

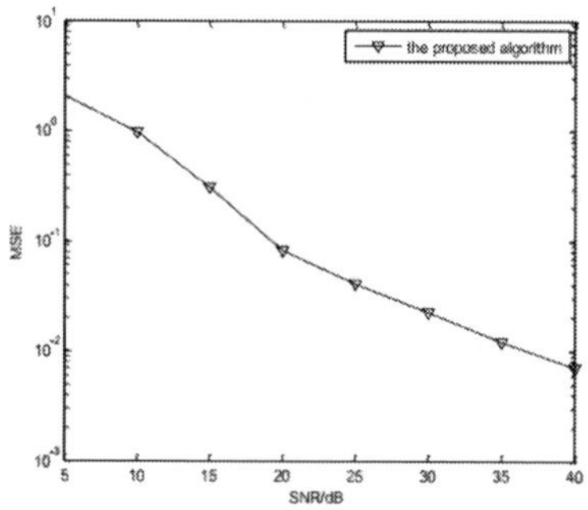

Fig.8 the algorithm performance in the underdetermined condition

Acknowledgment

This work is supported by National Nature Science Foundation of China (No. 61371169), Jiangsu Postdoctoral Research Funding Plan (No.1301013B). The authors wish to thank the anonymous reviewers for their valuable suggestions on improving this paper.

References

1. M. Amin, C. Wang, and A. Lindsey., IEEE Transactions on Signal Processing, 47(1999)
2. A. Belouchrani, M. Amin, Signal processing, 80, (2000).
3. Carlemalm, H. V. Poor, and A. logothetis, IEEE Journal on Selected areas in communications, 18, (2000).
4. A. Belouchrani, K. Abed-Meraim, J. F. Cardoso, and E. Moulines, IEEE Trans. Signal Processing, 45, (2000).
5. D. W. E. Schobben, and P. W. Sommen, IEEE Trans. Signal Process., 50, (2002).
6. H. Buchner, R. Aichner, and W. Kellermann, IEEE Transactions on Speech and Audio Processing, 13(2004).

7. L. De Lathauwer, SIAM Journal on Matrix Analysis and Applications, 28(2006).
8. J. B. Kruskal, Linear Algebra and its Applications, 18(1977).
9. Xijing Guo, Sebastian Miron, David Brie etc. IEEE Transactions on Signal Processing, 59 (2011).
10. Xiaofei Zhang, Jiangfeng Li, HanChen, and Dazhuan Xu, Wireless Pers. Communication, 67(2012).
11. R. Bro, N. D. Sidiropoulos, and G. B. Giannakis, International Workshop Independent Component Analysis and Blind Signal Separation, 1999.

Spectrum Assignment with Non-deterministic Bandwidth of Spectrum Holein Cognitive Radio Networks

Huang Jie , He Yuan, Zeng Xiaoping, Tan Xiaoheng and Miao Lijuan

College of Communication Engineering, Chongqing University, Chongqing, China

Abstract. The spectrum allocation for cognitive radio networks (CRNs) has received considerable studies under the assumption that the bandwidth of spectrum holes is static. However, in practice, the bandwidth of spectrum holes is time-varied due to primary user/secondary user (PU/SU) activity and mobility, which result in non-determinacy. This paper studies the spectrum allocation for CRNs with non-deterministic bandwidth of spectrum holes. We present a novel probability density function (PDF) model through order statistic to describe the non-deterministic bandwidth of spectrum holes and provide a bound to approximate it. After that, a statistical spectrum allocation model based on stochastic multiple knapsack problem (MKP) is established for spectrum allocation with non-deterministic bandwidth of spectrum holes. To reduce the computational complexity, we transform this stochastic programming probleminto a constant MKP though exploiting the properties of cumulative distribution function (CDF), which can be solved via MTHG algorithm by using auxiliary variable. Simulation results illustrate that the proposed statistical spectrum allocation algorithm can achieve better performances compared to the existing algorithms when the bandwidth of spectrum holes istime-varied.

Keywords-cognitive radio; time-varied; spectrum allocation; non-deterministic bandwidth of spectrum holes; stochastic programming.

1. Introduction

Radio spectrum is becoming one of the most important and scarcest resources in wireless communication system. However, because of the current spectrum allocation policies, the spectrum utilization efficiency in licensed spectrum is very low [1], which generates many non-continuous vacant frequency bands, referred to as spectrum holes [2]. Cognitive radio (CR) which allows secondary users (SUs) to opportunistically utilize the frequency spectrum originally assigned to licensed primary users (PUs) is a promising approach to alleviate spectrum scarcity.

Considerable achievementhave been made for spectrum allocation in CRNs by adopting game theory [3]-[4], graph theory[5]-[6] and linear programming[7]-[9], which assume that the parameters of radio environment are static. However, in practice, due to the imperfect spectrum sensing, transmission delay and time-varied spectrum environment etc, it is difficultforthe secondary network to have theaccuratereal-timeparametersin CRNs, which will make the parameters non-deterministic. Therefore, in fact, static spectrum allocationalgorithm without considering non-deterministic parameters may result in frequent spectrum collision and poor performance. Recently, dynamic resource allocation with non-deterministic parameters in CRNs has received considerable interest from academia, which mainly focus on non-deterministic channel gain and mutual interference [10]-[12]. [10] studies the resource allocation for CRNs under non-deterministic signal-to-interference-plus-noise ratio (SINR) and propose a power control scheme by using water filling algorithm and stochastic programming. On the other hand, [11] studies the distributed resourceallocation problem in CRNs by considering the non-deterministic channel gainand the authors propose arobustdistributedpowercontrol algorithm by applying secondorder cone programming (SOCP).[12] proposes a robust distributed uplink power allocation algorithm by using worst case robust optimization method, which consider channel gains from SUs to PUs' base station, and interferencecaused by PUs to the SUs' base station are non-deterministic. Inaddition,most existing works mainly focus on non-deterministic channel gain and mutual interference. However, due to PU/SU activity and mobility, spectrum environment will change frequently, which result in time-variedbandwidth of spectrum holes. Therefore, exploring a spectrum allocation algorithm by considering non-deterministic bandwidth of spectrum holes is necessary for future CRNs, which will reduce spectrum collision and greatlyimprove spectrum efficiency. To the best ofour knowledge, non-continuous spectrum allocation in CRNs by considering non-deterministic bandwidth of spectrum holeshas not beenstudied in previous works.

This paper studies the non-continuous spectrum allocation problem in CRNs where the bandwidth of spectrum holes isnon-deterministic due to PU/SU activity and mobility. We present a novel PDF model through order statistic to describe the non-deterministic bandwidth of spectrum holes and provide a bound to approximate it. This PDF model is different from existing researches which mainly focus on the time duration of spectrum holes[13][14]. After that, a statistical spectrum allocation model based on stochastic MKP is established for spectrum allocation with non-deterministic bandwidth of spectrum holes. To reduce the computational complexity, we transform this stochastic programming probleminto a constant MKP though exploiting the properties of CDF, which can be solved via MTHG algorithm by using auxiliary variable. Simulation results illustrate that the proposed statistical spectrum allocation algorithm can achieve better performances compared to the existing algorithms when the bandwidth of spectrum holes istime-varied.

The remainder of this paper is organized as follows. The layout problem is inSection II.InSection III, the statistical model for bandwidth of spectrum holes is presented. In Section IV, the statistical spectrum allocation algorithm based on stochastic MKP methodisproposed. Numerical results are provided in Section V to demonstrate theadvantages of the proposed schemes.We conclude this paper inSection VI.

2. Layout of Problem

In this CRNs system, the spectrum occupancy over frequency domain at a certain time is non-continuous, as depicted in Fig.1 The totally frequency band is T, and the number of spectrum holes is n. $x(i)$ and $y(i)$ are used to denote the start point of spectrum occupancy and the occupied bandwidth. The bandwidth of spectrum hole are denoted by b. To describe the non-deterministic bandwidth of spectrum holes, we establish a PDF model for bandwidth of spectrum holes and based on this,a statistical spectrum allocation algorithm is proposed.

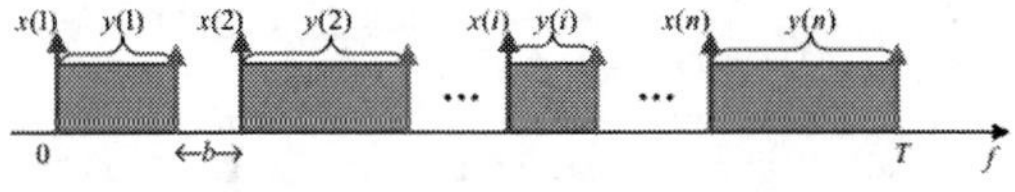

Figure 1. Spectrum occupancy over frequency domain

3. Statistical Modelfor Bandwidth ofSpectrumHoles

In this section, the non-deterministic bandwidth of spectrum holes has been formulated in a mathematic way. We consider the bandwidth of spectrum holes as a part of interval between two adjacent order statistics and represent it through a random variable. Then, base on the properties of two adjacent order statistics, the PDF of bandwidth of spectrum holes can be derived.

Theorem 1: The PDF of bandwidth of spectrum holes when $x(i)$ follows a uniform distribution $U(0,T)$ and $y(i)$ follows a conditional uniform distribution $U_y(0,x(i+1)-x(i))$ is given by

$$f_b(b)=\lambda E_1(\lambda b) \quad (1)$$

Where $\lambda=\frac{n}{T}$, and the upper and lower bounds for $f_b(b)$ is

$$\frac{1}{2}\lambda e^{-\lambda b}\ln(1+\frac{2}{\lambda b})<f_b(b)<\lambda e^{-\lambda b}\ln(1+\frac{1}{\lambda b}) \quad (2)$$

(This derivation method can also be used when $x(i)$ and $y(i)$ follow other distributions)

Proof: Let $z(i)=x(i+1)-x(i)$, where $z(i)$ represents the interval between two adjacent order statistics. Then the bandwidth of spectrum holes can be achieved by $b(i)=z(i)-y(i)$. According to [15], the joint probability distribution function of two order statistics can be obtained as

$$\begin{aligned} f_{x(i),x(j)}(u,v) &= \frac{n!}{(i-1)!(j-1-i)!(n-j)!}f_x(u)\cdot \\ & f_x(v)[F_x(u)]^{i-1}[F_x(v)-F_x(u)]^{j-1-i}\cdot \quad (3) \\ & [1-F_x(v)]^{n-j} \quad -\infty<u<v<\infty \end{aligned}$$

When $x(i)$ follows uniform distribution, $f_x(x)=\frac{1}{T}$, $F_x(x)=\frac{x}{T}$,the probability distribution function z can be expressed as

$$\begin{aligned} f_z(z) &= \int_{-\infty}^{+\infty} f_{x(i),x(i+1)}(u,u+z)du \\ &= \frac{1}{T}\frac{n!}{(i-1)!(n-i-1)!}(1-\frac{z}{T})^{n-1}\int_0^1 t^{i-1}(1-t)^{n-i-1}dt \end{aligned} \quad (4)$$

Substituting $B(\alpha,\beta)=\int_0^1 u^{\alpha-1}(1-u)^{\beta-1}du$ and $B(\alpha,\beta)=\frac{\Gamma(\alpha)\Gamma(\beta)}{\Gamma(\alpha+\beta)}$ in (4) yields

$$f_z(z)=\frac{n}{T}(1-\frac{z}{T})^{n-1}\approx\frac{n}{T}e^{-\frac{n}{T}z} \quad (5)$$

Since $y(i)$ follows a conditional uniform distribution, uniformly distributed in $(0, z)$, according to Bayes' theorem, the joint probability distribution function $f_{y,z}(y,z)$ can be obtain as

$$f_{y,z}(y,z)=\frac{1}{z}\cdot\frac{n}{T}e^{-\frac{n}{T}z} \quad (6)$$

The probability distribution function of b can be given as

$$f_b(b)=\int_{-\infty}^{+\infty} f_{y,z}(y,y+b)dy=\frac{n}{T}\int_b^T \frac{e^{-\frac{n}{T}u}}{u}du \quad (7)$$

Substitutingexponential integral function, $E_1(ax)=\int_x^{\infty}\frac{e^{-ax}}{x}dx$, in (7), result in

$$f_b(b)=\lambda[\mathrm{E}_1(\lambda b)-\mathrm{E}_1(n)] \quad (8)$$

Let $\lambda=\frac{n}{T}$, since the value of $\mathrm{E}_1(n)$ is minimal, (8) can be approximated as $\lambda\mathrm{E}_1(\lambda b)$. Based on the properties of exponential integralfunction[16], we obtain the upper and lower bounds for $f_b(b)$ is given as

$$\frac{1}{2}\lambda e^{-\lambda b}\ln(1+\frac{2}{\lambda b})<f_b(b)<\lambda e^{-\lambda b}\ln(1+\frac{1}{\lambda b}) \quad (9)$$

4. StatisticalSpectrumAllocationAlgorithmBased on StochasticMKP

4.1 Spectrum allocation model based on stochastic MKP

We assume that the CRNs consist of m SUs and n spectrum holes, where SUs'data rate requirements and the bandwidth of spectrum holes are indicated as $R=\{r_i,...,r_m\}$ and $C=\{c_1,...,c_n\}$ respectively. Among this, each SU i ($i=1,2,3,\cdots,m$) can utilize any spectrum hole j ($j=1,2,3,\cdots,n$), which means the non-continuous spectrum allocation problem can be transformed into a kind of MKP to achieve maximum transmission rate.

$$maximize \quad z=\sum_{i=1}^{m}\sum_{j=1}^{n} r_i x_{i,j} \quad (10)$$

$$s.t. \quad \sum_{i=1}^{m} b_i\cdot x_{i,j}\le c_j \qquad j\in N=\{1,...,n\} \quad (11)$$

$$\sum_{j=1}^{n} x_{i,j}\le 1 \qquad i\in M=\{1,...,m\} \quad (12)$$

$$x_{i,j}=0\ or\ 1 \qquad i\in M \quad j\in N \quad (13)$$

Let b_i be the necessary bandwidth that satisfieddata rate requirement r_i and $x_{i,j}=1$ represent SU i can allocated into spectrum hole j. Moreover, by adopting AMC scheme, SUs with different positions can achieve different data rate in same channel due to environment differences. According to G.J.Foschini's inference[17], the bandwidth needs for SU i on spectrum hole j can be given as

$$b_{i,j}=\frac{r_i}{\log_2(1+\frac{\gamma_{i,j}}{-\frac{2}{3}\ln\frac{P_b}{2}})} \quad (14)$$

Where P_b denotes maximum tolerable error rate and $\gamma_{i,j}$ indicates the SINR when SU i transmits on spectrum hole j. Substituting (14) into (11), we obtain a special MKP model, where each SU requires different bandwidth when allocated into different spectrum holes. Then (11) can be expressed as

$$\sum_{i=1}^{m}\frac{b_{i,j}}{ca_B}\cdot x_{i,j}\le\frac{c_j}{ch_B}\cdot ca_{num} \qquad j\in N=\{1,...,n\} \quad (15)$$

Where ca_B is the subcarrier bandwidth, ch_B denotes the channel bandwidth, ca_{num} indicates the number of subcarriers in each channel.

To consider non-deterministic bandwidth of spectrum holes, we take previous statistical model into this MKP model. Let $c_{j,t}$ be the available bandwidth of spectrum hole j detected bySUs' base station at time t and $C_j=\{c_{j,t-k+1},c_{j,t-k+2}...c_{j,t}\}$ represents the record of bandwidth of spectrum hole j in recent k times spectrum sensing. We can determine the parameter λ of the statistical model by applying point estimation method. The expectation of previous PDF can be written as

$$E(b)=\int_0^{\infty}\lambda bE_1(\lambda b)db=-\lambda\cdot(-\frac{\Gamma(2)}{2\lambda^2})=\frac{1}{2\lambda} \quad (16)$$

Hence, based on the method of moment[15], the PDF parameter λ_j of bandwidth of spectrum hole j can express as $1/2E(C_j)$.

We assume c_j is a random variable in (15) and follows the PDF (1), which transmits the MKP spectrum allocation model above into a stochastic programming model. The unknown random variable introduces uncertainty to (15) and makes the constraint quit complicated. Thus, to reduce the complexity, we tackle this issue by using chance-constraint programming method which finds the optimum solution under the condition that the constraints is feasible with probability greater than the threshold α. The constraint (15) above can be rewritten as

$$s.t.\ \Pr\{\sum_{i=1}^{m}\frac{b_{i,j}}{ca_B}\cdot x_{i,j}\le\frac{c_j}{ch_B}\cdot ca_{num}\}\ge\alpha_j \quad j\in N \quad (17)$$

This indicates the constraint must be feasible with the probability greater than α_j and this model maximizes the transmission rate by satisfying the constraint with high probability.

4.2 Simplification and solution

To reduce the computational complexity, we simplify the stochastic programming model above by the usage of the properties of CDF. For (17), since c_j follows the CDF $F_j(c)$, there must exist a K_j for each threshold α_j ($0 \le \alpha_j \le 1$) satisfies $\Pr\{K_j \le c_j\} = \alpha_j$. Therefore, (17) will be satisfied only if

$$\sum_{i=1}^{m} \frac{b_{i,j} \cdot ch_B}{ca_B \cdot ca_{num}} \cdot x_{i,j} \le K_j \quad (18)$$

Furthermore, as $\Pr\{K_j \le c_j\} = 1 - F_j(K_j)$, K_j can be represented by $F_j^{-1}(1-\alpha_j)$. We can transmit (17) as

$$\sum_{i=1}^{m} \frac{b_{i,j} \cdot ch_B}{ca_B \cdot ca_{num}} \cdot x_{i,j} \le F_j^{-1}(1-\alpha_j) \quad j \in N = \{1,...,n\} \quad (19)$$

Since the model proposed is a NP-hard problem, the existing algorithm obtains the solution by exhaustive search which will encounter complex computationswhen n and m have large values. Hence, we convert the model into a Generalized Assignment Problem (GAP) by using auxiliary variable and solve it via MTHG algorithm[17] which will get the suboptimal solution. Let

$$\frac{b_{i,j} \cdot ch_B}{ca_B \cdot ca_{num}} = w_{i,j},\ \tilde{c}_j = F_j^{-1}(1-\alpha_j),\ p_{i,j} = r_i,\ j \in N$$

and adding auxiliaryvariables $p_{i,n+1} = 0$, $\tilde{c}_{n+1} = m$ and $w_{i,n+1} = 1, i \in M$. Then, the model can be rewritten as

$$maximize \quad z = \sum_{i=1}^{m} \sum_{j=1}^{n} p_{i,j} x_{i,j} \quad (20)$$

$$s.t. \quad \sum_{i=1}^{m} w_{i,j} \cdot x_{i,j} \le \tilde{c}_j \quad j \in \tilde{N} = \{1,...,n,n+1\} \quad (21)$$

$$\sum_{j=1}^{n} x_{i,j} = 1 \qquad i \in M = \{1,...,m\} \quad (22)$$

$$x_{i,j} = 0 \; or \; 1 \qquad i \in M \quad j \in \tilde{N} \quad (23)$$

This GAP model means each SU must be allocated in a unique $\tilde{c}_j$ and $\tilde{c}_{n+1}$ contains the SUs which have not be allocated due to lack of spectrum resource. The GAP model can be solved by using MTHG algorithm, where iteratively consider all the unassigned SUs, and assign the SU i^* having the maximum difference between the largest and second largest $p_{i,j}$ to the spectrum hole j^* for which p_{i^*,j^*} is a maximum ($p_{i^*,j^*} = \max\{p_{i,j} : j \in \tilde{N}\}$).

5. Simulation Results

In this section, the performance of proposed statistical spectrum allocation algorithm has be studied thoughnumericalsimulations.We assume that all SUslocated around SUs' base station with the distance randomly ranged from 50 to 2000 m and the simulation parameters are set as TableI. In order to simulate the time-variedbandwidth of spectrum holes caused by PU/SU activity and mobility, we randomly generate spectrum holes with the bandwidth changed through change rate p (the probability of bandwidth of spectrum hole changed next moment) over the frequency band W. Then, we compare the performance of proposed statistical MKP spectrum allocation algorithm with static MKP spectrum allocation algorithm in [8].

TABLE I. SIMULATIONPARAMETERS

Parameters	**Values**
SU output power P_{max}	45dBm
Maximum tolerable error rate P_b	10^{-6}
SU number *SMUN*	30
Spectrum hole number *SH*	3~6
Noise power $N_0 B$	-111dBm
Chance programming threshold α	0.5
Bandwidth of spectrum hole change rate p	0.1~1
Total frequency band W	10MHZ

Fig. 2illustrates the CDF of SU spectrum efficiency using the two algorithmswith different situation ($p = 0.5$ and $p = 1$). As shown in Fig. 2, statistical MKP algorithm and static MKP algorithm occupy 61% and 52% respectively when the spectrum efficiency is large than 1.8, which means statistical MKPalgorithm has larger percentage of high spectrum efficiency than the static MKP algorithm. This case is becausestatistical MKP algorithm which takes the statistical properties of bandwidth of spectrum holes into consideration can achieve dynamic optimal and reduce bandwidth collisions caused by spectrum holes changes. Moreover, as p increased to 1, statistical MKP algorithm exceeds static MKP algorithm more, which means statistical MKP algorithm can better adapt to the scenario with frequency time-varied bandwidth of spectrum holes and achieve high spectrum efficiency.

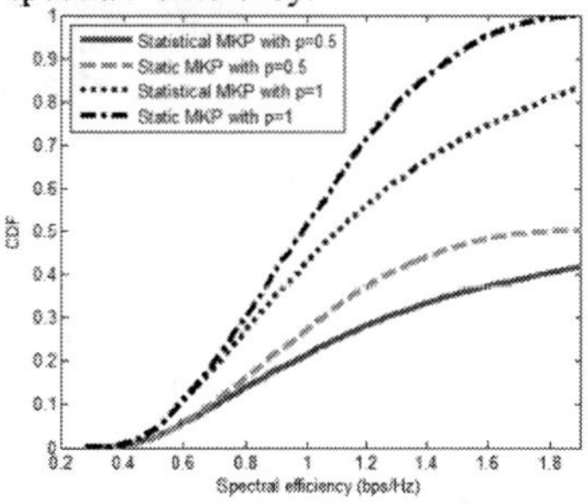

Figure 2. CDF of spectrum efficiency

In Fig.3, we evaluate the network utility with different bandwidth change rate p. The network utility is defined as[18]

$$U=\sum_{k=1}^{m}\ln R(\mathrm{k}) \quad (24)$$

Where $R(\mathrm{k})$ is the throughput for SU k. In theFig. 3, as p increased, the network utility is reduced due to bandwidth collision. The static MKP algorithm suffers lower networkutility thanstatistical MKP algorithm, because it only considers current static optimal and use constant allocation parameters, which may lead to more bandwidth collision. In addition, when the number of spectrum holes increase, the network utility of statistical MKP algorithm increase more than the static MKP algorithm, which means statistical MKP algorithm can obtain a good performance under the condition of non-continuous spectrum.

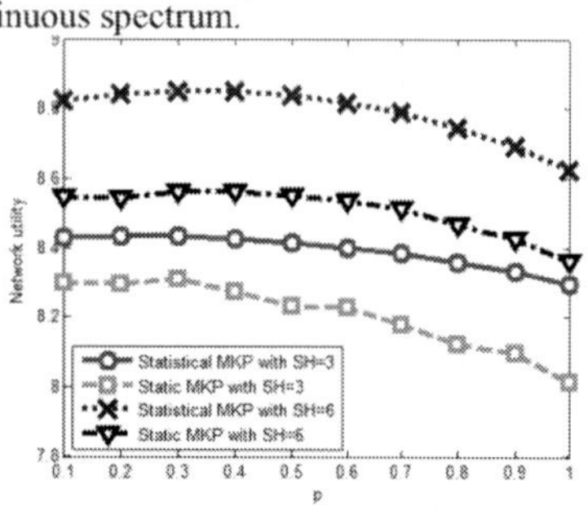

Figure 3. Network utility with different bandwidth of spectrum holes change rate p

Fig. 4 shows the average spectrum collision rate with different bandwidth change rate p. From Fig. 4, we can see that as p increase, the spectrum collision rate increased dramatically and statistical MKP algorithm achieve lower collision rate than the static MKP algorithm. It is becausestatistical MKP algorithm can relax the bandwidth collision through assigning spectrum resource with considering the statistical properties of spectrum holes. The result shows that statistical MKP algorithm can effectively reduce the spectrum collision when spectrum holes changes.

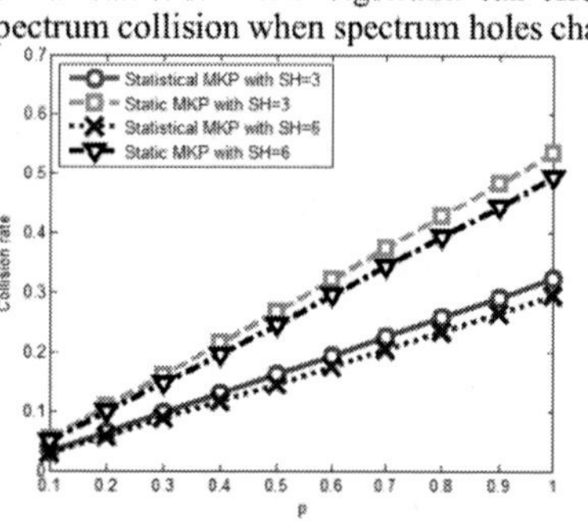

Figure 4. Average spectrum collision rate with different bandwidth of spectrum holes change rate p

6. Conclusion

In this paper, we study the problem of non-continuous spectrum allocation in CRNs where the bandwidth of spectrum holes is non-deterministic due to PU/SU activity and mobility.We present a novel PDF model through order statistic to describe the non-deterministic bandwidth of spectrum holes and provide a bound to approximate it.After that, a statistical spectrum allocation model based on stochastic MKP is established for spectrum allocation with non-deterministic bandwidth of spectrum holes. To reduce the computational complexity, we transform this stochastic programming probleminto a constant MKP though exploiting the properties of CDF, which can be solved via MTHG algorithm by using auxiliary variable. Simulation resultsverify that the proposedstatistical spectrum allocation algorithm can achieve better performances compared to the existing algorithms when the bandwidth of spectrum holes istime-varied.

Acknowledgment

ThisworkwassupportedbytheTrainingProgramoftheMajor ResearchPlanoftheNaturalScienceFoundationofChina(Grant No.91438104)andtheGraduateStudentsResearchProgramofC hongqing(GrantNo.CYS14005).

References

1. C. Kim, C. Leem, S. Kang,"Policy and technology of dynamic spectrum access in Korea,"2008 3rd International Conference on Cognitive Radio Oriented Wireless Networks and Communications, pp. 1-4, 2008, doi: 10.1109/CROWNCOM.2008.4562456.
2. M. Mchenry,E.Livsics,T.Nguyen,"XG dynamic spectrum access field test results," IEEE Communication Magazine,pp. 51-57,Jun. 2007.
3. J. Revathy,M. Senthil,"Resource allocation in next generation networks using game theory,"2014 International Conference on Information Communication and Embedded Systems (ICICES), pp. 1-4.
4. Z. Wu, P. Cheng, X. Wang,"Cooperative spectrum allocation for cognitive radio network: an evolutionary approach,"2011 IEEE International Conference on Communications (ICC), pp. 1-5.
5. Y. Li, L. Zhao, C. Wang,"Aggregation-based spectrum allocation algorithm in cognitive radio networks,"Network Operations and Management Symposium (NOMS), IEEE, pp. 506-509.
6. A. Plummer, S. Biswas,"Distributed spectrum assignment for cognitive networks with heterogeneous spectrum opportunities," Wireless Communications and Mobile Computing, pp. 1239-1253, Sep. 2011.
7. F. Wu, Y. Mao, S. Leng, "A carrier aggregation based resource allocation scheme for pervasive wireless networks,"2011 IEEE Ninth International Conference on Dependable, Autonomic and Secure Computing (DASC), IEEE, pp. 196-201.
8. C. Li, W. Liu, Q. Liu, "Spectrum aggregation based spectrum allocation for cognitive radio networks,"Wireless Communications and Networking Conference (WCNC), IEEE, pp. 1626-1631.
9. Y. Song, C. Zhang, Y. Fang,"Multiple multidimensional knapsack problem and its

applications in cognitive radio networks,"Military Communications Conference, MILCOM. pp. 1-7.

10. P. Setoodeh, S. Haykin,"Robust transmit power control for cognitive radio," Proc. of the IEEE,vol. 97, no.5, pp. 915-939, May. 2009.
11. S. Sun, W. Ni, Y.Zhu,"Robust power control in cognitive radio networks: a distributed way,"2011 IEEE International Conference on Communications (ICC), pp. 1-6.
12. S. Parsaeefard, A. Sharafat,"Robust distributed power control in cognitive radio networks," IEEE Transactions onMobile Computing, pp. 609-620, Apr. 2013.
13. J. Misic, V. Misic,"Probability distribution of spectral hole duration in cognitive networks,"INFOCOM, 2014 Proc. IEEE.pp. 2103-2111.
14. Y. Saleem, M. Rehmani,"Primary radio user activity models for cognitive radio networks: A survey," Journal of Network and Computer Applications, pp. 1-16, Mar. 2014.
15. G. Casella, R. Berger,"Statistical inference. Pacific Grove," CA: Duxbury, 2002.
16. M. Abramowitz,S. Irene,"Handbook of mathematical functions: with formulas, graphs, and mathematical tables," Courier Corporation, 1964.
17. S. Martello, P.Toth,"Knapsack problems: algorithms and computer implementations," John Wiley Sons, 1990.
18. Y. Wang, K. Pedersen, T. Sorensen,"Utility Maximization in LTE-Advanced Systems with Carrier Aggregation,"2011 IEEE 73rd Vehicular Technology Conference (VTC Spring), pp. 1 - 5.

Joint Interference Detection Method for DSSS Communications Based on the OMP Algorithm and CA-CFAR

Yongshun Zhang , Xin Jia and Ge Song

Equipment Academy, 101416 Beijing, China

Abstract. The existing direct sequence spread spectrum (DSSS) communications interference detection algorithms are confined to the high sampling rate. In order to solve this problem, algorithm for DSSS communications interference detection was designed based on compressive sensing (CS). First of all, the orthogonal matching pursuit (OMP) algorithm was applied to the interference detection in DSSS communications, the advantages and weaknesses of the algorithm were analyzed; Secondly, according to the weaknesses of the OMP algorithm, a joint interference detection method based on the OMP algorithm and cell average constant false alarm rate (CA-CFAR) was proposed. The theoretical analyze and computer simulation all proved the effectiveness of the new algorithm. The simulation results show that the new method not only could achieve the interference detection, but also could estimate the interference quantity effectively.

1 Introduction

Interference detection is an important part of communication anti-jamming technology; it provides the necessary information for the interference suppression. The existed interference detection algorithms could be divided into two parts, one is time-domain interference detection algorithm and the other is transform domain interference detection algorithm. The typical algorithm of time domain interference detection algorithm is the energy detection algorithm [1]. The transform domain interference detection algorithms are mainly based on the different domain characteristics of the signal, background noise and interference. The commonly transform domains include the frequency domain [2], the wavelet domain [3], the fractional Fourier transform domain [4], etc. These methods achieve interference detection successfully, but with the development of technologies, the signal bandwidth is growing, large bandwidth leads to a sharp increase of the sampling rate, the existing methods are confined to the high sampling rate.

Compressive sensing (CS) [5] is a new kind of signal and image processing technology. The technology breaks through the limitation of the Nyquist sampling theorem. This is of great significance in large bandwidth signal processing. At present, the study of signal processing based on CS are mainly focus on the precisely reconstruct of the original signal from compressed signal. However, it is not necessary in many cases, such as signal detection [6]. Due to the structure information of the signal are already included in the compressed signal, we could distinguish whether the interested signal exist just from the compressed data without reconstructing it correctly. Some scholars proposed the signal detection algorithm based on the CS [7, 8].

In this paper, we achieve the DSSS communications interference detection by using the OMP algorithm first. According to the deficiency existing in the algorithm, proposed a joint detection method based on the OMP algorithm and cell average constant false alarm rate (CA-CFAR). The theoretical basis of the method were analyzed, the effectiveness of the method was verified by computer simulation.

2 Compressive Sensing

There are three core issues in CS theory: signal sparse representation, measurement matrix design and signal reconstruction algorithm.

Signal sparse representation: for any signal $\mathbf{s} \in R^N$, if there is a transformation matrix $\mathbf{\Psi} = \{\psi_1, \psi_2, \psi_3, ..., \psi_N\}$, which satisfies (1), if there only K $(K \ll N)$ on-zero elements in $\mathbf{\alpha}$, so $\mathbf{s}$ can be sparse representation in $\mathbf{\Psi}$.

$$\mathbf{s} = \sum_{i=1}^{N} \psi_i \alpha_i \text{ , } \mathbf{s} = \mathbf{\Psi\alpha} \tag{1}$$

Measurement matrix design: we achieve the dimensionality reduction by construction the measurement matrix $\mathbf{\Phi}$. The process can be expressed in (2).

$$\mathbf{y} = \mathbf{\Phi s} = \mathbf{\Phi\Psi\alpha} = \mathbf{\Theta\alpha} \tag{2}$$

If the signal sparse degree is K , the length of the signal is N . Then the row numbers of the measurement matrix is $M \geq O(K \log_2(N/K))$.

Signal reconstruction: Candès proved that if the (sensing matrix) $\mathbf{\Theta}$ satisfies the restricted isometry property (RIP) [9], then we can reconstructed the original signal from the compressed data accurately by solving the optimal problem of l_0 norm.

$$\min_{\alpha} \|\boldsymbol{\alpha}\|_0 \quad s.\ t.\ \ \mathbf{y} = \mathbf{\Theta}\boldsymbol{\alpha} \tag{3}$$

In (3), $\|\bullet\|_p$ represents $p-norm$. Seeking for the sparse representation of a signal from the random sensing matrix is a non-deterministic polynomial problem [10]. A lot of algorithms could be used to solve the problem.

3 The DSSS communications interference detection BASED on CS

3.1 The Problem Description of the Interference Detection in DSSS Communications Based on CS

In DSSS communications interference detection process, the interested signal is the interference signal. Due to the DSSS signal is similar to the noise, the existence of the DSSS signal does not affect the interference detection seriously. The DSSS communications interference detection problem could be expressed to distinguish the two hypothetical questions as follows.

$$H_0:\ \mathbf{x} = \mathbf{s} + \mathbf{n} \tag{4}$$

$$H_1:\ \mathbf{x} = \mathbf{s} + \mathbf{j} + \mathbf{n} \tag{5}$$

In (4) and (5), $\mathbf{s}$ represents the DSSS signal, $\mathbf{j}$ represents the interference, $\mathbf{n}$ represents the white Gaussian noise.

If the interference can be sparse representation in $\mathbf{\Psi}_{jam}$, then the above equations could also be expressed in (6) and (7).

$$H_0:\ \mathbf{x} = \mathbf{s} + \mathbf{n} \tag{6}$$

$$H_1:\ \mathbf{x} = \mathbf{\Psi}_{jam}\boldsymbol{\alpha} + \mathbf{s} + \mathbf{n} \tag{7}$$

The compressed signal can be expressed as follows.

$$H_0:\ \mathbf{y} = \mathbf{\Phi}(\mathbf{s} + \mathbf{n}) \tag{8}$$

$$H_1:\ \mathbf{y} = \mathbf{\Phi}(\mathbf{\Psi}_{jam}\boldsymbol{\alpha} + \mathbf{s} + \mathbf{n}) \tag{9}$$

In (9), $\boldsymbol{\alpha}$ represents the coefficient vector, $\mathbf{\Phi}$ represents the measurement matrix. From (8) and (9), we know that when the interference does not exist, compressed signal projection values on $\mathbf{\Psi}_{jam}$ is small and distribute randomly. Only when the interference exists, big projection value appears. So we could achieve the interference detection by checking the projection values. Because only a few kinds of interference patterns in DSSS communications system could be sparse representation, we assume that the types of interference may exist in the DSSS communications are known in this paper, so we could get the interference sparse dictionary. This made the interference detection just like signal detection process. So the signal detection method based on the OMP algorithm [8] was used to achieve DSSS communications interference detection firstly.

3.2 Interference detection algorithm based on the OMP algorithm

The OMP [8] algorithm is the most commonly used greedy algorithm. Compared to the other reconstruction algorithm, it has low computational complexity. The description of the interference detection algorithm based on the OMP algorithm was as follows.

Input: measurement matrix $\mathbf{\Phi}$, interference sparse dictionary $\mathbf{\Psi}_{jam}$, iteration times K, threshold γ;

Initialization: initialization residual $\mathbf{r\_n} = \mathbf{y}$, incremental matrix $\mathbf{Aug\_t} = []$, iterations times $time = 1$, coefficient vector $\mathbf{aug\_y} = []$;

Output: detection result;

Cycle steps:

Step 1) Search the sensing matrix $\mathbf{\Theta} = \mathbf{\Phi}\mathbf{\Psi}_{jam}$ for the most relevant column vector with the residual, that is $Inp = \max_{i=1,2...N} |\langle\mathbf{r\_n}, \boldsymbol{\psi}_i\rangle|$;

Step 2) Update the incremental matrix $\mathbf{Aug\_t} = [\mathbf{Aug\_t}, \boldsymbol{\psi}_i]$;

Step 3) Calculate the coefficient vector $\mathbf{aug\_y}$ by using the least squares method, $\mathbf{aug\_y} = (\mathbf{Aug\_t}^T * \mathbf{Aug\_t})^{(-1)} * \mathbf{Aug\_t}^T * \mathbf{y}$;

Step 4) Update the residuals $\mathbf{r\_n} = \mathbf{y\text{-}Aug\_t*aug\_y}$;

Step 5) If $time < K$, $time = time + 1$, then return to step 1), else, execute step 6);

Step 6) If $\|\mathbf{aug\_y}\|_\infty > \gamma$, the interference exists, else, the interference does not exist.

4 The joint interference detection method based on the OMP and CA-CFAR

There three weaknesses in the interference detection algorithm based on the OMP algorithm.

(1) It is difficult to choose the iteration times in the algorithm, especially in the interference detection process;

(2) Due to the threshold changes with the signal-to-noise ratio (SNR), it is difficult to choose the right threshold in the algorithm.

(3) The hardware implemented of the OMP algorithm is still under study [11, 12].

In order to overcome the weakness of the algorithm based on the OMP algorithm, we proposed a joint interference detection algorithm based on the OMP algorithm and CA-CFAR.

4.1 Theoretical analyze of the new algorithm

The first step of the OMP algorithm is inner product searching. It was show in (10) and (11).

$$\hat{\mathbf{x}} = (\mathbf{\Phi}\mathbf{\Psi}_{jam})^H * (\mathbf{\Phi}\mathbf{\Psi}_{jam}\boldsymbol{\alpha}) \tag{10}$$

$$\hat{\mathbf{x}} = \mathbf{\Psi}_{jam}{}^H\mathbf{\Phi}^H * \mathbf{\Phi} * \mathbf{\Psi}_{jam} * \boldsymbol{\alpha} \tag{11}$$

The above expression can be expressed as follows.

$$\hat{\mathbf{x}} = \mathbf{\Psi}_{jam}{}^H * (\mathbf{\Phi}^H * \mathbf{\Phi}) * \mathbf{\Psi}_{jam} * \boldsymbol{\alpha} \tag{12}$$

$$\hat{\mathbf{x}} = \mathbf{\Psi}_{jam}{}^H * \mathbf{Phi} * \mathbf{\Psi}_{jam} * \boldsymbol{\alpha} \tag{13}$$

In (13), **Phi** is a normal matrix. As can be seen from the above equation, after the calculation, we got avector. Due to the characteristics of the matrix **Phi** and $\boldsymbol{\Psi}_{jam}$, $\hat{\mathbf{x}}$ is associated with the coefficient $\boldsymbol{\alpha}$. When there is a big sparse coefficient in $\boldsymbol{\alpha}$, there is a corresponding peak in $\hat{\mathbf{x}}$, so we also could achieve the interference detection by detecting the peak value in $\hat{\mathbf{x}}$. After the above operation, the random characteristics of the compressed data are overcome, we convert the compressed data to the same length as the original signal, the new signal contains the interference signal information, so the conventional detection method will suitable for the interference detection.

CA-CFAR [13] is a common method in radar signal detection. For a given expected average false alarm rate, the required threshold could be obtained by solving the following equations.

$$\bar{P}_{fa} = (1+\frac{T}{L})^{-L} \tag{14}$$

$$T = L(\bar{P}_{fa}^{-1/L} - 1) \tag{15}$$

In (14) and (15), $\bar{P}_{fa}$ represents the false alarm rate, T represents the threshold product factor, L represents the length of the lead and lag cell. The final detection threshold was showed as follows.

$$\hat{T}_h = \frac{T}{L}\sum_{i=1}^{L}\hat{\mathbf{x}}_i \tag{16}$$

In (16), $\hat{T}_h$ represents the threshold. When the peak value in is greater than the threshold, we believe that the existence of interference. When the peak is less than the threshold, we believe that there is no interference.

4.2 Algorithm description

According to the above description of the principle, the proposed algorithm could be described as follows:

Input: measurement matrix $\boldsymbol{\Phi}$, interference sparse dictionary $\boldsymbol{\Psi}_{jam}$, the false alarm rate P_{fa}, the lead and lag cells length L_c, the guard cell length L_g;

Initialization: initialization residual $\mathbf{r\_n} = \mathbf{y}$;

Output: detection result.

Steps:

Step 1) Calculate the product of the sensing matrix $\boldsymbol{\Theta} = \boldsymbol{\Phi}\boldsymbol{\Psi}_{jam}$ and the residual, namely $\hat{\mathbf{x}} = (\boldsymbol{\Phi}\boldsymbol{\Psi}_{jam})^H * \mathbf{r\_n} = \boldsymbol{\Theta}^H * \mathbf{r\_n}$;

Step 2) Using the matched filter to deal with the $\hat{\mathbf{x}}$, output signal is $\mathbf{r}$, $\mathbf{r} = \hat{\mathbf{x}} * \bar{\hat{\mathbf{x}}}$;

Step 3) Using the CA-CFAR detection method to deal with the output signal of the matched filter, we got to know whether the interference exist using the detection result.

5 Simulation Results and Analysis

We take the DSSS communications multi-tone interference detection as an example to verify the effectiveness of the algorithm. The DSSS signal parameters were set as follows: source rate was $100kb/s$, PN code was m sequence of 63, PN code rate was $6.3Mb/s$. The modulation scheme is binary phase shift keying (BPSK), the carrier frequency was $6.3MHz$. The multi-tone interference parameters were set as follows: the multi-tone interference frequency was $[5,8]MHz$, the unit is Hz. The sparse dictionary of the interference is Fourier orthogonal matrix. The CA-CFAR detection algorithm parameters were set as follows: the lead and lag cell is 50, the guard cell is 2. We use the Gaussian random matrix as the measurement matrix. The compression ratio is defined as $M = \text{actual sampling number/Nyquist sampling number}$. The compression ratio is 1/2 in this paper. Doing 2000 Monte-Carlo simulation for every experiment.

Fig. 1 shows the success rate of interference detection algorithm based on the OMP algorithm in the condition of 10dB noise, the threshold was set to 4.9 which was the optimal threshold obtained by multiple simulations, algorithm iteration number was set to 2. As can be seen from the figure, the algorithm could achieve the interference detection effectively. When the jammer-to-signal ratio (JSR) is more than -6dB, the success rate of interference detection was more than 0.95. This proves that with the prior information of the interference types and the number of iterations, the interference detection algorithm based on the OMP algorithm is feasible for the DSSS interference detection.

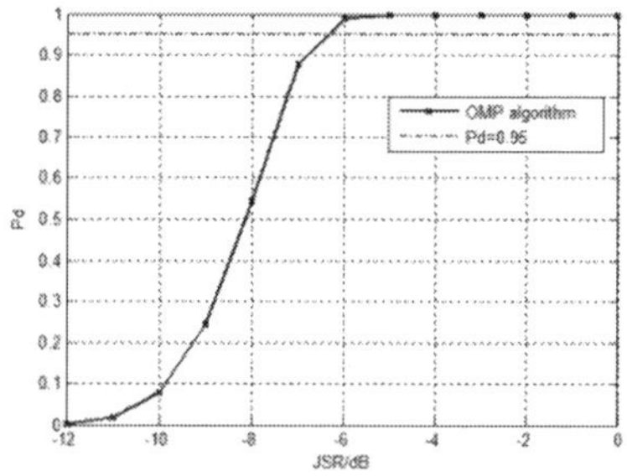

Figure 1. Success rate of interference detection algorithm based on the OMP algorithm (10dB noise)

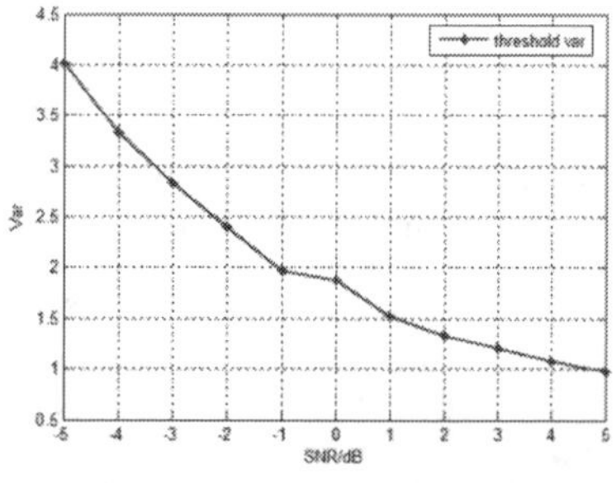

Figure 2. The threshold fluctuation with SNR

In the interference detection algorithm based on the OMP algorithm, the threshold is very important. Fig. 2 shows the threshold fluctuations of the algorithm. As can be seen from the figure, the threshold changes with the SNR, the fluctuations of the threshold is serious, this factors will confine the application of the algorithm. At this point, the interference detection algorithm based on the OMP algorithm was not applicable.

Fig. 3 shows the interference detection result using the interference detection algorithm based on the OMP algorithm and CA-CFAR. The JSR was set to -9dB, the SNR was set to 10dB, the false alarm rate was set to 10e-6. As can be seen from the figure, the algorithm works correctly in detecting the interference even through the interference power is lower than the signal. It proved that the algorithm is suitable for interference detection in DSSS communications.

CA-CFAR is an adaptive threshold detection algorithm, so the interference detection algorithm based on the OMP algorithm and CA-CFAR is essentially an adaptive threshold interference detection algorithm, the performance of the algorithm does not change over the threshold. Fig. 4 shows the performance of the algorithm based on the OMP algorithm and CA-CFAR in the condition of different SNR, the JSR was set to 10dB, the false alarm rate was set to 10e-6. As can be seen from the figure, the algorithm works pretty good. The simulation result proved the above analysis.

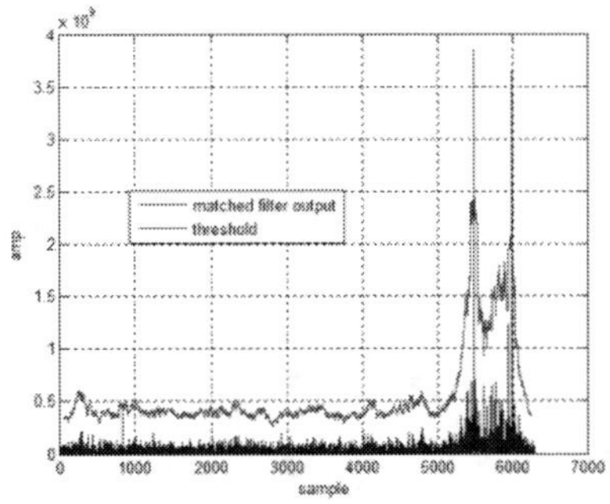

Figure 3. Interference detection result of the algorithm based on the OMP algorithm and CA-CFAR

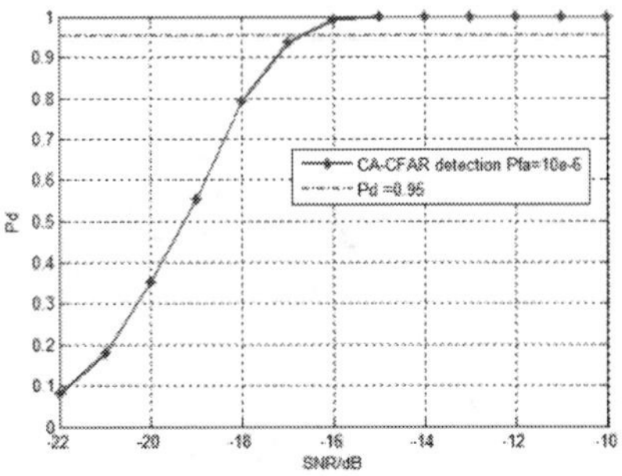

Figure 4. Success rate of interference detection in the condition of different SNR

Fig. 5 shows the comparison of the two algorithms. As can be seen from the figure, the algorithm based on the OMP algorithm and CA-CFAR got the similar performance with the algorithm based on the OMP algorithm when the false alarm rate was set to 10e-6. We also can see that the success rate of the interference detection in the new algorithm changes with the false alarm rate, with a higher success rate of interference detection, the false alarm rate also increased, so there should be a tradeoff between the false alarm rate and success rate of the interference detection in practical.

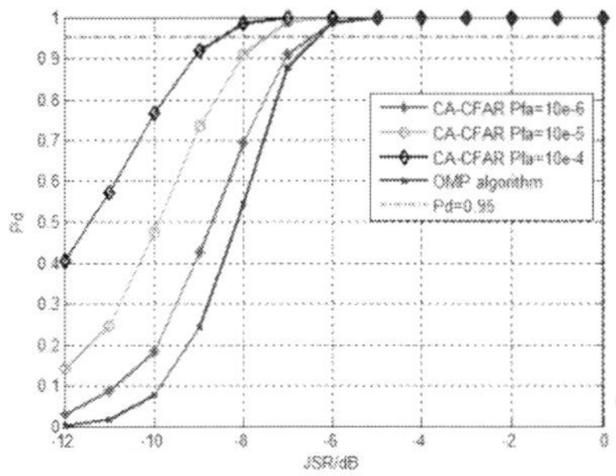

Figure 5. Success rate of the interference detection of the two algorithms

The interference detection algorithm based on the OMP algorithm and CA-CFAR not only can achieve the interference detection, but also can get the quantity information of the interference. Fig. 6 shows the quantity estimation results using the algorithm. As can be seen from the figure, when JSR was more than 6dB, the estimation result of the interference is close to the real value. According to Fig. 3, we can also see that, when the interference signal index positions in the sparse dictionary were in the same reference unit range, we may not estimate the quantity of the interference accurately.

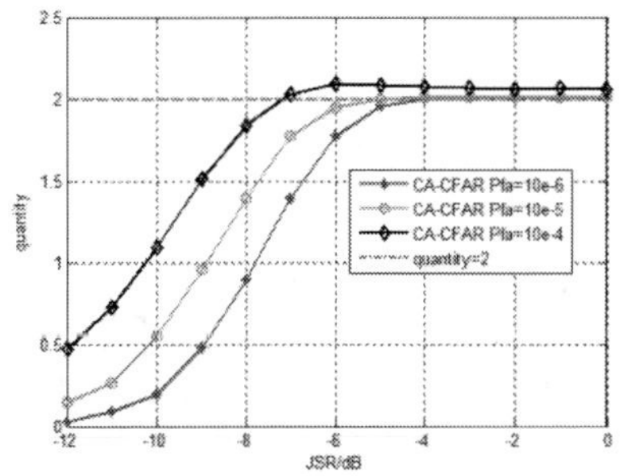

Figure 6. The estimation quantity of the interference

6 Conclusion

We proposed the joint compressive sensing interference detection algorithm for the DSSS communications based the OMP algorithm and CA-CFAR in this paper. With the prior information of the interference types that may exist in the DSSS communications system, the algorithm could achieve the interference detection effectively and it has a good robustness. Due to the algorithm achieves the DSSS communications interference detection by using the CS, the high sampling rate problem in the traditional interference detection algorithms was solved, also the new algorithm is useful to reduce the system costs. Interference detection process is the basis of the interference suppression method, this algorithm will help to achieve the DSSS communication interference suppression in compressed domain and we will carry out the research in this field in the future.

References

1. Y. Lee, D. Kang, and H. Park, Proc. 3rd International Conference on Anti-counterfeiting, Security, and Identification in Communication. (2009).
2. T. Yamamoto, K. Takeda, and F. Adachi, Proc. 2009 IEEEE 70th Vehicular Technology Conference Fall. (2009).
3. S. C. Olhed and G. Metikas IEEE Transactions on Signal Processing, **57**. (2009).
4. Y. Zhang, Q. Zhang, and S. Wu, 2008 International Conference on International Technology and Applications in Biomedicine. (2008).
5. X. Chu, M. C. Stamm, K. J. R. Liu IEEE Transactions on Information Forensics and Security, **10**. (2015).
6. J. Haupt and R. Nowak IEEE International Conference on Acoustics, Speech and Signal Processing. (2007).
7. M. F. Duarte, M. A. Davenport, M. B. Wakin. IEEE International Conference on Acoustics, Speech and Signal Processing. (2006).
8. L. Bing, F. Ping, M. Shengwei, Chinese journal of scientific instrument, **31**. (2010).
9. J. Castorena and C. D. Creusere, IEEE Transactions on Signal Processing, **62**. (2014).
10. A. Majumdar, R. K. Ward, IEEE Jounal on Emerging and Selected Topics in Circuits and Systems, **2**. (2012).
11. F. Lingzhi, M.S. thesis, South-Central University for Nationalities, Wuhan, (2013).
12. X. Chengyi and D. Panfeng, Journal of South-Central University for Nationalities (Nat. Sci. Edition), **32**. (2012).
13. M. A. Richards. New York: McGraw-Hill. (2005)

Frequency Estimation Using SAMP-SVD Based on Nyquist Folding Receiver

Jun Wang , Sujuan Chen , Kaili Jiang , and Bin Tang

School of Electronic Engineering, University of Electronic Science and Technology of China, Chengdu, Sichuan, P.R. China

Abstract. The Nyquist Folding Receiver (NYFR) is an efficient analogy-to-information (A2I) architecture that folds the broadband RF inputs by subsampling so that can sample with a low-speed ADC. The compressive sensing (CS) model of the NYFR can be built in the sparse environment to recover by traditional CS algorithms. This paper presents an improved algorithm SAMP-SVD based on sparsity adaptive matching pursuit (SAMP) and singular value decomposition (SVD) which is intended to deal with the problem caused by SAMP recovering—the inversion of matrix, avoiding the problem of whether matrix is singular and reducing the time of computation. A comparative analysis between the presented method and sparse reconstruction by separable approximation (SpaRSA) is discussed; simulation results show the algorithm can reconstruct the signal effectively and estimate the frequencies accurately.

1 Introduction

The bottleneck in reconnaissance receiver performance is in the analog-to-digital conversion (ADC) process, where sampling rate and accuracy in limitation [1], cannot matching to radar coverage. Given this discord, we need to improve the receiving system on the present device level, breaking the limitation of ADC.

There are currently several sub-sampling architectures: the random demodulator (RD) [2] and the modulated wideband converter (MWC) [3] as a practical implementation of compressive sensing (CS) [4], where the signals is sampled completely randomly for suppression of aliasing and guaranteed reconstruction. Another efficient analogy-to-information (A2I) architecture is called the NYFR [5] in 2006 presented, which folds the broadband radio frequency (RF) input by sub-sampling with a stream of short pulses. The modulation sampling induces a bandwidth widening that varies with the Nyquist zone (NZ, i.e. folding index number). It is a periodic non-uniform sampling, which is a tradeoff between folding and aliasing. The under-sampled signals then digitized by a narrowband ADC, which the same as CS techniques by using only a few non-zero coefficients in some suitable sparse representation, allowing recovery techniques in addition to traditional CS recovery methods such as reconstruction via l_1-minimization [6].

There are a number of classical approaches to recover information from signals, which are obtained as the solution to the constrained minimization problem in a CS setting, including l_1-minimization, Sparse Reconstruction by separable approximation (SpaRSA) [7], iterative greedy unfolding (IGU) [8] and matching pursuit (MP) [9], etc. l_1-minimization and SpaRSA are by means of optimization theory , the former tend to be slow for extremely large sampled data sets and, ill-suited for non-positive definite measurement matrix, if speed is not an issue, SpaRSA obtains higher resolution accuracy. MP and orthogonal matching pursuit (OMP) [10] techniques are the basis methods on pursuit algorithms, which enjoy highly efficient, low computational complexity and a restricted isometry property (RIP) [11]. stagewise orthogonal matching pursuit (StOMP) [12] reduces the complexity of the algorithm, but its accuracy not as good as OMP. regularized orthogonal matching pursuit (ROMP) [13] and compressive sampling matching pursuit (CoSaMP) [14] require the sparsity knowledge of signals. subspace pursuit (SP) [15] algorithm performance is similar to CoSaMP, and combination with OMP to get a powerful methods named SAMP [16].

This paper focuses on SAMP because of its efficiency and ease of implementation, and applies SVD to solve the least squares (LS) problem, for a high-accuracy frequency estimation under the condition of a low signal-to-noise ratio (SNR).

2 NYFR Sensing Matrix Model

CS builds upon the fundamental fact that signals can be represented by using only a few non-zero coefficients in a suitable basis or dictionary, projecting a high-dimensional signal onto a low-dimensional signal model, recovered such signals from very few measurements by optimization algorithm, to break through the traditional limits of sampling theory.

Assumptions about the analog RF input signals $x_a(t)$ of the NYFR architecture in a usual CS setting [8]; namely:

1) the input $x_a(t)$ can be sampled according to the Nyquist theorem, denoted as $x[n]$.

2) the discrete Fourier transform (DFT) $X[k]$ of $x[n]$ is sparse or highly compressible.

Note that the relationship between the NYFR architecture and the theory of CS can be presented as following in the assumptions setting. Then we represent $X[k]\in\mathbb{C}^N$ as a vector of s-sparse (i.e., the number of non-zero elements is s in the frequency domain), and consider the output measurement vector $y[n]\in\mathbb{C}^K$ after ADC sampling. Expressed the NYFR receiving model by underdetermined linear equations

$$y[n]=\phi x[n] \quad or \quad y[n]=\Phi X[k] \tag{1}$$

where $\phi\in\mathbb{C}^{K\times N}$ is measurement matrix and, $\Phi\in\mathbb{C}^{K\times N}$ is sensing matrix. Note that $N=K\cdot Z$, where Z denotes index number (or folding number).

Motivated by RD model, the direct measurement matrix (D-measurement matrix) is given by

$$\phi=diag(P_1,P_2,\cdots,P_K) \tag{2}$$

where $P_i=\left[p_{1+(i-1)Z},p_{2+(i-1)Z},\cdots,p_{Z+(i-1)Z}\right]$, $i=1,2,\cdots,K$, and $p_j=\exp\left(-j\left(2\pi kf_{s1}t_j+k_H\theta(t_j)\right)\right)$, $j=1,2,\cdots,N$, namely that the modulation sampling function is sampled by the Nyquist rate of the input $x_a(t)$.

A next consideration is signal reconstruction from $y[n]$ and ϕ by CS techniques. Actually, the NYFR signal reconstruction is dependent of k_H (namely, folding index number) estimated correctly and the synchronization of measurement matrix sampled with wideband-input signals under Nyquist theorem, from the mathematical point-of-view. Thus the sensitivity of folding index number estimation and the synchronization of sampling with inputs are important to estimate accurately, which show the deficiencies.

$$\begin{bmatrix} y[1]\\ y[2]\\ \vdots\\ y[K]\end{bmatrix}=\underbrace{\begin{bmatrix} I_K & I_K & \cdots & I_K\end{bmatrix}}_{R}\underbrace{\begin{bmatrix} I_K & & & \\ & e^{-j\theta(t)}I_K & & \\ & & \ddots & \\ & & & e^{-j(Z-1)\theta(t)}I_K\end{bmatrix}}_{S}\underbrace{\begin{bmatrix}\Psi_K & & & \\ & \Psi_K & & \\ & & \ddots & \\ & & & \Psi_K\end{bmatrix}}_{\Psi}\begin{bmatrix} X_1[k]\\ X_2[k]\\ \vdots\\ X_Z[k]\end{bmatrix} \tag{3}$$

Given the sensing matrix model without the estimation of folding index number as the reference [8]. The DFT of Nyquist rate sampled inputs with the length of N, partitioned into Z index numbers of length K each block matrix, meets $N=K\cdot Z$. Then induced modulation by diagonal matrix $S\in\mathbb{C}^{N\times N}$ which is the function of time on the inverse DFT (IDFT) block matrices $\Psi\in\mathbb{C}^{N\times N}$ of inputs with index numbers from zero to Z and, projected into baseband of length K also with matrix $R\in\mathbb{C}^{K\times N}$, where R, S and Ψ are partitioned as $X[k]$. This model is depicted by

where the measurement matrix is $\phi=RS\in\mathbb{C}^{K\times N}$, and the sensing matrix is $\Phi=RS\Psi\in\mathbb{C}^{K\times N}$, the block matrix of $X[k]$ is given by

$$X_i[k]=\begin{bmatrix} X[(i-1)K+1]\\ X[(i-1)K+2]\\ \vdots\\ X[iK]\end{bmatrix}\in C^{K\times 1},\ i=1,\ldots,Z \tag{4}$$

and the IDFT matrix of length K is

$$\Psi_K=\frac{1}{\sqrt{K}}\begin{bmatrix} 1 & 1 & 1 & & 1\\ 1 & \omega^1 & \omega^2 & \cdots & \omega^{K-1}\\ 1 & \omega^2 & \omega^4 & \cdots & \omega^{2(K-1)}\\ \cdots & \cdots & \cdots & \cdots & \cdots\\ 1 & \omega^{K-1} & \omega^{2(K-1)} & \cdots & \omega^{(K-1)(K-1)}\end{bmatrix} \tag{5}$$

where $\omega=e^{j2\pi/K}$. Nevertheless the deficiency on synchronization based on this model is the same as the model mentioned before, from the practical consideration.

3 Information Recovery

The OMP and the SP are the typical greedy algorithm of the CS algorithm, which can search for answers in the several regions of a solution space and jump out of local optimization at a greater probability, thus the global optimization can be found. The SAMP is the combination of the OMP look-ahead searching and the SP backtracking searching, modifying the support set (SS) with a fixed step length in iterative process, eliminating the poor atoms via new and the backtracking searching until residuals no longer shrinking.

The SAMP brings solution for blind signal recovery which does not require information of sparsity s of input signals as a prior, following greedy iteration principles through stage by stage to estimates the sparsity. The OMP and the SP are viewed as its special cases, namely that when $\hat{s}=1$, SAMP can be roughly regarded as OMP, which takes a few more iterations on more accuracy. In addition, when $\hat{s}=s$, SAMP becomes exactly SP. Thus, the stage is generally chosen one from $\hat{s}=1\sim s$. The improved SAMP algorithms which applied SVD on the basis of previous study for high-resolution reconstruction, called SAMP-SVD.

The following Table I shows the steps of SAMP algorithm [16].

Table 1. SAMP Steps

Input: Sampling matrix Φ, Sampling vector y ,step size d

Output: A s-sparse approximation $\hat{x}$ of the input signal

Initialization:

$\hat{x}=0$ {Trivial initialization}

$r_0=y$ {Trivial initialization}

$F_0=\varnothing$ {Empty finalist}

$I=d$ {Size of the finalist in the first stage}

$k=1$ {Iteration index}

$j=1$ {Stage index}

Repeat:

$S_k=Max\left(\left|\Phi^* r_{k-1}\right|,I\right)$ {Preliminary Test}

$F_k=F_{k-1}\cup S_k$ {Make Candidate List}

$\hat{x}_{F_k}=\arg\min\left\|y-\Phi_{F_k}x_{F_k}\right\|$ {Final test}

$r=y-\Phi_F\hat{x}_F$ {Compute Residue}

If halting condition true

then quit the iteration;

else if $\|r\|_2\geq\|r_{k-1}\|_2$

then {stage switching}

$j=j+1$ {Update the stage index}

$I=j\times d$ {Update the size of finalist}

else

$F_k=F$ { Update the finalist}

$r_k=r$ { Update the Residue}

$k=k+1$

end

Until halting condition true;

Output: $\hat{x}=\arg\min\left\|y-\Phi_F x\right\|$ {Prediction of non-coefficients}

An optimum solution is found by the minimum of residual to $\hat{x}_{F_k}=\arg\min\left\|y-\Phi_{F_k}x_{F_k}\right\|$, which is equivalent to a least squares problem as $\left(\Phi_{F_k}^H\Phi_{F_k}\right)x_{F_k}=\Phi_{F_k}^H y$ and, namely $\hat{x}_{F_k}=\left(\Phi_{F_k}^H\Phi_{F_k}\right)^{-1}\Phi_{F_k}^H y$. However, in solving the normal equation we are interested in the singularity of $\Phi_{F_k}^H\Phi_{F_k}$, unfortunately, the inverse of a nonsingular matrix costs a large computation and is difficult to implement in computer arithmetic. The SVD definitely point out the generalized inverse numerical solution of $\Phi_{F_k}^H\Phi_{F_k}$ which is avoided in the problem of singularity.

The solution of LS regular system by using SVD algorithm

$$\hat{x}_{F_k}=\sum_{i=1}^{l}u_i\left(u_i^H\Phi_{F_k}^H y/\sigma_i^2\right)\tag{6}$$

where u_i is the right singular vector, σ_i represents nonzero singular value and, the number of singular values is l which depends on $\Phi_{F_k}^H\Phi_{F_k}$.

And the following Table II shows the steps of SVD decomposition.

Table 2. SVD Decomposition Steps

Step 1:	the singularly valuable decomposition of Φ_{F_k}
	all the nonzero singular value is $\sigma_1,\sigma_2\cdots\sigma_s$
	the corresponding right singular vector $u_1,u_2\cdots u_s$
Step 2:	output the numerical solution $\hat{x}_{F_k}=\sum_{i=1}^{l}u_i\left(u_i^H\Phi_{F_k}^H y/\sigma_i^2\right)$

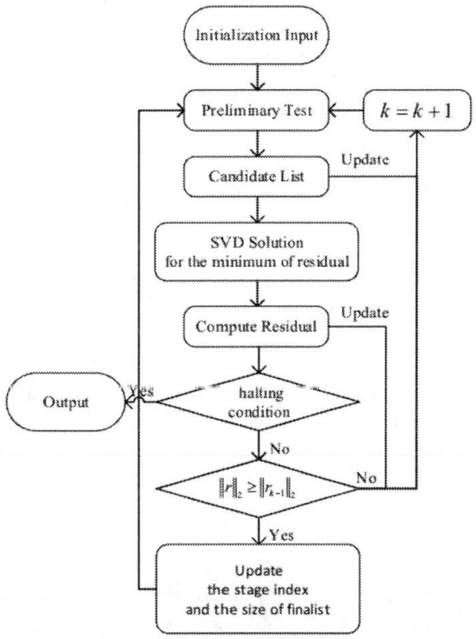

Fig. 1 The Flow Chart of SAMP-SVD Algorithm

where the output gets the numerical solutions from SVD algorithm to avoid the complexities of matrix inversion. SAMP-SVD obtains high-resolution reconstruction by combining the merits of SAMP and SVD, not requiring sparsity and singularity as prior. The flow chart of the SAMP-SVD algorithm is presented in Fig.1.

4 Results and Discussion

In this section, the performance of SAMP-SVD algorithm derived above is evaluated through simulation and compared with the SpaRSA that is one of traditional CS convex recovery algorithm, and we briefly discuss their strengths and weaknesses. The inputs can be written as fellows shown in (7)

$$x(n)=\sum_{m=1}^{N}A_m\cos(2\pi f_m n+\varphi_m)+w(n) \qquad (7)$$

where the A_m is the amplitude of the m th signal, f_m and φ_m represent the frequency and phase of input signal, and $w(n)$ is a vector of Gaussian white noise.

Example 1: At the first experiment, we will show the information recovery performance of the two matrices mentioned above. We consider the parameters of input signals are $f=5.36GHz$, $A=1$, $\varphi=\pi/6$ and assume the environment is without noise. For measurement matrix ϕ, we choose average sampling rate f_{s1} of 2GHz, and the modulated function $\theta(n)$ is written as $\sin(2\pi f_\theta n)$. As discussed earlier, in order to reconstruct more highly accuracy, we should give a true index zones. The NYFR sensing matrix Φ has the same modulated function as ϕ. Fig.2 shows the SAMP-SVD reconstruction spectrum obtained from both of two. Notice that there is an extra frequency at 6.65GHz in the first picture which use D-measurement matrix, but it is pure in the spectrum of second. As a consequence, we will use the NYFR sensing matrix in the following experiments.

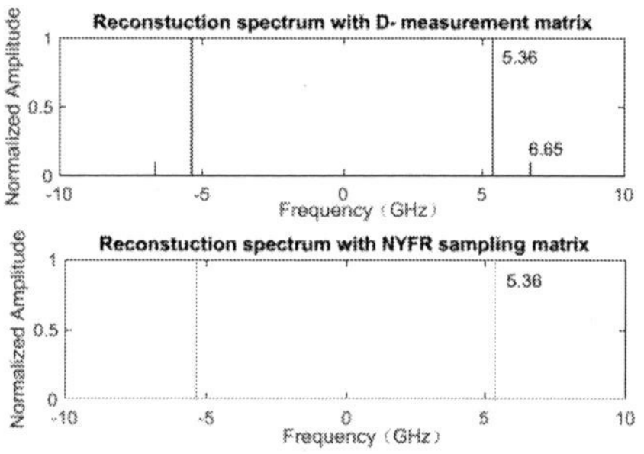

Fig. 2 Spectrum of reconstruction signal

Example 2: In this experiment, we assume that the input of NYFR consisting of monopulse (MP), linear frequency modulation (LFM) and binary phase shift keying (BPSK) signals at 2.1, 5.56, and 12.94GHz, and consider the amplitude of each signals are 1, the phase are $\pi/6$, $\pi/3$ and $\pi/2$, respectively. LFM signal is given a fixed-bandwidth of 5MHz, for the BPSK signal case, we choose random codes. Besides, the relative SNR is 10dB. Fig.3 shows the SAMP-SVD reconstruction magnitude spectrum of three input signals, in which algorithm uses the NYFR sensing matrix Φ. Notice that the modulated function $\theta(n)$ is written as $\sin(2\pi f_\theta n)$, where f_θ is 10MHz.

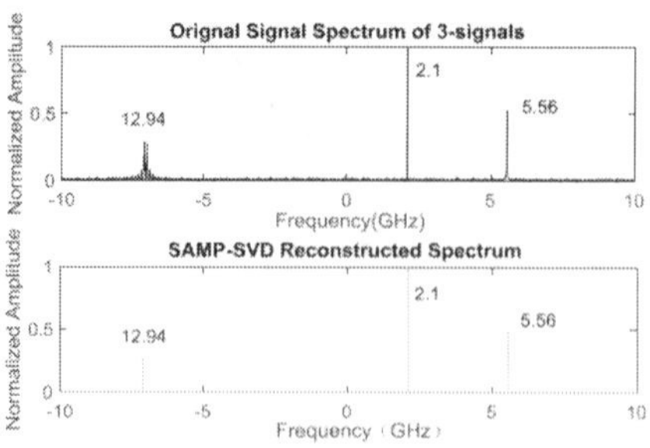

Fig. 3 Spectral of original and SAMP-SVD reconstruction signal

Comparing the spectrum of reconstruction signals with one of the original signals, we can see that SAMP-SVD algorithm recovers the 3-signal after the NYFR correctly. The accuracy of reconstruction is controlled by noise energy σ^2, comparing with residue r which is updated in each iteration, repeating this step until $r\le\sigma^2$. It can estimate all the frequencies by search for the peaks in the spectrum.

Example 3: Here we consider the situation almost the same with example 1, and we define normalized root mean square error (NRMSE) to evaluate the performance of the frequency estimation by using SpaRSA and SAMP-SVD, the results are based on 500 Monte Carlo simulations and shown in Fig.4, respectively.

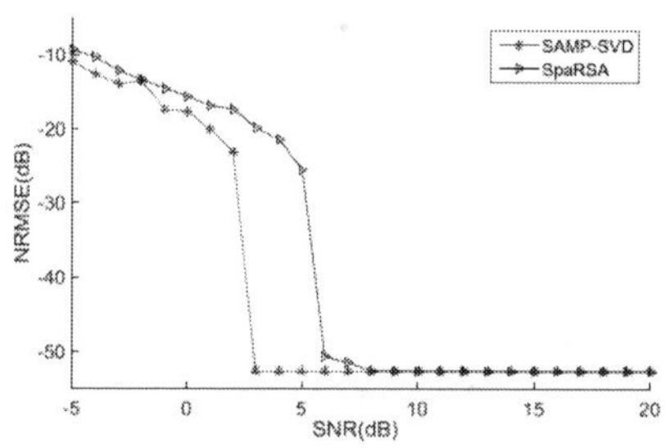

Fig. 4 NRMSE of frequency estimation (SAMP-SVD & SpaRSA)

The simulation shows the ability of the proposed algorithm on information reconstruction of inputs more faster than SpaRSA, because of SAMP-SVD without the process of matrix inversion, if time is a factor should be considered, the SAMP-SVD is a better choice. Fig.4 shows the NRMSE tends to a certain value when the SNR is greater than 8dB, and the performance of two algorithm is close to each other.

5 Conclusions

SAMP-SVD exhibits a tradeoff in computational complexity versus reconstruction performance if the RIP condition of measurement matrix is satisfied, contributes that not requiring sparsity of the receiving signals and singularity of sensing matrix as prior. And simulation shows that the algorithm achieves accurate frequency estimation than the existing SpaRSA in the low SNR environment. Future work will concentrate on wideband signals recovery and the effect of RIP on signal recovery based on the NYFR with information recovery methods of CS framework, promising for selecting the parameters and improving the structure of the NYFR.

References

1. R. H. Walden. "Analog-to-digital converter survey and analysis," *IEEE Journal on Selected Areas in Communications*, vol. 17(4), pp. 539-550, Apr 1999.
2. S. Kirolos, J. Laska, M. Wakin etc. "Analog-to-Information Conversion via Random Demodulation[C]," *Processing of the IEEE Dallas Circuits and Systems Workshop (DCAS)*. Dallas, Texas, 2006.
3. J. Laska, S. Kirolos, Y. Massoud etc. Random Sampling for Analog-to-Information Conversion of Wideband Signals[C]. Proceedings of the IEEE Dallas Circuits and Systems Workshop (DCAS'06). Dallas, Texas, 2006.
4. Cand`es, J. Romberg, and T. Tao, "Robust uncertainty principles: Exact signal reconstruction from highly incomplete frequency information," *IEEE Trans. on Information Theory*, vol. 52, pp. 489-509, Feb 2006.
5. G. Fudge, R. Bland, M. Chivers, S. Ravindran, J. Haupt, and P.Pace, "A Nyquist folding analog-to-information receiver," in *Proc. 42nd Asilomar Conf. on Signals, Compute,* Syst. (ACSSC), 2008, pp.541-545E.
6. J. Haupt and R. Nowak, "Signal reconstruction from noisy random projections," *IEEE Trans. Inf. Theory*, vol. 52, no. 9, pp. 4036-4048, Sept.2006.
7. S. J. Wright, R. D. Nowak, and M. A. T. Figueiredo, "Sparse reconstruction by separable approximation," *IEEE Trans. Signal Process.*, vol. 57, no. 7, pp. 2479-2493, Jul 2009.
8. R. Maleh, G. L. Fudge, F. A. Boyle, P. E. Pace, et al. "Analog-to-Information and the Nyquist Folding Receiver," *IEEE Journal on Emerging and Selected in Circuits and Systems*, 2012,2(3) . pp. 564-578.
9. S. G. Mallat and Z. Zhang, "Matching pursuits with time-frequency dictionaries," *IEEE Trans. Signal Process.*, vol. 41, no. 12, pp.3397-3415, Dec 1993.
10. Y. Pati, R. Rezaiifar, and P. Krishnaprasad, "Orthogonal matching pursuit: Recursive function approximation with application to wavelet decomposition," in Proc. 27th ACSSC, 1994, vol. 1, pp. 40-44.
11. Candes. E. J and Tao. T, "Decoding by linear programming [J]," *IEEE Transactions on Information Theory*, vol. 51, pp. 4203-4215, Dec 2005.
12. Donoho D. L, Tsaig. Y, Drori. I, et al. "Sparse solution of underdetermined linear equations by stage wise orthogonal matching pursuit [J]," *IEEE Transaction on Information Theory*, vol. 58, pp. 1094-1121, Feb 2012.
13. D. Needell and R. Vershynin. "Uniform uncertainty principle and signal recovery via regularized orthogonal matching pursuit," *Foundations of Computational Mathematies*, vol. 9, pp. 317-334, Mar 2009.
14. D. Needell, J. A. Tropp. "CosaMP: Iiterative Signal Recovery from Incomplete and Inaccurate Samples[J]," *Applied and Computational Harmonic Analysis*, vol. 26, pp. 301-320, Mar 2008.
15. Dai. W, Milenkovic. O, "Subspace pursuit for compressive sensing signal reconstruction," *IEEE Trans. Information Theory*, vol.55, pp. 2230-2249, May 2009.
16. Thong T D, Gan Lu, Nguyen. "Sparsity adaptive matching pursuit algorithm for practical compressed sensing [J]," in *Proc.* 2008 42nd *Asilomar Conf*, 2008, pp. 581-587.

Multi-windows Threshold-based Frame Synchronization in Variable Coded Modulation Service

Yong Sheng , Jinshu Chen and Xiaoming Fan

Department of Electronic Engineering, 100084 Tsinghua University, Beijing, China

Abstract. In this paper, an innovative and reliable frame synchronization scheme is proposed for second generation digital video broadcasting (DVB-S2) system in Variable Coded Modulation (VCM) service. Frame synchronization in VCM suffers many difficulties from large frequency offset, extremely low SNR and variable physical layer frame (PLFRAME) lengths. To cope with variable PLFRAME lengths, multi-windows are proposed and utilized. Multi-windows threshold-based search algorithm is presented with theoretical analysis to decrease the missed detection probability. Experimental results demonstrate that our proposed method is more robust and reliable than the existing approaches, especially in low signal to noise ratios (SNRs) and large frequency offsets.

1 Introduction

Second generation digital video broadcasting (DVB-S2) via satellites has been standardized for a decade. It draws much attention since its distinctive transmission performance could approach the Shannon limit. Compared with DVB, DVB-S2 provides Adaptive Coded Modulation (ACM) and Variable Coded Modulation (VCM), implying that channel coding and modulation could be changed on a frame-by-frame basis [1]. VCM makes full use of the channel by transmitting more information bits when signal-to-noise ratio (SNR) is high, e.g., uses 9/10 LDPC and 32APSK combination, and makes sure of receiver performance by throttling down the information rate when SNR is low, e.g., exploits 1/2 LDPC and QPSK combination. This flexible character results in a much higher total throughput than that obtained with constant coded modulation (CCM).

Even though VCM can provide an obvious improvement in achievable throughput over CCM, with nearly 1.7 times [2], it suffers many difficulties from frame synchronization, because of large frequency offset, extremely low SNR and variable physical frame lengths.

Many algorithms on frame synchronization have been researched and recommended. Based on the approximate maximum likelihood (ML) criterion, Choi and Lee proposed a differential detector in [3], known as CLD, which is insensitive to large frequency offset but has high complexity. NCPDI (Non-Coherent post detection integration) [4] and DPDI (Differential post detection integration) [5] are studied to decrease implementation complexity, while they are susceptible to unknown frequency offset. ETSI recommends a scheme in [6] to correlate on the SOF and PLSCODE differentially, which can work smoothly in large frequency offset and has comparatively low complexity.

After obtaining differential correlation, traditionally, decisions can be made simply by comparing correlation with a pre-set threshold [6]. Frame synchronization is deemed as achieved if the value is higher than the threshold. On account of inflexible threshold setting, peak detector is introduced in [7] to get rid of threshold. To strengthen the robustness of non-threshold scheme, double peak search algorithm is discussed in [8, 9].

There are, however, few investigations into frame synchronization in VCM application. Variable frame lengths make it hard to detect physical frame header (PLHEADER). How to ensure less frame synchronization drops in coding and modulation (CODMOD) switchovers is another issue to be solved. In addition, the performances of existing methods for frame synchronization are unsatisfactory at low SNRs. In this paper, multi-windows threshold-based search algorithm is developed to cope with these problems. Our simulation results indicate that the proposed approach can acquire satisfactory performance than existing methods.

The rest of this paper is organized as follows. In section 2, the structure and features of DVB-S2 physical frame are analysed. Then, VCM service is discussed in section 3, with description of the structure of VCM transmitter and receiver. Moreover, the complete frame synchronization schemes are proposed in detail in section 4, including differential detection module and multi-windows threshold-based algorithm. Some optimizations and comparisons of the scheme are discussed in section 5. Section 6 shows the performance of MATLAB simulation. Finally, conclusions are drawn in section 7.

2 DVB-S2 frame structure analysis

2.1 Physical layer frame structure

According to DVB-S2, Physical Layer Header (PLHEADER) occupies one slot (90 symbols) in Physical Layer (PL), consisting of Start of Frame (SOF) and Physical Layer Signalling Code (PLSC). SOF corresponds to 26-symbol fixed pseudo-random pattern (18D2E82), which assists in frame synchronization and carrier recovery. PLSC is a 64-symbol linear binary code, conveying seven bits of information (e.g., coding, modulation type, pilot insertion and frame length) by RM transformation. To make receiver synchronization easier and simpler, pilot blocks could insert every 16 slots. The pilot block is made up of 36 un-modulated symbols, which shall be utilized in section 5 to enhance performance. Fig. 1 shows the structure of PLFRAME.

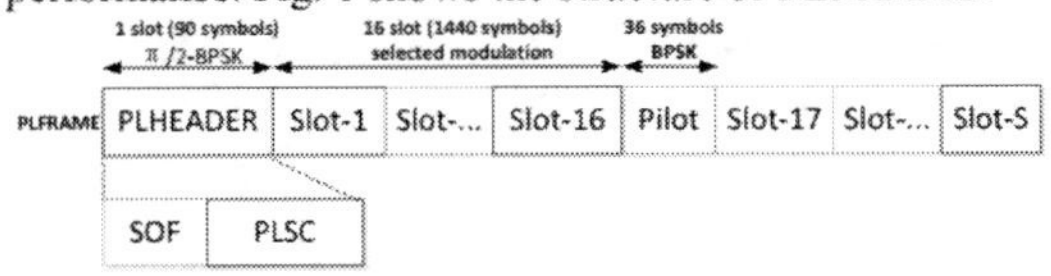

Figure 1. PLFRAME structure.

In VCM service, different frames can adopt optimal coding and modulation mode (CODMOD) to take advantage of dynamic link. Fig. 2 gives an example of variable CODMODs in continuous frames.

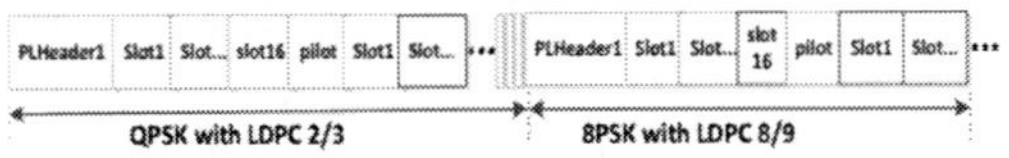

Figure 2. Example of VCM.

2.2 Variable PLFRAME lengths

According to [1], Forward Error Correction (FEC) Encoding is implemented by the concatenation of BCH outer code and LDPC (Low Density Parity Check) inner code. After FEC coding, the length of coded block shall be 64800 bits or 16200 bits, depending on the specific application field. When VCM is on service, FEC and modulation mode could vary in different frames, while they are kept constant in all slots of one frame. Modulation set contains QPSK, 8PSK, 16APSK and 32APSK. The length of PLFRAME can be obtained in table 1, which determines the width of window in section 4.

Table 1. Frame Parameters.

FECFRAME		PLFRAME			
Frame length (bits)	Mod type	Slot	Pilot block	Has Pilot	PLFRAME length (bauds)
64800	QPSK	360	22	Yes	33282
64800	QPSK	360	22	No	32490
64800	8PSK	240	14	Yes	22194
64800	8PSK	240	14	No	21690
64800	16APSK	180	11	Yes	16686
64800	16APSK	180	11	No	16290
64800	32APSK	144	8	Yes	13338
64800	32APSK	144	8	No	13050
16200	QPSK	90	5	Yes	8370
16200	QPSK	90	5	No	8190
16200	8PSK	60	3	Yes	5598
16200	8PSK	60	3	No	5490
16200	16APSK	45	2	Yes	4212
16200	16APSK	45	2	No	4140
16200	32APSK	36	2	Yes	3402
16200	32APSK	36	2	No	3330

3 VCM service design

Many satellite missions communicate in a highly dynamic environment, due to the variations in geometry, weather and interference [10]. To make best use of link margin when channel condition is good, high order coding and modulation (CODMOD) is preferred to obtain maximum throughput. In order to enable communication in poor channel, low order CODMOD is considered to guarantee the receiver performance. VCM service shall be designed to meet this requirement. Even though CODMOD could vary on frame basis, too many switchovers between CODMODs shall take down the overall performance contrarily for expensive overheads. According to [2], orbits are divided into a fixed number (e.g., 6) of sectors (geometric approach), and link budgets are computed in each sector according to local worst conditions.

In VCM service, the transmitter structure can be designed as Fig. 3. Several minutes before transmission mission, VCM system shall select certain sequence of CODMODs (e.g., 5 or 6) according to channel condition and link budget. During satellite passing, transmission system on the satellite shall follow presupposed CODMODs to process information frame on sector basis.

Receiver structure is showed in Fig. 4, where frame synchronization is in dashed box, which is the concern of this paper.

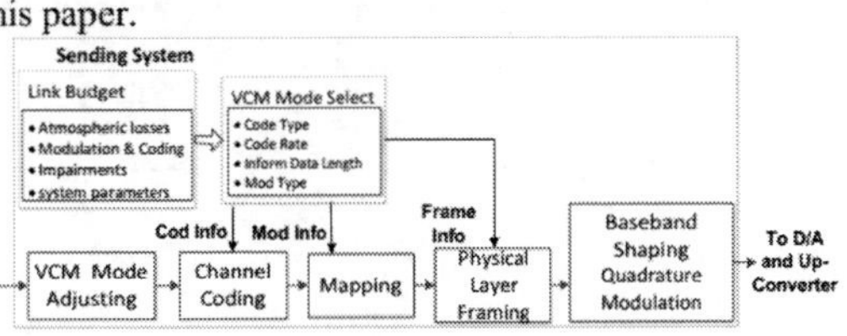

Figure 3. Transmitter structure of VCM.

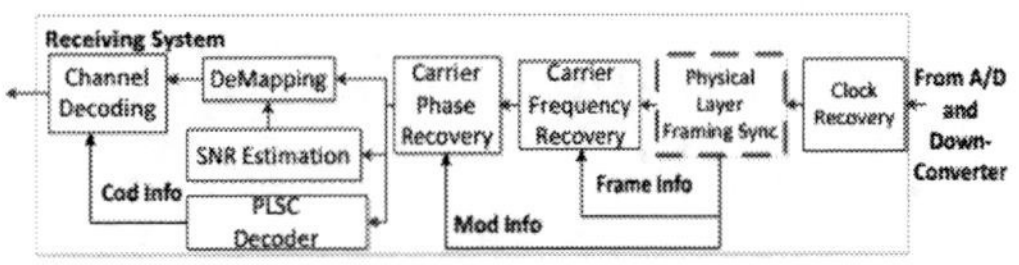

Figure 4. Receiver structure of VCM.

4 Proposed frame synchronization scheme

As shown in Fig. 4, after the steady convergence of receiver symbol timing, frame synchronization for PLHEADER shall be in place. Generally speaking, frame synchronization consists of acquisition and tracking. Frame acquisition is the process of calculating differential correlation, which is performed by differential detector. PLHEADER tracking is the phase to determine the accurate location of PLHEADER and modulation type in the current frame. To deal with variable frame lengths, multi-windows are utilized. An innovative and reliable method based on threshold is proposed to achieve satisfactory performance, especially in the situation of extremely low SNR and large frequency offset.

4.1 Differential detector

ETSI recommends a scheme in [6] to correlate on the SOF and PLSCODE differentially, which can work smoothly in the presence of large frequency offsets (20% of the symbol transmission rate) and low SNR (-2.35dB). The differential correlation Λ can be achieved in equation

$$\Lambda = \left| \sum_{k=1}^{25} r_k r_{k+1}^* C_{SOFk}^* + \sum_{m=1}^{45} R_{2m-1} R_{2m}^* C_{PLSCm}^* \right|, \quad (1)$$

where r denotes the received symbol of SOF, R is the received symbol of PLSC, C_{SOF} and C_{PLSC} are the corresponding differential coefficients obtained for SOF and PLSC respectively, and star mark represents complex conjugate. C_{SOF} and C_{PLSC} can be obtained by equation

$$C_{SOFk} = s_k s_{k+1}^* = \pm j, k = 1,2,...,24,25, \quad (2)$$

$$C_{PLSCm} = s_{2m-1} s_{2m}^* = \pm j, m = 1,2,...,44,45, \quad (3)$$

where s is the transmitted PLHEADER symbol.

The diagram of differential correlation circuit is shown in Fig. 5.

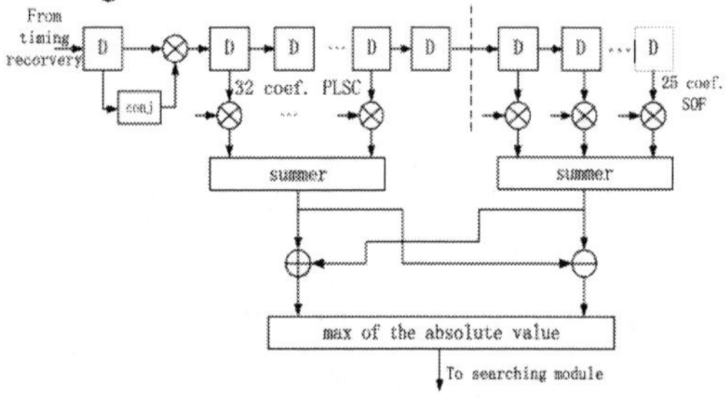

Figure 5. Differential detection circuit of the PLHEADER.

4.2 Multi-windows threshold-based search algorithm

The differential output is then sent to a peak search module to track the PLHEADER. The conventional peak search module simply compares the output of correlator with a predefined threshold [6]. In [7], peak search algorithm with one constant width window is raised to get rid of threshold setting. To decrease missed detection probability in poor channel condition, [8] suggests a double peak search method, choosing two maximum peaks in one window as the candidates. However, no matter one peak, double peaks or more, the missed detection probability is rather high when SNRs below -1dB. Moreover, to the best of our knowledge, none of the existing studies considers the VCM frame synchronization clearly and correctly, which shall be the crucial concern of this section.

As table 1 describes, 16 types of frame length are available in VCM system. Since there is one-to-one correspondence relationship between modulation and pilot status with frame length, frame length information can be achieved by comparing two neighbour PLHEADER positions. Without prior information about CODMODs in receiver, 16 types of search windows with corresponding widths shall be in consideration.

Without loss of generality, it is supposed that only 4 types of frame exist, QPSK, 8PSK, 16APSK and 32APSK with pilot insertion, and their corresponding widths are 13338, 16686, 22194 and 33282. As Fig. 6 shows, symbol differential correlations in 5 physical layer frames at SNR = -2.35dB are in hands. The five red bold lines represent the differential correlations of PLHEADER symbols. PHLEAHER symbol represent the first symbol of PLHEADER. Then, all of symbols whose differential correlations are above threshold 0.77 shall be in further investigation. In Fig. 7, any symbol with its differential correlation higher than 0.77 is recorded.

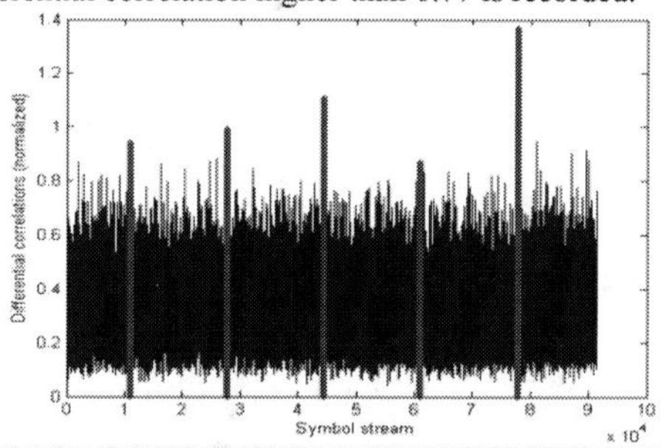

Figure 6. Symbol Differential correlations of 5 frames at SNR=-2.35dB.

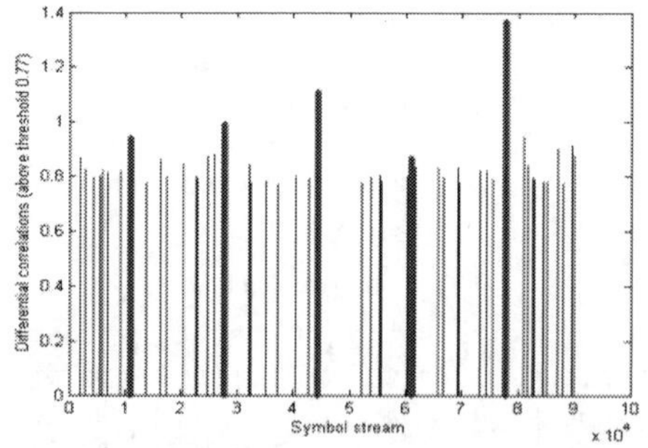

Figure 7. Symbol Differential correlations that are above 0.77.

Afterwards, position comparison is made within three successive search windows. As is shown in Fig. 8, the first search window corresponds to 32APSK with pilot insertion. Dashed boxes denote window inspection sections. If any two positions that relative to the left end of their own search window share the same value, the position is immediately deemed as aligned to the PLHEADER, and the search window shall slide to new place to start a new round of PLHEADER detection. If none can match in this kind of three successive windows, similar operations are carried on within next kind search window, which corresponds to 16APSK with pilot insertion. As is shown in Fig. 8, in the second type of search window, three positions relative to their own search window are quite same, indicating that PLHEDER is obtained at this position and the modulation of this PLFRAME is 16APSK.

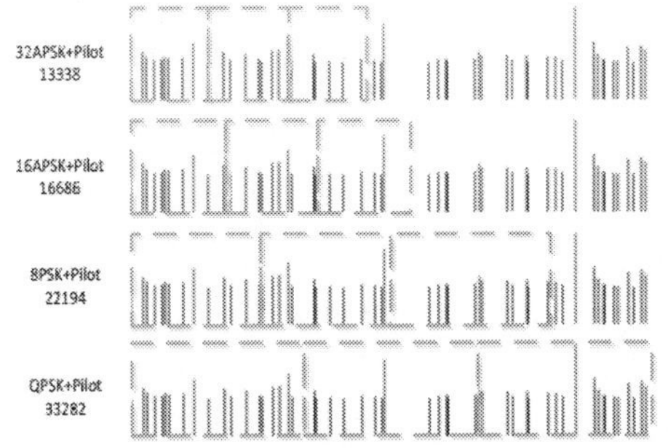

Figure 8. Multi-windows threshold-based algorithm processing on 5 frames.

Obviously, the value of threshold has a great influence on frame synchronization probability. In the next part, the relationship between threshold and frame synchronization probability shall be investigated.

Without noise interference, the differential correlation of PLHEADER symbol is 57——sum of PLSC differential correlation 32 and SOF differential correlation 25. Considering the poor condition with extremely low SNR = -2.35dB and large frequency offset of 20% of transmitting rate, some simulations over performance of differential correlation detector are conducted. In Fig. 9, PLHEADER symbol refers to the first symbol of PLHEADER, and non-PLHEADER symbols refer to other symbols except from PLHEADER symbol in a PLFRAME. All symbol differential correlations are normalized to 1 by dividing 57.

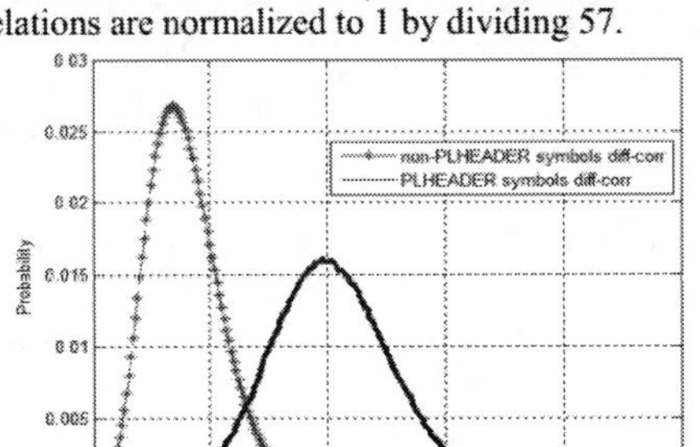

Figure 9. Symbol differential correlation distribution curve.

From Fig. 9, some hidden relationships between frame synchronization probability and false alarm probability with threshold could be discovered. Let T represent some certain threshold, $P_{header}(t)$ represents the blue solid line——PLHEADER symbol differential correlation distributions, and $P_{non\text{-}header}(t)$ denotes the dashed line with star marker —— non-PLHEADER symbol differential correlations distribution. P_+ represents the probability that PLHEADER symbol differential correlation is higher than threshold. And P_- refers to the probability of non-PLHEADER being below threshold.

$$P_+ = \int_T^{+\infty} P_{header}(t)dt \, , \tag{4}$$

$$P_- = \int_0^T P_{non-header}(t)dt \, . \tag{5}$$

If two candidates in three successive window share the same relative position in terms of their belonging windows, then this position can be deemed as PLHEADER. Supposing that differential correlations in different windows are independent identically distributed, the ideal synchronization rate can be get from the equation

$$P_{synched} = P_+^3 + C_3^2 P_+^2 (1 - P_+) \, . \tag{6}$$

Actually, the synched rate shall be lower than $P_{synched}$, for interfere of false PLHEADER symbols. If two relative positions of non-PLHEADER symbols exactly equal in continuous three frame, false synched PLHEADER occurs, leading to miss of true PLHEADER in that window. Considering the widest window is 33282, the number of false PLHEADER could be conducted in the following equation.

$$Numbers_{false-PLHEADER} = 33282 \times (P_-^3 + C_3^2 P_-^2 (1 - P_-)) \tag{7}$$

Even though P_- is rather small, i.e., e^{-4}, the number of false PLHEADER can be extraordinarily large after the multiplication of 33282.

Some conclusions can be drawn from Fig. 10 and Fig. 11. The higher threshold to be set, the less number of false PLHEADER can be detected incorrectly, and the lower true PLHEADER synched probability can be reached for more PLHEADER miss.

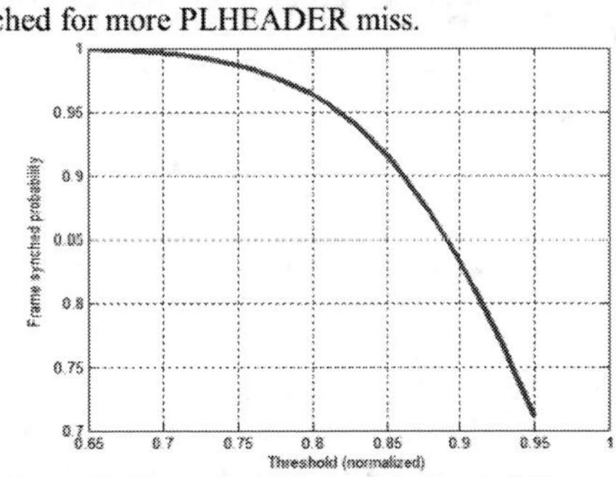

Figure 10. Theoretical frame synched probability versus threshold.

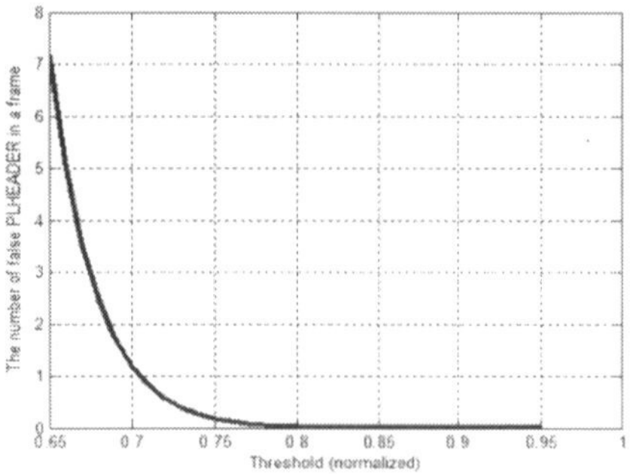

Figure 11. The number of false PLHEADER in a frame versus threshold.

5 Optimizations and simplifications

5.1 Optimizations of differential detector

The more features of the PLHEADER to be mined, the higher accuracy of the PLHEADER location can be achieved.

As described before, pilot blocks are inserted every 16 slots. Each pilot block is made up of 36 un-modulated symbols. To make full use of information in PLFRAME, differential detector can be more robust with the assistance of pilot block correlation. Pilot-symbol-assisted differential correlation is analysed in [9], with three pilot blocks participating in differential correlation in trade-off between complexity and performance. However, this is not applicable to VCM, whose pilot blocks could be less than 3. As can be seen in table 1, when information frame length is 16200 and modulation is 32APSK, there are only 2 pilot blocks in PLFRAME. To fit in with the various CODMOD in VCM, 2 pilot blocks are adopted in differential correlation as expressed in

$$\Lambda = \left| \sum_{k=1}^{25} r_k r_{k+1}^* C_{SOFk}^* + \sum_{m=1}^{45} R_{2m-1} R_{2m}^* C_{PLSCm}^* + \sum_{t=1}^{2}\sum_{h=1}^{35} p_{t,h} p_{t,h+1}^* C_{Pilott,h}^* \right|, \quad (8)$$

where r, R, C_{SOF} and C_{PLSC} have the same meanings with equation (1), (2), (3). p is the received symbol of pilot, t is the index of pilot blocks, h is the index of received symbol in a plot block, and star make represents complex conjugate.

C_{pilot} is the coefficient of pilot block symbol, which can be obtained as

$$C_{Pilott,h} = P_{t,h} P_{t,h+1} = \pm j, \pm 1 (t = 1,2, h = 1,2,...,35), \quad (9)$$

where P denotes transmitted pilot symbols.

To validate the effectiveness of pilot block, simulation experiments are conducted under large frequency about 20% of transmission rate and extremely low SNR (-2.35dB) in Additive White Gaussian Noise (AWGN) channel. After regularized, PLHEADER and non-PLHEADER differential correlation distribution are more concentrated with the help of pilot, as shown in Fig. 12. The simulation presents a conclusion that differential detector shall be more robust with 2 pilot blocks participating in.

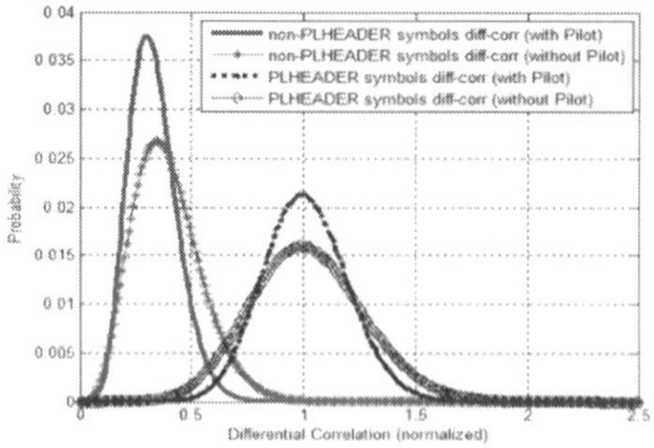

Figure 12. Differential correlation distribution in different detector.

5.2 Simplifications of proposed algorithms

In the proposed method, the set of window width shall cover all kinds of CODMODs, that is, frame synchronization controller shall search peak position in 16 windows in parallel. This also means that 16 suits of comparators and counters are in need, leading to high complexity. Some tricks can be done to simplify this nuisance in implementation.

First of all, there is a hidden fact that the length of FECREAME and pilot insertion usually remain constant, not varying from frame to frame. The length of the FECFRAME (64800 bits or 16200 bits) and pilot block insertion status depend on application area, meaning that we can get prior information about candidate windows type. Thus, in one satellite task, given the awareness of FECFRAME and pilot insertion status, only 4 types of window shall be considered, rather than 16.

Moreover, to balance transmission throughput and implementation complexity, VCM system usually switches CODMODs 4 or 5 times during one satellite mission. Every switchover only leads to 3 or 4 frames loss in proposed method, which is negligible compared with millions of data frame in a mission. And, these frame losses in switchover can be eliminated by inserting 3 or 4 dummy frames after switching to another CODMOD.

6 Simulation results

6.1 Threshold selecting in threshold-based algorithm

To study the effect of threshold on frame synchronization rate, MATLAB simulations are conducted at fixed SNR = -2.35dB and frequency offset of 20% of transmission rate. Results are shown in Fig. 13. In the first stage, frame synchronization probability rises with increase of threshold for less false PLHEADERs. However, frame synchronization probability goes down in the second stage, due to more PLHEADER miss in high threshold. In VCM service, extremely bad channel shall be in consideration, thus threshold can be selected in range from 0.76 to 0.78 to get optimal performance.

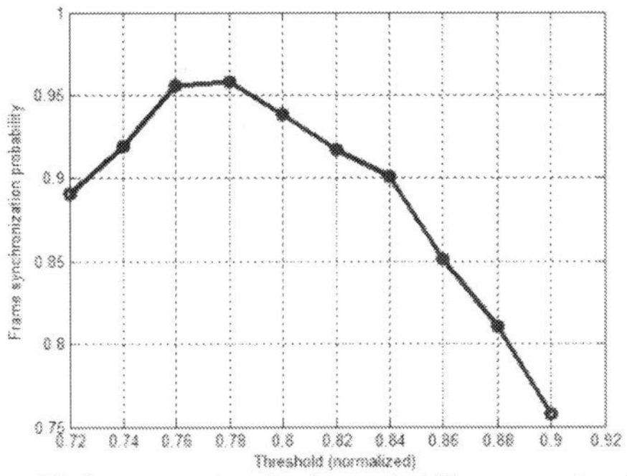

Figure 13. Frame synchronization probability versus threshold at SNR=-2.35dB.

6.2 Performance comparison of different algorithms

Experiments are conducted to investigate performance of proposed algorithm. With 20% frequency offset, Fig. 14 shows the performance comparison for different SNR ranging in between [-3, 1]. Threshold is set to 0.77 according to the result of Fig. 13 in threshold-based search algorithm. As Fig. 14 shows, peak search algorithm [7] and double peak search algorithm [8, 9] perform pretty well when SNRs are above -0.5 dB, and when channel condition turns to extremely bad, multi-windows threshold-based search algorithm shall prevail due to its novel scheme.

Actually, Multi-windows threshold-based search algorithm shares many similarities with double peak search algorithm in main procedure. The latter algorithm only considers two symbol locations of maximum differential correlation, while the former adds all locations whose differential correlations are above the threshold into inspection. When the channel is in good condition, e.g., the SNR is higher than -0.5dB, both can achieve ideal performance with frame synchronization probability equaling to 1. However, when SNR is lower than -0.5dB, and noise strongly interferes useful signal, making PLHEADER symbol differential correlation far behind from the maximum. In conclusion, One peak and double peak search algorithm can be thought as special cases of threshold-based algorithm when threshold are set rather high.

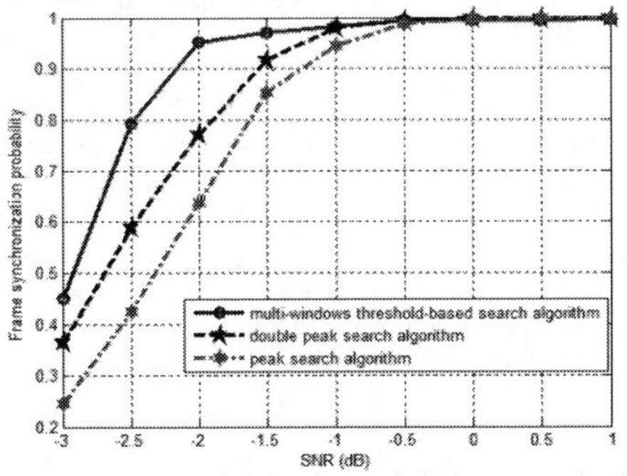

Figure 14. Frame synchronization probability versus threshold at SNR=-2.35dB.

7 Conclusions

Frame synchronization of DVB-S2 in VCM service is studied in this paper. Variability of frame coding and modulation makes it hard to achieve fast synchronization in VCM, thus, multi-windows are exploited. To decrease missed detection probability, threshold-based algorithm is proposed. Experimental results illustrate that our multi-windows threshold-based method is more robust and reliable than the existing methods, especially in the presence of large frequency offsets and low SNRs.

References

1. Digital video broadcasting (DVB); second generation framing structure, channel coding and modulation systems for broadcasting, interactive services, news gathering and other broadband satellite applications. (2005).
2. Estec ESA. VCM performance for LEO satellites at 26 GHz a realistic assessment. (2010).
3. Zae Yong Choi and Yong H Lee. IEEE Trans. Commun, **50**(7):1062–1065 (2002).
4. Giovanni E Corazza, Raffaella Pedone, and Marco Villanti. International Journal of Satellite Communications and Networking, **24**(2):185–201 (2006).
5. Pansoo Kim, Giovanni Emanuele Corazza, Raffaella Pedone, Marco Villanti, Dae-Ig Chang, and Deock-Gil Oh. In *Proc. IEEE Int. Conf. Wireless Communications and Networking*, pp. 1183–1187 (2007).
6. Digital video broadcasting (DVB) user guidelines for the second generation system for broadcasting, interactive services, news gathering and other broadband satellite applications (dvb-s2). (2005).
7. Feng-Wen Sun, Yimin Jiang, and Lin-Nan Lee. International journal of satellite communications and networking, **22**(3):319–339 (2004).
8. Li Q., Zeng X.Y., Wu c., et al, In *Proc. IEEE Int. Conf. Circuits and Systems*, pp. 956-959 (2008).
9. Yulong Zhang, Xubin Chen, Wenhua Fan, Jun Han, and Xiaoyang Zeng. In *Proc. IEEE Int. Conf. Wireless Telecommunications Symposium*, pp. 1–5 (2010).
10. Thomas A Sielicki, Jon Hamkins, and Denise Thorsen. In *Proc. IEEE Int. Conf. Aerospace*, pp. 1–7 (2013).

Channel Estimation on the (EW) RLS Algorithm Model of MIMO OFDM in Wireless Communication

Suzi Seroja Sarnin , Siti Maisurah Sulong and Habibah Hashim

Faculty of Electrical Engineering , UiTM Shah Alam, 40000, Malaysia

Abstract. This paper study the channel estimation based on the exponentially weighted (EW) RLS algorithm. The advantages of the proposed estimator arise from its implementation in the time-domain, whereas fewer channel parameters are required to be estimated compared with the frequency-domain channel coefficients. In addition, the matrix inversion operation required in LS and LMMSE estimators is avoided here by recursively updating the channel estimates. Therefore, the computational complexity is highly reduced compared to frequency-domain based channel estimation. Furthermore, the proposed estimator has good tracking capability due to exploiting the time correlation between the path gains. This paper provides analysis, evaluation and computer simulations in MATLAB. The performance is evaluated in terms of the MSE of the channel estimate and BER for different Doppler frequencies (correspond to different mobility speeds) and Monte Carlo simulations are performed and the MSE and BER performance versus SNR are obtained by averaging over 10000 channel realization. For comparisons, the BER performance is also presented for perfectly known channel at the receiver. In all the simulations, perfect synchronization between the transmitter and the receiver is assumed.

1 Introduction

OFDM is a multi-carrier modulation technique which is used to transmit high data rate data stream through wireless medium. MIMO-OFDM is a new wireless broadband technology which has great robustness against multipath fading and capability of high rate transmission and other channel impairments. MIMO-OFDM also has significant potential performance enhancements over existing wireless technology by concurrently exploiting the space, time and frequency domain. Channel estimation technique is introduced to improve accuracy of the received signal even though it is a challenging problem in any communication system especially when the channel is time-varying. These authors [1] proposed RLS channel estimator with improved decision-directed algorithm based on training sequences to eliminate interference. Performance of RLS channel estimation algorithm proved its efficiency in [2] by increasing the number of receive antenna even though RLS has high complexity. Due to the interference occur, original symbol that had been distorted during propagation need to be recovered back at the receiver [3]. A few authors [4] [5] also used different type of algorithm to combat interference and improve channel estimation accuracy. This part of the paper addresses the problem of channel estimation and tracking in MIMO OFDM systems based on adaptive filtering. The channel is assumed to be slow to moderate time-variant frequency-selective fading channel. The frequency selectivity of the channel arises as a result of relatively high transmission rates, where the transmission bandwidth exceeds the coherence bandwidth of the channel, thereby leading to significant changes of the channel within the band of the transmitted signal. The time-variant nature of the channel is considered to be significant within a frame of OFDM symbols. Hence, the need for channel tracking arises in order for the equalization to be effective.

2 Methodology

Adaptive channel estimation and tracking scheme for MIMO OFDM systems based on DD EW-RLS algorithm with time-domain training is presented. Channel estimation and then tracking are performed in time domain while channel equalization is implemented in frequency domain. The channel estimates are updated recursively upon receiving new training sequences. In general, we assume that training sequences of equal length n are employed, where n may take any value in the range $L \leq n \leq K - 1$. Further, assume that the channel remains fixed during the observation interval $0 \leq i \leq n$, where i denotes the recursion index. The estimation process is performed by recursively adjusting the time-varying tap-weight vector $\hat{h}_n$ of the transversal filter according to an instantaneous a *posteriori* estimation error $e(i)$, defined as the difference between the noisy received signal $y_q(i)$ and its estimate $\hat{y}_q(i)$ given by the adaptive filter output. Figure 1, illustrates how an adaptive filter is used to estimate unknown channel. The filter adaptation proceeds on a sample-by-sample basis, while the time update of the estimated channel proceeds on a block-by-block basis. Since the CIRs among the different transmit-receive antenna pairs are independent, channel estimation at each receive antenna is carried out independently.

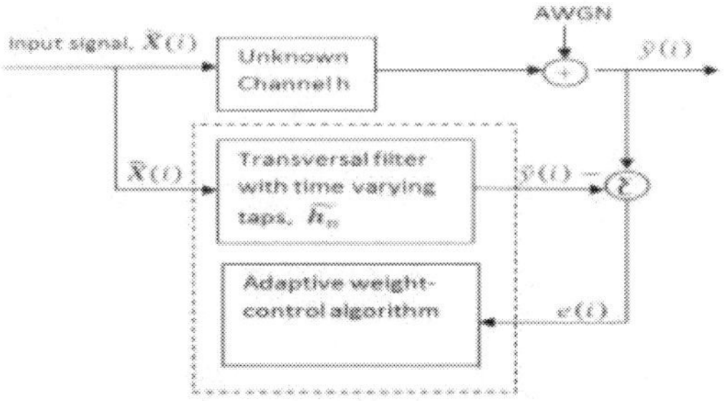

Figure. 1. Illustration of adaptive filter modelling of unknown channel.

Channel estimation and tracking based on adaptive algorithms is also considered. Specifically, the exponentially weighted recursive least squares (EW-RLS) and normalized least-mean square (NLMS) algorithms are used, jointly with decision-directed (DD) technique, for building the proposed estimators. First, the channel is estimated in time-domain followed by zero forcing equalization in frequency-domain. Then, the equalized data symbols are exploited as training sequences to train the estimator to blindly track the time-variations of the channel. Two estimation methods are proposed. In the first method, the CIRs are estimated in TD through exploiting preamble sequences inserted in TD. In the other method, the CIRs are obtained by substituting for the channel frequency responses (CFRs) by their corresponding representations as a FFT matrix multiplied by the CIRs. In such method, the pilot-symbols are inserted in FD on some pre-defined subcarriers. Inserting the training sequences in TD has the advantage of avoiding the problem of peak-to-average power ratio (PAPR) of the training sequences. The training symbols either as preambles or pilots are periodically transmitted to update the channel estimates.

By using typical baseband MIMO OFDM system, training sequences are demultiplexed in time –domain at the receiver. After removing guard interval (GI), the time domain K×1 received signal at the qth antenna during transmitting the mth OFDM symbol can be written as

$$\bar{y_q}(m) = \sum_{p=1}^{M_t} \bar{X_p}(m)\, h_{pq}(m) + v_q(m) \qquad (1)$$

Where $\mathbf{X}^{\sim}_p(m)$ is N x L Toeplitz matrix contains delayed versions of the input data sent from the pth transmit antenna. $\mathbf{h}_{pq}$ is the channel impulse response between the pth transmit and the qth receive antennas while $\mathbf{v}_q(m)$ is a K x 1 vector representing additive white Gaussian noise (AWGN) at the qth receive antenna with complex elements.

Data vectors are demodulated after assuming guard interval is larger than length of MIMO subchannels. ISI will not occur between OFDM symbols and received signals of the different subcarriers can be viewed as independent of each other. Then, frequency-domain received signal at the qth antenna over kth tone at the time m, $Y_q(m,k)$ can be expressed as

$$Y_q(m,k) = \sum_{p=1}^{M_t} H_{pq}(m,k) X_p(m,k) + V_q(m,k) \qquad (2)$$

where $X_p(m,k)$, $H_{pq}(m,k)$ and $V_q(m,k)$ denote the kth data sample of the transmitted OFDM symbol from the pth antenna, the channel coefficient of the kth tone between the pth transmit and the qth receive antennas, and the noise at the qth receive antenna on the kth tone, respectively.

3 Model System Settings

3.1 Initial Conditions

In RLS algorithm, there are two variables involved in the recursion that need initialization, namely, the channel under estimation $\hat{\mathbf{h}}_{(-1),q}$ and the inverse of the autocorrelation of the inputs $\mathbf{\Psi}(-1)$. In our simulations, the initial state conditions of these parameters are set as in Table 1, since our knowledge of these parameters is indistinct, a very high covariance matrix of $\mathbf{\Psi}(-1)$ is to be expected, and thus a high value to $\boldsymbol{\delta}$ is assigned.

Table 1. The initial conditions of the RLS algorithm parameters.

The parameter	Initial state
Channel under estimation , $\hat{\mathbf{h}}_{(-1),q}$	$0_{(M_t L\times 1)}$
Inverse of the autocorrelation matrices of the inputs, $\mathbf{\Psi}(-1)$	$\delta^{-1}\mathbf{I}_{M_t L}$
Regulation parameter (small positive constant),$\boldsymbol{\delta}$	1×10^{-6}

3.2 Simulation Parameters

We consider 2-transmit and 2-receive antennas (2×2) MIMO OFDM system, which can be straightforward extended to nay number of transmit and receive antennas. The simulated system parameters are given in Table 2. In the simulations, a forgetting factor of $\lambda = 0.995$ is used.

Table 2. System and Channel Parameters

Parameter	Value
Channel bandwidth (BW)	20MHz
Sampling frequency (f_s)	20MHz
FFT size (N_{FFT})	128
Number of subcarriers (K)	128

Subcarrier spacing (Δf)	147.06 kHz
Cyclic prefix length (N_{cp})	8 samples
Modulation type	QPSK
OFDM symbol duration (T_s)	6.8μs

4 Channel Estimation in Time-Domain(TD)

Training Schemes Design

Training sequences with constant magnitude are inserted in time-domain (i.e., after the IFFT modulators), thereby the problem of peak-to-average power ratio (PAPR) of the training sequences is avoided. Two training schemes are investigated:

Schemes I: The training sequences from the different antennas are arranged in a sequential manner (i.e., orthogonal in time) in a block or more, where each block has a length equivalent to the length of one OFDM symbol. The criterion for choosing the number of blocks required to transmit the training sequences depends on the number of subcarriers K, transmit antennas M_t, maximum channel length L, and the guard interval N_g. Assume that $N_g \geq L$, the inequality that shows the relation between these variables may be written as

$$K + N_g \geq M_t (L_{ts} + N_g) \quad (3)$$

$$L_{ts} \leq \frac{K + N_g(1 - M_t)}{M_t} \quad (4)$$

Where L_{ts} denotes the length of a training sequence.

Schemes II: The training sequences are simultaneously transmitted from the different antennas, where a number of OFDM symbols equal to the number of transmit antennas is used for training.

5 Results and Analysis

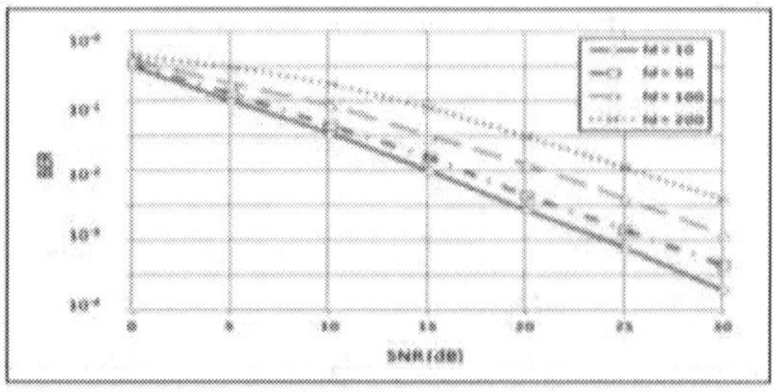

Figure. 2. MSE of channel estimates with time-orthogonal training sequences (Scheme I)

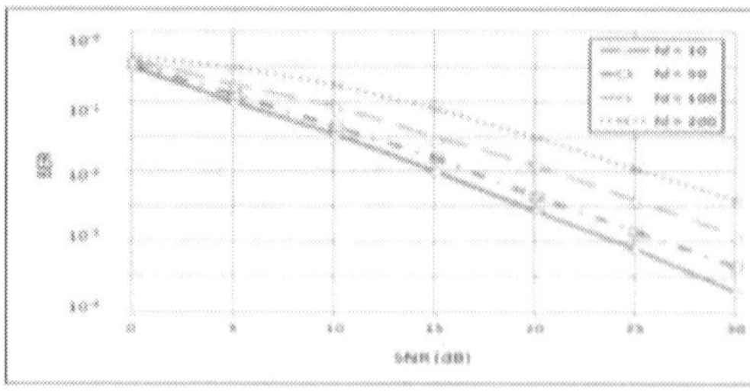

Figure. 3. The BER performance vs. SNR, with time-orthogonal transmitted training sequences (Scheme I)

Simultaneously Transmitted Training Sequences

The training sequences are simultaneously transmitted from the different antennas, where a number of OFDM symbols equal to the number of transmit antennas is needed for training. We refer to this training scenario as scheme II. For this case, the MSE of the channel estimates and the BER performance versus the SNR are shown in Figure 2 and 3, respectively.

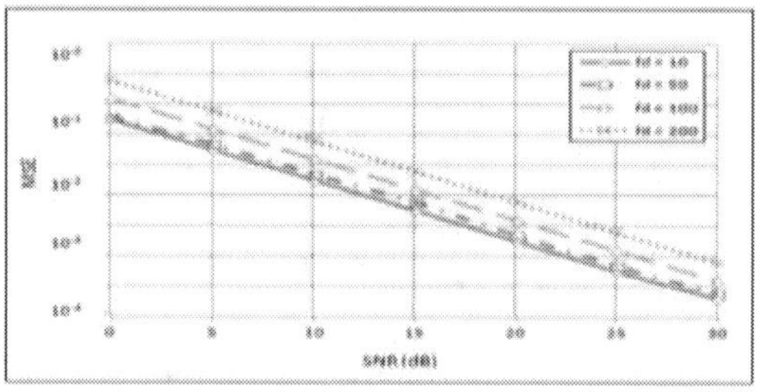

Figure. 4. MSE of channel estimates with simultaneously transmitted training sequences (Scheme II)

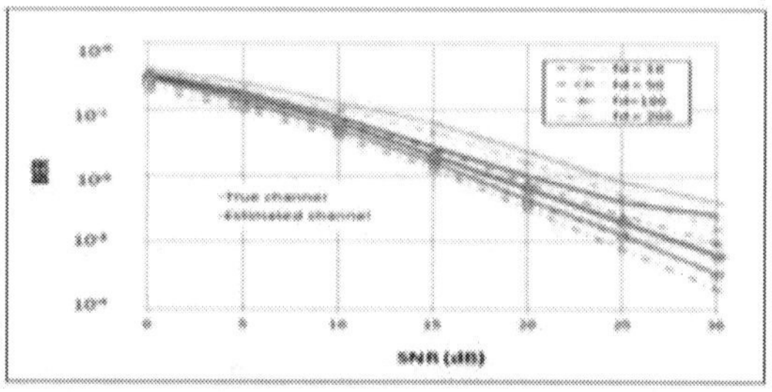

Figure. 5. The BER performance vs. SNR, with simultaneously transmitted training sequences (Scheme II)

From Figure 1 and 2, it can be seen that the proposed channel estimator with training scheme I has much better estimation performance than that with training scheme II. Quantitatively, for example, at a MSE of 10^{-3} a gain difference of 11.5 dB in favoring of training scheme I over training scheme II is obtained. It can also be noticed from the Figure 4 that the BER performance curves obtained with training scheme I are very close to those with perfect channel knowledge at the receiver. On the contrary, Figure 5 shows that the BER performance with training scheme II suffers some degradation due to the enhanced noise at the receiver. For example, to achieve a BER of 10^{-2} at f_d=100Hz, and increase in the SNR of about 2dB is required over that when training scheme I is used.

Channel Estimation in Frequency-domain

In case of the channel estimation is carried out in FD, pilot-symbols are multiplexed with the data symbols in each OFDM symbol where each eight subcarrier is used for training. The MSE of the channel estimates and the BER performances are shown in Figure 6 and 7, respectively.

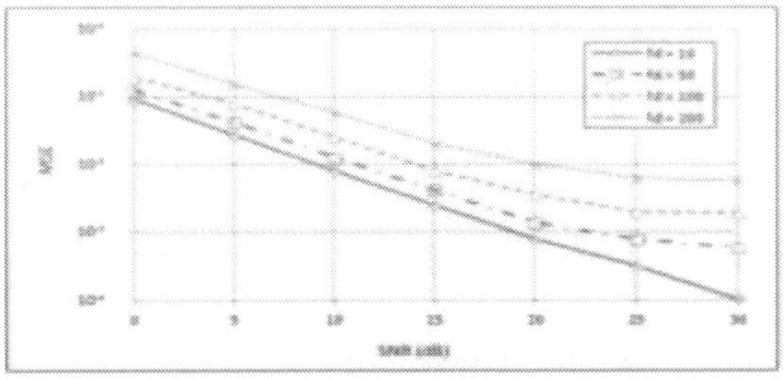

Figure. 6. MSE of channel estimates of the EW-RLS estimator in FD

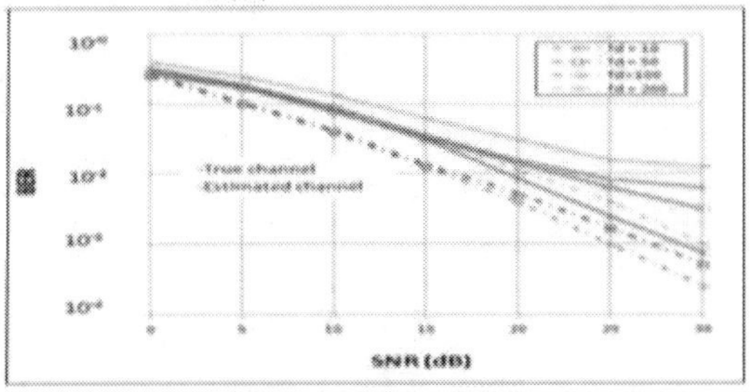

Figure. 7. The BER performance vs SNR, with EW-RLS channel estimation in FD

Comparing the result in Figure 6 and 7 with their corresponding Figure 2 and 3, obtained for EW-RLS channel estimator in TD, it can be noticed that at the lower SNR range and lower Doppler frequencies, more of 10dB SNR is required for the FD estimator to get the same MSE of the TD estimator. However, at the higher SNR and higher Doppler frequencies, an error floor has appeared in case of FD-based channel estimation, the situation which is not occurred in case of Td-based channel estimation. Similarly, it can be observed that the BER performance curves obtained for different Doppler frequencies with the TD channel estimation are very close to the ideal, while severe degradation in the BER performance has occurred when FD channel estimation is performed.

6 Conclusion

Simulation results show that the proposed estimator performance with the time-orthogonal training sequences outperforms that with simultaneously transmitted training sequences. In addition, one OFDM symbol duration is only required for training in case of time-orthogonal training sequences compared to two OFDM symbol durations in case of simultaneously transmitted training sequences. Simulation results also proved that the estimator performance, in terms of MSE and system BER, in TD is superior over that obtained in FD, for different Doppler frequencies. Channel estimation in time-domain is attractive over its counterpart in frequency-domain due to its lower computational complexity, accuracy, and effective channel impulse response tracking especially when the channel is time-varying.

References

1. M. Rana, "Performance comparison of LMS and RLS channel estimation algorithms for 4G MIMO OFDM systems," *... and Information Technology (ICCIT), 2011 14th ...*, no. Iccit, pp. 22–24, (2011)
2. J. Li, J. Ma, and S. Liu, "RLS channel estimation with superimposed training sequence in OFDM systems," *... Technology, 2008. ICCT 2008. 11th IEEE ...*, vol. 0, pp. 1–4, (2008)
3. B. Sklar, *Digital Communication Fundamentals and Application*, 2nd Editio. Prentice Hall, (2001)
4. D. Bhoyar, C. Dethe, and M. Mushrif, "Performance Comparison of Modified Variable Step Size Leaky LMS Algorithm for Channel Estimation in Noisy Environment," *arpublication.org*, vol. 1, no. 1, pp. 13–23,(2013)
5. T. Saikumar, B. Smitha, and P. S. Murthy, "CMA Channel Equalization Through An Adaptive MMSE Equalizer Based RLS Algorithm," *International Journal of Computer and Communication*, vol. 3, no. 1, pp. 65–70.

Performance Evaluation of Hadoop-based Large-scale Network Traffic Analysis Cluster

Ran Tao, Yuanyuan Qiao and Wenli Zhou

Beijing Key Laboratory of Network System Architecture and Convergence Beijing University of Posts and Telecommunications, Beijing 100876, China

Abstract. As Hadoop has gained popularity in big data era, it is widely used in various fields. The self-design and self-developed large-scale network traffic analysis cluster works well based on Hadoop, with off-line applications running on it to analyze the massive network traffic data. On purpose of scientifically and reasonably evaluating the performance of analysis cluster, we propose a performance evaluation system. Firstly, we set the execution times of three benchmark applications as the benchmark of the performance, and pick 40 metrics of customized statistical resource data. Then we identify the relationship between the resource data and the execution times by a statistic modeling analysis approach, which is composed of principal component analysis and multiple linear regression. After training models by historical data, we can predict the execution times by current resource data. Finally, we evaluate the performance of analysis cluster by the validated predicting of execution times. Experimental results show that the predicted execution times by trained models are within acceptable error range, and the evaluation results of performance are accurate and reliable.

1 Introduction

With the rapid development of cloud computing, Hadoop[1] as an advanced big data processing tool, has become the first choice for many researchers and companies. In order to analyze the massive traffic data efficiently, we developed a Hadoop-based Large-scale Network Traffic Analysis Cluster (HBLSNTAC). It consists of one master (running NameNode, ResourceManager), one backup master (also work as client access server, running SecondaryNameNode) and nine slaves (running DataNode, NodeManager, ApplicationMaster). HBLSNTAC stores massive network traffic data collected from operators of several cities in China in HDFS. Moreover, users run various off-line statistical analysis applications using the massive data on the cluster. These applications are aimed at analyzing the basic statistical characteristic of network traffic, such as the time distribution or geographical distribution of network behavior of mobile phone users.

The performance status would greatly affect the work efficiency of HBLSNTAC. A poorer performance means a lower work efficiency, which would slow down the whole analysis work schedule. If we can reliably evaluate performance of HBLSNTAC, the cluster manager could get an intuitive overview of performance status and then make timely adjustment to keep the cluster steady and in high efficiency. However, the performance of HBLSNTAC is affected by various factors, such as hardware device, software versions, Hadoop configuration parameters which are the main factor of performance, and so on. The total amount of Hadoop configuration parameters is excessive, let alone the complex interactive relationships among these parameters. So it is hard to accurately evaluate the performance of HBLSNTAC.

Since several aspects of factors would influence the performance of HBLSNTAC, it is very difficult for normal user to collect all the information of influenced factors. So normal user barely knows the actual performance of HBLSNTAC. Meanwhile, the execution time of the off-line analysis application is directly depended on the performance of HBLSNTAC. A poorer performance would cause a longer execution time. If normal user launches an application at improper timing, for example, when the performance is poor, it would end up lengthening the execution time of his own application, and further worsening the work efficiency of HBLSNTAC. Obviously, that is a dilemma we avoid to get in.

Thus, there is a pressing need for a performance evaluation system to evaluate the performance of HBLSNTAC. This paper aims to design and implement a performance evaluation system to HBLSNTAC. The main contributions of this paper are:

- We propose a performance evaluation system of HBLSNTAC, and the evaluation results are verified to be reliable. Thus we could have a quantitative understanding of performance and then guide user behaviors.

- Since the Hadoop configuration parameters are excessive and complex, we focus on the resource data, which is the essence of configuration parameters. And besides average value, we add three statistical dimensions of resource data, i.e., maximum, minimum and standard deviation.
- We apply principle component analysis and multiple linear regression in modeling phase. After training by historical data, the prediction of this model is relatively accurate with an average error less than 10%.

The remainder of this paper is organized as follows. Firstly, we introduce the related work in section 2. Then we propose the performance evaluation system and detailed introduce the design of the system in section 3. The experimental results and corresponding analysis are presented in section 4. Finally, we draw the conclusion and give the outlook of future work.

2 Related Work

Hadoop is a popular open source project. It has shown its significance in big data and cloud computing areas. Since increasing users and companies have applied Hadoop, the study and research of its performance has become more important and globalized.

Performance Optimization is a research focus for Hadoop cluster, because it could improve the work efficiency of cluster. Microsoft IT SES Enterprise Data Architect Team proposes various optimization techniques to maximize performance of Hadoop MapReduce jobs in [2]. Paper [3] proposes a statistic analysis method to identify the relationship among workload characteristics, configuration parameters and Hadoop cluster performance. They both optimize Hadoop configuration parameters in different situations, therefore shorten the execution times of applications to improve the efficiency.

Performance evaluation is another research hot topic. HiBench[4] is a representative and comprehensive benchmark suite for Hadoop, which contains ten applications, classified into four categories. It quantitatively evaluates and characterize the Hadoop development. Paper [5] proposes a performance evaluation for Hadoop cluster on virtual environment, it builds a Hadoop performance model and then examine several influence factors of cluster performance.

In summary, the existed researches of performance of Hadoop cluster mostly focus on Hadoop configuration parameters. However, the configuration parameters are excessive and complex. So in this paper, we study on resource data, which is the essence of Hadoop configuration parameters. Especially, we build a performance model and identify the relationship between resource data and execution times of applications, then we evaluate performance of HBLSNTAC by the validated performance model.

3 The Design of Performance Evaluation System

In this section, we introduce the design of performance evaluation system. The framework of system is presented in Figure 1. Firstly we set execution times of three benchmark applications as the performance benchmark of HBLSNTAC, and select total of 40 metrics of resource data, which cover the main resources of HBLSNTAC, such as CPU, memory, disk and network. Then we model historical data of execution times of three benchmark applications and resource data through principal component analysis and multiple linear regression. After that, we predict the execution times by the trained models. Finally, we evaluate the performance of HBLSNTAC based on the predicting of execution times.

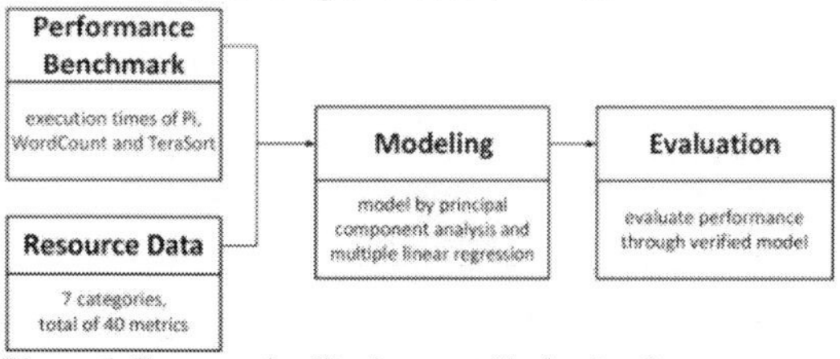

Figure 1. Framework of Performance Evaluation System

3.1 Performance Benchmark

Hadoop applications can be classified into four categories, CPU intensive, memory intensive, disk I/O intensive and network I/O intensive[6]. Since applications could never finish successfully unless there is sufficient memory resource, we exclude memory intensive applications. Moreover, Network I/O is not a bottleneck in our experiment, because the scale of cluster is small and all nodes are connected to the same switch. As a result, we select a CPU intensive application, i.e., Pi; a disk I/O intensive application, i.e., TeraSort; and a both CPU and disk I/O intensive application, i.e., WordCount. Furthermore, we use the execution time of three selected Hadoop benchmark applications, Pi, WordCount and TeraSort, as the performance benchmark of HBLSNTAC.

More importantly, the main function of HBLSNTAC is running off-line applications to analyze the basic statistical characteristic of network traffic. Both WordCount and TeraSort applications have statistical approaches as counting and sorting separately. So it makes them much more suitable to evaluate the actual performance of HBLSNTAC than other applications.

3.2 Statistical Resource Data

Hadoop has more than 200 configuration parameters[7], and the interactive relationships among parameters are complex. Although many parameter researches have proposed various parameter tuning strategies, but with the difference of Hadoop cluster environments, not all the strategies are suitable for any environment. So it is unachievable to propose a universal full-cover parameter tuning strategy for all environments. Thus we can't evaluate performance of HBLSNTAC through configuration parameters without that universal parameter tuning strategy.

Resource data of server, such as CPU, memory, disk and network, is the underlying data of Hadoop cluster. Application on Hadoop cluster would not be launched if not enough resources are allocated to it. Furthermore, most of configuration parameters are specified by resources. For instance, most of configuration parameters of YARN are used for allocations of CPU and memory resources. In other words, resource data is the essence of configuration parameters. The parameters tuning essentially means resources reallocation.

Thus, to conduct a more simplified and essential research of performance of HBLSNTAC, we study on resource data of server instead of configuration parameters. Based on the study of Hadoop resources, combining with accumulated management experience of HBLSNTAC, we select 7 categories of resource data, total of 40 metrics. The list of metrics is shown in Table 1.

Table 1. List of Resource Data Metrics

CPU	CPU utilization	× (maximum, minimum, average, standard deviation)
Memory	memory utilization	
Disk	disk utilization	
	disk I/O speed	
Network	upstream speed	
	downstream speed	
Open files	number of open files	
Load	custom server load	
HDFS	upload speed	
	download speed	

Since the resource of Disk, Network and HDFS, are subdivided into two smaller categories, we now have ten small categories. These ten small categories are the common resource data, which are studied frequently within Hadoop researches. Most of the researches only study the average value of resource data. It is generally known that average value does not reflect all of the statistical characteristics of the data. Therefore, for each small category, we add three statistical dimensions, maximum, minimum and standard deviation besides average. Then we have 40 metrics totally. Now, we not only study on the central tendency, but also study on the dispersion degree of resource data, which would improve the reliability of modeling analysis.

Here, the customized Load metric is used to describe the extent of a single server load exceeds the average server load of whole cluster, which would be expressed as $Load_i$. It is easy to understand that for metrics of CPU, Memory, Disk, Network and Open files, the larger the value is, the heavier the server load is. For server i, assume that R_n is the value of one of metrics, R_{avg} is the average value of the same metric of the cluster. Then we have L_n, i.e., the value of the customized load of a single metric, which could be computed by (1), and $Load_i$, i.e., the value of the customized server load of server i, which could be computed by (2).

$$L_n=\begin{cases} 0 & (R_n \le R_{avg}) \\ R_n \cdot (R_n - R_{avg}) & (R_n > R_{avg}) \end{cases} \quad (1)$$

$$Load_i = \sum_n L_n \quad (2)$$

3.3 Modeling Analysis

The purpose of modeling analysis is to figure out the relationship between performance benchmark, i.e., the execution times of three benchmark applications, and the recourse data. After training the model by historical data, we could predict the execution times by current resource data.

3.3.1 Principal Component Analysis

Principal Component Analysis (PCA) uses dimensionality reduction technique to make a set of possibly correlated original indicators into relatively fewer comprehensive and linearly uncorrelated indicators by linear combination, and retain most of the information of original target[8].

If we have the original data matrix as:

$$X=\begin{bmatrix} x_{11} & x_{12} & \cdots & x_{1n} \\ x_{21} & x_{22} & \cdots & x_{2n} \\ \cdots & \cdots & \cdots & \cdots \\ x_{m1} & x_{m2} & \cdots & x_{mn} \end{bmatrix}$$

where m is the number of sites, n is the number of variables in the index system, $x_{ij}(i=1,2,\cdots m; j=1,2,\cdots n)$ is the value of index variable j of site i.

(1) Standardization of original data

Usually the variables have different units of measurement (i.e., pounds, feet, rate, etc), that would cause different weight for each variable in the analysis. In order to make each variable to receive equal weight, it is necessary to standardize the original data.

(2) Normalized correlation matrix of data

$$R=(r_{ij})_{m\times n} \quad (3)$$

where r_{ij} is the correlation coefficient of index i and index j. The eigenvalues of matrix R are $\{\lambda_1, \lambda_1, \cdots \lambda_n\}$, and $\lambda_1 \ge \lambda_1 \ge \cdots \ge \lambda_n$.

(3) Calculation of variance contribution of each principal component

$$\alpha_i = \frac{\lambda_i}{\sum_{i=1}^{n} \lambda_i} \quad (4)$$

(4) Selection of principal components

Generally speaking, select number of principal components whose cumulative contribution reaches to rate of 85 ~ 100%, meaning these principal components retain 85 ~ 100% information of original data.

3.3.2 Multiple Linear Regression

Multiple Linear Regression (MLR) attempts to model the relationship between two or more explanatory variables and a response variable by fitting a linear equation to observed data. It is a statistical technique that uses several explanatory variables to predict the outcome of a response variable[9].

The general form of multiple linear regression is defined as:

$$y = \beta_0 + \beta_1 x_1 + \beta_2 x_2 + \cdots + \beta_n x_n + \varepsilon \qquad (5)$$

where y is the dependent or response variable, $x_i \, (i = 1,2,\cdots,n)$ are the independent or explanatory variables, $\beta_i \, (i = 0,2,\cdots,n)$ are the parameter of model and ε is the deviations of model.

3.4 Performance Evaluation

After verifying the trained model reliable, we could evaluate the performance of the HBLSNTAC by the prediction of the three execution times. We propose a score criterion by theses three execution times, then we select a typical practical application, whose execution time is a reflection of the practical performance of HBLSNTAC. By studying the relationship between the score and the execution time of practical application, we can find out several patterns between them. Finally we could evaluate the performance of HBLSNTAC by these patterns, thus to help us manager the HBLSNTAC in a better way.

4 Experiment

4.1 Experiment Setup

We setup a simple classical Hadoop cluster that consists of one master (running NameNode and ResourceManager) and three slaves (running DataNode, NodeManager and ApplicationMaster). All servers are connected to a 100Mbps Ethernet switch. Hardware and software configurations of the servers in this cluster are the same, as shown in Table 2.

Table 2. Configurations of Experimental Servers

Item	Value
Machine	Lenovo M4360
Processor	Pentium(R) E5800 @3.20GHz
Cores	2
Memory	4G
Hard Disk	500G
Network	100Mbps
Operation System	CentOS 6.5 x64
JVM	JDK 1.7.0_45
Hadoop	2.5.0

4.2 Modeling and Prediction

We run three benchmark applications for 25 times separately. The input of modeling is the resource data within 3 minutes before application get launched. And the output of modeling is the predictions of execution times of applications. It means that we would finally predict the execution time of an application by the resource data within 3 minutes before this application get launched.

4.2.1 Principal Component Analysis

We get total of 40 metrics of resource data, meaning 40 dimensions for data analysis, that is a quiet large number of dimensions. Moreover, there would be correlations between these metrics. Therefore, we would reduce dimensions from total 40 dimensions to much more fewer uncorrelated dimensions by Principal Component Analysis. The main results of three benchmark applications by Principal Component Analysis using SPSS software[10] are shown in Table 3, Table 4, Table 5 and Figure 2.

For Pi application, we can find out from Table 3 and Figure 2a that the cumulative contribution rate of first 6 Principal Components (PCs) is 98.422%, meaning they retain 98.422% information of original 40 metrics. Here, we have reduced dimensions from 40 metrics to 6 PCs, which we would express as $X_{Pi} = (x_1, x_2, x_3, x_4, x_5, x_6)$. Similarly, for WordCount application, the first 5 PCs expressed as $X_{WC} = (x_1, x_2, x_3, x_4, x_5)$ retain 97.713% information of original 40 metrics. And for TeraSort application, the first 5 PCs expressed as $X_{TS} = (x_1, x_2, x_3, x_4, x_5)$ retain 100.00% information of original 40 metrics. The component matrix would not be shown here due to space limitation.

4.2.2 Multiple Linear Regression

The purpose of MLR is to model the relationship between principal components and execution time of benchmark application, then make a prediction of execution time by a new set of principal components.

As for Pi application, the multiple linear regression equation of principal components as $X_{Pi} = (x_1, x_2, x_3, x_4, x_5, x_6)$ and execution time as y_{Pi} is as follow:

$$y_{Pi} = -0.632 + 0.713x_1 + 0.413x_2 + 0.527x_3 + 0.258x_4 + 0.179x_5 + 0.302x_6 \qquad (6)$$

As for WordCount application, the multiple linear regression equation of principal components as $X_{WC} = (x_1, x_2, x_3, x_4, x_5)$ and execution time as y_{WC} is as follow:

$$y_{WC} = -0.491 + 0.726x_1 + 0.619x_2 + 0.492x_3 + 0.325x_4 + 0.173x_5 \qquad (7)$$

Table 3. Total Variance Explained of Pi

Component	Initial Eigenvalues			Extraction Sums of Squared Loadings		
	Total	% of Variance	Cumulative %	Total	% of Variance	Cumulative %
1	14.636	36.590	36.590	14.636	36.590	36.590
2	9.452	23.630	60.220	9.452	23.630	60.220
3	6.694	16.735	76.955	6.694	16.735	76.955
4	3.950	9.875	86.830	3.950	9.875	86.830
5	3.010	7.525	94.354	3.010	7.525	94.354
6	1.627	4.068	98.422	1.627	4.068	98.422
7	.631	1.578	100.00			
…	…	…	…			

Table 4. Total Variance Explained of WordCount

Component	Initial Eigenvalues			Extraction Sums of Squared Loadings		
	Total	% of Variance	Cumulative %	Total	% of Variance	Cumulative %
1	16.980	42.451	42.451	16.980	42.451	42.451
2	11.518	28.794	71.245	11.518	28.794	71.245
3	5.000	12.500	83.745	5.000	12.500	83.745
4	3.557	8.893	92.638	3.557	8.893	92.638
5	2.030	5.075	97.713	2.030	5.075	97.713
6	.915	2.287	100.00			
…	…	…	…			

Table 5. Total Variance Explained of TeraSort

Component	Initial Eigenvalues			Extraction Sums of Squared Loadings		
	Total	% of Variance	Cumulative %	Total	% of Variance	Cumulative %
1	16.264	40.661	40.661	16.264	40.661	40.661
2	14.100	35.250	75.911	14.100	35.250	75.911
3	4.728	11.820	87.731	4.728	11.820	87.731
4	3.226	8.066	95.797	3.226	8.066	95.797
5	1.681	4.203	100.00	1.681	4.203	100.00
…	…	…	…			

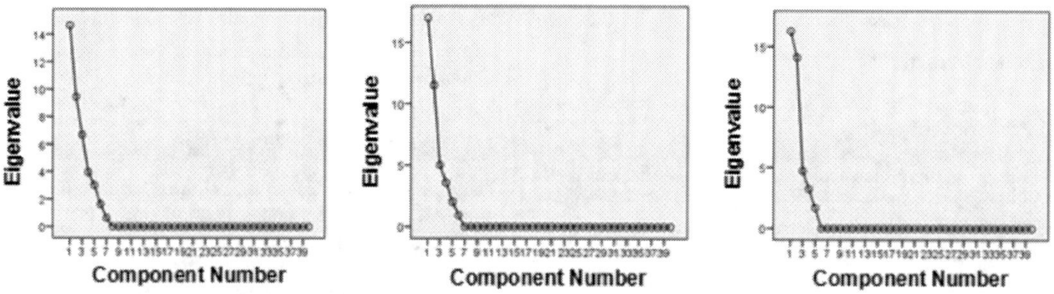

Figure 2. Scree Plots of Pi, WordCount and TeraSort (from left to right)

As for TeraSort application, the multiple linear regression equation of principal components as $X_{TS} = (x_1, x_2, x_3, x_4, x_5)$ and execution time as y_{TS} is as follow:

$$y_{TS} = -0.585 + 0.691x_1 + 0.543x_2 + 0.595x_3 + 0.362x_4 + 0.238x_5 \quad (8)$$

4.2.3 Prediction Verification

After training model by historical data, we need to verify the model by current data. We run three benchmark applications for 8 times separately to verify the prediction result of the trained models. The results are shown in Figure 3, Figure 4 and Figure 5. The relative error between predicted execution time and actual execution time of Pi, WordCount and TeraSort are 7.62%, 7.97% and 8.18% respectively. It means that these models can accurately predict the execution time of benchmark applications.

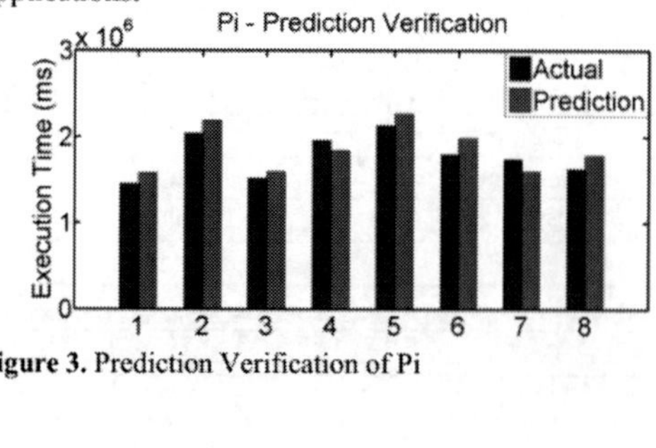

Figure 3. Prediction Verification of Pi

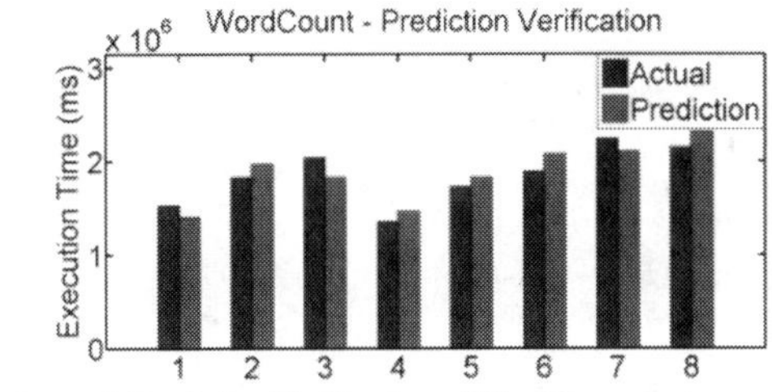

Figure 4. Prediction Verification of WordCount

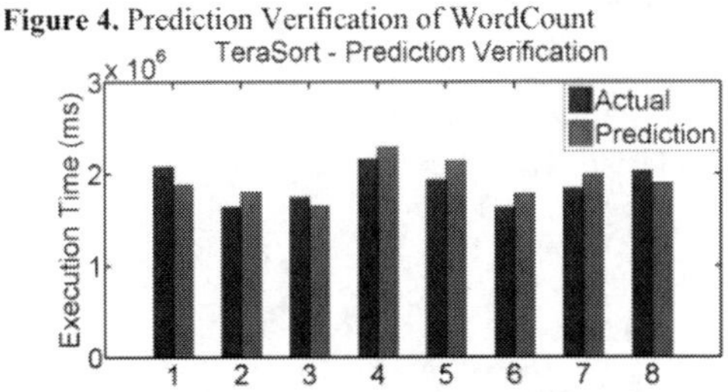

Figure 5. Prediction Verification of TeraSort

The trained models, whose prediction average errors are all less than 10%, are verified to be reliable. As a result, they can be used for performance evaluation of HBLSNTAC.

4.3 Performance Evaluation

We convert the prediction of each application into a centesimal score, then set the average of the three scores as the evaluation score of the performance of HBLSNTAC. The main function of HBLSNTAC is running off-line statistical applications. And a longer execution time of application means a poorer performance of HBLSNTAC. So we select a typical practical application, whose goal is to find out the top 3 visited websites of 6 time segments of the whole day for each user, and set its execution time as a reflection of the performance of HBLSNTAC. Then study the relationship between the evaluation score and execution time of this practical application.

4.3.1 One Single Practical Application

Relationship between evaluation score and execution time of one single practical application is shown in Figure 6.

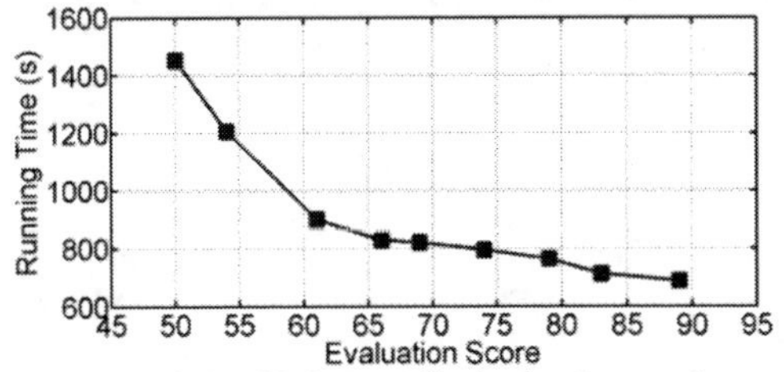

Figure 6. Relationship between Evaluation Score and execution time of One Single Practical Application

Roughly, the lower the score is, the longer the execution time is, thus the poorer the performance is, and vice versa. More closely, we can see that there is turning point near 60. When score is higher than 60, a decrease of 5 points of score would cause an increase of 50 seconds of execution time. When score is lower than 60, the same decrease of 5 point of score would cause an increase of 250 seconds of execution time, which is 5 times larger than the increase when score is higher than 60. It suggests that when the performance of HBLSNTAC reaches at a certain poor degree, such as 60 in this case, it is not a good choice to launch an application, because that would cause a much larger increase of execution time.

4.3.2 Multiple Practical Applications

Since the HBLSNTAC is running multiple application parallelly for most of the time, we further study the relationship between evaluation score and total execution time of multiple practical applications. We launch two practical application successively, and launch the second application before the first application completed. The circumstance before the first application launched is the same as one single practical application, so the evaluation score here is predicted before the second application launched. The result is shown in Figure 7.

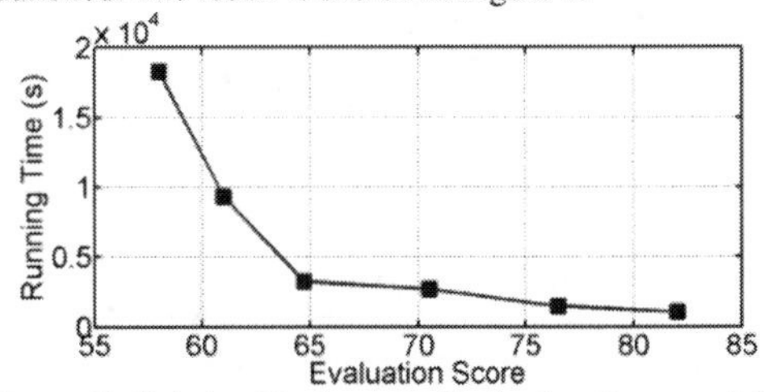

Figure 7. Relationship between Evaluation Score and Total execution time of Two Practical Applications

The trend of curve in Figure 7 is about the same as in Figure 6. However, comparing with Figure 6, the turning point is advanced from 60 to 65. It suggests that the greater the number of running applications is, the sooner that certain poor degree is coming. When score is higher than 65, a decrease of 5 points of score would cause an increase of 630 seconds of execution time. When score is lower than 65, the same decrease of 5 points of score would cause an increase of 11210 seconds of execution time, which is more than 17 times larger than the increase when score is higher than 65. Comparing with the circumstance of single practical application, the number of running applications doubles from 1 to 2, however, the increasing range is more than three times from an increase of 5 times to an increase of 17 times. It suggests that the increasing speed of total execution time, in order words, the worsening speed of performance, is more faster than the increasing speed of number of running applications.

4.3.3 Summary

The evaluation results provide a much more straightforward understanding of performance of HBLSNTAC. We used to only have a qualitative understanding of the performance, but with no idea of how exactly the patterns are. Now, we can get a quantitative understanding of performance with the clear evaluation scores. With accumulation of more experimental evaluation results, we can propose a comprehensive user guide strategy. It can guide using behavior of common users, such as suggest the proper timing for users to launch applications and forbid users from launching applications at improper timing, so that to keep HBLSNTAC in a continual good performance status.

5 Conclusion and Future Work

In this paper, we have proposed a performance evaluation system of HBLSNTAC. This system associates execution time of three benchmark applications and statistical resource data of servers, then model and analyze these two parts by principal component analysis and multiple linear regression. An evaluation score of performance of HBLSNTAC would be marked by the trained models. The experimental results show that the prediction of execution time of application is reliable with average error less than 10%. Furthermore, the evaluation score can quantitatively reflect the performance of HBLSNTAC, so it could be further used for user behavior guidance. The performance evaluation system is validated to be feasible and reliable.

As for the future work of next stage, firstly, with the expansion of application scenarios of HBLSNTAC, we would include more kinds of applications, such as machine learning and web searching, besides Pi, WordCount and TeraSort. Moreover, the modeling analysis approaches are both linear in this paper, we would change and choose nonlinear modeling approaches to analyze the nonlinear relationship between the execution time of application and resource data. Then study the pros and cons of linear and nonlinear modeling approaches.

Acknowledgement

This work is supported in part by the Fundamental Research Funds for the Central Universities(2015RC11) , 111 Project of China (B08004) and EU FP7 IRSES MobileCloud Project (Grant No. 612212).

References

1. T. White, Hadoop: The Definitive Guide. O'Reilly Media, Inc. p. 1-362 (2012)
2. Microsoft IT SES Enterprise Data Architect Team: Hadoop Job Optimization (Microsoft IT white paper), (2014)
3. Hailong Yang, Zhongzhi Luan, Wenjun Li, et al. Statistics-based Workload Modeling for MapReduce. Parallel and Distributed Processing Symposium Workshops & PhD Forum (IPDPSW). Shanghai, p. 2043 - 2051 (2012)
4. Shengsheng Huang, Jie Huang, Jinquan Dai, et al. The HiBench benchmark suite: Characterization of the MapReduce-based data analysis. Data Engineering Workshops (ICDEW). Long Beach, CA, p. 41-51(2010)
5. M. Ishii, J. Han, H. Makino. Design and Performance Evaluation for Hadoop Clusters on Virtualized Environment. International Conference on Information Networking (ICOIN). Bangkok, p. 244-249 (2013)
6. Jinsong Yin, Yuanyuan Qiao. Performance Modeling and Optimization of MapReduce Programs. Cloud Computing and Intelligence Systems (CCIS). Shenzhen, p. 180-186 (2014)
7. Hadoop Homepage: http://hadoop.apache.org/.
8. Moore, B. Principal component analysis in linear systems: Controllability, observability, and model reduction. Automatic Control. Vol. **26**, No. 1, p. 17-32 (1981)
9. Lu Li. An Internet of Things QoE evaluation method based on multiple linear regression analysis. Computer Science & Education (ICCSE). Cambridge, p. 925-928 (2015)
10. Zhang Liping, Pang Jie, Wang YongChao, et al. SPSS for Water Quality Assessment of Beijing Typical River Based on Principal Component Analysis. Digital Manufacturing and Automation (ICDMA). Changsha, p. 395-398 (2010)

The Design of Smart Home for the Elderly Based on ZigBee

Junhan He

School of Electronics and Information, Nantong University, 226019 Nantong, China

Abstract. In recent years, as the pace of people's life is speeding up, more young people working far from their parents, the elderly have to live alone. With the increase of age, they tend to forget things which may create safety accidents. Therefore, a smart home system for the elderly will have good application prospects. In this paper, a smart system based on ZigBee wireless network is proposed. This system uses wireless network for transmitting information between sensor nodes and uses embedded technology to analyse the data. Once an emergency occurs, the system will send an alarm to notify the elderly. At the same time, the young people can know the situation of the house through mobile phone

1 Introduction

In recent years, as the pace of people's life is speeding up, more young people working far from their parents, the elderly have to live alone. As people grow older, they will be likely to forget basic things like switching off lights. And also they may forget to switch off cylinder gas causing LPG gas leakage, or close the doors resulting in financial loss. In consideration of these circumstances, the smart home for the elderly is becoming definitely important. A rather simple solution to overcome these challenges is to install huge number of sensors to monitor the environment inside the house[1].

In a smart home[2], it needs various types of sensors to achieve the information of the living environment of the elderly, such as temperature, humidity, light, noise, human infrared, smoke. The collected data is real-time displayed on the LCD of embedded system, also the embedded system analyses environmental data, and control automatically the light, air conditioning, humidification, alarm and so on[3]. The selected sensors will be introduced in the third section.

Even though many intelligent wireless sensor networks developed for specific applications have been reported[4]. ZigBee technology is employed to transmit data. The ZigBee wireless communication module can realize receiving and transmitting functions by CC2530. The main objective of ZigBee is for battery-powered applications where lower data rate, low cost, and longer battery life are the key requirements[5]. Therefore, the ZigBee module based on wireless technology was used as an appropriate platform to transfer data.

The whole system uses ZigBee wireless sensor network and embedded Linux technology. The management applications are designed by the ARM processors and embedded Linux. Qt is used to design the interface of users. They can browse the data and control the whole system by touching screen which allows more convenience. Afterwards, the data is transmitted to IOS application via a Bluetooth module to display on the phone. Part 1 is mainly for background introduction; Part 2 is for system overview; Part 3 and 4 are for hardware and software implementation; Part 5 is for conclusion

2 System overview

As can be seen from the Figure 1, the whole system is composed of sensor nodes, coordinator, ARM and PC. These sensor nodes will be installed in the house, mainly including temperature and humidity sensors, human body infrared sensors, smoke sensors, doors and windows magnetic sensors. The data collected by each sensor is transmitted to the network coordinator using a wireless network.

The coordinator is responsible for the initiation and maintenance of the whole ZigBee network, and establishing the communication between the sensor nodes. The collected data sends through serial port to the embedded system according to a certain protocol format for a further processing.

DOI: 10.1051/matecconf/20165605016

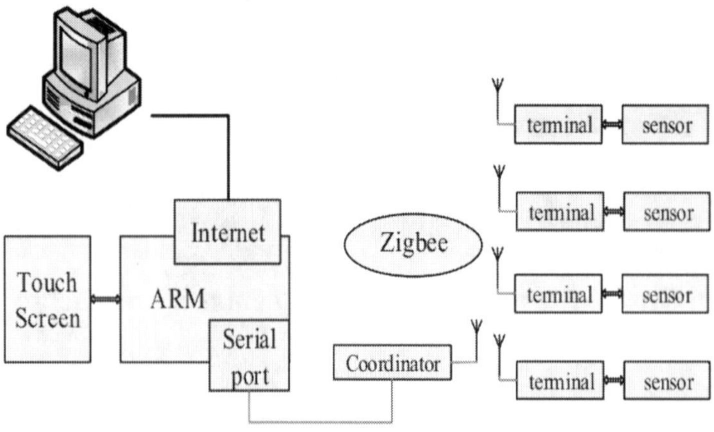

Figure 1. Block Diagram of the whole system

The embedded system collects the data from each sensor node. Each received value is processed. After that, the processed data will be in contrast with a standard threshold. When the measured data is greater than the threshold, a warning sound will be generated. At the same time, the data displays in the touch screen.

3 Hardware implementation

The smart home system mainly achieves the function of collecting and analysing data. It can collect and process the information of temperature, humidity, noise, smoke, the human body infrared and the door or window position. In order to achieve this goal, the next section mainly introduces the design of ZigBee communication modules and terminal node modules.

3.1 Magnetic contact switch

Figure 2 shows the design of the magnetic switch sensor. A voltage comparator LM393 is used in the circuit to prevent the error signal effectively caused by interference or false triggering. When the sensor is installed on the corresponding doors or windows, it can accurately detect whether a door or window is open or closed.

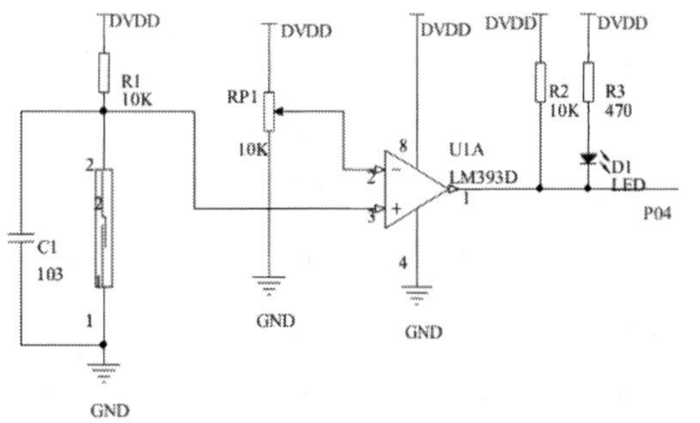

Figure 2. Magnetic switch sensor

3.2 Temperature and humidity sensor

The temperature and humidity can reflect the comfort degree of the living environment of the elderly. SHT10 is used in the system to detect the temperature and humidity. It can work in the range of -40~123℃. Its error range of measuring humidity is 4.5%RH. Its error range of measuring temperature is ±0.5℃.

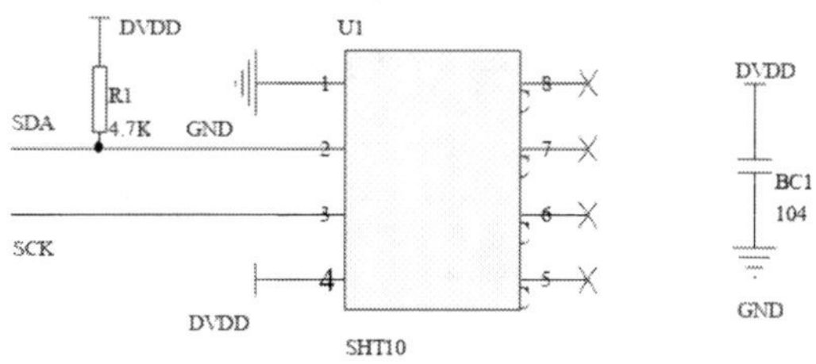

Figure 3. Circuit of temperature and humidity sensor

As shown in Figure 3, after power up, the sensor has a 11ms dormant period. It needs to carry on a time delay operation. In order to decouple filter, it needs increase a 100nf capacitance between the power and ground.

3.3 MQ-2 gas sensor

When MQ-2 is in a sensitive environment, the conductivity of the sensor increases with the increase of the concentration of sensitive gas in the air, and it is suitable for the detection of liquefied gas, methane, propane, methane, alcohol, smoke and so on. As shown in the Figure 4, the adjustable resistor is used to adjust the alarm threshold. After processing, the output signal is digital.

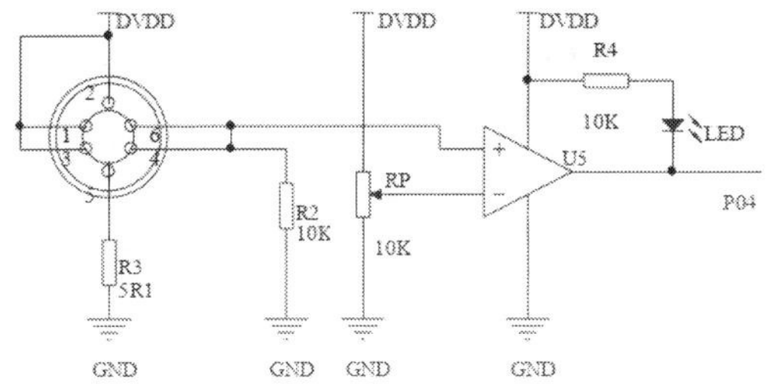

Figure 4. Circuit of gas sensor

3.4 Noise sensor

The noise sensor can detect environmental noise. When illegal person entering, it can detect footsteps, shattering glass, voice of conversation and then sending alarm information to inform the elderly. Due to the output of capacitor microphone is a weak signal, the system uses resistance capacitance coupling way to enlarge the signal. As shown in Figure 5, by using integrated operational amplifier LM358, the weak signal can be amplified 150 times.

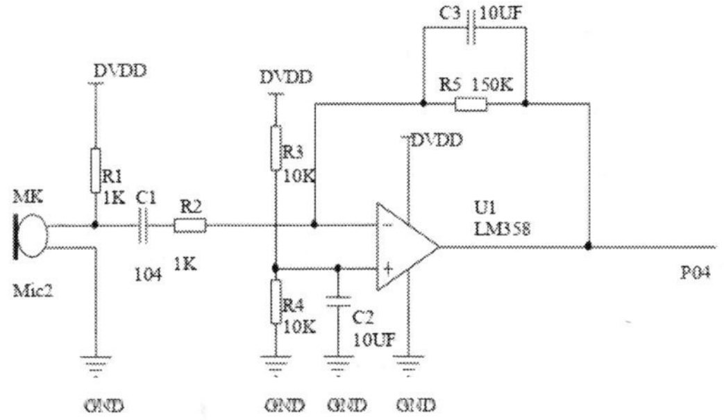

Figure 5. Circuit of noise sensor

4 Software implementation

4.1 Software system of sensor nodes

The software system of sensor nodes is mainly composed of two parts, the sensor nodes and ZigBee network communication. As shown in Figure 6, the data acquisition module is mainly responsible for sensor data collection. And the wireless communication module is mainly responsible for joining network and data sending and receiving.

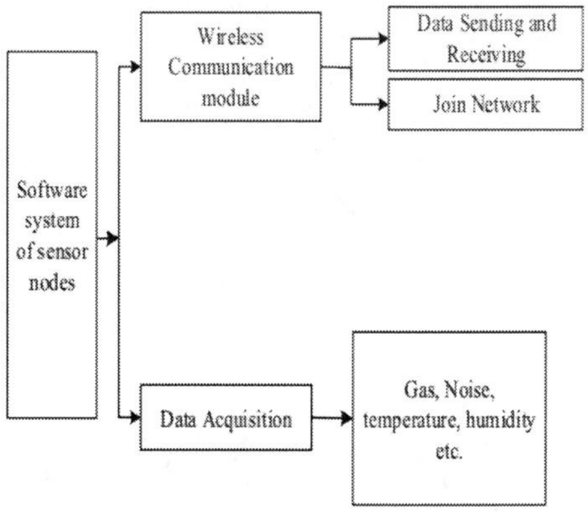

Figure 6. Software system of sensor nodes

4.2 Program design of ZigBee terminal nodes

The ZigBee terminal nodes mainly complete finding network, joining the network, data acquisition, sending and receiving control commands from the coordinator. The program flow chart of the ZigBee terminal node is shown in Figure 7.

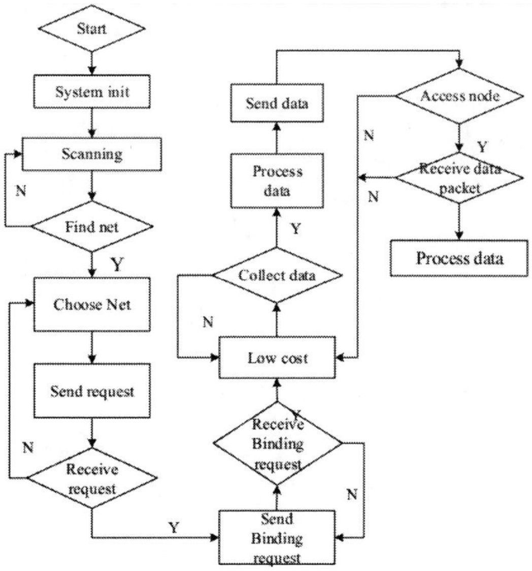

Figure 7. Flow chart of ZigBee terminal nodes

5 Result and conclusion

This system is a practical application. The stability of the system and the accuracy of the data determine the quality of the system. The designs of software system and hardware system are introduced in the front sections. The whole system is shown as Figure 8.

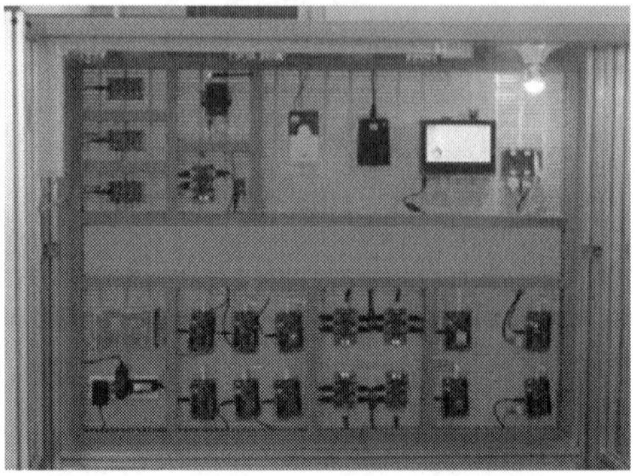

Figure 8. Structure of system

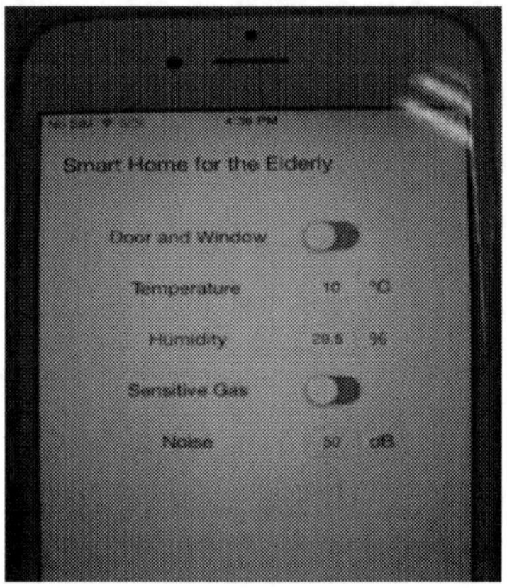

Figure 9. Data displays on the IOS device

By testing the entire system, wireless sensor nodes transmit data reliably. And the consumption of the system is low cost. The wireless transmission distance can meet the needs of the house environment. The embedded system runs stable, and interface is easy to use. The intelligent, low-cost, and low-power smart home system is realized. It has good economic efficiency and application value.

References

1. Dittmar A, Axisa F, Delhomme G, Gehin C, "New concepts and technologies in home care and ambulatory monitoring," Study Health Techno Inform, pp. 9-35(2004).
2. Ricquebourg, Vincent, David Menga, David Durand, Bruno Marhic, Laurent Delahoche, and Christophe Loge, "The smart home concept: our immediate future," In E-Learning in Industrial Electronics, 2006 1ST IEEE International Conference on, pp. 23-28(2006).
3. N. K. Suryadevara, S. C. Mukhopadhyay, R. K. Rayudu,and Y. M. Huang, "Sensor data fusion to determine wellness of an elderly in intelligent home monitoring environment," In Proc. IEEEI2MTC Conf., Graz, Austria, pp. 947–952(2012).
4. P. Tang, T. Venables, "Smart homes and telecare for independent living," J. Telemed. Telecare **6**, pp. 8–14(2000).
5. Rasika S. Ransing, Manita Rajput, "Smart Home for Elderly Care, based on Wireless Sensor Network," Nascent Technologies in the Engineering Field (ICNTE), 2015 International Conference on pp.1-5(2015).

Robotic Label Applicator: Design, Development and Visual Servoing Based Control

Chyi-Yeu Lin, Asheber T. Wagshum and Trung-Son Le

Department of Mechanical Engineering, National Taiwan University of Science and Technology, Taipei, Taiwan

Abstract. Use of robotic arms and computer vision in manufacture, and assembly process are getting more interest as flexible customization is becoming priority over mass production as frontier industry practice. In this paper an innovative label applicator as end of arm tooling (EOAT) capable of dispensing and applying label stickers of various dimensions to a product is designed, fabricated and tested. The system incorporates a label dispenser-applicator and had eye-in-hand camera system, attached to 6-dof robot arm can autonomously apply a label sticker to the target position on a randomly placed product. Employing multiple advantages from different knowledge basis, mechanism design and vision based automatic control, offers this system distinctive efficiency as well as flexibility to change in manufacturing and assembly process with time and cost saving.

1 Introduction

Advance in manufacturing and assembly line methods are made regularly as new and more efficient ways of achieving the goal of increased efficiency are exposed. The equipment used and arrangement of theirs plays a great role for overall productivity of the line. New applications in mature and emerging industries, in addition to growing interest in automation pose pressure to insure innovation of flexible technologies and systems that can accommodate the changes. This in turn further expands the market for robotized manufacture and assembly [1].

Advance in manufacturing and assembly line methods are made regularly as new and more efficient ways of achieving the goal of increased efficiency are exposed. The equipment used and arrangement of theirs plays a great role for overall productivity of the line. New applications in mature and emerging industries, in addition to growing interest in automation pose pressure to insure innovation of flexible technologies and systems that can accommodate the changes. This in turn further expands the market for robotized manufacture and assembly [1].

Automation may range from employing a simple ordinary sensor to sophisticated industrial robot and machineries. Its effectiveness and value is measured in its ability to provide increased productivity for the user [2]. The robotic label applicator proposed in this paper can lie within the range of automation class. The raising sales of industrial robots each year, signifies robotics have high number of presence and taking central role in manufacturing industries as dependable work force. In fact, it is the dominant automation means in selected industries such as automotive, electrical/electronic equipment's, transportation, chemical products, etc. and this high population of robots in industry floors encourages its use and exploitation of its flexibility potential for customized automations.

Manufacturers are putting a considerable amount of effort to distinguish themselves by employing efficient automation and optimum equipment in their factory floor to remain competitive in local and international markets[3, 4]. Robotic arms are in high number in industries thus adding more sound value to their end of arm tooling (EOAT) will significantly improve productivity. The more flexible and smarter EOAT that actually acts with intelligence rather than depending on the arm itself is emerging in high numbers on robotic arm assisted assembly. As a result, we are starting to see new category of end effectors; alike of the mechanism presented in this paper, and manufacturers, and customers are highly tapping from this development.

Over decades, there has been progressive effort to incorporate computer vision and industrial robotics applications. Most common tasks are inspection, part handling and assembling. One early comprehensive review can be found in [5] in which classic vision algorithms using connectivity analysis, size/shape, position/orientation descriptors were briefly collected and explained. These early algorithms try to simplify 3D pose estimation problem to 2D case by taking advantage of conveyor belt picking or small part placement error. A more recent review shows persisting researches to solve general 3D vision problem by introducing more sophisticated sensory systems including stereo cameras, combining mono-camera and range sensor, structured-light and mono-camera. In these systems, the cost for hardware and software development significantly increases. Regardless of quite a number of methods proposed, general part picking is still an unsolved challenge.

The vision algorithm used in this work has been a well-known problem in literature and was addressed by different terminologies. The early day literatures presented a bigger branch of research about structure and motion recovery [6, 7], where ours approach focuses on second, motion recovery subtopic. More recent literatures, Szeliski [8] describes a parametric feature-based alignment of two images in motion recovery with 6-DOF problems. It is well known that there is an ambiguity in translation solution by a scale factor [9, 10]. Zhang et al. proposed a solution for planar pattern [11] which required knowledge about the pattern's measurements as model-based approach. Our approach is a 2D motion recovery from parametric feature-based image alignment. This takes advantage of our current application, which is used on planar objects, and the depth of vision can be fixed for simplicity. This results in a linear relation between image space and task space mapping 2D motion between these two spaces.

The rest of this paper is organized as follows: Section II focuses on conceptual explanation on the mechanism employed and the proposed hardware mechanism designed. Section III will present system integration and vision based control of the label applicator EOAT. In Section IV, experimentation and result will be discussed and lastly conclusion is presented in Section V.

2 Mechanism design

In today's industries the demand and importance for product labeling and tagging is significantly high. This task is mostly done through a dedicated workstation, which is made for specific size, shape and position. Thus, when there is a need to change a product size and shape it necessitates a change in work flow setup and even the machineries mix and arrangements. Moreover manual labor is still used in some other industries. In this work a robotic mechanism for product labelling process presented, an intelligent product labeling and tagging.

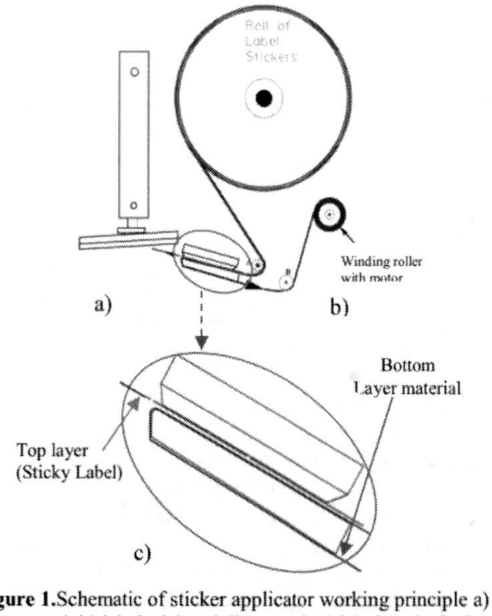

Figure 1. Schematic of sticker applicator working principle a) gripper unit b) label sticker delivery unit c) label peeler bed

2.1 Conceptual design

Product labeling in manufacturing processes or retail service have paramount importance and in many cases it is becoming crucial component. Usually this tagging or labeling is done through applying sticker on product surface or permanently engraved on the product body.

The two major steps, label supply and applying process, combined into one unit to form the operation principle of the robotic label applicator machine designed, manufactured and tested in this work, as schematically show in Figure 1. Thus label applicator system is modularly designed consisting of two basic subassemblies:

a) Label applicator module (Figure 1-a)and
b) Label supply module (Figure 1-b)

The label sticker is prepared in roll form, which is placed on passive roller on main frame of the system is made ready and supplied to applicator unit to paste to target position. The strip of the sticker label, label and underneath layer material, laid and pass over wedge shaped base (Figure 1-c magnified view) and other tensioning and guiding roller distributed along the path to deliver the sticker. The pull force applied by the winding roller (pull on under layer, blue strip, Figure 1b), powered by geared DC motor, will separate the upper sticky part (the label sticker, red strip, Figure 1b) from the underneath non sticky material. Once the sticker is ready the vacuum gripper is activated, signaled by the proximity sensor, to grab the label to apply to a target position. At each bend of the rollers, both driven or idlers, the contact between roller and strip of sticker paper must be insured for proper feeding without excessive slip [12]. Therefor the tension at the entry and exist of the roller will be different due to contact friction.

The distribution of idler and unwinding roller should consider avoidance of trough and wrinkle in the web and keep proper tension along web length [13]. However, the web length is relatively short such that most of this problem may not be a source of trouble in our design. The effect of roller speed, mass and consequently the force it induce can be neglected for the application considered in this work. Static analysis shows that tensions at entry and exit to non-driven roller (idlers A and B, Figure 1) can approximately relate as:

$$\frac{T_0}{T_i} = e^{\mu\beta} \quad (1)$$

Where μ is the coefficient of friction between roller surface stickers underneath player material and β is wrap angle with roller. Thus equation (1) is needs to be taken into consideration while selecting the actuator and controller tuning.

2.2 Mechanical design

Labels or tags on products have multiple of purpose and effects for the producer and consumers. Automotive, medical device, food and beverage, pharmaceutical, personal care, computer and consumer electronics are among predominant industries where automated assembly and product labeling is highly employed and continuously improved. In current changing demand of consumer, the product life cycles are not long enough to support customized machine tools. This condition then encourages robust automated solution, design of innovative robotic label applicator mechanism. The applicator presented here is coupled to robot arm as end of arm tooling (EOAT) equipment. Thus, the robot arm will autonomously seek for the target position to apply the label sticker guided by visual servoing control technique with vision camera fitted on this applicator as feedback source.

The main contribution of this paper lies in the innovative design development and implementation of robotic labeling mechanism that supply the label and paste to target surface. The basic features of the applicator allow flexibility in manufacturing and assembly process due to its self-contained compact autonomous features. Applying label sticker anywhere on the object can be done easy without a need for reconfiguration of the system.

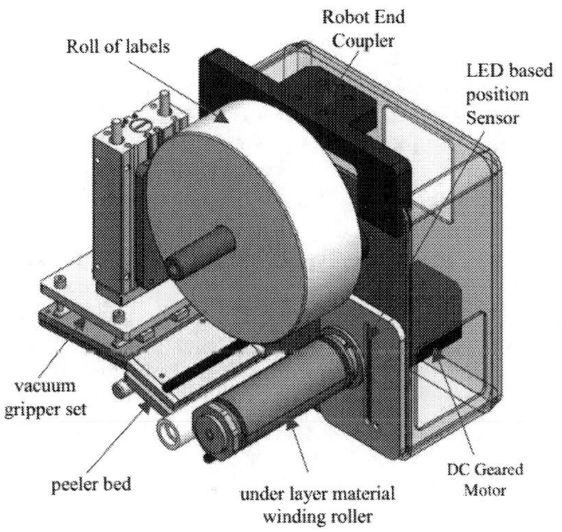

Figure 2. Label Applicator CAD model

To keep it light weight, the applicator most parts are designed for manufacture using rapid prototyping technology (3D printing) without losing structural strength required for its operation. As stated in the conceptual part the rotation of label dispensing roller, powered via DC geared motor, will separate the label from the underneath non sticky layer and makes it ready for the vacuum gripper. Table 1 shows the basic components of this robotic applicator.

The actuation unit includes a DC motor (output power 3.1W) with gearhead (total reduction ratio, 48:1), pneumatic piston and vacuum generator as can be seen in Figure 2. The controller regulates the DC motor to insure constant rate of label supply, which is achieved by feedback from line of LED array monitoring the diametrical size of underneath material collected by the winding motor.

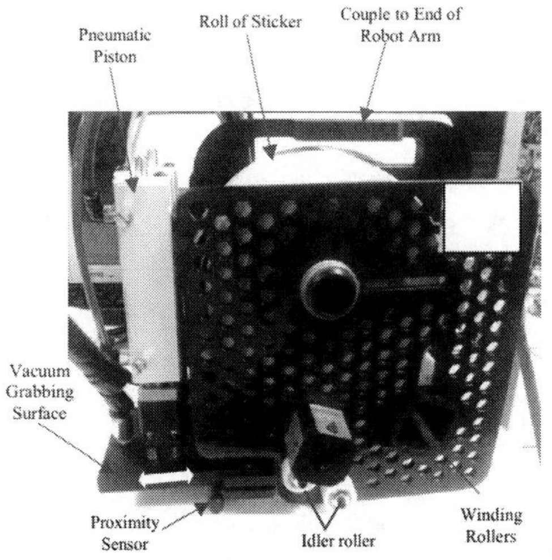

Figure 3. Complete Assembly of Label Applicator Prototype

Once grabbing surface hold the sticker it needs to move in the direction away from the peeler wedge edge to avoid interference while traveling down to the target area. The white thick arrow in Figure 3 shows the direction that sort of movement once the label is grabbed the grabbing surface moves 3 mm in the direction of the arrow to avoid collision with peeler bed.

Table 1. Applicator basic components.

Part Name	Description
Pneumatic gripper	Gripe surface + set of vacuum supply unit
Peeler Bed	Acrylic
Label rollers	Acrylic
DC Geared motor	IG32R 01004
Solenoid Valves	Standard Set
Vision Camera	Raspberry Pi Camera
Robot arm end coupler	Aluminium
Micro -Controller	Raspberry pi 2 model B V1.1
Proximity sensor	Standard

3 System integration and control

The embedded position, torque and proximity sensors (proximity sensor, position sensor) are shown in Figure 2&3. In typical control system diagram shown in Figure 4 the motor speed is controlled based on a feedback from line of proximity sensor LED distributes along the winding roller diameter monitoring the collected underneath layer material by the roller. Therefore, a constant speed of label deliver is achieved throughout the labeling process by controlling the speed of the driving motor based on this sensor feedback collected from the TMS32012810 based microprocessor.

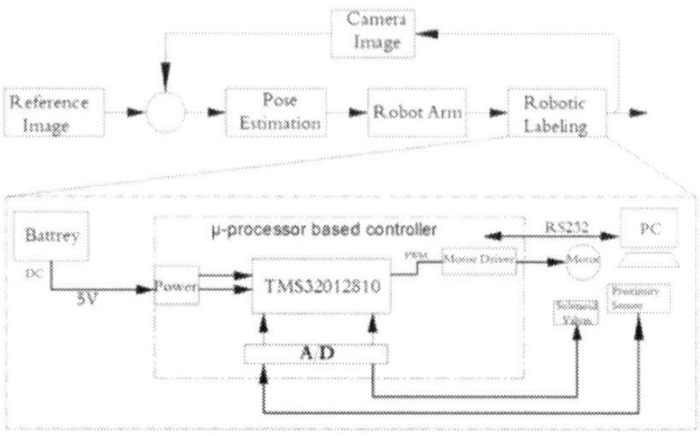

Figure 4. Control System Diagram

The operation of the system can be distinguished between the operation of the dispenser-applicator mechanism and of the vision-robot system. Therefore, the control involves two levels of scheme; the higher control stages where robot arm servoing is performed and labeling machine control in a second level (Figure 4). Raspberry Pi2 model Bv1.1 (BCM2836 ARM7 processor) was utilized as the main controller in the labelling machine and vision system control.

3.1 Vision System

A typical labeling task requires a target object to be placed at a known position and orientation, usually planar surface allowing a certain small error by an operator. This working condition provides a simplified assumption for vision and control processing. In particular, the placement error is restricted to a 2D motion and the known planar surface provides a fixed-depth pose estimation problem. Figure 5 shows the cooperation in control between main controller and vision module.

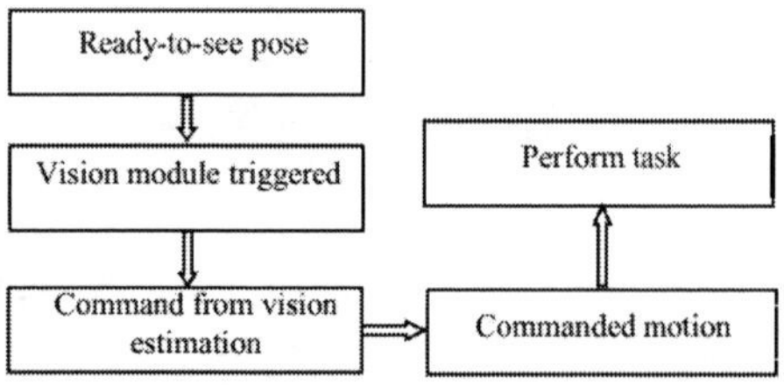

Figure 5. Vision based control operational flow

To perform the motion estimation in step 3, the algorithm starts with feature point detection which uses good feature to track [14]. This is followed by a matching using optical flow and a robust registration with Euclidean parametric model to reject outliers (Figure 6).

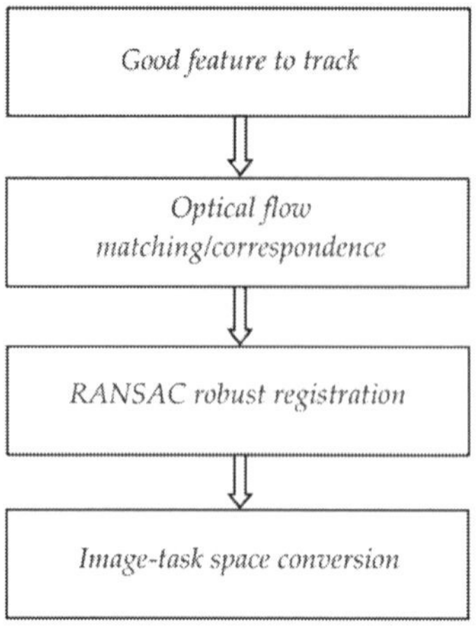

Figure 6. Motion estimation algorithm

The result motion solved in image-space is mapped to task space based on a linear relation $I = \alpha T$ where I, T are image-space and task-space measurements respectively; α is conversion scalar multi-disciplinary.

3.2 Dispenser-applicator Mechanism

The vision system operates according to Figure 5. At third step, the vision algorithm estimates the 2D motion to ready-for-tasking pose. After step5, straight-down motion, the controller triggers labeling process. It operates a dispensing motor and applicator gripper sequentially. The pneumatic gripper is fixed to a main frame on vertical guide, and moved to grasping direction simultaneously due to proximity sensor triggers the solenoid valves consequently the piston. Based on the information from the sensor the vacuum grasping surface on the pneumatic gripper will move to the bottom edge of the label sticker such that it sucks the sticker fully to its surface without damage. External compressed air supplied is distributed through the solenoid valves as depicted in Figure 7 to the two pneumatic pistons and vacuum grabbing surface.

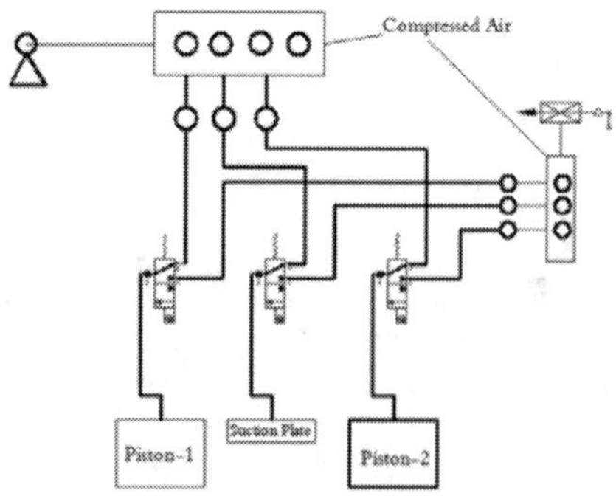

Figure 7. Schematics of the pneumatic circuits

4 Experiment and result

To validate the performance of the robotic label applicator system proposed in this paper prototype is developed and experiment has been conducted. Figure 8 show the experimental set up to test the performance of the applicator machine. It is attached to end of 6DOF robot arm with own separate controller. Vision processing is accomplished on Raspberry PI embedded computer via captured images from PI camera. Vision module is setup to communicate with PC using TCP/IP protocol.

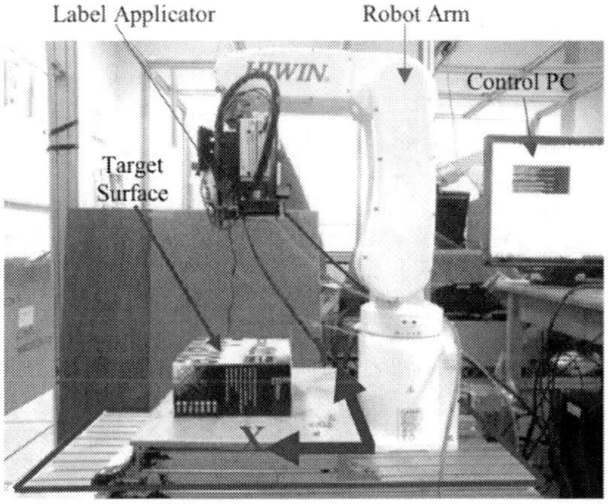

Figure 8. Experimental setup

The A4 sized chessboard template used in this test is shown in Figure 9a which is replaced for natural template in real-scenario targets due to the fact that some objects do not have feature points to be applied with current vision algorithm. The designed template provides the advantages for pose estimation of target pattern and the applied label by using straight-forward computer vision techniques as well as key-point features for vision module's algorithm. These pictures as can be seen from these captured images in Figure 9 b&c.

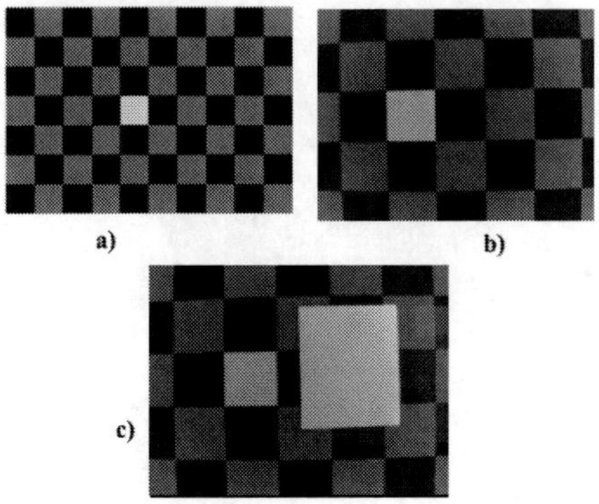

Figure 9. Template and reference pattern, reference pattern with label captured from PI camera

Experimental data collected for 30 repetitions shows some performance evaluation of the system (Figure10).

The overall control maintains a standard deviation of dT_x= -0.386, dT_y= 0.818, $d\theta$= 0.5 and mean value of -10.445, -6.325, 0.5 for case object is placed at T_X= 25 (mm), T_y= 5 (mm), θ= 3^0 deviated from reference position. Due to camera's field of view and vision algorithm, the object can be placed within an overall working range of T_X = 30mm, T_y = 30mm, and θ = 15^0.

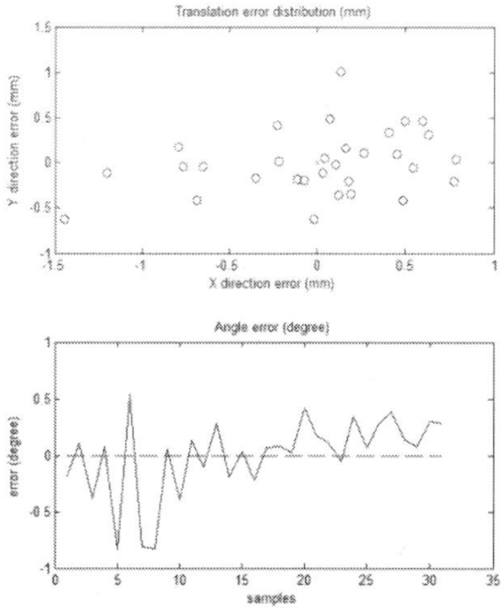

Figure 10. Performance error verification

Processing time of the system is also recorded. The graph shows 10 times of recording (Figure 10). Each record includes duration for vision processing, robot servoing and their total. Overall means are 1.7s, 2.27s and 3.98s for vision processing, robot servoing and total processing respectively. Among these records, instance 1, 7 and 10 show overshoots of processing time. This is the result of sampling-based algorithm where the correct solutions are not returned in first little iteration for these instances.

5 Conclusions

In this paper an innovative new robotic label stickers or tags dispensing and pasting system is reported. Intuitive and more natural steps involved in design of this mechanism therefore it is less complex and highly flexible to adapt changes in system arrangement or label size and type change. From systematic point of view, the paper proposes a robotic solution with functionality, structure, actuation, and sensor configuration and control strategies considered carefully. Successful experimentation on the system demonstrates that it's fairly efficient and accurate in applying a label sticker on randomly oriented object with certain features on its surface. Its contribution for wide range of industries is enormous in terms of time and cost saving in addition to its easily customizable to a change in manufacturing lines. The current robotic mechanism designed is tasted on flat surface however, with minor change on grabbing surface; one can makes it suitable for curved surface as well.

REFERENCES

1. C. M. P. Santos and M. J. Ferreira, "Control of an industrial desktop robot using computer vision and fuzzy rules," in *Industrial Electronics, 2005. ISIE 2005. Proceedings of the IEEE International Symposium on*, 2005, pp. 1297-1302 vol. 3.
2. F. Jovane, Y. Koren, and C. R. Boër, "Present and .Future of Flexible Automation: Towards New Paradigms," *CIRP Annals - Manufacturing Technology*, vol. 52, pp. 543-560, // 2003.
3. Å. Fast-Berglund, T. Fässberg, F. Hellman, A. Davidsson, and J. Stahre, "Relations between complexity, quality and cognitive automation in mixed-model assembly," *Journal of Manufacturing Systems*, vol. 32, pp. 449-455, 7// 2013.
4. A. G. Abad, K. Paynabar, and J. J. Jin, "Modeling and Analysis of Operator Effects on Process Quality and Throughput in Mixed Model Assembly Systems," *Journal of Manufacturing Science and Engineering*, vol. 133, pp. 021016-021016, 2011.
5. G. J. Agin, "Computer Vision Systems for Industrial Inspection and Assembly," *Computer*, vol. 13, pp. 11-20, 1980.
6. O. Faugeras, N. Deriche, and N. Navab, "From optical flow of lines to 3D motion and structure," in *Mustererkennung 1989*. vol. 219, H. Burkhardt, K. Höhne, and B. Neumann, Eds., ed: Springer Berlin Heidelberg, 1989, pp. 522-530.
7. G. Toscani and R. Deriche, "Cooperation between 3D Motion Estimation and Token Trackers," in *Multisensor Fusion for Computer Vision*. vol. 99, J. K. Aggarwal, Ed., ed: Springer Berlin Heidelberg, 1993, pp. 153-153.
8. R. Szeliski, *Computer Vision: Algorithms and Applications*, 2011 ed.: Springer, 2011.
9. P. Corke, *Robotics, Vision and Control: Fundamental Algorithms in MATLAB* vol. 2011, 2011.
10. N. Navab, R. Deriche, and O. D. Faugeras, "Recovering 3D motion and structure from stereo and 2D token tracking cooperation," in *Computer Vision, 1990. Proceedings, Third International Conference on*, 1990, pp. 513-516.
11. Z. Zhang, "A Flexible New Technique for Camera Calibration," *IEEE Trans. Pattern Anal. Mach. Intell.*, vol. 22, pp. 1330-1334, 2000.
12. H. Thanhtam, S. Hyeunhun, and L. Sangyoon, "Fuzzy control of the lateral position of a moving web in roll-to-roll processes," in *Robotics and Biomimetics, 2008. ROBIO 2008. IEEE International Conference on*, 2009, pp. 1637-1642.
13. S. Kee-Hyun and K. Soon-Oh, "The effect of tension on the lateral dynamics and control of a moving web," in *Industry Applications Conference, 2005. Fourtieth IAS Annual Meeting. Conference Record of the 2005*, 2005, pp. 529-536 Vol. 1.
14. J. Shi and C. Tomasi, "Good features to track," in *Computer Vision and Pattern Recognition, 1994. Proceedings CVPR '94., 1994 IEEE Computer Society Conference on*, 1994, pp. 593-600.

Examination of Color-Lighting Control System Using Colored Paper User Interface

Hiroto Aida[1], Kento Matsui[2], Soma Keisuke[2], Hiroki Murakami[2], Mistunori Miki[1]

[1]Department of Science and Engineering, Doshisha University, Japan
[2] Graduate School of Science and Engineering, Doshisha University, Japan

Abstract. In recent year, Full-Color LED Lighting that can be changed to various color such as red, green, blue has been appeared with development of LED Lighting. By Color-Lighting control, users affected such as concentrating and relaxing. Therefore, Color-lighting control will spread to various place such as home, offices, stations. However color-lighting control affected some disturbance such as daylight, display when Full-Color LED controlled indoors. Also, information devices control get difficult with information technology develop. I propose Color-Lighting Control System using Colored Paper User Interface(CLC/CPUI). The purpose of CLC/CPUI is that anyone can intuitively control Full-Color LED Lighting. CLC/CPUI uses colored paper as user interface by sensing the paper. CLC/CPUI realizes lighting color that user demanded to do feedback control. I conduct accuracy verification experiment of CLC/CPUI.

1 Introduction

LED lighting has advanced drastically to draw attention with the advent of high-brightness blue LED. Compared with fluorescent lamps, LED lamps are characterized by less power consumption, long life, and easier downsizing. They are intended to be introduced into offices and homes. Full color LED lamps can produce light in various colors (colored light) such as red, green, and blue. It has been found that controlling a full color LED lamp to vary the brightness and color shade of light, in particular its color temperature, affects user's work efficiency and psychological relaxation [1, 2, 3, 4].

There is a problem, however, of a disturbance effect by light sources such as daylight, display, and task light in controlling the colored light of an individual LED lamp indoors. The impact of disturbance precludes providing the colored light environment demanded by a user appropriately due to a difference between the colored light environment supposed to be provided by the system and the one actually provided. Therefore, it is required to search a colored light demanded by the user by sensing the current colored light environment and performing feedback control.

With the development of information technology, information devices are loaded with a variety of functions, which complicates the operation of such devices. It is the same with full color LED lighting. While it is possible to realize a variety of colors and brightness by using a full color LED lamp, it is necessary to change multiple parameters to control it. As a result of such complication, a user who is unaccustomed to the operation of information devices finds it reluctant to use such devices. A method using a piece of paper is thus suggested as an intuitive way of operating a system. Chand et al. proposed Jadoo, a system using paper UI. We conducted a human subject experiment of Jadoo with users unaccustomed to the operation of information devices [5]. Based on this study, it is estimated that using a piece of paper instead of a keyboard and a mouse enables users unaccustomed to operating information devices to operate a device and leads them to master its use. In this way, using a piece of paper as UI is effective in making the use of information devices intuitive and letting everyone use them easily.

In this study, we thus constructed a Color-Lighting Control System using Colored Paper User Interface (CLC/CPUI), which realizes a colored light a user demands by using colored paper as UI sensed by a vision sensor for feedback control of a full color LED lamp, for the purpose of enabling everyone to control a full color LED lamp in an intuitive manner.

The organization of this paper is as follows. Section 2 gives some remarks on colored light and colors. Section 3 describes the Color-Lighting Control System using Colored Paper User Interface. Section 4 describes the verification experiment for CLC/CPUI. Section 5 gives the conclusion.

2 Colored Light and Color

2.1 Light Color and Object Color

Colored light is light recognized to have color by human sensitivity. The color of colored light is called light color. Colored light is described by illuminance and chromaticity, which are used as indices in measurement in a colored light environment. Object color is color caused to appear by light reflected by the surface of an object or light permeating an object.

2.2 Illuminance

Illuminance is an indicator of brightness of an illuminated surface and represented by the unit of lux (lx). It has been found that appropriate illuminance differs by an individual's preference and the nature of work [6]. Japanese Industrial Standards (JIS) recommend working in different illuminance environments depending on the purpose of using a given space [7]. For instance, a high illuminance environment of 1, nm,. 000 lx or greater is recommended if precise work such as drafting or assembling is required. Medium illuminance of about 750 lx is recommended for a general office, meeting room, classroom, etc.

The above criteria, however, are based only on work on paper and do not consider the influence of a display in PC work or similar work. A high illuminance environment has a problem such as causing it difficult to see the screen when working with a display screen. Since brightness preference differs by individuals, it is considered necessary to vary an illuminance environment by an individual's preference and the nature of work rather than to use a constant illuminance environment in the same indoor space.

2.3 Chromaticity

Chromaticity is a measure of color used for object color and light color. While there are multiple ways to represent chromaticity, this experiment uses RGB, XYZ, and L*a*b* systems. The XYZ color system is a standard color coordinate system defined by CIE, and values of other color systems such as the XYZ or L*a*b* color system can be obtained by converting values of the XYZ color system. Since color recognition is based on human sensitivity, it is not possible to represent color only by the energy intensity of light. The XYZ color system, however, quantifies color, which is a sensible quantity, by numericizing the intensity of color sensation obtained by cones, color sensing organs, from the visible light spectrum. This numericized intensity of color sensation from the visible light spectrum is called a color matching function. A triple of values (X, Y, Z) representing a color can be obtained from color matching functions and a spectrum distribution of light. XYZ chromaticity is calculated by Formulae 1, 2 , and 3:

$$X = \sum_{\lambda=380}^{780} I(\lambda)\bar{x}(\lambda)d\lambda \quad (1)$$

$$Y = \sum_{\lambda=380}^{780} I(\lambda)\bar{y}(\lambda)d\lambda \quad (2)$$

$$Z = \sum_{\lambda=380}^{780} I(\lambda)\bar{z}(\lambda)d\lambda \quad (3)$$

λ : wavelength [nm] $I(\lambda)$: spectrum intensity

$\bar{x}(\lambda)$, $\bar{y}(\lambda)$, $\bar{z}(\lambda)$: color matching functions

Of the XYZ chromaticity coordinates, Y represents brightness, and X and Z represent a change in the shade of color. It is difficult to grasp the relation between these XYZ values and color, and the amount of a change in those values is not in proportion to that in perception. For this reason, a degree of difference between different colors (color difference) cannot be quantitatively represented using those values. Therefore, the L*a*b color system is used when color difference needs to be represented. The L*a*b color system is a chromaticity system obtained by conversion from XYZ values. L* value corresponds to brightness, a*, to a change in the shade of color from red to green, and b*, such a change from yellow to blue. The L*a*b* color system defines a standard color called white point. As (a*, b*)=(0,0) is applied for the white point color, L*a*b* values obtained by measurement are relative to the white point. If a different color is used as the white point, the chromaticity coordinate of the same color changes. For this reason, the standard light source D65 is used as the white point of the L*a*b* color system in this paper. The XYZ chromaticity coordinates are converted into the L*a*b* chromaticity coordinates by formulae 4, 5 and 6:

$$L^* = 116(Y/Y_n)^{\frac{1}{3}} - 16 \quad (4)$$

$$a^* = 500[(X/X_n)^{\frac{1}{3}} - (Y/Y_n)^{\frac{1}{3}}] \quad (5)$$

$$b^* = 200[(Y/Y_n)^{\frac{1}{3}} - (Z/Z_n)^{\frac{1}{3}}] \quad (6)$$

X, Y, Z : XYZ values

X_n, Y_n, Z_n : XYZ values of the white point

Since the L*a*b* color system is so constituted that a change in color in the color space is approximately uniform, color difference can be represented by taking the Euclidean distance between two points in the color space.

The color difference ΔE between Color 1 and Color 2 is calculated by Formula 7:

$$\Delta E = \sqrt{(L^* - L^{*\prime}) + (a^* - a^{*\prime}) + (b^* - b^{*\prime})} \quad (7)$$

L^*, a^*, b^* : color information of Color 1

$L^{*\prime}, a^{*\prime}, b^{*\prime}$: color informaiton of Color 2

JIS specifies the tolerance for color difference in industrial use in stages. The color difference distinguishable by visual observation is specified to be 3.2 or greater, and two colors are recognized as different if the color difference between them exceeds 25 [8]. In this experiment, based on these values of color difference ΔE, it is judged whether the target chromaticity space is realized or not. As each of primary colors is emitted by a full color LED light source, the chromaticity of a full color LED light source can be represented as the intensity of primary colors relative to the visible light spectrum.

2.4 Three Elements of Color

There are three measures of color as its three elements: hue, luminosity, and saturation. Any color is located somewhere in the color solid constituted by these three elements.

Hue is a measure of a particular shade of color such as red, green, or blue. A circle in which hues are arranged is called a color wheel. Luminosity is a measure of darkness or lightness of a color. Greater its luminosity is, lighter a color is. When its luminosity reaches 100%, a color becomes white. When its luminosity reaches 0%, a color becomes black. Whereas hue indicates a particular shade of color such as red, blue, or green, colors with the same hue are recognized as different if they differ in luminosity. In the L*a*b* color system used in this experiment, L* represents luminosity.

Saturation is a measure of the vividness of a color. Greater its saturation is, more vivid a color is. A color whose saturation is 0 is called an achromatic color, and white, black, and different shades of gray fall under achromatic colors. Fig. 1 shows the relation among three elements of color.

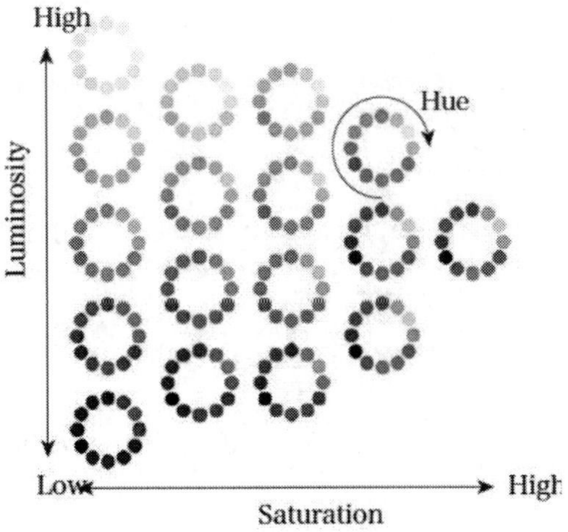

Figure 1. Three Elements of Color.

2.5 Color mixing

There are two ways of mixing colors: additive color mixing, a way of mixing light source colors, and subtractive color mixing, a way of mixing object colors. Additive color mixing results in a lighter color, while subtractive color mixing results in a darker color.

Additive Color Mixing

Additive color mixing refers to mixing of light source colors. Three colors of red, blue, and green are primary colors of light. If you mix these three colors, you get white. Fig. 2 shows three primary colors of light. In this way, mixing colors by additive color mixing results in a lighter color. Formulae 8, 9, and 10 for additive color mixing are given below:

$$R = \begin{cases} 255 & (R1 + R2 > 255) \\ R1 + R2 & (otherwise) \end{cases} \tag{8}$$

$$G = \begin{cases} 255 & (G1 + G2 > 255) \\ G1 + G2 & (otherwise) \end{cases} \tag{9}$$

$$B = \begin{cases} 255 & (B1 + B2 > 255) \\ B1 + B2 & (otherwise) \end{cases} \tag{10}$$

R, G, B : color information of a color resulting from additive color mixing
$R1, G1, B1$: color information of Color 1
$R2, G2, B2$: color information of Color 2

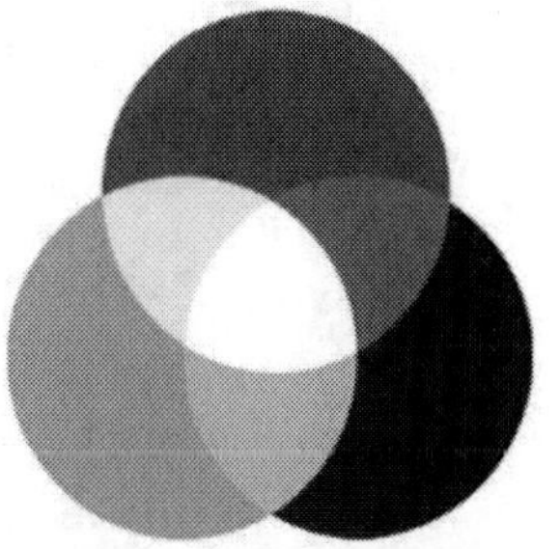

Figure 2. Three Primary Colors of Additive Color Mixing.

Subtractive Color Mixing

Subtractive color mixing refers to mixing of object colors. Magenta, cyan, and yellow are three primary colors of pigment. If you mix these three colors, you get black. Fig. 3 shows three primary colors. In this way, mixing colors by subtractive color mixing results in a darker color. Formulae 11, 12, and 13 for subtractive color mixing are given below:

$$R = \begin{cases} 0 & ((255 - R1) + (255 - R2) > 255) \\ 255 - ((255 - R1) + (255 - R2)) & (otherwise) \end{cases} \tag{11}$$

$$G = \begin{cases} 0 & ((255 - G1) + (255 - G2) > 255) \\ 255 - ((255 - G1) + (255 - G2)) & (otherwise) \end{cases} \tag{12}$$

$$B = \begin{cases} 0 & ((255 - B1) + (255 - B2) > 255) \\ 255 - ((255 - B1) + (255 - B2)) & (otherwise) \end{cases} \tag{13}$$

R, G, B : color information of a color resulting from subtractive color mixing
$R1, G1, B1$: color information of Color 1
$R2, G2, B2$: color information of Color 2

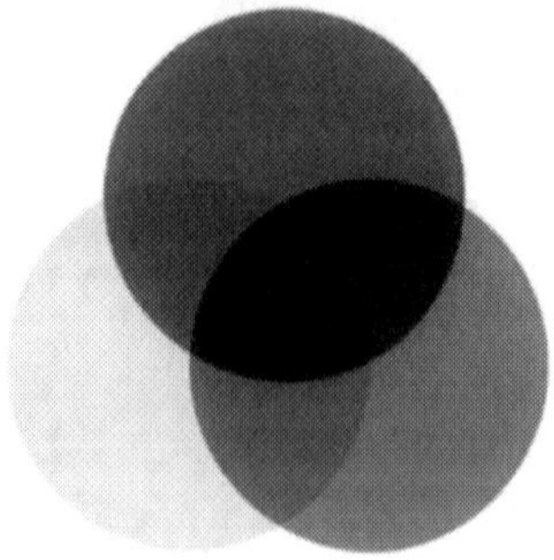

Figure 3. Three Primary Colors of Subtractive Color Mixing.

3 Color-Lighting Control System using Colored Paper User Interface

In this study, the Color-Lighting Control using Colored Paper User Interface (CLC/CPUI) is developed. The purpose of CLC/CPUI is to enable anyone to control full color LED lamp in an intuitive manner. A full color LED lamp is to be feedback-controlled by sensing a piece of colored paper as UI by a vision sensor to realize colored light demanded by a user. In CLC/CPUI, light color demanded by a user is determined by coloring a piece of paper serving as UI with a crayon or color pencil. In this way, CLC/CPUI has a feature that it is accessible even to a user unaccustomed to the operation of information device because of its UI using a piece of paper, a crayon, and a color pencil.

CLC/CPUI is implemented by connecting a vision sensor, a full color LED lamp, and a control computer to the same network. A piece of paper serving as UI is placed within the range of the vision sensor. The configuration diagram of CLC/CPUI is shown in Fig. 4.

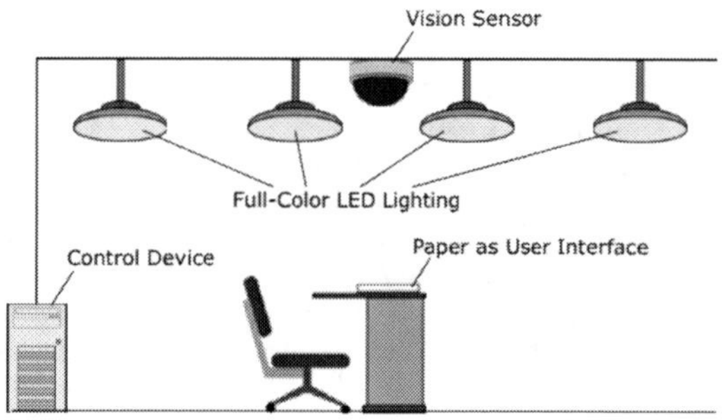

Figure 4. System Configuration of CLC/CPUI.

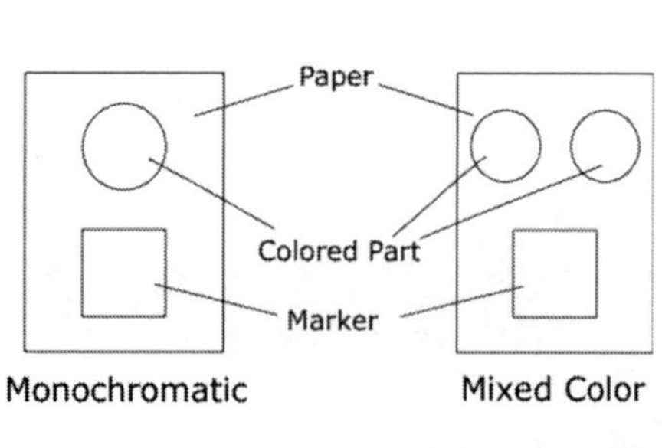

Figure 5. Paper as User Interface.

3.1 Paper UI

CLC/CPUI uses a piece of paper as UI. There are two types of UI: one used in monochromatic light control and one used in mixed color light control. Fig. 5 shows paper UI used in monochromatic light control and mixed color light control.

3.2 Method for Sensing UI in Monochromatic Light Control

A piece of paper to be used as UI has areas to be colored and sensed. A user colors the area to be colored with a crayon in the color he/she wants. The current colored light environment can be detected by sensing the RGB values of the sensing area with a vision sensor.

Shown below is the flow of sensing the current colored light environment with a vision sensor from a piece of paper used as UI.

1. Detect the sensing area of a piece of UI paper.
2. Obtain the RGB values of the sensing area.

By performing the above operations, the RGB values of the colored light environment demanded by a user and that of the current environment can be compared.

3.3 Method for Sensing UI in Mixed Color Light Control

A piece of paper to be used as UI has two areas to be colored and a sensing area. A user colors those areas with a crayon in the colors he/she wants.

Shown below is the flow of determining the color demanded by a user on the basis of the color of two areas to be colored.

1. Detect two colored areas on a piece of UI paper.
2. Obtain the RGB values of each colored area.
3. Mix two RGB values by subtractive color mixing.

By performing the above operations, the color information of colored light demanded by a user can be obtained

The sensing area is sensed in the same way as it is sensed when there is a single UI circle to detect the current colored light environment, which is compared with the environment that the user wants.

3.4 Controlling Algorithm of CLC/CPUI

CLC/CPUI is feedback-controlled on the basis of values of color information extracted from images transmitted regularly to the control computer by the vision sensor which senses UI paper. By repeating this feedback control process, the target colored light environment is to be provided. Equation 14 gives the objective function used for searching the target colored light in CLC/CPUI.

$$f_i = \sqrt{(L^* - L^{*\prime}) + (a^* - a^{*\prime}) + (b^* - b^{*\prime})} \quad (14)$$

L^*, a^*, b^* : color information of the target colored light

$L^{*\prime}$, $a^{*\prime}$, $b^{*\prime}$: color information of the current colored light

CLC/CPUI realizes the target colored light by controlling the objective function formalized as Equation 14 so as to minimize its value. The flow of CLC/CPUI control is shown below.

1. Turn on a LED lamp with the color set to white.
2. A user draws a circle on a piece of UI paper in any color to choose the target colored light.
3. Place the UI created in the range of the vision sensor.
4. The vision sensor obtains a UI image.
5. Analyze the image to obtain the RGB values of the color requested by the user.
6. Vary the luminance randomly in the neighborhood of ±20%.
7. The vision sensor obtains a UI image.
8. Analyze the image to obtain the RGB values of the current color environment.
9. Calculate the objective function on the basis of color information obtained in (8).
10. The vision sensor obtains a UI image.
11. Analyze the image to obtain the RGB values of the current color environment.
12. Calculate the objective function. If its value is better, then go to (6).
13. If the value of the objective function is worse, then revert the colored light to its previous state and go to (6).

The above operations constitute a single step of control. The element of LED lighting for each color is controlled. The colored light environment demanded by the user can be provided by repeated control.

4 Evaluation of CLC/CPUI

It is verified whether the target colored light environment is realized by using CLC/CPUI. In the verification experiment, an LED lamp is controlled by using CLC/CPUI toward the target colored light, and it is evaluated whether the target colored light is realized on the basis of chromaticity described earlier in Section 2.3.

4.1 Verification of the Precision of CLC/CPUI for Monochromatic Light

The verification experiment was conducted at the Intelligent Systems Design Laboratory, Kochikan Building, Doshisha University. The experiment was conducted by using 29 full color LED lamps controllable in four colors of red, blue, green and yellow made by SHARP, 1 iPhone 5s as a vision sensor, 1 control computer, and pieces of paper on which a circle in an arbitrary color is drawn. A single circle was printed on pieces of paper as UI. There are three colors of the circle: Color 1, Color 2, and Color 3. RGB values of each color is as follows: [254, 160, 186] for Color 1, [31, 80, 77] for Color 2, and [34, 38, 138] for Color 3. The target RGB values were the same as those for the painted color.

Fig. 6 gives the diagram of the experimental environment. Fig. 6 (a) is the plan of the experimental environment, and Fig. 6 (b) is the LED layout drawing. Fig. 7 shows a scene of the experiment.

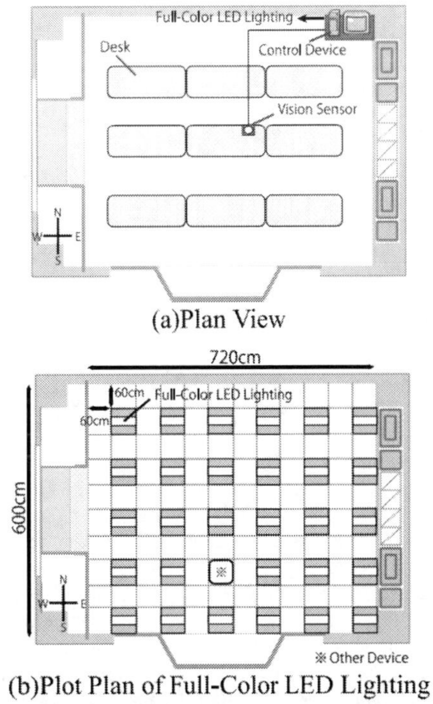

(a)Plan View

(b)Plot Plan of Full-Color LED Lighting

Figure 6. Experimental Environment

Figure 7. Experimental Landscape.

An experiment was conducted to verify that the target monochromatic light can be realized by using CLC/CPUI. There were three colors for a circle drawn on the piece of UI paper: Colors 1, 2, and 3.

The target colored light was set to have Color 1, 2, or 3. Fig. 8 shows the record of color difference under CLC/CPUI. In Fig. 8, the vertical axis represents the color difference between the target light environment and the current light environments, and the horizontal axis represents the number of control steps. The color difference after 300 steps was 21.7 with Color 1, 21.9 with Color 2, and 2.2 with Color 3.

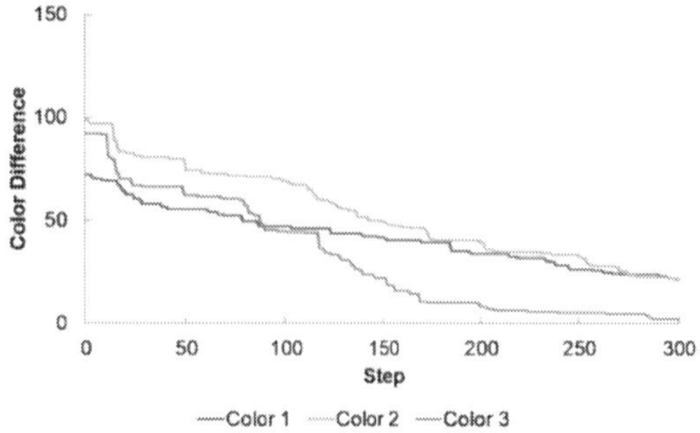

Figure 8. Result of CLC/CPUI for monochromatic light.

4.2 Verification of the Precision of CLC/CPUI for Mixed-Colored Light

The verification experiment was conducted at the Intelligent Systems Design Laboratory, Kochikan Building, Doshisha University. The experiment was conducted by using 29 full color LED lamps made by SHARP whose light sources in four colors of red, blue, green and yellow were controllable, 1 iPhone 5s as a vision sensor, 1 control computer, and pieces of paper on which a circle in an arbitrary color is drawn. Two circles were drawn on a single piece of UI paper with three input colors: Colors 4, 5, and 6. RGB values for Color 4 were [225, 58, 249], those for Color 5, [255, 250, 53], and those for Color 6, [93, 250, 253]. Two out these three colors are chosen to color circles on the piece of UI paper. Pieces of UI paper were created in three patterns. The result of mixing Colors 4 and 5 was named Pattern A, that of mixing Colors 5 and 6, Pattern B, and that of mixing Colors 4 and 6, Pattern C. Fig. 9 shows Patterns A, B, and C.

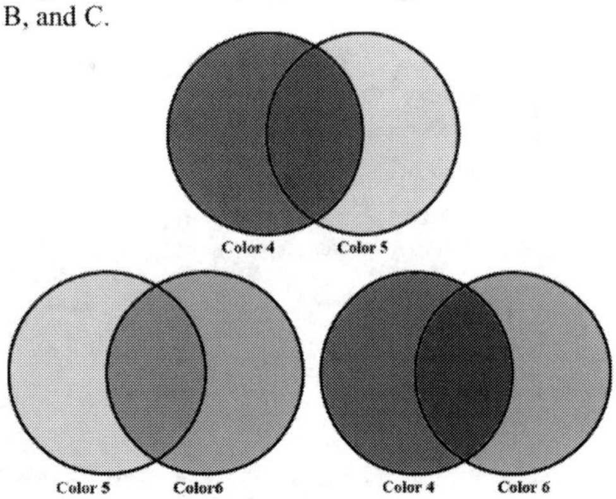

Figure 9. Mixed Color in three Patterns.

An experiment was conducted to verify that the target mixed-colored light can be realized by using CLC/CPUI. Fig. 10 shows the record of color difference when CLC/CPUI was controlled for the target colored light determined by three UI patterns. In Fig. 10, the vertical axis represents the color difference between the target light environment and the current light environment, and the horizontal axis represents the number of control steps. The color difference after 300 steps was 27.5 with Pattern A, 43.8 with Pattern B, and 0.3 with Pattern C.

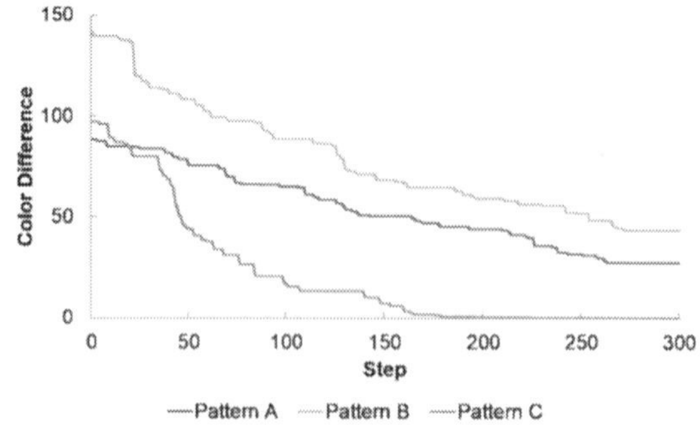

Figure 10. Result of CLC/CPUI for Mixed-Colored Light.

4.3 Verification of the Precision of CLC/CPUI for Mixed-Colored Light with Luminosity Fixed

This verification experiment was conducted in a dark room where any lamps other than the controlled lamp were turned off. The experiment was conducted by using 1 LED lamp Hue, made by Phillips, whose light sources in three colors of red, blue, and green were controllable, 1 iPhone 5s as a vision sensor, 1 control computer, and pieces of paper on which a circle in an arbitrary color is drawn. Two circles were drawn on the piece of UI paper, similarly, with three input colors: Colors 4, 5, and 6. RGB values for Color 4 were [225, 58, 249], those for Color 5, [255, 250, 53], and those for Color 6, [93, 250, 253]. Two out these three colors are chosen to color circles on a single piece of UI paper. Pieces of UI paper were created in three patterns. The result of mixing Colors 4 and 5 was named Pattern A, that of mixing Colors 5 and 6, Pattern B, and that of mixing Colors 4 and 6, Pattern C. Fig. 9 shows Patterns A, B, and C.

An experiment was conducted to verify that the target mixed-colored light can be realized by using CLC/CPUI when luminosity was fixed. Fig. 11 shows the record of color difference when CLC/CPUI was controlled for the target colored light determined by three UI patterns. In Fig. 11, the vertical axis represents the color difference between the target light environment and the current light environment, and the horizontal axis represents the number of control steps. The color difference after 300 steps was 21.9 with Pattern A, 28.7 with Pattern B, and 6.8 with Pattern C.

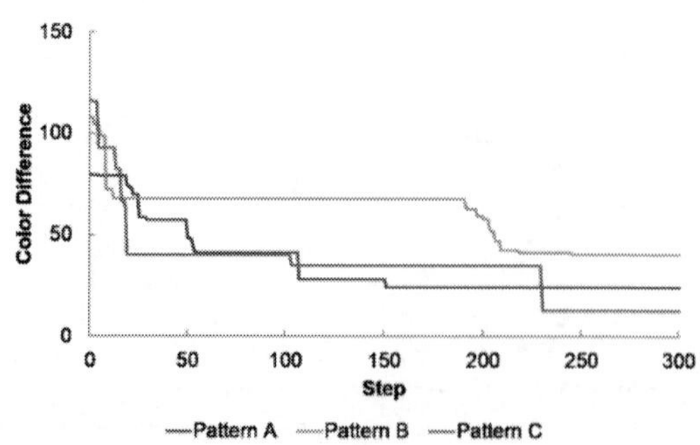

Figure 11. Result of CLC/CPUI for Mixed-Colored Light with Luminosity Fixed.

4.4 Discussion

It was found that, if a lamp is controlled by using CLC/CPUI for arbitrary colored light, colored light from the lamp approaches the target, both with monochromatic input color and mixed input color. According to the tolerance for color difference in industrial use prescribed in JIS, two colors are to be recognized as different if color difference between them exceeds 25.0. With the above result, when a single color was put into UI, color difference was 25.0 or less, whichever of piece of papers on which a circle was drawn in Color 1, 2, or 3 was used as UI. Therefore, the optimal solution is said to have been found in colored light environment search. On the other hand, when two colors were put into UI, there were cases where color difference was not 25.0 or less. The optimal solution failed to be found in colored light environment search. In this way, it is considered that there are both cases where the colored light environment demanded can be provided on the basis of color information of the target colored light and those where it cannot be provided. It is possible that this problem was caused by the inability to find the optimal solution due to a change in luminosity that was simultaneous with a change in the chromaticity of colored light.

CLC/CPUI is thus modified so that the luminosity of colored light does not change and then evaluated. As a result, it was found that controlling a lamp to produce arbitrary colored light by using modified CLC/CPUI which did not cause a change in luminosity resulted in colored light closer to the target. Even in this case, however, there was a case where color difference is not 25.0 or less, failing to find the optimal solution. Since lessening the number of parameters resulted in colored light closer to the target, a more efficient search for the optimal solution is considered to be possible by changing conditions.

5 Conclusion

This study, in order to realize an intuitive colored light control system, constructed a Color-Lighting Control System using Colored Paper User Inter-face (CLC/CPUI) and conducted verification experiments. CLC/CPUI attempts to bring colored light produced by a lamp closer to one demanded by a user through the feedback control using a vision sensor. Verification experiments for CLC/CPUI were conducted in three cases: for monochromatic light, mixed color light, and mixed color light with luminosity fixed. It was found that colored light from a lamp can be brought closer to the target by using CLC/CPUI in any of those cases.

It was found that, if a monochromatic color is input, colored light approached the target and converged within the tolerance. On the other hand, if a mixed color is input, colored light approached the target but sometimes failed to converge within the tolerance. It is possible that this problem was caused by the inability to find the optimal solution due to a change in luminosity that was simultaneous with a change in the chromaticity of colored light.

CLC/CPUI is thus modified so that the luminosity of colored light does not change simultaneously when its chromaticity is changed and then evaluated. Even in this case, colored light sometimes failed to fall in the tolerance. It became easier, however, to find the optimal solution and colored light was brought closer to the tolerance as luminosity was fixed and parameters were reduced to chromaticity only.

References

1. F. Obayashi, K. Tomita, Y. Hattori, etc. "A Study on Environmental Control Method to Improve Productivity of Office Workers -Development of an Illumination Control Method and its Experimental Evaluation," Human Interface Symposium 2006, No.1322, pp.151-156 (2006).
2. M. Miki, Y. Taniguchi, M. Yoshimi, "Effect of Preferred Illuminance and Color Temperature on Creative Works/Intellectual Productivity," Journal of the Illuminating Engineering Institute of Japan, 96[8], 437-441 (2012).
3. PHILIPS. ActiViva club.
4. Dong gi Lee, Hsin chen Chiang, Takaaki Koga, Kotaroh Hirate, "Basic Study on Discomfort Caused by LED Lighting Colors in Working Space," AIJ Journal of Technology and Design, Vol. 17, No.35, pp.201-204 (2011).
5. A. Chand, A. K. Dey "Jadoo: A Paper User Interface for Users Unfamiliar With Computers," Proc. CHI2016, pp.1625-1630 (2006).
6. P.R.Boyce, N.H.Eklund and S.N.Simpson, ``Individual Lighting Control: Task Performance, Mood, and Illuminance", Journal of the Illuminating Engineering Society, 131-142 (2000).
7. JIS. JISZ9110. 2010.
8. JIS. JISZ8721.1993.

Anti-rollover Control System Design of Articulated Off-road Vehicles

Shuai Wang[1], Yuxin Zhang[2], Chao Chen [1], and Huikai Liu [1]

[1]College of Mechanical Science and Engineering, Jilin University, 130022 Changchun, China
[2]State Key Laboratory of Automotive Simulation and Control, Jilin University, 130022 Changchun, China

Abstract. This paper designs an anti-rollover control system to improve articulated off-road vehicle rollover safety. This controller consists two level control which target to two different instability stages: 1) the active swing bridge adjustment control algorithm for the first stage instability; and 2) the active braking control algorithm for the second stage instability. The first level control is based on fuzzy PID control and second is based on switch control, they work together to prevent the articulated wheel loader from rollover accident. Multi-body dynamics model of a ZL50 articulated wheel loader is built by commercial software RecurDyn, and the MBD model and controller model were co-simulated under the same solver in RecurDyn/CoLink. The co-simulation results show that, during the high speed steering rollover process, these two level controller work efficiently to reduce the roll angle and improve the vehicle rollover stability.

1 Introduction

Articulated steering system is widely used in off-road and engineering vehicles due to its several advantages, such as small steering radius, good manoeuvring characteristics, simple structure, and low power consumption.

While the articulated steering vehicles became popular and widely used, the security problems are becoming increasingly prominent [1]. Compared to the vehicles with overall frame, the articulated engineering or off-road vehicles have a special body structure, which usually leads to an additional degree of freedom, and reduces vehicle lateral stiffness. Therefore, this kind of vehicles are easier to be caused rollover accidents which lead to casualties and economic loss [2].

In order to ensure the driver's life safety, international standard ISO3471 requirements that the engineering vehicles must be installed rollover protection structure (ROPS) which can protect the driver in rollover accidents passively [3], however this approach does not solve the root of the problem.

At present, the safety research about articulated off-road vehicles are mainly focused on passive rollover safety, rollover mechanism, rollover and steering stability analysis, etc. [4]. There are only a few researches on the active anti-rollover control system [5-7] and most of them are focused on commercial vehicles and SUVs.

D. J. M. Sampson et al developed an active anti-rollover control system for articulated commercial vehicles, and introduced the hardware structure and software development of the control system in detail [8]. Through controlling the active anti-roll bar by a hydraulic actuator, it can adjust the traction semi-trailer roll motion. The simulation results have shown that this control system can significantly improve traction semi-trailer stability, and reduce rollover accident probability.

P. Gaspar developed a reconfigurable control system for heavy vehicles, it combined active anti-roll bar control and active braking control [9]. At the same time, the optimal anti-rollover control strategy considered several roll dynamics parameters, such as roll angle, roll angular velocity, lateral angular velocity and lateral acceleration.

M. M. Islam built a 5 DOF single articulated trailer heavy vehicle model and a 7 DOF multi-trailer heavy vehicle model, and simulated them using TruckSim [10]. By combining trailer active steering control, active anti-roll bar control, and differential braking control, this control system can effectively improve low-speed maneuverability and high-speed cornering lateral stability. N. Daher investigated a yaw stability control system for articulated frame steering off-highway vehicles via novel steer-by-wire technology that they have recently developed [11], this work explores whether the new technology can also deliver increased safety as well, by providing active safety functions.

Y. X. Zhang established wheel loader dynamic rollover model and the rollover warning strategy model by MATLAB/Simulink [7]. Through the simulation results analysis, the effectiveness of warning strategy under high-speed transport and sharp steering was verified. What's more, a physical prototype test was performed using the remote control loader model vehicle.

In this paper, a fuzzy PID controller is designed to improve articulated off-road vehicle rollover safety, especially the research object is an articulated wheel loader. Section 2 introduces the design process of the active-anti rollover controller. Section 3 describes the multi-body dynamics modeling of the wheel loader. Section 4 sets up the co-simulation. Simulation results are shown in section 5. Finally, the conclusions are given in section 6.

2 Controller Design

The designed active anti-rollover control system is divided into two levels, they are used in different stages of the rollover process: 1) for the first stage instability of the loader, the active swing bridge adjustment control algorithm will be activated; and 2) for the second stage instability of the loader, the active braking control algorithm will be activated.

Fuzzy control has good noise suppression ability and good robustness, which is suitable for nonlinear systems. However, the inherent weakness of fuzzy control is that the steady state performance is poor, and it is easy to produce the limit oscillations near the equilibrium point. The traditional PID control has many advantages, but its fixed set of control parameters cannot adapt to environmental interference.

In order to improve the control effect, this paper developed a fuzzy PID controller, which takes advantages from both fuzzy control and traditional PID control. Based on fuzzy control, the fuzzy PID controller can tune the three key parameters K_P, K_I and K_D of the PID control online using fuzzy logic inference [12, 13].

As shown in Figure 1, a two dimensional fuzzy controller is used in this paper. The inputs are deviation *e* and its rate *ec*, and the outputs are the three key parameters of PID controller, K_P, K_I and K_D. In practice, in order to accelerate the control system response speed, the increment of these three parameters ΔK_P, ΔK_I and ΔK_D are used instead of directly tuning parameters K_P, K_I and K_D. Figure 1 shows the structure diagram of the fuzzy PID controller.

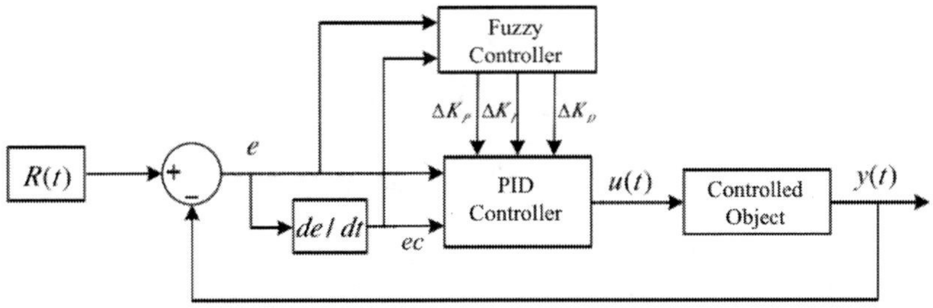

Figure 1. Structure diagram of the fuzzy PID controller.

For the PID controller part, its inputs K_P, K_I and K_D are calculated from the output of fuzzy controller ΔK_P, ΔK_I and ΔK_D and the previous system initial value K_{P0}, K_{I0} and K_{D0}, and then the control variable *u(t)* is generated and acted on the controlled object. The calculation formula of K_P, K_I and K_D are as follows:

$$K_P = K_{P0} + \Delta K_P \quad (1)$$

$$K_I = K_{I0} + \Delta K_I \quad (2)$$

$$K_D = K_{D0} + \Delta K_D \quad (3)$$

For the active swing bridge control, the feedback signal *y(t)* is roll angle $\varphi(t)$, while the given input is a fixed roll angle. Therefore, the roll angle deviation *e* of the fuzzy control is:

$$e = R(T) - \varphi(t) = -\varphi(t) \quad (4)$$

The change rate *ec* of the roll angle deviation is:

$$ec = de/dt = \frac{d(0-\varphi(t))}{dt} = -\omega(t) \quad (5)$$

The outputs of the fuzzy control are the correction parameters of the PID control Δk_p, Δk_i and Δk_d. The basic domains of *e*, *ec*, Δk_p, Δk_i and Δk_d are $[-e_{max}, e_{max}]$, $[-ec_{max}, ec_{max}]$, $[-\Delta k_{pmax}, \Delta k_{pmax}]$, $[-\Delta k_{imax}, \Delta k_{imax}]$, and $[-\Delta k_{dmax}, \Delta k_{dmax}]$, respectively. The linguistic variables of inputs and outputs are *E*, *EC*, K_P, K_I and K_D, respectively and their fuzzy discrete domains are $[-n_1, n_1]$, $[-n_2, n_2]$, $[-m_1, m_1]$, $[-m_2, m_2]$, and $[-m_3, m_3]$. The input quantization factors are:

$$K_e = n_1/e_{max} \quad (6)$$

$$K_{ec} = n_2/ec_{max} \quad (7)$$

The comparative factors of output variables are:

$$K_{u1} = \Delta k_{pmax}/m_1 \quad (8)$$

$$K_{u2} = \Delta k_{imax}/m_2 \quad (9)$$

$$K_{u3} = \Delta k_{dmax}/m_3 \quad (10)$$

The fuzzy subsets of the input and output variables are all seven: NB (Large negative), NM (Medium negative), NS (Small negative), ZO (zero), PS (Small positive), PM (Medium positive), PB (Large positive). The membership function of fuzzy subsets is:

$$\mu(x) = \begin{cases} \frac{x-a}{b-a}, & a < x < b \\ \frac{x-c}{b-c}, & b < x < c \end{cases} \quad (11)$$

where *a*, *b*, and *c* are slopes of membership functions.

Based on Mamdani's reasoning method, the fuzzy control rules of ΔK_P, ΔK_I and ΔK_D are shown in Table 1 to Table 3.

In this paper, weighted average method is used to realize the process of ambiguity resolution:

$$\Delta K_P = \frac{\sum_{i=1}^{49} \mu_{Pi}(\Delta K_P) \times \Delta K_{P_i}}{\sum_{i=1}^{49} \mu_{Pi}(\Delta K_P)} \quad (12)$$

where, $\mu_{Pi}(\Delta K_P) = \mu_{Pi}(E) \times \mu_{Pi}(EC)$. The processes of ambiguity resolution for ΔK_I and ΔK_D are similar.

Table 1. Fuzzy control rules of ΔK_P.

E \ EC	NB	NM	NS	ZO	PS	PM	PB
NB	PB	PB	PM	PM	PS	ZO	ZO
NM	PB	PM	PM	PS	PS	ZO	NS
NS	PM	PM	PS	PS	ZO	NS	NS
ZO	PM	PM	PS	ZO	NS	NM	NM
PS	PS	PS	ZO	NS	NS	NM	NM
PM	PS	ZO	NS	NS	NM	NM	NB
PB	ZO	ZO	NS	NM	NM	NB	NB

Table 2. Fuzzy control rules of ΔK_I.

E \ EC	NB	NM	NS	ZO	PS	PM	PB
NB	NB	NB	NM	NM	NS	ZO	ZO
NM	NB	NM	NM	NS	NS	ZO	PM
NS	NB	NM	NS	NS	ZO	PS	PM
ZO	NM	NM	NS	ZO	PS	PM	PM
PS	NM	NS	ZO	PS	PS	PM	PB
PM	NS	ZO	PS	PS	PM	PB	PB
PB	ZO	ZO	PS	PM	PM	PB	PB

Table 3. Fuzzy control rules of ΔK_D.

E \ EC	NB	NM	NS	ZO	PS	PM	PB
NB	PS	NS	NB	NB	NB	NM	PS
NM	PS	NS	NB	NM	NM	NS	PS
NS	ZO	NS	NM	NM	NS	NS	ZO
ZO	ZO	NS	NS	NS	NS	NS	ZO
PS	ZO	ZO	ZO	ZO	ZO	ZO	ZO
PM	PB	PS	PS	PS	PS	PS	PB
PB	PB	PM	PM	PM	PS	PS	PB

Finally, the Simulink model of the fuzzy PID controller is shown in Figure 2.

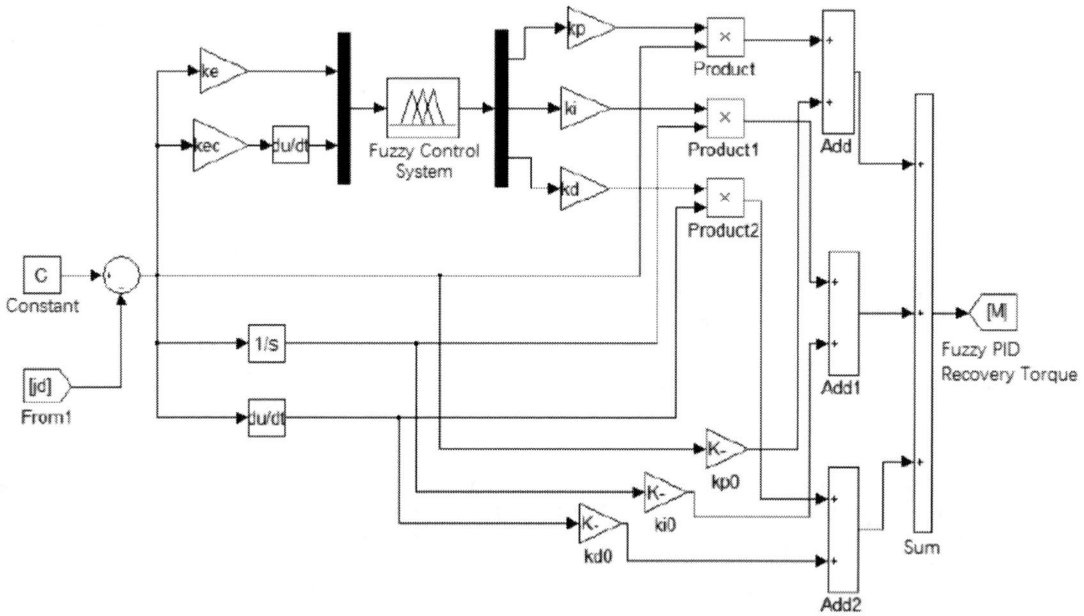

Figure 2. Simulink model of Fuzzy PID controller.

The two level Simulink control block diagrams are given below. Figure 3 shows the active swing bridge adjustment control algorithm, which can automatically tune the output torque by the fuzzy PID control. When the vehicle state detection module detects the vehicle is under the first stage instability risk, this controller will be triggered and provides a recovery torque to the swing bridge. Figure 4 shows the active braking control algorithm, which is based on switch control. When the vehicle state detection module detects the vehicle is under the second stage instability risk, this controller will be

 DOI: 10.1051/matecconf/20165606003

triggered and output a braking force to the brake pedal. Their work processes will be introduced in Section 4.

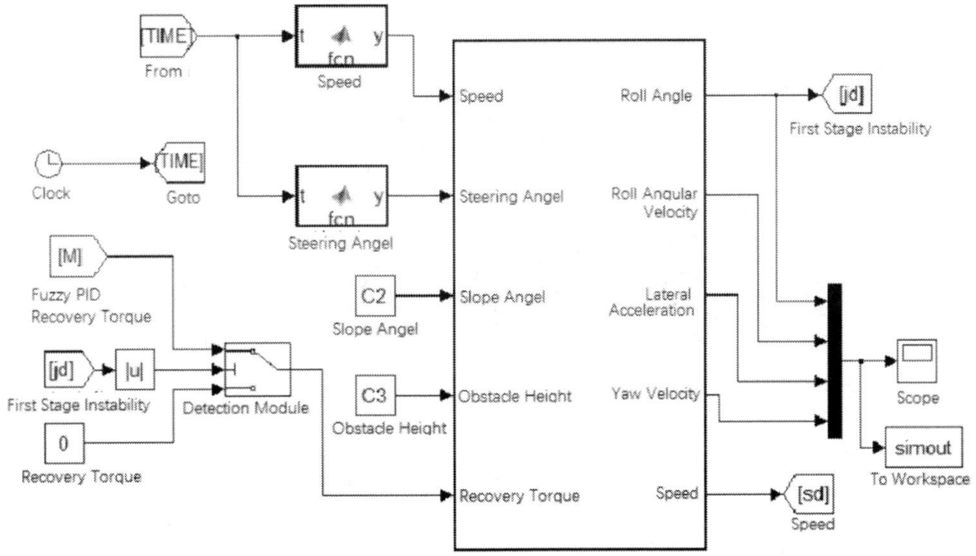

Figure 3. Active swing bridge adjustment control.

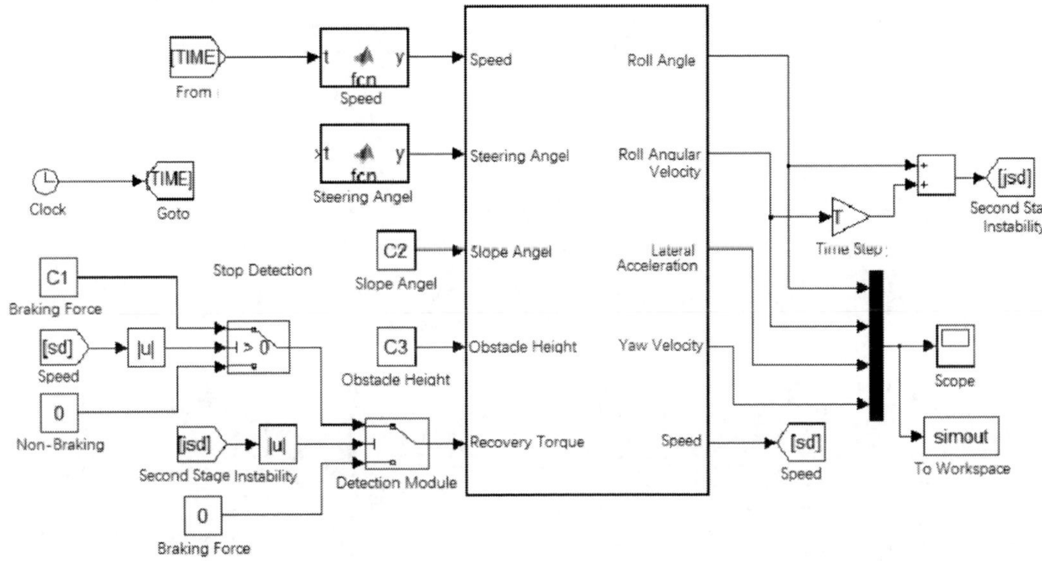

Figure 4. Active braking control.

3 Multi-body Dynamics Modelling

The multi-body dynamics model of a ZL50 articulated wheel loader is built by commercial software RecurDyn. This model consists of a front body, a rear body, and a rear axle. The front and rear bodies are linked together by a hydraulic articulated steering system; the rear body and rear axle are linked together by a swing bridge which makes them own relatively independent roll motion. For the tire, FIALA tire model provided by RecurDyn is used. Finally, the road and slope are modelled by Ground module in RecurDyn. After building all rigid models, tire model and road model, the constraints and drive forces between these components are added to the MBD model. The final MBD model is shown in Figure 5.

Figure 5. Multi-body dynamics model of a ZL50 wheel loader.

4 Simulation Setup

In order to verify the active anti-rollover control algorithm and analyse its effect during the first and second stage instability, a co-simulation was conducted based on the MBD model and the active anti-rollover control model via RecurDyn/CoLink.

Before co-simulating the MBD model and the control system, the input variables (PIN) output variables (POUT) were set up as follows:

PIN: Swing Bridge Recovery Torque, Braking Force;

POUT: *X*-axle speed, *Y*-axle speed, Yaw Velocity, *X*-axle acceleration, *Y*-axle acceleration, Roll Angular Velocity, Roll Angle, Lateral Angular Velocity, etc.

What's more, the driving speed and steering angle were set as constant into the model. Note that the driving speed is in the relative local coordinate system, which is calculated from *X*-axle speed, *Y*-axle speed and Yaw Velocity, and the lateral acceleration is also obtained by *X*-axle acceleration, *Y*-axle acceleration, Roll Angular Velocity, Roll Angle, and Lateral Angular Velocity. The final co-simulation model is shown in Figure 6.

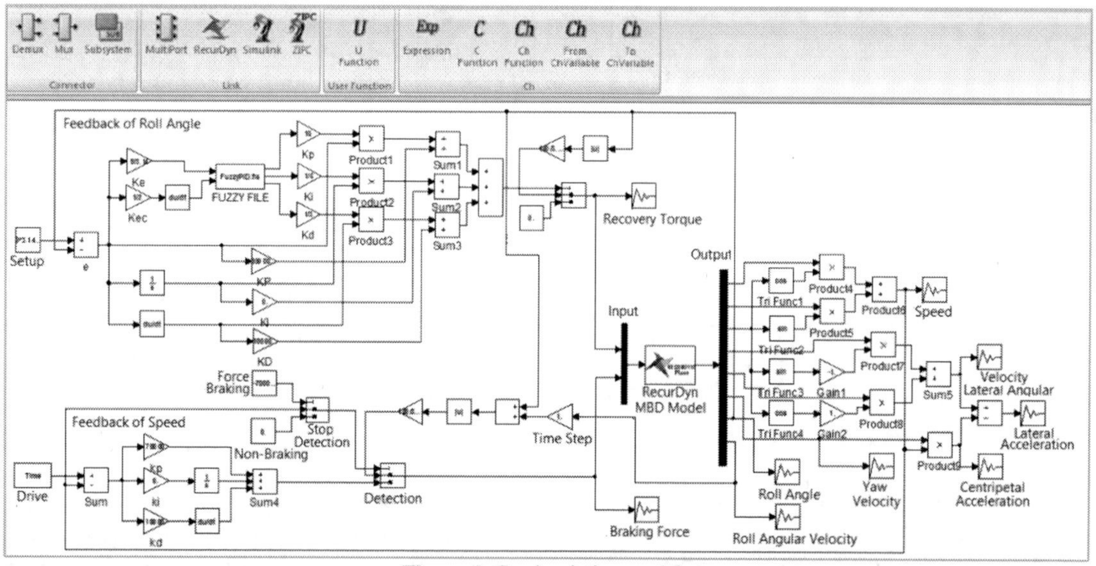

Figure 6. Co-simulation model.

5 Simulation Results

There are three different typical rollover situations for the wheel loader: tripped rollover, high speed steering rollover, and slope rollover. In order to verify the first and second instability anti-rollover control algorithm, we chose high speed steering rollover to conduct the simulation and analyse, which can cause rollover by adjusting the driving speed and steering angle. Figure 7 shows the rollover process by increasing the driving speed of the vehicle, Figure 7(a) shows the first instability (one front wheel leaves the ground) and Figure

7(b) shows the second instability (two wheel in one side of the vehicle all leave the ground).

(a) First stage instability

(b) Second stage instability

Figure 7. High speed steering rollover process of wheel loader.

When the wheel loader runs fast on straight road with a sudden steering, the first instability occurs when the steering angle reaches 30 degrees and the driving speed reaches 6 m/s. If the control system detects this situation, it will apply a recovery torque on the swing bridge. As shown in Figure 8 (a), a sudden steering angle is applied after 2 s straight drive: without control, the roll angle increases sharply and results in the first instability, after 6 s the whole loader is rollover; when the active control is activated, before the first instability occurs, as shown in Figure 8 (c), a swing bridge recovery torque works and forces the loader to stay stable.

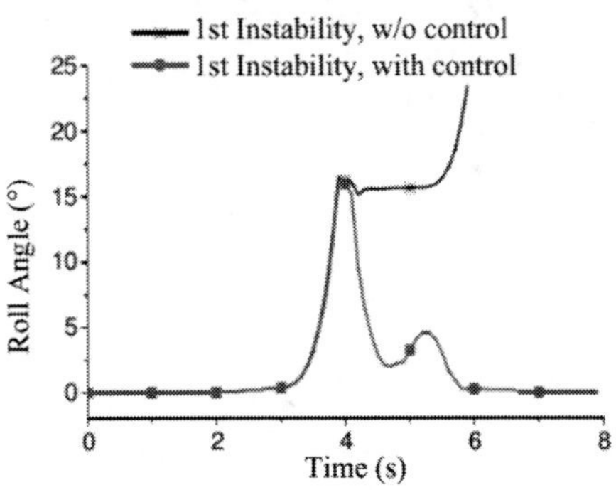

(a) Roll Angle

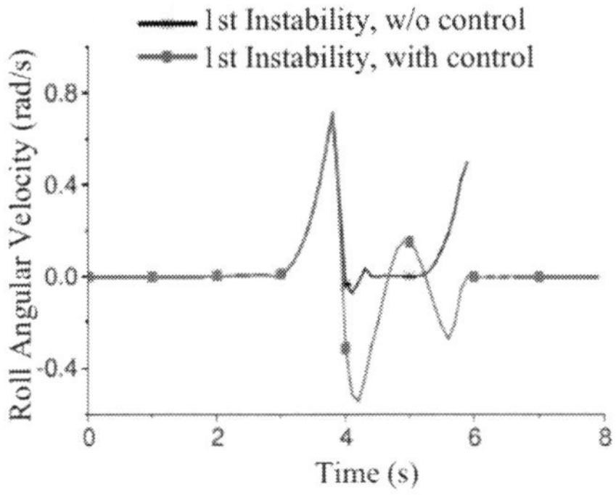

(b) Roll Angular Velocity

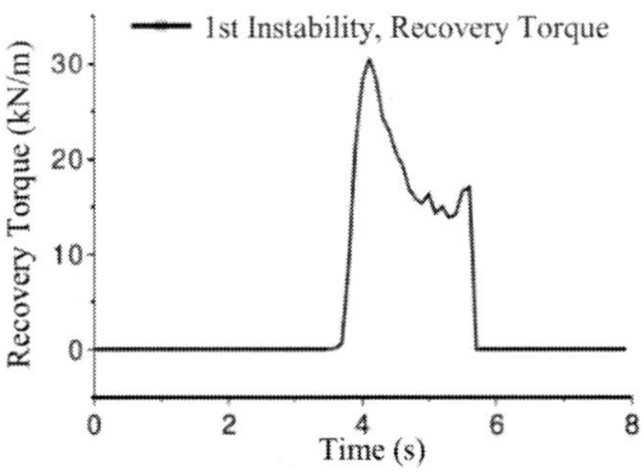

(c) Recovery Torque

Figure 8. The first stage instability control.

Similarly, when the driving speed continues to increase and reaches 8 m/s, the vehicle reaches to the limit of the second instability, then the control system applies a braking force and reduces the rollover possibility until the vehicle stops. Figure 9 (a) and (b) show the comparison between with and without active control when the vehicle is under the second instability. As shown in Figure 9 (a), without control, the roll angle increases dramatically after 4 s and finally rollover accident happens. When the active control is activated, the first instability also occurs at 4.4 s, and then the vehicle roll angle has a certain degree of reduction dues to the recovery torque, but it could not eliminate the rollover trend, and the second instability still occurs at 6.2 s. Finally, the active braking force is applied to reduce the roll angle and the vehicle recovers to safety condition finally.

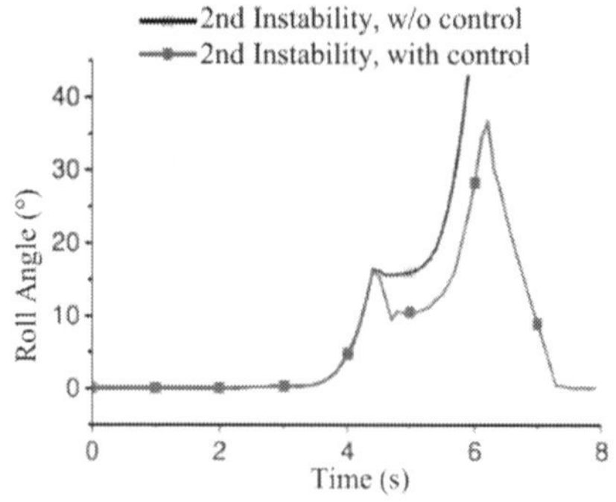

(a) Roll Angle

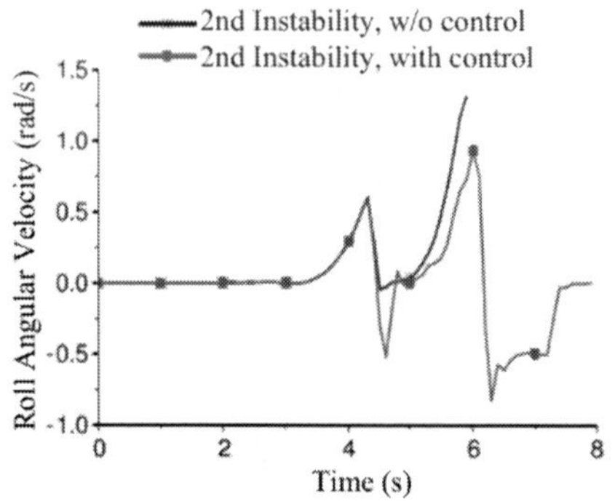

(b) Roll Angular Velocity

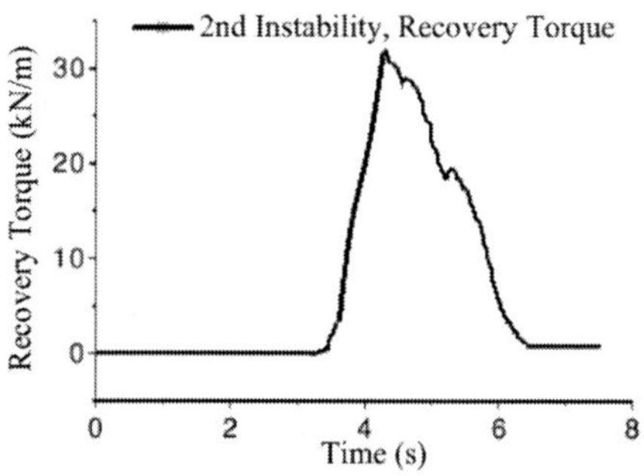

(c) Recovery Torque

Figure 9. The second stage instability control.

6 Conclusions

Articulated off-road vehicles are widely used in engineering construction and agricultural production, but most of them are suffered a high possibility of rollover. This paper took advantages from both fuzzy and PID control and developed an active anti-rollover control system. This controller consists two level control, which target to two different instability stages: 1) the active swing bridge adjustment control algorithm for the first stage instability; and 2) the active braking control algorithm for the second stage instability. They work together to prevent the articulated wheel loader from rollover accident.

The multi-body dynamics model of the wheel loader is built by commercial software RecurDyn, and then the MBD model and controller model are co-simulated under the same solver in RecurDyn/CoLink. The co-simulation results show that, during the high speed steering rollover process, the active swing bridge control system works well to prevent rollover when the steering angle reaches 30 degrees and the driving speed reaches 6 m/s. However, when the driving speed continues to increase and reaches 8 m/s, only the first level control is not enough to prevent rollover, and the active braking control is needed to reduce the accident possibility.

Acknowledgements

This study was supported by the National Natural Science Foundation of China (Grant No. 51175216).

References

1. G.M. Saltzman, M.H. Belzer. Nat. Inst. for Occu. Saf. and Hea. (2007).
2. M. Iida, H. Nakashima, H. Tomiyama. Comp. and Elec. in Agri., **76**(2): 277-283 (2011).
3. ISO 3471, *Earth-Moving Machinery-Roll-Over Protective Structures-Laboratory Tests and Performance Requirements*, (2008).
4. Y.X. Zhang, G.Q. Wang, C. Chen, J.N. Qu. Eng. Veh, **6**(42):56-63 (2011).
5. N.L. Azad, J. Mc Phee, A. Khajepour. *SAE Paper*, No.2005-01-3597 (2005).
6. S.J. Reynolds, W. Groves. Amer. J. of Prev. Med., **18**(4S): 63-69 (2000).
7. Y.X. Zhang. *Changchun: Jilin University*, 2012.
8. D.J.M. Sampson, G. Mckevitt, D. Cebon. Veh. Sys. Dyn., (**33**): 704-715 (2000).
9. P. Gaspar, I. Szaszi, J. Bokor. Cont. Eng. Pra., **13**(6): 699-711 (2005).
10. M.M. Islam. *Ontario: University of Ontario Institute of Technology* (2013).
11. N. Daher. Cont. Eng. Prac. (**45**): 46–5348(2015).
12. Y.T. Juang, Y.T. Chang, C.P. Huang. Info. Sci., **178**(5): 1325-1333 (2008).
13. J. Carvajal, G.R. Chen, H. Ogmen. Info. Sci., **123**(3): 249-270 (2000).

Control of AWD System for Vehicle Performance and Safety

Hojin Jung , Seibum Choi

KAIST, Department of Mechanical Engineering, 34141 Daejeon, Korea

Abstract. AWD (All-Wheel Drive) system transfers drive force to all wheels so that it can help vehicle escape low mu surface or climb hill more conveniently. Recently, AWD system for on road vehicle has become popular to improve vehicle driving performance. However, there has not been enough research of applying AWD system for vehicle stability especially for lateral movement. Compared with ESC (Electronic Stability Control), AWD system does not cause any inconveniences to the driver because it controls vehicle only by distributing front and rear drive torque, without using brake. By allowing slipping/locking of wet clutch inside the transfer case, AWD system can distribute different amount of torque between front and rear axle. This paper introduces modelling of AWD system and suggests the control of AWD system based on peak slip ratio and slip angle at which tyre saturates. Carsim based vehicle simulation results of AWD controller is presented.

1 Introduction

Compared with original 2WD vehicle which drives front wheels or rear wheels only, AWD system has a roll to distribute engine drive to torque to both front axle and rear axle. Usually AWD system was equipped with off-road vehicle that runs on rough road. However, with the growing of engine power, AWD system has become being equipped in high performance sedan.

Hardware configuration which is dealt with in this paper differs from center differential 4WD and selective 4WD because it distributes some portion of drive torque of main drive shaft which is directly linked to transmission output shaft to sub drive shaft by controlling wet clutch inside the transfer case. Actuator type for control varies from ball ramp control type by electromagnetic force to hydraulic piston control type by motor pump.

By utilizing advantages of active control system, AWD system can be used for both of longitudinal and lateral driving safety. In this regard it is important to find out the present vehicle state as well as road state. For example, vehicle climbs high slope hill, escapes from low mu surface and has lateral movement like double lane change or circular motion with high velocity. And research about model predictive control should be done including these various driving situations. Then it is expected that vehicle performances like driver's riding quality and steady state cornering are improved.

This paper suggests the control method of AWD system of wet clutch with hydraulic piston actuator type. Vehicle simulation environment was set based on simulink built transfer case system coupled with carsim software. Especially, possibility of lateral stability improvement by AWD system without intervention of ESC is discussed in detail as well as longitudinal performance.

2 System modelling

2.1 Driveline dynamic model

States of AWD vehicle change depending on the amount of torque which is delivered front or rear shaft. AWD system which is dealt in this paper can control the amount of torque which is transfered to front drive shaft in transient state, by using multi plate wet clutch. Thus simulation description should be available of variable distriburtion of front shaft torque depending on clutch engagement force especially between clutch slipping state and clutch lock up state. Driveline model from transmission output shaft to front and rear differential gear can be represened as follows, mathematical model is also as follows :

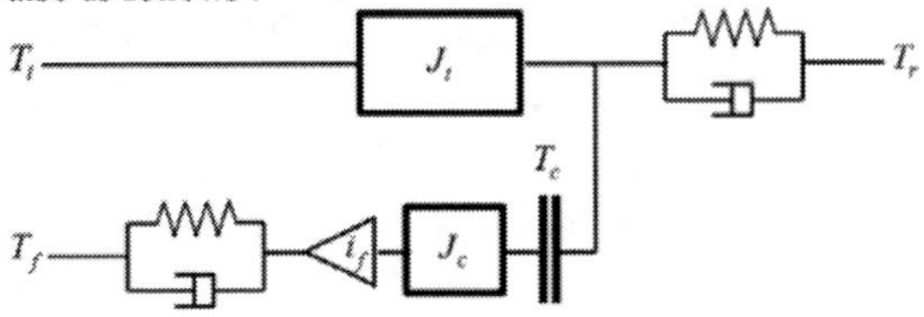

Figure 1. Schematic diagram of transfer case system.

$$T_t - T_c - T_r = J_t \dot{\omega}_t \tag{1}$$

$$T_c = f\left(F_c, \omega\right) \tag{2}$$

$$T_c - T_f / i_f = J_c \dot{\omega}_c \tag{3}$$

$$T_f = k_f\left(\theta_c - \theta_f\right) + b_f\left(\dot{\theta}_c - \dot{\theta}_f\right) \tag{4}$$

$$T_r = k_r\left(\theta_t - \theta_r\right) + b_r\left(\dot{\theta}_t - \dot{\theta}_r\right) \tag{5}$$

Where T_t is a transmission output torque, T_c is a clutch engagement torque, T_f and T_r are front and rear shaft torque each. J_t is a main drive shaft inertia, J_c is front drive shaft inertia, F_c is clutch engagement force given by hydraulic actuator. ω is a clutch slip angular velocity. Spring and damper system was added at each end of drive shaft to represent transfer torque given by distortion of shaft. Transmitted clutch torque is expressed as a function $f\,(F_c,\ \omega_{slip})$ and detailed expression is introduced in section 2.2.1.

2.2 Wet clutch model

Clutch is generally divided into dry clutch and wet clutch, depending on the kind of material and whether the oil is included or not. Dry clutch can be burned if slip duration prolonged. Wet clutch compensate the disadvantage of dry clutch, by absorbing the heat which is generated in friction plate. However, wet clutch is short of maximum allowable torque which friction plate endures. By overlapping multiple plates, maximum allowable torque can be enhanced.

2.2.1 Generalized Maxwell Slip (GMS) friction model

It is important to have a feasibility of friction model for the system which includes engagement force between two rolling plates to guarantee the validity of simulation. GMS model [1,2], which is well known for friction modelling, was opted for modelling transfer case system.

When clutch is fully lock up, differential equation of clutch transfer torque is as follows:

$$\frac{dM_i}{dt} = k_i\left(p, T_t\right)\cdot\omega \tag{6}$$

Where $k_i\,(p,\ T_t)$ is the function of clutch engagement force and transmission output torque. k_i can be mapped by comparing carsim internal 4WD model front shaft torque and developed transfer case system front shaft torque.

When clutch is slipping, differential equation of clutch transfer torque is as follows:

$$\frac{dM_i}{dt} = \mathrm{sgn}(\omega)C\left(\alpha_i - \frac{M_i}{s(\omega,p)}\right) \tag{7}$$

Where s(ω, p) is a stribeck equation about torque, which is expressed as follows:

$$s(\omega,p) = \mathrm{sgn}(\omega)\left(M_s + \left[M_c - M_s\right]e^{-\left|\frac{V_s(p)}{\omega}\right|}\right) \tag{8}$$

Wet clutch should consider viscous effect by the oil film between disc plates. Especially for wet clutch, drag torque takes quite some part of transmitted torque so that this should be modelled carefully to improve the accuracy of transient response [4]. Drag torque of wet clutch is originated from the shear stress between plates and can be modelled by fluid dynamic analysis [4-6]. Then, viscous friction torque equation is obtained as follows:

$$M_{vis} = 2\pi\int_{r_1}^{r_o}\frac{\mu\omega r^3}{h}\left(1 + 0.0012\,\mathrm{Re}_h^{0.94}\right)dr \tag{9}$$

Where r_1 is an clutch inner diameter, r_o is an effective diameter due to the centrifugal force, which is usually less than clutch outer radius. Re_h is a Reynolds number Automatic Transmission Fluid (ATF), which is written as follows:

$$\mathrm{Re}_h = \frac{\rho\omega r h}{\mu} \tag{10}$$

Where μ is a viscous coefficient of ATF, ρ is a density of ATF and h is a clearance length of clutch plate.

Finally, clutch transfer torque when slipping is as follows:

$$T_c = M_i + M_{vis} \tag{11}$$

To improve the accuracy of wet clutch engagement simulation, clutch experimental friction coefficient data which will be manufactured for vehicle implementation was used.

2.3 Lock up state representation by applying DOF (Degree Of Freedom) reduction model

GMS model which is introduced in section 2.2 matches well in normal driving simulation. However, complex or severe driving situation like rapid acceleration in low surface can cause model instability like chattering response of clutch transfer torque. Equation should be replaced when lock up state so that it was described by designing DOF reduction model. When lock up state, angular accelerations at both transmission output shaft and clutch driven shaft are same. Then, dynamic model of driveline is as follows:

$$\dot{\omega}_t = \dot{\omega}_c \tag{12}$$

$$T_t - T_f - T_r = \left(J_t + J_c\right)\dot{\omega}_t \tag{13}$$

$$T_f = k_f\left(\theta_c - \theta_f\right) + b_f\left(\dot{\theta}_c - \dot{\theta}_f\right) \tag{14}$$

$$T_r = k_r\left(\theta_t - \theta_r\right) + b_r\left(\dot{\theta}_t - \dot{\theta}_r\right) \tag{15}$$

2.4 State transition between clutch slipping and lock up

It is important to decide the current state whether clutch is slipping or locked up to improve the accuracy of AWD system simulation [3]. GMS model distinguish current state only by the limit of allowable static torque. However, for the case of simulation which include full vehicle model. Above condition was not enough to determine state transition accurately when clutch is engaged while input torque is rapidly increasing. Thus two conditions of static torque and clutch slip were used simultaneously to enhance the accuracy of state transition.

$$T_c \ge M_s \tag{16}$$

$$\omega_{slip} \le 0.05\,rad/s \tag{17}$$

3 Controller design

3.1. Vehicle state estimation

For the AWD system controller based on tyre model, current vehicle states should be estimated. Generally, the lowest wheel angular velocity value among 4 angular velocity sensors can be assumed to vehicle longitudinal velocity in 2WD. However, when clutch is engaged, vehicle longitudinal velocity of AWD cannot be assumed only with wheel sensors because all wheels become driven wheels. Thus vehicle longitudinal velocity should be estimated. In addition, vehicle lateral velocity should be estimated to know wheel side slip angle. By utilizing vehicle longitudinal dynamics, lateral dynamics, yaw dynamics and wheel dynamics, these vehicle states can be estimated by extended Kalman filter [7,8] using planar bicycle model shown in figure 2.

The equation of motion for the bicycle model shown in figure 2 can be categorized three dynamics.

Longitudinal dynamics is:

$$\dot{v}_x = \frac{1}{m}\Big\{F_{xf}\cos(\delta_f) - F_{yf}\sin(\delta_f) + F_{xr}\cdots \\ -\frac{1}{2}\rho_{air}C_d A v_x^2 - mg\sin(\theta) + v_y r\Big\} \tag{18}$$

Where v_x is the longitudinal velocity, m is the vehicle mass, F_{xf} is the front longitudinal axle force, F_{yf} is the front lateral axle force, δ_f is the tire steer angle, ρ_{air} is the air density, C_d is the drag coefficient, v_y is the lateral velocity, r is the yaw rate and θ is hill angle.

Lateral dynamics is:

$$\dot{v}_y = \frac{1}{m}\left\{F_{xf}\sin(\delta_f) + F_{yf}\cos(\delta_f) + F_{yr}\right\} + v_x r \tag{19}$$

Where F_{yr} is the rear lateral axle force.

Yaw dynamics is:

$$\dot{r} = \frac{1}{I_z}\left\{l_f\left(F_{xf}\sin(\delta_f) + F_{yf}\cos(\delta_f)\right) - l_r F_{yr}\right\} \tag{20}$$

Where I_z is the vehicle yaw moment of inertia, l_f is the distance from the center of mass to the front axle and l_r is the distance from the center of mass to the rear axle.

Another dynamics which force terms are related is wheel dynamics which is expressed as follows:

$$\dot{\omega} = \frac{1}{I_w}\left\{T_d - T_b - R_e F_{x,i} - R_e F_r\right\} \tag{21}$$

Where ω is the wheel angular velocity, I_w is the wheel inertia, T_d is the wheel drive torque, T_b is the wheel brake torque. In this paper, T_d is assumed to be known and does not consider braking situation. R_e is the wheel effective radius, F_r is the tire rolling resistance. $F_{x,i}$ is the each tire longitudinal force whose subscript i includes *fl* (front left), *fr* (front right), *rl* (rear left), *rr* (rear right) which is written throughout the paper.

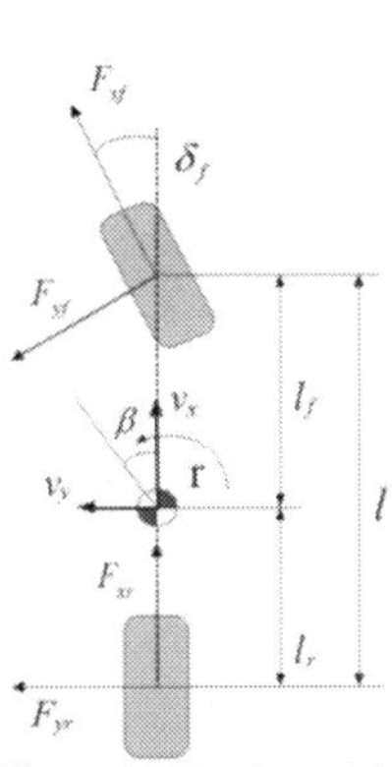

Figure 2. Bicycle model.

Using dynamics which are mentioned above, Kalman filter can be obtained. States of Kalman filter is defined as follows:

$$\mathbf{x}(t) = \left[v_x,\ v_y,\ r,\ \boldsymbol{\omega},\ \mathbf{F}\right]^T \tag{22}$$

Where

$$\boldsymbol{\omega} = \left[\omega_{fl},\ \omega_{fr},\ \omega_{rl},\ \omega_{rr}\right] \tag{23}$$

$$\mathbf{F} = \left[F_{xfl},\ F_{xfr},\ F_{xrl},\ F_{xrr},\ F_{yf},\ F_{yr}\right] \tag{24}$$

Tire longitudinal forces which are expressed in this paper are related as:

$$F_{xf} = F_{xfl} + F_{xfr},\ F_{xr} = F_{xrl} + F_{xrr}$$

The measurements signals from vehicle are as follows:

$$\mathbf{y}(t) = \left[a_x,\ a_y,\ r,\ \omega_{fl},\ \omega_{fr},\ \omega_{rl},\ \omega_{rr}\right] \tag{25}$$

Where a_x and a_y are related as:

$$a_x = \frac{1}{m}\Big\{F_{xf}\cos(\delta_f) + F_{yf}\sin(\delta_f) + F_{xr}\cdots \\ -\frac{1}{2}\rho_{air}C_d A v_x^2 - mg\sin(\theta)\Big\} \tag{26}$$

$$a_y = \frac{1}{m}\left\{F_{xf}\sin(\delta_f) + F_{yf}\cos(\delta_f) + F_{yr}\right\} \tag{27}$$

The above states and measurements are expressed as nonlinear function $f(\mathbf{x}(t),\mathbf{u}(t))$ and $h(\mathbf{x}(t),\mathbf{u}(t))$ as follows:

$$\dot{\mathbf{x}}(t) = f(\mathbf{x}(t),\mathbf{u}(t)) + w(t) \tag{28}$$

$$y(t) = h(\mathbf{x}(t),\mathbf{u}(t)) + v(t) \tag{29}$$

Estimated states are usually denoted as $\hat{\cdot}$. Then, the algorithms of discrete time extended kalman filter can be expressed as follows [10]:

$$\hat{\mathbf{x}}_k(-) = \hat{\mathbf{x}}_{k-1}(+) + \int_{t_{k-1}}^{t_k} f(\mathbf{x}(\tau),\mathbf{u}(\tau))d\tau \tag{30}$$

$$\mathbf{P}_k(-) = \mathbf{P}_{k-1}(+)\cdots \\ +\int_{t_{k-1}}^{t_k}\left\{\mathbf{F}(\tau)\mathbf{P}(\tau) + \mathbf{P}(\tau)\mathbf{F}^T(\tau) + \mathbf{L}\mathbf{L}^T\right\}d\tau \tag{31}$$

$$\mathbf{K}_k(-) = \mathbf{P}_k(-)\mathbf{H}_k^T\left[\mathbf{H}_k\mathbf{P}_k(-)\mathbf{H}_k^T + \mathbf{R}_k\right]^{-1} \tag{32}$$

$$\hat{\mathbf{x}}_k(+) = \hat{\mathbf{x}}_k(-) + \mathbf{K}_k\left[\mathbf{y}_k - h(\hat{\mathbf{x}}_k(-),\mathbf{u}(t))\right] \tag{33}$$

$$\mathbf{P}_k(+) = \left[\mathbf{I} - \mathbf{K}_k\mathbf{H}_k\right]\mathbf{P}_k(-) \tag{34}$$

Where F and H are the linearization of equation (29) and (30) around $\hat{\mathbf{x}}_k(-)$ at each time step. $\mathbf{P}$ is the error covariance and $\mathbf{L}$ is the sensor noise disturbance matrix.

Wheel slip ratio and side slip angle are obtained from the estimated states. Longitudinal slip ratio at each wheel in acceleration is given by:

$$\lambda_i = \frac{R_e \omega_i - \hat{v}_x}{R_e \omega_i} \tag{35}$$

Also side slip angle of front axle and rear axle which will be used in the AWD logic development are:

$$\alpha_f = \beta + \frac{l_f r}{\hat{v}_x} - \delta_f \tag{36}$$

$$\alpha_r = \beta - \frac{l_r r}{\hat{v}_x} \tag{37}$$

Where β is side slip angle which is defined as v_y / v_x.

3.2 AWD driving logic development

Certainly, AWD driving condition can be decided based on saturation point of tyre longitudinal and cornering stiffness. AWD system has a tendency to stabilize the vehicle both of longitudinal and lateral movements. There many approaches to develop vehicle control logic based on tyre stiffness. However, there has not been saturation point detection logic considering vehicle weight shifting and slip simultaneously. To solve this problem tyre lateral force function of two variables about vertical force and slip angle is presented here.

$$F_x = C_{x,1}\lambda_i + C_{x,2}\lambda_i^2 \tag{38}$$

$$F_y = C_{y,1}\left(1 - \frac{\Delta F_z}{F_{z,n}}\right)\alpha + C_{y,2}\alpha^2 \tag{39}$$

Where $F_{z,n}$ is nominal vertical force of normal load. ΔF_z is the difference between shifted vertical force and nominal vertical force.

Saturation point of slip ratio and slip angle can be calculated by differentiating equation (39) and (40).

$$\lambda_{sat} = -\frac{C_{x,1}}{2C_{x,2}} \tag{40}$$

$$\alpha_{sat} = -\frac{C_{y,1}}{2C_{y,2}\left(1 - \frac{\Delta F_z}{F_{z,n}}\right)} \tag{41}$$

Above saturation point is used in this paper to determine the engagement of wet clutch.

4 Simulation

To validadte the AWD system driving simulation by impementing developed transfer case plant model, carsim software was used. As shown in figure 2, simulink based transfer case model was coupled in carsim. Except for transfer case part carsim internal model was used. Simulation coupled with commercial sofware not only can verify the versatility of developed model but also help to easiliy grasp of change of vehicle state depending on various driving simulation. Vehicle driveline is based on front ship rear wheel drive (FR) type.

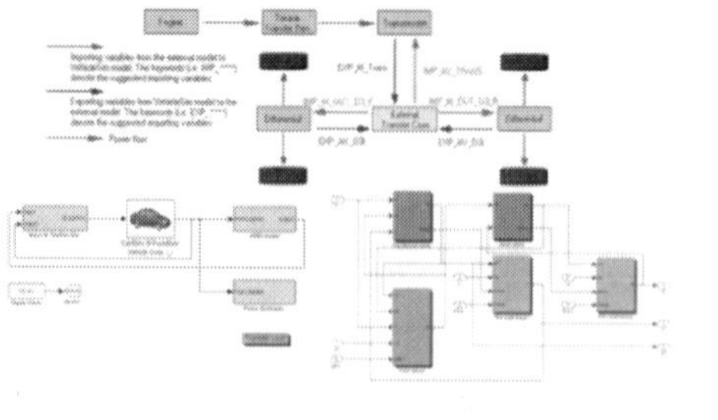

Figure 3. Transfer case simulink model coupled with carsim.

4.1 Accelerating in split mu situation

For the validation of longitudinal driving performance of AWD system, straight line with split mu simulation was conducted. Left wheel is driven on the low mu with friction coefficient of 0.2 and right wheel is driven on the high mu with friction coefficient of 0.85. While driving split mu with 30km/h, rapid full acceleration was conducted between 8-14 seconds to simulate the situation of high slip ratio. Based on the slip calculation logic and trigger condition which was mentioned in section 3.2 and 3.3, AWD system was activated. To exclude the influence of gear shifting, only 1st and 2nd gear shifting was allowed.

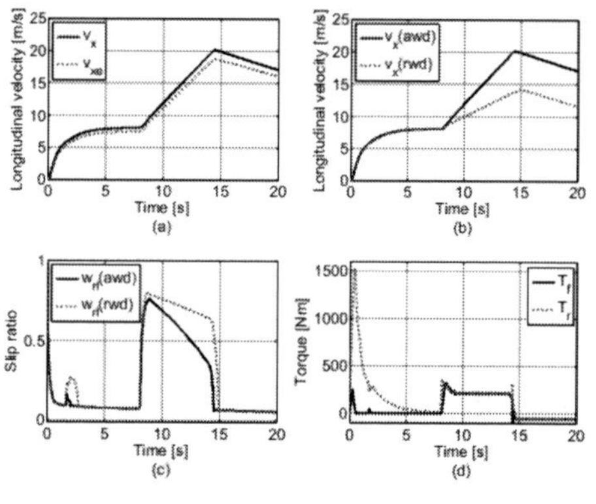

Figure 4. Simulation results of accelerating in split mu situation. (a) Comparison of longitudinal velocity between real value and estimated value (b) Comparison of longitudinal velocity between awd and rwd (c) Comparison of slip ratio between awd and rwd (d) front and rear shaft torque response in awd

Figure 4(a) shows the results of estimated awd longitudinal velocity. Although there appeared discrepancy between real value and estimated value, it can be used for slip ratio calculation. And figure 4(b) shows the results of longitudinal velocity between awd and rwd. Certainly, awd showed a better performance to escape the split mu road. Slip ratio reduction in awd compared to rwd can be verified in figure 4(c). Figure 4(d) shows the torque response of awd by the control logic.

4.2 Accelerating in steady state cornering situation

For the validation of lateral stability of AWD system, 500 feet radius circle lane simulation was conducted. While driving high mu of friction coefficient with 90km/h, rapid full acceleration was conducted between 15-25 seconds to simulate the situation of over-steering. Then, AWD system was activated based on same process which was mentioned in section 4.1. Maximum transmission gear was set to 5th.

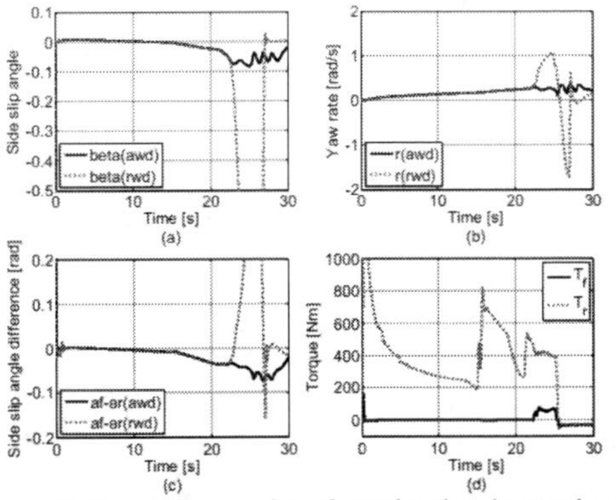

Figure 5. Simulation results of accelerating in steady state cornering. (a) Comparison of side slip angle between awd and rwd (b) Comparison of yaw rate between awd and rwd (c) Comparison of front and rear slip difference between awd and rwd (d) front and rear shaft torque response in awd

Figure 5(a) and 5(b) shows the results of side slip angle and yaw rate, which are principal indices of lateral stability, between awd and rwd. Certainly, rwd showed high side slip angle and yaw rate when tyre saturates. Figure 5(c) shows the results of front and rear axle side slip angle difference. Actually, awd showed slight side slip angle difference, which means vehicle moves similar to neutral steer. However, rwd showed an over-steering response when tyre force saturates. Figure 5(d) shows the torque response of awd by the control logic. It is noteworthy that small amount of torque distribution prevented over-steering response.

5 Conclusion

AWD control logic development for vehicle stability was conducted in this paper. For controller validation, transfer case modeling method was introduced and reliable simulation environment was built first, by implementing simulink transfer case as an external model in carsim software. Noticable reuslt of simulation is that upper level controller based on tyre force function of wheel slip ratio and slip angle helped tyre not to fall into saturation region, without intervention of ESC system. In addition, small amount of torque transfer in subdrive shaft stabilize the vehicle lateral movement efficiently. However, explanation of some principal vehicle states estimator is omitted by assuming they are measurable in this paper. Thus it should be furthur studied for practical application. Also, real time tyre stiffness and mu estimation based logic will improve the awd control logic for various driving scenario.

Acknowledgement

This work was supported by the National Research Foundation of Korea(NRF) grant funded by the Korea government(MSIP) (No. 2010-0028680).

This research was supported by the MSIP(Ministry of Science, ICT and Future Planning), Korea, under the C-ITRC(Convergence Information Technology Research Center) (IITP-2015-H8601-15-1005) supervised by the IITP(Institute for Information & communications Technology Promotion)

References

1. Agusmian Partogi Ompusunggu, Paul Sas, Hendrik Van Brussel, "Modeling and simulation of the engagement dynamics of a wet clutch system subjected to degradation: An application to condition monitoring and prognostics", Mechatronics (2013)
2. Vincent Lampaert, Farid Al-Bender, Jan Swevers, "A Generalized Maxwell-Slip Friction Model appropriate for Control Purposes", IEEE Physics and Control Proceedings. International Conference (2003)
3. Jiwon J. Oh, Seibum B. Choi, Jinsung Kim, "Driveline modeling and estimation of individual clutch torque during gear shifts for dual clutch transmission", Mechatronics (2014)
4. Hiromu HASHIMOTO, Sanae WADA and Yoshiaki MURAYAMA, "An Influence of Centrifugal Force on the Performance of Thermo-hydrodynamic Turbulent Thrust Bearings", Bulletin of JSME, **27**, No. 228 (1984)
5. Yiqing Yuan, Eysion A. Liu, James Hill, Qian Zou, "An Improved Hydrodynamic Model for Open Wet Transmission Clutches", ASME Journal of Fluids Engineering (2007)
6. Youhei Takagi, Yasunori Okano, Masatoshi Miyayaga and Nobuyuki Katayama, "Numerical and Physical Experiments on Drag Torque in a Wet Clutch", Japanese Society of Tribologists, pp. 242-248 (2012)
7. Laura R. Ray, "Nonlinear state and tire force estimation for advanced vehicle control", IEEE TRANSACTIONS ON CONTROL SYSTEMS TECHNOLOGY, **3**, NO.1 (1995)
8. Mooryong Choi, Jiwon J. Oh and Seibum Choi, "Linearized Recursive Least Square Methods for Real-time Identification of Tire-Road Friction Coefficient", IEEE TRANSACTIONS ON VEHICULAR TECHNOLOGY, **8**, (2013)

Active Head Motion Compensation of TMS Robotic System Using Neuro-Fuzzy Estimation

W.N. Wan Zakaria , Razali Tomari and Rafidah Ngadengon

Advanced Mechatronic Research Group (ADMIRE), Faculty of Electrical & Electronic Engineering, Universiti Tun Hussein Onn Malaysia, 86400 Parit Raja, Batu Pahat, Johor, Malaysia.

Abstract. Transcranial Magnetic Stimulation (TMS) allows neuroscientist to study human brain behaviour and also become an important technique for changing the activity of brain neurons and the functions they sub serve. However, conventional manual procedure and robotized TMS are currently unable to precisely position the TMS coil because of unconstrained subject's head movement and excessive contact force between the coil and subject's head. This paper addressed this challenge by proposing an adaptive neuro-fuzzy force control to enable low contact force with a moving target surface. A learning and adaption mechanism is included in the control scheme to improve position disturbance estimation. The results show the ability of the proposed force control scheme to compensate subject's head motions while maintaining desired contact force, thus allowing for more accurate and repeatable TMS procedures.

1 Introduction

Recent developments in the neuroscience area have heightened the need of robotized Transcranial Magnetic Stimulation (TMS) system to study human brain behaviour. In the TMS procedure, an electromagnetic coil is placed over the head, and a large but brief current pulse is produced and passed through the scalp [1]. The changing magnetic field which this generates induces a current in the conductive tissue of the head including the underlying cerebral cortex.

The cortical activity produced by TMS is critically dependent on coil location and orientation. Currently, the coil is positioned manually, which can be a problem for precise clinical evaluation and experiments of the TMS technique.

Although the TMS technique is very useful, it is not yet widely accepted because of the observed inconsistency of efficiency between subjects. TMS coils produce very dense and defined fields, which mean that a small movement of the coil position leads to a substantial change of the electromagnetic field delivered to a target nerve. It is due to the difficulty to indicate correct positioning with the current available stimulation systems which leads to poor repeatability [2][3].

This leads to applications of robotics in TMS to position the coil for efficient and reliable TMS procedures. Robot assistance in TMS procedure has been reported by a number of researchers [2-6]. These existing TMS robotic systems, which aim at replacing the hand and the arm of neurologist, are particularly focused on image-guided by means of navigation system to position the coil on subject's head.

Despite its safety and efficacy, current researchers are unable to address the force control as an important feature of the TMS procedure. Most of the research to date has tended to focus on image-guided and navigation system and neglected the study on contact force between the coil and the subject's head which is crucial. Excessive contact force between the coil and unconstrained subject's head may harm the subject.

During conventional hand-held procedure test, the following issues were observed and can be considered as challenges to the force-controlled TMS robot system implementation;

1. The applied force exerted by the operator along normal direction to the subject's head surface was found ranging from 6N to 10N in which the operator keeps the coil in close contact to the hot-spot position.
2. The subject's head geometrical, inertial and structural stiffness properties are totally unknown and vary from one subject to other subjects.
3. During the treatment, even the subject is in relax and conscious condition there is a higher possibility of involuntary head movement due to tremor, muscle contraction or relaxation involving to-and-from movement.

2 Human head-neck motion estimation

Human head-neck movement is extremely complex and is not completely understood. There is a considerable amount of literature published on this topic, however it is still impossible to model the head-neck system totally because of several factors namely; head-neck sheer complexity and lack of quantitative data on material properties, geometries and boundary conditions of individual tissues [7-11]. The outcome suggests that there is tremendous variation in the types of biomechanical models to approximate the human head-neck system [12]. This is due to the consideration of task, for example, slow voluntary movement is differed than movement resulting from an impact to the head or during walking. In addition, researchers also suggest those horizontal and vertical movement kinematics is coupled as the same muscles are involved in both types of movement [7] [12]. Typically, the movement is assumed to be a straight path with a single peak and bell-shaped velocity profile [13-15].

The problem is simplified from a task specification point of view in which force control is executed along the normal direction to the subject's head surface. Thus, it is necessary to establish the translational movement of the unconstrained head in relax and seated condition along this direction. A linear relationship between rotational and translational movement was established as described in following section.

2.1 Ellipsoidal shaped head geometry

An ellipsoidal shape was chosen to define the head geometry as it exhibits elevation as well as azimuth dependence [16]. To express the head geometry analytically, consider the head pitch movement (θ_p) around the rotation axis which is the centre of the ellipsoid as shown in Figure 1.

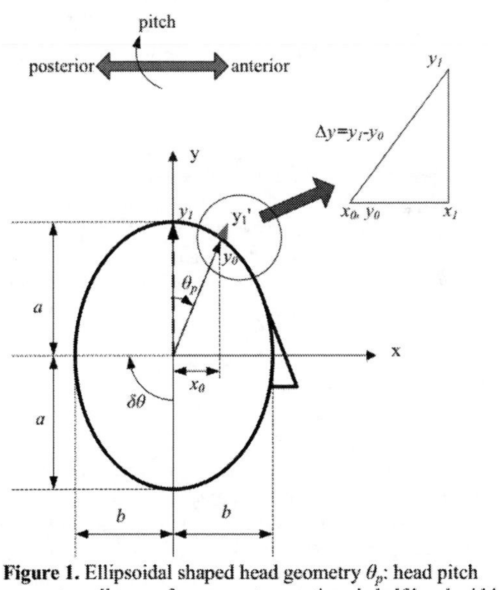

Figure 1. Ellipsoidal shaped head geometry θ_p: head pitch movement, a: distance from a centre rotation, b: half head width

As the head rotates about θ_p from start position y_1 to final position y_0, the translational distance can be expressed as $\Delta y = y_1 - y_0$. The following heuristic procedure is defined and yields a close approximation to determine Δy;

Find the point y_0,

$$y_0 = \frac{x_0}{tan\ \theta_p} \tag{1}$$

Find x_0 by using following ellipsoid equation;

$$\frac{x_0^2}{a^2} + \frac{y_0^2}{b^2} = 1 \tag{2}$$

Define the y_0 and gradient m as both y_1 and y_0 positions lie on the same straight line,

$$m = \frac{x_0}{y_0} = \frac{x_1}{y_1} = \frac{1}{tan\ \theta_p} \tag{3}$$

Combining equations (1), (2) And (3), thus x_0

$$x_0 = \sqrt{\frac{a^2 b^2}{\left(\frac{a}{tan\ \theta_p}\right)^2 + b^2}} \tag{4}$$

where, a is a distance from centre of rotation axis (approximately 30 mm inferior to the interaural axis [11] to the y_1 point and b is the half of human head width). Based on [8][11][17] works, a, b and θ_p can be defined as 160 mm, 95 mm and 10° respectively.

For further analysis, the head-neck [12] is related as following equation;

$$\frac{dx}{dt} = r.\frac{d\theta}{dt} \approx \vartheta = r.\omega \tag{5}$$

where r, (x,θ) and (υ,ω) are the vector directed from the axis of rotation, translational and angular position and translational and angular velocity respectively. To guide this analysis, the head movement parameters were estimated by considering various research studies. Ferman et al. [8] suggest that the subject head oscillate about 0.25 rad/s to 1.05 rad/s (amplitude about 10 degrees) in the horizontal, vertical or torsional plane during seated position while fixating target at optical infinity. This corresponds to 40-160mm/s translational velocities. This is almost consistent with the findings of [18] that investigated the amplitude of head rotation for the subject seated stationary in defined 'relax' and 'still' position with the head in a resting and facing forward. Their studies found the approximately similar oscillations of 0.5 rad/s to 1.5 rad/s in both conditions. These calculation parameters can be considered as motion disturbances to the force controller and will be used in following experimental programme.

3 Force Control Scheme

Figure 2 illustrates the block diagram of the force control architecture based on [19-22] as implemented in the TMS robotic system. Both internal position and external force control loops are performed in Cartesian coordinates.

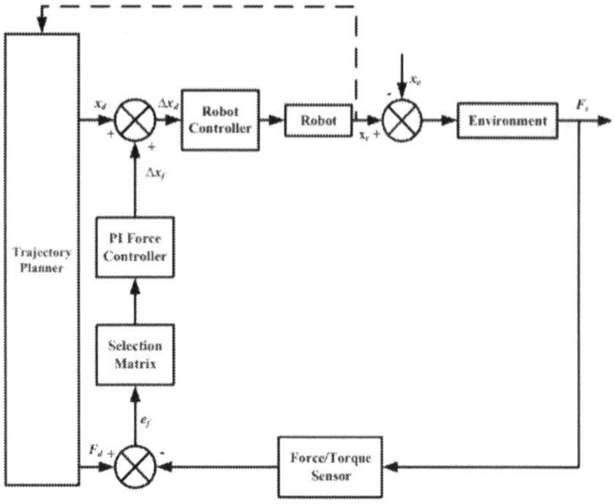

Figure 2. The external force control scheme

The key idea of this control strategy is that the force control loop generates the corrective position data from the force error, $e_f(t_k)$ through the Proportional-Integral (PI) force controller. The initial contact position $x_d(t_{k-1})$ is then modified by the force controller output $\Delta x_f(t_k)$. The resultant position demand $\Delta x_d(t_k)$ is forwarded to robot controller to determine the next desired position of the robot at every path update interval.

Prior to programming the robot to perform the TMS operation tasks, the issue relating to force control stability was first addressed. The implementation of a conventional PI force controller in TMS robotic system was found to be difficult. The preliminary test result suggests that unknown or uncertain environment stiffness and positional disturbance would drive the system into undesired behaviour. Thus, an intelligent approach is proposed to overcome these identified problems as discussed in the following section.

3.1. Adaptive Neuro-Fuzzy Force Controller based on MRAC

The Adaptive Neuro-Fuzzy Force Controller (ANFFC) structure is similar to that of a classical model reference adaptive control (MRAC) method as shown in Figure 3. The general idea of MRAC is to create a 'parameter adjustment' algorithm based on the model error ε to change the system response to match a desired model. The output of the parameter adjustment algorithm δ approaches unity as the system response tends to desired response.

The ANFFC consists of three main components: a reference model describing the desired behaviour of the force control scheme, a main Neuro-Fuzzy force controller (NFFC) that determines the adjustment to the position control loop and Neuro-Fuzzy learning and adaption mechanism (NFLA) that modifies the NFFC output.

The implementation of ANFFC is to compensate for any unknown or significant change in environment parameters that may lead to overshoot and undershoot error responses. This method behaves in a similar way to a human does, as we do not apply every control rule for particular situation but moderately adjusts the existing Neuro-Fuzzy Force Controller (NFFC) corresponding to the actual situation. The Neuro-Fuzzy [23-25] concept is applied to determine and generate the effective parameter adjustment algorithm.

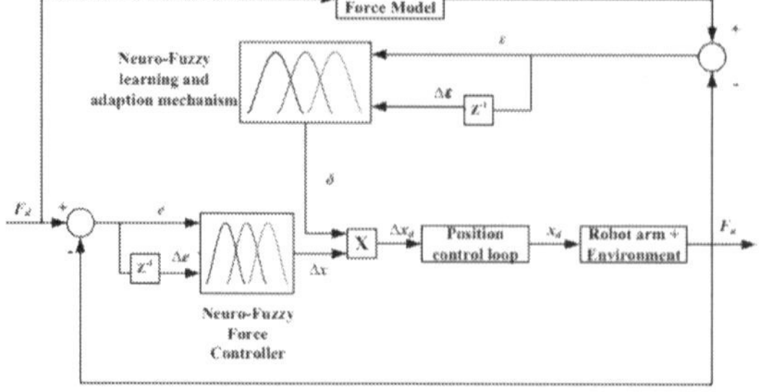

Figure 3. The Adaptive Neuro-Fuzzy Force Controller (ANFFC) structure

3.1.1 Reference Force Model

The reference model (equation 6) is chosen to generate the whole system desired behaviour in response to a force trajectory. In this research, the reference model is generated using the MATLAB® System Identification Toolbox™ software. Prior to that, a step force response test was conducted to identify the input-output data set of desired behaviour of force control system. The desired behaviour criteria are 0.2s rise time, overshoot of 10% with 5% steady-state error.

The input-output data was used to estimate a second-order linear system parameters and the model output is validated by using model-output plot where the identified model accurately fit the measured data of ~90%.

$$\frac{F_m}{F_d} = \frac{1}{s^2 + 7s + 100} \quad (6)$$

where damping ratio, ζ and natural frequency, ω are 0.35 and 10 respectively.

3.1.2 Design of Neuro-Fuzzy learning and adaption mechanism

The purpose of the NFLA is to learn and adapt to any changes in the environmental parameters and subsequently modify the NFFC output accordingly so that the overall system response is as close as possible to the desired reference model. The NFLA inputs are the difference between the reference model output and the measured contact force ε and the rate of error Δε. Its output is the parameter of factor adjustment δ which is used to modify the output of NFFC. The method is very useful in compensating any disturbance caused by environment position (moving object) which is a key requirement for the TMS robotic system. The Neuro–Fuzzy approach was applied to evaluate the degree of influence of the undesired response, and consequently, the adjustment algorithm generates a larger factor adjustment parameter for the larger undesired oscillation and unexpected overshoot/undershoot error. Similarly a smaller value is generated if the measured force is closed to the desired response.

The rule-base is created as follows;

If $\varepsilon(k)$ is ε_i and $\Delta\varepsilon(k)$ is $\Delta\varepsilon_i$, then output is δi

To begin with, the environment movement can be classified in terms of the sign of the model error ε:

1) **Negative** if the object moves towards the Stäubli robot

2) **Positive** if the object moves away from the Stäubli robot

Similarly the force controller behaviour can be defined by the rate of error sign as summarized in Table 1.

Table 1. A rule-base parameter determination

Δε sign	ε sign	Force controller response	Adjustment
Zero	Zero/Positive/ Negative	Satisfactory rate	No or small adjustment
Positive	Positive	Slow	Large adjustment
	Negative	Satisfactory rate	Small adjustment
Negative	Positive	Satisfactory rate	Small adjustment
	Negative	Slow	Large adjustment

4 Experimental Evaluation

This section describes how the force controller is able to accommodate a dynamic environment which is not stationary; namely a sequence of ramp inputs with/without parameter adjustment algorithms. Figure 4 illustrates the schematic diagram of ANNFC test set up.

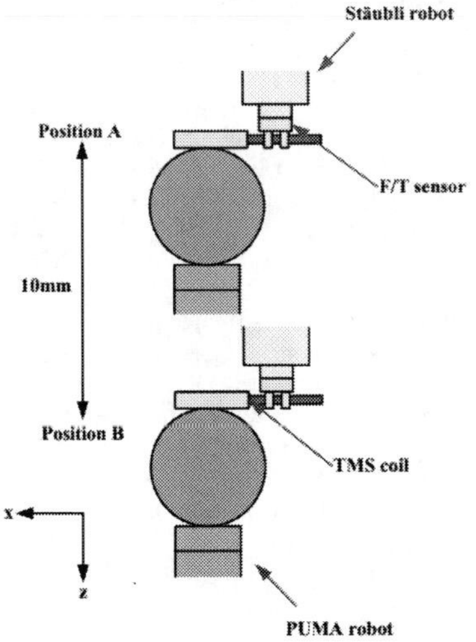

Figure 4. ANFFC experimental test set up

The Stäubli robot is first moved down at an approximate 10mm/sec until it makes contact with the ball. Once contact is made, the force control law is activated in which the robot was commanded to maintain 5N contact force in the z axis direction. Then, after approximately 7seconds the PUMA robot is moved about 10mm in towards direction (from position B to A) to the Stäubli robot with a speed approximately 2mm/sec. Another PUMA robot movement (from position A to B) was applied for the same travel distance. Note that, neither environment stiffness nor its position is known to the robot controller.

5 Results and Discussion

Figure 5 shows the results of force and position (along z axis) during this test. There was a significant difference between the conventional PI and the ANFFC controllers. It is apparent from the data in Figure 5, that the conventional PI force controller drives the manipulator into an unstable limit cycle. The problem can be attributed to the rapid changes in movement (either robot or environment) that leads to impulsive forces on the system. The form of the signal shows the PI controller was trying to compensate the impulsive force, but failed to do so. In this condition, it is desirable if the Stäubli robot can moves at velocity equal to the environment speed.

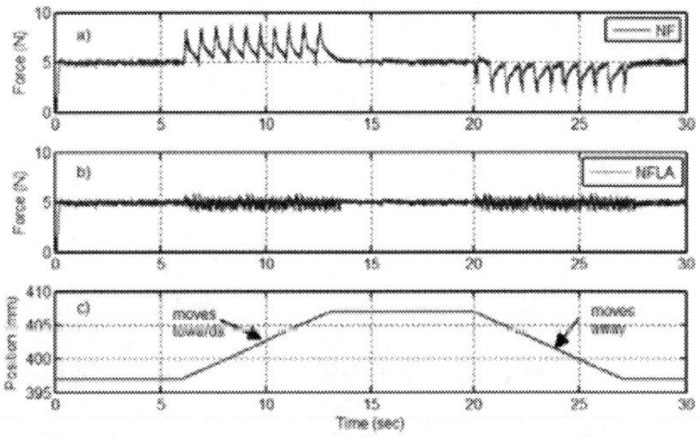

Figure 5. Test results of force response of (a) PI and (b) NFLA, and (c) Stäubli robot position

Consider the situation in which the environment moves towards the robot very rapidly, the large force error occurs before the robot controller can react to it due to small calculated incremental position. This results in the robot springing backward rapidly. Similar pattern is observed during away movement, if the environment movement speed is higher than resultant incremental position, the robot momentarily leaving contact with the environment. This inevitably results in repetitive bouncing action until the environment stop moving. Clearly, the dynamics of the force-controlled system are now considered to be extremely complex with many unknown independent factors. Thus the task of maintaining force response, as position changing with the conventional PI controller only is impossible.

6 Conclusion

The PI gain scheduling and Neuro-Fuzzy methods were successfully implemented in TMS robotic system and demonstrated improvement for the control system. Experimental results have shown the effectiveness of the proposed controller particularly for unknown/variable environment stiffness as well as in condition when the environment parameters and position are unknown and change substantially. In addition, the proposed method also employs a learning and adaptive mechanism to adjust a main loop Neuro-Fuzzy force controller online. The purpose of this mechanism is to learn the environment and adapt the controller accordingly. Adaption consists of comparing the model reference error and using the derivative of error as the input to the Neuro-Fuzzy algorithm. Any undesired response is immediately compensated with a suitable adjustment parameter according to the response behaviour.

Acknowledgement

This work was supported by Ministry of Education and Universiti Tun Hussein Onn Malaysia under Research Acculturation Grant Scheme Vot R032.

References

1. M. Hallett, 'Transcranial magnetic stimulation: A primer', Neuron, 55, (2), pp. 187-199. (2007)
2. L. Matthäus, A. Giese, D. Wertheimer, and A. Schweikard, 'Planning and analyzing robotized TMS using virtual reality', Studies in health technology and informatics, 119,pp. 373-378. (2006)
3. L. Matthäus, P. Trillenberg, and C. Bodensteiner, 'Robotized TMS for motion compensated navigated brain stimulation', International Journal of Computer Assisted Radiology and Surgery, 1, (SUPPL. 7), pp. 139-141. (2006)
4. M. Finke, T. Fadini, S. Kantelhardt, A. Giese, L. Matthaus, and A. Schweikard, 'Brain-Mapping using robotized TMS', International Conference of the IEEE Engineering in Medicine and Biology Society. Vancouver, CANADA, Aug 20-24. pp. 3929-3932. (2008)
5. J. Lancaster, S. Narayana, D. Wenzel, J. Luckemeyer, J. Roby, and P.Fox. 'Evaluation of an image-guided, robotically positioned transcranial magnetic stimulation system', Human Brain Mapping, 22, (4), pp. 329-340. (2004)
6. C. Lebosse, P. Renaud, B. Bayle, M. De Mathelin, O. Piccin, and J Foucher, 'A robotic system for automated image-guided Transcranial Magnetic Stimulation', IEEE/NIH Life Science Systems and Applications Workshop, LISA. Bethesda, MD, pp. 55-58. (2008)
7. BHannaford, W.S.Kim, S.H. Lee, and L Stark, 'Neurological control of head movements: Inverse modeling and electromyographic evidence', Mathematical Biosciences, 78, (2), pp. 159-178. (1986)
8. L. Ferman, H. Collewijn, T.C Jansen, and A. V. Van den Berg, 'Human gaze stability in the horizontal, vertical and torsional direction during voluntary head movements, evaluated with a three-dimensional scleral induction coil technique', Vision Research, 27, (5), pp. 811-828. (1987)
9. J. Winters, 'Biomechanical Modeling of the Human Head and Neck', in F.J.Richmond, B. W. P. a.(ed), Control of Head Movement. Oxford University Press. (1988)
10. A. Hagemann, K. Rohr, H.S. Stiehl, U. Spetzger, and J.M. Gilsbach, 'Biomechanical modeling of the human head for physically based, nonrigid image registration', Medical Imaging, IEEE Transactions on, 18, (10), pp. 875-884. (1999)
11. S. T. Moore, E. Hirasaki, T. Raphan, and B. Cohen, 'Instantaneous rotation axes during active head movements', Journal of Vestibular Research, 15, (2), pp. 73-80. (2005)
12. B. W. Peterson, and F. J. Richmond, 'Control of head movement'. Oxford University Press, USA. (1988)
13. R. Plamondon, 'A kinematic theory of rapid human movements - Part I. Movement representation and generation', Biological Cybernetics, 72, (4), pp. 295-307. (1995)
14. R. Plamondon, C. Feng, and A. Woch, 'A kinematic theory of rapid human movement. Part IV: A formal mathematical proof and new insights', Biological Cybernetics, 89, (2), pp. 126-138. (2003)
15. J. T. Cavanaugh, D. Goldvasser, C. A. McGibbon, and D. E. Krebs, 'Comparison of head- and body-velocity trajectories during locomotion among healthy and vestibulopathic subjects', Journal of Rehabilitation Research and Development, 42, (2), pp. 191-198. (2005)
16. R. O. Duda, C. Avendano, and V.R. Algazi, 'An adaptable ellipsoidal head model for the interaural time difference', IEEE International Conference onAcoustics, Speech, and Signal Processing, 15-19 Mar 1999. pp. 965-968 vol.2. (1999)
17. P. Stephen, 'Bodyspace: anthropometry, ergonomics, and the design of work'. Taylor & Francis. (1996)
18. R. Chan, D. K. Rogers, and D. I. McCloskey, 'Postural stability of the head in response to slowly imposed, small elastic loads', Neuroscience Letters, 214, (2–3), pp. 205-207. (1996)
19. J. De Schutter, 'A study of active compliant motion control methods for rigid manipulators based on a generic scheme', Robotics and Automation. Proceedings. 1987 IEEE International Conference on. Mar 1987. pp. 1060-1065. (1987)
20. J. De Schutter, and H. Van Brussel, 'Compliant Robot Motion I. A Formalism for specifying compliant motion tasks', International Journal of Robotics Research, 7, (4), pp. 3-17. (1988)
21. J. De Schutter, and H. Van Brussel, 'Compliant Robot Motion II. A Control Approach Based On External Control Loops', International Journal of Robotics Research, 7, (4), pp. 18-33. (1988)
22. E. Degoulange, and P. Dauchez, 'External force control of an industrial PUMA 560 robot', Journal of Robotic Systems, 11, (6), pp. 523-540. (1994)
23. J. S. R. Jang, 'ANFIS: adaptive-network-based fuzzy inference system', IEEE Transactions on Systems, Man and Cybernetics, 23, (3), pp. 665-685. (1993)
24. J. S. R. Jang, and C. T. Sun, 'Neuro-fuzzy modeling and control', Proceedings of the IEEE, 83, (3), pp. 378-406. (1995)
25. J. S. R. Jang, and C. T. Sun, and E. Mizutani, Neuro-fuzzy and soft computing: a computational approach to learning and machine intelligence. Prentice Hall. (1997)

Development of a Control and Vision Interface for an AR.Drone

Prasad Cheema*, Simon Luo* , and Peter Gibbens

School of AMME, The University of Sydney, 2006, Australia

* *Equal contributions*

Abstract. The AR.Drone is a remote controlled quadcopter which is low cost, and readily available for consumers. Therefore it represents a simple test-bed on which control and vision research may be conducted. However, interfacing with the AR.Drone can be a challenge for new researchers as the AR.Drone's application programming interface (API) is built on low-level, bit-wise, C instructions. Therefore, this paper will demonstrate the use of an additional layer of abstraction on the AR.Drone's API via the Robot Operating System (ROS). Using ROS, the construction of a high-level graphical user interface (GUI) will be demonstrated, with the explicit aim of assisting new researchers in developing simple control and vision algorithms to interface with the AR.Drone. The GUI, formally known as the Control and Vision Interface (CVI) is currently used to research and develop computer vision, simultaneous localisation and mapping (SLAM), and path planning algorithms by a number of postgraduate and undergraduate students at the school of Aeronautical, Mechanical, and Mechatronics Engineering (AMME) in The University of Sydney.

1 Introduction

Quadcopters have recently garnered a lot of interest in the aerospace community. This is predominantly due to the simple nature of their force calculations (only lift and thrust required), and the inexpensive nature of their construction which allows for the wide spread availability of this platform [1, 2]. This has ultimately lead to the production of the AR.Drone - a cheap quadcopter freely available for purchase by the public. As a result, the AR.Drone has received much interest to act as a research platform for robotic research in the fields of control and vision processing.

Traditional quadcopter research has involved using computer vision for autonomous flight in an unmapped environment, via simultaneous localization and mapping (SLAM) [3,4,5]. In the past, many researchers have relied on expensive aerial vehicles with computationally intensive sensors to accomplish SLAM [6]. However, recent research efforts have attempted to visually map the environment using low cost drones such as the Parrot AR.Drone by using efficient algorithms [7,8]. In addition to SLAM, a pertinent example may be seen by the Computer Vision Group at Tehcnische Universitat Munchen (TUM). This is most noted in their use of the AR.Drone as a platform for control algorithms and vision navigation research, as well as a tool for educational and outreach purposes. Their research mainly pertains to the use of monocular RGB-D cameras to implement SLAM for the purpose of scale aware autonomous navigation and dense visual odometry [9]. In their research the vision group at TUM have employed a PID controller successfully on the Parrot Drone [9]. Control techniques similar to those demonstrated by TUM have been shown in Altu & Taylor [10], as well as Bristeau et al. [11]. Furthermore, as researchers at Cornell University demonstrate, a particular class of drones known as micro aerial vehicles (MAV) have been able to be autonomously move in an indoor environment using algorithms based on feature detection. More specifically, canny edge detection has been used in conjunction with a probabilistic Hough transform to find the lines of a staircase. The MAV also uses the vanishing point (the furthest point until the image becomes a point) to navigate around a particular staircase [12].

Thus, there is a variety of research performed on MAV-type drones, but with the emergence of the AR.Drone a cheap and readily available research alternative to the traditional, more expensive drone models has been developed. It has been utilised in a variety of research, but due to the complexity of these research tasks it is often the case that the hand-over phase to new researchers is unreasonably long, and has a high associated learning curve. Thus, it is the purpose of this paper to try and develop a control and vision interface (CVI) in order to assist new researchers in control and navigation field, for interfacing with the AR.Drone.

2 Overview of the AR.Drone

The AR.Drone 2.0 is a remote controlled consumer quadcopter developed by Parrot. The body is made of a carbon fibre tube structure and high resistance plastic. A protection hull is made of Expanded Polypropylene (EPP) foam which is durable, light in weight, and recyclable. The propellers are powered by four brushless motors (35 000 rpm, 15W power), and on-board energy is provided by a Lithium polymer battery with a capacity of 1000 mAh. This allows a flight time of approximately 10 minutes. The internal processor of the Drone is a 468 MHz ARM9-processor and comes included with 128 Mb of RAM memory. The processor runs a minimalist version of Linux on board. Moreover an integrated 802.11g wireless card provides network connectivity with an external computing device via WiFi.

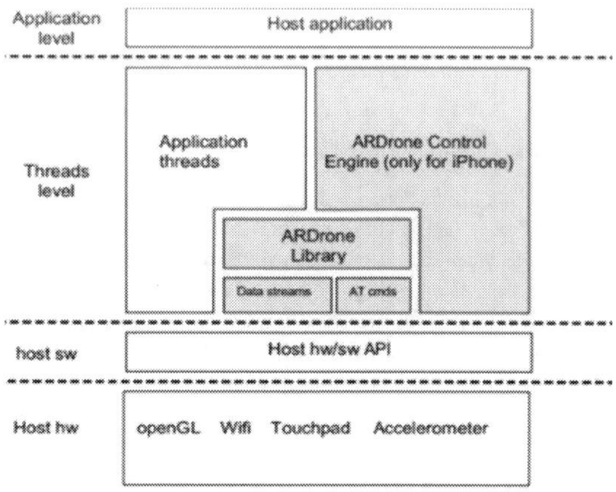

Figure 1. The layered architecture of the AR.Drone 2.0's SDK.

In regards to sensor technology and actuators, the AR.Drone 2.0 is equipped with a 3-axis accelerometer (BMA 150), and a 2-axis roll and pitch gyroscope (IDG-500) with an additional single axis gyroscope (Epson Toyocom XV-3500CB), which are of MEMS type. An ultra sound altimeter is also included for height calculations. Moreover the Drone has two cameras (forward and down facing). The forward facing camera covers a field of view 92°, has a resolution 640 x 360. According to literature it generally contains significant radial distortion that needs to be corrected through via calibration techniques. The second camera points down, covers a field of view of 64° and records at 60 fps. At any point in time only one video stream can be used to record video [9]. For movement the AR.Drone relies upon differential torque and thrust. By spinning each rotor with a different amount of angular velocity, the three basic motions of flight mechanics, yaw, pitch and roll can be achieved.

The Software Development Kit (SDK) that is used within AR.Drone is a low level C-based interface. Ultimately, it allows third party developers to create new software for the AR.Drone through the use of libraries which aid in vision processing and controllability of the drone. However, it must be noted that the SDK does not support the writing of one's own embedded software. That is to say, it offers no direct access to manipulate low level drone hardware. Figure 1 shows a layered architecture overview of the AR.Drone SDK. The yellow coloured portions of the diagram are those parts which the user has direct control over (for example the programmer will have the ability to read in data streams, as well as send data streams). Note that the AR.Drone Control Engine exists only specifically for the iOS devices. This allows the AR.Drone's Application Programming Interface (API) to be portable for such devices.

Table 1. Communication port streams used by the AR.Drone.

Port Number	Communication Type	Function
5554	UDP	Navdata stream sent from Drone to client
5555	TCP/IP	Image stream sent from Drone to client
5556	UDP	Controlling and configuring the Drone using commands
5559	TCP/IP	Transferring of critical system information

The SDK communicates through to an external PC via UDP and TCP connections for different purposes. These communication ports are outlined in Table 1.

Although very complete, the main drawback to the use of the SDK system is its low level nature. Instead of sending direct commands such as "move forward with a particular speed", it is necessary to send a complete bit sequence, with sequences using an unsigned 32 bit representation. This means that sending a complex command sequence to the drone can become very cumbersome. Moreover, working with the drone on a complex level, such in the case of SLAM, will involve an incredibly large amount of software development for even the simplest of SLAM algorithms to work. This is definitely counter-productive for the purpose of engineering research. If students are required to perform and learn large amounts of low level software programming it will slow down and perhaps even limit efforts in the use of drones for future research projects.

Hence, it was decided to implement the communication to the AR.Drone via the Robot Operating System (ROS) ARDrone autonomy driver. This driver provides a higher level abstraction to the SDK via the ROS system. It is still relatively low level in that it deals with Drone via C++, but the abstraction it provides allows the AR.Drone to move forward, for example, through a single velocity command rather than building up multiple bit streams. Of course understanding the manner in which ROS works to communicate with the drone does have an associated learning curve, however ROS is a well-documented and widely used environment for robotic research development that gaining familiarly for a ROS environment will be an ineffably easier task compared to learning to work with the even lower level (and poorly documented) SDK. A look into the ROS environment is explored in the following section.

3 Development of the Control and Vision Interface

This section of the paper explores the development of the control and vision interface (CVI) as a whole. It will first look at the C++ systems use build the CVI (through ROS and OpenCV). From there it shall explore how these components work together to generate a coherent computer system for researchers.

3.1. Understanding Robot Operating System

The Robot Operating System (ROS) is an open source platform used to aid communication with research robot. It is a `so called' operating system because it provides similar services to an OS. For example it includes multi-levelled hardware abstraction, low level device control, inter-process communications, and distributed package management. In addition to this, it is especially venerated within the robotics community due to its ample tools and libraries of algorithms, which allows for multi-layered communication with robots. There are several distinct advantages to the uses of ROS, which are briefly explored in the following sub-sections [13].

3.1.1 Distributed computation

ROS provides an abstraction system to help simplify distributed computations on robots. Distributed computation naturally arises throughout the use of multiple computers within the single robot, or when multiple robots attempt to cooperate and coordinate their efforts to achieve some overarching goal. Moreover, it is simpler to subdivide one's code into multiple blocks which can then co-operate together via inter-process communication (as is used in the development of the CVI's control and vision GUI).

3.1.2 A single, monolithic library

Although ROS contains some core libraries, more often than not the user will be required to download additional libraries to supplement their current library (or to use external code developed by other universities). The compilation and linkage of these libraries can cause significant issue if the user is not well acquainted with such aspects of software development. Moreover the linking of several external libraries will generally exponentially increase compilation time for the program.

3.2 Understanding OpenCV

In the construction of the CVI, the open computer vision (OpenCV) library was used for image processing. Briefly, OpenCV is an open source computer vision project supplied online by sourceforge. All libraries and systems of OpenCV are written in C and C++, and projects from OpenCV are able to run under Linux, Windows and the Mac OS X operating systems. Recently, there have been substantial efforts to port the OpenCV libraries to Python, Ruby, and MATLAB, although progression is generally slow, and support is very little. Due to OpenCV's dependency on the low level languages of C and C++, a strong focus of the OpenCV libraries is on computational efficiency and real time applications. It does this by looking at specific memory allocations on the processor level, and can in most cases utilize multi-core processing (a strong limitation of MATLAB for real time computer vision is that it can only use one core at most, unless an external parallel computing toolbox is installed). The main goal of OpenCV is to provide a simple-to-use computer vision infrastructure which will help people construct fairly sophisticated computer vision processes rapidly.

3.3 Development of Control and Vision Interface

Speaking from a high level perspective, the CVI development required a large degree of internal multi-threading. This is necessary because both of these applications (the control part - Control GUI, and the vision part - Vision GUI) require multiple infinite loops in order to function properly. For example a standard GUI thread will always require an infinite loop to physically show itself on the screen, and to perform checks on the buttons. The running of ROS would need its own infinite loop to check the global call back queue defined in the ROS API, and finally the "Video Thread" and "Video Processing Thread" both require their own threads. This is because the showing of the camera GUI showing, the video stream (and its processed version of itself) all require infinite loops in order to continue to compute image data and show it on screen. Thus the GUIs are in fact highly threaded to internalise computations and render them quick enough for a psuedo-real time environment to be used. A summary of this multi-threaded nature is shown in Figure 2.

In addition to multi-threaded code, the two GUIs (and a third process labelled as pseudo real-time graph which graphs state vector outputs) are able to communicate through Inter Process Communications (IPCs). However IPC is not the ideal option to communicate data between each GUI. If all the GUIs were internally multi-threaded into one another, then that would be most efficient. However threading all aspects into a single GUI would make the coding extremely difficult (lots of inherent multi-threading issues such as context switching and race conditions will arise), and therefore difficult to conceptually follow for new researchers in the field. Thus two (separate) processes were selected as the time optimal way to code up these applications.

Ultimately, the control GUI aims to facilitate control-based communication with the AR.Drone. It achieves this through the use of keyboard control, open access to navigation data, and button options to perform a pseudo real time plot of system states. In addition it provides the user with a link to a second GUI for video processing. The final, coded implementation of the control GUI is diagrammatically represented in Figure 3.

In Figure 3 it can be seen the user has the option to manually control the flight of the AR.Drone through simple keyboard controls as well as track the state variables defined in Table 2 in pseudo real-time.

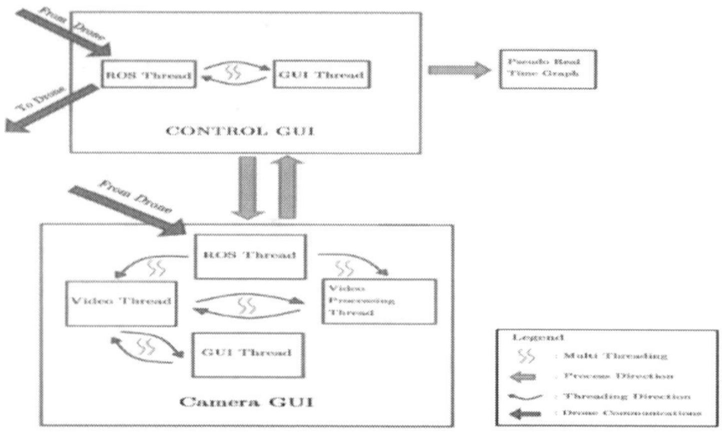

Figure 2. Demonstration of the Multi-Threaded Nature of the CVI back end.

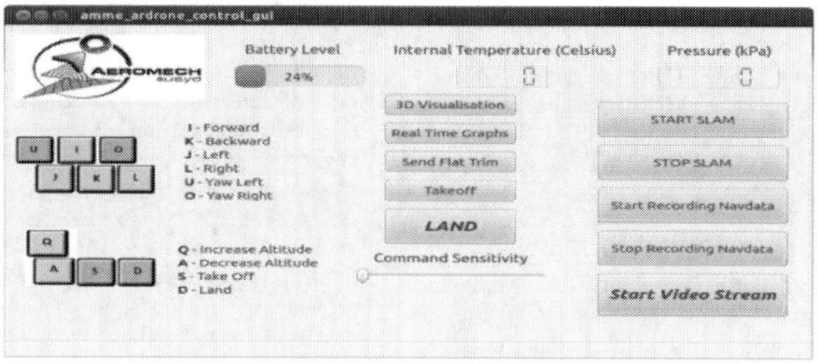

Figure 3. Final implementation of control GUI in its 'off' state.

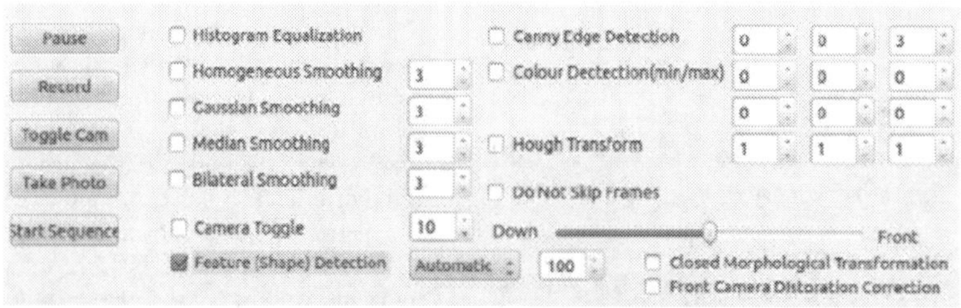

Figure 4. Final implementation of Camera GUI showcasing pseudo real-time edge detection.

Table 2. Summary of AR.Drone information and corresponding units extracted from the SDK via ROS.

Variable Recorded	Units
Timestamp	Second since Unix Epoch
Time between timestamps	Nanosecond
Rotation rates	Radians/second
Acceleration	Metres/second^2
Quaternions	Unitless
Euler Angles	Degrees
Altitude	Centimetres
Stream sequence number	Unitless

In addition to the ability to perform and track pseudo real-time control operations on the AR.Drone, a link can be directly set-up to the video steam GUI which can be initiated by clicking the *Start Video Stream* option shown in Figure 3. The completed state of the video GUI portion of the CVI interface is represented in Figure 4.

From Figure 4, it can be seen that the user is given the option to toggle between the front, and down facing cameras for the video stream. This is a consequence of the AR.Drone 2.0 only being able to show one video stream at one point in time via the ROS driver. Moreover, the GUI enables researchers to toggle between various image filtering techniques in pseudo real-time which is an invaluable experience for learning, as well as for research. In particular, Figure 4 shows an implementation of an in-house contour-based feature detection algorithm.

Although the CVI shown in Figures 3 and 4 seem extensive, obviously there are a multitude of other algorithms which can be implemented for various research purposes. Therefore, in order to accommodate the possibility of future research the code has been constructed in a modularised fashion, so that it essentially acts as API. Therefore future researchers need only use a few basic functions (the same ones used to create all the other buttons and features) within which new code can be written and dynamically integrated into the CVI. A few examples of the positive impacts of the CVI for future drone-based research is outlined in the following sections.

4 Application to Real Time Feature Detection

The framework for the Vision Node GUI has been adapted such that additional functions can be easily be added or removed into the video stream. The vision GUI has been extended to include real time feature detection. This is done by adding a feature detection algorithm to the video processing thread as shown in Figure 3. This algorithm includes a number of stages, first a Gaussian smoothing filter is applied to remove all the background noise. This is followed by Canny edge detection, which creates a binary image, which in turn places a pixel at regions with a large gradient change. This is then followed by the Suzuki algorithm [14] which is used to find closed contours in the image. These algorithms have been implemented in the CVI framework. The results are shown in sub figures 5 (a) and 5 (b).

Figure 5(a). The raw image from the AR.Drone.

Figure 5(b). The processed image with edge detection.

It can be seen in Figure 5 (b) that the feature detection algorithm converted the raw image into a binary image which consists of only the edges of the objects in the images. From there the aforementioned Suzuki algorithm detects a closed contour on the image. These closed contours are then matched against a database of features, in this case a triangle, rectangle and a pentagon. By implementing this multi-stage feature detection algorithm, it can be seen that the framework is robust enough to be able to support multi-stage algorithms.

Hence, the vision GUI has developed a pipeline which is highly effective in adding and removing computer vision algorithms. Even with algorithms which require high computational power such as Canny edge detection and contour detection algorithms. The framework is set up such that each critical section is isolated in its own thread allowing the software to run smoothly. When running in a computer with lower computational power, the frame work ensures that the video stream is running in real time by only retrieving the newest video frame.

5 Application to Monocular SLAM

Following the success of the feature detection, the CVI was extended to incorporate a SLAM framework. SLAM is a computational problem which requires user to develop a map of an unknown environment whilst keeping track of its location. The CVI has assisted in the development of SLAM by providing a simple API for mapping, control and video stream. The overall mapping and feature detection can be seen in Figure 6. The mapping feature uses the 3D visualisation tool Rviz provided by ROS. The feature detection algorithm is run in parallel with the SLAM algorithm in the vision node. The full SLAM algorithm has been implemented in the control node, with the features and landmarks detected through the vision GUI.

Figure 6. Map building and feature detection working for SLAM.

Ultimately, the effective use of multi-threading has allowed the control of the drone and SLAM algorithm to run simultaneously in the control node. Figure 6 shows that SLAM has been effectively implemented into the CVI framework. The mapping of the features during SLAM can be seen towards the bottom left corner of Figure 6, where the red dots indicate the features, green line showing the dead reckoning and the blue line indicating the prediction by SLAM. The implementation of SLAM has demonstrated an effective communication between the control and vision node and as well as the highly effective usage of using threads to ensure that all features in the CVI run in pseudo real-time.

6 Conclusion

In conclusion this paper has demonstrated the development and application of a control and vision interface (CVI) for use on a commercial quadcopter. Using the AR.Drone 2.0 as a simple, low-cost, readily available test-bed it is possible to research into a wide-range of control and vision-based applications. This paper has related how the use of a high level abstraction via OpenCV and ROS work intimately together to create a coherent and sufficiently abstracted interface to assist in user interaction with the AR.Drone. Ultimately this paper serves as an important introductory platform for any researchers new to the field of vision processing or controller design for quadcopter systems, as the basic structure and methodology behind interfacing with the quadcopter is explored here. Examples of the practicality of this approach have been demonstrated in the fields of real time feature detection and monocular SLAM.

Acknowledgements

The authors would like to thank the undergraduate students Andy Ho, and Trevor Wang, of The University of Sydney. They were able to assist in the appropriation of the CVI for the SLAM and feature detection applications explored in this paper.

References

1. M. Saska, T. Krajnik, J. Faigl, V. Vonasek, L. Preucil, IEEE IRS, 4808 – 4809 (2012)
2. M. Michael, The AR Drone LabVIEW Toolkit : A Software Framework for the Control of Low-Cost Quadrotor Aerial Robots (Tufts Univ. Ph. D. 2012)
3. S. Yang, S. Scherer, and A. Zell, IEEE ICRA, 5227 – 5232 (2014)
4. F. Sadat, S. A., K. Chutskoff, D. Jungic, J. Wawerla J., R. Vaughan, IEEE ICRA, 3870 – 3875 (2014)
5. G. H. Lee, F. Fraundorfer, and M. Pollefeys, IEEE ICRA, 3139 – 3144 (2011)
6. N. Karlsson, E. di Bernardo, J. Ostrowski, L. Goncalves, P. Pirjanian, and M. E. Munich, IEEE ICRA, (2005)
7. N. Dijkshroon, Simultaneous Localisation and Mapping with AR.Drone (Amsterdam Univ. Ph. D. 2012)
8. M. Blosch, S. Weiss, D. Scaramuzza, and R. Siegwart, IEEE ICRA, (2010)
9. J. Engel, J. Sturm, D. Cremers, J. of Rob. And Auto. Systems **62**, 1646 – 1656 (2014)
10. E. Altug, C. Taylor, IEEE ICM, 316 – 321 (2004)
11. P. J. Bristeau, F. Callou, D. Vissiere, and N. Petit, IFAC World Congress, 1477 – 1484 (2011)
12. C. Bills, J. Chen, and A. Saxena, IEEE ICRA, 5776 – 5783 (2011)
13. J. M. O'Kane, A Gentle Introduction to ROS, (Tech. Rep. 2014)
14. S. Suzuki, J. of Comp. Vis., Graphics, & Img. Proc. **30**, 32 – 46 (1985)

RHINO–an autonomous interactive surveillance robot for the needed ones: design and study case.

Michel Ochoa[1], Gabriel Aguiar[1] and Andrés Erazo[2]

[1]Universidad de las Fuerzas Armadas - ESPE, Department of Science of Energy and Mechanics, 170501Av. Gral. Rumiñahui, Ecuador
[2]Universidad de las Fuerzas Armadas - ESPE, Department of Electrical and Electronic Engineering, 170501 Av. Gral. Rumiñahui, Ecuador

Abstract. Healthcare services are in need of new tools and gadgets that could provide surveillance and people interaction of the needed ones with their surrounding environment. In order to address this problematic, an autonomous-interactive surveillance robot system is proposed in this research. RHINO (Robotic-Human Interface for the Needed Ones) was designed under particular normative and will provide commonly needed features in the healthcare field. One of the principal aspects of the device is its particular design which allows the patient to have some privacy under a surveilled environment and without causing any interference in their daily lives. Another main factor is the ability of notifying the care provider or the family of the patient with e-mail or text-sms notifications in case of abnormal activities; so that video access by provider or Skype calls by family relatives could be performed throughout RHINO.

1 Introduction

The surveillance of the needed ones and their interaction with people is a subtle topic for health care providers and families. The normative regulates the use of methods, technologies and applications in order to give best quality to these services. If we talk about surveillance, there are several different types of technology equipment that can be used for surveillance of the needed ones [1]. The most important are:

• Video-monitoring: it is possible to implement color audio-video communications in real-time for surveillance of the individual being monitored, but it is necessary that the care provider checks the video constantly.

• Health monitors: they are the size of a watch with a continuous pulse, skin, temperature and movement monitoring. Also, an alarm is transmitted in case of an abnormal activity.

• Fall detectors: there are wearable mechanisms that detect an impact greater than a certain threshold and send a notification.

In addition, there are hip protectors, pressure mats, door alerts, movement detectors, and others.

Limitations of these systems are the lack of autonomy and the need of the care provider to check constantly the equipment, and the patient. Robots can solve these problems, since they have the ability to interact with humans in their daily life. In a research in Osaka University, an English-speaker robot was used for interacting and teaching language to first and sixth level learners. The trial took place for the period of two weeks and the results proved that the children showed interest in interacting with the robot and their English-speaking abilities were increased during the trial time [2].

Robots are accepted entities in the medical fields and new methods to improve their functionality are developed; like ROS, a complete Open Source Operative System to control robots with all the tools and features a personal desktop could provide. Under this scope, the current research shows how an autonomous-interactive surveillance system could be incorporated into a robot for taking care of the needed ones. RHINO (Robotic-Human Interface for the Needed Ones) is a new technology, which follows the tendency of medical robots bringing services of care and interaction with patients.

2 Related Works

Nowadays, home accidents are a public health problem. Home accidents are responsible for the 5% of deaths in children aged 5 years or younger in Latin America [3]. However, the statistics don't show the actual number of people injured from home accidents, where we have between 200 and 1200 injuries per death. Ninety per cent of these accidents could have been prevented if there have been adult supervision [4].Home accidents are also a big problem between the elderly, researches indicate that a third of people aged 65 and older fall down each year, being half of them not just one but multiple falls. A lot of these falls are related with hospitalization, institutionalization, hip fractures, fear of falling, mortality and other consequences [5].

As a result of previous researches [2, 6], surveillance interactive robot has been proposed as a possible solution for helping, monitoring, and assisting people without the necessity of institutionalize them or to hire health care personnel. Companion robots can increase the positive mood in elderly living in nursing homes by decreasing levels of stress and loneliness, and increasing communication activity with others [6]. Currently, there are countless of companionship robots, four of the most relevant are PARO, PEARL, ICAT and AIBO [7]-[11](*Figure 1*).

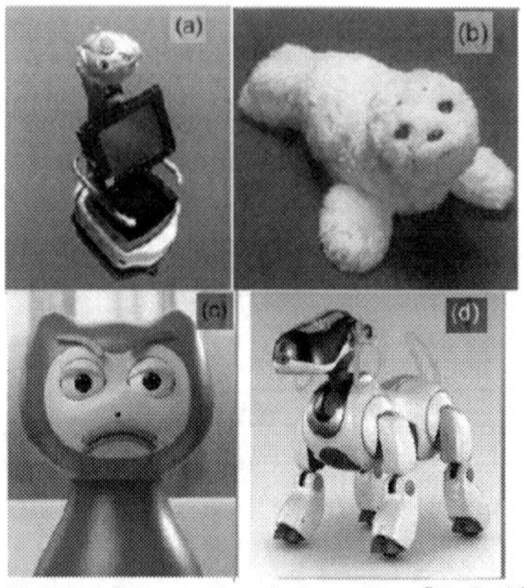

Figure 1. Companion and assisting robots [7, 8, 10, 11].

The most used configuration for smart robots, nomad scouts and assistive robots is the two-wheeled centered differential drive with third point of contact [12].

Some of the previous technologies described use tracking as their core system or component. Tracking gives to autonomous system the ability to interact with their environment. Human tracking algorithm [13]-[16] become the basis for human-robot interactions and their possible applications are uncountable [17].

Stationary results of the human tracking are reliable [18]-[20]. However, if the camera is tracking from a moving platform, those methods tend to fail. Choi et al indicates several approaches that aboard this problem [18]. In addition, the study proposes a general framework for tracking multiple people from a moving camera, performing data association to track people, and estimating the camera motion. Other alternatives, like adaptive HOG human tracker, upgraded by OpenCV libraries have proven their validity [18].

Choosing the camera for making the tracking cannot be taken lightly. A proper description of the use of stereo and monocular cameras for mapping in mobile platforms is developed at [21]. The use of RGB-D cameras such as the Microsoft Kinect, are good alternatives for capturing RGB images along with per-pixel depth information, mapping and recognition applications [21] [22]. Random errors of depth measurements are minimums when data is acquired within 1-3 meters of distance to the sensor [22].

Nowadays robots application had been improved by Internet standard communications. This communication techniques enable features as telecare and remote surveillance to a new level [23].

Most of autonomous surveillance systems are not capable of sending notifications, e-mails or text messages to the care providers. Such notification will inform anomalies detected in the surveillance. This tool will keep the care provider notified at all time, avoiding constant monitoring [24]. But a complete and limited autonomy from the surveillance system could raise ethical issues. In the work developed by A. Sharkey & N. Sharkey, they discuss main ethical concerns associated with robot care for the elderly. They explain that all autonomous surveillance systems should have their "off switch" [25].

The straightforward remote control applications used for accessing any system from a distance, not only reduce the autonomy of the system, they also allow the care provider to control the mobile robot and access the information collected from anywhere through the internet. The benefits of remote control application are that they provide a secure connection and environment, plus a register of all the times care providers have accessed to the surveillance system.

The work of care providers is very emotional and physical demanding. It can even affect the family members causing them burden and stress. The investigation held by J. Czaja & P. Rubert describe how technology can be used to provide support to care providers, increasing the quality of life for both, care providers and care recipients[17].

Technological developments focused on the care and surveillance of needed people, must consider fundamental care standards. Fundamental standards as person-centered care, dignity and respect, consent, safety, among others; should be of our concern when we perform innovation of any kind [26].

There are some organizations that monitors, review and regulate care services to make sure they meet this fundamental standard of quality and safety [27]. The independent regulator of health and social care in England and worldwide recognized, The Care Quality Commission (CQC), establishes a normative for using surveillance. This normative gives some guidelines for any technological developments in this field. In addition, there are some legal obligations under the Human Rights Act of 1998 and the Regulation of Investigatory Powers Act of 2000 (RIPA) that providers should be aware of [28].

The CQC advises that covert surveillance must be limited both in time and in purpose to remain lawful. Also, the provider must assure that only authorized people can access the information collected. The use of an autonomous and independent surveillance system, capable of detecting anomalies in the behavior of the individual being monitored, ensures the legal right of privacy of the individual. Furthermore, these systems must secure the information collected and guarantee that only authorized people access to this service. Finally, a record of when and why recorded information has been accessed could be generated by these systems. All of this will ensure the correct use of the critical information providers have to deal with.

3 RHINO Prototype

RHINO is an autonomous interactive surveillance system that allows the care provider to monitor the subject in real time(*Figure 2*). RHINO is a mobile robot that has two operating modes:

• Autonomous mode: the robot supervises the person being monitored using a computer vision algorithm and a pyroelectric infrared sensor. If an anomaly occurs, the robot could send a notification (mail or sms) to the care provider (*Figure* 3d).

• Control mode: the care provider could call the person been monitored using video calls and he could be able to control the robot from any place with internet access, allowing an interaction between them (*Figure* 3a and 3c).

Figure 2.RHINO prototype

The robot was armed on a two-wheeled centered differential drive with a third point of contact. The platform was armed using commercial elements, because is the first prototype of the RHINO models.

The robot uses two DC motors that are controlled by a microcontroller board that is connected to the master controller. The master node is an Intel NUC with Ubuntu 14 for its operating system. The motors are connected to an H-bridge for switching their spinning direction and each motor has an incremental encoder to control the displacement of each wheel during the movement of the robot.

The algorithm was implemented in ROS Indigo using node programming and communication by messages. The ROS node connects to the microcontroller board, and using serial communication, the master node sends a message telling the board that the robot should perform a movement or an action. For the control mode, a ROS node was used. This node allows the care provider to use a key of the keyboard storing it as a char variable and sent via serial port to the microcontroller board using the serial communication node of ROS (Rosserial_python). The variable will be validated in the microcontroller board, and depending of the validation, the robot will move forward, backward, right, left or stop(*Figure* 3a).

The autonomy of the RHINO's surveillance system is based on the detection and tracking of the individual being monitored (*Figure* 3b). Without this, RHINO will not be capable of monitoring the individual. In the previous section we discuss the advantages of using an RGB-D camera. RHINO is enhanced with a Microsoft Kinect camera and a human tracking method for mobile platforms. It also counts with an adaptive HOG human tracker, coded using OpenCV features.

Tracking will provide the position of the individual in an x, y and z system regarding the camera position, and therefore regarding the position of the robot. A secure zone is established to reduce the maximum possible intrusion of RHINO to the individual being monitored. If the individual is getting away from the robot around 2 meters, the robot will follow the subject; if not, RHINO will stay still monitoring the individual. A python subscriber receives the position data, process it and control the motor movement through serial communication.

Notifications to the care provider of an abnormal parameter, such as the lack of movement, is processed by a PIR sensor algorithm. The PIR sensor would be constantly sending high pulses while there is movement, and if the high pulses are not received by the microcontroller board, a flag via serial communication will be sent to a ROS node. After receiving the flag, the ROS node would subscribe to another node call pushing box which is connected to an internet cloud that can send notifications based on API calls. This cloud was previously configured with the service needed (e-mail, notify my android, prowl, etc.) and the scenario (sending a notification with the abnormal parameter, "lack of movement").

This autonomy is the key to provide the individual being monitored with a more truly privacy, without the need of a care provider always looking after the subject. And even more, all the connection activity is reported in a .txt document automatically. This is useful and important because this register is a key point of the care quality surveillance normative.

If the care provider or the family of the individual being monitored wants to communicate with the subject, they can use Skype to start a video call and RHINO will automatically answer it. To ensure privacy, it will be needed a password to establish the communication.

Skype's text, voice and video make it simple to connect people and stay in touch with them wherever you are. Boulos et al in their study of sociable technologies in health and health care education remarks the potential of Skype, and mention these technologies should not be underestimated. Features like automatic answering and passcode for accessing calls provide a valuable and secure use for care surveillance application. Also, the study affirms: *"Sociable interactive technologies have the potential to augment patients' and older peoples' sense of belonging to a supportive kindred community, thereby reducing their social isolation"* [29].

TeamViewer is the remote desktop that will control the autonomy of RHINO. The care provider can access RHINO through this application in his cell phone, tablet or PC. This provides the ability to turn off or on RHINO and access data recorded like PIR events and video. This is also the interface to control the robot from anywhere through the internet. For example, if the care provider calls the individual using Skype, he can control the movement of RHINO to interact with the individual in a more unique way.

DOI: 10.1051/matecconf/20165607003

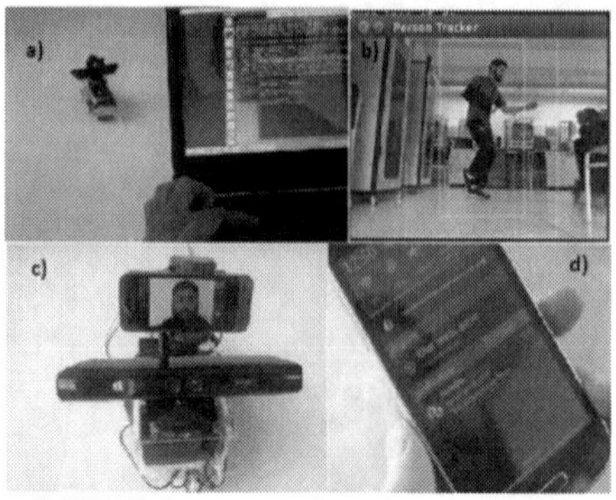

Figure 3:RHINO features; a) Control mode, b) Autonomous mode, c) RHINO interaction and d) Cell phone notification.(*https://www.youtube.com/channel/UC3qG3bW-FBaHf7Ol_UI4X3g*)

4 Results and Discussion

A Proof of Concept (POC) was conducted with potential users. These users have been people with someone that need surveillance and have experience with traditional surveillance systems, like doctors, medical interns and family of needed ones. The purpose of this POC was to identify how care provider's responds to this autonomous surveillance system. We describe the conceptual features to our potential users; and using video tools, we showed them the results obtained. A survey was conducted at the end of the POC (*Figure* 4).

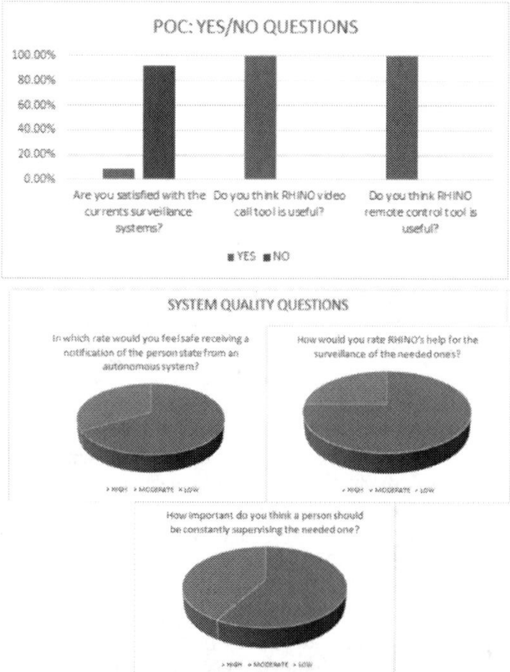

Figure 4:POC: Yes/No Questions and System Quality Questions.

The results obtained from this POC shows us that 92.31% of the respondents are not satisfied with the current and traditional surveillance system. It is considered that new systems should provide more liberty to care providers and individuals being monitored, without reducing the control and care. The results also revealed that 53.85% of the research participants considered really important a permanent surveillance of patients. However and in contrast, 46.15% of them presented a tendency to avoid permanent surveillance as an important factor. This second group could be related to the problematic of privacy and intrusive factor that our system deal with to provide a better service. Another important value to take into account is that more than 61.54% of the polled consider useful the alert notifies RHINO provided. In the same way and with almost the same percentage, the 69.23% of the participants approved and rated RHINO as a trusty autonomous surveillance system. In addition, RHINO was accepted as a more personalized way of surveillance for the loved ones. Two aspects of the system were checked by the respondents under this scope: video calls and teleoperation mode; and both tools were 100% accepted.

Under the same perspective, measures taken to ensure the security of the system, like the passcode for the video calls and the password for the remote control of RHINO, fulfill their purpose. The potential users considered that this aspect is important and was very well implemented.

The tracking algorithm has to deal with noises and occlusion from the Kinect. The uses of filters, like the Kalman filter, have reported good results; reducing noise and estimating with good accuracy human's motion detection in mobile robots [30] [31].Further investigations are needed to incorporate advanced features for medical and health care. One of the future improvement includes the implementation of a health monitor bracelet that will measure pulse, temperature and other important parameters and would transfer these data to the master node; which in turn will process this information and send a notification if there is any abnormality [1].

The approval RHINO received from care providers during the proof of concept demonstrated the growth of the acceptance of robots at healthcare services and facilities. This approval motivates us in developing new systems for care and surveillance for the needed ones. However, innovations not only should focus on the control of the patient and we expect that further versions of RHINO help patients regain their confidence in order to help them in a process of reintegration to the society. This will only be achieved throughout a more robust human robot interaction system, that could establish a truly bound of trust with the patient.

There has always been a constant debate between privacy rights and the need of complete surveillance [28]. Individuals that suffer from dementia or complete body paralysis must be under surveillance all the time. The only way these individuals could regain some of their privacy rights is changing the traditional surveillance methods to those who are autonomous. Surveillance made by a person is always intrusive; on the other hand, systems that doesn't require the control of a person to fulfill their main objective, give privacy and control care to individuals under surveillance [28].

4 Conclusions and Future Works

RHINO is the first prototype of RHINO models, which is why in our research we found some benefits and also some setbacks. ROS worked perfectly like a master node, allowing the care provider to control all nodes in the robot, and allowing the communications between them.

Also, the human tracking system, based on the Microsoft Kinect and OpenCV features, was able to complete the objective established for a noninvasive surveillance. Furthermore, the Node in charge of the lack of movement worked as expected, sending the right notification as needed. In parallel, the remote control provided by TeamViewer offers an exceptional communication between the care provider and the robot. In resume, this first prototype of the RHINO models proved that it is possible to develop surveillance systems that respect fundamental care standards and it is capable of performing care services.

Finally, the autonomy of RHINO accomplishes his main objective, reducing privacy invasion. POC results corroborate that this project respects the right of privacy of the individual being monitored to a greater degree than conventional methods. RHINO is the tool of the future, a health care monitoring system which respect your privacy and helps you when you are in need of your doctor, of your family or just of the attention of any of them.

References

1. F. G. Miskelly, *Assistive technology in elderly care*, Age Ageing, **30**, pp.455–458 (2001)
2. C. Trial, T. Hiranoet al, *Interactive Robots as Social Partners and Peer Tutors for childrens: A field trial*,HCI, **19**, pp.61–84 (2004)
3. A. Benavides-Lara & M. Vargas-Salas, *Original Mortality by accidental causes in children of 5 years and younger at the "Nacional de Niños" Hospital*, Acta Med. Costarric., **50**, pp.22–28 (2008)
4. Y. Benguigui, J.C Bossio& H. Fernández, *Operations Research on Integrated Management to Prevalent Childhood Illness*, OPS/OMS, pp.437-532 (2001)
5. C. a Reyes-Ortiz, S. Al Snih, & K. S. Markides, *Falls among elderly persons in Latin America and the Caribbean and among elderly Mexican-Americans*, RPSP/PAJPH, **17**, pp.362–9 (2005)
6. J. Broekens, M. Heerink, & H. Rosendal, *Assistive social robots in elderly care: a review*, Gerontechnology, **8**, pp.94-103(2009)
7. M. Pollack et al, *Pearl: A mobile robotic assistant for the elderly*, AAAI, pp.85-91, (2002)
8. K. Wada et al,*Effects of robot assisted activity to elderly people who stay at a health service facility for the aged*,IROS,**3**, pp.2847-2852 (2003)
9. K. Wadaet al, *Effects of robot-assisted activity for elderly people and nurses at a day service center*, IEEE, **92**, pp.1780–1788 (2004)
10. A. Van Breemen, X. Yan & B. Meerbeek, *iCat: an animated user-interface robot with personality*, AAMAS, pp.143-144 (2005)
11. M. Fujita, *AIBO: Toward the era of digital creatures*, IJRR, **20**, pp.781-794 (2001)
12. R. Siegwart& I. R. Nourbakhsh, *Introduction to Autonomous Mobile Robots*, p. 17, 30, 35 (2006)
13. O. Tuzel, F. Porikli, & P. Meer, *Human detection via classification on riemannian manifolds*, CVPR'07, pp.1-8 (2007)
14. V. Ferrari, M. Marin-Jimenez, & A. Zisserman, *Progressive search space reduction for human pose estimation*, CVPR'08, pp.1-8 (2008)
15. P. Felzenszwalb, R. Girshick et al, *Object detection with discriminatively trained part based models*, PAMI, 32, pp.1627-1645 (2010)
16. N. Dalal& B. Triggs, *Histograms of oriented gradients for human detection*, CVPR, **1**, pp. 886-893 (2005)
17. S. Czaja& M. Rubert, *Telecommunications technology as an aid to family caregivers of persons with dementia*, Psychosomatic medicine, **64**, pp.469-476 (2002)
18. W. Choi, C. Pantofaru& S. Savarese, *A general framework for tracking multiple people from a moving camera*, PAMI, **35**, pp.1577–1591 (2012)
19. M. D. Breitenstein, F. Reichlin et al, *Robust tracking-by-detection using a detector confidence particle filter*, ICCV, pp.1515-1522 (2009)
20. H. B. Shitrit, J. Berclaz et al,*Tracking multiple people under global appearance constraints* ICCV, pp.137-144 (2011)
21. P. Henry, M. Krainin et al, *RGB-D mapping: Using Kinect-style depth cameras for dense 3D modelling of indoor environments*, IJRR, **31**, pp.647–663 (2012)
22. K. Khoshelham& S. Elberink, *Accuracy and resolution of kinect depth data for indoor mapping applications*, Sensors, **12**, pp.1437-1454 (2012)
23. F. Grossi, V. Bianchi et al, *Internet-based home monitoring and control*, AAATE, pp.309-313 (2009)
24. O. Fuerst& T. Fuerst, U.S. Patent Application No. 11/198,588 (2005)
25. A. Sharkey& N. Sharkey, *Granny and the robots: ethical issues in robot care for the elderly*, Ethics and Information Technology, **14**, pp.27-40 (2012)
26. CQC,*Fundamental Standers*, http://www.cqc.org.uk/content/fundamental-standards, (accessed on December 2015).
27. CQC, About Us, http://www.cqc.org.uk/content/who-we-are, (accessed on December 2015).
28. CQC, *Using surveillance: Information for providers of health and social care on using surveillance to monitor services* (2015).
29. M. Boulos, N. Maged& S. Wheeler, *The emerging Web 2.0 social software: an enabling suite of sociable technologies in health and health care education*. HILJ, **24**, pp.2-23 (2007).
30. E. Machida, M. Cao et al, *Human motion tracking of mobile robot with Kinect 3D sensor*, SICE, pp.2207-2211 (2012)
31. S. Huang & J. Hong, *Moving object tracking system based on camshift and Kalman filter*,CECNet, pp.1423-1426 (2011)

MATEC Web of Conferences 56, 08001 (2016) DOI: 10.1051/matecconf/20165608001
ICCAE 2016

Modeling and Simulation for Nonlinear Pressurized Water Reactor Cores Using Gap Metric and Fuzzy Model with Transfer Function

Gang Li[1,2,3,4,5], Bin Liang[1,2,3,4,5], Xueqian Wang[1,2,3,4,5], Xiu Li[1,2,3,4,5] and Rongjian Liang[1,2,3,4,5]

[1]Tsinghua University, National Laboratory for Information Science and Technology, 100084 Beijing, China
[2]Tsinghua University, Graduate School at Shenzhen, 518055 Shenzhen, China
[3]Tsinghua University, Department of Automation, 100084 Beijing, China
[4]Shenzhen Key Lab of Space Robotic Technology and Telescience, 518055 Shenzhen, China
[5]Shenzhen Engineering Lab of Precision Geometry Measurement Technology, 518055 Shenzhen, China

Abstract. This investigation is to deal with the modeling issue for nonlinear pressurized water reactor cores. Nonlinearity measure with gap metric and T-S fuzzy modeling are exploited to build the fuzzy model of a nonlinear core which approximates this nonlinear core model. The gap metric of the core is proposed to quantify the core nonlinearity. The curve in the whole range of core power level that is solved adopting the gap metric is the core nonlinearity measure. In terms of the measure, six linearized core models at six power levels are selected as local models of the nonlinear core model. Based on the local models and the introduction of the triangle membership function, the core fuzzy model is achieved. The core fuzzy model and nonlinear core model are simulated. Simulation results show that the core fuzzy model can approximate and substitute the nonlinear core model.

1 Introduction

Nuclear energy as a kind of clean energy provides an incentive to sustainable development of nuclear power plants (NPPs) for electricity generation. Meanwhile, of research fields for NNPs, the modeling study on plants of NPPs such as reactor cores is always popular. The pressurized water reactor (PWR) cores in NPPs possess nonlinear characteristics, and parameters of them are varying at different operating conditions. Hence, cores are essentially time-varying complex systems so that modeling their models is difficult. It is necessary and meaningful to research the modeling issue of nonlinear cores.

Many researchers have utilized the point reactor core modeling method to construct PWR core models at the full power level. Kerlin et al.[1] modelled the PWR core of the H. B. Robinson NPP at the full power level which was also verified to be correct. PWR cores at the full power level were modeled by Edwards et al.[2-5]to design core power control systems.

However, such modeling for cores based on one power level cannot represent the dynamics of nonlinear core within the global ranges of power level. Stimulated by the considerations, the work in this paper is to model a nonlinear PWR core within the global ranges of power level by the use of gap metric to quantify the core nonlinearity measure and the T-S fuzzy principle. Finally, the T-S fuzzy model and nonlinear model of the core are simulated, and conclusions are drawn.

2 PWR core model

The point reactor core modeling as a traditional modelling method is adopted to model a PWR core, which is a lumped parameter method. For this modeling, a core is regarded as one point without any space profile, and parameters of the core only vary with time and have nothing to do with space positions.

Table 1. Main model parameters.

Parameter	Name
P_r	core power level
c_{ri}	ith group normalized precursor concentration
c_r	normalized precursor concentration
T_f	fuel average temperature
T_l	coolant inlet temperature
T_e	coolant outlet temperature
δrod	position variation of the control rod (fraction of core length)
δ	deviation of a parameter from initial steady-state value

According to the point reactor core modeling [1-3], the nonlinear PWR core is modelled adopting the point kinetics equations with six groups of delayed neutrons and reactivity feedbacks due to control rod movement and variations in fuel temperature and coolant temperature. Main model parameters are given in Table 1. The nonlinear core model is showed as Eqs.(1)-(6).

$$\frac{dP_r}{dt}=\frac{\rho-\beta}{\Lambda}P_r+\sum_{i=1}^{g}\frac{\beta_i c_{ri}}{\Lambda} \quad (1)$$

$$\frac{dc_{ri}}{dt}=\lambda_i n_r-\lambda_i c_{ri}, i=1,2,\ldots,g \quad (2)$$

$$\frac{dT_f}{dt}=\frac{f_f P_0}{\mu_f}P_r-\frac{\Omega}{\mu_f}T_f+\frac{\Omega}{2\mu_f}T_i+\frac{\Omega}{2\mu_f}T_e \quad (3)$$

$$\frac{dT_e}{dt}=\frac{(1-f_f)P_0}{\mu_c}n_r+\frac{\Omega}{\mu_c}T_f+\frac{2M-\Omega}{2\mu_c}T_i-\frac{2M+\Omega}{2\mu_c}T_e \quad (4)$$

$$\rho=\rho_{rod}+\alpha_f(T_f-T_{f0})+\frac{\alpha_c}{2}(T_i-T_{i0})+\frac{\alpha_c}{2}(T_e-T_{e0}) \quad (5)$$

$$\delta\rho_{rod}=\alpha_{rod}\delta rod \quad (6)$$

The small perturbation linearization methodology is utilized to linearize this nonlinear core model, and the linearized core model is expressed by Eqs.(7)-(11).

$$\frac{d\delta P_r}{dt}=-\frac{\beta}{\Lambda}\delta P_r+\sum_{i=1}^{g}\frac{\beta_i}{\Lambda}\delta c_{r_i}+\frac{P_{r0}}{\Lambda}\delta\rho \quad (7)$$

$$\frac{d\delta c_{ri}}{dt}=\lambda_i\delta n_r-\lambda_i\delta c_{ri}, i=1,\ldots,g \quad (8)$$

$$\frac{d\delta T_f}{dt}=\frac{f_f P_0}{\mu_f}\delta n_r-\frac{\Omega}{\mu_f}\delta T_f+\frac{\Omega}{2\mu_f}\delta T_i+\frac{\Omega}{2\mu_f}\delta T_e \quad (9)$$

$$\frac{d\delta T_e}{dt}=\frac{(1-f_f)P_0}{\mu_c}\delta n_r+\frac{\Omega}{\mu_c}\delta T_f+\frac{2M-\Omega}{2\mu_c}\delta T_i-\frac{2M+\Omega}{2\mu_c}\delta T_e \quad (10)$$

$$\delta\rho=\alpha_{rod}\delta rod+\alpha_f\delta T_f+\frac{\alpha_c}{2}\delta T_i+\frac{\alpha_c}{2}\delta T_e \quad (11)$$

One group delayed neutron model is utilized and the coolant inlet temperature is treated as a constant [2-4]. The transfer function and the state equation of the core are separately calculated and expressed by

$$G=\frac{\delta P_r}{\delta rod}=\frac{\sum_{i=0}^{3}a_i s^i}{\sum_{i=0}^{4}b_i s^i} \quad (12)$$

$$\begin{cases}\dot{\mathbf{x}}=A\mathbf{x}+Bu\\ y=C\mathbf{x}+Du\end{cases} \quad (13)$$

where u=δrod-the input; y=δP_r-the output; a_i (i=0,1,2,3)-numerator coefficients; b_i (i=0,1,2,3,4)-denominator coefficients; s-laplace operator; $\mathbf{x}=[x_1, x_2, x_3, x_4]^T=[\delta P_r, \delta c_r, \delta T_f, \delta T_e]^T$-the state matrix; A-the $R^{4\times4}$ system matrix; B-the $R^{4\times1}$ input matrix; C-the $R^{1\times4}$ output matrix; D-the zero matrix.

A transfer function of the core at a power level 10*i% is denoted by $G_{10*i\%}$ (i=1, . . . ,10). These transfer functions are calculated by using parameters from Ref. [6], in which the total primary heat output is 2200 MW, the primary coolant inlet temperature is 285°C, the primary coolant outlet temperature is 317°C, the primary coolant average pressure is 15.5 MPa, and the primary coolant mass flow is 12861.1 kg/s.

3 Nonlinearity measure with gap metric for core

The nonlinearity measure on the basis of the gap metric for the core is proposed to quantify the core nonlinearity. One group of linearized core models can be selected to approximate the nonlinear core in terms of distributing situations of the core nonlinearity measure in the while range of power level, and each linear model is regarded as a local model of the core.

According to the work [7, 8], the gap metric of the core is developed as follows to solve the core nonlinearity.

Let O_i (i=1,2) be two closed linear operators in a Hilbert space H_s. $H_s \times H_s$ is also a Hilbert space with inner product derived from H_s. The graph of O_i is denoted by $g_r(O_i)$. As O_i is linear, $g_r(O_i)$ is a subspace of the product Hilbert space $H_s \times H_s$.

Definition 3.1. The gap between two closed linear operators O_1 and O_2 is defined by the gap between $g_r(O_1)$ and $g_r(O_2)$:

$$\Delta(O_1,O_2)=\Delta(g_r(O_1),g_r(O_2)) \quad (14)$$

where $\Delta(\cdot)$is a metric.

Linear systems such as G of the core are viewed as mappings in a Hilbert space. Hence, G is treated as an operator.

Lemma 3.1. Frequency response systems are closed [7].

According to Lemma 3.1, G as a frequency response system is a closed linear operator in H_s. The gap metric in Definition 3.1 can be used to qualify the gap between two linearized systems of the core. So Eq.(15) is obtained by

$$\Delta(G_{10*i\%},G_{10*j\%})=\Delta(g_r(G_{10*i\%}),g_r(G_{10*j\%})) \quad (15)$$

where $\Delta(\cdot)\in[0\ 1]$; $i\neq j$; i,j=1,…,10.

Let H_∞ and H_2 be the standard Hardy spaces of functions, and let RH_∞ be the subspace of real rational functions in H_∞. $G_{10*i\%}$ is a 1×1 rational transfer matrix and has the normalized right coprime factorization as follows.

$$G_{10*i\%}=N_1M_1^{-1},\ G_{10*j\%}=N_2M_2^{-1} \quad (16)$$

with $M_1^*M_1+N_1^*N_1$=I and $M_2^*M_2+N_2^*N_2$=I.
where for i=1 or 2, M_i and N_i belong to RH_∞, $M_i(s)^*=M_i(-s)^T$, $N_i(s)^*=N_i(-s)^T$.

$g_r(G_{10*i\%})$ or $g_r(G_{10*j\%})$ is a closed subspace of $H_2 \times H_2$ that consists of all pairs (δrod, δP_r) such that $\delta P_r=G_{10*i\%}*\delta rod$ or $\delta P_r=G_{10*j\%}*\delta rod$. They are presented by

$$g_r(G_{10*i\%})=\begin{bmatrix}M_1\\N_1\end{bmatrix}H_2,\ g_r(G_{10*j\%})=\begin{bmatrix}M_2\\N_2\end{bmatrix}H_2 \quad (17)$$

And

$$\Delta(g_r(G_{10*i\%}),g_r(G_{10*j\%}))=\left\|\Pi_{g_r(G_{10*i\%})}-\Pi_{g_r(G_{10*j\%})}\right\| \quad (18)$$

where $\Pi_{(\cdot)}$ is the orthogonal projection onto ($\cdot$).

Lemma 3.2. The gap between $G_{10*i\%}$ and $G_{10*j\%}$ is calculated by

$$\begin{aligned}&\Delta(G_{10*i\%},G_{10*j\%})\\&=\left\|\Pi_{g_r(G_{10*i\%})}-\Pi_{g_r(G_{10*j\%})}\right\|\\&=\max\{\vec{\Delta}(G_{10*i\%},G_{10*j\%}),\vec{\Delta}(G_{10*j\%},G_{10*i\%})\}\\&=\max\{\inf_{q\in H_\infty}\left\|\begin{bmatrix}M_1\\N_1\end{bmatrix}-\begin{bmatrix}M_2\\N_2\end{bmatrix}q\right\|_\infty,\inf_{q\in H_\infty}\left\|\begin{bmatrix}M_2\\N_2\end{bmatrix}-\begin{bmatrix}M_1\\N_1\end{bmatrix}q\right\|_\infty\}\end{aligned} \quad (19)$$

where $\vec{\Delta}(\cdot)$ denotes a directed gap.

4 Fuzzy modeling

The T-S fuzzy method [9] can be used to handle modeling issues of nonlinear plants such as the core at varying operating conditions. According to this modeling, the power level operating range of the nonlinear core is partitioned into separate subregions, local dynamics in each region are represented by one linear model as a local model of this core; fuzzy rules are proposed adopting membership functions that are utilized to develop fuzzy sets of which the intersection of two ones is nonempty, the T-S fuzzy model that is the overall model to approximate the nonlinear core is then achieved via "blending" of all linear local models based on fuzzy rules.

4.1 Core local models based on nonlinearity measure

$G_{10\%}$ is chosen as a referenced model and a gap metric between $G_{10\%}$ and G_{Pr} for $P_r\in[0.1\ 1]$ is calculated by means of Section 3. This calculated gap metric is the core nonlinearity measure between 10% and P_r.

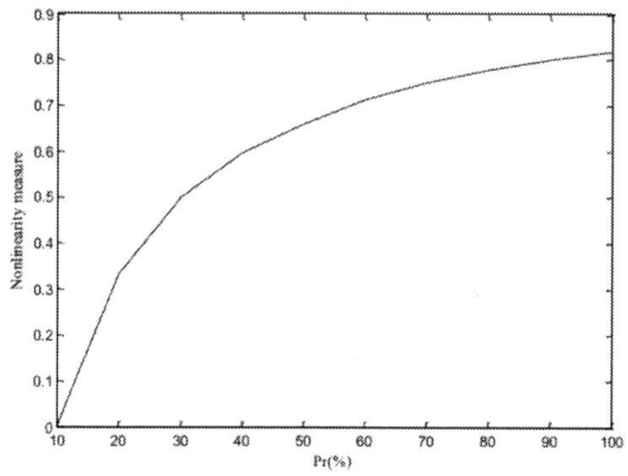

Figure 1. Nonlinearity measure of the core.

The nonlinearity measure curves of the core are obtained and shown in Fig.1, where the curve approximates a parabola. Comparing with two level areas [0.1 0.6] and (0.6 1.0], the curve change relatively rapidly in [0.1 0.6], and vary relatively gently in (0.6 1.0] in the form of an approximated straight line. These indicate that the core has the strong nonlinearity in [0.1 0.6] and the weak nonlinearity in (0.6 1.0]. Therefore, choose three linear models at 10%, 20%, 40%and 60% to substitute the nonlinear core in [0.1 0.6] with the strong nonlinearity, select two linear models at 80% and 100% to substitute the nonlinear core in (0.6 1.0] with the weak nonlinearity. Finally, linearized core models at five levels are selected as local models of the core to substitute the nonlinear core in (0 100%].

Let $G_1=G_{10\%}$, $G_2=G_{20\%}$, $G_3=G_{40\%}$, $G_4=G_{60\%}$, $G_5=G_{80\%}$, $G_6=G_{100\%}$. The state equation of G_i is represented by (A_i B_i C_i D_i) (i=1,…,6). Parameters of G_5 are enumerated: [a_3, a_2, a_1, a_0]=[5088 7021 1853 137.2], [b_4, b_3, b_2, b_1, b_0]=[1 401.4 692.2 389.9 43.2]. G_1, G_2, G_3, G_4 and G_5 are used to respectively denote dynamics of the nonlinear core located in the five power level ranges (0, 10%], [10%,

20%], [20%, 40%], [40%, 60%], [60%, 80%] and [80%, 100%] in turn.

4.2 Core fuzzy model

The triangle membership function is utilized to set up fuzzy rules. The membership function of the core fuzzy model at whole power levels is shown in Fig.2.

The fuzzy logic system (FLC) is of the following form.

Rule i: if P_r is M_i, then $Y_i = G_i$ (i=1,…,6)

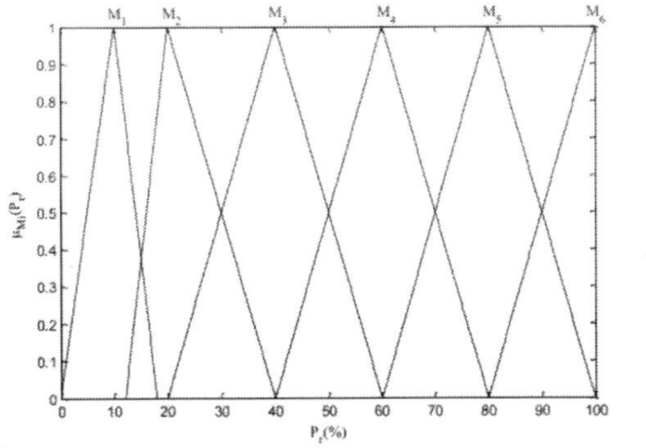

Figure 2. Membership function of core fuzzy model at whole power levels

where Rule i denotes the ith fuzzy rule; M_1, M_2, M_3, M_4, M_5 and M_6 respectively represent the fuzzy sets corresponding to the power levels 10%, 20%, 40%, 60%, 80% and 100%; $\mu_{Mi}(P_r)$ represents the membership when a power P_r belongs to the fuzzy set M_i, which is calculated in the light of a relationship between the membership and the core power level.

In the light of Fig.2, weight values q_i for the "blending" of G_i (i=1,…,6) is calculated as Eq. (20).

$$q_i = \frac{\mu_{M_i}(P_r)}{\sum_{i=1}^{6}\mu_{M_i}(P_r)}, i = 1,\ldots,6 \tag{20}$$

where $P_r \in (0\ 1]$, $q_i \in [0\ 1]$.

One value q_i indicates how the ith local model at a power level belongs to a core fuzzy model at this power level. Finally, the core fuzzy model is expressed by Eq.(21) or (22).

$$G_{FM} = \sum_{i=1}^{6} q_i G_i \tag{21}$$

$$\begin{cases} \dot{\mathbf{x}} = \sum_{i=1}^{6} q_i[A_i\mathbf{x} + B_i u] \\ y = \sum_{i=1}^{6} q_i C_i \mathbf{x} \end{cases} \tag{22}$$

5 Simulation

To verify effectiveness and correctness of this fuzzy modeling for the nonlinear core, the core fuzzy model (21) and the core nonlinear model are compared via simulations.

When the input δrod is taken as a 0.01 step, the fuzzy modeland the nonlinear model at ten typical power levels 10×i% (i=1,…,10) are simulated, and output responses of the fuzzy modeland nonlinear model at these levels are shown in Fig.3, which is namely responses of δP_r. In fig.3, the curve FMi (i=1,…,10) represents the output response of the fuzzy model at 10×i%, and the curve NLi (i=1,…,10) is the output response of the nonlinear model at 10×i%. From FMi and NLi, it can be observed that FMi approaches NLi. Similarly, the similar simulation results are able to be achieved for other power levels.

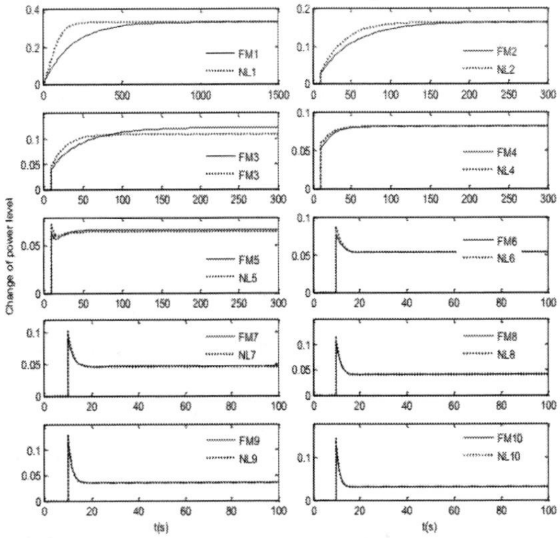

Figure 3. Output responses of fuzzy model and nonlinear model for the core at ten power levels

Hence, the core fuzzy model can approximate the core nonlinear model according to simulation results.

6 Conclusions

The work handles the modeling problem of nonlinear PWR cores. On the basis of modeling a nonlinear core and calculating the core nonlinear measure with the introduction of the gap metric, one group linear local models of the core is defined. In terms of local models and the triangle membership function, the T-S fuzzy model for the core is developed. From numerical simulations, it is obtained that the core fuzzy model can substitute the core nonlinear model. The proposed modeling principle as a reference method can be applied to model other reactor cores and nonlinear plants, and also applicable for stability analysis and control system design of nonlinear systems.

Acknowledgments

The authors would like to thank anonymous reviewers for their valuable comments. The work is funded by National High Technology Research and Development Program of

China (863 Program) (No.2015AAXX46611) and China Postdoctoral Science Foundation (No.20159200078).

References

1. T.W. Kerlin, E.M. Katz, J.G. Thakkar, J.E. Strange, Theoretical and experimental dynamic analysis of the H. B. Robinson nuclear plant, Nuclear Technology, **30**, 299-316 (1976).
2. R.M. Edwards, K.Y. Lee, M.A. Schultz, State feedback assisted classical control: an incremental approach to control modernization of existing and future nuclear reactors and power plants, Nuclear Technology, **92**, 167-185 (1990).
3. A. Ben-Abdennour, R.M. Edwards, K.Y. Lee, LQG/LTR robust control of nuclear reactors with improved temperature performance, IEEE Trans. on Nuclear Science, **39**, 2286-2294 (1992).
4. H. Arab-Alibeik, S. Setayeshi, Improved temperature control of a PWR nuclear reactor using an LQG/LTR based controller, IEEE Trans. on Nuclear Science, **50**, 211-218 (2003).
5. G.R. Ansarifar, H.R. Akhavan, Sliding mode control design for a PWR nuclear reactor using sliding mode observer during load following operation, Annals of Nuclear Energy, **75**, 611-619 (2015).
6. IAEA, *Directory of Nuclear Reactors Vol.IX*, 27-34, (International Atomic Energy Agency, 1971).
7. A.K. El-Sakkary, The gap metric: robustness of stabilization of feedback systems, IEEE Trans. Autom. Control, **AC-30**, 240 (1985).
8. T.T. Georgiou, On the computation of the gap metric, *Proc. Of 27th Conference on Decision and Control*, 1360 (1988).
9. K. Tanaka, M. Sugeno, Stability analysis and design of fuzzy control systems, Fuzzy Sets and Systems, **45**, 135-156 (1992).

Novel Laparoscopic Needle Holder

Han-Min Chien

China University of Technology, No. 56, Sec. 3, Xinglong Rd., Wunshan District, Taipei City 116, Taiwan

Abstract. This paper presents a new design of laparoscopic needle holder, which is able to reduce the laparoscopy operating time. Many reports showed that laparoscopic technique provide many benefits to patients, however, it also bring many disadvantages to surgeons. In contrast to conventional surgery, there are several disadvantages to this technique; these include restricted vision, image distortions, difficulty of hand-eye coordination and poor instrument design resulting to a higher probability of getting musculoskeletal disorders compared to conventional surgical procedures. The new design will be the one of solution to improve surgeons' quality of life.

1. Introduction

Laparoscopy (Figure 1), a kind of minimal access surgery technique, is used to reduce the traumatic effect of surgical operations. In contrast to conventional surgery, there are several advantages to this technique; these include decreasing surgical trauma, superiority of surgical care, patients experience less pain, patients can recover quickly and there are economic gains to the whole society. Laparoscopy however, also brings some disadvantages for surgeons [1]. For example surgeons are faced with restricted vision, image distortions, hand-eye coordination difficulty, poor instrument design and musculoskeletal disorder [2]. Surgeons usually complain that they have physical pain after performing laparoscopy [3, 4].

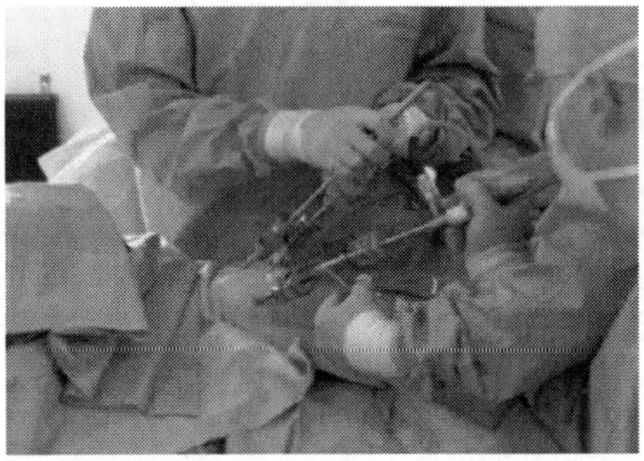

Figure 1. Picture of Performing Laparoscopy

In conventional surgery the surgeon has an unhindered view, which allows the surgeon to explore organs inside the body without a problem. For laparoscopy, an optical image system is needed to provide a view inside the abdomen. Therefore, the surgeon cannot see directly to control the movement of instruments and he needs to observe intra-abdominal information from the monitor [2].

The image systems, however, can only provide a two-dimensional (2D) enlarged image to the surgeon. Under this situation, the surgeon needs to imagine a normal visual (3D) image and measure monocular distance psychologically in order to identify the object and perform the task accurately [2, 5].

The hand-eye coordination is derived from a sequence of body sensing and control structures. First the image signals are transferred to the visual cortex via the retina. Then the visual cortex identifies the spatial position of the surgical instruments inside the patient's body by the signals sent from the retina. After that, the motor cortex generates stimulation signals to muscles in order to control the movement of arms. Therefore, the surgeon's action will depend on the transformation of visuo-motor signals, which were created during childhood development and enhanced by years of training or experience in their life [6].

The visuo-motor transformation and the real world do not match in the case of laparoscopy, which refers to surgeons do not visuo-motor transformation experience in normal life. Surgeons have been facing a huge challenge to deal with the mapping problem when they cannot look at their hands during the operation. The surgeon is often confused by the indirect method of view and control. Consequently, hand-eye coordination is inconsistent when the surgeon performs laparoscopy [5, 7, 8]. In addition, the surgeon must slow down the movement of the instrument in order to pay attention to the degree of enlargement (Figure 2). The disadvantages of visual problems have limited the adoption of laparoscopy [2, 9, 10].

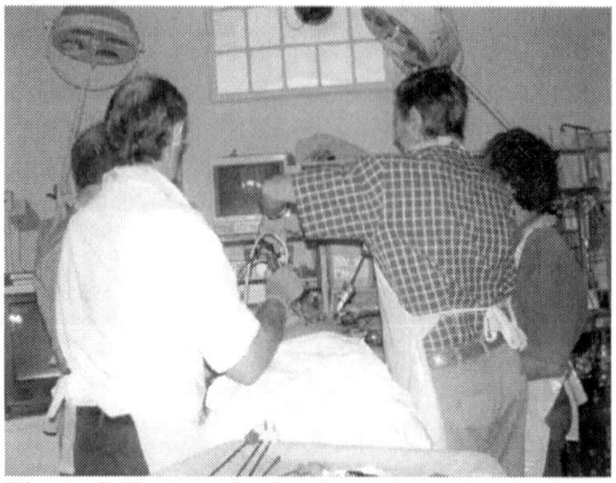

Figure 2. Posture during Laparoscopic surgery

A user-friendly interface instrument has present in this report, which is able to improve the quality of life of surgeons. The purpose of examination is to find if there are any differences between performing conventional laparoscopic instrument and new laparoscopic instrument. The results show the new concept for laparoscopic needle holder design, which reduces the operation time [2].

2. Observational Study

The problem is demonstrated by looking at a 2D image and attempting to do 3D measurements, for example the surgeon can estimate the distance between Point A and Point B (Figure 3) with the surgeon's eyes (3D).

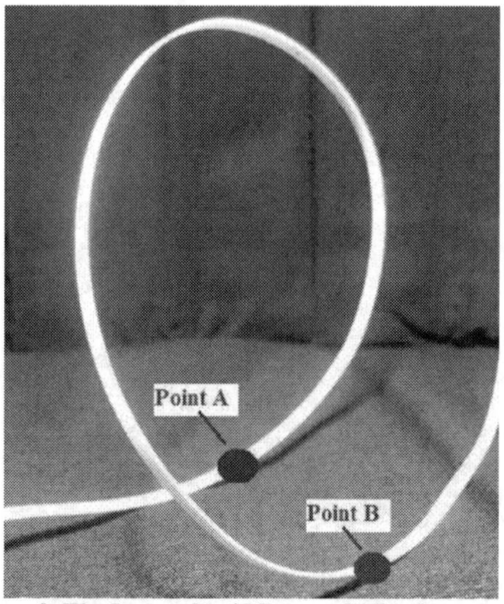

Figure 3. The Loop with 45 Degree Obliquity

However, it is not easy to guess the distance in front of the loop when the surgeon closes one of his/her eyes (2D). From another angle it seems that there is a complete loop from the front side of the loop (Figure 4), however, if the surgeon looks laterally then the two ends of the loop could be either quite close or widely separated.

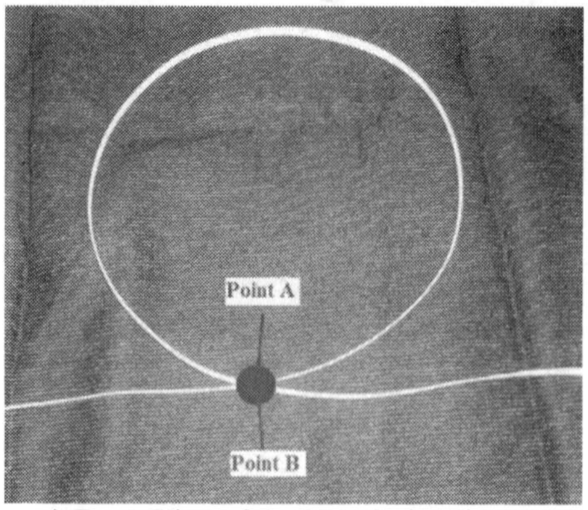

Figure 4. Front View of the Loop in 2D Picture

During the data collection for Investigation, it was found that surgeons move their arms in six DOF with four DOF vision. In this case, the surgeon finds it very difficult to make sure that the knot has been made, which could be the major problem with laparoscopic suturing. 3D imaging is one of the best ways to solve this problem. Otherwise it can only be solved though the surgeon's experience and increased operating time.

The aim here is to design a novel laparoscopic instrument to reduce the hand-eye coordination problem. The idea behind the proposition for a new needle holder is that if we cannot improve the surgeon's vision then we may need to limit the degree of movement required by the surgeon.

3. The New Concept of Design

In the new idea of laparoscopic needle holder, a "Thread Holder" is installed on the instrument stem, which is used to fix the suture thread behind the main jaw (Figure 5).

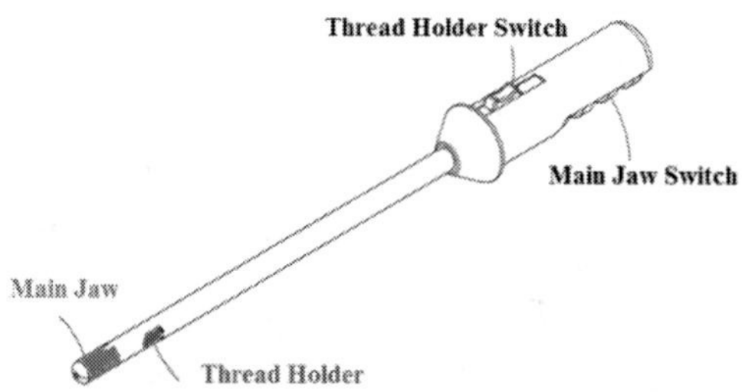

Figure 5. Proposed New Instrument Design

The structure of the new needle holder is shown in Figure 6. The instrument has to have as simple a construction as possible so that it may be sterilised repeatedly. For example, it can contain no electrical components. The handle will be similar to original laparoscopic instrument. The only difference is to install an innerspring switch on the top on the holder to control the "Thread Holder". The user can easily pull the switch with his thumb to open the "Thread Holder" or release to close it.

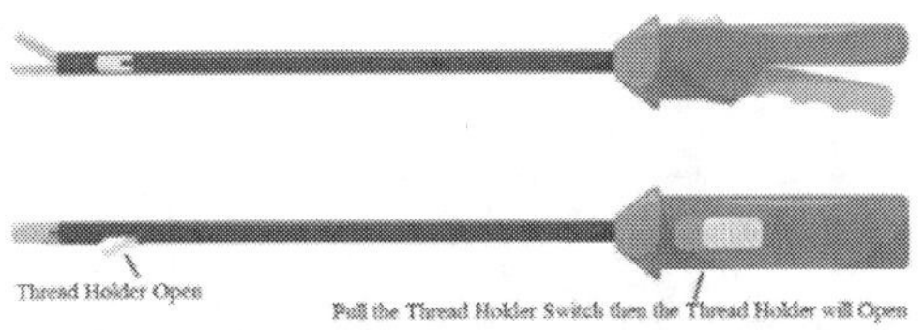

Figure 6. Demo of the New Laparoscopic Instrument Operating

The function of "Thread Holder" is to fix the position of the suture thread and let the surgeon make laparoscopic circular movements more simply and quickly (Figure 7). In this case, the suture line cannot be moved without turning the instrument, since the two ends have been held by the instrument jaw and the "Thread Holder".

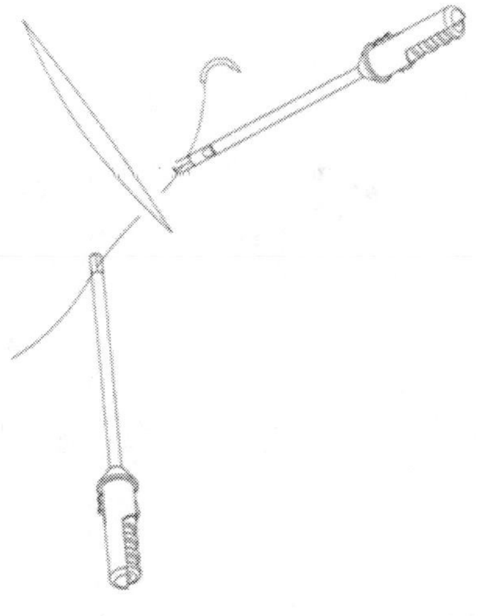

Figure 7. Operating of the New Instrument

The surgeon can mentally figure out the 3D position of the instruments by tactile sensation without 3D vision. The surgeon will finds it easier to determine the position of the two ends of the suture knot (Figure 8).

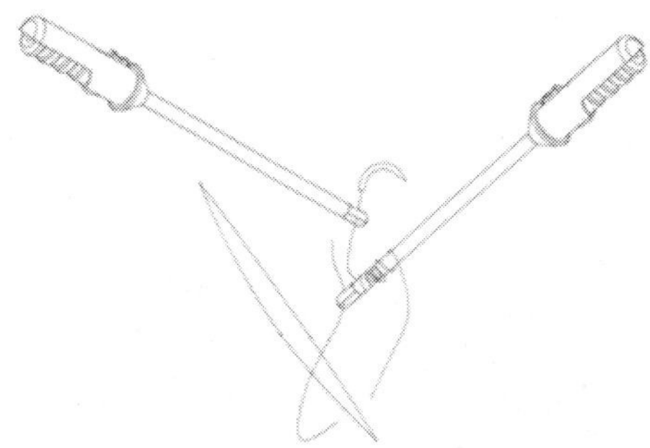

Figure 8. Suture Knot by New Instrument

4. Suturing Procedures by New Laparoscopic Instrument

In order to introduce the operation of the new instrument, the new procedure of laparoscopic suturing is described below step by step. This omits the needle stabbing steps.

Step 1. Use the left hand instrument to grasp the long end.

Step2. Place the new needle holder (right hand) under the suture thread. Then push the switch to open the "Thread Holder" (Figure 9).

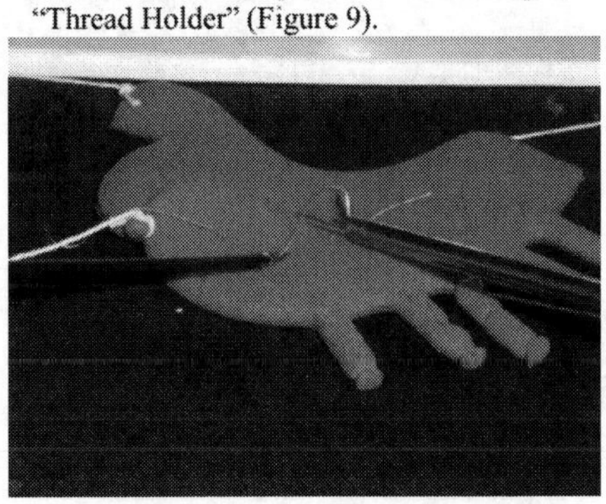

Figure 9. Place the new needle holder under the suture thread.

Step 3. Pick up the suture thread by the "Thread Holder" then pull the button to close the "Thread Holder". (Figure 10).

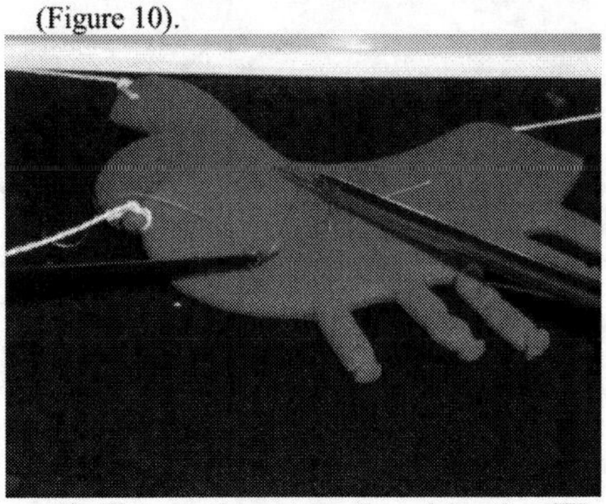

Figure 10. Pick up the suture thread by the "Thread Holder"

Step 4. Cycle the suture thread on the stem clockwise twice.

Step 5. Find the end of suture thread and pick it up by main jaw.

Step 6. Hold the end of suture thread by main jaw (Figure 11).

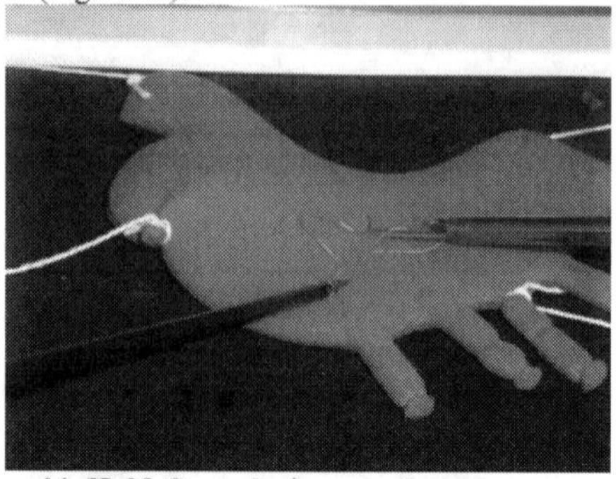

Figure 11. Hold the end of suture thread

Step 7. Push the switch to open the "Thread Holder" (Figure 12).

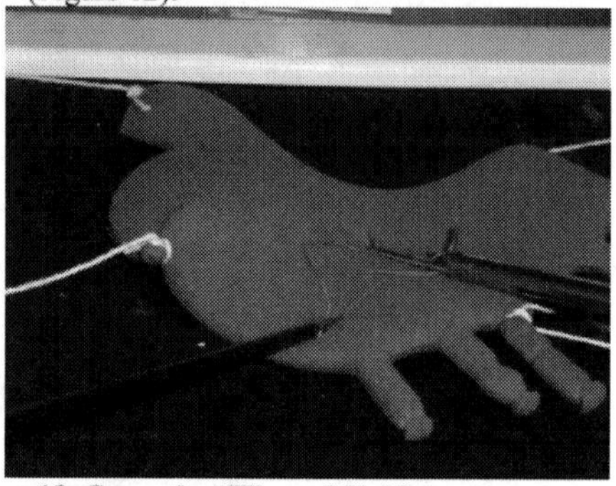

Figure 12. Open the "Thread Holder

Step 8. Pull both instruments back to tie it well and that will be a single knot.

Repeat these above steps for two more single knots. Turn the thread anti-clockwise in the first repeat cycle and turn the thread clockwise in the second repeat cycle. This process is based on normal suturing, where anti-clockwise and clockwise single-knots are made to increase the overall knot stability, and shows how easy it is to do this with the new tool.

5. Results

In order to measure operating time differences between original and novel laparoscopic instruments, 10 medical students were recruited and taught how to use the new needle holder. Three weeks training was provided to these students. Each participant was asked to perform laparoscopy sutures on a surgical practice kit using the original and the new instruments.

189.41 seconds were spent when these students tying a suture knot using the old instrument on average with the minimum 161.52 seconds and the maximum 210.55 seconds. However, they spent 146.5 seconds on average completing a laparoscopic suture with the new instrument, with the minimum 86 seconds and the maximum 176.53 seconds (Figure 13).

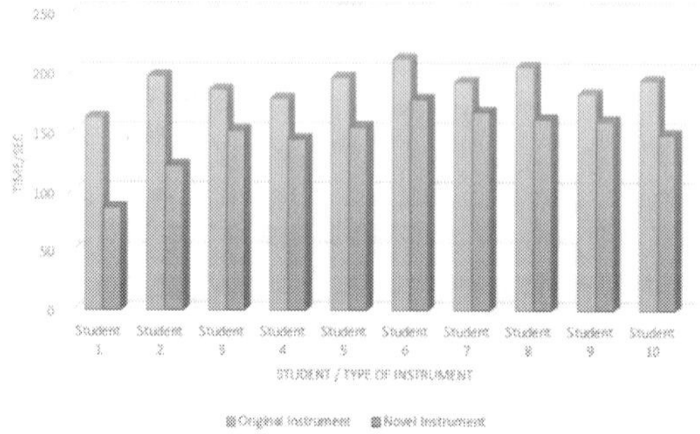

Figure 13. Students' Performances

This indicated that suturing using the updated instrument can reduce performance time by 22.7% compared with the original instrument (Figure 14). These differences between the old and new instrument suturing time were statistically significant ($P = 0.020$) by using a 5% significance level ($P \leq 0.05$) with the Mann-Whitney U-Test.

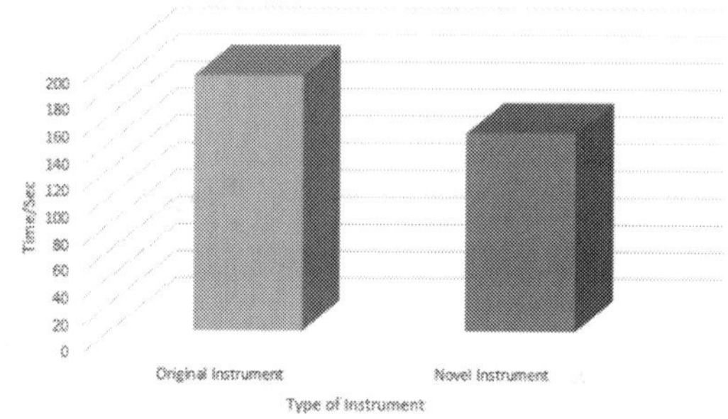

Figure 14. The Mean Value of Operating Time between Original and New Instruments

6. Conclusions

This research only considers the physical problem of the consultant surgeon. However, it is necessary to investigate team-based ergonomic problems by considering how the entire surgical team functions. This involves the participation of the assistant surgeon who is holding the camera and the nursing staff who transfer surgical instruments to surgeons. As a consequence, investigating the ergonomics of the surgical team may provide suggestions on how operating theatres should be designed to suit laparoscopic applications.

The major difficulty and time consuming part of laparoscopy is suturing. The problems of surgeons needing to recognise the position of the 3D thread and then to move their arms in the opposite direction as shown in a 3D image are surmounted with good co-ordination and training. 3D views of the environment are costly to achieve. A solution has been to design a new suture-friendly tool that reduces the need for the hand-eye co-ordination and finding the thread.

In conclusion, most medical engineering research is geared towards designing new equipment and technologies for the benefits of patients. Research by many groups has been carried out and the results demonstrated that most surgeons are not comfortable with laparoscopic instruments. It is important for researchers to find the ergonomic factors and disadvantages in operating with laparoscopic instruments. Ultimately, this will lead to improvements in terms of the quality of equipment for surgeons and other medical professionals.

References

1. T. Dohi, Computer Aided Surgery, "*The Proceeding of the 6th Asian-Pacific Conference on Medial and Biological Engineering 2005*", 24-27 April 2005. Tsukuba, Japan.
2. H.M. Chien, S. Ferguson and G. Dodds, A New Concept for Laparoscopic Instrument Design, *The Eighth Pan-Pacific Conference on Occupational Ergonomics (PPCOE 2007)*, 17-19 Oct, 2007, Bangkok, Thailand.
3. W.J. Lee, & Y.S. Chae, "Superficial Nerve Damage of Thumb of Laparoscopic Surgeon", *Surgical Laparoscopy Endoscopy & Percutaneous Techniques*, Vol. 11, No. 3, pp. 207-208, 2001.
4. N.T. Nguyen, H.S. Ho, W.D. Smith, C. Philipps, C. Lewis, R.M. De Vera, and R. Berguer, R., "An Ergonomic Evaluation of Surgeons' Axial Skeletal and Upper Extremity Movements During Laparoscopic and Open Surgery", *American Journal of Surgery*, Vol. 182, No. 6, pp. 720-724, 2001.
5. D.H. Birkett, L.G. Josephs and J. Este-McDonald, "A new 3-D laparoscope in gastrointestinal surgery", *Surgical Endoscopy*, Vol. 8, No. 12, pp. 1448-1451, 1994.
6. V. Sherman, L.S. Feldman, D. Stanbridge, R. Kazmi, and G.M. Fried, "Assessing the learning curve for the acquisition of laparoscopic skills on a virtual reality simulator", *Surgical Endoscopy*, Vol. 19, No. 5, pp. 678-682, 2005.
7. H.J. Chen & C.J. Lin, "The Investigation of Laparoscopic Instrument Movement Control", *Proceedings of the Internation MultiConference of Engineerings and Computer Scientists 2011*, Vol. II, IMECS 2001, 16-18 Mar, 2011, Hong Kong.
8. M.A. Van Veelen, J.J. Jakimowicz and G. Kazemier, "Improved Physical Ergonomics of Laparoscopic Surgery", *Minimally Invasive Therapy & Allied Technologies*, Vol. 13, No. 3, pp. 161-166, 2004.
9. M. Patkin, & L. Isabel, "Ergonomics, engineering and surgery of endosurgical dissection", *Journal of the Royal College of Surgeons of Edinburgh*, Vol. 40, No. 2, pp. 120-132, 1995.
10. M.A. Van Veelen, E.A.L. Nederlof, R.H.M. Goossens, C.J. Schot and J.J. Jakimowicz, "Ergonomic Problems Encountered by the Medical Team Related to Products Used for Minimally Invasive Surgery", *Surgical Endoscopy and Other Interventional Techniques*, Vol. 17, No. 7, pp. 1077-1081, 2003.

Comparisons of Feed-Forward and Multiple-Feedback Sigma-Delta Modulators for MEMS Accelerometers

Zhang Meimei , Quan Haiyang and Zhang Fuqiang

Beijing Microelectronics Technology Institute, 100076 Beijing, China

Abstract. This paper investigates two different architectures of a 5th order electro-mechanical sigma-delta modulator: a feed-forward (FF) architecture and a multiple-feedback (MF) architecture. And a comparison was performed in terms of stability and noise shaping ability, sensitivities to parameter variances due to fabrication tolerances and loop gain, and nonlinearity in feedback force. Both architectures were modeled in Simulink and investigated at system level. The results show that: a) both architectures are stable and achieve the similar noise floor level of -170dB within 250Hz in the ideal condition; b) both architectures have good ability in fabrication tolerance; c) the performance of the MF architecture will degrade heavily with the loop gain decreasing and become unstable if the loop gain beyond one optimal value, while the FF architecture is insensitive; d) the FF architecture controls the proof mass well and achieves better SNDR, whereas the MF has a 56dB degradation in consideration of nonlinearity in feedback force.

1 Introduction

MEMS (micro-electro-mechanical systems) capacitive sensing elements are commonly embedded in a closed loop, force feedback system to realize increased dynamic range, better linearity and wider bandwidth [1]. In particular, incorporated in a sigma delta modulator (SDM), the system can output direct digital signal, and avoid potential electro-static instability [2]. Earlier works showed that only using the MEMS sensing element as the loop filter, forming a second order electro-mechanical sigma delta modulator (EMSDM), suffers from relatively high quantization noise, idle tones and dead-zones [3]. To address these weaknesses, several research groups have designed high order EMSDMs, which cascading the sensing element with an electronic filter. The architecture of EMSDMs can be developed from conventional electronic SDMs, which can be sorted into single loop and multi stage noise shaping (MASH). Although MASH architectures have the considerable advantage of inherent stability and large input range, the severe drawback of the MASH is its sensitivity to component and parameter tolerances [4]. In contrast, the single loop architectures have been applied successfully to capacitive MEMS accelerometers [5, 6, 7, 8]. In this paper, the author discussed two typical architectures of the single loop, called the feed-forward (FF) and the multiple-feedback (MF), respectively. Simulation results showed that the FF has better performance in EMSDMs.

This paper is organized as follows: Section 2 describes the system level models of two architectures in general; Section 3 compares the FF and MF in terms of sensitivity to parameter variations and the loop gain, and nonlinearity in the feedback force; in Section 4 conclusions are drawn.

2 System level modeling

The Simulink models of the FF and MF are illustrated in Figure 1 and Figure 2, respectively. Traditionally, an EMSDM accelerometer system consists of the MEMS capacitive sensing element and the electrical interface circuit [9]. The sensing element, which is modeled as a mass-damping-spring system with a second order transfer function, converts acceleration into the displacement of the proof mass, which is reflected by the change of differential capacitances. The electrical interface circuit often includes the following blocks: the read-out circuit that outputs a voltage proportional to the displacement by measuring capacitor change; a phase compensator providing phase-lead compensation to stabilize the system (especially when the sensing element is under damped); an electronic loop filter that shapes quantization noise to high frequencies by oversampling; a quantizer outputting a digital bitstream, which decides the direction of the feedback electrostatic force to keep the displacement of the proof mass small.

To analyze both architectures mathematically, the quantizer is usually assumed as a simple gain (Kq) with additive white noise [10]. For small mass deflections, the pickoff and feedback circuits can also be modeled simply as gain constants Kxv and Kfb, respectively. Based on the above assumptions, the quantization noise transfer functions for the FF and the MF are given by:

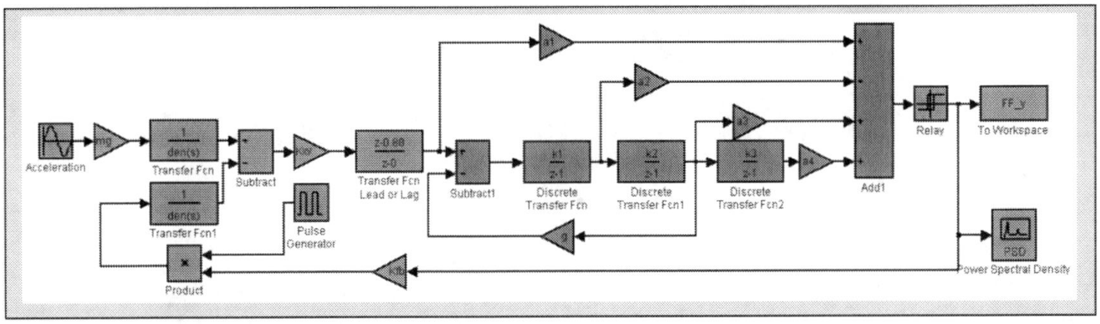

Figure 1. Simulink model of a FF 5th order EMSDM.

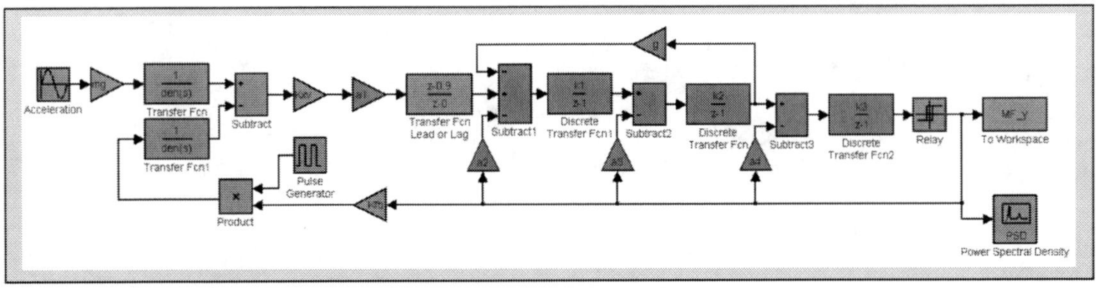

Figure 2. Simulink model of a MF 5th order EMSDM

$$QNTF\_FF = \frac{1}{1+K_qK_{fb}K_{xv}H_m(z)H_c(z)H_{FF\_e}(z)} \quad (1)$$

$$QNTF\_MF = \frac{1}{1+K_qH_{MF\_e1}(z)+K_qK_{fb}K_{xv}H_m(z)H_c(z)H_{MF\_e2}(z)} \quad (2)$$

where $H_m(z)$ is the transfer function of the sensing element in discrete-time domain [11]:

$$H_m(z) = \frac{K_m(z+\beta)}{(z-z_1)(z-z_2)} \quad (3)$$

$H_c(z)$ is a phase lead compensator:

$$H_c(z) = 1-\alpha z^{-1} \quad (4)$$

$H_{FF\_e}(z)$, $H_{MF\_e1}(z)$ and $H_{MF\_e2}(z)$, shown in (5) (6) (7), are the electronic filters of the FF and the MF, respectively.

$$H_{FF\_e}(z) = \frac{a_1\cdot\left[(z-1)^2+g\cdot k_1k_2\right](z-1)+a_2k_1\cdot(z-1)^2+a_3k_1k_2\cdot(z-1)+a_4k_1k_2k_3}{\left[(z-1)^2+g\cdot k_1k_2\right](z-1)} \quad (5)$$

$$H_{MF\_e1}(z) = \frac{a_4k_3\cdot\left[(z-1)^2+g\cdot k_1k_2\right]+a_3k_2k_3\cdot(z-1)+a_2k_1k_2k_3}{\left[(z-1)^2+g\cdot k_1k_2\right](z-1)} \quad (6)$$

$$H_{MF\_e2}(z) = \frac{a_1k_1k_2k_3}{\left[(z-1)^2+g\cdot k_1k_2\right](z-1)} \quad (7)$$

In order to make a comparison between the two architectures, a high performance accelerometer based on the bulk-machining technology and vacuum packaging, suffering from low Brown noise, is used for the system. Lumped parameters of the under-damped sensing element are listed in Table 1.

Table 1. Lumped parameters of the under-damped sensing element.

Parameter	Value	Unit
Mass	7.34e-6	kg
Damping coefficient	6e-3	N/m/s
Spring constant	33.8	N/m
Nominal electrode gap	2e-6	m
Area	9e-6	m^2
Brown noise	0.14	ug/sqrt(Hz)

A 62.5Hz sinusoidal input signal with normalized amplitude of -6dB of the full scale is used to avoid overloading, and the simulation results of power spectral density (PSD) are presented in Figure 3. It is clear that both architectures achieve excellent noise shaping ability, suppressing the noise floor below -170dB within 250Hz signal band. The estimated SNR values are 143.5dB and 143.1dB for the FF and the MF when the oversample rate (OSR) is 256.

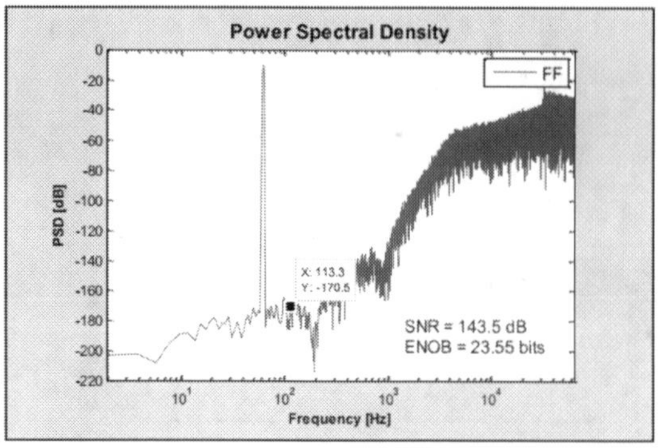

(a) FF

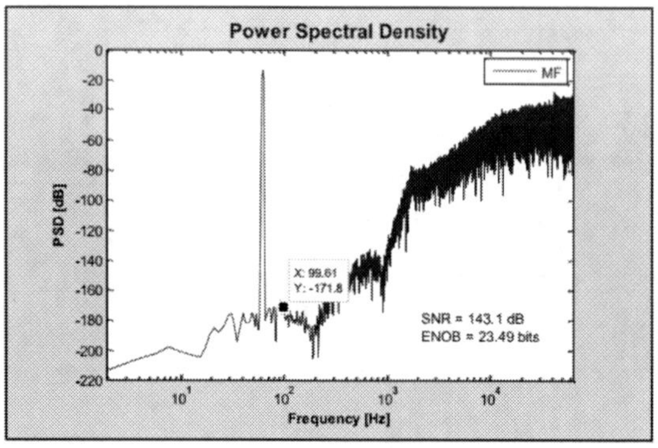

(b) MF

Figure 3. Noise shaping and SNR comparison for the FF and the MF.

3 Architectural advantages and disadvantages

In all-electronic SDMs, the two architectures give similar performance. An electromechanical SDM, however, includes the sensor element as part of the loop filter. In this section, we will make a comparison between the FF and the MF architectures with regard to some important practical issues. The results given here will be a useful reference for the designers to decide which one is their better choice.

3.1 Parameter sensitivity

Limited to current fabrication technique, the sensing element parameters are often varied by 10% of their ideal values, which may break the system stability easily. As a result, the sensitivity of the two architectures to the sensor parameter variations derived from fabrication tolerances is a significant comparison criterion. These deviations contribute to a change in the dynamic characteristic of the sensing element, affecting the noise-shaping ability of the whole system.

In Figure 4, the three lumped model parameters of the MEMS accelerometer: proof mass 'm', damping coefficient 'b', and spring constant 'k' are examined individually, to observe the parameter with the most critical effect. In this case, both of the two systems can be considered to be robust to parameter variation within the range of $\pm 10\%$, since the SNR only suffers from within 2 dB degradation.

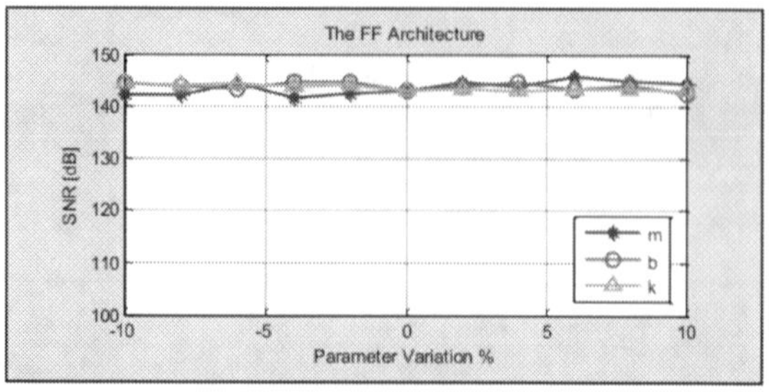

(a)FF

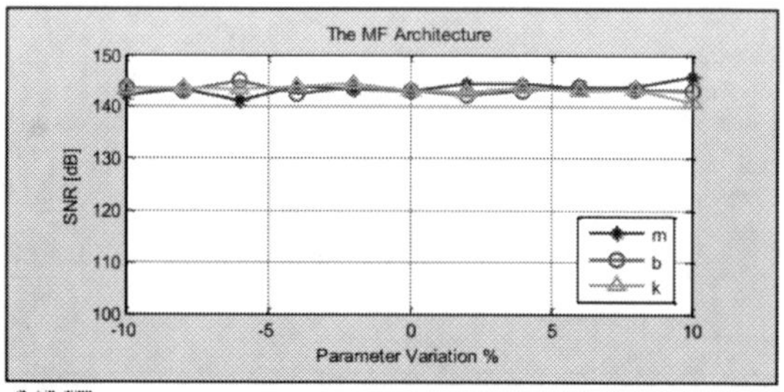

(b)MF

Figure 4. Parameter sensitivity analyse for the FF and the MF.

3.2 Sensitivity to the loop gain

Comparing equation (1) with (2), we can find that, in the FF, the mechanical element appears in series with the electrical filter and the quantizer. Since a 1-bit quantizer only outputs the sign of its input signal, this implies that any change in the loop gain independent of the frequency, such as Kfb, Kxv, will not affect the system stability. The MF, on the other hand, has an additional purely electrical signal path, which bypasses the mechanical element, other than their common electromechanical signal path. It contributes poles of QNTF dependent on the values of the feedback coefficients as well as the parameters of the sensor. Since the feedback coefficients do not track the variation in the sensor transfer function, there is uncertainty in the position of the poles of QNTF.

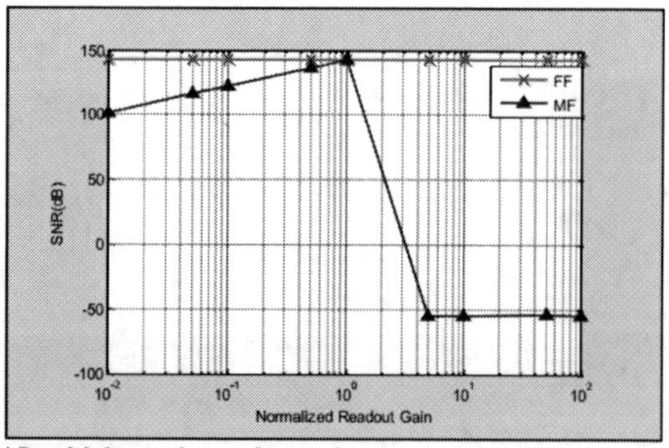

(a)Sensitivity to the readout gain

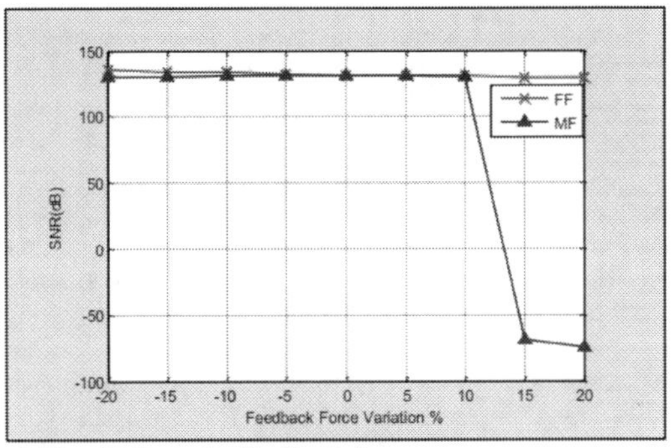

(b)Sensitivity to the feedback force

Figure 5. Sensitivity to the loop gain.

Figure 5 shows the changes of SNR in the FF and the MF with the normalized readout gain increasing from 10^{-2} to 10^{2} and the feedback force varying ±20%, respectively. Large changes in the loop gain has a negligible effect on the FF architecture performance, while the SNR of the MF degrades heavily, along with it decreases or increases. The tolerances on the feedback force means designers can improve the full range easily by applying a larger feedback voltage or longer feedback time. What's more, there is a possibility that the same FF architectures can be used with different sensors, while the MF is only designed for a special sensor.

3.3 Nonlinearity in the feedback force

In Figure 1 and Figure 2, the conversion of a feedback voltage to an electrostatic feedback force on the proof mass is assumed as a constant. Actually, it is nonlinear depending on the residual proof mass motion, and is given by:

$$
\begin{aligned}
K_{fb} &= \mathrm{sgn}(Y_{out})\frac{K_{fb0}}{\left[1+\mathrm{sgn}(Y_{out})(x/d_0)\right]^2} \\
&= \mathrm{sgn}(Y_{out})K_{fb0}\left[1+\mathrm{sgn}(Y_{out})\cdot 2(x/d_0)+3(x/d_0)^2+\mathrm{sgn}(Y_{out})\cdot 4(x/d_0)^3+\cdots\right]
\end{aligned} \tag{8}
$$

where $K_{fb0} = 2\varepsilon_0 A V_{fb}^{\ 2} / d_0^{\ 2}$. Equation (8) indicates that the electrostatic feedback force has higher harmonic content, which will cause harmonic distortion, resulting in a reduction in the SNDR.

To investigate the feedback force nonlinear effect on an EMSDM performance, system level models with the nonlinear electrostatic feedback force were simulated for the FF and MF once time. Assuming a normalized amplitude of -6dB of the full scale input signal with a 62.5Hz frequency, the maximum residual proof mass motion x of the FF and MF are about 1nm and 17nm, respectively; this is shown in Figure 6. The spectrum of the output bitstream Yout is plotted in Figure 7. Since even harmonics are canceled due to the time-averaging positive or negative feedback pulse, there is only an obvious third harmonic distortion peak at three times of the input signal frequency (assumed to be 187.5Hz). The SNDR drops about 56 dB in a 250Hz signal bandwidth in the MF, while it is too small that can be ignored in the FF.

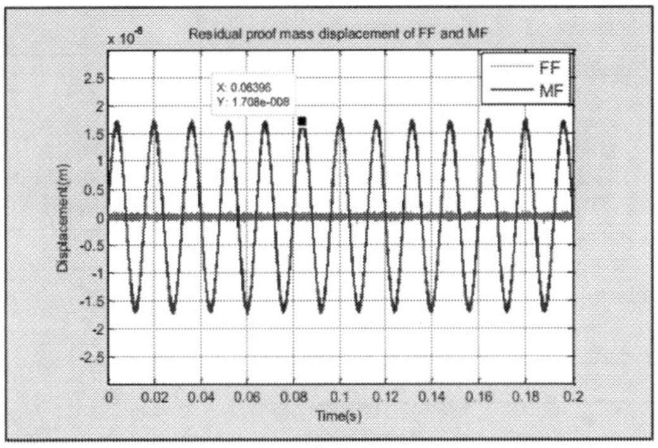

Figure 6. Residual proof mass displacement of a simulation run of the model in Figure 1 and Figure 2.

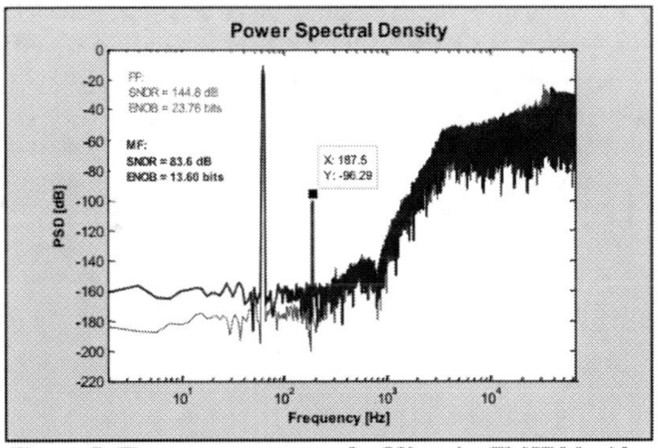

Figure 7. The output spectrum of a fifth-order EMSDM with a nonlinear force feedback.

This is not an accidental phenomenon. In purely electronic SDMs, the first integrator has much larger output amplitudes, which can be adjusted within a reasonable range by scaling the feedback coefficients a1, k1, than the FF. In EMSDMs, however, the gain of the sensing element cannot be changed, so what we can only do is changing the gain of Kxv to meet the stability requirement. Nevertheless, it is useless to narrow the displacement. Hence, we can conclude that, there exists more serious nonlinear, which means larger SNDR in the feedback force of the MF than the FF.

4 Conclusions

A comparative study was carried out between two 5th order EMSDMs for a MEMS capacitive accelerometer: a feed-forward (FF) architecture and a multiple-feedback (MF) architecture. Both architectures achieved over 140dB SNR (Signal to Noise Ratio) and suppressed a quantization noise floor level below -170dB with a bandwidth of 250Hz. Besides, the investigation confirmed their immunity to the sensing element parameter variations due to manufacture errors. The FF can be used with different sensors directly with its superiority of large tolerance on the loop gain, while the MF can only be designed for a special sensor because its performance degrades seriously as the loop gain changes. The former has slight third harmonic distortion, since it controls the displacement of the proof mass much smaller than the latter. As such, we believe that the FF architecture will be more popular with designers.

References

1. Dong, Yufeng, M. Kraft, and C. Gollasch. "A High performance accelerometer with fifth-order Sigma-Delta Modulator." *Journal of Micromechanics & Microengineering* 15.7(2005)
2. Henrion, W., et al. "Wide dynamic range direct accelerometer." *IEEE Solid-state Sensor & Actuator Workshop* IEEE. (1990)
3. Dong, Yufeng, M. Kraft, and W. Redman-White. "Higher Order Noise-Shaping Filters for High-Performance Micromachined Accelerometers." Instrumentation & Measurement IEEE Transactions on 56.5(2007)
4. Almutairi, Bader. Multi stage noise shaping (MASH) sigma delta modulator for capacitive MEMS inertial sensors. Diss. University of Southampton. (2015)
5. Y. Dong. Control systems for capacitive micromachined inertial sensors with high-order sigma-delta modulators. Diss. University of Southampton. (2006)
6. Zwahlen, P., et al. "Breakthrough in high performance inertial navigation grade Sigma-Delta MEMS accelerometer." *Position Location and Navigation Symposium (PLANS), 2012 IEEE/ION* IEEE. (2012)
7. Fu, Qiang, et al. "Design of high-performance fourth-order single-loop sigma-delta modulator for inertial sensor." Optoelectronics and Microelectronics (ICOM), 2012 International Conference on IEEE.(2012)
8. Zhang Feng, X. Jin, and L. Wang. "Modeling and test result of closed-loop MEMS accelerometer with wide dynamic range." *Microsystem Technologies* . (2015)
9. Wilcock, Reuben, and M. Kraft. "Genetic Algorithm for the Design of Electro-Mechanical Sigma Delta Modulator MEMS Sensors." Sensors11.10 (2011)
10. Norsworth, S.N.; Schreier, R.; Temes, C. *Delta-Sigma Data Converters, Theory, Design, and Simulation*; IEEE Press: Piscataway, NJ, USA. (1997)
11. Xuesong Jiang. Capacitive Position-Sensing Interface for Micromachined Inertial Sensors [D]. University of California at Berkeley. (2003)

Study on Wind Flow around Highway Embankments

Sun-Young Paek[1], Byung-Hee Nam[1] , Young-Moon Kim[2], Ki-Pyo You[2] and Jang-Youl You[3]

[1]*Department of Architecture Engineering, Chonbuk National University, Jeonju, Korea*

[2]*Department of Architecture Engineering, Chonbuk National University, Jeonju, Korea, Long-Span steel Frame system Research Center*

[3]*Department of Architecture Engineering Songwon University, Gwangju, Korea*

Abstract. The embankments established in conjunction with highway construction significantly influence ambient wind velocity in surrounding agricultural lands. Wind flow at levels lower than the highway embankment is blocked by the embankment so that the wind blows less here than in the surrounding regions thereby greatly affecting the agricultural crops. Various types of damage are incurred due to the road embankments. However, as yet, no study has been conducted on the micro-meteorological changes occurring around embankments. In this paper, a wind tunnel test was conducted to evaluate wind flow before and after construction of a highway embankment. The study showed that the wind speed was reduced by 40 to 60% after construction of the embankment as compared to the reference wind speed.

1 Introduction

The embankments required by road construction can affect air flows, thereby changing surrounding temperatures and incurring degradation to crops, such as fruit trees withering, crops being reduced, and flowering delayed. To prevent cold-weather damage resulting from the slowing down of wind flow around embankments, wind passages have been installed around embankments but no significant effects have been achieved. The cold-weather damage to fruit trees near to highway embankments is normally experienced in spring when the wind is blowing weakly. Blocked wind flows constitute a major cause of cold-weather damage to crops. The changes to wind flows can create various phenomena, depending on the geographical features, such as strong gusts of wind or reductions in wind speed due to the blocking effect. A large number of studies have been conducted on increased wind speed in mountainous terrains, valleys, and slopes. Jackson and Hunt[1] studied the effect of wind speed increases in low mountainous terrain, where flow separation phenomena did not occur. Bowen[2] studied the mean wind speed in simple two-dimensional (2D) mountainous terrain. Taylor and Lee[3] proposed an algorithm that predicted increases in wind speed at the top of simple mountainous terrains. Chock and Cochran[4] conducted a wind tunnel test to study the phenomenon of increasing wind speed on an island consisting of a complex terrain. They proposed an empirical model involving peak and mean wind speed increases that can be applied to real structure designs. Most of the studies were conducted in the upper parts of mountainous terrains where wind speed increases are common. However, few studies have been conducted on wind flow characteristics around the roads constructed through mountainous areas[5][6] .

Fruit trees planted in the bare ground, and grown on the lower reaches of sloping terrain, are particularly sensitive being affected by both temperature and wind speed. Thus, it is necessary to evaluate the effect on temperature and wind speed if road embankments are built in sloping areas. This study aimed to determine the changes in wind speed resulting from the construction of various types of road embankment. The changes in wind speed were determined by use of wind tunnel tests, using a scale reduction model and particle image velocimetry (PIV). The wind tunnel tests were conducted to determine the wind speed before and after the construction of an embankment in order to explain wind flow changes around highway embankment construction.

2 location and study methodology

An experimental area was selected in agricultural lands where a ground embankment was to be built for highway construction. Prior to the building of the embankment, wind blowing down from the mountain would follow a natural direction of flow. However, after the construction of the embankment this wind flow was significantly affected. A scale reduction ground model was constructed and a wind tunnel test was conducted to evaluate wind flow both before and after construction of the highway embankment.

3 Wind tunnel test

3.1. Test model

A wind speed measurement test was conducted with a 1/150 scale reduction model to investigate wind flow around the highway embankment slopes. The wind tunnel test model was constructed so as to include an area with a radius of 1km around the embankment. The wind tunnel ground model was made from styrofoam. The dimensions of the wind tunnel were 12m in width, 2.5m in height, and 40m in length. Figure 1 shows the ground model installed inside the wind tunnel. The test aimed to determine changes in wind speed relative to height differences in the surrounding area both before and after the construction of a highway embankment and when the wind was blowing at a constant speed. An omni-directional wind speed sensor was installed to measure wind speed around the embankment slope and the wind speed was measured using a multi-point anemometer (System 6242 Model 1560). Nineteen locations were selected in order to determine changes in wind speed around the embankment slope. The height difference between the ground and the embankment was larger on the south side of the embankment than on the north side. Because of this, anemometers were only installed on the south side. A wind speed test was conducted at five locations immediately below the embankment slope and 14 other locations surrounding the embankment.

(a) Before building the embankment wall

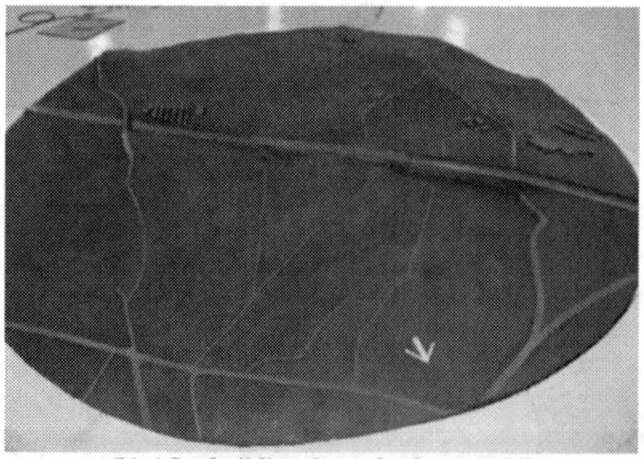

(b) After building the embankment wall

Figure 1. Ground model installed inside the wind tunnel

3.2 Result of wind speed measurement

The wind speed used for the wind speed measurement was 3 m/s, 5 m/s, and 7 m/s. This test aimed to detect changes in wind speed around the embankment slopes before and after the building of the embankment. Figure 2 shows the wind speed measurement locations and their vertical distances from the embankment top. The surrounding terrain below the embankment had a mean height difference of 8.5 m, with the left side showing the highest difference of 11 m, and the right side showing a 5.9 m height difference.

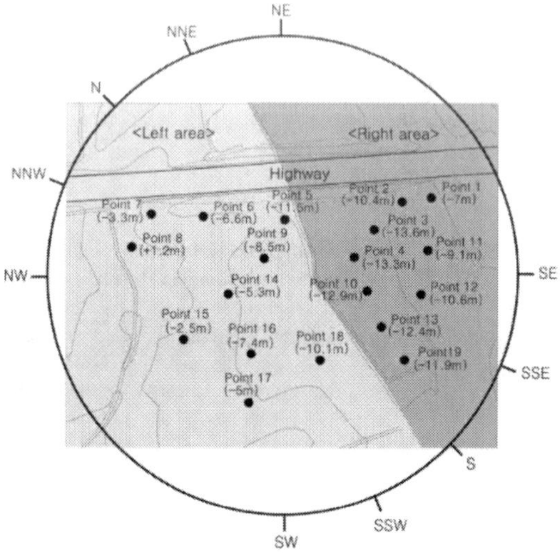

Figure 2. Measurement locations

The test results were analyzed using the concept of wind speed ratio as shown in Equation (1). Three different wind speeds (U_R) were set as reference values (3 m/s, 5 m/s, 7 m/s) in the test on the basis of embankment wall height. $(U_h)_i$ refers to the wind speed measured at each of 19 locations.

$$R_i = \frac{(U_h)_i}{U_R} \tag{1}$$

where $(U_h)_i$ is wind speed (m/s) at i location and U_R is the reference wind speed (m/s).

After the embankment wall was built, many locations experienced a reduced wind speed ratio. Figure 3 shows changes to the wind speed ratio before and after building the embankment wall at measurement locations 4 and 11, which are located in the same line as measurement location 15. In the legend for Figure 3, "before" means before building the embankment and "after" means after building the embankment. As shown in the figure, the wind speed ratio was reduced after building the embankment. For the center area (measurement locations 4, 5, 9, and 10) of the embankment wall, the wind ratio was reduced by 40% to 60% which was more than in the other measurement locations. These results revealed that the wind flow was affected by the building of the embankment wall.

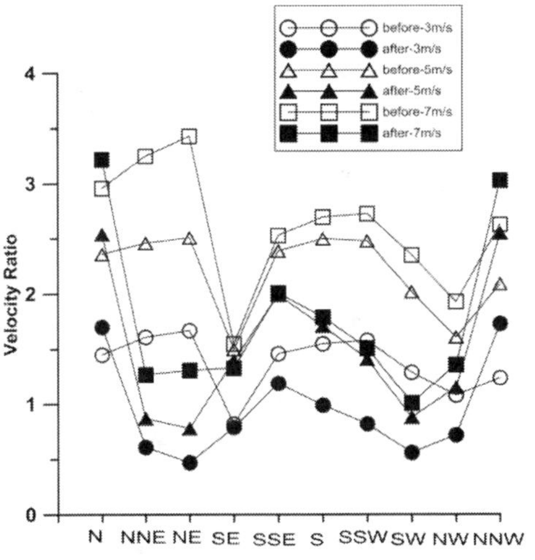

(a) Point 5

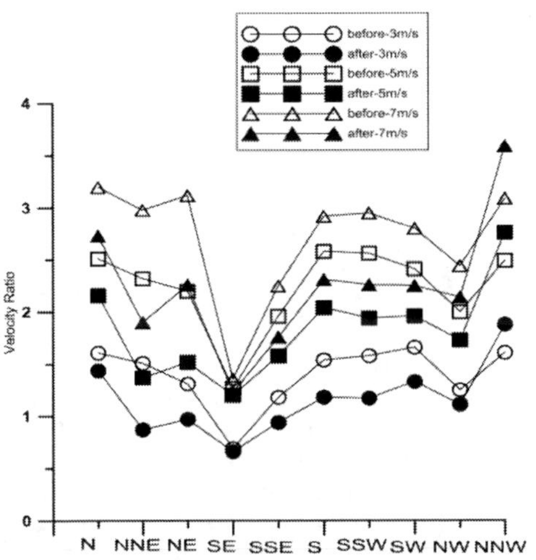

(b) Point 11

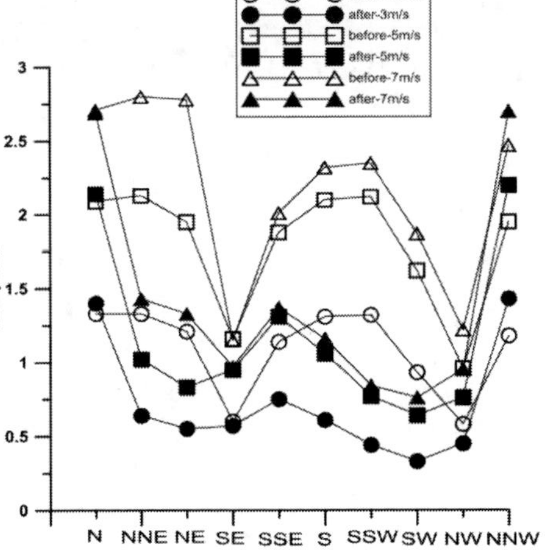

(c) Point 4

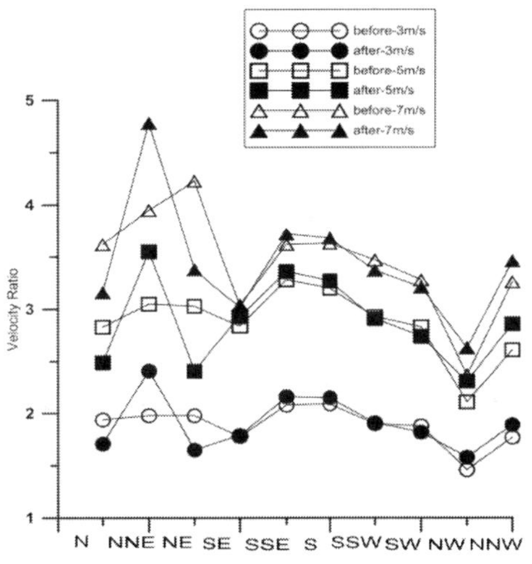

(d) Point 15

Figure 3. Changes in wind speed ratio before and after building the embankment at measurement locations according to changes in wind direction angle.

4 Experiment of particle image velocimetry (PIV)

In order to identify wind flow difference shown as relative wind speed vector magnitude before and after construction of the embankment, we used a PIV to interpret velocity distribution and wind profiles of the velocity field. When an embankment tunnel around the fruit farm was established at the center, we horizontally measured two sections (A and B) on the right and three sections (C, D, and E) on the left. Figure 4 shows the measurement locations.

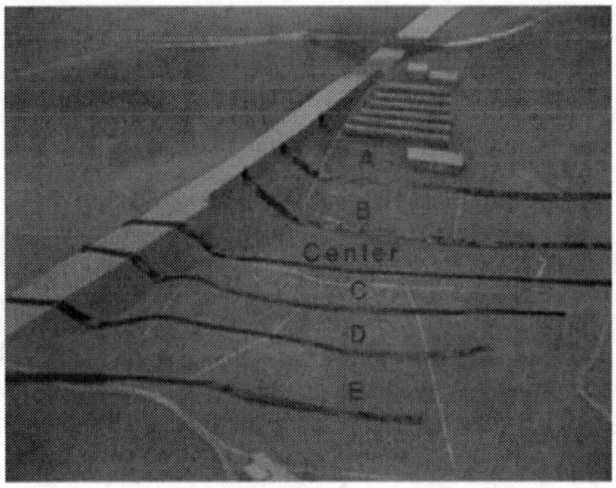

Figure 4. PIV locations for measurement

4.1. Analysis of measurement results

The PIV serves as a method for measuring flow velocity by tracking particles scattered with a laser light sheet and obtaining a velocity field on the sheet. We obtained images of 100 captures using cameras located in the five sections under the lower wind velocity of 1m/s or less in the wind tunnels, images that were used to obtain the average velocity field. Figures 5 and 6 show the velocity fields per measurement section before and after constructing the embankment. In the analysis images obtained through post-processing of the measured images, the wind velocity was relative and the speed was shown by colors. The velocity magnitude was shown by red > yellow > blue colors. The speed image velocity presented in the analysis of the experiment served as a relative value; the changes in the velocity field before and after constructing the embankment were presented as color changes. The sections where tunnels were placed in the embankment had greater topographical changes than the sections without tunnels, indicating that there were larger changes in the velocity field. The velocity field of the surrounding air current had been red-colored without the embankment, but became blue-colored after constructing the embankment. As for sections A, B, and C for measurement, greater topographical changes after the embankment indicated that the velocity field of the air current was sufficiently changed before and after the embankment.

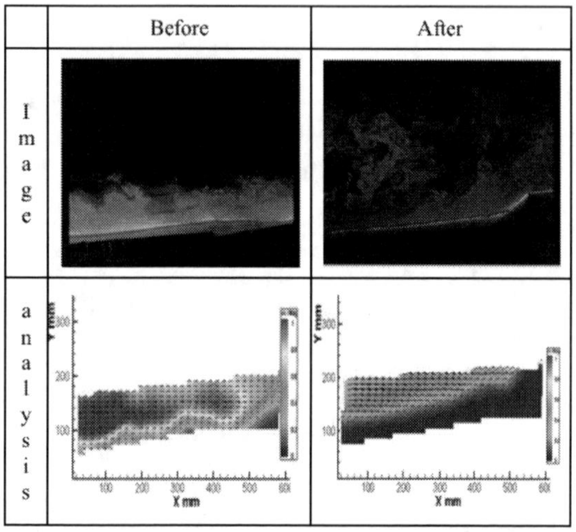

Figure 5. Images of measurement and analysis of measurement location (Center of embankment)

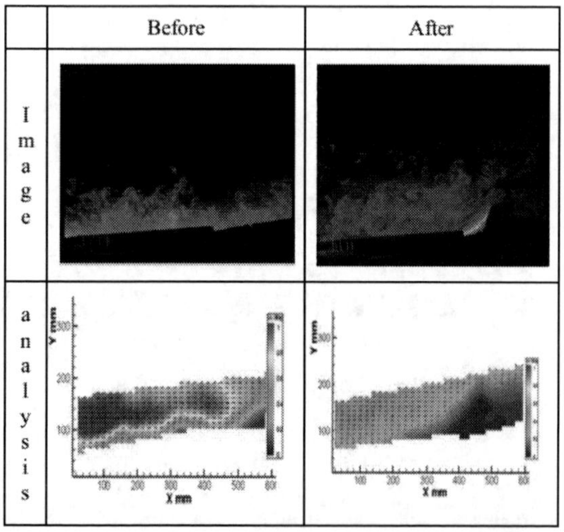

Figure 6. Images of measurement and analysis of measurement location (Section A)

5 Conclusion

The results of the present study on changes in wind speed around a highway embankment are as follows:

Building a highway embankment on sloping terrain affects the wind speed in the surrounding area. After building the highway embankment, the wind speed was reduced by up to 60%. The reduction in wind speed was greater at points some distance away from the embankment slope relative to areas immediately below the embankment slope. The reduction in wind speed was determined in relation to the height differences of the ground below the embankment slope. Changes in wind speed around the embankment showed that the embankment slope greatly affected the reduction in wind speed.

Acknowledgment

This work is supported by the National Research Foundation of Korea (NRF) grant funded by the Korea government. (MEST) (NO.2011-0028567).

References

1. P.S. Jackson, J.C.R. Hunt, *Turbulent flow over a low hill*, QJRMS, **101**, pp.929-955, (1975)
2. A.J. Bowen, *The prediction of mean wind speeds above simple 2-D hill shapes*, JWEIA, **15**, pp.259-270, (1983)
3. P.A.Tayor, R.J.Lee, *Simple guidelines for estimating wind speed variations due to small scale topographic features*, Climatological Bulletin (Canada), **18**(2), pp.3-32, (1984)
4. G.Y.K. Chock, L. Cochran, *Modeling of topographic wind speed effects in Hawaii*, JWEIA, **93**, pp.623-638, (2005)
5. A.S. Sergio, P. Santiago, L.G. Oscar, S.A. Angel , *Wind Tunnel Analysis of the Aerodynamic Loads on Rolling Stock over Railway Embankments: The Effect of Shelter Windbreaks*, Scientific World J, **2014** , pp.1-17,(2014)
6. Y.M. Kim, K.P. You, and J.Y. You, *Characteristics of wind velocity and temperature change near an escarpment-shaped road embankment*, The Scientific World J, **2014**, pp.1-13(2014)

Calculation of Electric Field at Ground Surface and ADSS Cable Prepared Hanging Point near EHV Power Transmission Tower

Bao-Qing Xu[1], Xiao-Qiang Ma[1], Sen Lv[1], Biao Li[1] and Lu Wang[2]

[1] Sanmenxia Power Supply Company, State Grid Henan Electric Power Company, Sanmenxia, China
[2]North China Electric Power University, Baoding, China

Abstract. A simplified model of the 750kV tower is established by CDEGS software which is based on the Method Of Moment. The power frequency electric field distribution on the ground is achieved by software calculation and field-measuring. The validity of the calculation is proved when compare the calculation and experiment results. The model also can be used to calculate the electric field in prepared hanging points on the tower. Results show that the electric field distribution on the ground surface around the tower and prepared hanging points are meet the standard by calculation and experiment.

1. Introduction

With the rapid development of the national power grid construction, there are many applications of ADSS (All Dielectric Self-Supporting) cables for electric power communication system. Most of the ADSS cables are used for replace the aging communication line. But the research on ADSS is still not perfect at home and abroad. There are problems on the sheath resistant to electric corrosion, designing of line construction, electrical fittings equipped method. It is very important to master and apply technology of ADSS cables.

According to [1]-[3], in the power frequency electric field environment of the 750 kV double circuit lines on the same tower even multi-circuit transmission line, the largest electric field near the house is 4 kV/m; In the general area, such as the public activities area, area that the line across the road, the electric field is limited to 7 kV/m; the electric field is limited to 10 kV/m when the transmission line across farmland to avoid the uncomfortable feeling caused by the discharge; In remote areas sparsely populated, the area of non-public activities area or person passing through occasionally, the limit value of the electric field can be relaxed to 12 kV/m.

Hanging points of ADSS cables is selected reasonably will ensure the cables in a small electric field environment. The position choice of ADSS cable hanging points is determined by the potential (12kV, 25kV) generally. But the researcher proposed that the main factor that affects the electric corrosion is the electric field strength at the cable hanging points. At present, electric field less than 10kV/m at the hanging points set as standard generally [2-4].

The effectiveness of CDEGS software based on Method of Moments has been verified in literature [4]. Charge Simulation Method was used in [5] to calculate the electric field generated by transmission lines, and the method is effective, but charge simulation method to calculate the electric field is more complicated in several cases such as consider the earth to non-perfect conductor, and taking into account the ground portion of the tower. Charge simulation method is generally used to analyze the current model. The frequency electric field of 500kV transmission line was calculated based on charge simulation method, and the influence of the electric field to the environment of tower was analyzed in literature [6]. Three-dimensional frequency electric field near high voltage transmission towers was calculated based on three-dimensional electric field computational method. Detailed distribution of three-dimensional frequency electric field was obtained in literature [7], and low field position fitting to ADSS cable suspension was achieved. The damage mechanism and countermeasures of ADSS optical cable in electric stress was analyzed detailedly in literature [3]

In this paper, CDEGS software is used to build the 750kV linear iron tower model. The distribution of the surface field strength of the tower is calculated, the correctness of the simulation was verified by experiment. And then the model is used to calculate the electric field strength of the prepared hanging points of cable in the tower to determine the reliability of the position.

2 Calculation principle

The method of moment is to transform the integral equation problem into a matrix equation. The numerical solution of the matrix equation is achieved by computer calculate[5]. Based on the numerical solution of the driving source distribution, the radiation field distribution

 DOI: 10.1051/matecconf/20165608005

and the wave impedance and other parameters can be calculated. The potential difference between two points on the conductor surface is determined by the leakage current. The electric potential difference between two points in the conductor section is generated by the self impedance and the axial current flowing through the conductor. The axial current on the conductor section can be expressed by the leakage current of each conductor section. The leakage current can be achieved by equations established by leakage currents as unknown quantity. The electric field near the transmission line can be obtained by leakage current[8-9].

3 Computational model

The 750kV double-circuit lines straight tower has been set as an example using CDEGS. The line voltage is 750kV and the phase conductor type is LGJ-400/50, 6-bundle transmission line is applied, and the spacing between intra-bundle conductors is 0.55m, right and left phase sequence are respectively as C/B/A and A/B/C. the ground wires type is JLB20A-150 and the diameter is 3.15mm, the span that in front of the tower and behind the tower are respectively 488m and 448m. Arc sag of transmission line left the tower is 13 meters, and arc sag of transmission line right the tower is 11 meters in figure 3. The earth is regarded as single uniform soil in the model, and the soil resistivity is 100Ω ·m. The two prepared hanging points of telecommunication optical fiber cables are respectively located at 12 meters and 11.46 meters high of the tower, as shown in figure 4.

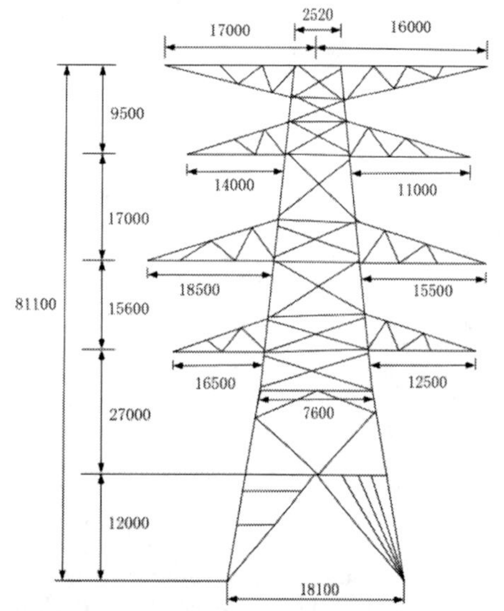

Figure 1. Tower geometry structure

The tower is needed to be simplified in the modeling process because of its complex structure, some factors such as the secondary metal components and insulator strings are ignored, the main metal frame is retained, and the platform of the main tower is equivalent to the square.Tower geometry structure is shown in figure 1, and simplified tower model is shown in figure 2. The wire arc sag is shown in figure 3.

Line voltage is change according to the law of symmetrical three-phase sine, and effective value is applied in calculation. The rated line voltage is 750kV, and calculated Voltage is 1.05 times greater than rated voltage when considering the actual operation. That is,

$$\dot{U}_A = \frac{750\times 1.05}{\sqrt{3}} kV$$
$$\dot{U}_B = \frac{750\times 1.05}{\sqrt{3}} \angle -120° kV \qquad (1)$$
$$\dot{U}_C = \frac{750\times 1.05}{\sqrt{3}} \angle 120° kV$$

Electric field of two prepared hanging points and the distribution of electric field on OA, OD and DF are calculated, these position shown in figure 4.

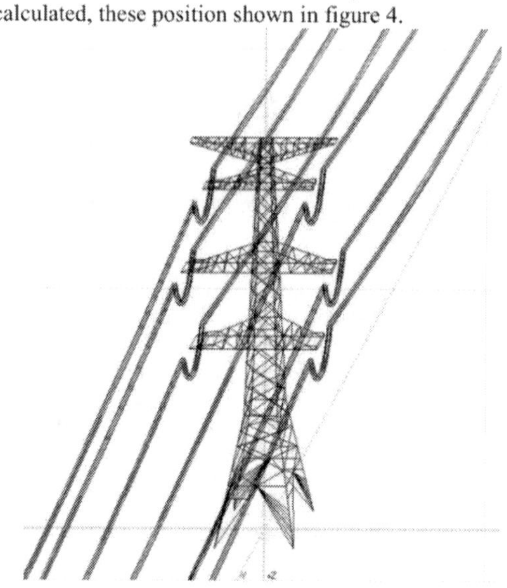

Figure 2. Simplified tower model

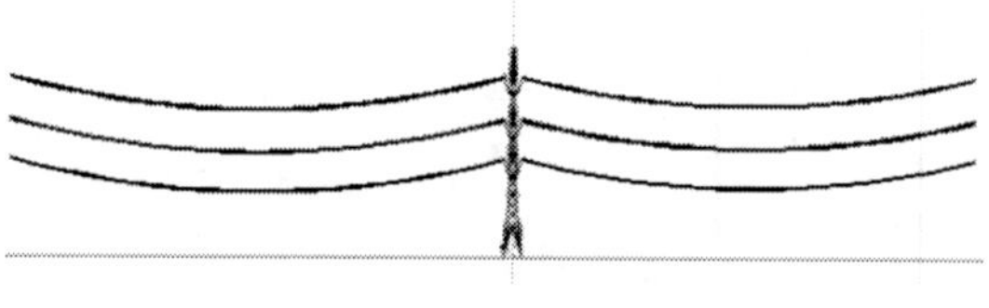

Figure 3. Left view of simplified tower model

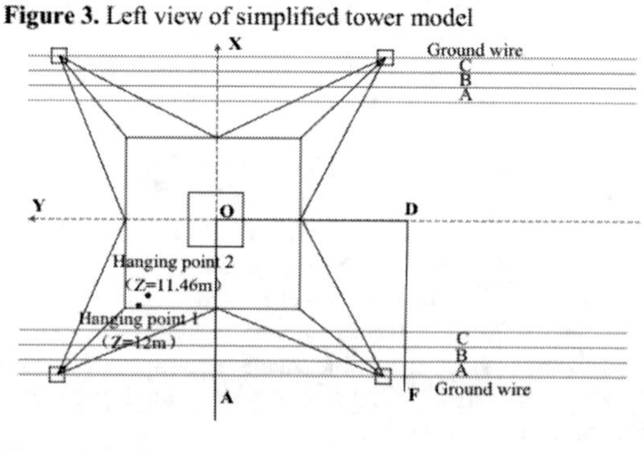

Figure 4. Position schematic diagram

The calculation results of ground surface field near the tower are shown in figure 5.

The calculation results of electric field of two prepared hanging points are shown in table 1.

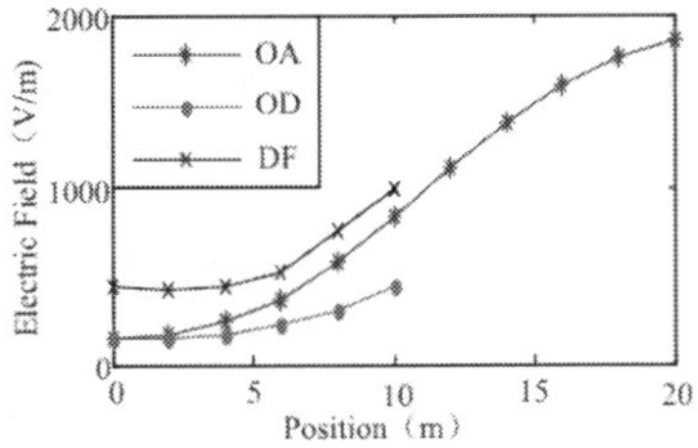

Figure 5. Calculation results of surface field

Table 1. Calculation of prepared hanging points

Prepared Hanging Point	Electric Field Intensity (kV/m)
hang point 1	7.37
hang point 2	6.43

4 Experimental measurement

Only the surface field strength can be measured in this experiment because of the tower charged operation. 750 kV tower center is selected to carry out measuring the electric field intensity under the transmission lines, as shown in figure 4.

Figure 6. Scene measure graph

EFA-300 electromagnetic field analyzer was applied in measurement, with temperature of 70C and humidity of 45%. The method of measurement refers to GBT 12720-1991 Power frequency electric field measurement and DLT 988-2005 Methods of measurement of power frequency electric field and magnetic field from high voltage overhead power transmission line and substation.

(1)The measurement of surface electric field: Schematic diagram of tower and transmission line is shown in figure 4. Measured height is 1.5 meter. The electric field intensity of the three lines around the iron tower that one measuring points per two meters are measured, which is OA(the tower center to 10 meters outside the tower)、OD(tower center to tower foot)、DF(center of tower lateral to tower foot) respectively. Test data and calculation data are both effective value.

(2) The measurement of electric field of two prepared hanging points: The electric field intensity of two pre hang points of communication optical cable were measured, as shown in figure 4, the electric field measuring probe was fixed to the insulating rod, the insulating rod was fixed to the edge of the tower by the operator. The specific position of the probe is determined by the laser range finder, and the distance of probe and tower is 10cm. Test data and calculation data are effective value.

The measurement results are shown in figure 7.

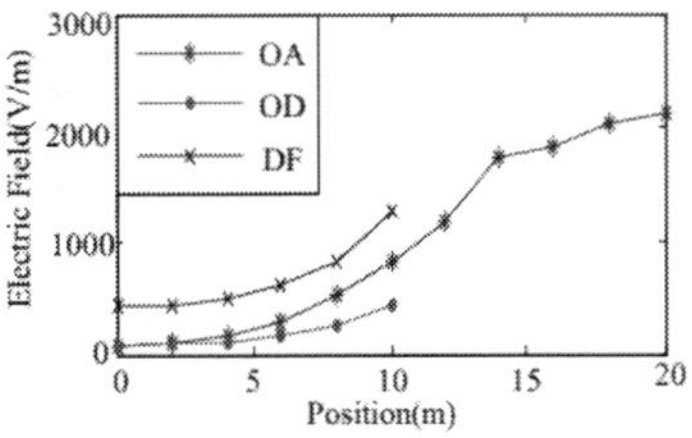

Figure 7. Measurement results of surface field

Table 2. Measurement results of prepared hanging points

Prepared Hanging Point	Electric Field Intensity (kV/m)
hang point 1	7.17
hang point 2	6.00

5 Analysis

5.1 Surface field strength comparison results

The comparison of OA direction between the measured results and the simulation results is shown in table 3.

The comparison of OD direction between the measured results and the simulation results is shown in table 4.

The comparison of DF direction between the measured results and the simulation results is shown in table 5.

The comparison of prepared hanging point between the measured results and the simulation results is shown in table 6.

Table 3. OA direction comparison results (V/m)

Observation position	Measurement	Simulation	Error%
0	81.3	99.5	22.2
2	140.2	118	13.2
4	166.2	185	11.3
6	279.1	185	13.5
8	518	525	1.3
10	815.8	795	2.5
12	1173	1080	7.9
14	1740	1350	22.4
16	1821	1570	13.7
18	2026	1730	14.6
20	2131	1840	13.6

Table 4. OD direction comparison results (V/m)

Observation position	0	2	4	6	8	10
measurement	81.38	89.83	108.3	153.3	263.1	434.7
simulation	99.5	105	127	181	278	412
error%	22.2	16.8	17.2	18.0	5.6	5.1

Table 5. DF direction results comparison(V/m)

Observation position	0	2	4	6	8	10
measurement	434.7	433.7	484.1	602.8	828.5	1250
simulation	445	435	453	517	756	1000
error%	2.4	0.3	6.4	14.2	8.7	20

Table 6. Comparison results of prepared hanging Points

Prepared Hanging Point	Hang Point 1 (kV/m)	Hang Point 2 (kV/m)
measurement	7.17	6.00
simulation	7.37	6.43
error%	2.7	7.2

According to the above tables,

(1)Surface field strength measured values and simulated values are both limited to 4kV/m within the safety limits.

(2)There is certain dispersion between the calculated values and the test values, but the trend is identical basically, and maximum error less-than 30%. Therefore, the proposed method is feasible, and the model and calculation process are correct. The model can be used to analyze the potential distribution of electric field near the tower, and calculate prepared hanging points of communication optical cable, thus the ability to analyze the problem of electrical corrosion of ADSS is enhanced.

(3)After analyzing, the main reason for the error is the simplification of the tower model. In particularly, the horizontal steel structure of the 12 meter platform and multi-stage platforms above are not given. These multi-layer horizontal steel structures is ignored in modeling process that have shielding effect on the electric field of transmission lines.

(4)The measuring point extends to 10 meters outside the tower along the OA direction, as seen from the table 2, after 10 meters the error is greater. This because there are a lot of trees outside the tower, and measurement probe cannot escape the trees that are difficult to ensure that the probe is in the vicinity of the 2.5 meter range to meet the measurement requirements. The ground surface of tower in a bad environment, the ground is uneven, and the gap is about 0.5-1m that is difficult to match the position coordinates of the actual measurement points with the calculation points.

5.2 Prepared hanging points calculation

By the above analysis of surface field strength, the method and modeling process are proved to be effective, and prepared hanging points are calculated. The position of two prepared hanging points are located in 12 meters high of the tower, and the distance between the position and the wire lowest point is 20 meters, therefore, the electric field of prepared hanging points can be measured manually. The results of simulation and experiment are shown in table 6, and the error less-than 10%. Prepared hanging point 1 and prepared hanging point 2 both meet the standard of ADSS communication optical cable that electric field strength less than 10kV/m. Therefore, the selection of the position of the prepared hanging point is reliable.

6 Conclusion

The validity and effectiveness of the model built by CDEGS are proved by the comparison between measured and simulated results. The model can be used to calculate and analysis the distribution of electric field intensity that tower of different phase sequence, phase angle, the position near transmission line which cannot be measured directly and ADSS cable hanging point. Meanwhile, the surface field strength of the 750kV tower is verified to meet the requirements of the national standard that less than 4kV/m; and the calculation of electric field intensity of the prepared hanging point of the ADSS communication cable of the tower that meet the requirements of the national standard that less than 10kV/m. The analysis results can provide a theoretical basis for the analysis of the distribution of the electric field strength around the tower, the safety protection of live working on the super high voltage tower and the safety of the ADSS cable.

References

1. G.-Z. Zhang, Y. Lu, B.-Q. Wan, X. Wu, Advances of power system and hydroelectric engineering **6**, 30 (2007)
2. Y.-F. Zhou, Z.-T. Cao, Telecommunications for electric power system **11**, 16 (2001)
3. D.-Q. Ding, Telecommunications for electric power system **9**, 1 (2001)
4. B. Zhang, X. Cui, T.-B. Lu, L. Li, Z.-B. Zhao, Power System Technology **27**, 5 (2003)
5. T.-B. Lu, K. Xiao, B. Zhang, High voltage technology 3, 24 (2001)
6. Y.-M. Li, Y.-Z. Fan, L.-W. Xu, Power system technology 3, 782 (2013)
7. X. Cui, F.-C. Wang, T.-B. Lu, C. Zhong, Telecommunications for electric power system **2**, 12 (2004)
8. Z.-B. Zhao, X. Cui, B. Zhang, L. Li, Proceedings of the CSEE **24**, 148 (2004)
9. G.-Z. Ni, S.-Y. Yang, J. Qiu, *Numerical calculation of engineering magnetic field* (China Machine Press, 2012)

Study on Tower Models for EHV Transmission Line

Bao-Qing Xu [1], Xiao-FeiZhang [1], Sen Lv [1], Rui-Xin Ding [1] and Lu Wang [2]

[1] Sanmenxia Power Supply Company, State Grid Henan Electric Power Company, Sanmenxia, China
[2] North China Electric Power University, Baoding, China

Abstract. Lightning outage accident is one of the main factors that threat seriously the safe and reliable operation of power system. So it is very important to establish reasonable transmission tower model and evaluate the impulse response characteristic of lightning wave traveling on the transmission tower properly for determining reliable lightning protection performance. With the help of Electromagnetic Transient Program (EMTP), six 500kV tower models are built. Aiming at one line to one transformer operating mode of 500kV substation, the intruding wave overvoltage under different tower models is calculated. The effect of tower model on intruding overvoltage has been studied. The results show that different tower models can result in great differences to the calculation results. Hence, reasonable selection of the tower model in the calculation of back- strike intruding wave is very important.

1 Introduction

500kV substation is the pivot of power system, and its performance is very important to power system. Lightning stroke in substation may lead to a lot of damage to electrical equipment. Hence, it is very important to calculate the over voltage of tower struck by lightning.

Tower model is important to considering lead-in line of the establishment of lightning invasion wave over voltage model. The lightning tower model was carried out by lots of researches [1-4]. But reflection on the process of the tower surge is still controversial, the fundamental reason is that the physical process of tower internal wave propagation is not clear. Especially with large span high tower widely used, lightning stoke opportunity, equivalent inductance of tower, overhead potential of tower and induced overvoltage are higher than common tower. At present, in the transmission line lightning protection analysis, three models, namely lumped inductance model, single-surge impedance model and multi-surge impedance model are selected for the tower simulation.

A 500kV substation is selected to calculate over voltage by using electromagnetic transient program (EMTP), combining substation incoming lines with thundering protection of transmission line and double circuit lines on same tower. The calculation method of tower surge impedance is introduced and six models, such as lumped inductance model, single-surge impedance model, lossy multi-surge impedance model and its simplified model, multi wave nondestructive model and Xi'an Jiao Tong university model are presented in paper. The effects of different models of substation lightning invasion wave are analyzed. Results show that different tower models can result in great differences to the calculation results. Hence, reasonable selection of the tower model in the calculation of back-strike intruding wave is very important.

2 Tower model

If the tower resistance can be ignored and the low precision is permitted, tower can be equivalent to one inductance, shown in figure 1(a), called lumped inductance model [5]. Formula of lumped inductance model is that:

$$L_{gt} = \tau Z_{gt} = \frac{h}{c} Z_{gt} \tag{1}$$

The surge impedance model is considered as a kind of transient wave process cross the tower. In the surge impedance models, tower over voltage is the results of superposition of the lightning over voltage and the reflected voltage wave from the tower bottom. The calculation principle of the single-surge impedance of tower is most regarded the tower as cone. Tour tower models provided by IEEE [6] are shown in figure.1 (b).

In figure.1 (b), shape1, shape2 and shape3 are the models used for general towers. The surge impedance of them is calculated by (2)-(4).

$$Z_{shape1} = 60\ln\frac{\sqrt{2(H^2 + r_3^2)}}{r_3} \tag{2}$$

$$Z_{shape2} = 60\ln\frac{2\sqrt{2}H}{r_2} - 60 \tag{3}$$

$$Z_{shape3} = \sqrt{\frac{\pi}{4}} \cdot 60[\ln \cot(\frac{1}{2}\tan^{-1}\frac{r_{avg}}{h_1 + h_2}) - \ln\sqrt{2}] \quad (4)$$

where, $r_{avg} = \dfrac{r_1 h_2 + r_2(h_1 + h_2) + r_3 h_1}{h_1 + h_2}$

The shape4 is the surge impedance model of tower and calculated by formula (5)

$$Z_{shape4} = \frac{Z_1 Z_2}{Z_1 + Z_2} \quad (5)$$

In the formula (5), Z1 is surge impedance of a vertical cylinder, Z2 is the weighted average of surge impedance of horizontal cylinder and vertical cylinder. Z1and Z2 are calculated from (6) and (7).

$$Z_1 = 60\ln\frac{2\sqrt{2}H}{r} - 60 \quad (6)$$

The shape4 is the surge impedance model of tower and calculated by formula (5)

$$Z_2 = (D \cdot 60\ln\frac{2H}{r} + HZ_1)/(H + D) \quad (7)$$

Lossy multi-surge impedance model is established based on the experiment. 1/70μs oblique wave current is injected into the top of the tower in the experiment, and the tower model and the calculation formula are obtained from the measurement results. The model consists of several lossless lines and several parallel connection of damping resistor R and damping inductance L, as shown in figure 1(c). The attenuation of shock wave is realized by resistance in the model, and the characteristic of high frequency attenuation is achieved by parallel connection of inductance and resistance [7-8]. For the 500kV tower, surge impedance is achieved by experiment that Zt1=Zt2=Zt3=ZT1=220Ω, Zt4=ZT2=150Ω. Damping resistor R and damping inductance L are calculated by (8)-(10), where vt denotes the propagation velocity of shock wave in the tower, and vt is 0.85 times than the speed of light.

$$R_i = \frac{-2Z_{T1}\ln\sqrt{\gamma_1}}{h_1 + h_2 + h_3} \cdot h_i \quad (i = 1,2,3) \quad (8)$$

$$R_4 = -2Z_{T2}\ln\gamma_2 \quad (9)$$

$$L_i = \alpha \cdot \tau \cdot R_i = \alpha \cdot (\frac{2H}{v_t}) \cdot R_i \quad (i = 1,2,3,4) \quad (10)$$

Takamitsu proposed simplified model based on lossy multi-surge impedance model, as shown in figure 1(d) [9]. Parameters of the model are the same with Zt1 and Zt2 in lossy multi-surge impedance model.

T.Hara proposed equivalent tower to lossless multi-surge impedance. The model will be subdivided into tower main materials、diagonal materials、and tower cross-arm, simulating with lossless circuit respectively, as shown in figure 1(e) [10-12] . The formula of main materials surge impedance is got by (11).

$$Z_{TK} = 60\ln(\frac{2\sqrt{2}h_k}{r_{ek}} - 2), (k = 1,2,3,4) \quad (11)$$

where,

$$r_{ek} = \begin{cases} \sqrt[8]{2} \cdot [\sqrt[3]{r_{TK} \cdot (r_B')^2}]^{1/4} \cdot [\sqrt[3]{R_{TK} \cdot (R_B')^2}]^{3/4} & (k = 1,2,3) \\ \sqrt[8]{2} \cdot [\sqrt[3]{r_{TK} \cdot (r_B)^2}]^{1/4} \cdot [\sqrt[3]{R_{TK} \cdot (R_B)^2}]^{3/4} & (k = 4) \end{cases} \quad (12)$$

The tower diagonal materials surge impedance is calculated by (13), and the length is assumed 1.5 times than the corresponding main material.

$$Z_{LK} = 9Z_{TK}(k = 1, 2, 3, 4) \quad (13)$$

The tower cross-arm can be treated as a normal horizontal conductor to calculate the surge impedance.

$$Z_{AK} = 60\ln(\frac{2h_k}{r_{Ak}}) \quad (k = 1,2,3,4) \quad (14)$$

Xi'an Jiao Tong University model is another multi-surge impedance model with the concept that surge impedance is different at different height based on vertical conductor. Same as lossy multi-surge impedance model, this model is divided into main support and ordinary support, and each part is assumed to be uniformly distributed. The surge impedance of each part is calculated by the calculation of tower geometrical dimension, as shown in figure 1(f) [13]. The formula of this model is same with ZTK and ZLK in lossless multi-surge impedance model.

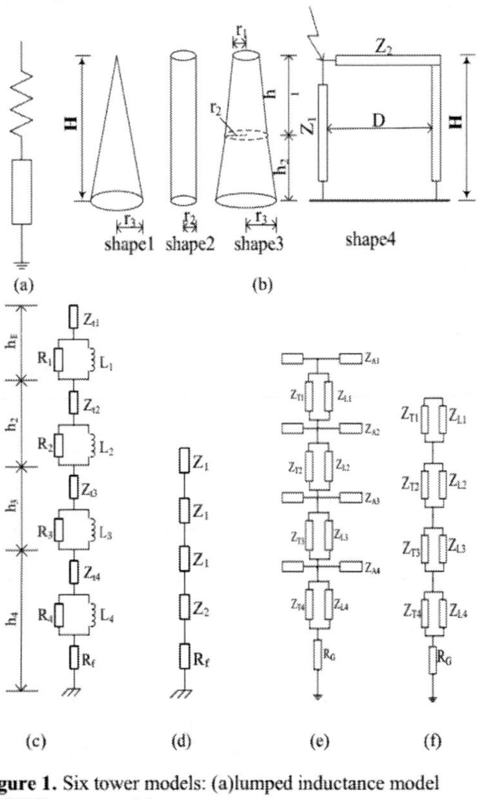

Figure 1. Six tower models: (a)lumped inductance model
(b)IEEE tower model
(c)lossy multi-surge impedance model
(d)simplified model
(e)lossless multi-surgeb Impedance
(f)Xi'an Jiao Tong University model.

3 Calculation conditions and results

3.1 Computational Condition

In this paper, one line to one transformer mode of 500kV was selected. Figure 2 is the equivalent circuit diagram. Overvoltage in six tower models was calculated respectively.

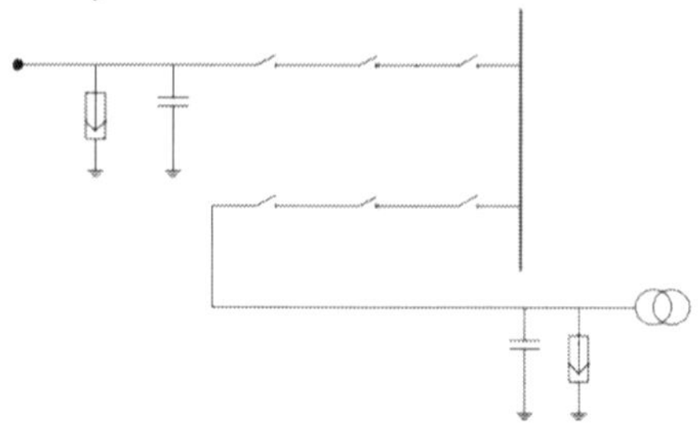

Figure 2. Equivalent circuit diagram of Substation.

The lightning current is selected as 2.6/50 μs, the amplitude is 216kA, and the lightning channel surge impedance is 300Ω. #2 is chosen to the lightning point, which located the range of 2km in substation incoming lines. Conductor type is 4×LGJ-400/50, lightning shield line type is LBGJ-120-40AC.

Arrester is installed at the entrance of the substation and beside the main transformer. The volt-ampere characteristics of arrester are shown in table 1.

Table 1. Arrester volt-ampere characteristics.

Arrester	rated voltage (kV)	volt-ampere characteristics					
		I(kA)	1	3	5	10	20
Line type	444	V(kV)	926	968	988	1030	1106
station type	420	V(kV)	876	916	934	974	1052

The lumped inductance is calculated by (1) that inductance value is 33.35μH. The single surge impedance is calculated to 106.4Ω by (4). Lossess multi-surge impedance model calculation results are shown in table 2.

Table 2. Calculation results of multi surge impedance.

K	1	2	3	4
Zi(Ω)	220	220	220	150
Ri(Ω)	4.589	9.906	10.051	16.736
Li(uH)	2.401	5.182	5.258	8.755

Zi in lossess multi-surge impedance model simplified model is same with table2. The calculation results of lossless multi-surge impedance model are shown in figure 3. The calculation results of Xi'an Jiao Tong University model are same with ZTK and ZLK in figure 3. The tower span is 500m, and the grounding resistance is 15Ω.

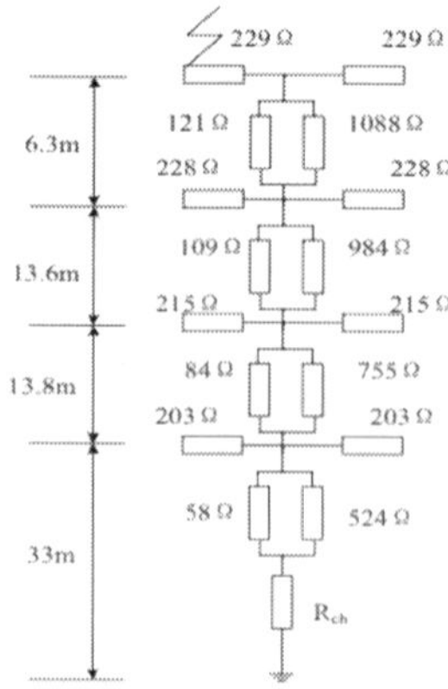

Figure 3. Lossless multi-surge impedance model.

All electrical facilities in the substation are simulated with the entrance capacitance. Capacitance values are shown in table 3.

Table 3. Electrical equipment inlet capacity.

Device Name	Symbol	Capacitance Value (pF)
Circuit breaker	D	330/150(Closing / Division)
Isolating switch	G	80/40(Closing / Division)
Current transformer	CT	700
capacitor voltage transformer	CVT	5000
transformer	B	5000

3.2 Computational Results

Six models are used to simulate the tower respectively, under one line to one transformer operation mode. Results of lightning intruding overvoltage in substation are shown in table 4.

Overvoltage is highest when using lossy multi-surge impedance model, the second is the simplified model, and then is lumped inductance model, single surge impedance model, Xi'an Jiao Tong University model, lossess multi-surge impedance model. Due to the resistance and inductance calculation formulas of lossy multi-surge impedance model are obtained by experiment using 1/70 μs oblique current wave, calculation results exist error. Overvoltage are different in capacitor voltage transformer with six models, lightning wave can enter secondary equipment through the capacitor voltage transformer in substation, so correction of tower model simulation have great influence to prevent substation lightning intruding overvoltage.

Arresters appear overvoltage first, followed by CVT and B, showing delay of wave process, as shown in figure 4. In arrester, overvoltage of lossy multi-surge impedance model is higer than lossess multi-surge impedance model, but the attenuation is faster, as shown in figure 5, showing characteristics of inductance and resistance branch.

Table 4. Lightning over voltage of substation equipment under six lightning tower models.

	Arrester(kV)	CVT(kV)	B(kV)
lumped inductance model	1123.1	1073.7	1119.7
single-surge impedance model	1060.5	1058.1	1104.6
lossy multi-surge impedance model	1099.6	1109.5	1253.3
simplified model	1095.7	1107.0	1250.8
lossess multi-surge impedance model	1022.6	1028.2	1095.8
Xi'an Jiao Tong university model	1056.7	1030.9	1090.4

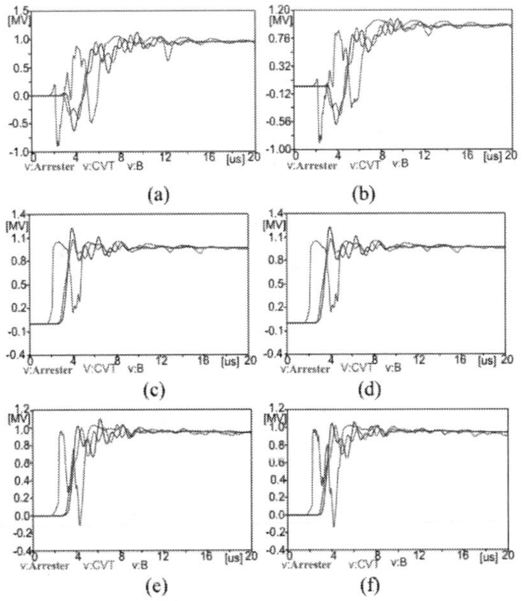

Figure 4. Calculation result of six models:(a)lumped inductance model
(b)single-surge impedance model
(c)lossy multi-surge impedance model
(d)simplified model
(e)lossess multi-surge impedance model
(f)Xi'an Jiao Tong university model.

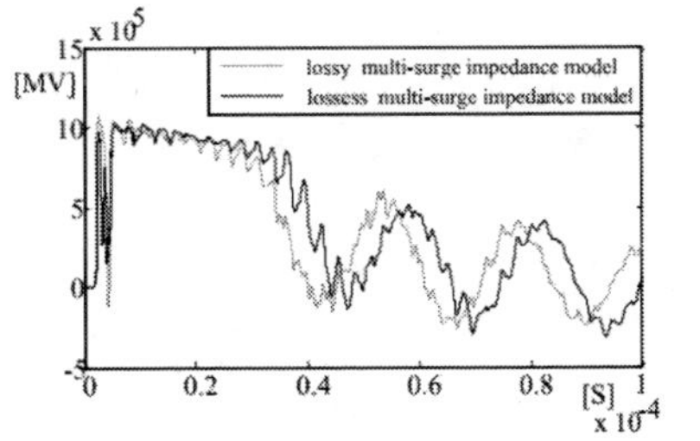

Figure 5. Arrester overvoltage of two models.

4 Conclusion

(1)Lumped inductance model is treated the tower as an equivalent inductance, and the potential of any point on the tower is same. Relative to the surge impedance model, lumped inductance model does not reflect lightning current propagation process cross the tower and the influence of reflection wave to each tower node potential when the lightning strikes the tower, that is the difference with the actual tower model.

(2)The wave propagation process of lightning current is considered in single surge impedance model, but the model does not take into account the structure of the tower itself, and the influence of the tower cross-arm, pillar and insulator flashover process are ignored.

(3)A hierarchical simulation of the tower is shown in lossy multi-surge impedance model, according to the actual height and cross-arm position on the tower. The parallel branch of the R and L reflects the attenuation process of the traveling wave transmission in the tower. But in the simulation results of the model, substation equipment overvoltage is the highest. One reason is the calculation formula of the model parameters directly using foreign calculation formula of damping resistance and damping inductance, obtained from experiment. This calculation is bound to lead to a certain error; on the other hand, lightning current also have an impact on the results. Due to the use of the lightning current waveform is domestic commonly used 2.6/50us double exponential waveform, which wave head is 2.6 time than the head wave of 1/70μs oblique wave current. The simplified model is obtained from lossy multi-surge impedance model, and the simulation results are close, so the 2.6/50us lightning current will have a same impact on lossy multi-surge impedance simplified model. Therefore, it's necessary to test lossy multi-surge impedance model of tower aiming at the domestic commonly using 2.6/50us double exponential lightning wave.

(4) Results of lossess multi-surge impedance model are close to Xi'an Jiao tong University model, both the two models segmenting the tower according to the concept that surge impedance are different at the different height of vertical conductor of the tower. But the results of overvoltage of the model of Xi'an Jiao Tong University are higher than the lossess multi-surge impedance model. The reason is the tower cross-arm model is not considered in Xi'an Jiao Tong University model, and propagation time of electromagnetic wave on the tower cross-arm is ignored. Therefore, the overall structure of the tower can be reflected in lossess multi-surge impedance model, and is often applied in the actual transmission line tower modeling. More tower section number, in calculation, the model is more precise, but the corresponding modeling will be more complex, so selecting the number of segment need to consider the accuracy and complexity of the modeling integrally.

(5)Lossy multi-surge impedance model and lossess multi-surge impedance model are both considered the height of tower and the position of tower cross-arm. Considering the attenuation of traveling wave in the transmission process of tower is the advantages of lossy multi-surge impedance model, but the consideration of the tower cross-arm surge impedance is the advantages of lossess multi-surge impedance model, so the two models can be combined in later simulation.

References

1. A. Ametani, Y. Kasai, J. Sawada, IEE Proc-Gener Transm Distrib **141**, 339 (1955)
2. B. Yoshihiro, IEEE Transactions on Power Delivery **14**, 630 (1999)
3. B. Yoshihiro, I. Masaru, IEEE Transactions on Power Delivery **15**, 1010 (2000)
4. Zhang Y.-J., W.-X. Si-Ma, Z.-J. Zhang, High Voltage Engineering **32**, 93 (2006)
5. Y. Xu, Y.-Z. An, Shaanxi Electric Power **3**, 37 (2011)
6. DL/T620-1997. Overvoltage Protection and Insulation Coordination for AC Electrical Installations, (1997)
7. D.-J. Wang, H. Zhou, J.-M. Chen, Power System Technology **31**, 11(2007)
8. W.-H. Wu, X.-L. Cao, China Water Resources and Hydropower Publishing House, (2012)
9. Z.-X. Yuan, H. Li, L. Xiang, High Voltage Engineering **34**, 867 (2008)
10. Takamitsu Ito, Toshiaki Ueda, IEEE Transactions on Power Delivery **18**, 545 (2003)
11. Y. Zhang, Y.-D. Gao, B. Du, W. Shi, JOURNALOF XI'AN JIAOTONG UNIVERSITY **38**, 365 (2004)
12. W.-W. Huang, M.-J. Xiao, X. Yang, High Voltage Apparatus **51**, 166 (2015)
13. X.-H. Zhang, Electrical and Electronic Engineering (2014)

Multilayer PV-storage Microgrids Algorithm for the Dispatch of Distributed Network

Ping Yang[1], Yu-jia Zhang[1], Zhi-rong Xu[1], Xu Yin[1], Shao-xiong Zhou[2], Jin-yong Lei[3]

[1]School of Electric Power, South China University of Technology, Guangzhou 510640, China.
[2]Guangdong Intework Energy Technology Co, Ltd , Guangzhou,511458, China.
[3]Electric Power Research Institute, CSG, Guangzhou 510080, China.

Abstract. In recent years, due to the support of our country, PV-storage microgrid develops rapidly. However, the flexible network operation modes of PV-storage microgrid change flexibly and the operating characteristics with a large amout of sources is highly complicated. Based on the existing microgrid coordinate control methods, this paper proposes multilayer PV-storage microgrid algorithm for fitting dispatch of distributed network, which achieves maximum output of renewable energy when meeting the scheduling requirements of network, by building PV-storage microgrid type dynamic simulation system in a variety of conditions in PSCAD. Simulation results show that the heuristic algorithm proposed can achieve microgrid stable operation and satisfy the demands of the dispatch in distributed network.

1 Introduction

As severe environmental problems such as pollution and depletion of fossil fuels appeared, developing clean energy becomes more and more important. Solar energy, wind energy and other renewable energy power generation has achieved rapid development and wide application[1]. At the same time, as industrial civilization developing rapidly, problems and challenge toward EHV and long-distance power transmission appeared while the electricity demand increased and the size of traditional power grid expanded. Accordingly, microgrid has received attention from domestic and foreign researchers.

Microgrid is a small-scale power grid that can work connected to the utility grid or in island mode. The capacity of microgrid is from kilowatts to megawatts so the energy generated of the microgrid can be consumed locally[2,3]. Most of the power generators in microgrid, whose kinds of battery contain lithium batteries, flow batteries and other hybrid energy storage devices as power storage unit, includes photovoltaic generator, wind power generator, fuel cell, micro turbine, diesel generator and hydropower generator use renewable energy[3,4].

The most important characteristic of the microgrid is that it controls and dispatches the power sources and loads uniformly through the use of rational allocation of distributed power source and advanced control technology. It helps the independent generators and distribution system work as Cells[5]. Microgrid can either be connected with network or run independently. If we dispatch power source and the load effectively, it is of benefit to meet the users' requirements of power quality and power reliability. Moreover, it can reduce the negative impact of the high penetration of renewable energy generation on the distribution network[6-7].

Microgrid has a variety of distributed power types and operating modes so the coordinated control algorithm of distributed power is quite complex[8-9]. Under the premise of guaranteeing the power quality, the mainly technical difficulties for large-scale application of PV-storage microgrid is efficiently taking advantage of each microsource generation, reliably controlling multiple types of microsources in PV-storage microgrid, and effectively meeting the energy balance and output requirements of user[10-12]. In order to solve those problems, this paper has researched the operation of microgrid system network.

The control between micro-grid and distribution grid are the key of safe and stable operation of microgrid. Document [13] put forward a network operation control technique which estimated whether the frequency value and power value met the requirement of disconnection by acquiring frequency and power value from the point of common coupling(PCC). If the value met the disconnecting requirement, the connected point would be cut off. However, this technique did not fully consider the inner condition of microgrid, and it cannot make photovoltaic reach maximum output. Document [14] proposes another control mode. This mode controlled the super capacitor to amend the discharge status of battery through detecting the SOC (State of Charge). It makes use of central control devices, energy control device and load control device to control the energy exchange between utility grid and microgrid and to solve the energy exchange problems, but this technique require demanding coordinate control.

Each existing microgrid coordinated control algorithm has its own advantages and disadvantages. However, both of them are difficult to be applied widely. Based on existing micro-grid coordinated research, this paper proposed a multilayer PV-storage micro-grid algorithm with upper scheduling. Having fully considered the multiple operating states of photovoltaic power, this algorithm can achieve maximum output of renewable energy while meeting the distribution scheduling requirements and thus, provided an effective solution for microgrid application.

2 PV-storage microgrid system designs

As distributed photovoltaic power generation developing rapidly, the PV-storage microgrid with photovoltaic microsource generation and energy storage system have received attention. In order to achieve power supply reliability, power quality and economical benefit, PV-storage microgrid has a lot of different types of connection and operation. Therefore, when studying the coordinated control algorithms of PV-storage type microgrid, it is necessary to analyze the design of PV-storage type microgrid system.

2.1 Typical structure of PV-storage microgrid

The typical topology structure of PV-storage microgrid system is shown in Fig.1. Photovoltaic generators connect to microgrid common bus through a contact line with storage system. When the microgrid system runs in grid-connected mode, photovoltaic generators run under maximum or extreme power output and storage batteries

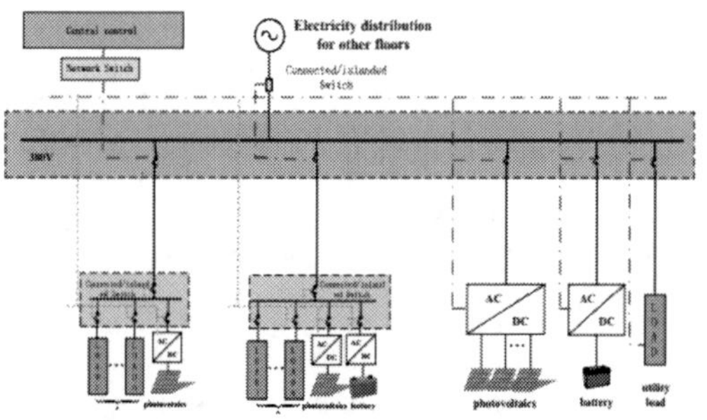

Fig. 1. PV-storage type microgrid schematic diagram

smooth the fluctuations of output power. When an islanding event occurs, microgrid system disconnects form the utility grid and runs in islanded operation mode. Storage batteries work under the V/f control as main power source of the microgrid system to support its voltage and frequency stability. Meanwhile, photovoltaic generators work under maximum or extreme power output to supply power to load in system. These two different operation modes show the flexibility and complexity of microgrid.

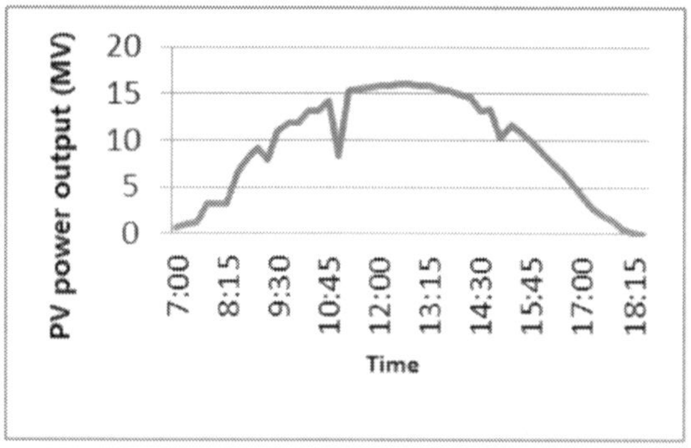

Fig. 2. The characteristic of photovoltaic generation with time changing

2.2 Microsources function orientation and access principle

The power supply of PV-storage microgrid includes photovoltaic generation system and storage system. Photovoltaic generator converts solar energy into electrical energy which is the main power source of the PV-storage microgrid besides the utility grid. The output power of photovoltaic generation depends on illumination intensity and temperature. Because illumination intensity and temperature of the day change as the weather conditions and time change, output of photovoltaic generation will show randomness, disturbance, intermittence and other characteristics, as Fig.2 shown.

Therefore, accessing photovoltaic generation should obey the following three principles:

(1) The capacity of photovoltaic generation should not be too large. Photovoltaic generation is the main power supply of microgrid, but it requires complicated control and can only be used during daytime. Accessing excessive capacity will bring negative effect to the utility grid.

(2)Accessing photovoltaic generation requires adjustable power supply for better control of the microgrid.

(3) Photovoltaic generators should always be able to work under maximum power output for effectively control in order to improve stability.

3 Hierarchical control idea of PV-storage microgrid

Illumination, temperature and other ambient conditions will greatly influence photovoltaic generation bringing significant vibration and intermittent problems. Energy storage system can change the status of charge and the output power through appropriate control, achieving power complement with photovoltaic generation. However, because of the cost and the limited storage capacity of battery, the storage system can only serve as a short-term energy supplement. Master-slave control, equal control and hierarchical control are three main control methods. For master-slave control, main storage power supply determines the coordinated control of other microsources supply so it requires demanded storage reliability. For equal control, photovoltaic generation and energy storage system equally share power shortage to maintain the energy balance and to meet the output requirement. But by using this method, photovoltaic generation cannot always work under extreme power output. Also, this method requires frequent charging and discharging to the storage system, which will shorten the lifespan of the energy storage system. Therefore, multilayer control becomes the mainstream and has engaged on many demonstration engineering projects.

Aiming at satisfying the demand of large-scale application, this paper uses microgrid coordinated control algorithm based on layered structure to divide the control of microgrid into three layers according to different response speed, time scales and communication needs[15], as Fig.3 shown. The first layer is local control layer. It controls photovoltaic generation, energy storage system and other underlying single device, response and process quickly and does not rely on communication. The second layer is centralized control layer. It maintains voltage and frequency stability of PV-storage microgrid, and keeps the PV-storage type microgrid security and stable. Compared with local control layer, it relies on communication and has a long response time. The third layer is distribution scheduling layer.

It coordinates control between multiple microgrid and dispatches each microgrid output power according to distribution network power requirements and constraints (minimum power loss, lowest economic cost, etc.), which needs several minutes to response and requires high communication reliability[16].

Coordinated control algorithm of centralized control layer is shown in Fig 4. Its main functions include:

(1) Acquiring and processing data from the underlying individual device, monitoring the distributed units and the controllable load.

(2) Achieving short-term and ultra-short-term prediction of photovoltaic generation and loads based on illumination, temperature, historical data and other information.

(3) Acting as "PQ" source when the microgrid operates at grid connected mode and optimizing each power output according to power supply and load forecasting, distribution scheduling.

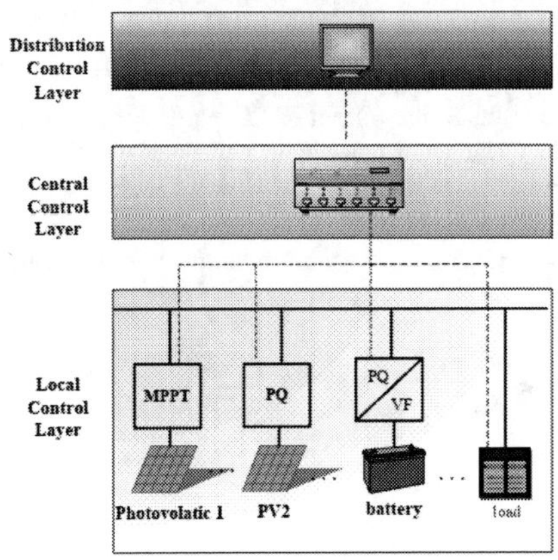

Fig.3. Microgrid coordinated control algorithm

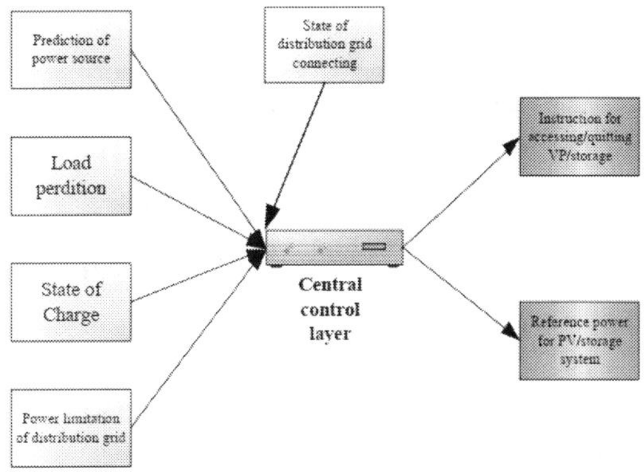

Fig.4. Microgrid Centralized control layer

4 Grid-connected operation algorithm of PV-storage microgrid

In grid-connected operation mode, the voltage and frequency of the microgrid are supported by the utility distribution network. The main objective of control algorithms is to maximum the output power of photovoltaic generation and to keep the power in tie line within requirement of distribution network scheduling.

The microgrid working under grid-connected operation mode should satisfy the following constraints:

$$\begin{cases} P_{net} = \sum P_{pv} + \sum P_{bat} - \sum P_{load} - P_{loss} \\ P_{net} \le P_{dpt} \\ f_{\min} \le f \le f_{\max} \\ U_{\min} \le U \le U_{\max} \end{cases} \tag{1}$$

Where:

P_{net} is the active power in tie line, threat microgrid transports power to distribution network as positive;

P_{pv} is the active power in photovoltaic generation;

P_{bat} is the active power in storage system, threat discharge as positive and charge as negative;

P_{load} is the active power consumed by loads;

P_{loss} is the internal active power loss in PV-storage microgrid;

P_{dpt} is the active power set by distribution scheduling layer, threat transporting from microgrid to distribution network as positive;

$f_{\max}$, $f_{\min}$ are upper and lower limitation of frequency;

$U_{\max}$, $U_{\min}$ are upper and lower limitation of voltage.

Coordinated control algorithm of PV-storage microgrid in grid-connected mode is shown as Fig.5-6. In Fig. 5, $\Delta P = P_{dpt} - P_{net}$ is the difference between active power set by distribution scheduling layer and actual active power in tie line, By setting P_{ofset1} (positive) and P_{ofset2} (negative) reasonably, the algorithm can achieve "hysteresis control" and avoid the fluctuation of value of active power in tie line which may lead the control system act frequently. When ΔP is in hysteresis, the algorithm will then modify the charge and discharge instruction of the central control layer according to the storage charge and discharge power limitation, as shown in the formula (2) :

$$\begin{cases} P_{chi} = \min(P_{batChSeti}, P_{batChLmti}) \\ P_{dhi} = \max(P_{batChSeti}, P_{batDhLmti}) \end{cases} \quad (2)$$

Where,

P_{chi} is the corrected charging power of storage system i ;

$P_{batChSeti}$ is the charging power of storage system i set by centralized control layer;

$P_{batChLmti}$ is the charging power limit of storage system i , related to SOC state;

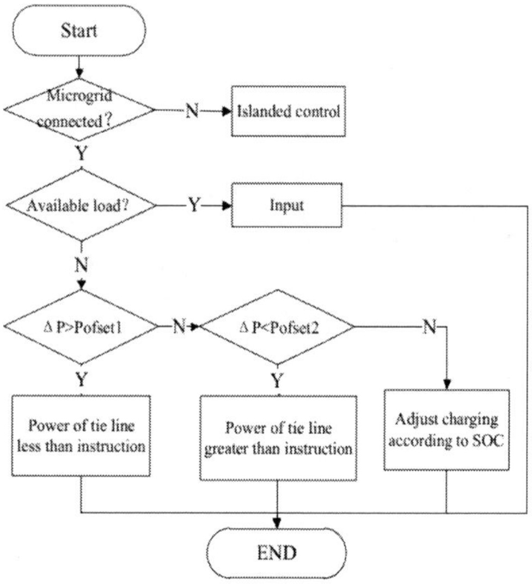

Fig.5. Control algorithm of grid-connected state

P_{dhi} is the corrected discharge power of storage system i ;

$P_{batDhSeti}$ is the discharge power of storage system i set by centralized control layer;

$P_{batDhLmti}$ is the storage system i discharge power limit, related to SOC state.

Then we are going analyze two modes of coordinated control algorithm when working in grid-connected mode:

A. When tie-line power is less than scheduling instruction

(1) If energy charging storage exists, reduce energy storage system charging power as following formula:

$$P_{batChSeti} = \min(0, P_{batRli} + \Delta P) \quad (3)$$

Where, P_{batRli} is the actual active power of storage system i . This situation generally occurs when the tie line power on the last control period cycle greater than scheduling instruction, and charge the storage system.

(2) If the charging storage cannot meet the requirement, connected photovoltaic generators into the grid, If the photovoltaic system operating mode is MPPT, put photovoltaic power that meet the requirement of the formula (4) into the grid in sequence.

$$\begin{cases} P_{mpPr\,j} = \max\{P_{mpPr\,i}\} \\ P_{mpPr\,i} < \Delta P \end{cases} \quad (4)$$

Where, $P_{mpPr\,j}$ is the predictive value of the MPPT active power that will be put in;

j is the number of power supply to be put in;

$P_{mpPr\,i}$ is the predictive value of the MPPT active power that can be put in;

If the operating mode of photovoltaic system is PQ mode, then start all photovoltaic generators and set the PQ output of photovoltaic system according to power distribution principle, as the following formula (5):

$$P_{pqSeti} = \min(P_{pqPri}, \frac{P_{pqPri}}{\sum P_{pqPri}} \times \Delta P) \quad (5)$$

Where, P_{pqSeti} is the setting active value of PQ photovoltaic power to be put in; P_{pqPri} is the predictive active value of PQ photovoltaic power to be put in.

(3) If the whole system was in the condition that the storage system has been charged and the photovoltaic generation system was running on full output power but the system still failed to meet the needs of active distribution scheduling, we can increase the active power output of PQ system as follows:

$$\begin{cases} P_{pqAdi} = \min(P_{pqRmi}, \frac{P_{pqRmi}}{\sum P_{pqRmi}} \times \Delta P) \\ P_{pqRmi} = P_{pqPri} - P_{pqRli} \\ P_{pqSeti} = P_{pqRli} - P_{pqAdi} \end{cases} \quad (6)$$

Where, P_{pqAdi} is the increasing active power instruction of PV system under PQ operation mode, P_{pqRli} is the actual active output of PQ photovoltaic system.

(4)If the photovoltaic generation system was fully utilized but still failed to meet the needs of active distribution scheduling, discharged the available energy storage system, its discharging power is set as following formula (7):

$$P_{batSeti} = \min(P_{batDhLmti}, P_{batRli} + \frac{P_{batDhLmti} - P_{batRli}}{\sum(P_{batDhLmti} - P_{batRli})} \times \Delta P) \quad (7)$$

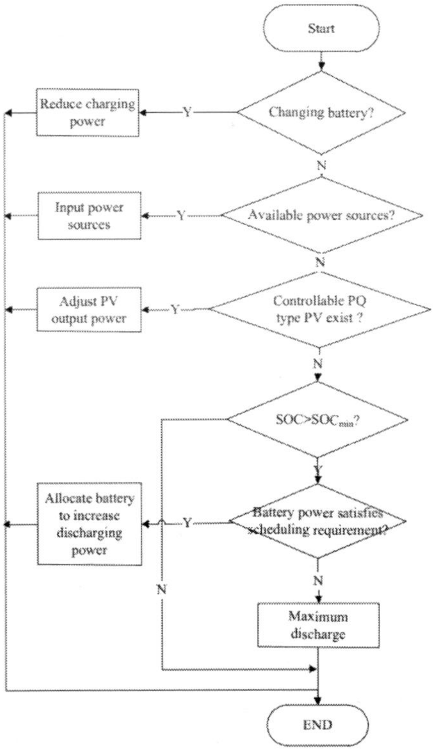

Fig.6. Tie line power is less than scheduling instructions

B. When tie-line power is greater than scheduling instruction

(1) If energy discharging storage exists, reduce energy storage system discharging power as following formula:

$$P_{batDhSeti} = \max(0, P_{batRli} + \Delta P) \tag{8}$$

This situation generally occurs when the tie line power on the last control period cycle less than scheduling instruction, and discharge the storage system.

(2)By following the principle of avoiding illumination abandoning control, calculate rechargeable restore power and the sum of PQ active power to determine whether removal MPPT and photovoltaic power operation or not, calculation as follows:

$$P_{sumCtrl} = \sum P_{pqRli} + \sum (P_{batRli} - P_{batChLmti}) \tag{9}$$

(3) Distribution scheduling layer typically want the active output power of PV-storage microgrid to be restricted in scheduling limits. So when removing MPPT photovoltaic systems, remove the nearest value of scheduling active power vacancies, as following formula :

$$\begin{cases} k = j, \left|P_{mpRlj} + \Delta P\right| = P_{\min} \\ P_{\min} = \min\{\left|P_{mpRli} + \Delta P\right|\}, j = 1,2\ldots n \end{cases} \tag{10}$$

Where, P_{mpRli} is active output power of MPPT photovoltaic system needed to be removed. k is the number of MPPT photovoltaic generator nearest to the value of scheduling active power vacancies.

(4) If the whole system was in the condition that the storage system has been discharged and the photovoltaic generation system was cut off but still failed to meet the needs of active distribution scheduling, reduced the active power output of PQ system as following formula (11).

$$P_{pqSeti} = P_{pqRli} + \max(-P_{pqRli}, \frac{P_{pqRli}}{\sum P_{pqRli}} \times \Delta P) \tag{11}$$

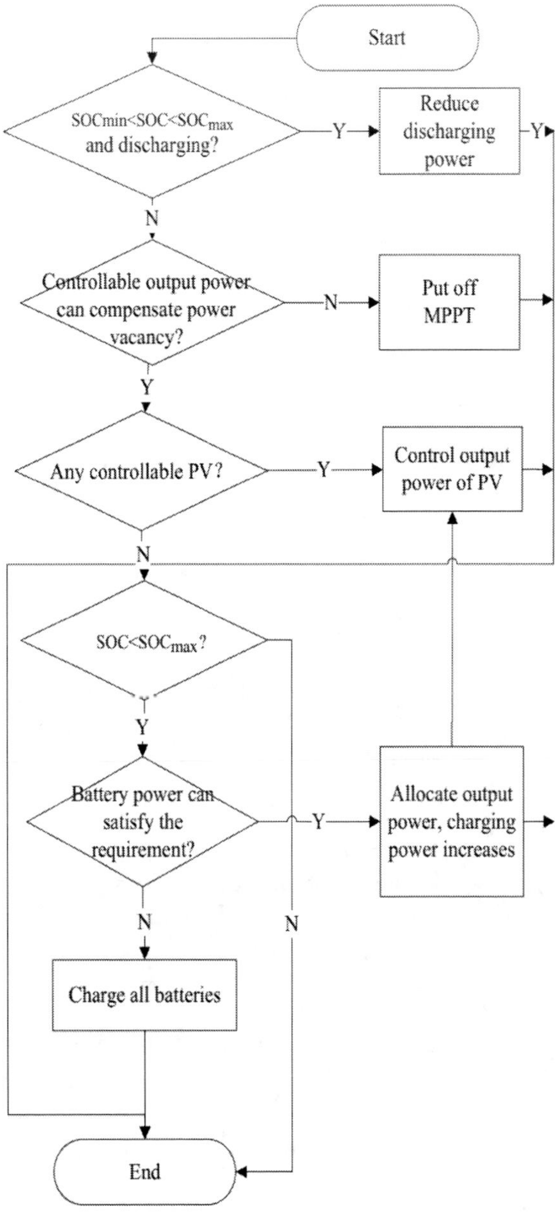

Fig.7. Tie line power is greater than scheduling

(5) If the output power of photovoltaic generation system was reduced but still failed to meet the needs of

DOI: 10.1051/matecconf/20165608007

active distribution scheduling, charged the available energy storage system, its charging power was set as following formula.

$$P_{batSeti} = \max(P_{batChLmti}, \frac{P_{batRli} - P_{batChLmti}}{\sum(P_{batRli} - P_{batChLmti})} \times \Delta P) \tag{12}$$

5 Simulation verification

The test circuit topology of control algorithm of microgrid in grid-connected operation mode is shown in Fig.8. The system consists of photovoltaic, energy storage and load. The photovoltaic system consists of one MPPT type photovoltaic power source whose rated power is 30kW and two PQ type photovoltaic power sources whose rated power is 30kW. The energy storage system consists of one main battery and one auxiliary battery, the rated power capacity of the main battery is 90kW, the rated power capacity of the auxiliary battery is 30kW. There are three static loads whose rated power capacity is 50kW.

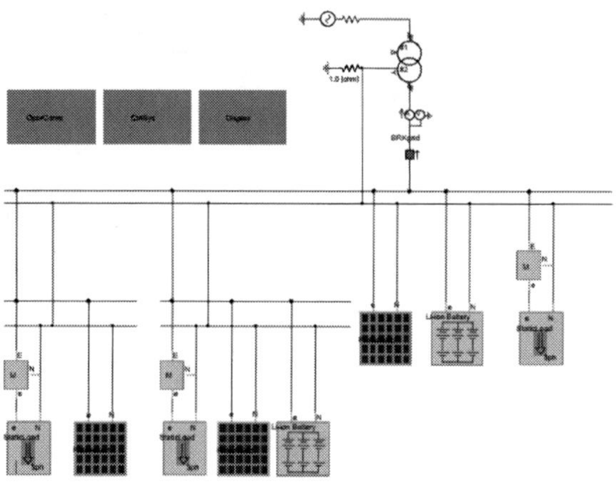

Fig.8. Microgrid Topology

Condition: change both distribution scheduling, illumination and load

Initial condition:

Illumination: 0-5s: 700 W/m^2, 5-10s: 900 W/m^2, 10-15s: 1000 W/m^2, 15-20s: 600 W/m^2, 20-25s: 700 W/m^2, 25-30s: 900 W/m^2, 30-35s: 1000 W/m^2, 35-50s: 600 W/m^2;

Temperature: 25 $°C$;

Scheduling value: 1-5s: 30kW; 5-10s: 10kW; 10-20s: -150kW; 20-25s: 30kW; 25-30s: 75kW; 30-35s: 110kW; 35-40s: 75kW; 40-45s: 0kW.

PV 1: type: MPPT, capacity: 30kW;

PV 2: type: limited power operation, capacity: 30kW;

PV 3: type: limited power operation, capacity: 30kW;

Main battery: type: VF, capacity: 90kW, SOC: 65%;

Auxiliary battery: type: PQ, capacity: 30kW, SOC: 60%;

Load 1: 0-5s: 50kW; 5-10s: 0kW; 10-20s: 15kW; 20-30s: 50kW; 30-40s: 40kW; 40-45s: 20kW;

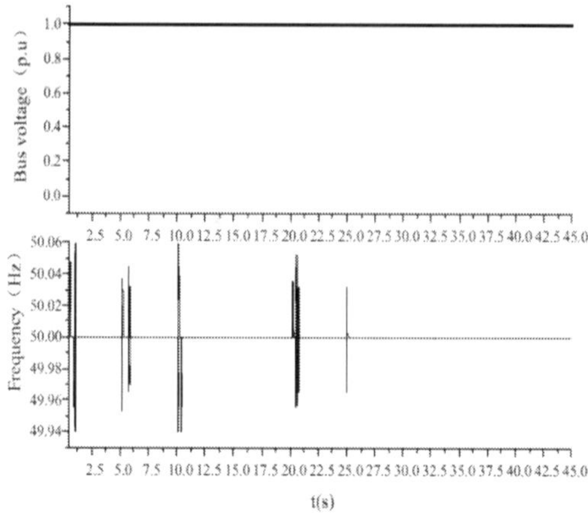

(a) Bus bar voltage and frequency change

Load 2: 0-5s: 50kW; 5-10s: 0kW; 10-20s: 15kW; 20-30s: 50kW; 30-40s: 40kW; 40-45s: 20kW;

Load 3: 15kW static load;

SOC fixed value: SOCmin1: 10, SOCmin2: 30, SOCmax2: 80, SOCmax1: 90;

Simulation time 20s represents 1 hour.

The initial condition of the system:

Load capacity: 105kW

PV capacity: 60kW

Power on tie line: -50kW.

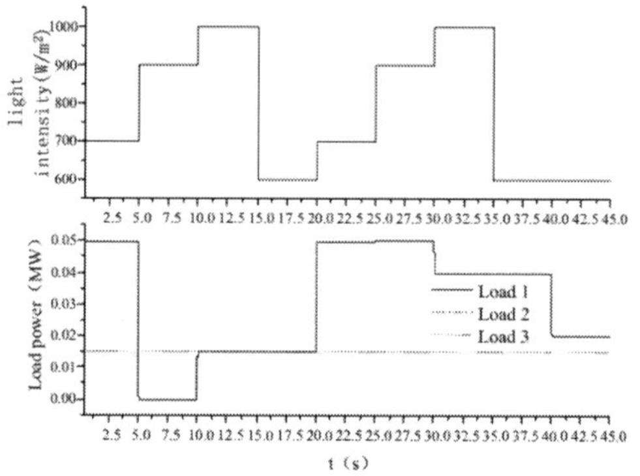

(b) Illumination intensity and load change

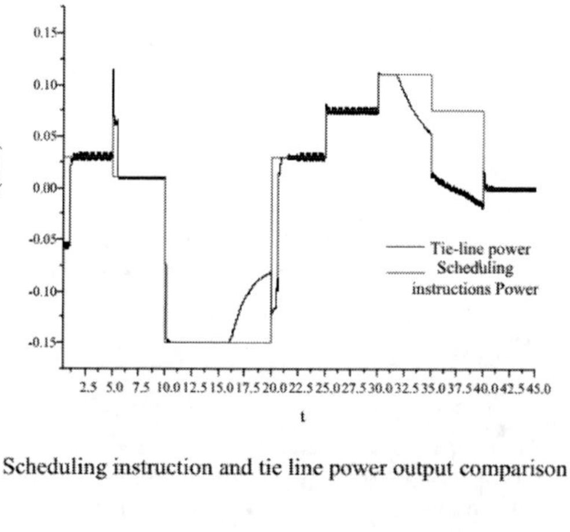

(c) Scheduling instruction and tie line power output comparison

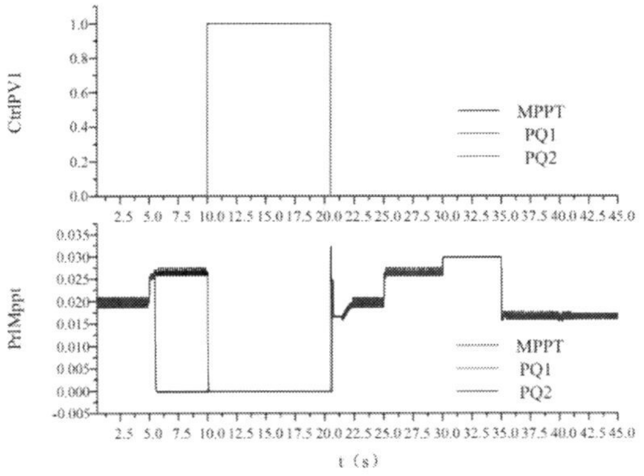

(d) PV control signal and output change

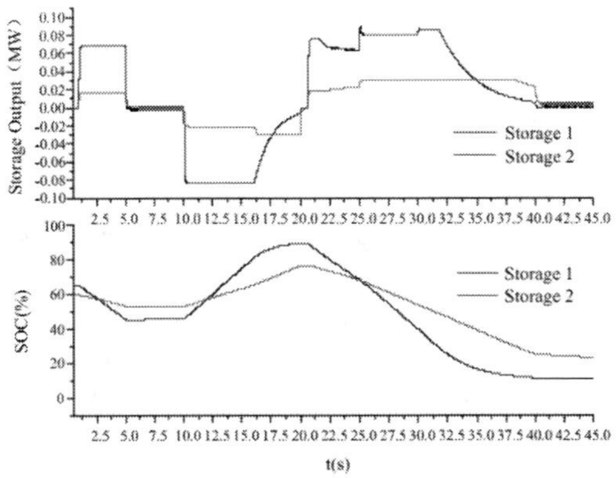

(e) Storage output and SOC change

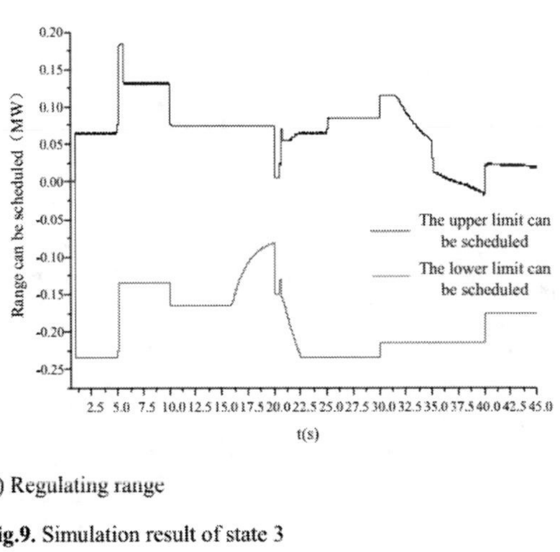

(f) Regulating range

Fig.9. Simulation result of state 3

1s later, the distribution scheduling layer sent down instruction of 30kW. The power of illumination is 700 W/m^2 . Under the control of grid-connected algorithm, the two batteries started to discharge, whose discharge power are 69.3kW and 16.7kW. Three photovoltaic power supplies do not work. At this time, the tie line power P_{net} is 30kW, which can follow the scheduling instructions of distribution network and the adjustable power ranges -235-64kW.

5s later, the distribution scheduling layer sent down instruction of 10kW. The power of illumination turns into 900 W/m^2 and the capacity of load turns into 15kW. The output power of MPPT photovoltaic power supply is 26.35kW. Under the control of grid-connected algorithm, the two The two storage batteries begin to charge to absorb fluctuation power produced by MPPT photovoltaic power source. At this time, the tie line power P_{net} is 10kW, which can follow the scheduling instructions of distribution network and the adjustable power ranges -135-131kW.

10s later the distribution scheduling layer sent down instruction of -150kW. The power of illumination turn into 1000 W/m^2 and the capacity of load turn into 15kW. Under the control of grid-connected algorithm, the MPPT photovoltaic power sources are cut off. The charging power of the two batteries increases to -83.1kW and -21.6kW. The PQ type photovoltaic power source does not work. At this time, the tie line power P_{net} is -145kW, which can follow the scheduling instructions of distribution network and the adjustable power ranges -165-74.8kW.

The simulation results show that under coordinated algorithm control, PV-storage microgrid can operate safely and stably in each condition.

6 Conclusions

China has paid great attention on renewable energy and the intelligent grid was developing rapidly. It is of great importance to develop microgrid. The control of PV-storage microgrid is the key of the application of distributed photovoltaic generation as well as the construction of intelligent grid. PV-storage microgrid contains varieties of power sources and complicated working state so that a reliable and efficient way of control is needed to guarantee the safety, stability and economic benefit of the microgrid.

This paper proposes a grid-connected working model of PV-storage microgrid and puts forward a heuristic algorism of multilayer PV-storage microgrid according to the coordinated control of microgrid. This algorism can satisfy the demand of scheduling and enable maximum power output of renewable energy. The results of the simulation show that this algorism enables the microgrid working efficiently and satisfies the requirement of the distributed network dispatches.

Acknowledgments

This work was supported by the national high technology research and development program (863 Program, 2014AA052001); Guangdong Province science and technology plan project (2012B040303005); Electric Power Research Institute, CSG, science and technology project(SEPRI-K143003); China Southern Power Grid, CSG, science and technology project(K-KY2014-009).

References

1. ZHANG Bo-quan, YANG Yi-min, "Status and trend of wind/photovoltaic power development", Electric Power, 2006,06:65-69.
2. GAO Jun-yan, MA Xiu-fan, "Risk Analysis on Influence of Distributed Renewable Power Supply on the Investment Plan of Distribution Network", Power System and Clean Energy, 2008(12):8-12.
3. EI Wei, SHENG Kun, KONG Li, QI Zhi-ping, "Impact and Improvement of Distributed Generation on Distribution Network Voltage Quality", Proceedings of the CSEE, 2008,13:152-157.
4. CHENG Miao-miao, "Optimal Capacity of Energy-Storing Section in PV/Wind Hybrid System", Electrotechnical Application, 2006,06:87-90.
5. Liserre M., Sauter T., Hung J. Y. ,"Future Energy Systems: Integrating Renewable Energy Sources into the Smart Power Grid Through Industrial Electronics ", Industrial Electronics Magazine, IEEE, 2010, 4(1):18-37.
6. Zeng J, Wu J, Liu J F, et al., "An agent-based approach to renewable energy management in eco-building", Sustainable Energy Technologies, 2008. ICSET 2008. IEEE International Conference onIEEE, 2008:46-50.
7. Chang J, Jia S Y., "Windy-Solar Power Generation System Based On Multi-Agent System", Machine Learning and Cybernetics, 2008 International Conference on2008:2446 - 2449.
8. XIAO Zhao-xia , "Control and Operation Characteristic Analysis of a MicroGrid", Tianjin University, 2009.
9. "2014 Distributed generation grid-connected innovation development seminar[Z]", EPTC, 2015:2015.
10. R. H. Lasseter, "MicroGrids," in Power Engineering Society Winter Meeting, 2002. IEEE: IEEE, 2002, pp. 305-308 vol.1.
11. N. Hatziargyriou, H. Asano, R. Iravani, and C. Marnay, "Microgrids," IEEE POWER & ENERGY MAGAZINE, vol. 5, pp. 78-94, 2007.
12. M. Ross, R. Hidalgo, C. Abbey, and G. Joos, "Energy storage system scheduling for an isolated microgrid," IET RENEWABLE POWER GENERATION, vol. 5, pp. 117-123, 2011.
13. PANG Xue-yue, LIAO Yi, ZHANG Xue-yan, "Microgrid grid-connected operation control approach and system[P]", Guangdong, CN103208819A, 2013-07-17.
14. WU Xing, ZHAO Yang-jie, ZHU Yi-xin, "An approach for grid-connected and off-net microgrid and energy control [P]", Shanxi: CN103647274A, 2014-03-19.
15. DING Fei, "Transient Modeling and Simulation of the LV Micro-grid with Multiple Distributed Energy Sources and Energy Storage Elements", Tianjin University, 2010.
16. Adhikari S. Fangxing L., "Coordinated V-f and P-Q Control of Solar Photovoltaic Generators With MPPT and Battery Storage in Microgrids", Smart Grid, IEEE Transactions on, 2014, 5(3):1270-12.

AUTHOR INDEX

AUTHOR INDEX